AF692820

Sustainable Development Goals: A Pragmatic Approach

Sustainable Development Goals: A Pragmatic Approach

Editors

Idiano D'Adamo
Massimo Gastaldi
Manoj Kumar Nallapaneni

MDPI

Basel • Beijing • Wuhan • Barcelona • Belgrade • Novi Sad • Cluj • Manchester

Editors

Idiano D'Adamo
Sapienza University of Rome
Rome
Italy

Massimo Gastaldi
University of L'Aquila
L'Aquila
Italy

Manoj Kumar Nallapaneni
City University of Hong Kong
Kowloon
Hong Kong

Editorial Office
MDPI AG
Grosspeteranlage 5
4052 Basel, Switzerland

This is a reprint of articles from the Special Issue published online in the open access journal *Sustainability* (ISSN 2071-1050) (available at: https://www.mdpi.com/journal/sustainability/special_issues/3EDH89176K).

For citation purposes, cite each article independently as indicated on the article page online and as indicated below:

Lastname, A.A.; Lastname, B.B. Article Title. *Journal Name* **Year**, *Volume Number*, Page Range.

ISBN 978-3-7258-2579-0 (Hbk)
ISBN 978-3-7258-2580-6 (PDF)
doi.org/10.3390/books978-3-7258-2580-6

Cover image courtesy of Idiano D'Adamo
Vasto, located in the Abruzzo Region, is a city in which to experience the beauty of nature.

© 2024 by the authors. Articles in this book are Open Access and distributed under the Creative Commons Attribution (CC BY) license. The book as a whole is distributed by MDPI under the terms and conditions of the Creative Commons Attribution-NonCommercial-NoDerivs (CC BY-NC-ND) license.

Contents

About the Editors

Idiano D'Adamo

Idiano D'Adamo is an Associate Professor of Management Engineering at the Department of Computer, Control, and Management Engineering of Sapienza University of Rome, where he teaches Business Management, Economics of Technology, Management and Economics, and Management of Energy Sources and Services. Idiano was among world's top 0.1% of scientists in 2023 and among the world's top 2% of scientists for the fifth consecutive year (Elsevier Repository Data). He received a master's of Science in Management Engineering in 2008 and his PhD in Electrical and Information Engineering in 2012. In August 2015, he obtained the Elsevier Atlas Award with a work published in *Renewable and Sustainable Energy Reviews*. He collaborates continuously with several journals as an Editor in Chief (*Sustainability*), Subject Editor (*Sustainable Production and Consumption*), and Associate Editor (*Environment, Development and Sustainability, Global Journal of Flexible Systems Management, Frontiers in Sustainability, Sustainable Development, Supply Chain Analytics*, and the *Decision Analytics Journal*). During his academic career, Idiano D'Adamo published 165 papers in the Scopus database, reaching an h-index of 50. He has participated in scientific research projects (PRIN2022 'Protecting the Environment: Advances in Circular Economy (PEACE)', PNRR-PE 'Made-in-Italy circolare e sostenibile', Horizon 2020 'Star ProBio', Life 'Force of the Future'), has collaborated with relevant national institutes (MATTM, CNBBSV, SVIMEZ), and has written for major national newspapers (Formiche, il Messaggero). His current research interests are the bioeconomy, circular economy, green energy, sustainability, and waste management.

Massimo Gastaldi

Massimo Gastaldi is now a Full Professor in Managerial Engineering at University of L'Aquila, Department of Industrial and Information Engineering and Economics, where he teaches Applied Economics in Engineering Science, Analysis of Financial Systems, and Circular Economy and Sustainability Management. Since 1993, he has contributed to several publications and edited books while working for the National Research Council (C.N.R.) and at the University of Rome "Tor Vergata". In 1995, he was a Visiting Professor at the California State University, San Marcos, USA, and in 1997 he worked at Colorado University, Builder, USA. His current research interests are regulation and public utilities, green supply chain management, sustainability and the circular economy, finance, renewable energy, and real options. He is the responsible manager of private and public research projects. He is the past Editor in Chief of the *International Journal of Management and Network Economics* published by Inderscience. He is author of more than 200 papers published in peer-reviewed books, journals, and conference proceedings.

Manoj Kumar Nallapaneni

Dr. Nallapaneni is a transdisciplinary energy and sustainability engineer with a PhD in the Digital Circular Economy and Circular Power System from the School of Energy and Environment, City University of Hong Kong. He has obtained two master's degrees, one in Renewable Energy Technologies from Karunya University, India, and the other in Environmental Economics from Annamalai University, India. He holds a bachelor's degree in Electrical and Electronics Engineering from GITAM University. Before joining the CityU, he worked as a Research Fellow at Universiti Malaysia Pahang, Malaysia, on a project which focused on using solar photovoltaics as urban and rural infrastructure. Prior to this, he worked as an Assistant Professor in the Department of Electrical and Electronics Engineering at the Bharat Institute of Engineering and Technology, Hyderabad, India, and Energy Engineer at Atiode Solar Systems Limited, Benin City, Nigeria. Dr. Nallapaneni's research works focus on the topics of simulation, experimental, real-time empirical and location-, or climate-specific studies mainly focused on building sustainable and resilient systems (that are decentralized, networked, and centralized) across critical infrastructure sectors by adopting nexus thinking and systems innovation, with an emphasis on circular business models and digitalization. He has worked on key sustainability challenges that include the design, performance modeling, and analysis of a wide range of clean energy and environmental systems, the food–energy–water–waste nexus, industrial symbiosis, waste valorization and material passports, carbon accounting, and pricing. He possess an interdisciplinary skill set that includes performance analytics, techno-economics, life cycle assessments, resilience assessments, leveraging digital innovations (blockchain, IoT, smart contracts, and AI), business model innovation, and nexus systems design, with excellent conceptualization skills.

Preface

The most promising visions of the future are those in which young people are trusted and experts pass on their knowledge to them. However, respect is essential. We hope for a society built on peace, and a more humane and fraternal Europe with a pragmatic approach to sustainability is our future horizon.

This Special Issue was made possible by the effort, talent, and support of the Managing Editor, Ms Milica Milosev.

Idiano D'Adamo, Massimo Gastaldi, and Manoj Kumar Nallapaneni
Editors

sustainability

MDPI

Editorial

Europe Moves toward Pragmatic Sustainability: A More Human and Fraternal Approach

Idiano D'Adamo [1,*], Massimo Gastaldi [2] and Manoj Kumar Nallapaneni [3,4,5]

1 Department of Computer, Control and Management Engineering, Sapienza University of Rome, 00185 Rome, Italy
2 Department of Industrial and Information Engineering and Economics, University of L'Aquila, 67100 L'Aquila, Italy; massimo.gastaldi@univaq.it
3 School of Energy and Environment, City University of Hong Kong, Kowloon, Hong Kong; nallapanenichow@gmail.com
4 Swiss School of Business and Management Geneva, Avenue des Morgines 12, 1213 Genève, Switzerland
5 Center for Circular Economy Skills, HICCER—Hariterde International Council of Circular Economy Research, Palakkad 678631, Kerala, India
* Correspondence: idiano.dadamo@uniroma1.it

Citation: D'Adamo, I.; Gastaldi, M.; Nallapaneni, M.K. Europe Moves toward Pragmatic Sustainability: A More Human and Fraternal Approach. *Sustainability* **2024**, *16*, 6161. https://doi.org/10.3390/su16146161

Received: 12 July 2024
Accepted: 16 July 2024
Published: 18 July 2024

Copyright: © 2024 by the authors. Licensee MDPI, Basel, Switzerland. This article is an open access article distributed under the terms and conditions of the Creative Commons Attribution (CC BY) license (https://creativecommons.org/licenses/by/4.0/).

"Sustainable by passion and deliberate choice, not mere interest" This is the first message that young people should read, as should more mature adults. Sustainability had its beginnings in environmental protection, which was being defaced with the idea that resilience would always lead nature to regenerate. However, humankind has opted for the overuse of resources, and while scepticism about climate change still exists, it is clear that urgent action is needed. However, it is incumbent to be realistic and implement actions that protect the environment and not seek momentary consensus by making promises that cannot be kept. Nature must be protected, but so must the needs of humankind, and therefore, social welfare must also be preserved and economic opportunities generated. Hence, the 1987 Brundtland Report calls for respect for future generations by allocating at least the same resources to them as we currently have. In this context, the concept of sustainability shifts from an environmental view to a more comprehensive one that includes economic and social spheres. Is it then simple to find this balanced point? The answer is that with a purely ideological approach, disconnected from reality, we are far from improving our existing situation. Therefore, the need to find a pragmatic approach emerges. As such, what is derived from thought is subjected to the scrutiny of experience and an attempt is made to find an optimal point not only in an abstract model but also in a concrete one. Theories or words are not enough; instead, real actions and deeds are needed.

Pragmatic sustainability is a model of sustainability that contemplates the three dimensions of environmental, economic, and social factors and does not stop at enunciating a concept but ensures it is implemented in practice and is made concrete after providing analyses that support it. Clearly, an optimal point does not always derive from the point of maximum utility for each of the three dimensions but from a proper balance in which one or more of the three dimensions could be more highly valued. But who does this depend on? Stakeholders. As a function of what? The balance of ecosystems. So, the balance between humans and nature drives decision-making and a humane and fraternal approach to life needs to be put back at the centre of the agenda. A pragmatic view, therefore, values the great challenge of sustainability: overcoming personal selfishness to protect ecosystems and achieve the triple aim of economic performance, environmental protection, and social progress [1].

Pragmatic sustainability is based on the concept of altruism, just as the morning sentinels baptised by St. John Paul II defended "life at every moment of its earthly development" to make it "ever more habitable for all". In the 2003 post-synodal exhortation

"Ecclesia in Europa", Europe is seen not in opposition or competition with the world community but in harmony with it. It is pointed out how the increasing openness of peoples to one another and the reconciliation between those who had long been enemies allowed for recognition, collaboration and exchanges of every order. In this way, "a culture, indeed a European consciousness, is created, which should especially among young people, grow the feeling of fraternity and the will to share".

The European Coal and Steel Community came into effect in 1952, bringing together Germany, France, Italy, the Netherlands, Belgium and Luxembourg. The goal was to prevent any country alone from making weapons of war for use against others. Unfortunately, even today, there are armed conflicts that take the lives of innocent people and children. The European Union is seeing its borders expand but also shrink due to the United Kingdom voting to leave. Europe aims to be peaceful, united and prosperous; it should also promote the entry of new countries. The goal is to be able to compete with two global giants, such as the U.S. and China.

The European Green Deal has the potential to generate a competitive advantage, saving our planet but also creating new jobs. It is a set of policy initiatives involving states, civil society, and private actors in the fight against climate change [2]. In this innovative framework, digital technologies can enable the achievement of these policies, but they are also characterised by barriers [3]. However, it is crucial to combine sustainability with the needs of households, businesses and citizens without generating new dependencies or destroying entire production chains.

Forward-looking choices toward the future include ex-ante spatial impact assessments and sustainability analyses that demonstrate the relevant environmental, economic and social benefits. These are adequately outlined in the opinion "A Just Transition for All EU Regions" by Abruzzo Region President Marco Marsilio (rapporteur) and Stefano Cianciotta (expert). This opinion, which was voted on in July 2024 by the Coter Commission of the European Committee of the Regions, emphasises the relevance of technology neutrality, the role of storage, and the security of energy networks. This pragmatic approach to sustainable energy is a topic proposed in the literature [4], where it is emphasised that this approach is at the core of engineering activities that focus on technical needs to achieve energy sustainability and less on the role of economics, politics and other non-technical factors. Similarly, interdisciplinary collaboration, pragmatism, reliance on youth, and altruism are believed to be the resources that can support the realisation of a sustainable community within a university setting, where the phenomena of sustainable washing should be avoided [5]. This is a change resulting from the Sustainable Development Goals (SDGs), which concern all universities at the European level [6].

The topic of sustainability requires further analysis, as some aspects of this concept are not yet well understood [7,8]. Similarly, other analyses show a concentration of research and development on the SDGs in developed countries rather than in developing and underdeveloped countries [9]. The measurement of the SDGs plays a key role, and some analyses propose assessing their status [10], while others focus on comparing certain countries [11] and finally, approaches are also proposed at the level of each individual nation [1].

This Special Issue, titled "Sustainable Development Goals: A Pragmatic Approach", aims to counter non-doing, i.e., day-to-day approaches that deny climate change or ideological approaches that do not allow for community-useful projects and infrastructure. Stakeholder engagement toward sustainable development is as basic as the balance between ecosystems.

The current context sees the Suez Canal as the needle in the reconciliation of new international balances, and a necessary ecological and digital transition cannot materialise by creating dependencies on raw materials from Asian territories. In this context, territorial specificities must be strengthened, poles of excellence must be generated, programs for the exchange of resources and skills must be fostered, and support must be given to the industrial system to modernise and be competitive. In this framework, public administration is called upon to equip itself with new professionalism, to reduce unproductive delays related to the Not In My Term Of Office (NIMTO) phenomenon, and include citizens in decision-making processes to reduce the Not In My Back Yard (NIMBY) phenomenon. The goal of European funding programs is to foster a just transition by focusing on circular, digital and green investments with continuous monitoring of the proper use of public money. Europe must offer its countries and citizens a model of development that is sustainable but also inclusive and resilient. It is crucial for an area to avoid a diaspora of talent and graduates and to be attractive to acquire new knowledge.

Europe must be ambitious and be a spokesperson for a change in geopolitical balances. Pope Francis speaks of an "open sore in our humanity" that has seen the deaths of men, women and children in the Mediterranean Sea trying to reach Europe. The Mattei Plan has become part of the G7 agenda and envisions a cooperative approach between Europe and Africa. This Plan concerns not only cultural cooperation but also the implementation of investments in energy and infrastructure. Figure 1 shows that the same picture can be painted in different colours. The most promising future visions are those in which the respect of young people is given for more experienced profiles, and there is trust that more mature people for the new generation will come from a variety of different backgrounds. A more humane and fraternal Europe with a pragmatic approach to sustainability is our future horizon.

Figure 1. Vasto (CH). City located in the Adriatic Sea in the Abruzzo Region, known as the green region of Europe.

Author Contributions: Conceptualisation, I.D., M.G. and M.K.N.; writing—original draft preparation, I.D., M.G. and M.K.N.; writing—review and editing, I.D., M.G. and M.K.N.; supervision, I.D. All authors have read and agreed to the published version of the manuscript.

Funding: This study was carried out within the PEACE (Protecting the Environment: Advances in Circular Economy) study, which received funding from the "Fund for the National Research Program and Projects of Significant National Interest (PRIN)", Investment M4.C2.1.1-D.D. 104.02-02-2022, 2022ZFBMA4, funded by the European Union-Next Generation EU. This manuscript reflects only the authors' views and opinions, and the authors can be considered solely responsible for them.

Data Availability Statement: The original contributions presented in the study are included in the article; further inquiries can be directed to the corresponding author.

Acknowledgments: This study was carried out within the PEACE (Protecting the Environment: Advances in Circular Economy) study, which received funding from the "Fund for the National

Research Program and Projects of Significant National Interest (PRIN)", Investment M4.C2.1.1-D.D. 104.02-02-2022, 2022ZFBMA4, funded by the European Union-Next Generation EU. This manuscript reflects only the authors' views and opinions, and the authors can be considered solely responsible for them.

Conflicts of Interest: The authors declare no conflicts of interest.

References

1. D'Adamo, I.; Di Carlo, C.; Gastaldi, M.; Rossi, E.N.; Uricchio, A.F. Economic Performance, Environmental Protection and Social Progress: A Cluster Analysis Comparison towards Sustainable Development. *Sustainability* **2024**, *16*, 5049. [CrossRef]
2. Vela Almeida, D.; Kolinjivadi, V.; Ferrando, T.; Roy, B.; Herrera, H.; Vecchione Gonçalves, M.; Van Hecken, G. The "Greening" of Empire: The European Green Deal as the EU first agenda. *Polit. Geogr.* **2023**, *105*, 102925. [CrossRef]
3. Sharma, R.; Lopes de Sousa Jabbour, A.B.; Jain, V.; Shishodia, A. The role of digital technologies to unleash a green recovery: Pathways and pitfalls to achieve the European Green Deal. *J. Enterp. Inf. Manag.* **2022**, *35*, 266–294. [CrossRef]
4. Rosen, M.A. Energy Sustainability: A Pragmatic Approach and Illustrations. *Sustainability* **2009**, *1*, 55–80. [CrossRef]
5. Biancardi, A.; Colasante, A.; D'Adamo, I.; Daraio, C.; Gastaldi, M.; Uricchio, A.F. Strategies for developing sustainable communities in higher education institutions. *Sci. Rep.* **2023**, *13*, 20596. [CrossRef] [PubMed]
6. Pactwa, K.; Woźniak, J.; Jach, K.; Brdulak, A. Including the social responsibility of universities and sustainable development goals in the strategic plans of universities in Europe. *Sustain. Dev.* **2024**, *in press*. [CrossRef]
7. Yamaguchi, N.U.; Bernardino, E.G.; Ferreira, M.E.C.; de Lima, B.P.; Pascotini, M.R.; Yamaguchi, M.U. Sustainable development goals: A bibliometric analysis of literature reviews. *Environ. Sci. Pollut. Res.* **2023**, *30*, 5502–5515. [CrossRef] [PubMed]
8. Giannetti, B.F.; Agostinho, F.; Almeida, C.M.V.B.; Liu, G.; Contreras, L.E.V.; Vandecasteele, C.; Coscieme, L.; Sutton, P.; Poveda, C. Insights on the United Nations Sustainable Development Goals scope: Are they aligned with a 'strong' sustainable development? *J. Clean. Prod.* **2020**, *252*, 119574. [CrossRef]
9. Mishra, M.; Desul, S.; Santos, C.A.G.; Mishra, S.K.; Kamal, A.H.M.; Goswami, S.; Kalumba, A.M.; Biswal, R.; da Silva, R.M.; dos Santos, C.A.C.; et al. A bibliometric analysis of sustainable development goals (SDGs): A review of progress, challenges, and opportunities. *Environ. Dev. Sustain.* **2023**, *26*, 11101–11143. [CrossRef] [PubMed]
10. Halkos, G.; Gkampoura, E.-C. Where do we stand on the 17 Sustainable Development Goals? An overview on progress. *Econ. Anal. Policy* **2021**, *70*, 94–122. [CrossRef]
11. Nakhle, P.; Stamos, I.; Proietti, P.; Siragusa, A. Environmental monitoring in European regions using the sustainable development goals (SDG) framework. *Environ. Sustain. Indic.* **2024**, *21*, 100332. [CrossRef]

Disclaimer/Publisher's Note: The statements, opinions and data contained in all publications are solely those of the individual author(s) and contributor(s) and not of MDPI and/or the editor(s). MDPI and/or the editor(s) disclaim responsibility for any injury to people or property resulting from any ideas, methods, instructions or products referred to in the content.

sustainability

MDPI

Article

Economic Performance, Environmental Protection and Social Progress: A Cluster Analysis Comparison towards Sustainable Development

Idiano D'Adamo [1,*], Cristina Di Carlo [2], Massimo Gastaldi [3], Edouard Nicolas Rossi [2] and Antonio Felice Uricchio [4,5]

1 Department of Computer, Control and Management Engineering, Sapienza University of Rome, 00185 Rome, Italy
2 Sapienza University of Rome, 00185 Rome, Italy; cridicarlo03@gmail.com (C.D.C.); edouard.rossi97@gmail.com (E.N.R.)
3 Department of Industrial and Information Engineering and Economics, University of L'Aquila, 67100 L'Aquila, Italy; massimo.gastaldi@univaq.it
4 President of the Governmental Agency for the Evaluation of Universities and Research System (ANVUR), 00153 Rome, Italy; antonio.uricchio@anvur.it
5 University of Bari Aldo Moro, 70121 Bari, Italy
* Correspondence: idiano.dadamo@uniroma1.it

Abstract: Sustainable development monitoring reveals the extent to which local and national territories are progressing towards sustainability goals. This study considered 105 indicators associated with the Equitable and Sustainable Wellbeing (BES) framework and 139 indicators associated with the Sustainable Development Goals (SDGs), using multicriteria decision analysis (MCDA) and cluster analysis to compare regional performance across the Italian territory in 2022. At the SDG level, Lombardia exhibited the highest performance, while the provinces of Trento and Bolzano led at the BES level. The results were further analyzed with respect to geographic macro-areas and the three dimensions of sustainability, via separate cluster analyses comparing the BES and SDG results. Northeast regions emerged as the top performers, and comparable performance was shown by regions in the center and northwest. The development of a sustainable innovation model, alongside territorial cooperation and synergy between regional specificities, may generate competitive advantages, especially when combined with resources and skills with an international profile.

Keywords: cluster analysis; equitable and sustainable wellbeing; indicators; Italy; multicriteria decision analysis; sustainable development goals

Citation: D'Adamo, I.; Di Carlo, C.; Gastaldi, M.; Rossi, E.N.; Uricchio, A.F. Economic Performance, Environmental Protection and Social Progress: A Cluster Analysis Comparison towards Sustainable Development. *Sustainability* **2024**, *16*, 5049. https://doi.org/10.3390/su16125049

Academic Editor: Cristina Raluca Gh. Popescu

Received: 1 May 2024
Revised: 30 May 2024
Accepted: 7 June 2024
Published: 13 June 2024

Copyright: © 2024 by the authors. Licensee MDPI, Basel, Switzerland. This article is an open access article distributed under the terms and conditions of the Creative Commons Attribution (CC BY) license (https://creativecommons.org/licenses/by/4.0/).

1. Introduction

1.1. The Concept of Sustainability

The topic of sustainability requires in-depth exploration, as some aspects of the concept are not yet well understood [1,2]. The definition offered by the Brundtland Report in 1987 stressed the need to consider future generations, moving beyond a concept of sustainability anchored in a purely environmental perspective [3]. Thus, sustainability may be considered to encompass not only environmental protection but also the balancing of social welfare and economic opportunities. In this vein, Saint John Paul II, on World Youth Day 2000 in Rome, urged young people to be "sentinels of the morning". Importantly, the concept of pragmatic sustainability may be contrasted with that of ideological sustainability, as Europe's goal of climate neutrality by 2050 requires solutions that meet the needs of various stakeholders while also involving young people and fostering a sense of brotherhood among peoples [4].

1.2. Development Goals towards Sustainability

Compared to the last publication of the development goals towards sustainability (accessed on 19 April 2023), the number of published papers on the topic has increased to 31,757 (as of 15 April 2024). A keyword search was run on the Web of Science (WoS) using the search terms "SDGs" and/or "Sustainable Development Goals". The data covered the past 4.5 years (2020–2024), and a 14% growth in the number of publications over this period was evident. Notably, 62.2% of the papers were published with open access, representing an increase of 60% with respect to the previous set.

- Table A1 lists the top 10 WoS categories, with "environmental sciences" leading, followed by "green sustainable science technology".
- Table A2 indicates that approximately 83% of the work originated in the top 10 countries, with China contributing approximately 18%, followed by the USA (12.9%) and the UK (10.3%). South American countries were absent from the ranking, while India, Spain, Australia, Italy, Germany, South Africa and Canada were included.
- Table A3 presents the analysis by WoS index, showing the Science Citation Index Expanded (57.4%) as the most significant, followed by the Social Sciences Citation Index (39.4%).
- Table A4 describes the distribution of papers across the individual SDGs. The results indicated a shift from the previous ranking, where SDG 3 prevailed, followed by SDG 2 and SDG 1. In the new ranking, SDG 13 led, followed by SDG 3 and SDG 11; SDGs 14, 15 and 16 received the least attention.

1.3. The Role of SDG and BES Indicators

In 2015, all United Nations Member States adopted the 2030 Agenda for Sustainable Development. The reviewed literature analyzed the SDGs with respect to different goals, including food consumption [5], medical waste management [6], education institutions [7], tourism [8], digitalization [9], the circular economy [10], smart cities [11], fashion [12] and space [13]. Core research topics included the SDGs, Agenda 2030, climate change and sustainability indicators [14]. Since the launch of the SDGs, some environmental and social goals have shown significant improvement [15]. In this regard, the use of indicators to measure performance at the territorial level is crucial [16–18]. Globally defined targets are not easily transmitted to the national level [19], and the maintenance of ecosystem services and promotion of well-being in low-income countries must be made a priority [20].

SDG calculations benefit businesses by encouraging managers to adopt green solutions [21]. Additionally, they allow politicians and the public to measure performance at the territorial level [22]. Consequently, techniques for measuring and evaluating the SDGs are crucial for various categories of stakeholders [23]. Importantly, social welfare must be measured alongside SDG indicators [24], and several European countries have proposed specific tools for this purpose [25]. Italy, for instance, considers Equitable and Sustainable Wellbeing (BES) indicators across 12 dimensions [26]. It is considered a virtuous example [27], as it was the first OECD country to introduce supplementary measures to GDP at the economic planning stage [28].

Scholars have suggested regional evaluations incorporating both BES and SDG metrics [29] while stressing the importance of preserving the uniqueness of these indices [30]. The pursuit of equitable, sustainable development and a high standard of living requires the identification of SDG and BES overlap [31]. However, the aggregation of information (which may or may not be related) is a complex challenge [32], and composite indicators may be needed for this purpose [33]. The effectiveness of composite indicators relies on the weighting system used, which can be implicit (i.e., equally weighted) or explicit (i.e., based on expert judgment) [34]. Equal weighting methods are often preferred due to their simplicity and immediacy [35].

In the literature, the Italian territory has been analyzed with respect to the SDGs [22,31,36]. As both the SDGs [4] and the BES [37] have been classified for the Italian territory, this paper aims to address a gap in the literature by providing aggregate analyses of BES and SDG indicators (using MCDA and cluster analysis) to compare the territorial performance of Italian regions in 2022. The work not only provides a global comparison of the two sets of indicators but also develops the analysis at the macro-geographical level and for the three dimensions of sustainability (i.e., economic, environmental, social), generating a ranking for the different Italian regions.

This paper is structured as follows: Section 2 provides a review of the literature; Section 3 defines the methods used; Section 4 presents the results; Section 5 discusses the implications of the analyses; and Section 6 offers a brief conclusion.

2. Literature Analysis

The present literature analysis aimed at identifying the main topics and trends of authors and countries relative to the SDGs. Given the vast amount of literature on the SDGs, spanning both STEM (i.e., science, technology, engineering, mathematics) and non-STEM disciplines, the review adopted a high-level approach. As indicated in Section 1.2, WoS yielded a total of 31,757 papers, which was too extensive for this work and for analysis using bibliometric methods. Therefore, we employed a novel approach, conducting three separate searches on Scopus (accessed on 25 April 2024): one for publications related to economic sustainability from 2020–2024 (Section 2.1), one for publications related to social sustainability from 2020–2024 (Section 2.2) and one for publications related to environmental sustainability from 2020–2024 (Section 2.3). Finally, Section 2.4 proposes the results of a general search of papers related to the entire field of sustainability.

2.1. Economic Sustainability

We commenced with a search on Scopus for publications on economic sustainability. The search was conducted using keywords such as "SDGs", "Sustainable Development Goals", and "economic sustainability", focused on the economic pillar of sustainability. The search was limited to articles written in the English language. Details of the advanced search are provided in Appendix A (Table A5). The search yielded 3940 documents. As depicted in Figure A1, there was a steady increase in scientific production from 2020 to 2023, with a decrease in 2024 due to the availability of only partial results. In terms of the most relevant sources (Table 1), Sustainability had the highest number of publications, followed by the Journal of Cleaner Production and Environmental Science and Pollution Research. Figure A2 shows the most relevant countries in terms of production, with the top five as follows: China (656), Italy (219), India (201), Spain (187) and the USA (164).

Table 1. Three top journals: economic sustainability.

Journal	Number of Published Articles
Sustainability	539
Journal of Cleaner Production	225
Environmental Science and Pollution Research	119

2.2. Environmental Sustainability

For the advanced Scopus search of environmental sustainability publications, details can again be found in Appendix A (Table A5). The 5162 retrieved papers focused on the environmental pillar of sustainability and were limited to articles written in English. As shown in Figure A3, there was a positive trend from 2020 to 2023, similar to the economic case. Table 2 shows that the most relevant sources were Sustainability, the Journal of Cleaner Production and Environmental Science and Pollution Research. The most relevant countries in terms of production (Figure A4) were again China (799), Italy (312), India (299), the USA (273) and Spain (238).

Table 2. Three top journals: environmental sustainability.

Journal	Number of Published Articles
Sustainability	597
Journal of Cleaner Production	303
Environmental Science and Pollution Research	130

2.3. Social Sustainability

In total, 3601 publications were related to social sustainability. Again, the search was limited to articles written in English and details of the advanced Scopus search can be found in Appendix A (Table A5). As in the previous cases, annual scientific production registered a positive trend, as seen in Figure A5. This time, the most relevant sources were Sustainability and the Journal of Cleaner Production, with Sustainable Development occupying the third position (Table 3). Figure A6 shows the most relevant countries in terms of production: China (404), Spain (238), Italy (233), the USA (211) and the UK (167).

Table 3. Three top journals: social sustainability.

Journal	Number of Published Articles
Sustainability	544
Journal of Cleaner Production	175
Sustainable Development	58

2.4. Global SDG Literature Analysis

Analyzing the current literature regarding the three pillars of sustainability, we found positive trends in academic research across these dimensions. At this point, we found it natural to conduct only a global assessment. Specifically, we conducted an advanced search on Scopus (Table A5) for articles related to the "SDGs" and "Sustainable Development Goals" from 2023 to 2024, which retrieved 12,080 articles in the English language. The most relevant sources were Sustainability, the Journal of Cleaner Production and Environmental Science and Pollution Research (Table 4).

Table 4. Three top journals: environmental sustainability.

Journal	Number of Published Articles
Sustainability	787
Journal of Cleaner Production	311
Environmental Science and Pollution Research	260

Figure 1 highlights the authors of the corresponding articles, with SCP and MCP representing intra-country and inter-country collaboration, respectively. China accounted for 2418 articles, India for 887, the USA for 562, the UK for 455 and Italy for 436.

Figure 2 presents the co-occurrence network of our search results. A co-occurrence network is a graphical representation of relationships between entities based on their co-occurrence in a set of documents. In the context of a literature review, this involves identifying and analyzing the frequency with which terms (e.g., keywords, concepts, authors) appear together in documents [38]. In our network, some important links were notable, including a strong connection between "sustainable development" and "Sustainable Development Goals". Indeed, in the modern academic landscape of sustainability, the SDGs are fundamental goals for future progress. Additionally, the network displayed a triangle formed by the SDGs, sustainable development and the word "human". In this regard, we must in fact affirm that humans are the centerpiece of sustainable development. The continuous search for sustainable solutions makes us not only actors but also observers of the changes we seek. Finally, as seen throughout this literature review, China has positioned itself at the forefront of sustainable research and development.

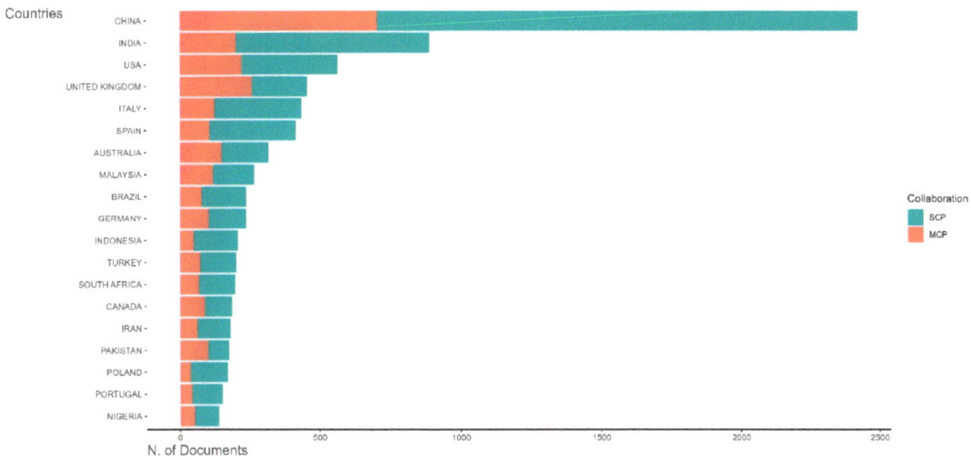

Figure 1. Top countries: literature analysis.

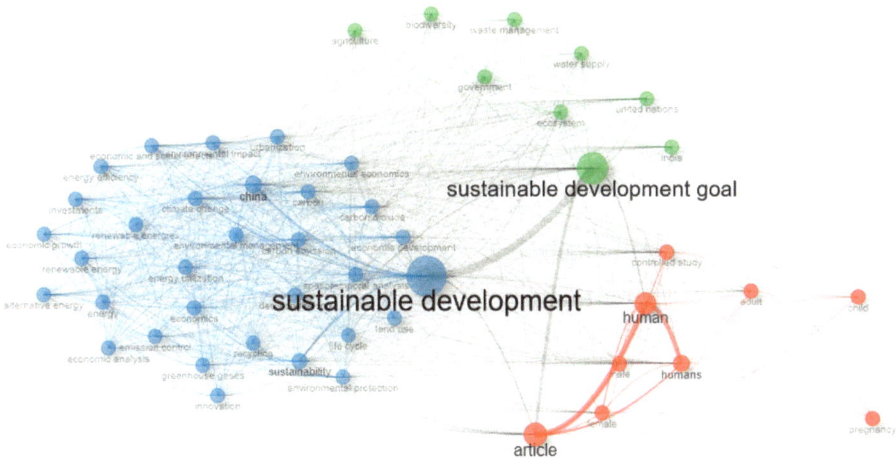

Figure 2. Co-occurrence network.

In light of these topics, we conducted an analysis to determine the top five authors (Table 5) and top five affiliations (Table 6) based on document counts in Scopus from 2023 to 2024, using an advanced search: "sustainable AND development OR sustainable AND development AND goal" (accessed on 21 May 2024). To align with the scope of the present research, we restricted the analysis to the Italian territory. The results showed that, globally, Adetunji Charles O. from Edo University Iyamho authored 69 papers over the period, followed by two authors from the University of Sharjah: Olabi Abdul Ghani and Abdelkareem Mohammad Ali. At the Italian level, the most productive authors were D'Adamo Idiano from Sapienza University of Rome, followed by Gastaldi Massimo from University of L'Aquila and Appolloni Andrea from Tor Vergata University of Rome.

Table 5. Top five authors.

International		Italy	
Author	**Number of Published Articles**	**Author**	**Number of Published Articles**
Adetunji, C.O.	69	D'Adamo I.	33
Olabi, A.G.	66	Gastaldi M.	18
Abdelkareem, M.A.	64	Appolloni A.	17
Adebayo, T.S.	58	Kraus, S.	16
Leal Filho, W. and Guo H.	54	Valeri, M.	15

Table 6. Top five affiliations.

International		Italy	
Affiliation	**Number of Published Articles**	**Affiliation**	**Number of Published Articles**
Chinese Academy of Sciences	1981	Sapienza Università di Roma	334
Ministry of Education of the People's Republic of China	1352	Politecnico di Milano	305
University of Chinese Academy of Sciences	1002	Consiglio Nazionale delle Ricerche	304
Tsinghua University	717	Alma Mater Studorium Università di Bologna	269
University of Johannesburg	681	Università degli Studi di Napoli Federico II	256

3. Materials and Methods

The present study compared Italian regional performance with respect to BES and SDG indicators using MCDA. This methodology, which is well known in the literature [39–41], is able to synthesize a large multiplicity of data, even considering data of different types and belonging to different scenarios characterized by contrasting objectives. The strength of the technique lies in its ability to create a composite indicator to rank different alternatives, thereby providing support for policymakers. In more detail, the method represents a mathematical combination of a set of elementary indicators representing the different components of a multidimensional concept to be measured. The matrix $X = \{x_{ij}\}$ of the original data is formed by n rows (regions) and m columns (indicators). Of note, the present analysis considered 21 regions, since Trentino Alto Adige is divided into two macro-areas, the provinces of Trento and Bolzano, and m columns, relating to either the j indicators of the BES (with j = 1, …, m = 105) or the SDG indicators (with j = 1, …, m = 139). Once the data matrix was constructed, it could be normalized by obtaining the matrix $Y = \left\{y_{ijt}\right\}$cont $= 2022$. For region I, the composite indicator was determined according to the following formula:

$$IC_{it} = f(y_{i1t}, y_{i2t}, y_{i3t}, \ldots, y_{imt}; w_1, w_2, w_3, \ldots, w_m)$$

with f representing a linear or non-linear aggregation function and e w_j ($j = 1, \ldots m = 105$ for BES; or $j = 1, \ldots m = 139$ for SDG) representing the weight of the single indicator j. The construction of the composite indicator was ensured through a dynamic process. First, elementary BES and SDG indicators were selected; then, these indicators were normalized and finally aggregated, determining the sustainable performance of each region. The min–max normalization method was used to create the composite indicator by determining values between 0 (worst performance) and 1 (best performance) and aggregating the results using the arithmetic mean (of note, all indicators had the same weight) [4,37]. In this way, two composite indicators were constructed: the first related to BES indicators and the second related to SDG indicators for the 21 regions in the year 2022. These indicators, having a common dimensionless range of variation (0–1), were thus fully comparable. Data for the analysis were taken from ISTAT's official website and related to all indicators for which there were available data at the regional level during the study period [42].

4. Results

The results of this paper refer to the calculation of regional performance against the SDGs (considering 139 indicators), compared to the BES (considering 105 indicators), for the year 2022. Section 4.1 reports the results of the baseline scenario and then breaks them down into functions of the three dimensions of sustainability (Section 4.2). Finally, a cluster analysis is proposed (Section 4.3).

4.1. The SDG–BES Comparison for Italian Regions

Within the 0–1 value normalization approach, a score of 1 is indicative of excellent performance. In the present analysis, no region produced a 1. Thus, the SDG and BES rankings showed different leading regions (Table 7). For the colored maps, the average value for Trentino Alto Adige was considered (0.587 for SDG and 0.732 for BES), considering the two provinces of Trento and Bolzano (Figure 3).

Table 7. SDG–BES comparison across Italian regions.

	Sustainable Development Goal			Equitable and Sustainable Well-Being	
1	Lombardia	0.626	**1**	Provincia Autonoma di Trento	0.740
2	Provincia Autonoma di Trento	0.612	**2**	Provincia Autonoma di Bolzano	0.724
3	Emilia-Romagna	0.607	**3**	Friuli-Venezia Giulia	0.609
4	Toscana	0.580	**4**	Valle d'Aosta	0.593
5	Friuli-Venezia Giulia	0.580	**5**	Emilia-Romagna	0.580
6	Veneto	0.569	**6**	Veneto	0.570
7	Provincia Autonoma di Bolzano	0.561	**7**	Lombardia	0.565
8	Valle d'Aosta	0.556	**8**	Umbria	0.565
9	Umbria	0.555	**9**	Toscana	0.564
10	Piemonte	0.538	**10**	Lazio	0.552
11	Lazio	0.538	**11**	Marche	0.541
12	Marche	0.527	**12**	Piemonte	0.535
13	Liguria	0.521		Italy	0.518
	Italy	0.503	**13**	Liguria	0.515
14	Abruzzo	0.487	**14**	Abruzzo	0.484
15	Molise	0.446	**15**	Sardegna	0.477
16	Sardegna	0.439	**16**	Molise	0.453
17	Basilicata	0.402	**17**	Basilicata	0.419
18	Puglia	0.389	**18**	Puglia	0.371
19	Campania	0.354	**19**	Calabria	0.358
20	Sicilia	0.341	**20**	Campania	0.335
21	Calabria	0.328	**21**	Sicilia	0.332

According to the SDG, Lombardia excelled with a score of 0.626 (far from the theoretical maximum of 1), followed by the province of Trento with 0.612 and Emilia-Romagna with 0.607. Thirteen regions were above the national average (0.503), with the first position being 0.123 away from the benchmark. The remaining eight were below the national average, with the lowest ranked, Calabria, only 0.175 away.

All eight regions below the national average were located in the south, led by Abruzzo with a score of 0.487. In contrast, regions in the north and center were above the benchmark. Among those in the center, Tuscany scored highest (0.580) while Marche scored lowest (0.527). This result was nonetheless better than that of the lowest-performing northern region, Liguria (0.521). Table A6 shows the percentage change in each region's score compared to each other region. The maximum variation was 91%, between the first (Lombardia) and last ranked (Calabria).

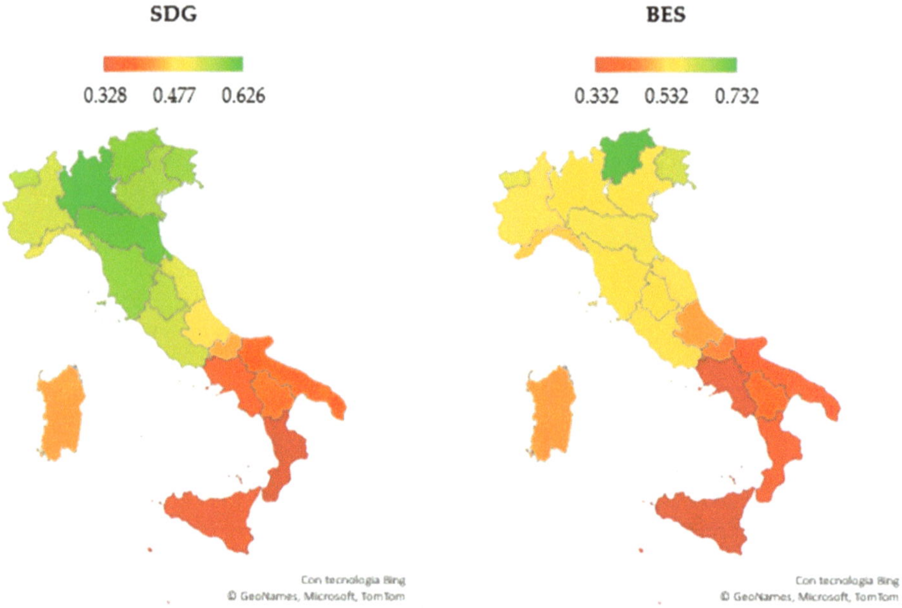

Figure 3. Map of Italy in 2022.

This difference between the first and last positions was 0.298, which was less marked than that of the BES (0.408). In alignment with previous research [37], the province of Trento (0.740) emerged in the top position, followed closely by the province of Bolzano (0.724). Much more significant was the distance with Friuli-Venezia Giulia (0.609). The following differences were also evident:

- The national average was 0.518 for the BES, slightly higher than the 0.503 for the SDGs; also for the BES, Liguria, in addition to the southern regions, fell below the national average.
- The negative changes from the BES ranking mainly concerned Lombardia and Toscana, which lost six and five positions, respectively.
- Positive changes from the BES ranking mainly concerned the provinces of Bolzano and Valle d'Aosta, which gained five and four positions, respectively.
- There was a change of two positions shown by four regions (Emilia-Romagna, Friuli-Venezia Giulia, Piemonte and Calabria), while five regions had the same position in the two rankings (Veneto, Liguria, Abruzzo, Basilicata and Puglia).

The numerical variation between the SDGs and the BES (Figure A7) showed that for nine regions there was a higher value, with Lombardy showing a positive delta of 0.061. However, while the difference for Sardegna (−0.038) was small, it was much more marked for the provinces of Trento (−0.128) and Bolzano (−0.163). This figure may be explained by the very significant BES performance of these regions, which was not, however, negative. In fact, the province of Trento ranked second in the SDG ranking.

4.2. Sustainability in Its Three Dimensions (i.e., Economic, Environmental, Social)

A useful analysis involved aggregating and disaggregating the results. The aggregation step considered the three Italian macro-areas (north, center and south) (Figure 4). In this analysis, no significant differences emerged between the two sets of indicators, although the delta between the north and center reduced when considering the SDGs due to the reduction in overall value. For instance, while Lombardia and Emilia-Romagna registered the most significant growth, the provinces of Trento and Bolzano followed the opposite direction. Thus, in the SDG ranking, the north (0.575) prevailed over the center (0.550). The difference with the south was much more pronounced, with the south recording the lowest-performing value (0.398), similar to that of the BES and on par with the central regions.

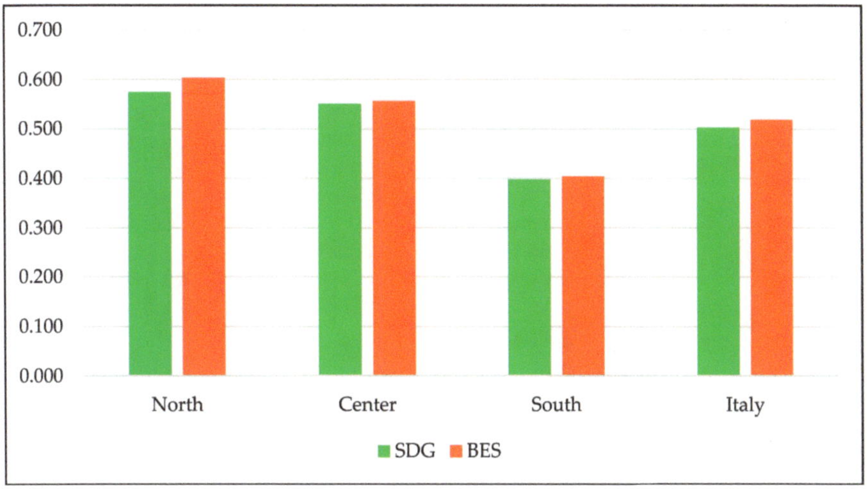

Figure 4. SDG and BES comparison at the macro-geographical level.

In the disaggregation step, the results were broken down according to the three dimensions of sustainability (Table 8). In accordance with the literature, the following classification was used [43,44]:

- Economic dimension: SDGs 7, 8, 9, 11 and 12.
- Social dimension: SDGs 1, 2, 3, 4, 5, 10, 16 and 17.
- Environmental dimension: SDGs 6, 13, 14 and 15 (SDG 14 was not included due to data unavailability).

Table 8. The three dimensions of sustainability: SDG side.

	Social Dimension			Environmental Dimension			Economic Dimension	
1	Emilia-Romagna	0.641	**1**	Provincia Autonoma di Trento	0.704	**1**	Lombardia	0.649
2	Lombardia	0.635	**2**	Provincia Aut onoma di Bolzano	0.653	**2**	Provincia Autonomadi Bolzano	0.618
3	Provincia Autonoma di Trento	0.622	**3**	Valle d'Aosta	0.645	**3**	Provincia Autonoma di Trento	0.574
4	Toscana	0.611	**4**	Toscana	0.637	**4**	Veneto	0.544
5	Umbria	0.609	**5**	Sardegna	0.627	**5**	Emilia-Romagna	0.539
6	Friuli-Venezia Giulia	0.606	**6**	Liguria	0.605	**6**	Friuli-Venezia Giulia	0.525
7	Valle d'Aosta	0.587	**7**	Basilicata	0.598	**7**	Lazio	0.523
8	Veneto	0.585	**8**	Lazio	0.584	**8**	Toscana	0.502
9	Marche	0.569	**9**	Abruzzo	0.581	**9**	Piemonte	0.490
10	Piemonte	0.563	**10**	Friuli-Venezia Giulia	0.570	**10**	Valle d'Aosta	0.472
11	Lazio	0.540	**11**	Emilia-Romagna	0.570	**11**	Liguria	0.463
12	Liguria	0.540	**12**	Molise	0.549		Italy	0.454
13	Provincia Autonoma di Bolzano	0.527		Italy	0.548	**12**	Marche	0.441
14	Abruzzo	0.523	**13**	Umbria	0.535	**13**	Umbria	0.440
	Italy	0.521	**14**	Veneto	0.519	**14**	Basilicata	0.395
15	Molise	0.478	**15**	Piemonte	0.499	**15**	Abruzzo	0.392
16	Sardegna	0.449	**16**	Marche	0.499	**16**	Sardegna	0.380
17	Puglia	0.412	**17**	Puglia	0.475	**17**	Molise	0.355
18	Basilicata	0.387	**18**	Campania	0.470	**18**	Campania	0.351
19	Sicilia	0.363	**19**	Sicilia	0.412	**19**	Puglia	0.322
20	Calabria	0.350	**20**	Lombardia	0.408	**20**	Sicilia	0.280
21	Campania	0.345	**21**	Calabria	0.364	**21**	Calabria	0.275

Lombardia confirmed its leadership in the economic dimension (0.649), took second place in the social dimension (0.635, only 0.006 lower than its first position) and performed poorly in the environmental dimension (0.408), ranking only 20th, below the national average. Emilia-Romagna led in the social dimension, while the province of Trento led in the environmental dimension. Regions in the north were above the national average in the economic and social dimensions. However, in the environmental dimension, Lombardia, Piemonte and Veneto ranked below the national average. Central regions ranked above the national average only in the social dimension, and Marche and Umbria ranked below average in both the environmental and economic dimensions. Southern regions showed slightly different results: all were below the national average in the economic dimension, while, in the social dimension, Abruzzo was just above, with a score of 0.523 (compared to the benchmark of 0.521). In the environmental dimension, Molise held a higher position with 0.549 (compared to the benchmark of 0.548), and Abruzzo rose to ninth place, exceeding the national average by 0.033.

To compare the three dimensions of sustainability related to the SDGs, we integrated BES indicators into the same dimensions (Table 9) as follows:

- Economic dimension: (i) economic well-being and (ii) innovation, research and creativity.
- Social dimension: (i) health, (ii) education and training, (iii) work and life time balance, (iv) social relationships, (v) policy and institutions, (vi) safety, (vii) subjective well-being, and (viii) quality of services.
- Environmental dimension: (i) landscape and cultural heritage and (ii) environment.

Table 9. The three dimensions of sustainability: BES side.

	Social Dimension			Environmental Dimension			Economic Dimension	
1	Provincia Autonoma di Trento	0.776	**1**	Provincia Autonoma di Bolzano	0.674	**1**	Provincia Autonoma di Trento	0.720
2	Provincia Autonoma di Bolzano	0.742	**2**	Provincia Autonoma di Trento	0.592	**2**	Provincia Autonoma di Bolzano	0.715
3	Friuli-Venezia Giulia	0.620	**3**	Friuli-Venezia Giulia	0.581	**3**	Lazio	0.676
4	Valle d'Aosta	0.604	**4**	Umbria	0.563	**4**	Lombardia	0.668
5	Emilia-Romagna	0.596	**5**	Toscana	0.557	**5**	Friuli-Venezia Giulia	0.654
6	Lombardia	0.576	**6**	Valle d'Aosta	0.530	**6**	Emilia-Romagna	0.651
7	Veneto	0.573	**7**	Piemonte	0.527	**7**	Umbria	0.639
8	Toscana	0.567	**8**	Marche	0.524	**8**	Veneto	0.639
9	Lazio	0.559	**9**	Molise	0.510	**9**	Marche	0.605
10	Umbria	0.557		Italy	0.495	**10**	Toscana	0.603
11	Marche	0.545	**10**	Abruzzo	0.490	**11**	Valle d'Aosta	0.583
12	Liguria	0.545	**11**	Veneto	0.487		Italy	0.550
13	Piemonte	0.543	**12**	Lombardia	0.484	**12**	Liguria	0.517
	Italy	0.523	**13**	Basilicata	0.480	**13**	Piemonte	0.514
14	Abruzzo	0.503	**14**	Sardegna	0.477	**14**	Basilicata	0.512
15	Sardegna	0.479	**15**	Emilia-Romagna	0.466	**15**	Sardegna	0.467
16	Molise	0.453	**16**	Liguria	0.451	**16**	Puglia	0.423
17	Basilicata	0.387	**17**	Lazio	0.448	**17**	Molise	0.421
18	Puglia	0.363	**18**	Puglia	0.406	**18**	Calabria	0.419
19	Calabria	0.353	**19**	Calabria	0.397	**19**	Abruzzo	0.419
20	Campania	0.332	**20**	Sicilia	0.378	**20**	Sicilia	0.365
21	Sicilia	0.322	**21**	Campania	0.368	**21**	Campania	0.333

The province of Trento confirmed its leadership in the economic and social dimensions, with scores of 0.720 and 0.776, respectively. It was followed in both cases by the province of Bolzano, with close scores of 0.715 and 0.742, respectively. For the environmental dimension, first place was obtained by the province of Bolzano (0.674), followed by the province of Trento, whose score was only 0.082 lower. Of note, in the social dimension, northern regions consistently scored above the national average. However, this was not the case in the other two dimensions. In the environmental dimension, Veneto, Lombardia, Emilia-Romagna and Liguria fell below the national average, as did Liguria and Piemonte in the economic dimension. Among the central regions, all were above the national average in the social and economic dimensions. Particularly in the economic dimension, Lazio took third place with a score of 0.676. However, in the environmental dimension, it was the only region below the national average, holding the 17th position with a score of 0.448 (0.047 away from the benchmark). Southern regions occupied the final positions in the ranking across all dimensions. In the social dimension, they were below the national average, led by Abruzzo, which ranked just below the average with a score of 0.503 (0.020 from the benchmark). In the economic dimension, Basilicata scored highest among these regions with a score of 0.512, which was still below the national average by 0.038. In the environmental dimension, Molise ranked just above the national average with a score of 0.510, exceeding the benchmark by 0.015. Abruzzo, similar to its performance in the social dimension, ranked just below the national average with a score of 0.490 (0.005 below the benchmark). Finally, in both types of indicators, the social dimension had the fewest regions below the national average. Conversely, the environmental dimension for the BES and the economic dimension for the SDGs had the least number of regions above the national average.

4.3. Clustering Sustainability Pillars

In this subsection, we present our comparative analysis of the SDG and BES indices, focusing on the social, environmental and economic dimensions. Our aim was to examine the relationship between these indices using scatterplots and Spearman correlations to gain insight into their interplay across Italian regions and their differences in capturing the multifaceted aspects of sustainability and well-being (Figure A8). In general, the Spearman correlation coefficient of 0.92 indicated a significant correlation between the two classifications.

4.3.1. Social Assessment

Figure 5 assesses the social aspect of sustainable development in the form of a scatterplot, with the SDG index on the x-axis and the BES index on the y-axis. Red lines indicate the respective medians of 0.540 for the SDG and 0.545 for the BES. The Spearman correlation coefficient of 0.84 indicated a strong linear correlation between the two indices. Here, we refer to the upper right quadrant, defined by high scores on both the SDG and the BES indices, as the UP quadrant. Notably, eight Italian territories fell into this quadrant: the province of Trento, Friuli-Venezia Giulia, Valle d'Aosta, Emilia-Romagna, Veneto, Toscana, Lombardia and Umbria. Marche did not belong in this quadrant as its score (0.545) fell exactly on the BES median, similar to Lazio and Liguria, with scores of 0.540.

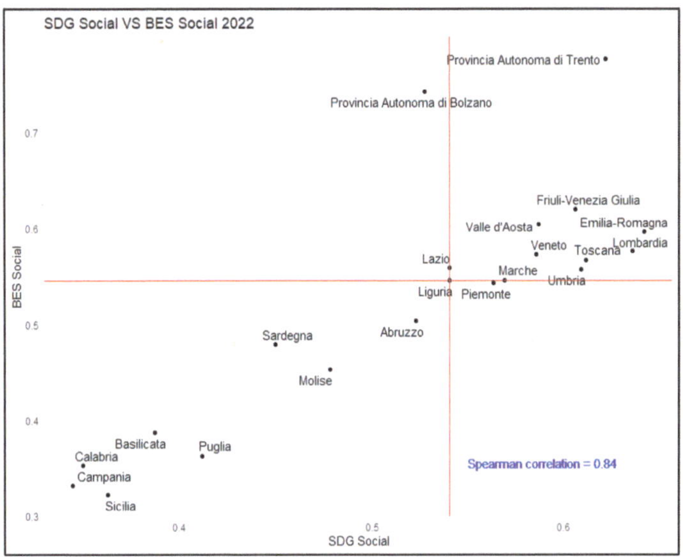

Figure 5. SDG and BES: social dimension.

4.3.2. Environmental Assessment

Figure 6 examines the environmental aspect of sustainable development in the form of a scatterplot. The median for the SDG index was 0.559, while that of the BES index was 0.488. The Spearman correlation coefficient of 0.56 was moderate, suggesting that the two indices weighed environmental aspects differently. In particular, the regions of Sardegna, Liguria, Basilicata, Lazio and Emilia-Romagna (which, despite appearing to be on the median, actually scored 0.570) scored highly on the SDG index for their environmental performance. However, they fell below the median for the BES index. This discrepancy suggests that, while these regions may excel in certain environmental indicators highlighted by the SDG framework, they may not perform as well when considering the broader range of environmental factors included in the BES index. Conversely, the four regions of Umbria, Piemonte, Marche and Molise (but not Veneto, which scored 0.48) scored highly on the BES index but fell below the SDG median. Six regions in the UP quadrant registered high environmental scores on both indices: the province of Bolzano, the province of Trento, Toscana, Valle d'Aosta, Abruzzo and Friuli-Venezia Giulia. This highlights the remarkable environmental performance of these regions, according to both indices.

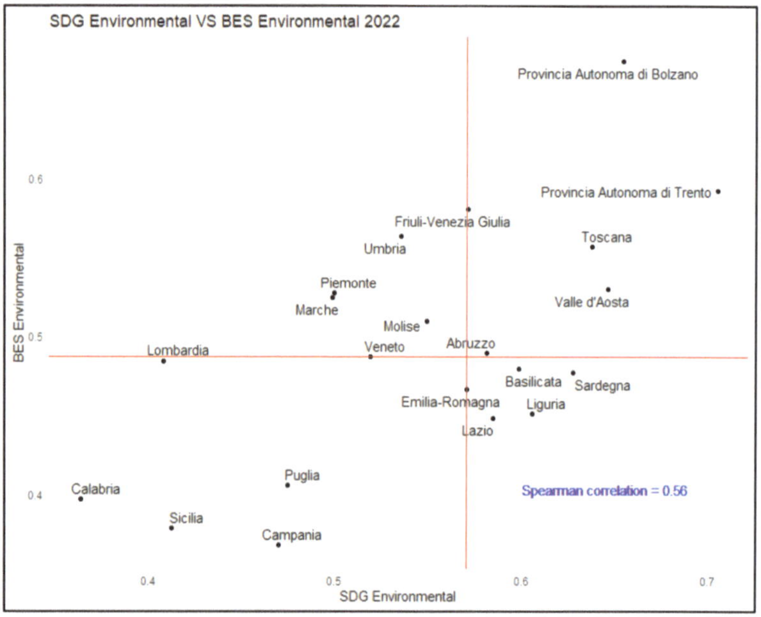

Figure 6. SDG and BES: environmental dimension.

4.3.3. Economic Assessment

Figure 7 presents a comparative analysis of the economic aspect of the SDG and BES indices. The median for the SDG economic index was 0.458, while that for the BES index was 0.566. The Spearman correlation coefficient of 0.9 indicated a strong positive correlation between the two indices, representing the highest among the three dimensions. This suggested significant convergence in the assessment of economic aspects, with both indices broadly agreeing on the economic performance of the respective regions. Nine regions fell in the UP quadrant, demonstrating high economic scores on both indices: the province of Trento, the province of Bolzano, Lombardia, Lazio, Friuli-Venezia Giulia, Emilia-Romagna, Veneto, Toscana and Valle d'Aosta. Notably, Valle d'Aosta, while aligned with the BES median, actually exceeded the median score with a value of 0.583. This convergence highlights the consistency in the assessment of economic performance across regions, providing valuable insight into areas of economic strength and potential strategies for enhancing economic sustainability.

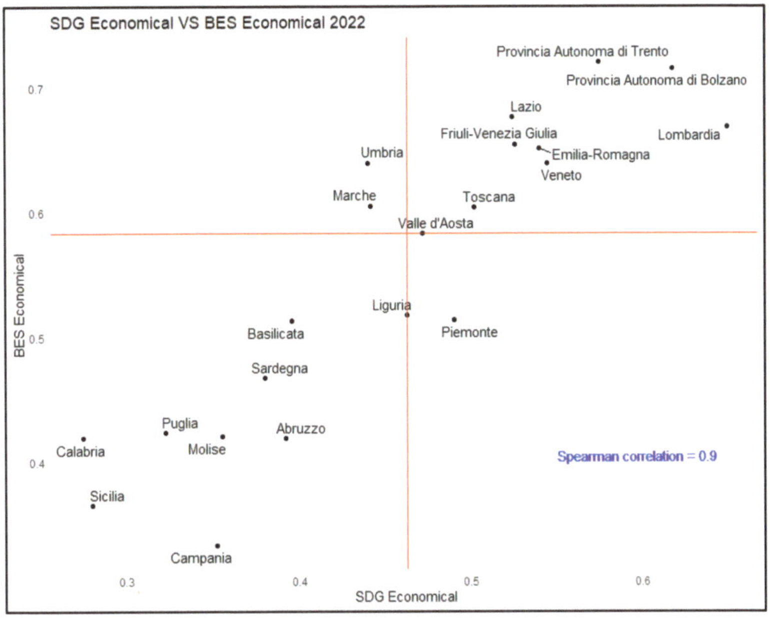

Figure 7. SDG and BES: economic dimension.

4.3.4. Mapping Italian Regional Performance

Finally, we summarized the results of our previous analysis by mapping Italian regions according to their relative scores in the scatterplots. In more detail, we assigned a score to each region based on its presence in the UP quadrant: 1 if present once, 2 if present twice and 3 if present three times. Notably, the provinces of Bolzano and Trento were treated as a single entity, Trentino-Alto Adige.

Figure 8 shows the results of this clustering analysis:

- Four regions scored 3: Trentino-Alto Adige, Toscana, Friuli-Venezia Giulia and Valle d'Aosta.
- Three regions scored 2: Emilia-Romagna, Lombardia and Veneto.
- Three regions scored 1: Umbria, Lazio and Abruzzo.

The analysis also revealed that three regions narrowly missed being ranked with a score of 1: Marche, Liguria and Piemonte, due to their slightly below-average scores in the social dimension. The findings also revealed interesting patterns in regional sustainability performance across Italy. In particular, three of the four regions scoring 3 were located in the north, with Trentino-Alto Adige and Friuli-Venezia Giulia all falling in the northeast. All regions scoring 2 were also in northern Italy. In central Italy, only Toscana scored 3 and two regions scored 2. Conversely, in southern Italy, only Abruzzo scored 1, while all other southern regions scored 0. Overall, northeastern regions emerged as the top performers, with all scoring 3 or 2. Southern Italy, however, showed lower performance across the board. These observations underline regional disparities in sustainability performance, highlighting potential areas for targeted interventions and policy initiatives to promote more equitable and balanced development across Italy.

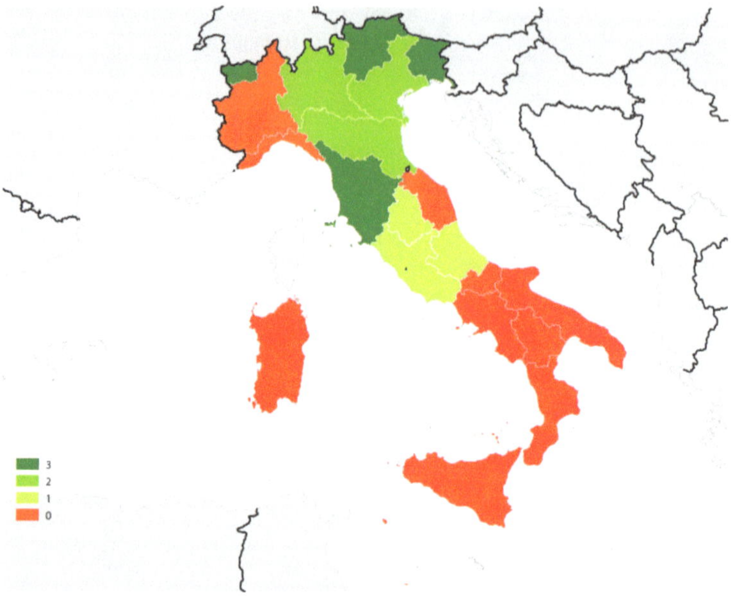

Figure 8. Clustering sustainability pillars in Italian regions.

5. Discussion

While the topic of sustainability has gained significant traction in recent years, some stakeholders remain focused solely on their own benefits. Consequently, phenomena such as green economy rebound, circular economy rebound and greenwashing have emerged [45–48], necessitating appropriate management to prevent a loss of public confidence. Public involvement is crucial for sustainability efforts, as highlighted by the prominence of the word "human" in the co-occurrence network (Figure 2). Some authors have called for new regulatory approaches and business models, emphasizing that "progress is too slow" and societal value creation remains underutilized [49].

The SDGs have gained fundamental relevance within the scientific community and civil society. Previously considered niche, the urgent need to address climate change has elevated these goals to a central focus. In particular, this urgency has underscored the need to develop innovative ideas and concepts to support SDG achievement, prompting the development of a new section within sustainability [50]. Thus, a vision of a sustainable community requires interdisciplinary contributions from various perspectives [51–53].

Local and global analyses often have different scopes of analysis, stakeholder categories may have different interests, and indicators sometimes provide competing information. This paper built on a review of the literature [29–31], showing that the BES and SDG indices are complementary while providing distinct insights. The results obtained from the analysis must now be integrated with the existing literature [37].

The first methodological contribution of this research was the creation of scatterplots ranking alternatives based on the two sets of indicators across the three dimensions of sustainability. The interplay between economic well-being and the SDGs is not only an Italian priority but also a European one [54].

From this analysis, a second consideration emerged, this time of a managerial nature. The cluster analysis made it possible to redefine the geographical structure of Italy, showing that it cannot be divided into north, center and south according to SDG and BES indicators. In fact, the data revealed that the northeast significantly outperformed the northwest (0.586 vs. 0.560 at the SDG level and 0.645 vs. 0.552 at the BES level). Additionally, values in the center were close to those registered in the northwest. Interestingly, among the southern regions, Abruzzo performed similarly to the central regions, reducing the overall value by only 0.012 in terms of the SDGs and 0.015 with respect to the BES. This suggests that central regions have great potential, demonstrated by Abruzzo's strong performance in the environmental dimension, Lazio's in the economic dimension and Umbria's in the social dimension, combined with Toscana's strong performance across all three dimensions. Although Marche narrowly missed the mark in the social dimension, it could still make a fundamental contribution. This shows that the realization of a sustainable community in these regions, facilitated by the exchange of services and products, may generate a competitive macro-area. This requires further data monitoring, which is already showing growth in these regions and the northeast's superior performance [4].

The "Made in Italy" brand aims to integrate regional disparities, rather than highlight them, to produce unique brands that are globally competitive [37]. However, this integration cannot overlook existing disparities, particularly in southern Italy. The south's potential, while significant, has yet to be fully and efficiently harnessed. Promoting the south will be essential for achieving balanced regional development in Italy and upholding the Italian pillars of sustainability. In addition, promoting technological innovation in the south may contribute to what many consider the fourth pillar of sustainability. By leveraging each region's unique strengths and fostering nationwide collaboration, Italy may advance towards holistic sustainability and enhance its global competitiveness under the "Made in Italy" banner. The involvement of new generations, alongside the experience of older generations, will be crucial for building sustainable community models based on skills and resources [47].

Finally, we must highlight a third implication, which is political. The use of European funds should not focus solely on individual territories but incorporate a future vision incorporating points of interconnection and uniting the Adriatic with the Tyrrhenian to achieve significant logistical advantages. In central Italy, cohesion between national and local governments may provide political stability and a comprehensive perspective, thereby supporting the interception of European funds and promoting green, circular and digital projects. Important initiatives include those of Abruzzo, Marche and Umbria, along with their respective "confindustries" universities and industrial development companies. These entities have collaborated in the Hamu (Hub Abruzzo Marche Umbria) project, experimenting with ecosystem building and value generation in central Italy. These territories should foster the degree of attractiveness to their own talents and those from other countries. A further collaborative effort involves the financial institutions of Lazio, Abruzzo, Marche and Umbria, which have signed a partnership agreement on European Funds in Rome.

Central Italy currently represents a model of sustainable innovation that should aspire to emulate the performance of northeastern regions. A divided and fragmented Italy hindered by ideological visions has no future. We must therefore pursue a pragmatic vision that recognizes the great challenge of sustainability: overcoming personal selfishness to protect ecosystems and achieve the triple goal of economic performance, environmental protection and social progress. Indicators allow decision-makers and the public to monitor the performance of individual territories towards this goal. While the SDG and BES indicators share some criteria, their rankings reveal critical differences, indicating that their outputs are complementary, rather than redundant. This highlights the importance of developing new tools to integrate these rankings.

Limitations of the present work include the time period of reference, which could be extended in future research. In this regard, close monitoring of the relevant data will be necessary to assess regional performance in light of interregional policies. In addition, it may be useful to study the relationships between these data and those related to the implementation of sustainability goals by universities in their respective territories. Such research may also open up a social perspective, exploring how these indicators might influence young people's choice of university and base for skills training. Further analysis could evaluate the present results concerning culture and income readiness, providing assessments at the macro-geographical level.

The SDG–BES pairing promotes ethical sustainability, engaging individuals' religious and philosophical beliefs to facilitate an ecological consciousness that may restore the human–nature relationship.

6. Conclusions

The great challenge of sustainability is to overcome personal selfishness, as this is crucial for protecting ecosystems and achieving the threefold goal of economic performance, environmental protection and social progress. Indicators allow decision-makers and the public to monitor the performance of individual territories towards this goal. Although the SDG and BES indicators share some criteria, their rankings also highlight some differences. Consequently, their outputs are complementary, rather than redundant, emphasizing the need for tools capable of integrating the different rankings.

In the present study, a cluster analysis was conducted to differentiate the various territorial realities. The regions of Trentino-Alto Adige, Toscana, Friuli-Venezia Giulia and Valle d'Aosta scored highly across all three dimensions of sustainability. Emilia-Romagna, Lombardia and Veneto performed positively in two of the three dimensions and Umbria, Lazio and Abruzzo achieved similar results in one dimension. Regions outside these clusters have gaps that require strengthening. However, the present work did not aim to highlight territorial differences but attempted to suggest actions to enhance sustainability contributions from all regions.

In this direction, reintroducing the "Made in Italy" concept may foster the development of an innovative, sustainable model based on territorial cooperation and related synergies, thereby maximizing the use of resources and skills to enhance global competition.

The present findings showed strong performances by regions in the northeast and center of Italy. To overcome the north–south divide, some southern regions must improve their performance, and the present analysis indicated that this is starting to happen. There are many challenges ahead, but with a stable political climate and proactive decision-making, the dream of a more sustainable country may become a reality. Present issues must be addressed with foresight to ensure that benefits are generated for a wide range of stakeholder categories.

Author Contributions: Conceptualization, I.D., C.D.C., M.G., E.N.R. and A.F.U.; methodology, I.D., C.D.C., M.G., E.N.R. and A.F.U.; data curation, I.D., C.D.C., M.G., E.N.R. and A.F.U.; writing—original draft preparation, I.D., C.D.C., M.G., E.N.R. and A.F.U.; writing—review and editing, I.D., C.D.C., M.G., E.N.R. and A.F.U.; supervision, I.D. All authors have read and agreed to the published version of the manuscript.

Funding: This research received no external funding.

Institutional Review Board Statement: Not applicable.

Informed Consent Statement: Not applicable.

Data Availability Statement: The original contributions presented in the study are included in the article, further inquiries can be directed to the corresponding author.

Acknowledgments: This study was carried out within the PEACE (Protecting the Environment: Advances in Circular Economy) study, which received funding from the "Fondo per il Programma Nazionale di Ricerca e Progetti di Rilevante Interesse Nazionale (PRIN)" Investimento M4.C2.1.1-D.D. 104.02-02-2022, 2022ZFBMA4 funded by the European Union-Next Generation EU. This manuscript reflects only the authors' views and opinions, and the authors can be considered solely responsible for them.

Conflicts of Interest: The authors declare no conflicts of interest.

Appendix A

Table A1. Trend of published articles per Web of Science categories (top 10 categories).

Category	Number of Published Articles
Environmental sciences	9554
Green sustainable science technology	7153
Environmental studies	6023
Energy fuels	1892
Public environmental occupational health	1684
Economics	1519
Engineering environmental	1483
Management	1305
Business	1290
Education educational research	1180

Table A2. Trend of published papers per country (top 10 countries).

Country	Number of Published Articles
China	5706
USA	4085
UK	3256
India	2737
Spain	2248
Australia	2081
Italy	1823
Germany	1703
South Africa	1350
Canada	1332

Table A3. Trend of published papers per Web of Science index.

WoS Index	Number of Published Articles
Science Citation Index Expanded (SCI-EXPANDED)	18,246
Social Sciences Citation Index (SSCI)	12,516
Emerging Sources Citation Index (ESCI)	6877
Conference Proceedings Citation Index-Science (CPCI-S)	1033
Book Citation Index—Social Sciences & Humanities (BKCI-SSH)	465

Table A4. Trend of published papers per SDG.

SDG	Description	Number of Published Articles
13	Climate action	5927
3	Good health and well-being	4495
11	Sustainable cities and communities	3632
15	Life on land	2685
12	Responsible consumption and production	2295
1	No poverty	2125
6	Clean water and sanitation	1799
2	Zero hunger	1789
9	Industry, innovation and infrastructure	1671
7	Affordable and clean energy	1591
4	Quality education	1278
10	Reduced inequality	723

Table A4. *Cont.*

SDG	Description	Number of Published Articles
8	Decent work and economic growth	709
14	Life below water	690
5	Gender equality	519
16	Peace and justice strong institutions	216

Table A5. Advanced search.

Section	Query String
2.1 (Economic)	TITLE-ABS-KEY ((sustainable AND development AND goals OR sdgs) AND economic AND sustainability) AND PUBYEAR > 2019 AND PUBYEAR < 2025 AND (LIMIT-TO (DOCTYPE , "ar")) AND (LIMIT-TO (LANGUAGE , "English")
2.2 (Environmental)	TITLE-ABS-KEY ((sustainable AND development AND goals OR sdgs) AND environmental AND sustainability) AND PUBYEAR > 2019 AND PUBYEAR < 2025 AND (LIMIT-TO (DOCTYPE , "ar")) AND (LIMIT-TO (LANGUAGE , "English"))
2.3 (Social)	TITLE-ABS-KEY ((sustainable AND development AND goals OR sdgs) AND social AND sustainability) AND PUBYEAR > 2019 AND PUBYEAR < 2025 AND (LIMIT-TO (DOCTYPE , "ar")) AND (LIMIT-TO (LANGUAGE , "English"))
2.4 (Global)	TITLE-ABS-KEY (sdgs OR sustainable AND development AND goals) AND PUBYEAR > 2022 AND PUBYEAR < 2025 AND (LIMIT-TO (LANGUAGE , "English")) AND (LIMIT-TO (DOCTYPE , "ar"))

Table A6. Percentage variation between each region's SDG score compared with every other region. The following acronyms are used: R1: Lombardia; R2: Provincia Autonoma di Trento; R3: Emilia-Romagna; R4: Toscana; R5: Friuli-Venezia Giulia; R6: Veneto; R7: Provincia Autonoma di Bolzano; R8: Valle d'Aosta; R9: Umbria; R10: Piemonte; R11: Lazio; R12: Marche; R13: Liguria; R14: Abruzzo; R15: Molise; R16: Sardegna; R17: Basilicata; R18: Puglia; R19: Campania; R20: Sicilia; R21: Calabria.

	R1	R2	R3	R4	R5	R6	R7	R8	R9	R10	R11	R12	R13	R14	R15	R16	R17	R18	R19	R20	R21
R1																					
R2	2%																				
R3	3%	1%																			
R4	8%	5%	5%																		
R5	8%	6%	5%	0%																	
R6	10%	8%	7%	2%	2%																
R7	12%	9%	8%	3%	3%	1%															
R8	13%	10%	9%	4%	4%	2%	1%														
R9	13%	10%	9%	5%	5%	3%	1%	0%													
R10	16%	14%	13%	8%	8%	6%	4%	3%	3%												
R11	16%	14%	13%	8%	8%	6%	4%	3%	3%	0%											
R12	19%	16%	15%	10%	10%	8%	6%	5%	5%	2%	2%										
R13	25%	22%	21%	15%	15%	13%	12%	11%	10%	7%	7%	5%									
R14	29%	26%	25%	19%	19%	17%	15%	14%	14%	10%	10%	8%	3%								
R15	41%	37%	36%	30%	30%	28%	26%	25%	24%	21%	21%	18%	13%	9%							
R16	43%	39%	38%	32%	32%	30%	28%	27%	26%	23%	23%	20%	15%	11%	2%						
R17	56%	52%	51%	45%	44%	42%	406	39%	38%	34%	34%	31%	25%	21%	11%	9%					
R18	61%	58%	56%	49%	49%	46%	44%	43%	43%	38%	38%	36%	29%	25%	15%	13%	3%				

Table A6. *Cont.*

	R1	R2	R3	R4	R5	R6	R7	R8	R9	R10	R11	R12	R13	R14	R15	R16	R17	R18	R19	R20	R21
R19	77%	73%	72%	64%	64%	61%	59%	57%	57%	52%	52%	49%	42%	38%	26%	24%	14%	10%			
R20	84%	80%	78%	70%	70%	67%	65%	63%	63%	58%	58%	55%	47%	43%	31%	29%	18%	14%	4%		
R21	91%	86%	85%	77%	77%	73%	71%	69%	69%	64%	64%	61%	53%	48%	36%	34%	22%	18%	8%	4%	

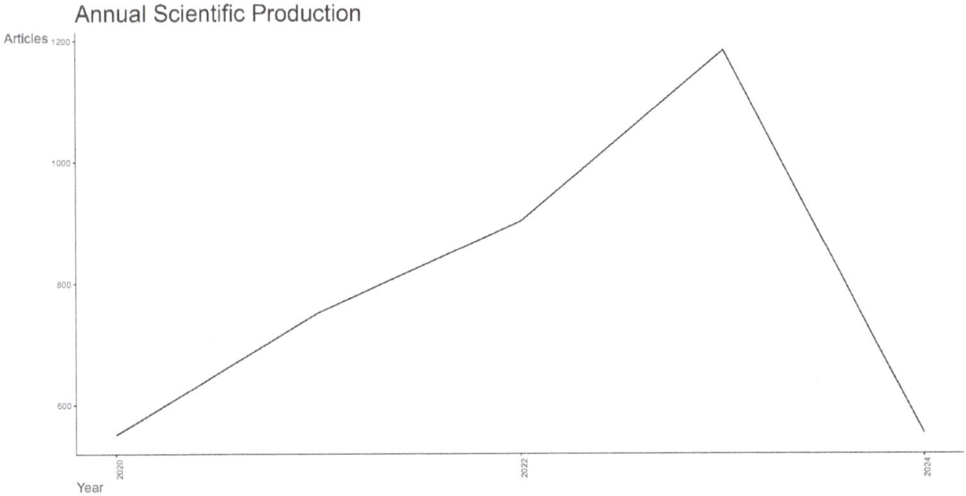

Figure A1. Annual scientific production: economic sustainability.

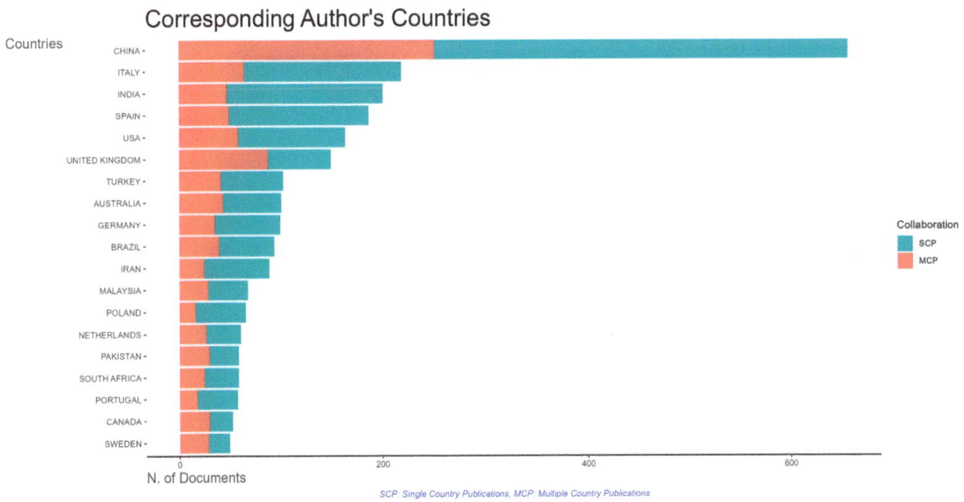

Figure A2. Corresponding authors' countries: economic sustainability.

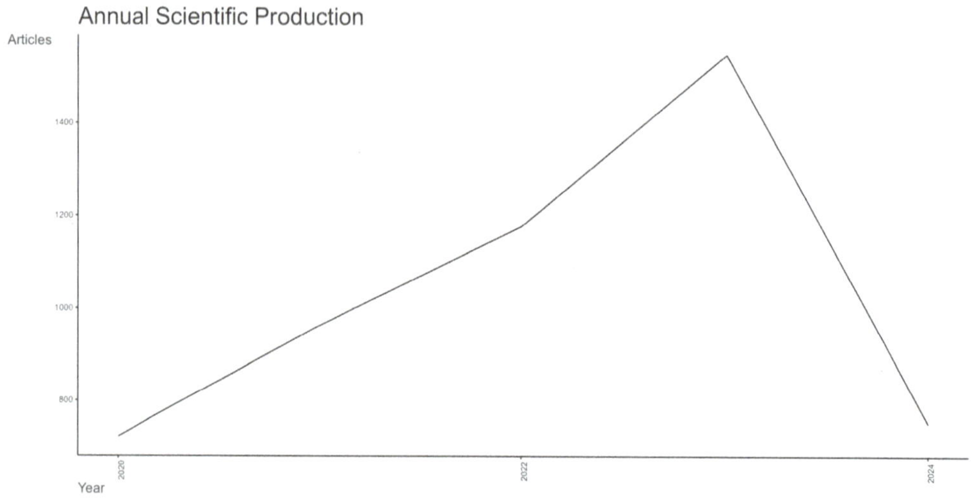

Figure A3. Annual scientific production: environmental sustainability.

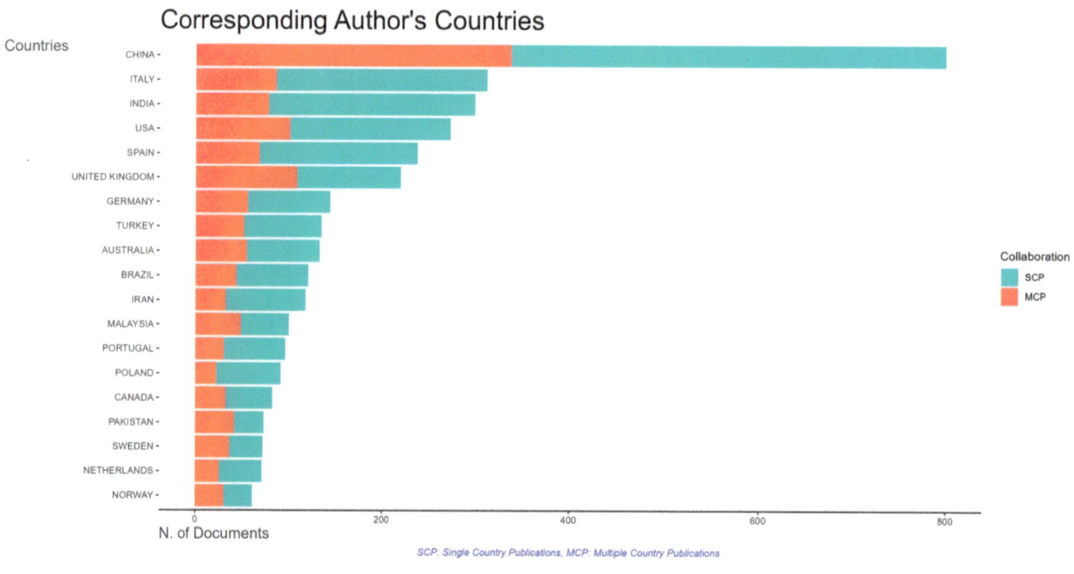

Figure A4. Corresponding authors' countries: environmental sustainability.

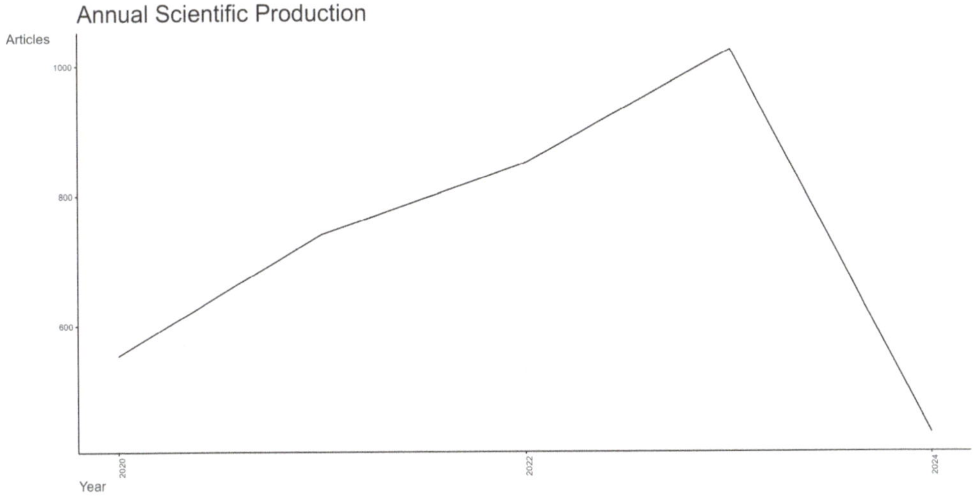

Figure A5. Annual scientific production: social sustainability.

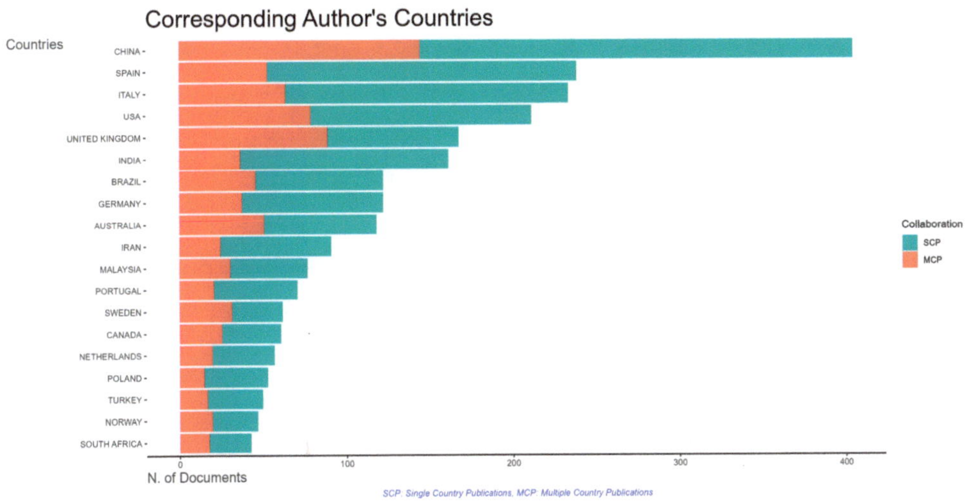

Figure A6. Corresponding authors' countries: social sustainability.

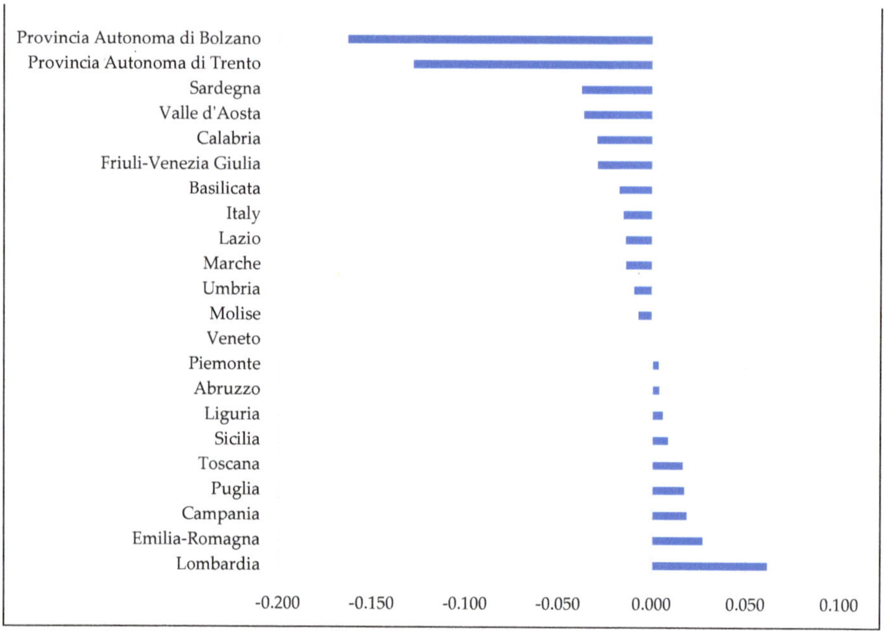

Figure A7. Numerical variation between SDG and BES in 2022.

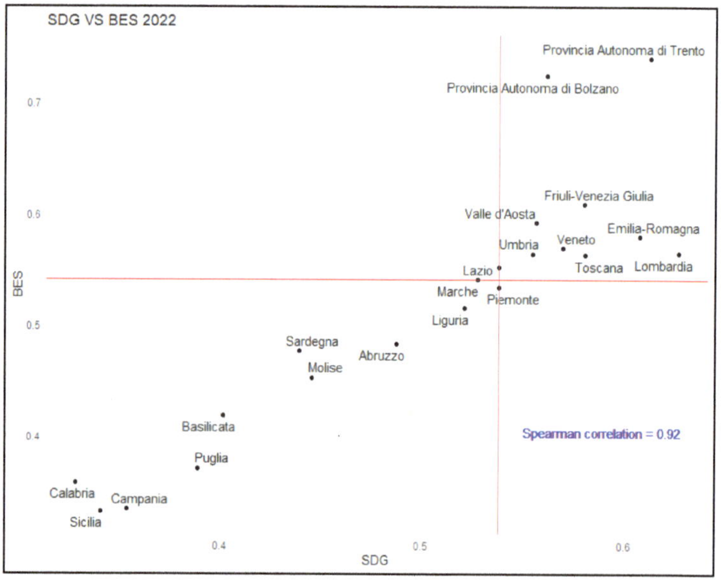

Figure A8. SDG and BES: overall dimension.

References

1. Yamaguchi, N.U.; Bernardino, E.G.; Ferreira, M.E.C.; de Lima, B.P.; Pascotini, M.R.; Yamaguchi, M.U. Sustainable development goals: A bibliometric analysis of literature reviews. *Environ. Sci. Pollut. Res.* **2023**, *30*, 5502–5515. [CrossRef] [PubMed]
2. Giannetti, B.F.; Agostinho, F.; Almeida, C.M.V.B.; Liu, G.; Contreras, L.E.V.; Vandecasteele, C.; Coscieme, L.; Sutton, P.; Poveda, C. Insights on the United Nations Sustainable Development Goals scope: Are they aligned with a 'strong'sustainable development? *J. Clean. Prod.* **2020**, *252*, 119574. [CrossRef]

3. World Commission on Environment and Development. *Our Common Future*; Oxford University Press: Oxford, UK, 1987; ISBN 019282080X.
4. D'Adamo, I.; Gastaldi, M. Monitoring Performance of Sustainable Development Goals in the Italian Regions. *Sustainability* **2023**, *15*, 14094. [CrossRef]
5. Mensah, K.; Wieck, C.; Rudloff, B. Sustainable food consumption and Sustainable Development Goal 12: Conceptual challenges for monitoring and implementation. *Sustain. Dev.* **2024**, *32*, 1109–1119. [CrossRef]
6. Olabi, A.G.; Shehata, N.; Obaideen, K.; Sayed, E.T.; Mahmoud, M.; AlMallahi, M.N.; Abdelkareem, M.A. COVID-19: Medical Waste Management, Impact on Sustainable Development Goals, and Bibliometric Analysis. *Chem. Eng. Technol.* **2024**, *47*, 4–19. [CrossRef]
7. Serafini, P.G.; de Moura, J.M.; de Almeida, M.R.; de Rezende, J.F.D. Sustainable Development Goals in Higher Education Institutions: A systematic literature review. *J. Clean. Prod.* **2022**, *370*, 133473. [CrossRef]
8. Buhalis, D.; Leung, X.Y.; Fan, D.; Darcy, S.; Chen, G.; Xu, F.; Wei-Han Tan, G.; Nunkoo, R.; Farmaki, A. Editorial: Tourism 2030 and the contribution to the sustainable development goals: The tourism review viewpoint. *Tour. Rev.* **2023**, *78*, 293–313. [CrossRef]
9. Parmentola, A.; Petrillo, A.; Tutore, I.; De Felice, F. Is blockchain able to enhance environmental sustainability? A systematic review and research agenda from the perspective of Sustainable Development Goals (SDGs). *Bus. Strateg. Environ.* **2022**, *31*, 194–217. [CrossRef]
10. Vann Yaroson, E.; Chowdhury, S.; Mangla, S.K.; Dey, P.; Chan, F.T.S.; Roux, M. A systematic literature review exploring and linking circular economy and sustainable development goals in the past three decades (1991–2022). *Int. J. Prod. Res.* **2024**, *62*, 1399–1433. [CrossRef]
11. Sharifi, A.; Allam, Z.; Bibri, S.E.; Khavarian-Garmsir, A.R. Smart cities and sustainable development goals (SDGs): A systematic literature review of co-benefits and trade-offs. *Cities* **2024**, *146*, 104659. [CrossRef]
12. Pelikánová, R.M.; Němečková, T.; MacGregor, R.K. CSR Statements in International and Czech Luxury Fashion Industry at the Onset and during the COVID-19 Pandemic—Slowing Down the Fast Fashion Business? *Sustainability* **2021**, *13*, 3715. [CrossRef]
13. Wood, D.; Rathnasabapathy, M.; Stober, K.J.; Menon, P. Challenges and progress in applying space technology in support of the sustainable development goals. *Acta Astronaut.* **2024**, *219*, 678–692. [CrossRef]
14. Mishra, M.; Desul, S.; Santos, C.A.G.; Mishra, S.K.; Kamal, A.H.M.; Goswami, S.; Kalumba, A.M.; Biswal, R.; da Silva, R.M.; dos Santos, C.A.C.; et al. A bibliometric analysis of sustainable development goals (SDGs): A review of progress, challenges, and opportunities. *Environ. Dev. Sustain.* **2023**, *26*, 11101–11143. [CrossRef] [PubMed]
15. Ordonez-Ponce, E. Exploring the Impact of the Sustainable Development Goals on Sustainability Trends. *Sustainability* **2023**, *15*, 16647. [CrossRef]
16. Akande, A.; Cabral, P.; Gomes, P.; Casteleyn, S. The Lisbon ranking for smart sustainable cities in Europe. *Sustain. Cities Soc.* **2019**, *44*, 475–487. [CrossRef]
17. Hametner, M.; Kostetckaia, M. Frontrunners and laggards: How fast are the EU member states progressing towards the sustainable development goals? *Ecol. Econ.* **2020**, *177*, 106775. [CrossRef]
18. Kostetckaia, M.; Hametner, M. How Sustainable Development Goals interlinkages influence European Union countries' progress towards the 2030 Agenda. *Sustain. Dev.* **2022**, *30*, 916–926. [CrossRef]
19. Hickmann, T.; Biermann, F.; Spinazzola, M.; Ballard, C.; Bogers, M.; Forestier, O.; Kalfagianni, A.; Kim, R.E.; Montesano, F.S.; Peek, T.; et al. Success factors of global goal-setting for sustainable development: Learning from the Millennium Development Goals. *Sustain. Dev.* **2023**, *31*, 1214–1225. [CrossRef]
20. Henderson, K.; Loreau, M. A model of Sustainable Development Goals: Challenges and opportunities in promoting human well-being and environmental sustainability. *Ecol. Model.* **2023**, *475*, 110164. [CrossRef]
21. Perevoznic, F.M.; Dragomir, V.D. Achieving the 2030 Agenda: Mapping the Landscape of Corporate Sustainability Goals and Policies in the European Union. *Sustainability* **2024**, *16*, 2971. [CrossRef]
22. D'Adamo, I.; Gastaldi, M.; Ioppolo, G.; Morone, P. An analysis of Sustainable Development Goals in Italian cities: Performance measurements and policy implications. *Land Use policy* **2022**, *120*, 106278. [CrossRef]
23. Siddiqui, A.; Altekar, S.; Kautish, P.; Fulzele, S.; Kulkarni, N.; Siddiqui, M.; Bashir, M.F. Review of measurement of sustainable development goals: A comprehensive bibliometric and visualized analysis. *Environ. Sci. Pollut. Res.* **2023**, *30*, 91761–91779. [CrossRef] [PubMed]
24. Ciommi, M.; Gigliarano, C.; Emili, A.; Taralli, S.; Chelli, F.M. A new class of composite indicators for measuring well-being at the local level: An application to the Equitable and Sustainable Well-being (BES) of the Italian Provinces. *Ecol. Indic.* **2017**, *76*, 281–296. [CrossRef]
25. Bacchini, F.; Baldazzi, B.; Di Biagio, L. The evolution of composite indices of well-being: An application to Italy. *Ecol. Indic.* **2020**, *117*, 106603. [CrossRef]
26. Davino, C.; Dolce, P.; Taralli, S.; Vinzi, V.E. A Quantile Composite-Indicator Approach for the Measurement of Equitable and Sustainable Well-Being: A Case Study of the Italian Provinces. *Soc. Indic. Res.* **2018**, *136*, 999–1029. [CrossRef]
27. Ermini, B.; Santolini, R.; Ciommi, M. Equitable and sustainable well-being in Italian municipalities: Do women in politics make the difference? *Socioecon. Plann. Sci.* **2023**, *90*, 101741. [CrossRef]
28. Dello Strologo, A.; D'Andrassi, E.; Paoloni, N.; Mattei, G. Italy versus Other European Countries: Sustainable Development Goals, Policies and Future Hypothetical Results. *Sustainability* **2021**, *13*, 3417. [CrossRef]

29. Tebala, D.; Marino, D. A Synthetic Indicator BES-SDGs to Describe Italian Well-Being. *New Metrop. Perspect.* **2021**, *178*, 1862–1871.
30. Guarini, E.; Mori, E.; Zuffada, E. Localizing the Sustainable Development Goals: A managerial perspective. *J. Public Budg. Account. Financ. Manag.* **2022**, *34*, 583–601. [CrossRef]
31. Richiedei, A.; Pezzagno, M. Territorializing and Monitoring of Sustainable Development Goals in Italy: An Overview. *Sustainability* **2022**, *14*, 3056. [CrossRef]
32. Ricciolini, E.; Tiralti, A.; Paolotti, L.; Rocchi, L.; Boggia, A. Sustainable development according to 2030 agenda in European Union countries: Evidence of the enlargement policy. *Sustain. Dev.* **2023**, *32*, 1894–1912. [CrossRef]
33. Ruiz, F.; Cabello, J.M. MRP-PCI: A Multiple Reference Point Based Partially Compensatory Composite Indicator for Sustainability Assessment. *Sustainability* **2021**, *13*, 1261. [CrossRef]
34. Mazziotta, M.; Pareto, A. Weighting in composite indices construction: The case of the Mazziotta-Pareto index. *Riv. Ital. di Econ. Demogr. e Stat.* **2022**, *76*, 1–10.
35. Gan, X.; Fernandez, I.C.; Guo, J.; Wilson, M.; Zhao, Y.; Zhou, B.; Wu, J. When to use what: Methods for weighting and aggregating sustainability indicators. *Ecol. Indic.* **2017**, *81*, 491–502. [CrossRef]
36. Puertas, R.; Marti, L. Regional analysis of the sustainable development of two Mediterranean countries: Spain and Italy. *Sustain. Dev.* **2023**, *31*, 797–811. [CrossRef]
37. D'Adamo, I.; Di Carlo, C.; Gastaldi, M.; Uricchio, A.F. Equitable and sustainable well-being indicators: A study of Italian regional disparities towards sustainable development. *Sustain. Dev.* **2024**. [CrossRef]
38. Booth, A.; Sutton, A.; Papaioannou, D. *Systematic Approaches to a Successful Literature Review*; SAGE Publications Ltd.: Thousand Oaks, CA, USA, 2016.
39. D'Adamo, I.; Gastaldi, M. Sustainable Development Goals: A Regional Overview Based on Multi-Criteria Decision Analysis. *Sustainability* **2022**, *14*, 9779. [CrossRef]
40. Sousa, M.; Almeida, M.F.; Calili, R. Multiple Criteria Decision Making for the Achievement of the UN Sustainable Development Goals: A Systematic Literature Review and a Research Agenda. *Sustainability* **2021**, *13*, 4129. [CrossRef]
41. Baffo, I.; Leonardi, M.; Bossone, B.; Camarda, M.E.; D'Alberti, V.; Travaglioni, M. A decision support system for measuring and evaluating solutions for sustainable development. *Sustain. Futur.* **2023**, *5*, 100109. [CrossRef]
42. ISTAT Benessere e Sostenibilità. Available online: https://www.istat.it/it/benessere-e-sostenibilit%C3%A0 (accessed on 3 December 2023).
43. Costanza, R.; Daly, L.; Fioramonti, L.; Giovannini, E.; Kubiszewski, I.; Mortensen, L.F.; Pickett, K.E.; Ragnarsdottir, K.V.; De Vogli, R.; Wilkinson, R. Modelling and measuring sustainable wellbeing in connection with the UN Sustainable Development Goals. *Ecol. Econ.* **2016**, *130*, 350–355. [CrossRef]
44. Kettunen, M.; Boywer, C.; Vaculova, L.; Charveriat, C. *Sustainable Development Goals and the EU: Uncovering the Nexus between External and Internal Policies*; Think2030 Discuss. Pap. IEEP Brussels; Institute of European Environmental Policy: Brussels, Belgium, 2018.
45. Fan, W.; Anser, M.K.; Nasir, M.H.; Nazar, R. Uncertainty in firm innovation scheme and impact of green fiscal policy; Economic recovery of Chinese firms in the post-Covid-19 era. *Econ. Anal. Policy* **2023**, *78*, 1424–1439. [CrossRef]
46. Ferrante, M.; Vitti, M.; Facchini, F.; Sassanelli, C. Mapping the relations between the circular economy rebound effects dimensions: A systematic literature review. *J. Clean. Prod.* **2024**, *456*, 142399. [CrossRef]
47. Biancardi, A.; Colasante, A.; D'Adamo, I.; Daraio, C.; Gastaldi, M.; Uricchio, A.F. Strategies for developing sustainable communities in higher education institutions. *Sci. Rep.* **2023**, *13*, 20596. [CrossRef] [PubMed]
48. de Freitas Netto, S.V.; Sobral, M.F.F.; Ribeiro, A.R.B.; Soares, G.R.D.L. Concepts and forms of greenwashing: A systematic review. *Environ. Sci. Eur.* **2020**, *32*, 19. [CrossRef]
49. Van Tulder, R.; van Mil, E. *Principles of Sustainable Business: Frameworks for Corporate Action on the SDGs*; Routledge: New York, NY, USA, 2022; ISBN 1003098355.
50. Ali, S.M.; Appolloni, A.; Cavallaro, F.; D'Adamo, I.; Di Vaio, A.; Ferella, F.; Gastaldi, M.; Ikram, M.; Kumar, N.M.; Martin, M.A.; et al. Development Goals towards Sustainability. *Sustainability* **2023**, *15*, 9443. [CrossRef]
51. Suguna, M.; Sreenivasan, A.; Ravi, L.; Devarajan, M.; Suresh, M.; Almazyad, A.S.; Xiong, G.; Ali, I.; Mohamed, A.W. Entrepreneurial education and its role in fostering sustainable communities. *Sci. Rep.* **2024**, *14*, 7588. [CrossRef] [PubMed]
52. Lennon, B.; Dunphy, N. Sustaining energetic communities: Energy citizenship and participation in an age of upheaval and transition. *Sci. Rep.* **2024**, *14*, 3267. [CrossRef] [PubMed]
53. Skaloumpakas, P.; Sarmas, E.; Rachmanidis, M.; Marinakis, V. Reshaping the energy landscape of Crete through renewable energy valleys. *Sci. Rep.* **2024**, *14*, 8038. [CrossRef]
54. Cook, D.; Davíðsdóttir, B. An appraisal of interlinkages between macro-economic indicators of economic well-being and the sustainable development goals. *Ecol. Econ.* **2021**, *184*, 106996. [CrossRef]

Disclaimer/Publisher's Note: The statements, opinions and data contained in all publications are solely those of the individual author(s) and contributor(s) and not of MDPI and/or the editor(s). MDPI and/or the editor(s) disclaim responsibility for any injury to people or property resulting from any ideas, methods, instructions or products referred to in the content.

sustainability

MDPI

Article

Monitoring the Performance of Sustainable Development Goals in the Italian Regions

Idiano D'Adamo [1,*] and Massimo Gastaldi [2]

1 Department of Computer, Control and Management Engineering, Sapienza University of Rome, Via Ariosto 25, 00185 Rome, Italy
2 Department of Industrial and Information Engineering and Economics, University of L'Aquila, Piazzale E. Pontieri 1, 67100 L'Aquila, Italy; massimo.gastaldi@univaq.it
* Correspondence: idiano.dadamo@uniroma1.it

Abstract: The Sustainable Development Goals (SDGs) are a shared agenda among countries but also a rallying point for forward-looking policy dialogue. Analysis and monitoring of the SDGs are decisive steps in evaluating possible corrective actions. This paper aims to reach two research objectives: (i) providing methodological insights for the application of multicriteria decision analysis (MCDA) in the evaluation of the SDGs and (ii) emphasizing the relevance of monitoring the outcomes of the SDGs by evaluating the Italian regions. For the first objective, an online survey among twenty academics is used, while for the second, an MCDA is proposed that compares the temporal performance of a sustainability score for each Italian region. The results, based on 27 targets, show that in 2021 the northern regions showed better performance, with the province of Trento topping the list. This is followed by Valle d'Aosta and the province of Bolzano, confirming the trio that emerged in the previous year. A very interesting fact is the growth of the central regions, which overall tend to reach a value close to that of the northern regions. In particular, Toscana, Marche and Lazio stand out for a good performance. It is also confirmed that the southern regions occupy the last places in the ranking with the only exception of Abruzzo. The implications of this paper suggest collaboration between different regions in order to achieve a social community in which resources and skills can be enhanced.

Keywords: Italy; monitoring; multicriteria decision analysis; performance indicators; sustainable development goals; sustainability

Citation: D'Adamo, I.; Gastaldi, M. Monitoring the Performance of Sustainable Development Goals in the Italian Regions. *Sustainability* **2023**, *15*, 14094. https://doi.org/10.3390/su151914094

Academic Editor: Cristina Raluca Gh. Popescu

Received: 28 August 2023
Revised: 15 September 2023
Accepted: 21 September 2023
Published: 22 September 2023

Copyright: © 2023 by the authors. Licensee MDPI, Basel, Switzerland. This article is an open access article distributed under the terms and conditions of the Creative Commons Attribution (CC BY) license (https://creativecommons.org/licenses/by/4.0/).

1. Introduction

The topic of Sustainable Development Goals (SDGs) shows a very growing trend in the literature. A major problem in assessing the SDGs is the multiplicity of aspects to be considered and the different scales [1], but also the relationships among the indicators [2]. While, in fact, SDG 5 (Gender Equality) has the most relationships, unlike SDG 7 (Affordable and Clean Energy) [3], the literature has placed more emphasis on SDG 3 (Good Health and Well-Being) followed by SDG 2 (Zero Hunger) and SDG 1 (No Poverty), and less on SDG 14 (Life Below Water) and SDG 16 (Peace and Justice Strong Institutions) [4]. Thus, to achieve the 2030 Agenda, synergies among the goals must be exploited and suitable trade-offs should be identified [5].

Climate change, health and global governance puts the SDGs at the center of research [6], but nevertheless, care must be taken to avoid sustainable washing phenomena by favoring a pragmatic approach [4]. Markets are characterized by new economic models, as it has been shown that gross domestic product does not capture the full economic dimension of sustainability, and an overcoming of the selfish view that may characterize some stakeholders emerges in importance [7]. Another key aspect is cooperation to support economic exchanges, foster technological innovation and develop a global culture of sustainable development [8]. One issue to be resolved is the structural distances between actors and the consequences of their acts, and policy has the task of reducing and eliminating them [9]. Thus, it is necessary to think of the SDGs as a shared agenda among different countries, but they are also a proposed meeting point for forward-looking policy and programmatic dialogue. Indeed, there is a problem of fragmentation in global sustainability governance [10]. Therefore, improvements in green technological innovation can be identified [11], but a balance between the digital economy and industrial eco-efficiency needs to be identified [12]. Similarly, there is also a need for policies to encourage sustainable monetary supply [13] through public–private partnership [14].

In addition, analyses on the SDGs require appropriate policies because otherwise their achievement is at risk [15] and it is necessary to assess which policies are most appropriate [16] by fostering a global partnership between developed and developing countries [17]. Some authors emphasize how achieving the SDGs requires each city government to demonstrate its support toward these goals and congruity with global sustainability [18]. Along this direction, cities' progress toward SDGs can be proposed according to seven directions (economy, environment, human, operations, organization, society and staff). In particular, the criteria considered most relevant are sustainable policy and green public investments, and rationing public spending and political stability also show important relevance [19]. In addition, public administrators can put data-driven policy targets into their government programs [20].

The relationship between indicators on the SDGs and policy implications is an essential aspect of their achievement [21], and the literature highlights the key role of indicators in monitoring progress. Therefore, it is necessary to measure and monitor progress toward the SDGs [22,23]. It is useful to create a sustainable development index and dashboard to track countries' development in relation to the SDGs [24]. Analyses can be conducted at various levels, concerning a single country [25], as a comparison of several countries [23] or arriving at a global view [26]. Thus, the use of statistical methods supports assessments of sustainability goals [27,28], and the analyses need to integrate different information [29,30]. According to the literature, ranking various options is likely to raise awareness and accountability for achieving the SDGs [31,32]. The use of multicriteria decision analysis (MCDA) can aggregate different data to compare several territories [33]. The MCDA is used to measure the performance of Italian regions, and monitoring their performance is crucial to the achievement of the SDGs [24]. This paper covers two research objectives (ROs):

RO1—The first is to provide methodological insights for the application of MCDA in the evaluation of the SDGs.

RO2—The second is to emphasize the relevance of monitoring the outcomes of the SDGs by evaluating the Italian regions.

The methodological implications of RO1 can be used in other MCDAs to be implemented in different geographical contexts. RO2 can provide policy-makers with information regarding the roles and degrees of satisfaction of the different SDGs. This will help indicate where to focus future investments so as to foster sustainable development in different regions. This approach needs to be applied from both local and national perspectives, fostering interregional collaborations for sustainability.

The paper is organized as follows: after this introduction, the methodology and dataset used are presented in Section 2, where the online survey among academics and the MCDA are proposed. The results for the two ROs are proposed in Section 3, and discussion are presented in Section 4. Section 5 shows the conclusions.

2. Materials and Methods

This section consists of an analysis of an online survey among academics to assess insights on MCDA versus the SDGs (Section 2.1—RO1) and a description of MCDA and related data used in this paper (Section 2.2—RO2).

2.1. Survey with Academic Experts

Expert surveys are a method used in studies described in the literature to enhance their backgrounds in order to provide useful implications for the development of the topic examined [25,26]. In this paper, we want to consider what approaches may be most useful for conducting MCDA. Responses can be scored from 0% to 100%. The content of the email sent to experts described the purpose of the paper, the time to complete the questionnaire (maximum 5 min), the methodology used, and that only the first 20 responses would be included. In addition, it was specified that anonymity would be guaranteed. Other works in the literature use samples with the same number of observations [19,27]. The list of profiles to contact was chosen by looking at the Scopus database [28] and identifying academics with at least ten years of experience and expertise on SDG issues [19].

These authors were identified from those who published several papers with the word SDG in the title, abstract or keywords. Approximately one hundred emails were forwarded, and among the twenty responses obtained (Table 1), five of them were received from women (25%). In addition, eleven of these twenty experts work in Europe. The survey was conducted in August 2023.

Table 1. List of academic experts.

Number	Role	Country	Years of Experience
E1	Professor	France	16
E2	Professor	Spain	11
E3	Professor	India	14
E4	Professor	Sweden	13
E5	Professor	Italy	11
E6	Professor	Spain	12
E7	Professor	Italy	13
E8	Professor	Germany	15
E9	Professor	Italy	18
E10	Professor	India	20
E11	Professor	China	18
E12	Professor	Canada	12
E13	Professor	Australia	14
E14	Professor	Bangladesh	11
E15	Professor	Cyprus	19
E16	Professor	Greece	14
E17	Professor	China	16
E18	Professor	USA	18
E19	Professor	Poland	11
E20	Professor	Brazil	12

Before being sent, the questions were submitted to the attention of two colleagues who have opened discussions on the special issues on SDG themes and have more than ten years of experience each. Some suggestions were implemented. This initial survey was done through a video call lasting about half an hour. Table 2 presents the nine questions submitted to the twenty academics who participated in the online survey. Experts could provide comments.

Table 2. List of questions.

Number	Question
Q1	How useful do you think MCDA is as a method for assessing the SDGs?
Q2	How congruous do you think the number of public indicators on the SDGs is?
Q3	How robust do you think an indicator composed of 27 targets is for assessing individual local realities?
We proceed below to evaluate the approach related to values and weights, where it is specified that answers should be provided for a small sample of targets (about 30).	
Q4	How valid do you think the 0–1 range approach is (0 = weak performance; 1 = strong performance)?
Q5	How valid do you think the maximum value approach is as a benchmark (1 = strong performance)?
Q6	How valid do you think equal weight among SDGs (EWG) scenario is?
Q7	How valid do you think equal weight among indicators (EWI) scenario is?
Finally, two final questions are provided on the possible decomposition of the results obtained.	
Q8	How useful do you think it is to break down the final data according to the three dimensions of sustainability?
Q9	How useful do you think it is to break down the final figure according to geographic macro-areas?

2.2. Multicriteria Decision Analysis

MCDA synthesizes a large multiplicity of data, also considering their different nature and belonging to different concepts. The method is widely used in the literature to make comparisons between different geographical realities, and the initial objective is to arrive at a ranking among different alternatives in order to provide insights for the decision-maker [29]. Thus, we can identify three precise stages:

- The first step is to assign suitable criteria to achieve this goal;
- The second step is to assign weights and values to these criteria;
- The third step is to aggregate the different information obtained, calculating the sustainability score of each alternative.

One of the major limitations is the lack of dynamicity, such that evaluation of the interactions between the different variables does not occur.

MCDA falls under the field of operations research, and the final results can be decomposed to understand which criteria affect the others the most. In addition, strengths and weaknesses can be ranked for each alternative, and how the trend varies over the years can be measured. The alternatives are represented by the geographic realities, which in this paper are the twenty regions of Italy (considering, for the Trentino Alto Adige region, the decomposition into the two provinces of Bolzano and Trento)—Figure 1.

The accuracy of these analyses depends on their data reliability, and for this reason, the data released by ASviS (Italian Alliance for Sustainable Development) are used, an approach already used in the literature [30,31], given the active role that ASviS plays in providing recent and reliable data.

This paper aims to monitor the data comparing the 2021 and 2022 ASviS reports [31,32], where 26 targets were confirmed, while two targets were not re-proposed: (i) target 4.1 (a)—By 2030, reduce the number of students who do not reach the sufficient level of numerical proficiency (18–19 years old) below the 15% quota and (ii) target 4.1 (b)—By 2030, reduce the number of students who do not reach the sufficient level of literacy proficiency (18–19 years old) below the 15% quota. Instead, this paper, in accordance with the new report released by ASviS for 2022, considers a new target [33]: by 2027 to reach at least 33% of placement in early childhood education services (3–36 months). Specifically, the latest report shows values for 27 targets attributable to 16 of the 17 SDGs (SDG 17 is absent)—Table 3.

Figure 1. The map of Italy.

Table 3. List of criteria.

SDG	Target	Unit
SDG 1	Target 1.2—By 2030 reduce the number of people at risk of poverty or social exclusion by 16% compared to 2020	%
SDG 2	Target 2.4 (a)—By 2030 reduce the use of fertilizer distributed in non-organic agriculture by 20% compared to 2020	quintals per ha
SDG 2	Target 2.4 (b)—By 2030 achieve the 25% share of UAA invested by organic crops	%
SDG 3	Target 3.4—By 2025 reduce the probability of dying from no communicable diseases by 25% compared to 2013	%
SDG 3	Target 3.6—By 2030 reduce road traffic injuries by half compared to 2019	per 10,000 population
SDG 4	Target 4.1—By 2030 reduce early exit from education and training (18–24 years old) below the 9% rate	%
SDG 4	Target 4.2—By 2027 achieve at least 33% of places in early childhood education services (3–36 months)	%
SDG 4	Target 4.3—By 2030 to reach the 50% share of college graduates (30–34 years old)	%
SDG 5	Target 5.5—By 2030 to halve the gender employment gap compared to 2020	females/males * 100
SDG 6	Target 6.3—By 2027 ensure high or good ecological quality status for all surface water bodies	%
SDG 6	Target 6.4—By 2030 achieve 90% efficiency share of drinking water distribution networks	%
SDG 7	Target 7.2—By 2030 achieve at least 45% share of energy from renewable sources	%
SDG 7	Target 7.3—By 2030 to reduce final energy consumption by at least 20% compared to 2020	ktoe per 10,000 population
SDG 8	Target 8.5—By 2030 achieve 78% share of the employment rate (20–64 years old)	%
SDG 8	Target 8.6—By 2030 reduce the share of NEETs to below 9% (15–29 years old)	%
SDG 9	Target 9.5—By 2030 achieve the share of 3% of GDP devoted to research and development	%
SDG 9	Target 9.c—By 2026 ensure that all households have coverage to the Gigabit network	%
SDG 10	Target 10.4—By 2030 reduce net income inequality (S80/S20) to levels observed in the best of European countries	s80/s20
SDG 11	Target 11.2—By 2030 increase public transport seat-km per inhabitant offered by 26% compared to 2004	places-Km per inhabitant
SDG 11	Target 11.6—By 2030 reduce PM10 exceedances to below 3 days per year	days

Table 3. *Cont.*

SDG	Target	Unit
SDG 12	Target 12.4—By 2030 reduce the share of municipal waste generated per capita by 26% compared to 2004	kg/inhab.* year
SDG 13	Target 13.2—By 2030 reduce emissions of CO_2 and other climate-altering gases by 55% from 1990 levels	ton CO_2 equivalent per capita
SDG 14	Target 14.5—By 2030 achieve 30% share of marine protected areas	%
SDG 15	Target 15.3—By 2050 achieve zero increase in annual land consumption	ha per 100,000 population
SDG 15	Target 15.5—By 2030 achieve 30% share of terrestrial protected areas	%
SDG 16	Target 16.3—By 2030 achieve zero overcrowding in penal institutions	%
SDG 16	Target 16.7—By 2026 reduce the average duration of civil proceedings by 40 percent compared to 2019	days

In particular, it emerges that SDG 4 is the one most represented, with 3 targets, while there are 2 targets for the following SDGS: 2, 3, 6, 7, 8, 9, 11, 15 and 16. The remaining 6 targets refer to SDGs 1, 5, 10, 12, 13 and 14, while SDG 17 is absent.

Another observation concerns the reference year related to the target data. In fact, for 12 targets, it is 2021 (targets 1.2, 2.4 (a), 2.4 (b), 3.6, 4.1, 4.3, 5.5, 8.5, 8.6, 15.3, 16.3 and 16.7), while for 10 other targets, it is earlier than 2021 (targets 3.4, 4.2, 6.3, 6.4, 9.5, 11.2, 11.6, 12.4, 14.5 and 15.3). Finally, for 5 targets, the Italian figure is 2021, while that of individual regions is before 2021 (targets 7.2, 7.3, 9.c, 10.4, 13.2). It should also be pointed out that the latest available data are used.

The product between a value and the weights associated with the various criteria was the basis for the overall sustainability indicator, which was calculated for all alternatives. Having completed the first stage of the multicriteria analysis, we proceeded to examine whether the data are homogeneous with each other or not, and it emerged that they are all comparable. Thus, the more populous regions have data that can be compared with the less populous ones. Particular mention should be made of target 14.5, whose value is not proposed for six regions (Piemonte, Valle d'Aosta, Lombardia, the province of Bolzano, the province of Trento and Umbria). A number of 26 targets is therefore considered for these six regions. Instead, when the value 0 is proposed, it is considered as input data.

The second step was to assign weights and values to the criteria for the different alternatives. A value of 1 was assigned to the most relevant performance and 0 to the least relevant performance. In addition, an intermediate value was assigned to the other alternatives using the interpolation method. The choice of the normalized method is the one that emerges most from the indications from the literature [7]. In addition, this choice was verified before presenting RO2 results by examining what emerged from RO1.

Table 4 proposes all the values assigned to the 21 alternatives. Two assumptions are considered: (i) Italy's data for target 11.6 is estimated as a function of population and (ii) the maximum value for target 7.2 is considered to be 100%.

Table 4. List of values. IT = Italy; A1 = Piemonte; A2 = Valle d'Aosta; A3 = Liguria; A4 = Lombardia; A5 = Province Bolzano; A6 = Province Trento; A7 = Veneto; A8 = Friuli Venezia Giulia; A9 = Emilia Romagna; A10 = Toscana; A11 = Umbria; A12 = Marche; A13 = Lazio; A14 = Abruzzo; A15 = Molise; A16 = Campania; A17 = Puglia; A18 = Basilicata; A19 = Calabria; A20 = Sicilia; A21 = Sardegna.

Target	IT	A1	A2	A3	A4	A5	A6	A7	A8	A9	A10
1.2	0.617	0.794	0.938	0.706	0.851	1.000	0.886	0.828	0.848	0.970	0.821
2.4 (a)	0.634	0.521	1.000	0.676	0.000	0.972	0.958	0.028	0.479	0.183	0.831
2.4 (b)	0.449	0.102	0.000	0.335	0.087	0.122	0.198	0.111	0.219	0.455	0.968
3.4	0.583	0.528	0.833	0.583	0.722	0.917	1.000	0.889	0.667	0.833	0.667
3.6	0.636	0.717	0.854	0.000	0.655	0.574	0.809	0.650	0.730	0.377	0.280
4.1	0.625	0.721	0.522	0.610	0.728	0.610	0.912	0.875	0.926	0.831	0.743
4.2	0.491	0.600	0.897	0.627	0.591	0.370	0.815	0.609	0.721	0.900	0.806

Table 4. *Cont.*

Target	IT	A1	A2	A3	A4	A5	A6	A7	A8	A9	A10
4.3	0.559	0.596	0.615	0.522	0.839	0.429	0.988	0.807	0.509	0.981	0.696
5.5	0.546	0.747	1.000	0.694	0.755	0.794	0.850	0.677	0.744	0.772	0.766
6.3	0.384	0.503	0.874	0.707	0.260	0.933	0.826	0.340	0.388	0.248	0.298
6.4	0.406	0.585	1.000	0.448	0.770	0.857	0.648	0.439	0.296	0.728	0.382
7.2	0.107	0.137	1.000	0.000	0.076	0.660	0.432	0.116	0.146	0.044	0.114
7.3	0.388	0.313	0.013	0.650	0.313	0.181	0.163	0.231	0.031	0.000	0.456
8.5	0.581	0.808	0.866	0.751	0.866	1.000	0.895	0.840	0.888	0.927	0.831
8.6	0.574	0.743	0.796	0.726	0.778	1.000	0.813	0.974	0.874	0.922	0.800
9.5	0.529	1.000	0.000	0.529	0.471	0.176	0.588	0.471	0.647	0.882	0.588
9.c	0.920	0.697	0.127	0.984	0.636	0.000	0.352	0.376	0.434	0.592	0.516
10.4	0.368	0.737	0.921	0.447	0.632	0.921	0.816	0.789	0.868	0.816	0.737
11.2	0.371	0.373	0.018	0.328	1.000	0.309	0.354	0.457	0.330	0.217	0.206
11.6	0.367	0.022	0.944	0.944	0.000	0.967	0.900	0.022	0.578	0.167	0.622
12.4	0.512	0.515	0.095	0.393	0.576	0.597	0.522	0.549	0.478	0.000	0.180
13.2	0.519	0.432	0.247	0.494	0.531	0.457	0.469	0.259	0.667	0.284	0.617
14.5	0.370			0.130				0.000	0.109	0.000	1.000
15.3	0.730	0.597	0.803	1.000	0.790	0.777	0.853	0.617	0.813	0.590	0.820
15.5	0.331	0.203	0.436	0.124	0.150	0.556	0.853	0.128	0.192	0.094	0.165
16.3	0.442	0.518	0.917	0.252	0.085	0.000	1.000	0.222	0.128	0.403	0.598
16.7	0.609	0.919	1.000	0.853	0.834	0.953	0.946	0.797	0.951	0.842	0.733

Target	A11	A12	A13	A14	A15	A16	A17	A18	A19	A20	A21
1.2	0.851	0.908	0.600	0.425	0.428	0.000	0.438	0.440	0.226	0.167	0.408
2.4 (a)	0.775	0.746	0.732	0.789	0.944	0.704	0.775	0.915	0.845	0.887	0.944
2.4 (b)	0.446	0.685	0.653	0.347	0.143	0.510	0.592	0.714	1.000	0.630	0.297
3.4	0.833	0.972	0.444	0.639	0.472	0.000	0.611	0.556	0.417	0.306	0.444
3.6	0.730	0.439	0.480	0.765	1.000	0.957	0.604	0.911	0.987	0.760	0.809
4.1	0.676	0.978	0.882	0.971	1.000	0.353	0.265	0.919	0.529	0.000	0.588
4.2	1.000	0.606	0.736	0.436	0.324	0.000	0.261	0.318	0.027	0.045	0.597
4.3	1.000	0.689	0.776	0.578	0.950	0.211	0.081	0.429	0.236	0.000	0.248
5.5	0.738	0.652	0.621	0.379	0.203	0.000	0.067	0.223	0.070	0.022	0.632
6.3	0.000	0.344	0.342	0.346	0.010	0.301	0.011	0.025	1.000	0.571	0.507
6.4	0.030	0.648	0.075	0.000	0.299	0.301	0.313	0.313	0.319	0.152	0.131
7.2	0.175	0.125	0.036	0.218	0.348	0.136	0.123	0.485	0.391	0.068	0.217
7.3	0.188	0.694	0.794	0.538	0.613	1.000	0.625	0.619	0.913	0.950	0.725
8.5	0.792	0.780	0.636	0.562	0.364	0.016	0.192	0.390	0.032	0.000	0.399
8.6	0.743	0.883	0.639	0.661	0.374	0.096	0.248	0.483	0.122	0.000	0.552
9.5	0.235	0.235	0.824	0.235	0.294	0.412	0.176	0.059	0.000	0.176	0.176
9.c	0.392	0.115	1.000	0.268	0.033	0.840	0.455	0.178	0.150	0.577	0.225
10.4	0.842	1.000	0.395	0.789	0.789	0.000	0.395	0.842	0.289	0.079	0.368
11.2	0.114	0.147	0.508	0.158	0.000	0.120	0.130	0.048	0.099	0.106	0.234
11.6	0.422	0.678	0.144	0.811	1.000	0.133	0.889	0.900	0.911	0.844	0.578
12.4	0.451	0.468	0.502	0.627	0.922	0.641	0.573	1.000	0.878	0.661	0.661
13.2	0.432	0.728	0.778	0.704	0.111	1.000	0.296	0.185	0.654	0.617	0.000
14.5		0.000	0.109	0.261	0.000	0.435	0.217	0.000	0.196	0.478	0.413
15.3	0.657	0.780	0.850	0.000	0.480	0.800	0.663	0.617	0.937	0.750	0.710
15.5	0.218	0.297	0.410	1.000	0.000	0.906	0.455	0.665	0.571	0.342	0.083
16.3	0.472	0.550	0.448	0.470	0.298	0.397	0.062	0.647	0.565	0.677	0.943
16.7	0.558	0.735	0.612	0.719	0.500	0.371	0.362	0.000	0.161	0.340	0.458

Once the values were obtained, we proceeded to consider the weights to be assigned to the criteria. Two approaches (EWG, Equal Weights Goals, and EWI, Equal Weights Indicators) are proposed in the literature [24], and this choice emerged from the result related to the question proposed in RO1. Thus, also for this methodological stage, the choices that would be implemented in RO2 would depend on what emerged in RO1. However, it is also useful to show what the literature proposes on the topic. The choice of not giving more emphasis to some goals and indicators is proposed in several works [34,35].

This practice appears to be widespread because it is based on the logic of assigning the same impact to the individual indicators analyzed [36,37]. However, approaches with different weights can also be used [38].

Finally, during the third step, all data were aggregated and a sustainability score for each alternative was obtained by the product between a row vector (referred to as the values of the criteria) and a column vector (referred to as the weights of the criteria). The results were also broken down according to the three main geographic macro-areas of Italy:

- North—Valle d'Aosta, Piemonte, Lombardia, Liguria, Trentino Alto Adige, Veneto, Friuli Venezia Giulia and Emilia Romagna.
- Center—Toscana, Umbria, Marche and Lazio.
- South—Abruzzo, Molise, Campania, Puglia, Basilicata, Calabria, Sardegna and Sicilia.

3. Results

This section is divided as follows: results related to RO1 are proposed in Section 3.1 and those related to RO2 are examined in the following subsections. In fact, first, the sustainability score in the base scenario is proposed (Section 3.2), then its value is monitored over time (Section 3.3), and finally, an alternative scenario is considered (Section 3.4).

3.1. Methodological Insights for the Application of MCDA in the Evaluation of the SDGs

The responses from the different experts are collected and proposed in Table 5. It should be pointed out that in order to ensure anonymity, there is no correspondence between the numbers of the experts in Tables 1 and 5. In order to give statistical significance to the results obtained, a Kruskal–Wallis test was conducted. The comparison of the questions determines that the hypothesis H_0 was rejected ($\chi 2 = 146.68$, $p < 0.001$). This suggested that certain groups' average ranks were indeed not equal. The post hoc Dunn's test using a Bonferroni corrected alpha of 0.0014 indicated that the mean ranks of the following pairs are significantly different: Q1–Q2, Q1–Q3, Q1–Q5, Q1–Q6, Q1–Q8, Q2–Q5, Q2–Q6, Q3–Q4, Q3–Q7, Q4–Q5, Q4–Q6, Q4–Q8, Q5–Q7, Q5–Q9, Q6–Q7, Q6–Q9, Q7–Q8 and Q8–Q9.

Table 5. Survey online according to 20 academic experts—all data are in percentages.

	E1	E2	E3	E4	E5	E6	E7	E8	E9	E10	E11	E12	E13	E14	E15	E16	E17	E18	E19	E20
Q1	95	100	100	95	100	100	95	100	100	95	100	95	100	95	100	100	100	100	90	90
Q2	90	90	90	85	90	95	90	90	85	85	85	85	90	90	85	85	90	90	80	80
Q3	70	80	85	75	85	85	75	90	80	70	70	65	85	75	75	70	85	85	65	70
Q4	95	100	95	100	95	95	90	95	100	90	95	95	95	90	95	95	95	100	90	90
Q5	70	80	70	75	65	70	65	70	70	65	70	65	65	60	65	65	70	70	65	65
Q6	75	70	70	65	65	60	60	65	60	65	70	70	65	60	60	60	65	70	70	70
Q7	90	95	95	90	90	90	85	90	90	85	90	90	90	85	90	95	95	95	85	80
Q8	75	65	65	70	75	65	75	75	90	75	80	60	75	65	85	80	85	90	65	75
Q9	95	90	90	90	85	85	90	85	90	90	95	90	95	85	90	90	85	90	85	90

The next step was to aggregate all responses, giving the experts equal weighting. It is worth mentioning that the experts could assign a percentage value between 0% and 100%—Table 6.

Table 6. Average values of questions.

Number	Question	Percentage Average
Q1	How useful do you think MCDA is as a method for assessing the SDGs?	97.50
Q2	How congruous do you think the number of public indicators on the SDGs is?	87.50
Q3	How robust do you think an indicator composed of 27 targets is for assessing individual local realities?	77.00
Q4	How valid do you think the 0–1 range approach is (0 = weak performance; 1 = strong performance)?	94.75
Q5	How valid do you think the maximum value approach is as a benchmark (1 = strong performance)?	68.00
Q6	How valid do you think equal weight among SDGs (EWG) scenario is?	65.75
Q7	How valid do you think equal weight among indicators (EWI) scenario is?	89.75
Q8	How useful do you think it is to break down the final data according to the three dimensions of sustainability?	74.50
Q9	How useful do you think it is to break down the final figure according to geographic macro-areas?	89.25

The responses to question Q1 show a percentage of 97.5%, with more than half of the experts assigning the maximum value. The result is not surprising given the wide use of MCDA. A limitation of the paper also emerges from this question, as some experts pointed out that such an approach is not the only useful way to compare different methodologies in order to monitor and analyze SDG values.

However, one of the problems encountered when using these methodologies is the availability of data, and the Q2 figure should be interpreted in this direction. A relevant value emerges, of 87.5%; however, a comment that has come from several experts should be highlighted. For example, Eurostat is a useful tool that compares different countries; however, then, the national figure cannot always be broken down to a more detailed local level.

In this direction, we then investigated whether an indicator consisting of 27 targets (Table 3) would be suitable for evaluating a local reality. The Q3 value is 77%. It is useful to underline that having specified individual local realities led to providing a higher value to this question.

The analyses of questions Q4 and Q5 aimed to evaluate the method to be assigned to the criteria values. Absolute values should tend to be normalized in order to be made homogeneous. In these questions, a choice toward the 0–1 approach emerges (94.75% vs. 68%) because it is believed that, just as the emphasis is placed on the best performance, likewise the weakest performance should be penalized. Clearly, the 0–1 approach tends to accentuate differences in terms of normalized values.

Instead, questions Q6 and Q7 evaluate the method related to the assignment of weights. In these questions, too, a clear choice toward the EWI scenario emerges (89.75% vs. 65.75%) because with a small number of criteria, individual SDGs might themselves be underrepresented. Thus, both methods are considered suitable with a larger sample of criteria, since being alternatives may or may not confirm the results obtained.

Finally, the last two questions, Q8 and Q9, aim to provide insights into the breakdown of sustainability scores at the size and geographic area level. The subdivision by economic, environmental and social dimensions is considered very useful; however, like the previous observation, it is noted that its significance loses significance (74.5%) when the sample number is small. In contrast, geographic data are crucial (89.25%) when conducting analyses at the level of individual territorial realities.

Consequently, it follows from this analysis that the MCDA of this paper will use the following assumptions:

- Value will be assigned through the 0–1 approach.
- Weight will be assigned through the EWI approach.
- Results will be decomposed only at the geographical area level.

3.2. *Sustainability Score for Italian Regions—Baseline Scenario*

The main result of a quantitative analysis is to provide numbers, and in the case of this paper, the sustainable performance of the Italian regions is indicated. Another key step is to be able to compare these results with those of the previous year. This approach allows the performance of alternatives to be monitored and improvements or worsening to be evaluated.

The number of criteria examined is small compared to other studies [39], but nevertheless, it still plays its role of providing new information to stakeholders. Pragmatic sustainability calls for evaluating the performance of alternatives based on real data, which could be affected by virtuous policies. Within the EWI scenario, the different values were normalized, which made it possible to homogenize criteria characterized by different units of measurement (Table 3).

This subsection shows the sustainability score in the EWI scenario, in which a value trending toward 1 indicates excellent sustainability performance. All alternatives were compared with each other, but also, with respect to the Italian average, it was considered in addition to the 21 alternatives. It is worth noting that Figure 2 proposes the average value for the Trentino Alto Adige region (that is composed of the two provinces of Bolzano and Trento, and the same will also be repeated in the monitoring phase).

Province of Trento	0.725
Valle d'Aosta	0.643
Province of Bolzano	0.620
Toscana	0.602
Marche	0.588
Lazio	0.557
Piemonte	0.555
Friuli Venezia Giulia	0.543
Lombardia	0.538
Liguria	0.538
Umbria	0.530
Emilia Romagna	0.521
Abruzzo	0.507
Italy	**0.505**
Veneto	0.485
Basilicata	0.477
Calabria	0.464
Sardegna	0.457
Molise	0.441
Campania	0.394
Sicilia	0.378
Puglia	0.366

Figure 2. Sustainability score in EWI scenario.

The results see the province of Trento excel, with a score of 0.725, which is a high performance and shows a significant difference from that of Valle d'Aosta, which follows with 0.643, and the province of Bolzano, with 0.620. The difference is thus 0.082 with the second, and it increases to 0.220 compared to the national benchmark. There are 13 alternatives placed above the national average (0.505), with Abruzzo placed at 0.507, while the last position, occupied by Puglia, is 0.139 away from the national benchmark. These results show how northern regions occupy the top positions, while southern regions are placed at the bottom of the ranking. It is interesting to note the performance of the three central regions (Toscana, Marche and Lazio), which are behind the first three northern regions mentioned above.

In order to understand the results obtained, we identify which alternatives perform better or worse in the individual targets, and the corresponding concentration indicator (Table 7 and Figure 3).

Table 7. The best and worst alternative for each target audience and the value of indicator concentration for each target.

Target	Max	Min	Indicator Concentration
1.2	Province of Bolzano	Campania	0.644
2.4 (a)	Valle d'Aosta	Lombardia	0.700
2.4 (b)	Calabria	Valle d'Aosta	0.410
3.4	Province of Trento	Campania	0.635
3.6	Molise	Liguria	0.671
4.1	Molise	Sicilia	0.697
4.2	Umbria	Campania	0.538
4.3	Umbria	Sicilia	0.580
5.5	Valle d'Aosta	Campania	0.543
6.3	Calabria	Umbria	0.421
6.4	Valle d'Aosta	Abruzzo	0.416
7.2	Valle d'Aosta	Liguria	0.240
7.3	Campania	Emilia Romagna	0.476
8.5	Province of Bolzano	Sicilia	0.611
8.6	Province of Bolzano	Sicilia	0.630
9.5	Piemonte	Valle d'Aosta, Calabria	0.389
9.c	Lazio	Province of Bolzano	0.426
10.4	Marche	Campania	0.642
11.2	Lombardia	Molise	0.250
11.6	Molise	Lombardia	0.594
12.4	Basilicata	Emilia Romagna	0.538
13.2	Campania	Sardegna	0.474
14.5	Toscana	Veneto, Emilia Romagna, Marche, Molise, Basilicata	0.223
15.3	Liguria	Abruzzo	0.710
15.5	Abruzzo	Molise	0.374
16.3	Province of Trento	Province of Bolzano	0.460
16.7	Valle d'Aosta	Basilicata	0.650

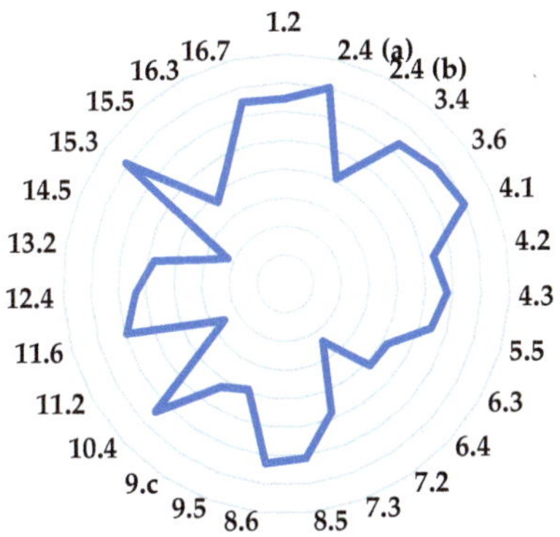

Figure 3. Indicator concentration.

The province of Trento confirms its leadership in targets 3.4 and 16.3, and while it loses the first position in target 4.1, it still maintains a high value, whereas its reduction is much more significant in target 9.5. The province of Bolzano retains leadership in targets 8.5 and

8.6, and gains it in target 1.2, replacing Valle d'Aosta, which nevertheless confirms it in targets 2.4 (b), 5.5, 6.4, 7.2 and 16.7. Most of the different alternatives present a highest value in the different targets; the only ones never to reach this target are Veneto, Friuli Venezia Giulia, Emilia Romagna, Puglia, Sicilia and Sardegna. However, the data in Table 7 show how for some targets of the same SDG, the alternatives are able to achieve diametrically opposite results. This is the case for Valle d'Aosta for target 2.4 (a) and 2.4 (b), Lombardia and Molise for targets 11.2 and 11.6, and Abruzzo for targets 15.3 and 15.5.

Focusing on the alternatives occupying the first three positions, however, it should be pointed out that they too show weak performance in the following cases: the province of Trento presents a value below 0.2 only in targets 2.4 (b) and 7.3; the province of Bolzano, in addition to the two mentioned above, is also weak in target 9.5, but especially occupies the last position in targets 9.c and 16.3 and finally Valle d'Aosta is last in targets 2.4 (b) and 9.5 and below 0.2 in targets 7.3, 11.2 and 12.4. Thus, it emerges that the province of Trento prevails over Valle d'Aosta despite having fewer first positions in the targets, as it has a higher value in the other criteria.

Among the northern regions, Veneto is the only one that is below the national average, presenting a performance below 0.2 in targets 2.4 (a), 2.4 (b), 7.2, 11.6, 14.5 and 15.5, and, as highlighted above, does not excel in any target. As noted earlier, the central regions show very interesting performances. Primacy is verified in target 14.5 (Toscana), targets 4.2 and 4.3 (Umbria), target 10.4 (Marche) and target 9.c (Lazio). In contrast, the situation tends to change considerably when analyzing the southern ones, where only Abruzzo is above the national average. Abruzzo is first in target 15.5 and it has very high performance in target 4.1 and high performance in target 11.6. On the other hand, it is last in targets 6.4 and 15.3 and is also weak in target 11.2. The other seven regions occupy the last positions in the ranking. This aspect clearly denotes a strong criticality, which, however, is also proclaimed at the European level, where targeted funding programs are allocated precisely for southern Italy.

The indicator concentration shows that where this value is high, many alternatives have significant performances, and this reduces the advantages of those occupying the top positions over the other alternatives. Similarly, when the value of the concentration index is low, the alternatives occupying the top positions will have much more significant numerical values than the alternatives that have values close to the concentration index. High values occur for targets 15.3 and 2.4 (a), while low values occur for targets 14.5, 7.2 and 11.2.

3.3. *Sustainability Score Monitoring for Italian Regions*

An additional element of the analysis was to compare the sustainability score in the EWI scenario for the two reference years (Figure 4 and Table 8). Results from the 2022 ASviS report were obtained for this paper and can be compared with what is reported in the literature regarding the 2021 ASviS report [24].

The two maps show no significant color differences, while the analysis of the values shows that numerical changes have occurred. As noted earlier, the very low target concentration index, 9.5, in the previous year led the province of Trento to have a very high value. This explains the 0.060 reduction. The other alternative that marks a reduction is Veneto. As for increases, there are very important changes for the central regions Lazio (0.058) and Toscana (0.053). At the ranking level, the most significant change is in Lazio, which climbs seven positions, while Lombardia and Liguria lose three positions. These variations also determine that the performance of the three macro-areas tends to increase when comparing years (Figure 5). While the increase in the northern regions is minimal (+0.004), the increase in the southern regions is appreciable (+0.022), wherein a very important role is played by Sardegna and Puglia. However, it is the central regions, as shown above, that mark the most significant increases, with +0.041. However, these data confirm the existence of gaps between regions with different performances. It is worth noting that the new figure for the central regions (0.569) is close to that of the northern regions (0.574). These values are significantly different from that of the southern regions (0.436).

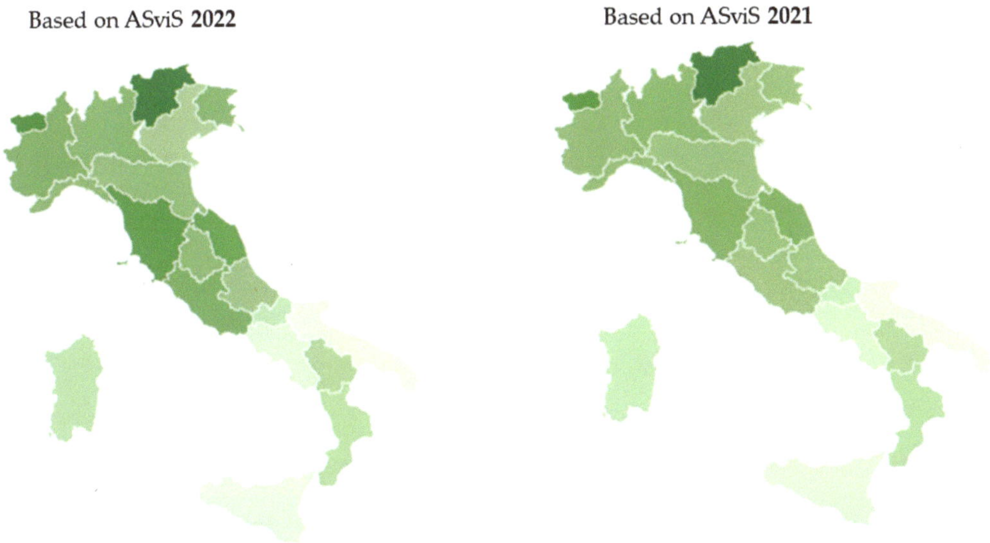

Figure 4. Monitoring of SDGs in Italian regions in the period 2021–2022.

Table 8. Delta sustainability score in the period 2021–2022.

	ASviS 2021	Delta ASviS 2022–ASviS 2021	Delta Ranking (2022 vs. 2021)
Province of Trento	0.785	−0.060	0
Valle d'Aosta	0.628	0.015	0
Province of Bolzano	0.612	0.009	0
Toscana	0.548	0.053	+1
Marche	0.556	0.032	−1
Lazio	0.498	0.058	+7
Piemonte	0.528	0.027	+1
Friuli Venezia Giulia	0.514	0.029	+1
Lombardia	0.536	0.002	−3
Liguria	0.529	0.009	−3
Umbria	0.508	0.021	−1
Emilia Romagna	0.502	0.019	0
Abruzzo	0.504	0.003	−2
Italia	0.494	0.011	
Veneto	0.495	−0.010	0
Basilicata	0.460	0.017	0
Calabria	0.446	0.018	0
Sardegna	0.418	0.040	0
Molise	0.411	0.030	0
Campania	0.380	0.014	0
Sicilia	0.358	0.020	0
Puglia	0.333	0.033	0

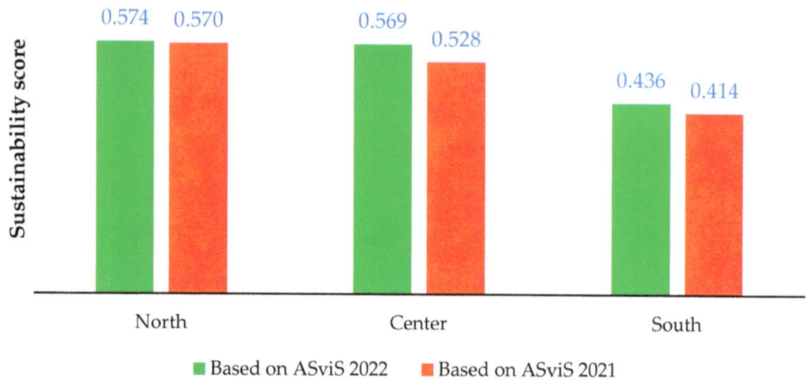

Figure 5. Sustainability score in EWI scenario—a comparison among macro-areas.

3.4. Sustainability Score for Italian Regions—Alternative Scenario

Finally, in order to give robustness to the results obtained, we proceeded to consider an alternative scenario. Specifically, the EWG scenario was chosen, in which the same criteria values were considered, but we evaluated the aggregation of targets within the SDGs. This inevitably tended to mediate the effects for targets that were a larger sample within the SDGs (Table 9 and Figure 6). For the six regions that do not consider target 14.5, the contribution from 15 SDGs is assessed.

Table 9. Delta sustainability score in EWG scenario in 2022.

	EWG Scenario	Delta EWG-EWI	Delta Ranking EWG-EWI
Province of Trento	0.784	0.059	0
Province of Bolzano	0.712	0.091	+1
Valle d'Aosta	0.704	0.062	−1
Toscana	0.661	0.060	0
Marche	0.642	0.053	0
Piemonte	0.607	0.052	+1
Lombardia	0.602	0.064	+2
Friuli Venezia Giulia	0.593	0.050	0
Lazio	0.590	0.033	−3
Liguria	0.581	0.043	0
Umbria	0.580	0.050	0
Abruzzo	0.557	0.050	+1
Italia	0.545	0.040	
Emilia Romagna	0.530	0.010	−1
Basilicata	0.526	0.049	+1
Calabria	0.518	0.054	+1
Veneto	0.516	0.031	−2
Sardegna	0.497	0.040	0
Molise	0.476	0.035	0
Campania	0.437	0.043	0
Sicilia	0.434	0.056	0
Puglia	0.413	0.047	0

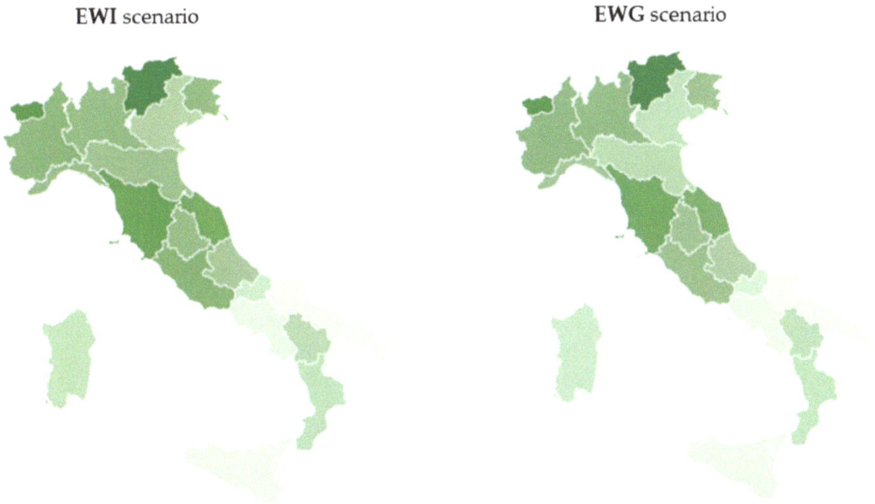

Figure 6. A comparison of the sustainability score between EWI and EWG scenarios.

It can be seen that EWG values are higher than EWI values due to the different normalization approaches to criteria. This alternative scenario demonstrates how much a chosen method influences the results obtained in this paper. The EWG scenario, which tends to be more synthetic than EWI, leads to different measurements. The results of this EWG scenario sees several ranking positions change, and 11 alternatives keep the same position. In the top part of the ranking, the province of Bolzano overtakes Valle d'Aosta, and Lombardia gains two positions, while Lazio loses three. As for the final part of the ranking, no changes are noted. Further relevant data include that twelve of the thirteen alternatives maintain a value above the national average; Emilia Romagna is the exception.

Furthermore, we proceed to aggregate the data at the macro-area level (Figure 7), and since the value of the national average in the EWG scenario is greater than that in the EWI scenario (0.545 vs. 0.505), the three individual values turn out to be greater. The difference between the areas does not tend to change significantly, as the values vary in the range of 0.047–0.053. The northern regions are slightly ahead of the central regions (0.627 vs. 0.618), while the difference is much more pronounced than in the southern regions (0.482).

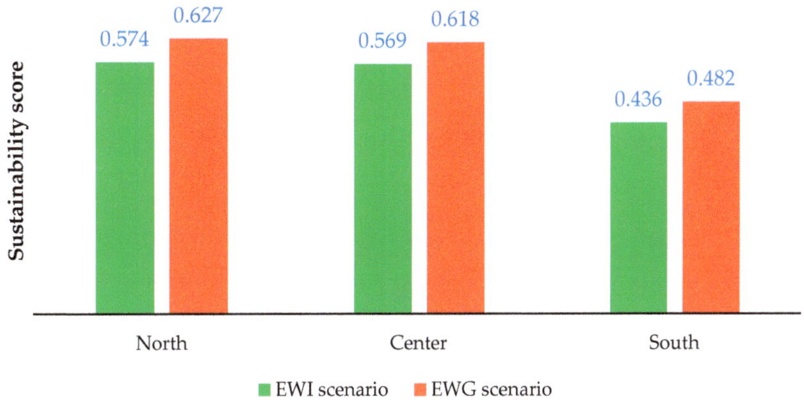

Figure 7. Sustainability score in both EWI and EWG scenarios—a comparison among macro-areas.

4. Discussion

The issue of sustainability has entered the agenda of many governments [28], and many citizens are involved and interested in applying a social welfare model. The challenge is complex and takes time to meet. This does not imply that there is no urgency, but a goal must be set. Some authors have pointed out that any paper that talks about the SDGs must specify what goal it aims to achieve [4]. Academic works must be projected to real problems in order to support policy-makers. This paper aims to focus its attention on the strategic role of monitoring the SDGs. A key step is to highlight critical issues that emerge over time in order to implement possible corrective measures.

This paper has limitations related to the number of criteria that are available, but first and foremost we need to be grateful to those who make it possible to have this data available. In this context, ASviS supports the monitoring of SDGs [30,31]. This research proposes not only to compare individual alternatives with the national benchmark but also to make comparisons, since the normalized 0–1 method determines that the value achieved does not depend only on one's own performance. However, it is desirable to invest in obtaining more available data and that these data are updated. In fact, the other limitation of the research is that not all data are available for 2021. In addition, the same literature proposes several approaches to managing data related to the SDGs [39,40].

This paper confirms the north–south divide in Italy already proposed in the literature [24], and its comparison with the existing literature makes it possible to highlight the decisive role of monitoring. Where the number of targets is greater, more information can be extrapolated [19]. Nonetheless, this paper provides managerial and policy implications. It shows how sustainability can succeed where different strategic choices and funding policies have failed to close the gap between different areas of Italy. The Next Generation EU (NGEU) in Italy was changed because some projects would not be completed on time. Sustainability suggests not using approaches in which investments are concentrated in a limited period, but are spread out over time and allow for the diffusion of skills and resources [41].

This paper considers 16 of the 17 SDGs (SDG 17 (Partnerships to achieve the SDG) is not considered), and so the approaches used obviously change according to specific contexts. The basic idea would be to invest in the sectors that will drive the economy of the future, where there will be demand, in order to have adequate supply. Likewise, this supply must have a strong national character in order not to be subjected to geopolitical risks and foreign dependence. In this direction, the new name of the Ministry of Enterprises and Made in Italy clearly indicates this direction. Thus, the choices of the future are based on a policy direction in which there is a tendency to produce within one's own country, in which research and innovation occur, in which health skills are valued, and in which social approaches that counter forms of selfishness are developed.

Stakeholder engagement is therefore essential, with input also required from all citizens, and in particular there is a need to invest in the youngest of them, the target audience of the NGEU [29]. Businesses are called upon to revise their strategies and public administrations to change to be able to capture the dynamic aspects coming from the external context [4].

The basic question is as follows. Economic theories have taught us that a euro today is worth more than a euro tomorrow, but how do we quantify this tomorrow; how much is the opportunity cost of capital for projects that are different from each other and affect not only the Italian context but have a global vocation? The challenge, therefore, calls not only for germinating national resources but also for building bridges of freedom and democracy with other countries to move toward a common goal. Is this the hope with which Europe was born? Sustainability is, thus, not only the 2050 goal for climate neutrality but requires a pragmatic approach in which solutions are explored that give opportunities to future generations and create a brotherhood among peoples [29].

Green and circular resources call for thinking with a perspective that is not only focused on the short term but looks to the future, in which citizens will be involved to make them feel part of the change. Likewise, there is a need to overcome parochialism in order to bring out collaborations between territorial realities, since in the struggle between the small ones, in the end, those who emerge and win are not them, but they instead come out even weaker. Finally, where young people are trained in the different trades, it is necessary to retain these backgrounds. However, their demands are modified, as the abandonment of stress, the idea of being part of a group, and the possibility of having recognition for achievements determine insights that all public and private actors should reflect on.

Achieving the SDGs is done by giving confidence to young people, creating partnerships with more experienced profiles, fostering an opening of the university world to the real world, and allowing younger people to be directed to the profiles required by the market. However, it is also crucial to create and foster the concept of a community that does not only look at its own backyard but shares it with its neighbors in order to be globally competitive and attractive.

5. Conclusions

This paper makes a contribution to the pragmatic view of sustainability since, starting from an objective dataset, it tries to aggregate the data to provide different information to stakeholders. This research aims at the achievement of the different proposed targets and, thus, overall, pays attention to all SDGs, with the exception of SDG 17, which was not included in the source database.

The methodological contribution of this paper is about a very established methodology such as MCDA, in which insights are provided to be applied to the attribution of values and weights. It emerges that the range 0–1 is considered correct in order to bias the strongest and weakest performances of the examined alternatives. Moreover, in a context where the number of criteria is small, the EWI method appears to be more appropriate than the EWG method. The motivation is mathematical since where there is already a contained data availability, the additional aggregation step risks losing some of the information that emerges from the individual criteria.

The pragmatic contribution of this paper confirms the north–south divide in Italy, in which northern regions excel in the sustainability indicator, while seven of the southern ones close the ranking list. In this direction, the result of Abruzzo is very important, which aspires to have a performance equal to that of the central regions. It is precisely these regions, and in particular, Toscana, Lazio and Marche, that are on the edge of the podium occupied by three alternatives belonging to northern Italy. The province of Trento, confirming its leadership, is followed by Valle d'Aosta and the province of Bolzano. On the other hand, if these regions travel with a green card toward sustainability goals, Campania, Sicilia and Puglia take a red card.

This paper has two limitations, namely the number of criteria examined and the time period covered, which highlight how useful it is for public decision-makers to invest in this aspect. In addition, it would also be useful to analyze the impact that could be exerted by stakeholder engagement from the perspective of territorial collaboration between different regions. However, in order to make the right choices, it is necessary to have data and to assess their trends over time through appropriate monitoring of the SDGs.

It is the time for action, for proposing ideas and solutions, and not the time to stop at sterile no's and maintain one's own interests, in order to build the Europe of the future with a key role played by the realities of the Mediterranean, which will play a key role in a global economy oriented toward sustainability.

Author Contributions: Conceptualization, I.D. and M.G.; methodology, I.D. and M.G.; data curation, I.D. and M.G.; writing—original draft preparation, I.D. and M.G.; writing—review and editing, I.D. and M.G.; supervision, I.D. All authors have read and agreed to the published version of the manuscript.

Funding: This study was carried out within the PEACE (Protecting the Environment: Advances in Circular Economy) and received funding from the "Fondo per il Programma Nazionale di Ricerca e Progetti di Rilevante Interesse Nazionale (PRIN)" Investimento 1.1-D.D. 104.02-02-2022, 2022ZFBMA4. This manuscript reflects only the authors' views and opinions, and they can be considered responsible for them.

Institutional Review Board Statement: Not applicable.

Informed Consent Statement: Informed consent was obtained from all the participants.

Data Availability Statement: Not applicable.

Conflicts of Interest: The authors declare no conflict of interest.

References

1. Krishankumar, R.; Mishra, A.R.; Ravichandran, K.S.; Peng, X.; Zavadskas, E.K.; Cavallaro, F.; Mardani, A. A Group Decision Framework for Renewable Energy Source Selection under Interval-Valued Probabilistic linguistic Term Set. *Energies* **2020**, *13*, 986. [CrossRef]
2. Bali Swain, R.; Ranganathan, S. Modeling interlinkages between sustainable development goals using network analysis. *World Dev.* **2021**, *138*, 105136. [CrossRef]
3. Kuc-Czarnecka, M.; Markowicz, I.; Sompolska-Rzechuła, A. SDGs implementation, their synergies, and trade-offs in EU countries—Sensitivity analysis-based approach. *Ecol. Indic.* **2023**, *146*, 109888. [CrossRef]
4. Ali, S.M.; Appolloni, A.; Cavallaro, F.; D'Adamo, I.; Di Vaio, A.; Ferella, F.; Gastaldi, M.; Ikram, M.; Kumar, N.M.; Martin, M.A.; et al. Development Goals towards Sustainability. *Sustainability* **2023**, *15*, 9443. [CrossRef]
5. Kostetckaia, M.; Hametner, M. How Sustainable Development Goals interlinkages influence European Union countries' progress towards the 2030 Agenda. *Sustain. Dev.* **2022**, *30*, 916–926. [CrossRef]
6. Sianes, A.; Vega-Muñoz, A.; Tirado-Valencia, P.; Ariza-Montes, A. Impact of the Sustainable Development Goals on the academic research agenda. A scientometric analysis. *PLoS ONE* **2022**, *17*, e0265409. [CrossRef]
7. D'Adamo, I.; Gastaldi, M.; Morone, P. Economic sustainable development goals: Assessments and perspectives in Europe. *J. Clean. Prod.* **2022**, *354*, 131730. [CrossRef]
8. Zhao, W.; Yin, C.; Hua, T.; Meadows, M.E.; Li, Y.; Liu, Y.; Cherubini, F.; Pereira, P.; Fu, B. Achieving the Sustainable Development Goals in the post-pandemic era. *Humanit. Soc. Sci. Commun.* **2022**, *9*, 258. [CrossRef]
9. Bonnedahl, K.J.; Heikkurinen, P.; Paavola, J. Strongly sustainable development goals: Overcoming distances constraining responsible action. *Environ. Sci. Policy* **2022**, *129*, 150–158. [CrossRef]
10. Bogers, M.; Biermann, F.; Kalfagianni, A.; Kim, R.E.; Treep, J.; de Vos, M.G. The impact of the Sustainable Development Goals on a network of 276 international organizations. *Glob. Environ. Chang.* **2022**, *76*, 102567. [CrossRef]
11. Li, Z.; Huang, Z.; Su, Y. New media environment, environmental regulation and corporate green technology innovation: Evidence from China. *Energy Econ.* **2023**, *119*, 106545. [CrossRef]
12. Liu, L.; Liu, M. How does the digital economy affect industrial eco-efficiency? Empirical evidence from China. *Data Sci. Financ. Econ.* **2022**, *2*, 371–390. [CrossRef]
13. Prah, G.J. Innovation and economic performance: The role of financial development. *Quant. Financ. Econ* **2022**, *6*, 696–721. [CrossRef]
14. Masuda, H.; Kawakubo, S.; Okitasari, M.; Morita, K. Exploring the role of local governments as intermediaries to facilitate partnerships for the Sustainable Development Goals. *Sustain. Cities Soc.* **2022**, *82*, 103883. [CrossRef]
15. Fuldauer, L.I.; Thacker, S.; Haggis, R.A.; Fuso-Nerini, F.; Nicholls, R.J.; Hall, J.W. Targeting climate adaptation to safeguard and advance the Sustainable Development Goals. *Nat. Commun.* **2022**, *13*, 3579. [CrossRef]
16. Biermann, F.; Hickmann, T.; Sénit, C.A.; Beisheim, M.; Bernstein, S.; Chasek, P.; Grob, L.; Kim, R.E.; Kotzé, L.J.; Nilsson, M.; et al. Scientific evidence on the political impact of the Sustainable Development Goals. *Nat. Sustain.* **2022**, *5*, 795–800. [CrossRef]
17. Halkos, G.; Gkampoura, E.-C. Where do we stand on the 17 Sustainable Development Goals? An overview on progress. *Econ. Anal. Policy* **2021**, *70*, 94–122. [CrossRef]
18. Leavesley, A.; Trundle, A.; Oke, C. Cities and the SDGs: Realities and possibilities of local engagement in global frameworks. *Ambio* **2022**, *51*, 1416–1432. [CrossRef]
19. D'Adamo, I.; Gastaldi, M.; Ioppolo, G.; Morone, P. An analysis of Sustainable Development Goals in Italian cities: Performance measurements and policy implications. *Land Use Policy* **2022**, *120*, 106278. [CrossRef]
20. Beccarello, M.; Di Foggia, G. Sustainable Development Goals Data-Driven Local Policy: Focus on SDG 11 and SDG 12. *Adm. Sci.* **2022**, *12*, 167. [CrossRef]
21. Miola, A.; Schiltz, F. Measuring sustainable development goals performance: How to monitor policy action in the 2030 Agenda implementation? *Ecol. Econ.* **2019**, *164*, 106373. [CrossRef] [PubMed]
22. De Neve, J.-E.; Sachs, J.D. The SDGs and human well-being: A global analysis of synergies, trade-offs, and regional differences. *Sci. Rep.* **2020**, *10*, 15113. [CrossRef]

23. Huan, Y.; Liang, T.; Li, H.; Zhang, C. A systematic method for assessing progress of achieving sustainable development goals: A case study of 15 countries. *Sci. Total Environ.* **2021**, *752*, 141875. [CrossRef]
24. D'Adamo, I.; Gastaldi, M. Sustainable Development Goals: A Regional Overview Based on Multi-Criteria Decision Analysis. *Sustainability* **2022**, *14*, 9779. [CrossRef]
25. Chand, M. Strategic assessment and mitigation of risks in sustainable manufacturing systems. *Sustain. Oper. Comput.* **2021**, *2*, 206–213. [CrossRef]
26. Sarker, M.R.; Moktadir, M.A.; Santibanez-Gonzalez, E.D.R. Social Sustainability Challenges Towards Flexible Supply Chain Management: Post-COVID-19 Perspective. *Glob. J. Flex. Syst. Manag.* **2021**, *22*, 199–218. [CrossRef]
27. Pasalari, H.; Nodehi, R.N.; Mahvi, A.H.; Yaghmaeian, K.; Charrahi, Z. Landfill site selection using a hybrid system of AHP-Fuzzy in GIS environment: A case study in Shiraz city, Iran. *MethodsX* **2019**, *6*, 1454–1466. [CrossRef] [PubMed]
28. Alfirević, N.; Malešević Perović, L.; Mihaljević Kosor, M. Productivity and Impact of Sustainable Development Goals (SDGs)-Related Academic Research: A Bibliometric Analysis. *Sustainability* **2023**, *15*, 7434. [CrossRef]
29. Anselmi, D.; D'Adamo, I.; Gastaldi, M.; Lombardi, G.V. A comparison of economic, environmental and social performance of European countries: A sustainable development goal index. *Environ. Dev. Sustain.* **2023**. [CrossRef]
30. Richiedei, A.; Pezzagno, M. Territorializing and Monitoring of Sustainable Development Goals in Italy: An Overview. *Sustainability* **2022**, *14*, 3056. [CrossRef]
31. Mazziotta, M.; Pareto, A. Measuring Well-Being Over Time: The Adjusted Mazziotta–Pareto Index Versus Other Non-compensatory Indices. *Soc. Indic. Res.* **2018**, *136*, 967–976. [CrossRef]
32. ASviS. Report Territories 2021. Available online: https://asvis.it/rapporto-territori-2021/ (accessed on 28 June 2022).
33. ASviS. Report Territories 2022. Available online: https://asvis.it/rapporto-territori-2022/ (accessed on 8 August 2023).
34. Marti, L.; Puertas, R. Assessment of sustainability using a synthetic index. *Environ. Impact Assess. Rev.* **2020**, *84*, 106375. [CrossRef]
35. Sachs, J.; Schmidt-Traub, G.; Kroll, C.; Lafortune, G.; Fuller, G. *SDG Index and Dashboards Report 2018*; Bertelsmann Stiftung and Sustainable Development Solutions Network (SDSN): New York, NY, USA, 2018.
36. Gan, X.; Fernandez, I.C.; Guo, J.; Wilson, M.; Zhao, Y.; Zhou, B.; Wu, J. When to use what: Methods for weighting and aggregating sustainability indicators. *Ecol. Indic.* **2017**, *81*, 491–502. [CrossRef]
37. Roszkowska, E.; Filipowicz-Chomko, M. Measuring sustainable development in the education area using multi-criteria methods: A case study. *Cent. Eur. J. Oper. Res.* **2020**, *28*, 1219–1241. [CrossRef]
38. Guijarro, F.; Poyatos, J.A. Designing a Sustainable Development Goal Index through a Goal Programming Model: The Case of EU-28 Countries. *Sustainability* **2018**, *10*, 3167. [CrossRef]
39. Alaimo, L.S.; Ciacci, A.; Ivaldi, E. Measuring Sustainable Development by Non-aggregative Approach. *Soc. Indic. Res.* **2021**, *157*, 101–122. [CrossRef]
40. Anderson, C.C.; Denich, M.; Warchold, A.; Kropp, J.P.; Pradhan, P. A systems model of SDG target influence on the 2030 Agenda for Sustainable Development. *Sustain. Sci.* **2022**, *17*, 1459–1472. [CrossRef]
41. De La Torre, E.M.; Perez-Encinas, A.; Gomez-Mediavilla, G. Fostering Sustainability through Mobility Knowledge, Skills, and Attitudes. *Sustainability* **2022**, *14*, 1349. [CrossRef]

Disclaimer/Publisher's Note: The statements, opinions and data contained in all publications are solely those of the individual author(s) and contributor(s) and not of MDPI and/or the editor(s). MDPI and/or the editor(s) disclaim responsibility for any injury to people or property resulting from any ideas, methods, instructions or products referred to in the content.

sustainability

MDPI

Article

Trainee Teachers' Shift towards Sustainable Actions in Their Daily Routine

Asheena Singh-Pillay [1,*] and Jayaluxmi Naidoo [2]

1 Science and Technology Education Cluster, School of Education, University of KwaZulu-Natal (UKZN), Durban 4041, South Africa
2 Mathematic Education Cluster, School of Education, University of KwaZulu-Natal (UKZN), Durban 4041, South Africa; naidooj2@ukzn.ac.za
* Correspondence: pillaya5@ukzn.ac.za

Abstract: The argument made in this study is that sustainability is a system of beliefs that extends beyond policy and the classroom to transform actions and societies into pro-sustainable behaviours. Therefore, it is crucial to equip trainee teachers to embrace their role as catalysts for change in driving Sustainable development goals, sustainable thinking, systems thinking, wise consumption, sustainable competencies, and sustainable action in their daily lives. The lack of practice-led research on incorporating Education for Sustainable Development (ESD) into trainee teachers' training programmes is noted in the literature. This qualitative practice-led research explored trainee mathematics and technology teachers' (TMTTs) experiences learning about sustainability. Kolb's experiential theory framed this study theoretically. The data were collated from twenty purposively selected TMTTs via semi-structured interviews and reflective diaries. All ethical protocols were observed. The analysis highlighted the core experiences TMTTs gained in learning about sustainability, transformative learning, design thinking, wise consumption, agency, and sustainable teaching strategies. Theoretically, the findings emphasise the importance of using an ESD perspective and experiential learning methods to integrate sustainability education into mathematics and technology teaching. The results suggest that when TMTTs are encouraged to actively engage with sustainability issues in their communities, informed decisions are made about their future roles as teachers, the teaching strategies they intend to use, and the type of learning they aim to foster in their learners.

Keywords: education for sustainable development; sustainable development goals; trainee teachers; system thinking; wise use

Citation: Singh-Pillay, A.; Naidoo, J. Trainee Teachers' Shift towards Sustainable Actions in Their Daily Routine. *Sustainability* **2024**, *16*, 8933. https://doi.org/10.3390/su16208933

Academic Editors: Manoj Kumar Nallapaneni, Idiano D'Adamo and Massimo Gastaldi

Received: 30 July 2024
Revised: 11 October 2024
Accepted: 14 October 2024
Published: 15 October 2024

Copyright: © 2024 by the authors. Licensee MDPI, Basel, Switzerland. This article is an open access article distributed under the terms and conditions of the Creative Commons Attribution (CC BY) license (https://creativecommons.org/licenses/by/4.0/).

1. Introduction

Sustainable development is a buzzword that has gained significance. It involves the responsible utilization of natural resources to meet current needs while safeguarding the ability of future generations to meet their own needs. Countries globally are trying to advance the 17 Sustainable development goals in their social, economic, and educational terrains. Globally, documents such as the Decade of Education for Sustainable Development, which was extended from 2004 to 2014 [1], UNESCO's Global Action Programme: Sustainability began with teachers [2], and Agenda 2030 emphasise the need to incorporate education on sustainable development into the curriculum for teacher training programmes [3]. Education for sustainable development (ESD) refers to education systems that privilege sustainable development across disciplines and education mandates [4]. The 17 SDGs foreground inclusive, peaceful societies that collaborate to fight inequalities and encourage economic growth while caring for the planet [5]. South Africa's educational policies, including the White Paper on Education and Training [6], the National Curriculum Statement [7], and the National Development Plan [8], are all geared towards achieving the Sustainable Development Goals (SDGs) outlined in Agenda 2030. Furthermore, there is a growing consensus that universities in South Africa must take a leading role in addressing the SDGs through various initiatives and projects, and that incorporating Education for Sustainable Development (ESD) into their curricula is crucial in tackling the escalating sustainability challenges.

These calls have impacted the traditional role of HEIs as knowledge producers [9]. Consequently, HEIs are expected to provide leadership and sustainable, innovative solutions via their research [10] to bring about social transformation [11]. HEIs are construed as catalytic tools for achieving the 17 SDGs via research, curriculum development, learning outcomes, student engagement, faculty development, and community engagement. All 17 SGDs intrinsically connect to education [12]. For example, when people have access to quality teaching and learning (SDG4), they are better educated and have access to jobs, which invariably reduces unemployment, poverty (SDG1), hunger (SDG2), and inequalities (SDG5), contributing to economic growth (SDG8) and stimulating good health and well-being (SGD 3); people live in peace (SDG16); people afford clean water and sanitation (SDG6) and energy (SDG7), leading to sustainable cities and communities (SDG9).

Moreover, these calls for integrating ESD into the curriculum have catapulted teachers and teacher educators to the centre stage as brokers of ESD curriculum transformation, propagators of systems thinking, sustainable thinking, an ethic of care towards sustainability issues, sustainable actions, student agencies and developers of sustainable competencies, collaborations and partnerships. The calls for integrating ESD into the curriculum also have implications for student engagement in terms of empowerment to be champions of ESD in their teaching and to take up their role as critical citizens at a community level [13]. Sinakou et al. [14] argued that trainee teachers should be provided with an opportunity to experience learning and practice sustainable thinking and behaviour, not via exposure to the things taught in teaching but through participating in the learning itself. Thus, Refs. [15,16] emphasise that academics have a crucial responsibility to make intentional decisions in their curriculum design and teaching practices to promote and instil sustainable thinking and values in their students.

This paper asserts that teachers and teacher educators must assume a pivotal role in promoting ESD and the SDGs, thereby cultivating a mindset of sustainability among their students. This means they must be reflexive practitioners who disrupt, rethink, and re-envision the curriculum and espouse pedagogies that support the integration of ESD, activating student engagement and collaboration to promote sustainable thinking and actions in students without being prescriptive. By transforming the curriculum and pedagogy, TMTTs can develop a profound understanding of sustainability, foster a deep sense of environmental stewardship, and acquire the skills to tackle complex sustainability challenges in their local communities. This empowering experience enables them to integrate ESD into their teaching practices, thereby catalysing a ripple effect that inspires schools and communities to adopt sustainable practices and demand systemic change.

We are teacher educators at a teacher training university in South Africa, the locale of this research. We lecture to the same cohort of students training to be specialist Technology and Mathematics teachers. The majority of the participant were from previously disadvantaged communities and received funding for their studies. They were between the ages of 19 and 23. In their technology lectures on the processing of polymers (which deals with the chemistry of processing polymers (Crude oil is a monomer/building block of polymers)), students drew from mathematics to perform calculations, audits, and visual representations. The approved module content and assessment design for the polymers module were theoretical and lecture-based. They did not allow TMTTs to reflect on sustainability issues and link theory to contextual problems (2.4 million tons of plastic waste is disposed of recklessly in South Africa) in order to engage TMTTs in experiential learning, and initiate their citizenship and environment stewardship. The module content and assessments were realigned to the global and national call to integrate ESD/SDGs into the curriculum and to contextualise the activities and tasks. Contextualised learning is an innovative approach that bridges the gap between academic learning and real-life experiences, allowing students to apply their knowledge to tackle authentic sustainability challenges in their local community. By doing so, it fosters a deeper understanding of complex issues, encourages creative problem-solving, and helps students develop a richer appreciation for their daily lives and the world around them. The polymers module was used to pilot a project. Assessment activities were reconstructed in the module to connect to the community so that learners could learn from actual sustainability issues. The literature is replete with studies on climate change [17], as well as the pivotal role that rural universities can play in spearheading sustainable development initiatives [18]. Furthermore, a plethora of research has delved into the multifaceted challenges that higher education leaders confront in their quest to drive sustainable development and create a more resilient future, including the goal of quality education in South Africa [19], perspectives of academic staff on embracing SGDs [20], challenges of involving students in SDG-related practices [21], and challenges in meeting SDGs in Southern Africa [22]. Few studies emphasise ESD integration across curriculum and pedagogy [23]. Moreover, there is a scarcity of empirical studies on the practicalities of integrating ESD into the higher education curriculum and examining students' firsthand experiences of learning about ESD. This paper reports on a study that explored trainee mathematics and technology teachers' (TMTTs) experiences in learning about sustainability and addresses the gaps identified in the literature.

While existing research, such as [24], focuses on quantitative monitoring of sustainable development progress using established indicators, there remains a critical gap in understanding the qualitative experiences of implementing sustainability education in teacher education programmes, particularly in developing contexts and specific subject areas.

This study addresses multiple gaps in the current literature. First, while many studies emphasise macro-level sustainability indicators and outcomes, our research uniquely explores the practical challenges and opportunities of integrating ESD into specific subject areas—mathematics and technology teacher training. Second, in contrast to research conducted in developed nations using top-down monitoring approaches, this study provides valuable insights from the South African context, specifically examining the experiences of students from previously disadvantaged communities. Third, unlike existing studies that focus on policy analysis or theoretical frameworks, our research employs a practice-led approach to understand how trainee teachers engage with and conceptualise sustainability in their learning journey.

The findings of this study are significant as they not only illuminate the crucial role of universities in advancing SDGs through teaching and learning but also provide empirical evidence of the practicalities, challenges, and opportunities of integrating ESD into the higher education curriculum. By focusing on the intersection of subject-specific pedagogy (mathematics and technology) with sustainability education, this research offers unique insights into how disciplinary knowledge can be leveraged to develop sustainability competencies. Furthermore, by examining the experiences of trainee teachers who will be future drivers of the SDGs, this study contributes to our understanding of how to effectively prepare educators to champion sustainability in diverse educational contexts.

2. Literature Review

This literature review focuses on the role of universities and teachers in promoting Education for Sustainable Development (ESD) and implementing Sustainable Development Goals (SDGs). The review is structured to address three key areas that are crucial to understanding the context and importance of our study:

Universities and SDGs: this section explores how higher education institutions are uniquely positioned to drive sustainability initiatives through research, education, and community engagement.

Teachers' views and attitudes towards ESD: here, we examine existing research on how teachers view and approach ESD, highlighting the importance of positive attitudes and proper training in effectively implementing sustainability education.

Relevant pedagogical approaches for ESD integration: the final section reviews current pedagogical strategies that are considered most effective for teaching ESD principles.

By examining these areas, we seek to establish the theoretical foundation for our study on pre-service teachers' experiences in learning about sustainability. This review will help contextualise our research questions and highlight the gaps in current knowledge that our study aims to address.

2.1. Universities and SDGs

Universities' pivotal role in fostering sustainability extends beyond traditional research and education to encompass cultural transformation and community building [25]. While the existing literature emphasises universities' capacity to leverage diverse faculties and forge strategic collaborations [25,26], emerging perspectives suggest a more student-centred and community-oriented approach. This evolving paradigm advocates for institutions to amplify student voices and cultivate a heightened sense of collective responsibility. By reorienting academic assignments and research initiatives to prioritise widespread societal impact over individual achievement, universities can nurture a culture that transcends personal interests in favour of communal benefit [26]. This approach aligns with and enhances the existing framework of universities as catalysts for sustainable development, where the focus extends beyond environmental stewardship to encompass social cohesion and stress reduction within academic communities. Moreover, this perspective complements the current emphasis on developing transformative leaders [27] by suggesting that true transformation begins with fostering a culture of empathy, mutual responsibility, and community-minded thinking. This cultural shift could significantly enhance universities' effectiveness in implementing SDGs and ESD principles, as it addresses not just the academic and research aspects but also the fundamental social and psychological barriers to sustainability. The integration of such community-building approaches with existing pedagogical strategies could create a more holistic and impactful framework for sustainability education in higher institutions [28,29].

2.2. Teachers' Views and Attitudes towards ESD

Corney and Reid [30] argue that the effectiveness of teachers in incorporating sustainability into their teaching hinges on their expertise and perspectives on subject matter and teaching methods. They propose that fostering a positive attitude towards sustainability among teachers could lead to greater engagement with Education for Sustainable Development (ESD). Tomas et al. [31] suggest that teachers with favourable attitudes towards the environment demonstrate a strong self-efficacy in promoting ESD. Similarly, ref. [32] discovered that effectively teaching ESD concepts hinges on educators possessing a genuine enthusiasm for and commitment to the underlying principles of ESD. Howlet et al. [33] (2016) believe that teachers' understanding and awareness of ESD are shaped by their attitudes to ESD. Gan and Gal [34] note that teachers need proper training to infuse ESD into their lessons. Therefore, there is a consensus that ESD should be integrated into initial teacher training programs.

Research by [35] indicates that teachers believe integrating ESD into the curriculum would enhance their ability to teach about sustainable development (SD), particularly emphasising its importance in early childhood education. However, ref. [36] found that teachers acknowledge the need for further education on environmental issues. Burgener and Barth [37] assert that teachers' views of ESD impact their teaching. Additionally, ref. [35] reveals that teachers feel ESD knowledge, skills, and attitudes are not adequately addressed in science education. Redman et al. [36] also identify a gap between ESD learning outcomes and what is being taught, suggesting that integrating ESD would improve the relevance of science education. This highlights the need for scientific educators to understand ESD more deeply.

Burgener and Barth [37] noted that many teachers lack the specialised knowledge, skills, and pedagogical expertise required to effectively incorporate ESD into their teaching, thereby hindering the adoption of transformative and impactful learning approaches. They propose transformative learning, utilising real-life challenges and examples. Pedagogical Content Knowledge (PCK) should encompass subject content, teaching methods, ESD knowledge, and competencies [37]. Maidau et al. [38] emphasise the need for ESD-specific Pedagogical Content Knowledge (ESD PCK), which includes knowing the content and appropriate understanding of sustainable development issues and pedagogical techniques for integrating ESD into teaching. They advocate for science, technology and mathematics teachers to connect content to societal challenges associated with ESD, thus aligning with ESD PCK. Furthermore, they stress the importance of guiding science, technology and mathematics teachers in integrating ESD and change-oriented teaching experiences into scientific instruction, addressing real-life and community concerns.

2.3. *Relevant Pedagogical Approaches for ESD Integration*

According to [39], teachers must embody both subject-matter knowledge and pedagogical sophistication to teach the principles of ESD and cultivate a profound appreciation of sustainable development concepts among their students. The prevailing consensus among scholars highlights learner-centred approaches, experiential and participatory methods [40], praxis-oriented and place-based instruction, and interdisciplinary and inquiry-based techniques [41]. These teaching methodologies encourage behavioural shifts and alter individuals' perspectives and responses towards processes and knowledge. Biasutti [42] asserts that these approaches are preferred because they foster critical thinking skills among learners and are learner-centred [43]. ESD emphasises cultivating cooperative skills, such as decision-making, critical thinking, and envisioning future scenarios, necessitating a re-evaluation of current educational delivery methods [44]. It is imperative to reshape teacher training towards sustainability and embrace more systemic learning and instructional models [45], departing from conventional teaching paradigms [31]. Teachers recognise that teaching ESD principles requires specific learner-centred, action-oriented pedagogies [46]. Such pedagogical strategies motivate learners to engage in actions promoting sustainable development.

2.4. *Kolb's Experiential Learning Theory*

This research was framed by Kolb's Experiential Learning Theory (ELT) [47]. It centred on the experiences of TMTTs as they learned about sustainability within the polymers module. ELT was deemed suitable for several reasons: Firstly, it outlines how shifts in experiences lead to learning, particularly emphasising the journey towards sustainability. Secondly, it acknowledges that TMTTs engage in a holistic learning process. Thirdly, it facilitates transformative learning experiences, crucial for sustainability education, when all four stages of ELT are experienced. Lastly, ELT has been proven effective in teaching sustainability in consumption.

ELT is a cyclic process with concrete experience (CE) and abstract conceptualisation (AC) for understanding experience and reflective observation (RO) and active experimentation (AE) for transforming experiences. When TMTTs undertook their task in their community, they traversed all the stages of ELT: (1) encountering a concrete experience, (2) reflecting on and observing that experience, (3) integrating reflections to conceptualise interventions, and (4) applying new ideas through active experimentation.

3. Materials and Methods

3.1. *General Background*

The interpretive paradigm steered this study. A vital feature of the interpretative paradigm is that it aims to describe and make sense of the phenomenon explored from the participants' experiences, opinions and perspectives [48]. Aligned with the interpretive paradigm, a qualitative research approach was used as it captures participants' lived experiences from their perspectives [49].

The study was conducted at a teacher training institution in Kwa Zulu Natal in South Africa. The data were collected from 20 TMTTs who specialised in technology and mathematics and were registered in 2018 for the polymers module. All ethical protocols concerning informed consent and voluntary participation were adhered to. A purposive sampling technique was used. The identifying criteria for selection were that participants had to specialise in technology and mathematics and had to be registered for the processing of the polymer module. All 20 TMTTs (10 males and 10 females) teachers lived in university residences around the campus. Participants self-selected four groups to work in for the task-based activity. Each group had 10 participants, and they were assigned pseudonyms A and B, respectively. All participants were workshopped on participatory action research (PAR) as it was used to generate data for their tasks. The workshop focused on teaching the participants about the core principles and key features and stages of Participatory Action Research so that they could apply this methodology to collect data for their own research tasks or projects. Interactive methods, such as small group discussions and hands-on exercises, were used to engage the participants, provide practical experience with PAR techniques, and equip participants with the skills and knowledge to conduct PAR in their own communities or organisations, fostering a more democratic and action-oriented approach to research and problem-solving.

The participants had to perform the following audit tasks at their residences and on campus by inspecting garbage contents. Students were provided with disposable gloves and masks. A waste audit is an exciting opportunity to investigate the types and quantities of materials thrown away on campus daily. The task was designed due to South Africa's massive problem with plastic consumption and disposal [13] and it allowed for reflection and critique of consumption habits, hopefully resulting in a change of behaviour. This task was linked to the polymers module as participants engaged with content on the classification, chemical properties, structure, and recycling of plastics. The participants had to share their findings with the residence students they worked with and develop a joint solution to address consumption patterns.

3.2. Instruments

The data were collected via individual interviews and reflective journals. The semi-structured interviews were audio recorded and were of thirty-minute duration. The interview questions focused on consumption patterns, disposal, recycling, ESD, ESD behaviour, changing how coffee and takeout are served, environmental degradation, citizenship responsibility, and agency. TMTTs recorded their experiences engaging in project-based learning and ESD in the reflective journals. The protocol of the reflective journal was for the participants to record their experiences as listed above.

3.3. Data Analysis

The interview transcripts and reflective journals were labelled from 1 to 20 for analysis before data analysis could commence. The transcripts of the interviews were sent to TMTTs for member checking to establish that their responses were captured accurately [49]. Member checking enhances the trustworthiness and validity of the data, ensuring that the findings accurately reflect the experiences and perspectives of the participants. Both data sets were analysed thematically.

Multiple readings of the transcripts were completed to identify similarities and divergences before coding could begin. This was completed individually by each author and compared thereafter. Codes were positioned alongside each other before they could be regrouped into themes. Table 1 reflects the themes and codes. The themes were also subjected to member checking.

Table 1. Categories and codes for interview questions.

Interview Question	Themes	Codes
What is your Experience in Learning about sustainability using PAR in communities?	Habits shape behaviour	Application of theory to solve contextual issues, fossil fuels, human activities, poor habits, repeated habits, progression of bad habits, emissions and global warming
	Default daily practices	Awareness of using natural resources, wise consumption, minimalist lifestyle, changing daily habits, one step at a time, ESD a way of life from early years
	Pedagogies for ESD	Consciousness, critical reflection, sustainable actions, learning for sustainability

4. Findings and Discussion

The audit findings are presented first, followed by the three experiences TMTTs encountered in learning about sustainability.

4.1. Audit of Residence Garbage

As mentioned earlier, TMTTs had to audit the garbage at their residences for over one week (five days). Table 2 highlights the items disposed of by students residing at the campus residences into garbage packets. The figures displayed in Table 2 are an average over five days.

Table 2. Audit results.

ITEM	Group A	Group B	Category of Plastic	Recycling Status
Coffee cups	**1500**	**1200**	Thin plastic lining	4
Polystyrene take-out containers	**1000**	**950**	Polystyrene	5
Plastic cutlery	**2500**	**1600**	Bisphenol	7
Shopping bags	**300**	**200**	Low-density polyethene	4
Used/soiled serviettes	**600**	**700**	Biodegradable	Can be composted
Disposal razor	**80**	**60**	High-density polyethylene handle	2
Water bottles	**500**	**400**	High-density polyethylene	2
Uneaten food	**yes**	**yes**	Organic	Can be composted
Cider/beer bottles	**700**	**890**	Glass	Can be recycled

The items listed in Table 2 indicate trainee teachers' (TTs) waste separation behaviour. TTs did not separate/segregate their litter or consider whether it could be recycled or composted. They also did not consider their consumption habits or focus on reducing the waste they produce. Bins are provided around the residences for different categories of waste, such as paper, glass, plastics, and composting. Coffee cups, polystyrene take-out containers, plastic cutlery, disposable shavers, water bottles and shopping bags are all non-biodegradable. TTs seem oblivious that they cannot decompose. These items are difficult to recycle and most likely end up in the landfill close to campus. The failure to separate/segregate the litter produced means it will end up in the landfill in the same mixed assortment. Moreover, the mixing of waste complicates the handling and separation of recyclables. As well as the reduction of waste volume. Uneaten food, soiled serviettes and liquid from coffee cups and water bottles can mix and decompose, releasing harmful gases/emissions into the atmosphere and runoffs into the soil. TTs seem oblivious to the impact of their waste separation behaviour (or lack thereof) on the environment and natural resources. The audits indicate that TTs need to be empowered to become more sustainable, conscious of their consumption behaviour, and aware of the natural resources used to produce these waste items. The TMTTs have been driving the change campaign on campus to reduce the use of plastics and drive pro-sustainable behaviour among students.

The audits conducted by TMTTs underpin their experiences in learning about sustainability. These experiences are discussed next.

4.2. Habits Shape Behaviour

During the audit activity, TMTTs connected the theory from the polymer modules with the activity of sorting/setting garbage, as is visible in the excerpts below.

> *Engaging in the audit was eye and mind-opening for me . . .I could apply the content from polymers to understand more about our garbage and its impact on us and the world. We need to inculcate the behaviour of sorting waste as early as possible—from toddlers.. it starts at home. It must be reinforced at nursery school, primary school, and high school . . . we have to see the bigger picture when applying theory to solve a contextual problem about consumption, waste produced, natural resources, fossil fuels, used and its impact on the environment.* (P7 interview)

> *These tasks that we have been doing as part of the module have extended our application of learning, and these kinds of activities and tasks are invaluable in learning about sustainability. Doing this garbage sorting task has made me realise that all of us are equally responsible for our actions and its impact on the environment and resources.* (P12, Interview)

> *The waste or garbage we produce is a human activity that we can control; our actions of using products made from non-renewable resources impact the availability of non-renewable resources and contribute to global warming and the greenhouse effect, I am applying what was discussed in our lectures to understand the contents of the garbage bags. I look back at students' poor garbage disposal habits and look forward to how I can change PSTs behaviour. These are educated citizens, yet their behaviour and habits on plastic use and disposal are shocking.* (P19 reflective diary)

Participants applied the content from the polymer modules to complete the audit. In the process, they see teaching as a contextualised activity to solve local problems, such as the reckless disposal of plastic and an opportunity to engage in critical thinking and reflection. They linked pre-service teachers' indiscriminate waste disposal behaviours to old, well-established, hard-to-beak habits that are so well ingrained and entrenched that they persist despite all the sustainability awareness campaigns to reduce consumption and environmental impact. This finding resonates with that of [50], which asserts that habits are the basis for daily actions and become barriers to change. What comes to the fore is that students' waste disposal behaviour is linked to habits that subvert knowledge of sustainability awareness campaigns or learning about pro-sustainable actions. Verplanken [51] noted that habits cue our actions and responses and supersede our intentions and learnings. Similarly, ref. [52] indicated that sustainable actions can only occur if habits change.

TMTTs also linked the students' poor waste management and excessive use and disposal of plastic to the earth's non-renewable resources. TMTTs could see the bigger picture and discern the link between human activity, consumption actions, non-renewable resources, and their environmental impact. The audit conscientised TMTTs. It was an eye-opener to them about their role in resolving environmental issues and promoting sustainable behaviour, actions, and sustainable competencies as early as possible in children. This finding confirms that a challenge in TT curricula is helping students unlearn habits and transform their actions. Furthermore, despite the many active learning approaches included in pre-service teacher education curricula, competencies for ESD still need to be developed. It also made them realise that all pre-service teachers should be responsible for sustainability issues, behaviours, and actions and have sustainability competencies. This means that sustainability should not be the responsibility relegated to some pre-service teachers only. The findings of this study align with existing research [53], reaffirming the critical role of ESD in fostering a sustainable mindset in trainee teachers, enabling them to promote environmental care and sustainable behaviours in communities.

4.3. Default Daily Practices

TMTTs reflect deeply on what they can do to bring about a change in students' habits, behaviour and actions concerning waste disposal, as can be gathered from the excerpts below:

> *How do we get students to change their ill-formed habits? It is not just what we learn in classes; we have a bigger responsibility. We are all training to be teachers. We have to drive sustainable actions and develop sustainable competencies amongst the student cohort on campus. We have to care about the world we live in and how our actions impact the ecosystems and natural resources.* (P1 interview)

> *It is our responsibility to drive change amongst our fellow pre-service teachers. We have in-depth knowledge of plastics and should share it so these students can also change their habits of sorting their waste and consuming plastic products and think about the depletion of natural resources.* (P15, reflective diary)

> *We are always learning and encouraged to be change agents. We are enlightened with polymer content knowledge, skills, and sustainability competencies. We must empower students to adopt sustainable lifestyles, making conscious choices about consumption and renewable and non-renewable resources and its environmental impact—in short, they need to be introduced to systems thinking.* (P2, reflective diary)

From the above testimonies, it is evident that TMTTs have been reflecting on their experiences and feel empowered with knowledge, skills, and sustainable competencies, such as critical thinking, to take on their role as change agents in influencing students' daily practice regarding waste disposal, poor consumption habits, wise consumerism, and awareness of natural resources, which are not limitless. Learning about ESD has changed TMTTs' outlook on the environment and the need for systems thinking when addressing poor environmental behaviour. TMTTs underscore the importance of ESD in fostering awareness about pro-sustainable behaviour. The positive impact of integrating an ESD approach to the polymer module is evident in the above testimonies. TMTTs take up their role as drivers of ESD by seeking to transform students' waste disposal and consumption habits. This finding concurs with that of [54], which noted the positive impact of teaching students about environmental education at tertiary institutions. Kopnina [27] asserts that teaching students about environmental concerns is crucial, as they are our future.

4.4. Pedagogies for ESD

TMTTs recognise, first, that learning extends beyond that classroom and, second, the need for activities and tasks that allow learners to apply theory to solve contextualised problems in their communities, as in the excerpts below:

> *I have learned that traditional pedagogies are not suited for ESD activities. To learn about ESD, students must be allowed to develop awareness and think critically to solve a contextual problem in their community. The community should be the backdrop for the setting of tasks.* (P13, interview)

> *Engaging in the polymer module and the audit activities has made me realise that teaching learners about ESD requires transformative pedagogies, like project-based learning, action research, and pedagogies that allow students to collaborate, solve problems, think critically and be reflexive of their actions.* (P16, interview)

> *I think everyone must learn about ESD. I plan to integrate ESD in all sections of technology and math, for example, in the section on electricity, I will integrate sustainable use of resources, wise consumption, discuss illegal electricity connections, social justice, learners cannot be taught a solution to an environment challenge, they have to evaluate, think critically to solve the problem.* (P11, reflective diary)

The testimonies above confirm that learning about ESD requires active learning tasks in community settings to bridge learning with community action and transform habits, mindsets and sustainability competencies. TMTTs understand that addressing sustainability issues requires employing contextualised learning within a community context. In other words, the pedagogy and the learning environment of the task must allow for critical thinking, collaboration, consciousness-raising, reflection, change in values, habits, and behaviour and a heightened sense of care for the environment and its resources [15]. It must allow them to reflect and critique their actions and perspectives. TMTTs are drawing on their experience and involvement in a community-based audit task to mould the pedagogies they plan to use in their classrooms. Such pedagogies bode well for sustainability issues. ESD pedagogies are collaborative, situated, and social processes.

5. Conclusions

This research explored South African Trainee Mathematics Technology Teachers' (TMTTs) experience of learning about sustainability during the polymer module. The module espoused SDG 4, which underscores quality teaching and learning, while the audit task embraced SDG 12 (responsible consumption) and SDG 11 (sustainable communities). Our findings reveal that to effectively pursue sustainability goals, universities must undergo a fundamental shift in their approach, particularly in how they engage with and empower students.

Universities serve as catalysts for change, driving the development of sustainable communities through education, research, and community engagement, as highlighted by SDG 11. However, our research suggests that this role needs to evolve beyond traditional approaches. To truly pursue the goal of sustainability, universities should accomplish the following:

- Lead cultural change by actively listening to student voices and perspectives;
- Increase students' sense of responsibility towards others;
- Design assignments that maximise community impact.

This paper emphasises the importance of appropriate pedagogies for teaching ESD, such as active learning tasks in community settings. By engaging pre-service teachers in these practical, community-oriented experiences, universities can directly contribute to sustainable community development while preparing future educators.

The research argues for the need to empower trainee teachers as agents of change in driving sustainable thinking, systems thinking, wise consumption, sustainable competencies, and sustainable action in their daily lives. This empowerment aligns with the university's role in fostering sustainable communities. By equipping trainee teachers with these skills and perspectives, universities extend their impact beyond campus boundaries.

Our findings indicate that when trainee teachers engage in community-focused sustainability initiatives, they develop the following traits:

- A deeper understanding of sustainability challenges;
- Enhanced problem-solving skills in real-world contexts;
- A stronger sense of community responsibility.

As a result of this approach to building sustainable communities through teacher training education, these teachers will be better prepared to undertake the following tasks:

- Adapt their teaching methods to incorporate sustainability principles;
- Influence not just students, but also parents and the wider community;
- Contribute to a broader cultural shift towards sustainability.

This strategic approach to empowering trainee teachers about sustainability creates a long-term, widespread impact, amplifying the university's role in community sustainability. It aligns with the growing recognition that solutions should accrue benefits to many, not just a few, and that building community and overcoming individualistic approaches are crucial for sustainable development.

Our research suggests several key implications:

- Policymakers should prioritise solutions that benefit the broader community;
- Universities should redesign curricula to emphasise community impact;
- Stress reduction and community building should be integrated into sustainability education.

Ultimately, this approach helps develop environmentally conscious, socially responsible, and action-oriented future generations capable of addressing the complex sustainability challenges our world faces. Through this process, universities become true catalysts for sustainable community development, fulfilling their educational mission while actively contributing to global sustainability goals.

Author Contributions: Conceptualization, A.S.-P.; Methodology, A.S.-P.; Formal analysis, J.N.; Investigation, J.N.; Writing—original draft, A.S.-P. and J.N. All authors have read and agreed to the published version of the manuscript.

Funding: This research received no external funding.

Institutional Review Board Statement: HSSREC 00003472/21.

Informed Consent Statement: Informed consent was obtained from all subjects involved in the study.

Data Availability Statement: All data are available from the corresponding author upon reasonable request.

Conflicts of Interest: The authors declare no conflicts of interest.

References

1. McGrath, S.; Powell, L. Skills for Sustainable Development. *Int. J. Educ. Dev.* **2016**, *50*, 12–19. [CrossRef]
2. United Nations Educational, Scientific and Cultural Organization (UNESCO). *EFA Global Monitoring Report for 2005. Education for All: The Quality Imperative*; Final Report; UNESCO: Paris, France, 2005.
3. United Nations Educational, Scientific and Cultural Organization (UNESCO). *UNESCO Roadmap for Implementing the Global Action Programme on Education for Sustainable Development*; UNESCO: Paris, France, 2014.
4. United Nations Educational, Scientific and Cultural Organization (UNESCO). Transforming Our World: The 2030 Agenda for Sustainable Development. 2015. Available online: https://documents-dds-ny.un.org/doc/UNDOC/GEN/N15/291/89/PDF/N1529189.pdf?OpenElement (accessed on 7 February 2023).
5. Garcia-Meca, E.; Lopez-Iturriaga, F.; Tejerina-Gaite, F. Institutional investors on boards: Does their behavior influence corporate finance? *J. Bus. Ethics* **2017**, *146*, 365–382. [CrossRef]
6. Republic of South Africa Department of Basic Education. *The White Paper on Education and Training*; Government Printer: Pretoria, South Africa, 1995. Available online: https://www.gov.za/documents/white-paper-education-and-training (accessed on 9 June 2023).
7. Department of Basic Education. Curriculum Assessment Policy Statements (CAPS): Civil Technology. 2011. Available online: https://www.education.gov.za/Portals/0/Documents/CIVIL%20TECHNOLOGY%20FET.pdf (accessed on 9 June 2024).
8. PricewaterhouseCoopers. Make It Your Business: Engaging with the Sustainable Development Goals. 2016. Available online: https://www.pwc.co.za/en/assets/pdf/sustainable-development-goals-south-africa-focus.pdf (accessed on 9 February 2024).
9. Zaléniené, I.; Pereira, P. Geography and sustainability higher education for sustainability: A global perspective. *Geogr. Sustain.* **2021**, *2*, 99–106.
10. Too, L.; Bajracharya, B. Sustainable campus: Engaging the community in sustainability. *Int. J. Sustain. High. Educ.* **2015**, *16*, 57–71. [CrossRef]
11. Rosak-Szyrocka, J.; Apostu, S.A.; Ali Turi, J.; Tanveer, A. University 4.0 sustainable development in the way of Society 5.0. *Sustainability* **2022**, *14*, 16043. [CrossRef]
12. Pasara, M.T.; Dunga, S.H. Trade creation and diversion effects in the tripartite region: A gravity approach. *Int. J. Econ. Financ. Stud.* **2020**, *12*, 305–320. [CrossRef]
13. Adeniran, A.A.; Shakantu, W. The health and environmental impact of plastic waste disposal in South African townships: A review. *Int. J. Environ. Res. Public Health* **2022**, *19*, 20779. [CrossRef] [PubMed]
14. Sinakou, E.; Donche, V.; Boeve-de Pauw, J.; Van Petegem, P. Designing powerful learning environments in education for sustainable development: A conceptual framework. *Sustainability* **2019**, *11*, 15994. [CrossRef]
15. Kalsoom, Q.; Khanam, A.; Quraishi, U. Sustainability consciousness of pre-service teachers in Pakistan. *Int. J. Sustain. High. Educ.* **2017**, *18*, 1090–1107. [CrossRef]
16. Leal Filho, W. "Sustainability 2.0" a new age of sustainable development in higher education. *Int. J. Sustain. High. Educ.* **2015**, *16*, 57–71. [CrossRef]

17. Chersich, M.F.; Wright, C.Y. Climate change adaptation in South Africa: A case study on the role of the health sector. *Glob. Health* **2019**, *15*, 22. [CrossRef] [PubMed]
18. Uleanya, C. Rural University Education for Sustainable Development in South Africa: A Review. *Afr. Educ. Rev.* **2022**, *19*, 160–176. [CrossRef]
19. Pramjeeth, S.; Nupen, D.; Jagernath, J. Challenges Impacting Higher Education Leaders in Achieving the Sustainable Development Goal of Quality Education in South Africa. *Afr. J. Inter/Multidiscip. Stud.* **2023**, *5*, 1–13. [CrossRef]
20. Nhamo, G.; Chapungu, L. Seven years of embracing the sustainable development goals: Perspectives from University of South Africa's academic staff. *Front. Educ.* **2024**, *9*, 1354916. [CrossRef]
21. Mawonde, A.; Togo, M. Challenges of involving students in campus SDGs-related practices in an OdeL context: The case of the University of South Africa (Unisa). *Int. J. Sustain. High. Educ.* **2021**, *22*, 1487–1502. [CrossRef]
22. Mutanga, S.S.; Skhosana, F.; Mateyisi, M.; Thenga, H.; Naidoo, S.; Lumsden, T.; Ramoelo, A.; Nangombe, S.S. Environmental challenges to meeting sustainable development goals in Southern Africa. In *Sustainability of Southern African Ecosystems under Global Change: Science for Management and Policy Interventions*; von Maltitz, G.P., Midgley, G.F., Brümmer, C., Veitch, J., Rötter, R.P., Viehberg, F.A., Veste, M., Eds.; Springer: Cham, Switzerland, 2024; Volume 248, p. 1. [CrossRef]
23. Wu, K.; Liao, C.; Tseng, M.; Chiu, K. Multi-attribute approach to sustainable supply chain management under uncertainty. *Ind. Manag. Data Syst.* **2016**, *116*, 777–800. [CrossRef]
24. D'Adamo, I.; Di Carlo, C.; Gastaldi, M.; Rossi, E.N.; Uricchio, A.F. Economic Performance, Environmental Protection and Social Progress: A Cluster Analysis Comparison towards Sustainable Development. *Sustainability* **2024**, *16*, 5049. [CrossRef]
25. El-Jardali, F.; Ataya, N.; Fadlallah, R. Changing roles of universities in the era of SDGs: Rising up to the global challenge through institutionalising partnerships with governments and communities. *Health Res. Policy Syst.* **2018**, *16*, 38. [CrossRef]
26. Bhowmik, J.; Selim, S.; Huq, S. The role of universities in achieving the Sustainable Development Goals. In *CSD-ULAB and ICCCAD Policy Brief*; ULAB: Dhaka, Bangladesh, 2018.
27. Kopnina, H. Teaching sustainable development goals in The Netherlands: A critical approach. *Environ. Educ. Res.* **2018**, *24*, 1268–1283. [CrossRef]
28. Togo, M. Students as agents of social change: Student initiatives at Rhodes University, South Africa. *S. Afr. J. Environ. Educ.* **2009**, *26*, 232–242.
29. Trencher, G.; Broto, V.; Takagi, T.; Sprigings, Z.; Nishida, Y.; Yarime, M. Innovative policy policies to advance building energy efficiency and retrofitting: Approaches, impacts and challenges in ten C40 cities. *Environ. Sci. Policy* **2016**, *66*, 353–365. [CrossRef]
30. Corney, G.; Reid, A. Student teachers' learning about subject matter and pedagogy in education for sustainable development. *Environ. Educ. Res.* **2007**, *13*, 33–54. [CrossRef]
31. Tomas, L.; Girgenti, S.; Jackson, C. Pre-service teachers' attitudes toward education for sustainability and its relevance to their learning: Implications for pedagogical practice. *Environ. Educ. Res.* **2017**, *23*, 324–347. [CrossRef]
32. Chunteng, L. Survey of primary and secondary school teachers' teaching competence for environmental education in Xicheng District of Beijing. *Chin. Educ. Soc.* **2004**, *37*, 39–44. [CrossRef]
33. Howlett, C.; Ferreira, J.A.; Blomfield, J. Teaching sustainable development in higher education: Building critical, reflective thinkers through an interdisciplinary approach. *Int. J. Sustain. High. Educ.* **2016**, *17*, 305–321. [CrossRef]
34. Gan, D.; Gal, A. Self-efficacy for promoting EfS among pre-service teachers in Israel. *Environ. Educ. Res.* **2018**, *24*, 1062–1075. [CrossRef]
35. Qablan, A. Building capacities of educators and trainers. In *Issues and Trends in Education for Sustainable Development Education on the Move*; Leicht, A., Heiss, J., Byum, W.J., Eds.; UNESCO Publishing: Paris, France, 2018.
36. Redman, E.; Wiek, A.; Redman, A. Continuing professional development in sustainability education for K-12 teachers: Principles, programmes, applications, outlook. *J. Educ. Sustain. Dev.* **2018**, *12*, 59–80. [CrossRef]
37. Burgener, L.; Barth, M. Sustainability competencies in teacher education: Making teacher education count in everyday school practice. *J. Clean. Prod.* **2018**, *174*, 821–826. [CrossRef]
38. Maidou, A.; Plakisti, K.; Polatoglou, H.M. Knowledge, perceptions, and attitudes on education for sustainable development of pre-service early childhood teachers in Greece. *World J. Educ.* **2019**, *9*, 1–15. [CrossRef]
39. Barth, M. *Implementing Sustainability in Higher Education: Learning in an Age of Transformation*; Routledge: London, UK, 2015.
40. Lysgaard, J.A.; Simovska, V. The significance of 'participation' as an educational ideal in education for sustainable development and health education in schools. *Environ. Educ. Res.* **2016**, *22*, 613–630. [CrossRef]
41. Saka, A.; Sahintűrk, A. Attitudes of prospective forest engineers and primary school teachers towards a sustainable environment. *Pol. J. Environ. Stud.* **2013**, *22*, 1533–1557.
42. Biasutti, M. An intensive programme on education for sustainable development: The participants' experience. *Environ. Educ. Res.* **2015**, *21*, 734–752. [CrossRef]
43. United Nations Educational, Scientific and Cultural Organization (UNESCO). *Education for Sustainable Development in Action*; UNESCO: Paris, France, 2012.
44. United Nations Educational; Scientific and Cultural Organization (UNESCO). *United Nations Conference on Environment and Development*; UNESCO: Geneva, Switzerland, 1992.
45. Sterling, S. Transformative learning and sustainability: Sketching the conceptual ground. *Learn. Teach. High. Educ.* **2011**, *5*, 17–33.

46. Rieckmann, M. Learning to transform the world: Key competencies in education for sustainable development. In *Issues and Trends in Education for Sustainable Development Education on the Move*; ALeicht, A., Heiss, J., Byum, W.J., Eds.; UNESCO Publishing: Paris, France, 2018.
47. Kolb, D.A. *Experiential Learning: Experience as the Source of Learning and Development*, 2nd ed.; FT Press: Upper Saddle River, NJ, USA, 2015.
48. Cohen, L.; Manion, L.; Morrison, K. *Research Methods in Education*, 8th ed.; Routledge: London, UK, 2018.
49. Moser, A.; Korstjens, L. Series: Practical guidance to qualitative research. Part 3: Sampling, data collection and analysis. *Eur. J. Gen. Pract.* **2018**, *24*, 9–18. [CrossRef]
50. Wood, W.; Rünger, D. Psychology of habit. *Annu. Rev. Psychol.* **2016**, *67*, 289–314. [CrossRef]
51. Verplanken, B. Introduction. In *The psychology of Habit: Theory, Mechanisms, Change, and Contexts*; Verplanken, B., Ed.; Springer: Cham, Switzerland, 2018; pp. 1–10.
52. Schultz, P.W. Conservation means behavior. *Conserv. Biol.* **2011**, *25*, 1080–1083. [CrossRef]
53. Laurie, R.; Tarumi, Y.N.; McKeown, R.; Hopkins, H. Contributions of education for sustainable development (ESD) to quality education: A synthesis of Research. *Cent. Environ. Educ.* **2016**, *10*, 226–242. [CrossRef]
54. Brennan, M. Struggles for Teacher Education in the age of the Anthropocene. *J. Educ.* **2017**, *69*, 43–65.

Disclaimer/Publisher's Note: The statements, opinions and data contained in all publications are solely those of the individual author(s) and contributor(s) and not of MDPI and/or the editor(s). MDPI and/or the editor(s) disclaim responsibility for any injury to people or property resulting from any ideas, methods, instructions or products referred to in the content.

sustainability

Article

The Paradox of Progress towards SDG7: Governance Quality and Energy Poverty Dynamics in Pakistan

Rongbing Liu [1], Afifa Qadeer [1], Junqi Liu [1], Suleman Sarwar [2,*] and Muhammad Wasim Hussan [1,*]

1 School of Finance & Trade, Liaoning University, Shenyang 110036, China; rbllnu@163.com (R.L.); afi.kadeer@gmail.com (A.Q.); jqllnu@yeah.net (J.L.)
2 Department of Finance and Economics, College of Business, University of Jeddah, Jeddah 23218, Saudi Arabia
* Correspondence: ch.sulemansarwar@gmail.com (S.S.); mwasimhassan@outlook.com (M.W.H.)

Abstract: This study investigates the multidimensional aspects of energy poverty in Pakistan from 2000 to 2022, specifically evaluating the direct, indirect, and total effects of socioeconomic and environmental factors. We employed Partial Least Squares Structural Equation Modeling (PLS-SEM) to examine the impacts of income, population, governance quality, energy intensity, fuel prices, and renewable energy consumption on energy poverty. The study further contributes by examining the mediating role of governance quality and developing the World Governance Indicators (WGI) Index. The findings indicate significant negative effects of energy intensity and renewable energy consumption on energy poverty. Conversely, population growth and income levels demonstrate positive effects, contradicting conventional economic development and energy access assumptions. Governance quality establishes direct and indirect effects that mediate most relationships between independent variables and energy poverty. Bootstrapping analysis confirms the significance of governance quality as a mediator. The model describes significant energy poverty variance with robust predictive relevance. This study emphasizes the need to adopt a comprehensive strategy to decrease Pakistan's energy poverty by articulating socioeconomic, environmental, and governance factors. Our findings offer valuable information for policymakers to achieve UN Sustainable Development Goal 7, embarking on governance reforms, promoting sustainable growth, and enforcing investments in energy efficiency and renewable sources as Pakistan approaches the 2030 SDG 7 deadline.

Keywords: energy poverty; Sustainable Development Goals (SDG7); governance quality; WGI index

Citation: Liu, R.; Qadeer, A.; Liu, J.; Sarwar, S.; Hussan, M.W. The Paradox of Progress towards SDG7: Governance Quality and Energy Poverty Dynamics in Pakistan. *Sustainability* **2024**, *16*, 8291. https://doi.org/10.3390/su16198291

Academic Editors: Idiano D'Adamo, Massimo Gastaldi and Manoj Kumar Nallapaneni

Received: 28 July 2024
Revised: 17 September 2024
Accepted: 18 September 2024
Published: 24 September 2024

Copyright: © 2024 by the authors. Licensee MDPI, Basel, Switzerland. This article is an open access article distributed under the terms and conditions of the Creative Commons Attribution (CC BY) license (https://creativecommons.org/licenses/by/4.0/).

1. Introduction

A major global problem, energy poverty impacts approximately a billion individuals who are without access to affordable and secure electricity. Energy is the essential force behind contemporary society, fueling various sectors such as industries, residences, and transportation networks [1,2]. Nevertheless, as the worldwide population grows and economies expand, the demand for energy increases, resulting in a complex set of challenges known as the world energy conundrum [3]. This complex issue involves aspects of sustainability, accessibility, affordability, and environmental effects among the primary urgent concerns of the 21st century. The core of the global energy challenge is the conflict between the high energy demands of continuous population growth and the reduction of the harmful environmental effects of energy generation and consumption [4]. Despite their historical status as primary sources of energy, fossil fuel extraction, burning, and greenhouse gas emissions, which include coal, oil, and natural gas, lead to environmental devastation and global warming. Consequently, there is a collective need to switch to greener, more environmentally friendly energy sources like geothermal, hydropower, solar, and wind power [5]. Developing countries need help with basic access to power and clean energy, whereas developed economies have dependable energy facilities and services. Energy poverty jeopardizes public health, impedes economic growth, and maintains social inequity [6,7].

Furthermore, the global energy problem connects to geopolitical conflicts and economic inequalities. Energy reliance on a few oil-producing countries leads to inadequate global supply systems, and clashes over limited fossil fuel supplies frequently cause fuel crises and instability in global politics [8,9]. The shift towards renewable energy prospects to decentralize energy generation, diminish reliance on imported fuels, and promote energy self-sufficiency at the national and local levels. However, there are obstacles to the shift to sustainable energy. Despite significant progress, renewable energy systems encounter obstacles, such as intermittency, storage constraints, and initial expenses [10]. Moreover, the deeply rooted vested interests in the fossil fuel sector, combined with the lack of political drive and regulatory obstacles, can hinder the advancement of renewable energy implementation. A comprehensive and synchronized strategy is necessary, which incorporates technical advancements, governance changes, and active public involvement to overcome these difficulties [11].

The world attempts to address the problems of the global energy dilemma. The United Nations Framework Convention on Climate Change (UNFCCC) Conference of the Parties (COP) meetings [12] and the Sustainable Development Goals (SDGs) are at the center of these attempts [13]. All member countries of the United Nations accepted the Sustainable Development Goals in 2015, which offer an extensive plan to tackle the most important global issues. The Sustainable Development Goals (SDGs) offer a framework for nations to organize and set priorities for their initiatives to attain a sustainable energy future [14,15]. The temporal distribution of Sustainable Development Goals (SDGs) characterizes the change in research emphasis towards various SDGs. SDG-related publications reveal a consistent 14% increase during the 2020–2024 timeframe. Significantly, 62.2% of these publications are available with open access, indicating a remarkable 60% growth. This tendency suggests an increasing scholarly interest in Sustainable Development Goals (SDGs). The latest research [16] shows SDG 13 (Climate Action) as the primary area of concern. SDG 3, which emphasizes Good Health and Well-being, ranks second. SDG 11, which focuses on Sustainable Cities and Communities, is in third position, while SDG 7 is ranked 10th with 1591 published papers. SDG 7 highlights energy as a primary catalyst for achieving sustainable development and enhancing human welfare. It underscores the pressing necessity to tackle energy poverty, given that more than 700 million individuals globally still do not have access to power, and almost 2.4 billion depend on ineffective and environmentally harmful cooking techniques [17]. The objective is based on the shift towards sustainable energy sources, such as solar, wind, and hydropower, which are crucial for mitigating carbon emissions, addressing climate change, and maintaining the environment [18,19]. Fulfilling SDG 7 necessitates both the growth of renewable energy sources and substantial enhancements in energy efficiency, which can reduce energy use and promote environmentally friendly consumption and production habits [20].

Furthermore, Sustainable Development Goal 7 aims to foster inclusiveness by tackling inequalities in energy access, especially for excluded groups and women who bear a disproportionate burden of energy poverty. The objective further emphasizes the need for international collaborations and financial support to finance the shift towards sustainable energy generation systems. Sustainable Development Goal 7 promotes economic growth, social fairness, and environmental protection by ensuring affordable and sustainable energy access. It aligns energy programs with the overarching objectives of poverty reduction and climate adaptation.

The recent COP 28 aims to drive progress towards a more equitable, sustainable, and adaptable global energy future through global collaboration in renewable energy implementation, investment for clean energy initiatives in underdeveloped nations, technology advancement, and capacity-building activities [21,22].

In Pakistan, energy poverty is a profoundly ingrained problem that presents major obstacles to social progress, economic growth, and environmental sustainability [23,24]. Pakistan is under immense pressure to meet the energy requirements of its population, which exceeds 220 million, with a developing economy [25,26]. One of the most apparent indicators of Pakistan's energy poverty is the frequent power outages and load shedding that many households and businesses encounter. Additionally, unstable electricity supplies exacerbate the disproportionate negative impact on senior citizens, women, and young people, among other vulnerable groups [27]. Apart from the scarcity of electricity, one major obstacle in Pakistan is to obtain renewable energy. According to a World Bank estimation, 60% of people use conventional biomass-based fuels like wood, agricultural waste, and animal excrement, which degrades the environment and causes domestic air pollution and respiratory ailments, i.e., allergies and asthma [28].

Pakistan's energy poverty exacerbates environmental and socioeconomic challenges and hinders ecologically conscious progress. Dependence on fossil fuels, especially imported gas and oil, burdens the nation's financial resources and exacerbates the issue of global warming, polluting the environment and greenhouse gas emissions resulting in energy poverty and environmental destruction. Pakistan's energy poverty must be addressed with a multidimensional, holistic, and comprehensive strategy that incorporates social efforts, funding for infrastructure, policy reforms, and technological advancements [29].

Governance plays a crucial role in reducing energy poverty, which occurs when households cannot access modern energy services necessary for necessities such as cooking, heating, lighting, and communication [30,31]. Effective governance, which includes strong institutions, openness, accountability, and adherence to the rule of law, substantially impacts energy policy, infrastructure development, and resource allocation and affects the extent and severity of energy poverty. Good governance also fosters an environment conducive to innovation and adaptation in the energy sector, facilitates research and development, supports the adoption of new technologies, and promotes sustainable energy solutions through investment in renewable energy sources, energy storage technologies, and smart grid systems [32–34]. Additionally, governance supports the development of innovative business models, such as micro grids and pay-as-you-go systems that can provide affordable energy access to low-income households. Furthermore, a sound regulatory environment, underpinned by good governance, is critical for ensuring that energy markets function efficiently and equitably [35,36]. Effective regulation promotes competition, attracts private investment, and protects consumers from manipulation by setting fair tariffs, ensuring service quality, and preventing anticompetitive practices. Strong regulatory bodies, free from political interference, oversee the energy sector and enforce regulations impartially [33]. Effective governance determines the development, execution, and oversight of programs to reduce energy poverty. It is well-defined and consistent with energy policies that tackle energy supply and demand issues. Governance affects resource allocation, especially in rural or low-income areas. Effective resource allocation programs, like India's Deen Dayal Upadhyaya Gramme Jyoti Yojana (DDUGJY), improve rural electrification as they possess strong governance frameworks [37]. Through governmental supervision and efficient policy formulation, India substantially reduced energy poverty by 30% in six states, extending the provision of electricity to all income and social groups from 2015 to 2018 [38]. Governance requires establishing institutions that oversee energy markets, encourage competition, and ensure affordable access to energy to prevent monopolistic practices and pricing structures. Effective governance promotes a competitive and transparent energy market environment and facilitates private sector investment in energy infrastructure, especially in regions with limited government capabilities.

Pakistan is unlikely to attain the Sustainable Development Goal 7 (SDG 7) targets by 2030, despite worldwide attempts and efforts established by the United Nations General Assembly in 2015 [39]. However, Pakistan still faces significant challenges to overcome energy poverty. Pakistan's energy distribution issue stems from a complex interplay of socioeconomic disparities, governance failures, and geopolitical challenges. Income inequality influences energy access and affordability in Pakistan. While urban areas and high-income communities often have better energy infrastructure, many low-income households, especially in remote areas, need more access to the grid entirely [40]. This disparity propagates energy poverty. Due to income inequality, rich households can afford backup power systems (like generators), while poorer households are more likely to experience frequent outages and high tariffs.

Another major barrier to Pakistan's efficient energy distribution is corruption. Within the energy sector, corruption is rampant at all levels, from misallocating resources to inflating project costs and mishandling money, using illicit connections to steal electricity, which burdens an already overburdened system [41,42]. Special interests have maintained control over the energy industry at the policy level due to ineffective administration and a lack of political will to pursue reforms. It gets worse when decision-makers and political connections profit from subsidized energy pricing.

The geopolitical context of Pakistan exacerbates the country's energy distribution issue [43]. The nation is vulnerable to supply chain disruptions and global price changes due to its reliance on imported energy, particularly gas and oil. Potential energy cooperation, such as the Turkmenistan–Afghanistan–Pakistan–India (TAPI) gas pipeline, is impacted by geographical issues [44]. Cross-border energy projects face obstacles by political instability in the wider region, which further limits Pakistan's capacity to diversify its energy supplies. Currently, a significant portion of the population still lacks accessibility to renewable energy, while the energy infrastructure is confronted with challenges related to energy demand [23,45]. Only six years remain until the 2030 deadline, so prompt action and meticulous preparation are needed to bridge this gap effectively.

Based on our extensive analysis and the consideration of SDG7 standpoints, our study offers significant contributions to energy poverty studies by addressing several gaps and integrating a comprehensive analytical framework, specifically in the context of developing countries like Pakistan. We conduct a holistic analysis of energy poverty, assimilating socioeconomic, energy-related issues, and governance aspects, with a detailed understanding of energy poverty dynamics. Our study contributes to existing literature through the following aspects. Firstly, our study explores the direct and indirect relationships between energy poverty and socioeconomic factors, energy intensity, fuel prices, and renewable energy consumption from 2000 to 2022 which provides a longitudinal perspective [46,47] of energy poverty dynamics in Pakistan. The study aims to explain the connection between income levels and energy poverty, with a challenge of economic development and access to energy simultaneously. Moreover, our study examines various energy sources and their impacts on energy poverty and adds to the literature on energy shifts in emerging economies as affordable and renewable energy alleviates energy poverty [48,49]. This holistic approach provides a nuanced understanding of how these variables interplay, which may be overlooked in other studies [50–53], focusing on individual aspects in isolation and not integrating the interplay between socioeconomic factors, fuel prices, and governance quality.

Secondly, by incorporating governance quality as a mediating factor, our research adds a crucial dimension to the analysis. The mediating role of governance quality is a relatively novel aspect of energy poverty research. Governance quality can influence the effectiveness of policies aimed at alleviating energy poverty and can mediate the impact of socioeconomic and environmental factors. Previous studies [54–57] investigated the determinants of energy poverty but have overlooked the important role of governance quality. This aspect needs to be explored more in existing literature, making this study innovative in highlighting the importance of governance.

Thirdly, this study employs Partial Least Squares Structural Equation Modeling (PLS-SEM) to construct an extensive and effective model. PLS-SEM allows for the analysis of complex relationships between observed and latent variables, providing a more accurate and comprehensive understanding of the factors influencing energy poverty. Many previous studies have used simpler statistical methods, which may not capture the full complexity of these relationships. For instance, studies [53,58–61] focused on specific indicators or regions. Still, they did not utilize PLS-SEM to analyze their data, potentially overlooking intricate interdependencies. Furthermore, by measuring the effect sizes of different factors, we offer a data-based framework for defining policy priorities in resource-constrained environments, as recommended by [62].

Fourthly, Pakistan faces distinctive socioeconomic and energy challenges similar to numerous emerging nations. The study's emphasis on Pakistan will yield context-specific insights essential for developing targeted interventions, an aspect inevitably omitted in broader and more generalized studies. Our analysis evaluates Pakistan's progress in achieving SDG7 by 2030, integrating academic research with global development goals, as [63] in their emphasis on sustainable development objectives. Our study links theoretical concepts and practical policy inferences, making valuable contributions to academic literature and policy development. That validates the [20,64] discussion on the practical challenges of achieving SDG7. Ultimately, it assists the nation's efforts to achieve the United Nations Sustainable Development Goals (SDGs), namely SDG 7, and offers realistic policy recommendations.

Based on the contribution of this study following are some objectives. First, to investigate the direct and indirect relationships between energy poverty and socioeconomic factors (income level, population), energy intensity, fuel prices (oil, coal, natural gas, LNG), and renewable energy consumption in Pakistan. The second objective is to determine the mediating role of governance quality in the relationships between socioeconomic and environmental factors and energy poverty. The third objective of our study is to construct an effective and extensive PLS-SEM model that elucidates the intricate paths by which these variables influence energy poverty in Pakistan.

This study aims to assess the intricate interaction among socioeconomic, environmental, and governance elements that impact energy poverty in Pakistan. It will use PLS-SEM modeling to offer policymakers a detailed understanding of these factors and suggest practical approaches for attaining SDG7 goals by 2030.

2. Literature Review

Access to renewable energy sources is crucial to alleviate energy poverty and enhance the quality of life [58]. Numerous studies have highlighted the issues related to traditional biomass-based fuels like wood and manure as fuel causes domestic air pollution and respiratory illnesses, especially in underdeveloped nations where women and children are more likely to suffer from these conditions [28,65]. The availability of clean fuels, especially liquefied petroleum gas (LPG) and electricity, has been associated with reduced levels of energy poverty and strengthened socioeconomic parameters [66]. The relationship between income level, population, energy intensity, and renewable energy in relation to clean energy is complex, as it reflects the economic, demographic, and energy traits of the country [4,67,68].

The level of income is a crucial factor in determining energy poverty, as income significantly impacts households' aptitude to buy contemporary energy products and services [61]. Energy costs disproportionately affect low-income households, resulting in energy scarcity and restricted access to other necessities [69,70]. Research has continuously demonstrated a negative relationship between energy poverty and income level, with greater incomes being linked to improved accessibility to dependable and environmentally friendly energy sources [71]. Rapid population growth in another reason for energy poverty, particularly in nations with inadequate resources and facilities. Fast urbanization and a growing population strain energy infrastructures, causing erratic service delivery and higher levels of energy poverty [72]. Studies reveal that population growth has a detrimental effect on people's ability to access electricity, especially in rural and remote areas where the growth of infrastructure is inadequate [73].

The term "energy intensity" refers to the quantity of energy required to generate a single unit of economic activity and is also significantly correlated with energy poverty. High energy intensity can exacerbate energy poverty, which increases energy costs and adverse ecological effects, which indicates inadequacies in energy consumption [48]. Ref. [74] documented that controlling energy poverty to achieve sustainable development objectives necessitates a combination of energy-saving measures and a decrease in energy intensity. One effective approach to reduce energy poverty and mitigate the impacts of climate change is to enhance the proportion of renewable energy sources [48]. Renewable energy sources, such as hydro, wind, and solar power, offer environmentally friendly and long-lasting replacements for conventional fossil fuels [68]. Literature has demonstrated how renewable energy improves affordability, accessibility to energy, and environmental sustainability [75]. The interaction between income level, population dynamics, energy intensity, and renewable energy substantially impacts the availability of clean energy in Pakistan and provides a foundation for formulating the first hypothesis of this research.

H1: *Energy poverty in Pakistan is significantly influenced by the nexus of socioeconomic, energy intensity, and environmental factors.*

Effective governance significantly impacts the results of energy access, affordability, and sustainability. Robust governance structures and policies are linked to enhanced energy service provision, infrastructure advancement, and regulatory structures, resulting in increased energy accessibility and decreased poverty rates [30,31]. Effective governance standards, such as rational energy regulations, regulatory structures, and financial incentives for investment, reduce risks associated with energy poverty and facilitate renewable energy consumption [33]. Governance quality influences energy strategy, investments, and environmental regulations, which in turn influences energy intensity and the adoption of renewable energy. Transparent and liable governance structures facilitate sustainable energy shifts through venture capital investment, technological advancement, and public involvement in renewable energy initiatives [76,77].

According to [32,63] empirical research, governance quality mediates the relationship between socioeconomic determinants, energy availability, and poverty. Countries with well-functioning governance structures show a more pronounced connection between income levels, population dynamics, energy intensity, renewable energy consumption, and energy poverty. In countries with robust governance, energy policies are often inclusive and well-targeted, ensuring that vulnerable populations receive the necessary support. Strong governance in Germany has led to effective energy policies promoting renewable energy and energy efficiency, significantly reducing energy poverty rates [78]. Conversely, in countries with weak governance, corruption may influence energy policies, resulting in misallocation of resources and insufficient support for people with low incomes. In Nigeria, poor governance has contributed to widespread energy poverty despite the country's abundant energy resources [79].

The energy governance in Morocco enabled significant private investments in renewable energy, such as the Noor Ouarzazate Solar Complex, which is among the largest solar power plants [80]. The rigorous regulatory systems assured the provision of affordable energy, therefore mitigating energy poverty. Bangladesh has achieved significant progress in alleviating energy poverty by implementing effective governance measures that have enabled the widespread adoption of Solar Home Systems (SHS) in rural regions [81]. The Infrastructure Development Company Limited (IDCOL) initiative has successfully established energy supply to more than 4 million households. Strong governmental control, useful financial incentives, and a mechanism that enabled cost recovery while enhancing access to sustainable energy are all part of the governance structure supporting this initiative. Botswana effectively controlled corruption through prudent governance practices to enhance its energy industry [82]. In contrast, Nigeria, although a major global oil producer, experiences substantial energy poverty mostly due to corrupt governance systems [83]. Despite the considerable potential income from natural resources, the mismanagement of oil earnings, power sector reforms, and infrastructural endeavors have resulted in the persistence of energy poverty among large population segments. This literature is consistent with our second hypothesis; hence, on the basis of the mentioned studies, we developed our second hypothesis.

H2: *Governance quality enhances the impact of socioeconomic and environmental factors to reduce energy poverty, particularly through income levels, population dynamics, energy intensity, and renewable energy consumption.*

Governance quality includes the authority's efficacy, accountability, and integrity to run the economy. Numerous scholarly investigations have emphasized the noteworthy influence of governance quality on positive development consequences, such as energy poverty mitigation, economic development, and environmental durability [74]. Strong governance structures are necessary to effectively address energy poverty and advance sustainable energy accessibility. Transparent, efficient, and responsible governance systems enable the creation and execution of energy policies, ensure fair resource allocation, and ensure the availability of clean and affordable energy services [84,85]. Studies have documented that countries with better governance quality experience less energy poverty because they have stronger enforcement of regulations, consistent policies, and improved infrastructure [33]. Effective governance frameworks enable collaborative efforts among stakeholders, leading to improved energy accessibility, reduced energy costs, and enhanced service reliability [86]. In Pakistan, the impact of governance quality on energy poverty has been recognized as a crucial factor influencing the efficiency and progress of the energy sector and the results of poverty [29]. Inadequate governance structures, bureaucratic inefficient and bottleneck processes, and corruption have impeded the progression of energy sector transformations and infrastructural development, resulting in energy poverty and socioeconomic inequalities [87].

Research findings of [53] suggest that improving governance quality might directly reduce energy poverty by enhancing the implementation of policies, allocation of resources, and collaboration of stakeholders. Improved performance of governing authorities, enhanced accountability and transparency, and combating corrupt behavior are needed to achieve sustainable energy access and reduce energy poverty in Pakistan [74]. The mentioned direct impact of effective governance gives us the base to formulate our third hypothesis.

H3: *Governance quality has a direct negative effect on energy poverty in Pakistan.*

Literature shows [46,50,55,88] that the advancement made towards achieving the goals of SDG 7 has been a combination of positive and negative outcomes. Even if there have been tremendous advancements in certain areas, there remain numerous obstacles to overcome and gaps to close before 2030, with access to reliable, affordable, and environmentally friendly energy [89]. Therefore, there is a clear need for more focused research in Pakistan. This research gap underscores the importance of further investigation to understand this correlation within Pakistan's distinct socioeconomic landscape.

3. Research Methodology

3.1. Composition of WGI Index

This study examines the relationship between governance indicators and energy poverty in Pakistan. The World Governance Indicators (WGI) play an essential role in the analysis. These indicators comprise five dimensions: Political Stability and Absence of Violence/Terrorism, Government Effectiveness, Regulatory Quality, Rule of Law, and Control of Corruption. They offer a thorough evaluation of a country's governance performance. We initially collected the World Governance Indicators (WGI) data for Pakistan from 2000 to 2022. The dataset includes annual estimates for the five WGI variables, with higher values indicating stronger governance performance. However, separate analyses of these variables may result in potential multicollinearity problems and mislead the interpretation of the findings [90,91].

To address this potential issue, we applied factor analysis techniques to integrate the information from the five WGI variables into a unified composite index. Our approach is meticulous, and we performed a common factor analysis in Table 1, and results show that a single latent factor can explain a substantial portion of the variation in the WGI variables by following the [92] approach. The factor loadings demonstrate the strong correlations between the variables and the latent factor, ranging from 0.94 to 0.98.

Table 1. Common factor analysis of governance.

Variables	Notation	Factor Loadings
Political stability and absence of violence/terrorism	PS	0.94
Government effectiveness	GE	0.96
Regulatory quality	RQ	0.95
Rule of law	RL	0.98
Control of corruption	CC	0.97

Our study uses principal component analysis (PCA) to derive a unique factor based on the outcomes of the common factor analysis. This factor represents the shared variation across the five WGI variables and validates our index composition that consolidated the governance information into a single measure.

Furthermore, we employ dynamic factor analysis, an expanded version of factor analysis that includes time-series data, to model the fluctuations in the fundamental component across time Table 2. This technique examines the potential connection between consecutive data points and the interdependence among distinct parts of the data. Additionally enhances our understanding of how the notion of governance, as measured by the WGI variables, evolves over time.

Table 2. Dynamic factor analysis results of governance.

Variables	Notation	Factor Loading	Proportion Variance	Cumulative Variance
Political stability and absence of violence/terrorism	PS	0.94	0.176	0.176
Government effectiveness	GE	0.96	0.184	0.36
Regulatory quality	RQ	0.95	0.18	0.54
Rule of law	RL	0.98	0.192	0.732
Control of corruption	CC	0.97	0.188	0.92

The dynamic factor analysis indicates that the latent factor gradually declines over time, with occasional variations. However, the factor loadings exhibit a consistent level of stability, demonstrating that the connections between the observable WGI variables and the latent factor do not experience substantial changes across the studied timeframe.

At first, the composite index using the factor scores exhibited negative values for all years. It is crucial to understand that negative numbers do not always indicate insufficient effectiveness in governance; rather, negative values illustrate that Pakistan's performance is below the global average, which is set at 0 for the WGI variables [91]. To clarify the index values, we modified the index scale to range from 0 to 100, where 50 equates to the worldwide average, as recommended by [93,94]. The rescaling method entailed standardizing the index values and implementing a linear transformation to translate the normalized values to the target range. The rescaled composite index presents a more precise measure, where values above 50 reflect performance exceeding the worldwide average, and values below 50 indicate a decline in performance from the global average. Subsequently, the study uses a rescaled index for further research to investigate its relationship with energy poverty in Pakistan.

Through factor analysis techniques, we combined the five WGI variables into a single composite index [95] to represent the core governance paradigm in a more concise, parsimonious, and interpretable way while maintaining essential data on Pakistan's governance performance over time. This approach resolves possible problems related to multicollinearity and makes it significantly easier to include governance indicators in our investigation of energy poverty in Pakistan. The approach is mathematically represented by the Equation (1).

The study aims to offer a more precise and succinct assessment of the state of governance by employing dynamic factor analysis as discussed by [96,97], to facilitate stakeholders and policymakers to develop a more comprehensive understanding of Pakistan's governance initiatives.

$$S(I,J,T) = \{Sijt\},\ i = 1,\ j = 1,\ 2,\ldots,5,\ t = 1,\ 2,\ldots.,T \tag{1}$$

Our study establishes the matrix Z of $S(I,J,T)$ which includes the elements $Sijt$ where i represents Pakistan, j represents the five WGI variables, and t corresponds to time. This matrix contains temporal data on the WGI variables over time for Pakistan, as represented by Equation (1). As indicated by Equation (2), we decompose the correlation matrix Z of $S(I,J,T)$ into three smaller components and combine them. The three components of this equation are

$$X = X(I)\rho^{\wedge *} + X(IT) + X(T)^{\wedge *} \tag{2}$$

The static matrix $X(I)^{\wedge *}$ represents Pakistan, illustrating the intertemporal correlation between the WGI variables. The temporal dimension does not impact this static matrix. The dynamic difference matrix $X(IT)$ integrates the WGI variables and time. The average dynamic change matrix, represented as $X(T)^{\wedge *}$, displays changes across time while excluding the country-specific component.

Equation (2) is simplified into Equation (3) in the following way:

$$X = X(I)^{\wedge *} + X(IT) + X(T)^{\wedge *} = X(T) + X(T)^{\wedge *} \tag{3}$$

$X(T)$ is the principal component analysis average correlation matrix across years, derived through the transformation of panel data into a cross-sectional matrix. $X(T)^{\wedge *}$ refers to a linear regression model extract that captures fluctuations across time [96,97].

As indicated in Equation (4), we use $X(T)$ to compute the solution and identify the primary components from the correlation matrix to reduce dimensionality.

$$X(T) = \frac{1}{T}\sum\nolimits_{t=1^{\wedge}}^{T} X(t) \tag{4}$$

This Equation represents the average correlation matrix calculated across years and displays the temporal fluctuations of the WGI variables. Data standardization is crucial to mitigate the impact of various scales and units of measurement between 2000 and 2022. For this uniformity, the data are comparable and suitable for analysis [98,99]. Subsequently, we conducted a dynamic factor analysis to explore the results for each WGI variable. The dynamic factor analysis results validate that dimensional reduction parallels expectations. Furthermore, we calculated the cumulative variance contribution rate, followed by [100], and noticed it is close to 100%. It suggests that, as shown in Table 1, the significance of interpretation for the five WGI variables is approximately in line with the predicted level.

3.2. Partial Least Squares Structural Equation Modeling

This study employs Partial Least Squares Structural Equation Modeling (PLS-SEM) to examine the complex, intricate connections among socioeconomic determinants, energy-related variables, governance quality, and energy poverty in Pakistan used by [101,102] and recommended by [103,104]. The selection of PLS-SEM is due to its ability to handle complex models that involve several constructs, indicators, and interactions. Additionally, it is known for its robustness and ability to handle non-normal data and small sample sizes [105]. Our approach aligns with previous studies [60,106] on energy poverty. Given the specific attributes of our data and the complexity of our model, (PLS-SEM) is very suitable for this research.

3.3. Data and Variables

The study employs annual data from 2000 to 2022 for Pakistan. The dependent variable is energy poverty, measured by access to clean fuels and technology for cooking, following the approach of [107]. Independent and mediating variables include a wide range of socioeconomic, environmental, and energy-related variables. The variables encompassed in the dataset are listed in Table 3:

Table 3. Variables, units of measurement, and data sources.

Variable	Notation	Measure	Source
Energy Poverty	EP	The proportion of the population with access to clean fuels and technology for cooking	Word development indicators
Income level	Income	GDP per capita (constant 2015 US$)	Word development indicators
Population	Pop	Total Population	Word development indicators
Governance Quality	WGII	Index	Word development indicators
Energy intensity	EI	Mega joules per USD (constant 2017 PPP GDP)	Word development indicators
Oil Price	OP	US dollars per barrel	BP statistical review
Coal Price	CP	US dollar per tonne.	Federal reserve economic data
Natural Gas Price	NGP	US dollars per million Btu	BP statistical review
LNG Price	LNG	US dollars per million Btu	Federal reserve economic data
Renewable energy consumption	Renew	% of total final energy consumption	Word development indicators

All variables are log-transformed to tackle putative non-linearity and enhance the model fit, an approach [108] utilized in their energy poverty study analysis.

3.4. Model Specification

Our study employs the PLS-SEM methodology to pursue our research objectives and test our hypotheses. This approach enables us to investigate the intricate relationships between socioeconomic determinants, energy-related variables, governance quality, and energy poverty in Pakistan. Table 4 depicts the variables specification model.

Table 4. Model specification and variables.

Endogenous Variable	Exogenous Variables	Mediating Variable
The proportion of the population with access to clean fuels and technology for cooking (representing Energy Poverty)	Income level Total Population Energy intensity Oil Price Coal Price Natural Gas Price LNG Price Renewable energy consumption	Governance Quality (WGII)

The WGI index acts as both an exogenous variable and as a mediator, following an identical strategy to [35,109] in their studies. The structural relationships in our model are illustrated in Equations (5)–(8), which incorporate direct effects, mediating effects, indirect effects, and total effects, following the integrated modeling methodology of [110,111].

3.4.1. Direct Effects on Energy Poverty

$$EP = \beta_1 Income + \beta_2 Pop + \beta_3 WGII + \beta_4 EI + \beta_5 OP + \beta_6 CP + \beta_7 NGP + \beta_8 LNG + \beta_9 Renew + \varepsilon_1 \quad (5)$$

This Equation captures the core of our model, specifically targeting our primary research objective to determine the factors of energy poverty in Pakistan. Each β coefficient signifies the direct effect of an independent variable on energy poverty (*EP*). It enables us to test our hypotheses concerning the impacts of income level (*Income*), total population (*Pop*), governance quality (*WGII*), energy intensity (*EI*), energy prices (*OP*, *CP*, *NGP*, *LNG*), and renewable energy consumption (*Renew*) on energy poverty (*EP*).

3.4.2. Effects on Governance Quality (Mediator)

$$WGII = \gamma_1 Income + \gamma_2 Pop + \gamma_3 EI + \gamma_4 OP + \gamma_5 CP + \gamma_6 NGP + \gamma_7 LNG + \gamma_8 Renew + \varepsilon_2 \quad (6)$$

This Equation models the factors that influence governance quality specifically emphasizing our secondary objective to observe the role of governance in energy poverty dynamics. It enables us to test hypotheses regarding the impact of different factors on governance quality, which subsequently affects energy poverty.

3.4.3. Indirect Effects

$$IE\_i = \gamma_i \times \beta_4 (for\ i = 1\ to\ 8) \quad (7)$$

This Equation assesses the indirect effects of every exogenous variable on energy poverty by considering governance quality. It supports us in addressing our research objective by grasping the mediating role of governance in the relationship between different components and energy poverty and examining our hypotheses regarding indirect effects.

3.4.4. Total Effects

$$TE\_i = \beta\_i + IE\_i (for\ i = 1\ to\ 8, \text{excluding } \beta_4) \quad (8)$$

This Equation comprehensively incorporates the direct and indirect effects and provides a thorough understanding of each variable's total effect on energy poverty. It fulfills our objective to evaluate the overall influence of several factors on energy poverty, implemented both directly and through a governance-mediated path. In Equations (5)–(8), β_1 to β_9 and γ_1 to γ_8 represents path coefficients and ε_1 and ε_2 denote the error terms.

These Equations together allow us to test our hypotheses regarding the direct, indirect, and total effects of several socioeconomic and energy-related factors on energy poverty in Pakistan. By addressing governance quality as both a direct factor and a mediator, we can investigate the complex connections between institutional factors and energy poverty, confronting the multifaceted nature of this issue in developing countries.

4. Results and Discussion

4.1. Descriptive Analysis

Our comprehensive descriptive analysis includes the central tendency, dispersion, and distribution shape measures and reveals a significant finding. This method is similar to the approach employed by [112], who presented comprehensive descriptive statistics in their study of energy poverty in Nigeria. We observe a rising trend of energy poverty, which aligns with the results of [61] in their examination of Indian households. This alignment with previous studies reinforces the validity of our findings. Table 5 displays the descriptive statistics for all variables used in the study.

Table 5. Descriptive statistics of key variables.

Variable	Mean	Std. Dev	Max	Min	Skewness	Kurtosis
EP	3.59	0.27	3.96	3.17	−0.02	−1.21
Income	7.17	0.16	7.44	6.96	0.34	−1.15
Pop	19.07	0.14	19.28	18.85	−0.01	−1.22
WGII	3.50	0.10	3.68	3.31	−0.09	−0.78
EI	1.50	0.09	1.63	1.36	−0.15	−1.14
OP	4.09	0.52	4.72	3.20	−0.45	−1.08
CP	4.46	0.62	5.96	3.41	0.55	0.30
NGP	1.91	0.54	3.19	1.07	0.58	−0.51
LNG	2.21	0.57	3.51	1.49	0.77	−0.35
Renew	3.85	0.05	3.94	3.74	−0.36	−0.83

The descriptive statistics indicate considerable variability in the variables during the study period. Energy poverty demonstrates an increasing tendency (mean = 3.59, SD = 0.27), indicating a long-standing concern in Pakistan. Income level (mean = 7.17, SD = 0.16) and total population (mean = 19.07, SD = 0.14) demonstrate consistent growth, indicating Pakistan's economic and demographic shifts. The WGI index with (mean = 3.50, SD = 0.10) demonstrates minor fluctuations, suggesting variations in governance quality. Energy intensity indicates a declining trend (mean = 1.50, SD = 0.09), suggesting progress in energy efficiency. Energy prices, specifically oil (mean = 4.09, SD = 0.52) and coal (mean = 4.46, SD = 0.62), indicate significant volatility reflecting global market dynamics. The skewness and kurtosis values suggest that many variables are almost normally distributed, which supports the rationale for using log-transformed data in our analysis.

4.2. Measurement Model Assessment

Since our model uses single-item measurements, standard assessments such as internal consistency reliability (Cronbach's alpha, composite reliability) and convergent validity are not pertinent. We analyze the collinearity between indicators employing Variance Inflation Factor (VIF) values in Table 6, as proposed by [103,113].

Table 6. Variance inflation factor (VIF) values for multicollinearity.

Variable	VIF
Income	12.34
Pop	15.21
WGII	1.53
EI	8.76
OP	5.87
CP	6.42
NGP	9.15
LNG	8.93
Renew	3.21

VIF values above 5 suggest potential collinearity issues [114], Figure 1.

Various variables in our model exceed the specified threshold, notably income level (VIF = 12.34), total population (VIF = 15.21), and natural gas price (VIF = 9.15). Our results indicate the presence of multicollinearity, specifically for total population and income level, similar to the findings of [115,116]. It validates the presence of multicollinearity and justifies using PLS-SEM (Python 3.12.4), which is more robust to such problems than traditional regression methods. The high VIF values indicate that these variables share considerable variance, which is to be anticipated due to the interrelated nature of socioeconomic and energy dynamics. However, this multicollinearity does not affect the predictive power of the model.

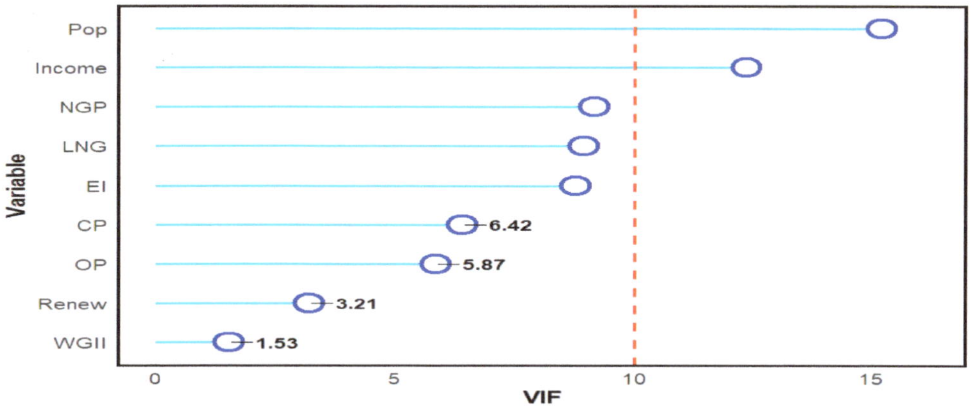

Figure 1. Variance inflation factor (VIF) values for multicollinearity assessment.

4.3. Structural Model Assessment

The structural model assessment examines the relationships between exogenous variables, the mediator (World Governance Indicators Index), and the endogenous variable (Energy Poverty). The model reveals important pathways and the relative significance of different variables, offering insights into the complex and multifaceted interplay of socioeconomic and governance factors in influencing energy access and affordability.

4.4. Path Coefficients and Significance

The study examines the structural model by analyzing the path coefficients, their statistical significance, and their relevance. Table 7 presents these results.

Table 7. Path coefficients and statistical significance of the model.

Path	Coefficient	t-Value	*p*-Value	Significance
Income → EP	0.312	4.567	<0.001	***
Pop → EP	0.485	6.892	<0.001	***
WGII → EP	−0.079	2.765	0.006	**
EI → EP	−0.176	3.124	0.002	**
OP → EP	0.087	2.023	0.043	*
CP → EP	0.098	2.11	0.035	*
NGP → EP	0.046	1.364	0.173	ns
LNG → EP	0.035	1.079	0.281	ns
Renew → EP	−0.157	2.982	0.003	**

Note: *** $p < 0.001$, ** $p < 0.01$, * $p < 0.05$, ns = not significant.

The results reveal that the majority of the hypothesized relationships are statistically significant. Income level ($\beta = 0.312$, $\rho < 0.001$) and total population ($\beta = 0.485$, $\rho < 0.001$) have a strong positive impact on energy poverty and support our hypotheses and are aligned with the findings reported by [67,117]. Results indicate that as population and income levels increase, the challenge of energy poverty becomes more severe, potentially because of higher increased energy consumption and potential imbalances in energy distribution. The WGI index indicates a significant negative effect ($\beta = -0.079$, $\rho = 0.006$), underscoring the crucial role of effective and efficient governance in mitigating energy poverty. This finding is consistent with the existing literature [30] highlighting the impact of governance quality on energy accessibility and affordability. Establishing high governance quality, signified by robust public services and minimal corruption, is crucial for enacting policies that efficiently tackle energy poverty [31]. Oil and coal prices show modest but significant positive effects and demonstrate that a rise in these prices intensifies energy poverty. Rising fossil fuel prices increase energy costs, reducing real purchasing power. Ref. [118] study indicates that this tendency is associated with increased energy poverty during the global energy crisis, especially among low-income households. The findings underscore the pressing necessity for policies to tackle the volatility of fossil fuel costs and advocate for sustainable energy alternatives to alleviate energy poverty [48].

Energy intensity ($\beta = -0.176$, $\rho = 0.002$) and renewable energy consumption ($\beta = -0.157$, $\rho = 0.003$) display significant negative effects. The negative impact of energy intensity and renewable energy consumption on energy poverty align with the findings of [119,120], respectively, hence validating our hypotheses (Figure 2). This specifies that developments in energy efficiency and expanded usage of renewable energy can significantly help in mitigating energy poverty. Surprisingly, the effect of natural gas and LNG prices is not statistically significant, contrary to our expectations. One plausible factor may be Pakistan's convoluted natural gas and LNG pricing structures. Various elements, like government rules, subsidies, or contract lengths, can influence the pricing mechanism and may obscure the direct effect of global price variations on the variable under investigation. Our finding aligns with prior research that underscores the intricacies of energy pricing in developing countries [121,122]. However, it contradicts studies [123] proposing a more direct correlation between fuel prices and domestic energy usage.

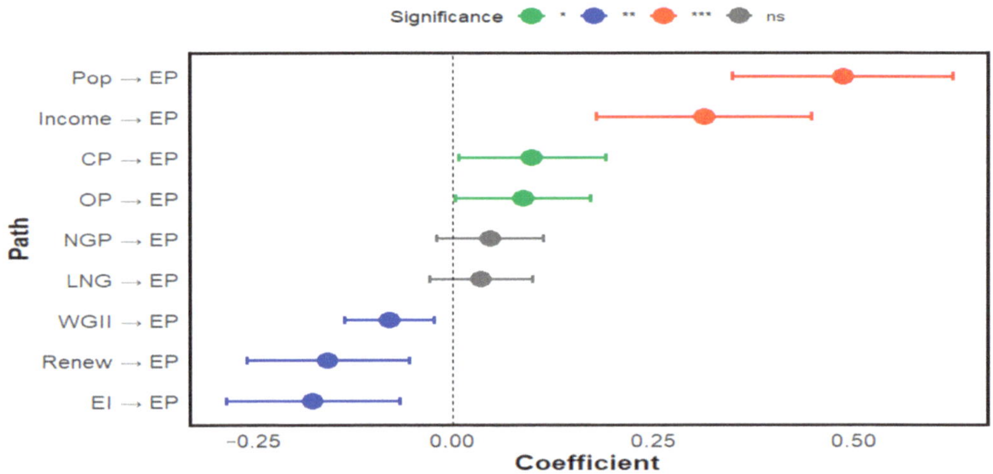

Figure 2. Path coefficients and statistical significance in the structural model. *** $p < 0.001$, ** $p < 0.01$, * $p < 0.05$, ns = not significant.

Another aspect to consider is the minor contribution of natural gas and LNG to household energy consumption in Pakistan. The complex correlation between energy prices and residential energy use is shaped by various socio-economic, political, and cultural elements that might differ in different contextual regions.

4.5. Model's Explanatory and Predictive Power

To measure the model's explanatory power, we assess the R-squared $(\mathcal{R}^2)$ value Table 8 for the endogenous construct (energy poverty). To assess predictive relevance, we employ the Stone–Geisser Q^2 value.

Table 8. R-squared and q-squared values for model fit and predictive accuracy.

Endogenous Construct	R^2	R^2 Adjusted	Q^2
EP	0.968	0.955	0.942

Our model's high $\mathcal{R}^2$ value (0.968) implies that the model describes 96.8% of the variance in energy poverty, which signifies considerable explanatory power. The adjusted $\mathcal{R}^2$ of 0.955 provides strong evidence, even considering the number of predictors. Ref. [124] states that $\mathcal{R}^2$ values of 0.75, 0.50, and 0.25 are categorized as substantial, moderate, and weak, respectively. Our $\mathcal{R}^2$ value surpasses the threshold required for substantial explanatory power. These results are similar to the findings of [125] in their extensive energy poverty study.

The Q^2 value of 0.942 is considerably greater than zero, representing the strong predictive significance of the model. Q^2 values above 0, 0.25, and 0.50, defined by [103], indicate the PLS path model's small, medium, and large predictive importance. Our Q^2 results clearly show that the model has a large predictive relevance and significance for energy poverty.

4.6. Effect Sizes

We employ the f^2 effect sizes Table 9 to assess each predictor's practical relevance, a method commonly employed in energy poverty research, as demonstrated by [126]. According to [127] guidelines, f^2 values of 0.02, 0.15, and 0.35 signify small, medium, and large effects, respectively. Figure 3 demonstrates the graphical representation of effect size.

Table 9. Effect sizes (f^2) for key predictors in the model.

Path	f^2	Effect
Income → EP	0.287	Medium
Pop → EP	0.452	Large
WGII → EP	0.058	Small
EI → EP	0.124	Small
OP → EP	0.063	Small
CP → EP	0.072	Small
NGP → EP	0.034	Small
LNG → EP	0.026	Small
Renew → EP	0.116	Small

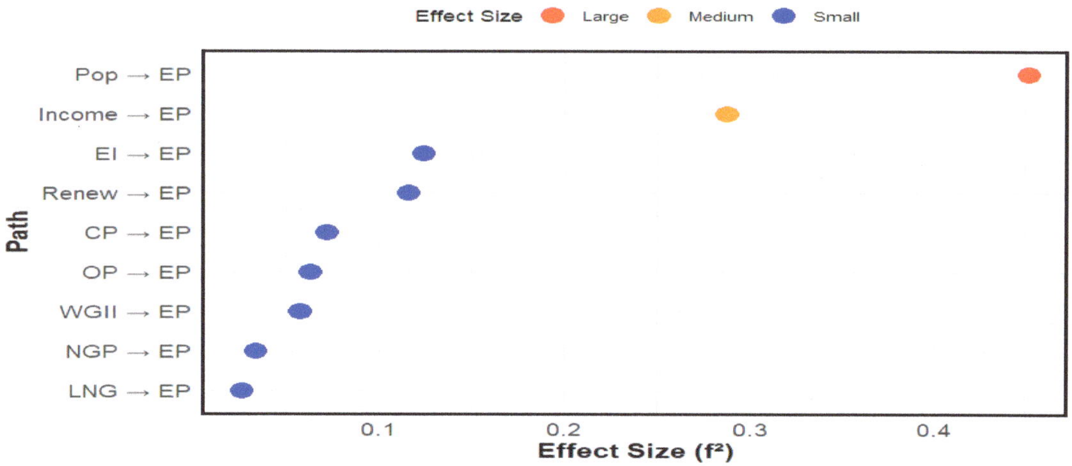

Figure 3. Effect size (f^2) assessment for practical relevance of predictors.

Our effect size total population analysis confirms a significantly large effect size (f^2 = 0.452), demonstrates its strong practical significance to determine the energy poverty. Income level has a medium effect (f^2 = 0.287), while the left over variables show small effect sizes on energy poverty. The effect sizes offer further supportive data for the path coefficients to prioritize components to handle energy poverty.

4.7. Mediation Analysis

To evaluate the mediating role of governance quality (WGI index), we performed a mediation investigation employing the approach proposed by [128]. Table 10 and Figure 4 displays the indirect effects and their statistical significance.

Table 10. Indirect effects with statistical significance in the model.

Mediation Path	Indirect Effect	t-Value	*p*-Value
Income → WGII → EP	−0.028	2.345	0.019
Pop → WGII → EP	−0.035	2.567	0.010
EI → WGII → EP	−0.015	2.123	0.034
OP → WGII → EP	−0.007	1.876	0.061
CP → WGII → EP	−0.008	1.945	0.052
Renew → WGII → EP	−0.013	2.234	0.026

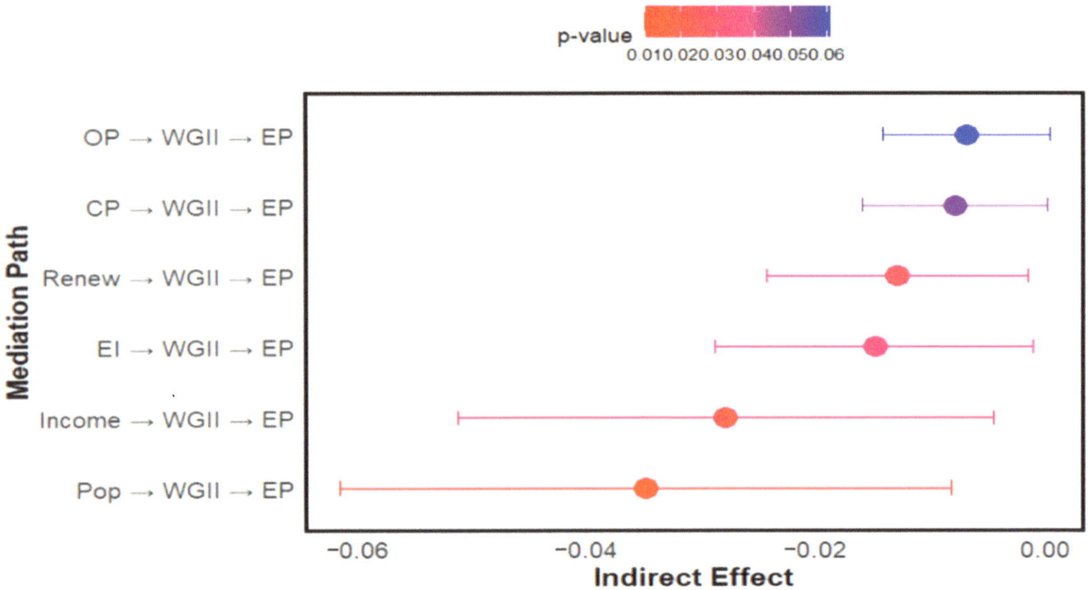

Figure 4. Mediation analysis: Indirect effects of governance quality (WGII) on energy poverty.

The findings demonstrate the significant mediating effects of governance quality on the connections between most independent variables and energy poverty, except for oil and coal prices. These findings are consistent with the results of [36,129], noted the key role of governance in energy poverty evolution. It affirms our hypothesis of the governance quality as a mediator. The negative indirect effects indicate that advancements in governance quality can boost the influence of additional contributing variables to reduce energy poverty. The research conducted by [83,84] emphasizes that effective governance enhances the provision of regulatory rules, essential for providing fair and equal access to energy resources. Strong governance structures enable countries to establish a conducive environment for the efficient implementation of policies targeting the reduction of energy poverty, optimizing the effectiveness of supplementary interventions. Figure 5 presents the structure model assessment of the direct and mediating effects of variables.

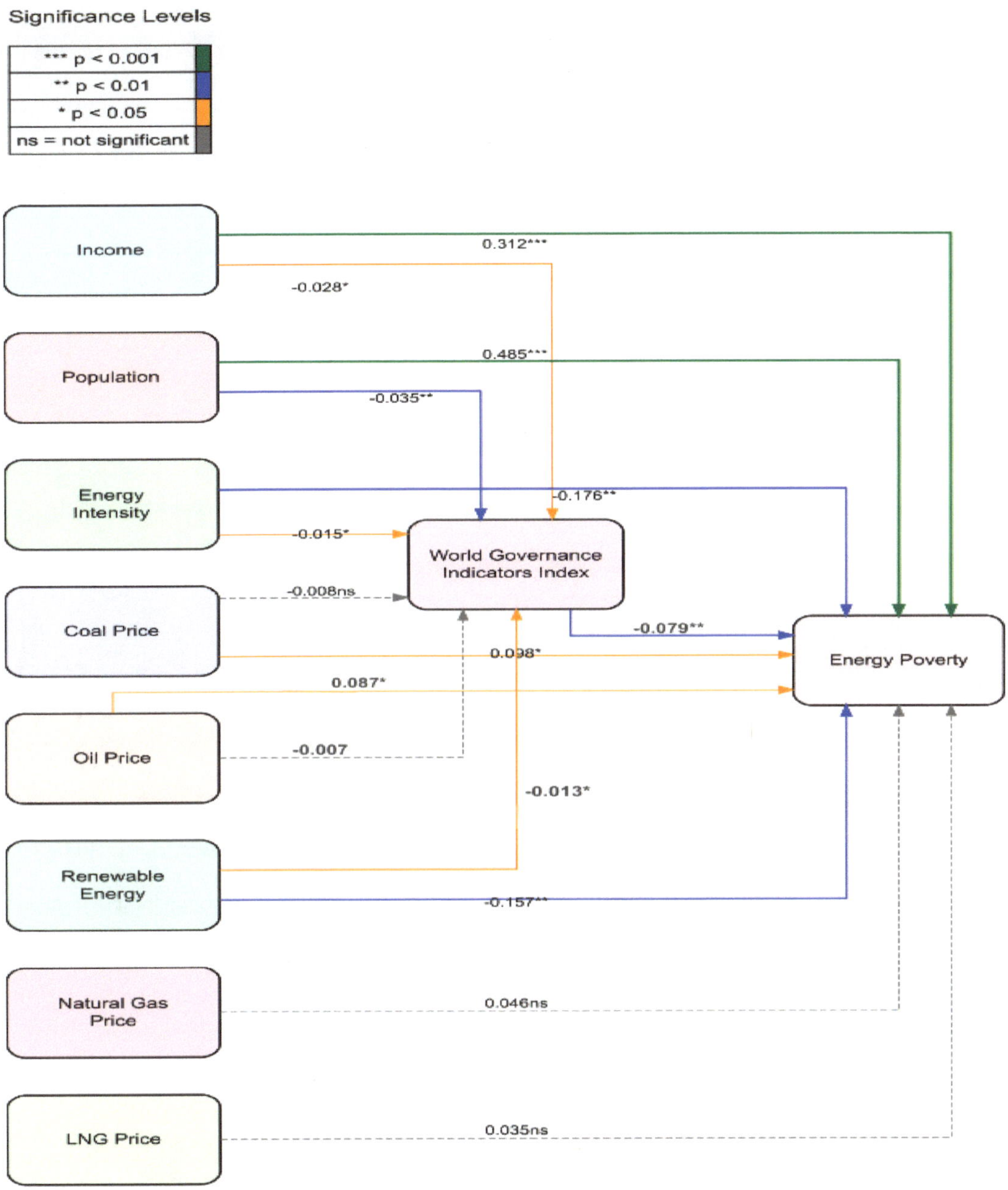

Figure 5. Structural model assessment.

4.8. Bootstrapping for Mediation

To strengthen the validity of the mediation results and address potential limitations associated with normal theory techniques in small sample sizes, we conducted bootstrapping with 5000 subsamples, a method also utilized by [130] in their poverty governance research. Table 11 and Figure 6 presents the bootstrap estimates and confidence intervals for the indirect effects.

Table 11. Bootstrap estimates and confidence intervals for indirect effects.

Mediation Path	Indirect Effect	95% CI Lower	95% CI Upper
Income → WGII → EP	−0.028	−0.052	−0.004
Pop → WGII → EP	−0.035	−0.064	−0.006
EI → WGII → EP	−0.015	−0.029	−0.001
OP → WGII → EP	−0.007	−0.015	0.001
CP → WGII → EP	−0.008	−0.017	0.001
Renew → WGII → EP	−0.013	−0.026	−0.001

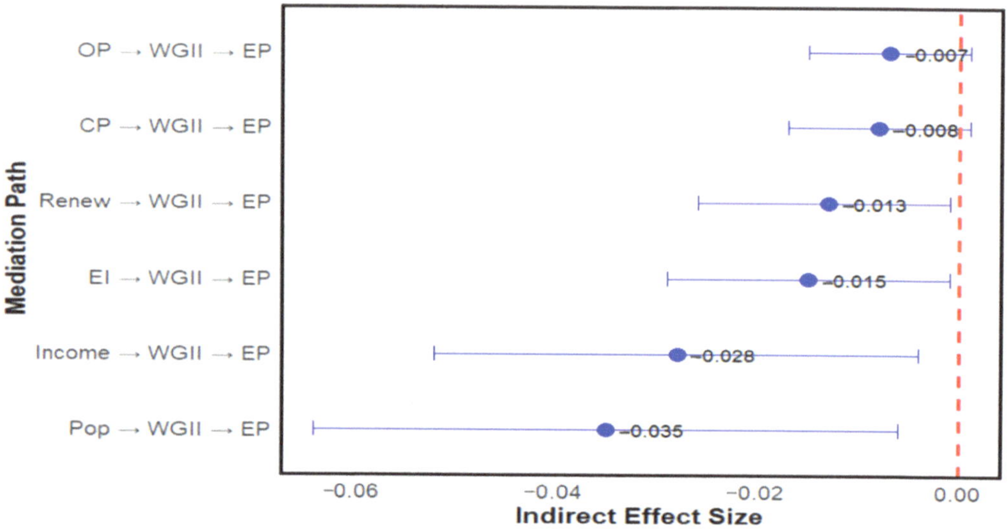

Figure 6. Bootstrap estimates for validation of mediation path.

Our rigorous methodology verifies the significance of the mediating effects for income level, total population, energy intensity, and renewable energy consumption, as their confidence intervals do not contain zero. Furthermore, our study presents persuasive evidence to support our mediation hypothesis, emphasizing the significance of governance quality in influencing the energy poverty trends in Pakistan.

4.9. Total Effects Analysis

To obtain a thorough comprehension of the influence of each factor on energy poverty, we compute the total effects, which include both direct and indirect effects (refer to Table 12 and Figure 7).

Table 12. Total effects of governance and socioeconomic factors on energy poverty.

Variable	Direct Effect	Indirect Effect	Total Effect
Income	0.312	−0.028	0.284
Pop	0.485	−0.035	0.450
WGII	−0.079	-	−0.079
EI	−0.176	−0.015	−0.191
OP	0.087	−0.007	0.080
CP	0.098	−0.008	0.090
NGP	0.046	-	0.046
LNG	0.035	-	0.035
Renew	−0.157	−0.013	−0.170

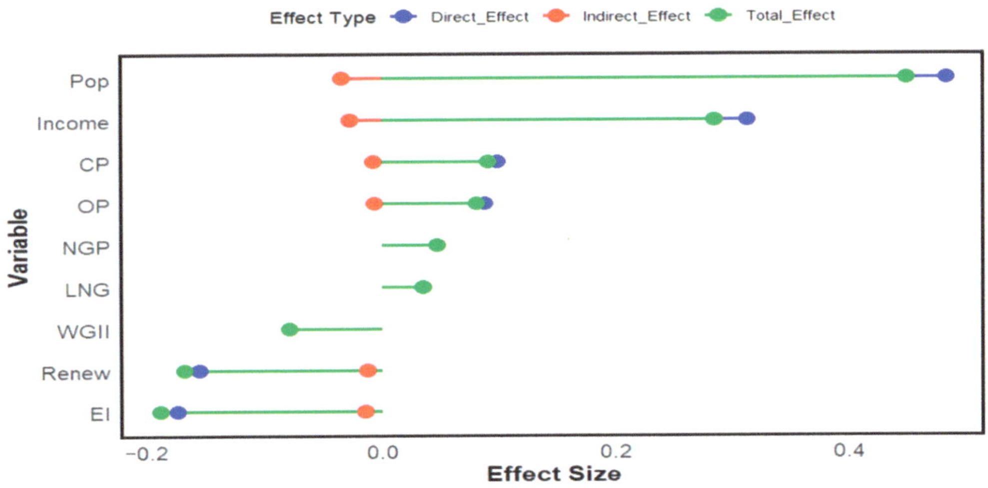

Figure 7. Total effects, including direct and indirect influences on energy poverty.

Our total effects examination indicates that the total population has the highest overall impact on energy poverty (0.450), followed by income level (0.284), which aligns with the findings of [88,131] on energy poverty. The noteworthy negative total effects of energy intensity (−0.191) and renewable energy (−0.170) corroborate the conclusions of [132], respectively highlighting the significant potential of renewable energy to combat energy poverty. Energy communities, as localized collectives striving for sustainable energy goals, have a vital role in promoting multiple Sustainable Development Goals (SDGs), namely SDG 7 (Affordable and Clean Energy), SDG 11 (Sustainable Cities and Communities), and SDG 13 (Climate Action) [16]. By facilitating the local communal ownership and administration of renewable energy resources, these communities may directly supply cost-effective and sustainable electricity to those deemed highest at risk of experiencing energy poverty. This decentralized strategy enables local production and consumption of energy, generally at a reduced cost compared to conventional energy suppliers, alleviating the financial strain on low-income people [133]. By offering individuals and communities more autonomy over their energy sources, energy communities advance energy justice, guaranteeing everyone access to cost-effective and sustainable energy and substantially reducing energy poverty. Moreover, they strengthen the ability to withstand high energy expenses and promote active participation from the local community, increasing energy availability for underprivileged communities.

The WGI index, although its direct impact is limited, plays a crucial role in mediating the effects of other variables, emphasizing the vital role of governance in tackling energy poverty. This comprehensive analysis shows strong evidence in support of our hypotheses. It provides valuable insights into the complicated mechanism of energy poverty in Pakistan, which is consistent with and improves the findings of recent literature in this study domain.

5. Conclusions and Policy Implications

5.1. Conclusions

Our findings emphasize the crucial role of governance quality as a mediator between different socioeconomic and energy-related factors and energy poverty in Pakistan, especially relevant given the impending 2030 deadline for SDG7, which strives to provide universal access to affordable, consistent, and sustainable energy. This study validates that income level and population growth positively and substantially impact energy poverty. Our results validate the "energy ladder" theory [134] but specify that developing countries like Pakistan move at different paces. This complication highlights the challenges of SDG7 goals by 2030 and suggests that economic development alone will not be enough to achieve this goal. The study highlights the negative relationships between energy intensity and renewable energy consumption with energy poverty and affirms the energy transition theory, which suggests that advanced technology and transition to cleaner energy sources can help reduce energy poverty [135]. The governance quality (WGI index) significantly mediates the relationship between most independent variables and energy poverty. It validates the institutional theory in energy policy [59]. This assertion indicates that the effectiveness of energy poverty elevation strategies depends on the efficacy of governance and institutions. Our research conclusion provides insight into why Pakistan, despite its efforts, might need help to meet SDG7 targets and emphasizes the importance of institutional reforms alongside technical solutions, aligning with the results of [136]. The disparity effects of various energy prices on energy poverty (significant for oil and coal, non-significant for natural gas and LNG) underscore the intricate relationship between global energy markets and regional energy poverty characteristics. The volatility of oil and coal prices indicates that Pakistan's energy mix and pricing structure may require modification to comply more effectively with SDG7 targets and protect vulnerable populations from energy poverty [45,87].

5.2. Policy Implications

To address the persistent obstacles towards SDG7 objectives, policymakers in Pakistan must give precedence to unified national energy initiatives with this goal. It entails establishing clear, time-bound targets to enhance energy accessibility, efficiency, and renewable energy adoption that coincide with the 2030 deadline [60,137].

Our counterintuitive finding of income level highlights the need for more reasonable economic growth policies, such as the introduction of subsidies on solar panels or progressive energy pricing mechanisms to ensure energy access for all societal groups, complying with the "leave no one behind" premise of the SDGs [4,61]. The negative effect of high energy intensity on energy poverty urges policymakers to prioritize programs in both industrial and residential sectors that improve energy efficiency, such as the implementing rules, appliance standards, and the adoption of industrial energy management technologies mediating role of governance quality indicates that effective governance is essential for the reduction of energy poverty in Pakistan. It is crucial to increase the autonomy of regulatory institutions to conduct energy market regulation, enforce energy efficiency standards, and implement transparent tariff regulations without political interference. Transparency through implementing rigorous audits and imposing penalties for electricity theft is essential to combat corruption [30,32]. Establishing regional autonomous organizations is also necessary to cater effectively to the varied requirements of each province in Pakistan. Provinces can assume more accountability for energy management and customize solutions to meet their unique needs when energy distribution is decentralized. The diverse effects of energy prices emphasize the necessity for effective, nuanced, sophisticated price policies. Pakistan's experience building wind and solar farms with the private sector has demonstrated that public–private partnerships (PPPs) can result in more effective project delivery [138]. These collaborations ought to be promoted, with the government helping to share risk and giving investors assurances. The government should also finance large-scale renewable energy projects by issuing green bonds, as green bonds have the potential to appeal to both local and global investors concerned with sustainable development. International organizations, including the Green Climate Fund (GCF), have already provided Pakistan with climate financing [139]. Reducing dependence on imported fossil fuels can be achieved by facilitating the development of solar, wind, and hydroelectric infrastructure by implementing well-designed programs that increase access to these funds.

The Pakistani government has established ambitious renewable energy objectives to increase the proportion of renewable energy sources in the national energy mix to 60% by 2030. The Gharo Wind Corridor and the Quaid-e-Azam Solar Park are noteworthy projects that the Alternative Energy Development Board (AEDB) has already successfully supported [140]. Nevertheless, further endeavor is required to expand these to achieve the objectives set for 2030. In order to promote energy efficiency in businesses, households, and agriculture, the government has launched programs such as the National Energy Efficiency and Conservation Act of 2016. Nevertheless, there has been uneven execution, and more work is required to guarantee broad adoption.

Although this study focuses on the 2030 deadline (SDG 7), sustained efforts in critical areas are necessary to achieve long-lasting gains for long-term sustainability after 2030. First, investing in renewable energy infrastructure, including grid modernization and energy storage, can help manage intermittent renewable sources and improve system resilience. Furthermore, supporting decentralized energy systems like micro-grids might improve rural communities' access to electricity while ensuring long-term dependability and autonomy. Secondly, governance reforms must be institutionalized to safeguard against future political instability or corruption, thereby ensuring transparency and accountability. In conclusion, it is essential to prioritize energy efficiency, climate resilience, and capacity building to guarantee that the progress achieved by 2030 continues to develop and meet future energy requirements while accommodating technological advancements and climate change challenges. The formulation of these policies ought to emphasize SDG7 goals and a balance between affordability and sustainability [107,120].

5.3. Limitations and Future Research

Despite the significant insights on governance and energy poverty provided by this study, it is important to recognize its substantial limitations. A fundamental limitation is the availability of data. While the current study relies on annual data, using higher-frequency data such as quarterly or monthly will allow for a more precise analysis of short-term fluctuations in energy availability and governance effectiveness. Moreover national-level approach may overlook geography or demography variations in energy poverty. Furthermore, future studies should explore the importance of private sector participation, which has been important in successfully executing renewable energy initiatives in other developing countries. The integration of private sector investment models can offer substantial insights into the financing of sustainable energy development. Furthermore, future research should consider more precise factors, such as technological advancements and the impact of foreign aid on the development of energy infrastructure.

This study offers a static view of energy poverty factors. However, the potential of future research using dynamic panel models to observe the evolution of these interactions over time, especially in the lead-up to the 2030 SDG timeframe, is another vision to advance our understanding. Furthermore strengthening this quantitative analysis with qualitative studies, like case studies or interviews with policymakers and contributed groups, can offer profound insights into the causes driving energy poverty and the obstacles in meeting SDG7 in Pakistan.

Our study enhances the comprehension of energy poverty dynamics in Pakistan through a complete model that takes into account various socioeconomic, energy-related, and governance challenges. The findings provide a foundation for evidence-based policy-making. The ultimate goal is to boost sustainable development and alleviate energy poverty in Pakistan by achieving the SDG7 2030 target. The constant challenges underlined in this study emphasize the urgency of swift action and focused interventions to ensure Pakistan can fulfill its SDG7 commitments. Future studies addressing the limitations mentioned earlier will enhance our knowledge of this important problem and bolster more effective interventions to reduce energy poverty and achieve universal energy access by 2030.

Author Contributions: Conceptualization, R.L.; Methodology, M.W.H.; Software, M.W.H.; Formal analysis, S.S.; Investigation, S.S.; Data curation, A.Q.; Writing—original draft, R.L., A.Q. and M.W.H.; Writing—review & editing, S.S.; Supervision, J.L. All authors have read and agreed to the published version of the manuscript.

Funding: This research received no external funding.

Institutional Review Board Statement: Not applicable.

Informed Consent Statement: Not applicable.

Data Availability Statement: The raw data supporting the conclusions of this article will be made available by the authors on request.

Conflicts of Interest: The authors declare no conflict of interest.

References

1. Garzón Baquero, J.E.; Bellon Monsalve, D. From Fossil Fuel Energy to Hydrogen Energy: Transformation of Fossil Fuel Energy Economies into Hydrogen Economies through Social Entrepreneurship. *Int. J. Hydrogen Energy* **2024**, *54*, 574–585. [CrossRef]
2. Madurai Elavarasan, R.; Nadarajah, M.; Pugazhendhi, R.; Sinha, A.; Gangatharan, S.; Chiaramonti, D.; Abou Houran, M. The Untold Subtlety of Energy Consumption and Its Influence on Policy Drive towards Sustainable Development Goal 7. *Appl. Energy* **2023**, *334*, 120698. [CrossRef]
3. Rehman, A.; Ma, H.; Ozturk, I.; Ulucak, R. Sustainable Development and Pollution: The Effects of CO2 Emission on Population Growth, Food Production, Economic Development, and Energy Consumption in Pakistan. *Environ. Sci. Pollut. Res.* **2022**, *29*, 17319–17330. [CrossRef] [PubMed]
4. Khan, I.; Hou, F.; Irfan, M.; Zakari, A.; Le, H.P. Does Energy Trilemma a Driver of Economic Growth? The Roles of Energy Use, Population Growth, and Financial Development. *Renew. Sustain. Energy Rev.* **2021**, *146*, 111157. [CrossRef]
5. Rahman, A.; Farrok, O.; Haque, M.M. Environmental Impact of Renewable Energy Source Based Electrical Power Plants: Solar, Wind, Hydroelectric, Biomass, Geothermal, Tidal, Ocean, and Osmotic. *Renew. Sustain. Energy Rev.* **2022**, *161*, 112279. [CrossRef]
6. Nsafon, B.E.K.; Same, N.N.; Yakub, A.O.; Chaulagain, D.; Kumar, N.M.; Huh, J.-S. The Justice and Policy Implications of Clean Energy Transition in Africa. *Front. Environ. Sci.* **2023**, *11*, 1089391. [CrossRef]
7. Kod'ousková, H.; Ilavská, A.; Stašáková, T.; David, D.; Osička, J. Energy Transition for the Rich and Energy Poverty for the Rest? Mapping and Explaining District Heating Transition, Energy Poverty, and Vulnerability in Czechia. *Energy Res. Soc. Sci.* **2023**, *100*, 103128. [CrossRef]
8. Yuan, H.; Zhao, L.; Umair, M. Crude Oil Security in a Turbulent World: China's Geopolitical Dilemmas and Opportunities. *Extr. Ind. Soc.* **2023**, *16*, 101334. [CrossRef]
9. Hafezi, R.; Souhankar, A. Energy Security in a Resource-Rich Economy: Case of Iran. In *The Handbook of Energy Policy*; Springer Nature Singapore: Singapore, 2023; pp. 97–127. [CrossRef]
10. Qadeer, A.; Hussan, M.W.; Aziz, G.; Waheed, R.; Sarwar, S. Emerging Trends of Green Hydrogen and Sustainable Environment in the Case of Australia. *Environ. Sci. Pollut. Res.* **2023**, *30*, 115788–115804. [CrossRef]
11. Salman, M.; Zha, D.; Wang, G. Assessment of Energy Poverty Convergence: A Global Analysis. *Energy* **2022**, *255*, 124579. [CrossRef]
12. Zaidan, E.; Antoine Ibrahim, I. Achieving Energy Justice: The Role of Supervisory and Compliance Mechanisms in Global Frameworks and the International Community. *Energy Strategy Rev.* **2024**, *52*, 101335. [CrossRef]
13. Minas, A.M.; García-Freites, S.; Walsh, C.; Mukoro, V.; Aberilla, J.M.; April, A.; Kuriakose, J.; Gaete-Morales, C.; Gallego-Schmid, A.; Mander, S. Advancing Sustainable Development Goals through Energy Access: Lessons from the Global South. *Renew. Sustain. Energy Rev.* **2024**, *199*, 114457. [CrossRef]
14. Painuly, P.K.; Tyagi, R.; Vishwakarma, S.; Khare, S.K.; Haghighi, M. Energy Supply Using Nexus Approach for Attaining Sustainable Development Goal 7. *Afford. Clean. Energy* **2021**, 562–573. [CrossRef]
15. D'Adamo, I.; Gastaldi, M.; Morone, P. Economic Sustainable Development Goals: Assessments and Perspectives in Europe. *J. Clean. Prod.* **2022**, *354*, 131730. [CrossRef]
16. D'Adamo, I.; Di Carlo, C.; Gastaldi, M.; Rossi, E.N.; Uricchio, A.F. Economic Performance, Environmental Protection and Social Progress: A Cluster Analysis Comparison towards Sustainable Development. *Sustainability* **2024**, *16*, 5049. [CrossRef]
17. Pachauri, S.; Poblete-Cazenave, M.; Aktas, A.; Gidden, M.J. Access to Clean Cooking Services in Energy and Emission Scenarios after COVID-19. *Nat. Energy* **2021**, *6*, 1067–1076. [CrossRef]
18. Boeri, A.; Gianfrate, V.; Boulanger, S.O.M.; Massari, M. Future Design Approaches for Energy Poverty: Users Profiling and Services for No-Vulnerable Condition. *Energy* **2020**, *13*, 2115. [CrossRef]
19. Ceglia, F.; Marrasso, E.; Samanta, S.; Sasso, M. Addressing Energy Poverty in the Energy Community: Assessment of Energy, Environmental, Economic, and Social Benefits for an Italian Residential Case Study. *Sustainability* **2022**, *14*, 15077. [CrossRef]
20. Manko, K.; Watkins, T.A. Microfinance and SDG 7: Financial Impact Channels for Mitigating Energy Poverty. *Dev. Pract.* **2022**, *32*, 1036–1048. [CrossRef]
21. Chopra, R.; Rehman, M.A.; Yadav, A.; Bhardwaj, S. Revisiting the EKC Framework Concerning COP-28 Carbon Neutrality Management: Evidence from Top-5 Carbon Embittering Countries. *J. Environ. Manag.* **2024**, *356*, 120690. [CrossRef]
22. Jiang, T.; He, X.; Su, B.; Havea, P.H.; Wei, K.; Kundzewicz, Z.W.; Liu, D. COP 28: Challenge of Coping with Climate Crisis. *Innovation* **2024**, *5*, 100559. [CrossRef] [PubMed]
23. Wang, Z.; Amin, A.; Chandio, A.A.; Shah, A.H.; Ullah, M.I. Dynamical Assessment of Multi-Dimensional Energy Poverty at the National and Sub-National Levels in Pakistan. *Energy Effic.* **2024**, *17*, 18. [CrossRef]
24. Batool, K.; Zhao, Z.-Y.; Nureen, N.; Irfan, M. Assessing and Prioritizing Biogas Barriers to Alleviate Energy Poverty in Pakistan: An Integrated AHP and G-TOPSIS Model. *Environ. Sci. Pollut. Res.* **2023**, *30*, 94669–94693. [CrossRef]

25. Khan, N.U.; Zhongyi, P.; Ullah, A.; Mumtaz, M. A Comprehensive Evaluation of Sustainable Mineral Resources Governance in Pakistan: An Analysis of Challenges and Reforms. *Resour. Policy* **2024**, *88*, 104383. [CrossRef]
26. Raza, M.Y.; Lin, B. Coal Efficiency, Carbon Reduction, and Future Policy Perspective in Pakistan's Economic Growth: A Decomposition and Decoupling Approach. *Front. Energy Res.* **2023**, *11*, 1275221. [CrossRef]
27. Hussain, S.; Xuetong, W.; Maqbool, R. Understanding the Power Disruption and Its Impact on Community Development: An Empirical Case of Pakistan. *Sustain. Energy Technol. Assess.* **2023**, *55*, 102922. [CrossRef]
28. Imran, M.; Ozcatalbas, O. Determinants of Household Cooking Fuels and Their Impact on Women's Health in Rural Pakistan. *Environ. Sci. Pollut. Res.* **2020**, *27*, 23849–23861. [CrossRef]
29. Jan, M.Z.; Ullah, K.; Abbas, F.; Khalid, H.A.; Bajwa, T.M. Barriers to the Adoption of Social Welfare Measures in the Electricity Tariff Structure of Developing Countries: A Case of Pakistan. *Energy Policy* **2023**, *179*, 113648. [CrossRef]
30. Bakhsh, S.; Zhang, W.; Ali, K.; Oláh, J. Strategy towards Sustainable Energy Transition: The Effect of Environmental Governance, Economic Complexity and Geopolitics. *Energy Strategy Rev.* **2024**, *52*, 101330. [CrossRef]
31. Du, J.; Shen, Z.; Song, M.; Vardanyan, M. The Role of Green Financing in Facilitating Renewable Energy Transition in China: Perspectives from Energy Governance, Environmental Regulation, and Market Reforms. *Energy Econ.* **2023**, *120*, 106595. [CrossRef]
32. Agostino, M. Extreme Weather Events and Firms' Energy Practices. The Role of Country Governance. *Energy Policy* **2024**, *192*, 114235. [CrossRef]
33. Acheampong, A.O.; Opoku, E.E.O.; Dogah, K.E. The Political Economy of Energy Transition: The Role of Globalization and Governance in the Adoption of Clean Cooking Fuels and Technologies. *Technol. Forecast. Soc. Chang.* **2023**, *186*, 122156. [CrossRef]
34. Iqbal, K.M.J.; Akhtar, N.; Amir, S.; Khan, M.I.; Shah, A.A.; Tariq, M.A.U.R.; Ullah, W. Multi-Variable Governance Index Modeling of Government's Policies, Legal and Institutional Strategies, and Management for Climate Compatible and Sustainable Agriculture Development. *Sustainability* **2022**, *14*, 11763. [CrossRef]
35. AlShiab, M.S.I.; Al-Malkawi, H.-A.N.; Lahrech, A. Revisiting the relationship between governance quality and economic growth. *Int. J. Econ. Financ. Issues* **2020**, *10*, 54–63. [CrossRef]
36. Shahbaz, M.; Wang, J.; Dong, K.; Zhao, J. The Impact of Digital Economy on Energy Transition across the Globe: The Mediating Role of Government Governance. *Renew. Sustain. Energy Rev.* **2022**, *166*, 112620. [CrossRef]
37. Pelz, S.; Chindarkar, N.; Urpelainen, J. Energy Access for Marginalized Communities: Evidence from Rural North India, 2015–2018. *World Dev.* **2021**, *137*, 105204. [CrossRef]
38. Acharya, R.H.; Sadath, A.C. Achievements and Challenges of Energy Poverty Alleviation Policies: Evidence from the Select States in India. *J. Public Aff.* **2023**, *23*, e2839. [CrossRef]
39. UNG Assembly. United Nations General Assembly Resolution A. *Antarct. Int. Law.* **2015**, *15900*, 1–35.
40. Masuku, B. Rethinking South Africa's Household Energy Poverty through the Lens of off-Grid Energy Transition. *Dev. S. Afr.* **2024**, *41*, 467–489. [CrossRef]
41. Cummins, M.; Gillanders, R. Greasing the Turbines? Corruption and Access to Electricity in Africa. *Energy Policy* **2020**, *137*, 111188. [CrossRef]
42. Sovacool, B.K. Clean, Low-Carbon but Corrupt? Examining Corruption Risks and Solutions for the Renewable Energy Sector in Mexico, Malaysia, Kenya and South Africa. *Energy Strategy Rev.* **2021**, *38*, 100723. [CrossRef]
43. Khurshid, N. Does the Causality between Environmental Sustainability, Non-Renewable Energy Consumption, Geopolitical Risks, and Trade Liberalization Matter for Pakistan? Evidence from VECM Analysis. *Heliyon* **2023**, *9*, e21444. [CrossRef] [PubMed]
44. Ali, Y.; Ahmad, M.; Sabir, M.; Shah, S.A. Regional Development through Energy Infrastructure: A Comparison and Optimization of Iran-Pakistan-India (IPI) & Turkmenistan-Afghanistan-Pakistan-India (TAPI) Gas Pipelines. *Oper. Res. Eng. Sci. Theory Appl.* **2021**, *4*, 82–106. [CrossRef]
45. Awan, A.; Bilgili, F.; Rahut, D.B. Household Fuel Choices and Consumption Intensity in Pakistan: Evidence from HIES Data 2001–2019. *Environ. Sci. Pollut. Res.* **2023**, *30*, 1–16. [CrossRef] [PubMed]
46. Xia, S.; Yang, Y.; Qian, X.; Xu, X. Spatiotemporal Interaction and Socioeconomic Determinants of Rural Energy Poverty in China. *Int. J. Environ. Res. Public Health* **2022**, *19*, 10851. [CrossRef]
47. Tundys, B.; Bretyn, A.; Urbaniak, M. Energy Poverty and Sustainable Economic Development: An Exploration of Correlations and Interdependencies in European Countries. *Energy* **2021**, *14*, 7640. [CrossRef]
48. Zhao, J.; Dong, K.; Dong, X.; Shahbaz, M. How Renewable Energy Alleviate Energy Poverty? A Global Analysis. *Renew. Energy* **2022**, *186*, 299–311. [CrossRef]
49. Mahalik, M.K.; Mallick, H.; Padhan, H. Do Educational Levels Influence the Environmental Quality? The Role of Renewable and Non-Renewable Energy Demand in Selected BRICS Countries with a New Policy Perspective. *Renew. Energy* **2021**, *164*, 419–432. [CrossRef]
50. Katoch, O.R.; Sharma, R.; Parihar, S.; Nawaz, A. Energy Poverty and Its Impacts on Health and Education: A Systematic Review. *Int. J. Energy Sect. Manag.* **2024**, *18*, 411–431. [CrossRef]
51. Nussbaumer, P.; Bazilian, M.; Modi, V. Measuring Energy Poverty: Focusing on What Matters. *Renew. Sustain. Energy Rev.* **2012**, *16*, 231–243. [CrossRef]
52. Pachauri, S.; Spreng, D. Measuring and Monitoring Energy Poverty. *Energy Policy* **2011**, *39*, 7497–7504. [CrossRef]

53. Qurat-ul-Ann, A.-R.; Mirza, F.M. Multidimensional Energy Poverty in Pakistan: Empirical Evidence from Household Level Micro Data. *Soc. Indic. Res.* **2021**, *155*, 211–258. [CrossRef]
54. Sharif, M.; Khan, F.N. Unveiling the Implications of Energy Poverty for Educational Attainments in Pakistan: A Multidimensional Analysis. *Int. J. Energy Econ. Policy* **2023**, *13*, 472–483. [CrossRef]
55. Sovacool, B.K. The Political Economy of Energy Poverty: A Review of Key Challenges. *Energy Sustain. Dev.* **2012**, *16*, 272–282. [CrossRef]
56. Awan, A.B.; Khan, Z.A. Recent Progress in Renewable Energy–Remedy of Energy Crisis in Pakistan. *Renew. Sustain. Energy Rev.* **2014**, *33*, 236–253. [CrossRef]
57. Rasool, Y.; Zaidi, S.A.H.; Zafar, M.W. Determinants of Carbon Emissions in Pakistan's Transport Sector. *Environ. Sci. Pollut. Res.* **2019**, *26*, 22907–22921. [CrossRef]
58. Byaro, M.; Mmbaga, N.F.; Mafwolo, G. Tackling Energy Poverty: Do Clean Fuels for Cooking and Access to Electricity Improve or Worsen Health Outcomes in Sub-Saharan Africa? *World Dev. Sustain.* **2024**, *4*, 100125. [CrossRef]
59. Dragomir, V.D.; Dumitru, M.; Perevoznic, F.M. Carbon Reduction and Energy Transition Targets of the Largest European Companies: An Empirical Study Based on Institutional Theory. *Clean. Prod. Lett.* **2023**, *4*, 100039. [CrossRef]
60. Phoumin, H.; Kimura, F. The Impacts of Energy Insecurity on Household Welfare in Cambodia: Empirical Evidence and Policy Implications. *Econ. Model.* **2019**, *82*, 35–41. [CrossRef]
61. Sadath, A.C.; Acharya, R.H. Assessing the Extent and Intensity of Energy Poverty Using Multidimensional Energy Poverty Index: Empirical Evidence from Households in India. *Energy Policy* **2017**, *102*, 540–550. [CrossRef]
62. Primc, K.; Slabe-Erker, R.; Majcen, B. Energy Poverty: A Macrolevel Perspective. *Sustain. Dev.* **2019**, *27*, 982–989. [CrossRef]
63. Sy, S.A.; Mokaddem, L. Energy Poverty in Developing Countries: A Review of the Concept and Its Measurements. *Energy Res. Soc. Sci.* **2022**, *89*, 102562. [CrossRef]
64. Andrews-Speed, P. Applying Institutional Theory to the Low-Carbon Energy Transition. *Energy Res. Soc. Sci.* **2016**, *13*, 216–225. [CrossRef]
65. Smith, K.R.; Bruce, N.; Balakrishnan, K.; Adair-Rohani, H.; Balmes, J.; Chafe, Z.; Dherani, M.; Hosgood, H.D.; Mehta, S.; Pope, D.; et al. Millions Dead: How Do We Know and What Does It Mean? Methods Used in the Comparative Risk Assessment of Household Air Pollution. *Annu. Rev. Public Health* **2014**, *35*, 185–206. [CrossRef] [PubMed]
66. Ghosh, P.; Chatterjee, V.; Paul, A.; Ghosh, D.; Husain, Z. Reducing Energy Poverty: How to Empower Women and Switch to Clean Fuel in India? *Energy Res. Soc. Sci.* **2024**, *110*, 103444. [CrossRef]
67. Carfora, A.; Passaro, R.; Scandurra, G.; Thomas, A. The Causal Nexus between Income and Energy Poverty in EU Member States. *Energy* **2022**, *15*, 2822.
68. Rahman, A.; Murad, S.M.W.; Mohsin, A.K.M.; Wang, X. Does Renewable Energy Proactively Contribute to Mitigating Carbon Emissions in Major Fossil Fuels Consuming Countries? *J. Clean. Prod.* **2024**, *452*, 142113. [CrossRef]
69. Jones, A.; Nock, D.; Samaras, C.; Qiu, Y.; Xing, B. Climate Change Impacts on Future Residential Electricity Consumption and Energy Burden: A Case Study in Phoenix, Arizona. *Energy Policy* **2023**, *183*, 113811. [CrossRef]
70. Heltberg, R. Fuel Switching: Evidence from Eight Developing Countries. *Energy Econ.* **2004**, *26*, 869–887. [CrossRef]
71. Igawa, M.; Managi, S. Energy Poverty and Income Inequality: An Economic Analysis of 37 Countries. *Appl. Energy* **2022**, *306*, 118076. [CrossRef]
72. Perera, A.T.D.; Javanroodi, K.; Nik, V.M. Climate Resilient Interconnected Infrastructure: Co-Optimization of Energy Systems and Urban Morphology. *Appl. Energy* **2021**, *285*, 116430. [CrossRef]
73. Acheampong, A.O.; Erdiaw-Kwasie, M.O.; Abunyewah, M. Does Energy Accessibility Improve Human Development? Evidence from Energy-Poor Regions. *Energy Econ.* **2021**, *96*, 105165. [CrossRef]
74. Wu, B.; Liu, S.; Wang, J.; Tahir, S.; Patwary, A.K. Assessing the Mechanism of Energy Efficiency and Energy Poverty Alleviation Based on Environmental Regulation Policy Measures. *Environ. Sci. Pollut. Res.* **2021**, *28*, 40858–40870. [CrossRef] [PubMed]
75. Singh, S.; Ru, J. Accessibility, Affordability, and Efficiency of Clean Energy: A Review and Research Agenda. *Environ. Sci. Pollut. Res.* **2022**, *29*, 18333–18347. [CrossRef]
76. UNEP Falling Clean Energy Costs Provide Opportunity to Boost Climate Action in COVID-19 Recovery Packages. Available online: https://www.unep.org/news-and-stories/press-release/falling-clean-energy-costs-provide-opportunity-boost-climate-action (accessed on 9 July 2024).
77. Wahlund, M.; Palm, J. The Role of Energy Democracy and Energy Citizenship for Participatory Energy Transitions: A Comprehensive Review. *Energy Res. Soc. Sci.* **2022**, *87*, 102482. [CrossRef]
78. Bouzarovski, S.; Haarstad, H. Rescaling Low-carbon Transformations: Towards a Relational Ontology. *Trans. Inst. Br. Geogr.* **2019**, *44*, 256–269. [CrossRef]
79. Akinyemi, R.; Ojagbemi, A.; Akinyemi, J.; Salami, A.; Olopade, F.; Farombi, T.; Nweke, M.; Uvere, E.; Aridegbe, M.; Balogun, J.; et al. Gender Differential in Inclination to Donate Brain for Research among Nigerians: The IBADAN Brain Bank Project. *Cell Tissue Bank.* **2019**, *20*, 297–306. [CrossRef]
80. Terrapon-Pfaff, J.; Ersoy, S.R.; Fink, T.; Amroune, S.; Jamea, E.M.; Zgou, H.; Viebahn, P. Localizing the Water-Energy Nexus: The Relationship between Solar Thermal Power Plants and Future Developments in Local Water Demand. *Sustainability* **2020**, *13*, 108. [CrossRef]
81. Jones, N.; Warren, P. Innovation and Distribution of Solar Home Systems in Bangladesh. *Clim. Dev.* **2021**, *13*, 386–398. [CrossRef]

82. Saad, R.; Plazas-Niño, F.; Cannone, C.; Yeganyan, R.; Howells, M.; Luscombe, H. Long-Term Energy System Modelling for a Clean Energy Transition and Improved Energy Security in Botswana's Energy Sector Using the Open-Source Energy Modelling System. *Climate* **2024**, *12*, 88. [CrossRef]
83. Kaitah, A.; Frimpong, F.B. Understanding differential effects of energy governance constraints in nigeria: A transitional justice perspective. *WIT Trans. Ecol. Environ.* **2023**, *261*, 25–33. [CrossRef]
84. Sinha, A.; Bekiros, S.; Hussain, N.; Nguyen, D.K.; Khan, S.A. How Social Imbalance and Governance Quality Shape Policy Directives for Energy Transition in the OECD Countries? *Energy Econ.* **2023**, *120*, 106642. [CrossRef]
85. World Bank. More People Have Access to Electricity Than Ever Before, but World Is Falling Short of Sustainable Energy Goals. Available online: https://www.worldbank.org/en/news/press-release/2019/05/22/tracking-sdg7-the-energy-progress-report-2019 (accessed on 9 July 2024).
86. UNDP. *SDG 7: Affordable and Clean Energy*; UN, ESCAP: Bangkok, Thailand, 2023.
87. Aized, T.; Rehman, S.M.S.; Sumair, M. Pakistan Energy Situation, Policy, and Issues. In *Recent Advances in Renewable Energy Technologies*; Elsevier: Amsterdam, The Netherlands, 2021; pp. 387–428. [CrossRef]
88. Faiella, I.; Lavecchia, L. Energy Poverty Indicators. In *Urban Fuel Poverty*; Elsevier: Amsterdam, The Netherlands, 2019; pp. 127–141. [CrossRef]
89. Jayachandran, M.; Gatla, R.K.; Rao, K.P.; Rao, G.S.; Mohammed, S.; Milyani, A.H.; Azhari, A.A.; Kalaiarasy, C.; Geetha, S. Challenges in Achieving Sustainable Development Goal 7: Affordable and Clean Energy in Light of Nascent Technologies. *Sustain. Energy Technol. Assess.* **2022**, *53*, 102692. [CrossRef]
90. Kaufmann, D.; Kraay, A.; Mastruzzi, M. The Worldwide Governance Indicators: Methodology and Analytical Issues. *Hague J. Rule Law.* **2011**, *3*, 220–246. [CrossRef]
91. Abegaz, M.B.; Debela, K.L.; Hundie, R.M. The Effect of Governance on Entrepreneurship: From All Income Economies Perspective. *J. Innov. Entrep.* **2023**, *12*, 1. [CrossRef]
92. Qin, L.; Aziz, G.; Hussan, M.W.; Qadeer, A.; Sarwar, S. Empirical Evidence of Fintech and Green Environment: Using the Green Finance as a Mediating Variable. *Int. Rev. Econ. Financ.* **2024**, *89*, 33–49. [CrossRef]
93. He, Y.-T.; Wang, X.-J.; He, H.; Zhai, J.; Wang, B.-S. Moving-Average Based Index to Timely Evaluate the Current Epidemic Situation after COVID-19 Outbreak. *MedRxiv* **2020**, 2020–2023. [CrossRef]
94. Gu, Y.; Qin, X.; Wang, Z.; Zhang, C.; Guo, S. Correction to: Global Justice Index Report. *Chin. Political Sci. Rev.* **2020**, *5*. [CrossRef]
95. Fernando, M.; Samita, S.; Abeynayake, R. Modified Factor Analysis to Construct Composite Indices: Illustration on Urbanization Index. *Trop. Agric. Res.* **2012**, *23*, 327. [CrossRef]
96. Stock, J.H.; Watson, M.W. Generalized Shrinkage Methods for Forecasting Using Many Predictors. *J. Bus. Econ. Stat.* **2012**, *30*, 481–493. [CrossRef]
97. Bai, J.; Ng, S. Determining the Number of Factors in Approximate Factor Models. *Econometrica* **2002**, *70*, 191–221. [CrossRef]
98. Munda, G.; Nardo, M. On the Methodological Foundations of Composite Indicators Used for Ranking Countries. *Ispra Italy Jt. Res. Cent. Eur. Communities* **2003**, 1–19.
99. Nardo, M.; Saisana, M.; Saltelli, A.; Tarantola, S.; Hoffman, A.; Giovannini, E. *Handbook on Constructing Composite Indicators: Methodology and User Guide*; OECD Publishing: Paris, France, 2005.
100. Greco, S.; Ishizaka, A.; Tasiou, M.; Torrisi, G. On the Methodological Framework of Composite Indices: A Review of the Issues of Weighting, Aggregation, and Robustness. *Soc. Indic. Res.* **2019**, *141*, 61–94. [CrossRef]
101. Gyimah, J.; Batasuma, S.; Yao, X.; Wauk, G. The Adoption of Renewable Energy towards Environmental Sustainability: Evidence from Partial Least Square Structural Equation Modelling (PLS-SEM). *PLoS ONE* **2024**, *19*, e0299727. [CrossRef]
102. Messie Pondie, T.; Engwali, F.D. Energy Poverty in Light of the Climate Emergency in Sub-Saharan Africa: Impact and Transmission Channels. *Nat. Resour. Forum* **2024**. [CrossRef]
103. Hair, J.F.; Risher, J.J.; Sarstedt, M.; Ringle, C.M. When to Use and How to Report the Results of PLS-SEM. *Eur. Bus. Rev.* **2019**, *31*, 2–24. [CrossRef]
104. Vishnoi, S.K.; Mathur, S.; Bagga, T.; Singhal, A.; Rawal, P.; Sharma, S.; Yadav, R. Construct Modelling, Statistical Analysis and Empirical Validation Using PLS-SEM: A Step-by-Step Guide of the Analysis Procedure. *Int. J. Data Anal. Tech. Strateg.* **2024**, *16*, 162–180. [CrossRef]
105. Sarstedt, M.; Ringle, C.M.; Hair, J.F. Treating Unobserved Heterogeneity in PLS-SEM: A Multi-Method Approach. *Partial. Least Sq. Path Model. Basic. Concepts Methodol. Issues Appl.* **2017**, 197–217. [CrossRef]
106. Batool, K.; Zhao, Z.-Y.; Irfan, M.; Żywiołek, J. Assessing the Role of Sustainable Strategies in Alleviating Energy Poverty: An Environmental Sustainability Paradigm. *Environ. Sci. Pollut. Res.* **2023**, *30*, 67109–67130. [CrossRef]
107. Khattak, S.I.; Ahmad, M.; Khan, Z.U.; Khan, A. Exploring the Impact of Innovation, Renewable Energy Consumption, and Income on CO2 Emissions: New Evidence from the BRICS Economies. *Environ. Sci. Pollut. Res.* **2020**, *27*, 13866–13881. [CrossRef]
108. Zhang, D.; Li, J.; Han, P. A Multidimensional Measure of Energy Poverty in China and Its Impacts on Health: An Empirical Study Based on the China Family Panel Studies. *Energy Policy* **2019**, *131*, 72–81. [CrossRef]
109. Sophia Yuliantini, L.; Nurmandi, A. The The Impact of the E-Government Development Index (EGDI) on the Worldwide Governance Indicator (WGI) in European Union Countries. *Policy Gov. Rev.* **2023**, *7*, 140. [CrossRef]
110. Tai, A.; Lin, S. Integrated Multiple Mediation Analysis: A Robustness-specificity Trade-off in Causal Structure. *Stat. Med.* **2021**, *40*, 4541–4567. [CrossRef] [PubMed]

111. Cheung, M.W.-L. Synthesizing Indirect Effects in Mediation Models with Meta-Analytic Methods. *Alcohol Alcohol.* **2022**, *57*, 5–15. [CrossRef]
112. Ogwumike, F.O.; Ozughalu, U.M. Analysis of Energy Poverty and Its Implications for Sustainable Development in Nigeria. *Environ. Dev. Econ.* **2016**, *21*, 273–290. [CrossRef]
113. Fotheringham, A.S.; Oshan, T.M. Geographically Weighted Regression and Multicollinearity: Dispelling the Myth. *J. Geogr. Syst.* **2016**, *18*, 303–329. [CrossRef]
114. Kyriazos, T.; Poga, M. Dealing with Multicollinearity in Factor Analysis: The Problem, Detections, and Solutions. *Open J. Stat.* **2023**, *13*, 404–424. [CrossRef]
115. Månsson, K.; Kibria, B.M.G.; Shukur, G.; Sjölander, P. On the Estimation of the Co2 Emission, Economic Growth and Energy Consumption Nexus Using Dynamic OLS in the Presence of Multicollinearity. *Sustainability* **2018**, *10*, 1315. [CrossRef]
116. Midi, H.; Sarkar, S.K.; Rana, S. Collinearity Diagnostics of Binary Logistic Regression Model. *J. Interdiscip. Math.* **2010**, *13*, 253–267. [CrossRef]
117. Antepara, I.; Papada, L.; Gouveia, J.P.; Katsoulakos, N.; Kaliampakos, D. Improving Energy Poverty Measurement in Southern European Regions through Equivalization of Modeled Energy Costs. *Sustainability* **2020**, *12*, 5721. [CrossRef]
118. Causa, O.; Soldani, E.; Luu, N. A Cost-of-Living Squeeze? Distributional Implications of Rising Inflation. *Public Sect. Econ.* **2023**, *47*, 431–460. [CrossRef]
119. Mlaskawa, J. Renewable Energy and Poverty in Sustainable Development of the European Union. *Probl. Ekorozwoju* **2022**, *17*, 110–123. [CrossRef]
120. Balogun, A.; Oloja-Ojabo, E.D. Impact of Renewable Energy Consumption and Public Capital Expenditure on Poverty Reduction in Nigeria. *Int. J. Bus. Manag. Econ.* **2023**, *4*, 212–230. [CrossRef]
121. FathollahZadeh Aghdam, R.; Ahmad, N.; Naveed, A.; Berenjforoush Azar, B. On the Relationship between Energy and Development: A Comprehensive Note on Causation and Correlation. *Energy Strategy Rev.* **2023**, *46*, 101034. [CrossRef]
122. Calì, M.; Cantore, N.; Iacovone, L.; Pereira-López, M.; Presidente, G. Too Much Energy The Perverse Effect of Low Fuel Prices on Firms. *J. Environ. Econ. Manag.* **2022**, *111*, 102587. [CrossRef]
123. Solarin, S.A. Towards Sustainable Development in Developing Countries: Aggregate and Disaggregate Analysis of Energy Intensity and the Role of Fossil Fuel Subsidies. *Sustain. Prod. Consum.* **2020**, *24*, 254–265. [CrossRef]
124. Bari, T.J.; Hallager, D.W.; Tøndevold, N.; Karbo, T.; Hansen, L.V.; Dahl, B.; Gehrchen, M. Moderate Interrater and Substantial Intrarater Reproducibility of the Roussouly Classification System in Patients with Adult Spinal Deformity. *Spine Deform.* **2019**, *7*, 312–318. [CrossRef]
125. Zhao, Y.; Shuai, J.; Wang, C.; Shuai, C.; Cheng, X.; Wang, Y.; Zhang, Z.; Ding, L.; Zhu, Y.; Zhou, N. Do the Photovoltaic Poverty Alleviation Programs Alleviate Local Energy Poverty?—Empirical Evidence of 9 Counties in Rural China. *Energy* **2023**, *263*, 125973. [CrossRef]
126. Giger, E.; Gall, H.C. Effect Size Analysis. In Proceedings of the 2013 1st International Workshop on Data Analysis Patterns in Software Engineering (DAPSE), San Francisco, CA, USA, 21 May 2013; pp. 11–13. [CrossRef]
127. Cohen, J. *Statistical Power Analysis for the Behavioral Sciences*; Routledge: London, UK, 2013.
128. Zhao, X.; Lynch, J.G.; Chen, Q. Reconsidering Baron and Kenny: Myths and Truths about Mediation Analysis. *J. Consum. Res.* **2010**, *37*, 197–206. [CrossRef]
129. Nguyen, C.P.; Su, T.D. The Influences of Government Spending on Energy Poverty: Evidence from Developing Countries. *Energy* **2022**, *238*, 121785. [CrossRef]
130. Abdulkarim, U.F.; Nurudeen, S.O.; Faruk, B.U. Moderating Role of Government Intervention on the Relationship between Interest Rate and SMES Performance; Implication on Poverty Alleviation in Zamfara State. *Acad. Account. Financ. Stud. J.* **2022**, *26*, 1–19.
131. Hong, M.P.; Wang, K.-T.; Khudoykulov, K.; Trung, L.M.; Ngo, T.Q.; Nguyen, T.T.H. Assessing Multidimensional Energy Poverty and Its Economic Impact on N11 Countries: Mediating Role of Energy Efficiency. *Front. Energy Res.* **2022**, *10*, 900449. [CrossRef]
132. Murshed, M. An Empirical Analysis of the Non-Linear Impacts of ICT-Trade Openness on Renewable Energy Transition, Energy Efficiency, Clean Cooking Fuel Access and Environmental Sustainability in South Asia. *Environ. Sci. Pollut. Res.* **2020**, *27*, 36254–36281. [CrossRef]
133. Minuto, F.D.; Lanzini, A. Energy-Sharing Mechanisms for Energy Community Members under Different Asset Ownership Schemes and User Demand Profiles. *Renew. Sustain. Energy Rev.* **2022**, *168*, 112859. [CrossRef]
134. Hosier, R.H.; Dowd, J. Household Fuel Choice in Zimbabwe: An Empirical Test of the Energy Ladder Hypothesis. *Resour. Energy* **1987**, *9*, 347–361. [CrossRef]
135. Manjon, M.-J.; Merino, A.; Cairns, I. Business as Not Usual: A Systematic Literature Review of Social Entrepreneurship, Social Innovation, and Energy Poverty to Accelerate the Just Energy Transition. *Energy Res. Soc. Sci.* **2022**, *90*, 102624. [CrossRef]
136. Xu, D.; Abbasi, K.R.; Hussain, K.; Albaker, A.; Almulhim, A.I.; Alvarado, R. Analyzing the Factors Contribute to Achieving Sustainable Development Goals in Pakistan: A Novel Policy Framework. *Energy Strategy Rev.* **2023**, *45*, 101050. [CrossRef]
137. Mehmood, A.; Moktadir, M.A.; Zhou, J.; Ren, J.; Zhang, L. Energy Sustainability Blueprint: A Critical Analysis of Macro and Micro Influential Factors in Pakistan. *Energy Sustain. Dev.* **2023**, *77*, 101311. [CrossRef]
138. Chunling, L.; Memon, J.A.; Thanh, T.L.; Ali, M.; Kirikkaleli, D. The Impact of Public-Private Partnership Investment in Energy and Technological Innovation on Ecological Footprint: The Case of Pakistan. *Sustainability* **2021**, *13*, 10085. [CrossRef]

139. Masud, M.A.K.; Sahara, J.; Kabir, M.H. A Relationship between Climate Finance and Climate Risk: Evidence from the South Asian Region. *Climate* **2023**, *11*, 119. [CrossRef]
140. Xin, Y.; Bin Dost, M.K.; Akram, H.; Watto, W.A. Analyzing Pakistan's Renewable Energy Potential: A Review of the Country's Energy Policy, Its Challenges, and Recommendations. *Sustainability* **2022**, *14*, 16123. [CrossRef]

Disclaimer/Publisher's Note: The statements, opinions and data contained in all publications are solely those of the individual author(s) and contributor(s) and not of MDPI and/or the editor(s). MDPI and/or the editor(s) disclaim responsibility for any injury to people or property resulting from any ideas, methods, instructions or products referred to in the content.

sustainability

MDPI

Article

The Feminisation of Poverty in European Union Countries—Myth or Reality?

Joanna Perzyńska [1,*] and Małgorzata Klaudia Guzowska [2]

1 Department of Mathematical Applications in Economics, Faculty of Economics, West Pomeranian University of Technology, 71-270 Szczecin, Poland
2 Department of Econometrics Statistics, Institute of Economics and Finance, University of Szczecin, 71-101 Szczecin, Poland; malgorzata.guzowska@usz.edu.pl
* Correspondence: joanna.perzynska@zut.edu.pl

Abstract: The feminisation of poverty is a complex, multidimensional phenomenon related to gender inequality in various aspects of life. Women are disproportionately affected by the gender pay gap, unequal intra-household resource distribution, unpaid domestic work, caregiving responsibilities, single motherhood, employment or educational barriers, violence, gender discrimination, and period poverty. Combating poverty and inequality are among the main goals of the 2030 Agenda for Sustainable Development adopted by the United Nations General Assembly on 25 September 2015, and their great importance is further highlighted in the preamble of the resolution outlining the plan to transform our world by 2030. This study uses SDG indicators from the Eurostat database to assess the feminisation of poverty in the EU-27 member states in 2020 based on selected diagnostic characteristics mainly related to SDG 1 and SDG 5; they are also related to other goals due to the cross-cutting nature of the topic. The characteristics were transformed to reflect gender gaps and afterwards unitised with a veto threshold indicating gender balance. These were then used to calculate a synthetic taxonomic measure, allowing for linear ordering and classification of countries based on the feminisation of poverty levels. The study confirmed significant feminisation of poverty in the EU-27, with a few aspects showing masculinisation. High feminisation of poverty was observed in both emerging and advanced economies.

Keywords: feminisation of poverty; gender gap; multivariate comparative analysis; Sustainable Development Goals; SDG indicators

Citation: Perzyńska, J.; Guzowska, M.K. The Feminisation of Poverty in European Union Countries—Myth or Reality? *Sustainability* **2024**, *16*, 7594. https://doi.org/10.3390/su16177594

Academic Editors: Manoj Kumar Nallapaneni, Idiano D'Adamo and Massimo Gastaldi

Received: 30 July 2024
Revised: 29 August 2024
Accepted: 30 August 2024
Published: 2 September 2024

Copyright: © 2024 by the authors. Licensee MDPI, Basel, Switzerland. This article is an open access article distributed under the terms and conditions of the Creative Commons Attribution (CC BY) license (https://creativecommons.org/licenses/by/4.0/).

1. Introduction

The lack of a universally accepted definition of poverty has led to its examination from diverse theoretical perspectives. Codes A.L. [1] outlines several core approaches, including subsistence (material poverty), basic necessities, and relational deprivation. When delineating poverty in terms of subsistence, it is fundamentally tied to the essential requirements for human survival, encompassing basic nutritional and physical needs. The basic needs perspective includes clothing, education, transportation, and other necessities beyond nutritional and physical demands. Relational deprivation pertains to an individual's social sphere and their interactions with the environment. Sen and Grusky et al. [2,3] define poverty not merely as a lack of income but as the inability to lead a meaningful life due to a shortage of economic resources, thus providing a multifaceted perspective that includes social and political dimensions.

Similarly, the concept of sustainability should be approached in a multifaceted way, taking into account its three dimensions: environmental, economic, and social [4]. At the same time, the implementation of this concept should be supported by analyses so that it becomes pragmatic sustainability in real life [5,6]. The Sustainable Development Goals (SDGs) of Agenda 2030 [7,8], emphasise tackling gender poverty and promoting gender equality [9]. Achieving SDG 1, "No poverty", means addressing the root causes of gender-based poverty to end poverty for all, including women and girls. The implementation of SDG 5, "Gender equality", is to ensure the achievement of gender equality and the empowerment of all women and girls, focusing on equal access to education, economic opportunities, and participation in decision-making. Due to the multifaceted nature of sustainability, references to both also appear in other Sustainable Development Goals. Agenda 2030 recognises that gender equality and women's empowerment are integral to all SDGs. Efforts to combat gender poverty include measures such as equal pay, quality education and healthcare access, women's empowerment in decision-making, and eliminating gender-based violence.

Both women and men experience gender-based inequalities, but it is mostly women who face poverty because of it. The term "feminisation of poverty", which refers to the increased representation of women among the impoverished or economically disadvantaged, was originally coined by Diana Pearce [10]. In her 1978 work, "The Feminization of Poverty: Women, Work, and Welfare", Pearce highlighted that in the United States, two-thirds of those living in poverty were women aged 16 and above [11]. This phenomenon primarily relates to the overrepresentation of women and children in lower socioeconomic status groups compared to men of the same socioeconomic status. The causes of feminisation of poverty include factors such as family and household structures, employment opportunities, violence, access to education, the impact of climate change, economic disparities affecting women (femonomics), and health disparities. Persistent traditional stereotypes regarding women continue to be ingrained in many cultures, limiting income opportunities and community participation for many women. Today, the concept of the feminisation of poverty has evolved into a broader and more complex idea. It must be examined in the context of the social inequalities that women experience. Women not only earn lower wages and pensions than men but also face higher levels of unemployment (being not employed but available for work and actively seeking work), including long-term unemployment (for 12 months or more), the challenges of single motherhood, and the burdens of household responsibilities [12].

Among the most common forms of female poverty, literature studies include:

- Decision-making power—women often experience decision-making poverty, meaning that they do not have an equal say in decisions about their lives, whether family, social, or professional [13–15];
- Disparate income (gender pay gap)—income inequality, often referred to as the gender pay gap, signifies that women receive lower earnings than men in identical or similarly qualified positions. This disparity in income stands as a primary factor contributing to the heightened risk of poverty among women. Insufficient income hinders women from acquiring assets and translating their financial resources into social and economic standing. Moreover, higher income not only provides increased opportunities for skill development but also leads to higher earnings [16–19];
- Energy poverty—means lack of access to essential energy, which can affect women's quality of life and health, especially in low-income countries and in households where the woman is the head of the family [20–22];
- Lack of assets—a primary element exacerbating poverty among women is their restricted ability, potential, and authority to reach and manage productive resources, which encompass land, employment, human capital assets, such as education and health, and social capital assets, such as engagement at various levels [23,24], legal entitlements, and safeguards [25];
- Time poverty—time is a component integrated into the concept of poverty because it represents a fundamental resource that is frequently distributed unevenly among individuals, particularly when other resources are inadequate. Women unquestionably experience a greater scarcity of time compared to men [26,27]. Research reveals that in addition to their paid employment, women are also heavily engaged in reproductive and unpaid tasks (domestic work) [28–30]. The distribution of time between men and

women within the household and the broader economy is a significant gender-related issue in the ongoing discussion on time scarcity. Time poverty can be understood in terms of the insufficient availability of time for rest and sleep. The more time dedicated to paid or unpaid work, the less time remains for other activities like relaxation and enjoyment;

- Capability deprivations—lack of access to opportunities, for example, educational and vocational, can affect women's opportunities for development and their ability to improve their social and economic status [31–33];
- Deprivation of health outcomes—lack of access to health care, health education, and adequate living conditions can result in poverty in health outcomes and increase women's risk of health problems, particularly in underdeveloped countries [34];
- Period (menstrual) poverty—limited or no access to menstrual products, sanitation and hygiene facilities, and education and awareness to manage menstrual health [35–37]. This can affect women's health and dignity, and lower their involvement in work or education [38];
- Social and cultural exclusions—refers to a situation where women may be excluded from social and cultural participation due to stereotyping, discrimination, and gender inequality. This can affect their access to resources and opportunities [39].

An analysis of work from recent years shows that the COVID-19 pandemic has significantly exacerbated the feminisation of poverty, in terms of most of the categories mentioned earlier. The economic downturn caused by the pandemic disproportionately affected women, particularly in low-income countries and vulnerable sectors like the service industry, where women are overrepresented [40]. Additionally, the pandemic intensified social inequalities, leading to increased job losses among women, a rise in unpaid domestic work, and greater barriers to accessing healthcare and education [41]. These factors have collectively deepened the gender disparities in poverty, reinforcing the economic and social vulnerabilities of women worldwide [42].

The feminisation of poverty is measured by analysing gender differences in the context of poverty and economic inequality. There are several indicators and approaches used to measure the feminisation of poverty. Poverty based on income compares the incomes of women and men, paying attention to the gender earnings gap. If women have significantly lower incomes than men, this can be considered a manifestation of the feminisation of poverty [43,44]. Using official poverty definitions, poverty rates for women and men can be calculated and compared. If poverty rates are higher among women, then we can talk about the feminisation of poverty [45–47]. The Gini Index measures income inequality within a given population. A higher Gini coefficient means greater income inequality. Comparing the Gini index for men and women can reveal income inequality between the genders [48–51]. Analysing employment and unemployment rates between genders can provide information about differences in access to work and labour market conditions [52–54]. Education is an important factor in combating poverty. Analysing educational indicators such as educational attainment and access to education can help understand the extent to which women have access to tools to avoid poverty [55,56]. Analysis of health indicators and access to healthcare highlights possible health inequalities that may impact women's risk of poverty [57–59].

Based on the above indicators, the feminisation of poverty is measured primarily by three international indicators that provide comprehensive analysis and data on gender inequality in the context of poverty and the role of women in society. The Gender Development Index (GDI) is an indicator developed by the United Nations Development Program (UNDP) that combines the traditional Human Development Index (HDI) with a focus on gender gaps. The GDI takes into account not only economic indicators such as per capita income and educational attainment but also gender disparities in these areas. The GDI helps understand how women and men participate in social and economic development processes [60–63]. The Gender Empowerment Measure (GEM) is another indicator developed by UNDP. It measures women's level of autonomy and influence in society. The

GEM assesses gender inequalities in women's participation in public life, politics, and the economy. This helps to understand the extent to which women have the opportunity to participate and make decisions in these areas [60,64,65]. The Human Poverty Index (HPI) is a United Nations index that takes into account various aspects of poverty such as health, education, and standard of living. In the context of the feminisation of poverty, the HPI also examines gender inequalities in these areas. This allows us to understand how gender disparities affect poverty levels in a given population [66–68].

A literature review on the feminisation of poverty in European Union (EU) countries reveals a complex and multifaceted issue. The term "feminisation of poverty" refers to the disproportionate representation of women among the world's poor, often resulting from gender-based inequalities and discrimination. In the context of EU countries, this phenomenon manifests itself in various ways, and research has explored these dynamics extensively. This literature review provides an overview of key findings and examples from EU member states to illustrate the feminisation of poverty. Economic disparities and the gender pay gap are persistent problems in all EU member states. Women tend to earn less than men, resulting in lower lifetime earnings and reduced economic security. Research by the European Institute for Gender Equality (EIGE) highlights the gender pay gap as a key contributor to the feminisation of poverty in the EU [69]. According to Eurostat, for the economy as a whole, in 2022, women's gross hourly earnings were on average 12.7% below those of men in the European Union and 13.2 % in the euro area [70]. The problem of the Gender Wage Gap in EU countries has been addressed by many authors [71–74]. Who, in addition to analysing the gender pay situation in individual EU countries, also study the impact of eliminating the gender pay gap on income and poverty [19]. Women in the EU are more likely to engage in part-time and precarious employment, which often leads to financial vulnerability. Studies have shown that part-time work can limit access to social security benefits and career advancement, thereby increasing the risk of poverty [75–78]. Single-parent households, predominantly led by women, face higher poverty rates. The lack of support and affordable childcare services can make it difficult for single mothers to escape poverty. A study around the EU (and OECD countries) explores the experiences of single parents, especially single mothers living in poverty, and their challenges in accessing social services [79–84]. Women often bear the brunt of unpaid care work, such as taking care of children, the elderly, or family members with disabilities. This can limit their participation in the labour market and their ability to earn a living. An analysis of the unequal distribution of care work in EU countries can be found in [85–88]. Intersectionality plays a crucial role in understanding the feminisation of poverty. Immigrant and minority women may face compounded forms of discrimination that contribute to their higher poverty rates [89–91]. The effectiveness of social welfare policies in addressing the feminisation of poverty varies among EU member states. Research on the impact of welfare policies on gender equality in different EU countries is discussed by Sánchez-Lopez and de Paz Báñez [92], McDevitt [93], or Ostner and Lewis [94]. Within the analyses of the Social Welfare Policies of the EU countries, the problem of poverty among the elderly, including women, is also frequently raised [95–98].

This paper adopts the concept of the feminisation of poverty being a multidimensional phenomenon that is related to gender inequality occurring in various aspects of life. The purpose of this study was to examine the presence of the feminisation of poverty in European Union member states and to compare its level in these countries based on selected SDG indicators. Variables from the Eurostat database that directly represent poverty, as well as those that can influence it, were selected for the study. To measure the level of feminisation of poverty, the use of transformed gender gaps was proposed. The specific objectives of the study were to rank countries according to the level of feminisation of poverty and to create classes of countries similar in terms of their level of feminisation of poverty. The empirical study carried out allowed the following questions to be answered: (1) Is there a feminisation of poverty in EU member states? (2) By which diagnostic characteristics does the feminisation of poverty occur? (3) Are there diagnostic characteristics that contribute to the masculinisation of poverty or because of which there is a gender balance in terms of poverty? (5) Does the level of feminisation of poverty vary across EU member states? (6) Which countries have the highest and the lowest levels of feminisation of poverty? (7) Which countries are similar with regard to the level of feminisation of poverty?

2. Materials and Methods

The empirical study covered the 27 member states of the European Union (EU-27). It used statistics showing the selected SDG indicators for 2020 by sex from the database of the statistical office of the European Union, Eurostat. The selection of primary data took into account both variables representing the level of poverty directly (e.g., the percentage of people at risk of income poverty or experiencing severe material deprivation) and those that may contribute to or counteract it (e.g., access to education, employment and percentage of inactive people due to caring responsibilities). Indicators on health or access to water and sanitation that relate to period poverty were also included in the study [99,100]. The primary variables collected were related to the Agenda 2030 Goals: SDG 1 "No poverty", SDG 3 "Good health and well-being", SDG 4 "Quality education", SDG 5 "Gender equality", SDG 6 "Clean water and sanitation", SDG 8 "Decent work and economic growth"; they were also related to other goals due to the cross-cutting nature of the topic. In order to select variables with the highest possible diagnostic values, the primary data set was subjected to selection according to statistical criteria [101]. To eliminate variables with low variability and those highly correlated with others, cut-off values of Pearson's linear correlation coefficient and classical coefficient of variation of 0.7 and 10%, respectively, were adopted [102]. Due to the sufficiently high coefficient of variation values, Pearson's coefficient proved to be the main statistical criterion for selection. Two primary variables, SDG 1.10 "Persons at risk of poverty or social exclusion" and SDG 11.11 "Severe housing deprivation rate", highly correlated with several other variables and were finally excluded from the set of collected sustainability indicators. However, this did not result in a significant loss of information, as the removed variable SDG 1.10 is a combination of indicators SDG 1.20, SDG 1.31, and SDG 1.40, and they remained in the final set.

Finally, 15 diagnostic features were selected, the values of which for the study objects (EU-27 member states) by sex were collected in matrices:

$$X = [x_{ij}]_{27\times 15}, \quad (1)$$

$$Y = [y_{ij}]_{27\times 15}, \quad (2)$$

where:

i—object number (i = 1, 2, …, 27),
j—diagnostic feature number (j = 1, 2, …, 15),
X—matrix of values of all diagnostic features for women and for all objects,
x_{ij}—the value of the j-th diagnostic feature for women and for the i-th object,
Y—matrix of values of all diagnostic features for men and for all objects,
y_{ij}—the value of the j-th diagnostic feature for men and for the i-th object.

The columns of the X and Y matrices contain the values of the diagnostic features for all objects by sex, i.e., X_j and Y_j represent the j-th diagnostic feature for women and men respectively.

Table 1 summarises selected descriptive characteristics of diagnostic features by sex. Most of the diagnostic features (for j = 1, 2, …, 7, 11, 12, 13, 15) are poverty stimulants, which are positively correlated with the level of the complex phenomenon of poverty. The remaining four variables (for j = 8, 9, 10, 14) are destimulants of poverty, i.e., are negatively correlated with the level of poverty. Most of the diagnostic features are characterised by right-handed asymmetry, while left-handed asymmetric are the variables X_9, X_{14}, and Y_9, which are destimulants of poverty. Comparing the descriptive parameters of the features for men and women, it can be seen that the median of most of the diagnostic features has larger values for women. The largest difference of 22.3 pp. is for the stimulant representing inactive population due to caring responsibilities (j = 11), followed by 11.1 pp. for the destimulant of tertiary educational attainment (j = 8). A smaller median value for women applies to only four features (numbered j = 4, 7, 14, 15), the largest difference (7.6 pp.), to women's disadvantage, is for the destimulant representing employment rate (j = 14). Similar relationships apply to the other parameters of the diagnostic features for men and women—the mean and the extremes.

Table 1. Selected descriptive characteristics of diagnostic features by sex.

j	Diagnostic Feature (in %)	X_j					Y_j				
		Mean	Median	Min	Max	Skewness	Mean	Median	Min	Max	Skewness
1	SDG 1.20 People at risk of income poverty after social transfers	17.1	16.7	11.7	25.8	0.54	15.3	15.0	7.2	22.1	0.10
2	SDG 1.31 Severe material and social deprivation rate	6.7	4.7	1.4	26.5	2.17	6.1	4.3	1.5	24.0	2.20
3	SDG 1.40 People living in households with very low work intensity	7.6	7.6	3.6	13.0	0.36	7.2	7.2	3.9	11.9	0.30
4	SDG 1.41 In work at-risk-of-poverty rate	6.9	7.3	3.0	11.4	0.08	8.6	8.1	3.3	18.1	0.78
5	SDG 1.50 Housing cost overburden rate	7.6	6.3	1.8	34.2	3.21	6.9	5.2	2.0	32.3	3.51
6	SDG 3.60 Self-reported unmet need for medical examination and care	2.7	1.7	0.0	14.6	2.35	1.9	1.4	0.0	11.1	2.70
7	SDG 4.10 Early leavers from education and training	7.0	6.3	2.0	16.6	1.03	10.3	10.0	2.4	20.2	0.27
8	SDG 4.20 Tertiary educational attainment	35.0	37.7	17.5	46.7	−0.32	27.3	26.6	15.0	40.2	−0.06
9	SDG 4.31 Participation in early childhood education	90.4	92.0	71.5	100	−1.05	90.2	91.6	71.0	100.0	−1.03
10	SDG 4.60 Adult participation in learning	11.3	8.7	1.0	35.5	1.35	8.8	7.4	1.0	23.0	0.96
11	SDG 5.40 Inactive population due to caring responsibilities	32.5	30.4	7.5	66.6	0.32	12.3	8.1	1.5	41.4	1.44
12	SDG 6.10 Population having neither a bath, nor a shower, nor indoor flushing toilet in their household	1.9	0.4	0.0	21.3	3.76	2.0	0.4	0.0	21.1	3.51
13	SDG 8.20 Young people neither in employment nor in education and training	16.9	17.1	7.7	29.3	0.34	11.1	10.6	5.0	21.0	0.79
14	SDG 8.30 Employment rate	68.5	70.6	48.7	77.4	−1.24	78.3	78.2	68.1	87.2	−0.11
15	SDG 8.40 Long-term unemployment rate	2.4	1.8	0.6	13.5	3.46	2.2	1.9	0.5	8.2	2.30

In the empirical research carried out, the pattern-free method of multivariate comparative analysis was used to assess the level of feminisation of poverty in the EU-27 member states. The research procedure followed six steps.

Step 1. Determination of the matrix:

$$V = [v_{ij}]_{27\times 15}, \tag{3}$$

with elements representing gender gaps (cf. [103]):

$$v_{ij} = \begin{cases} x_{ij} - y_{ij} & \text{for stimulants} \\ y_{ij} - x_{ij} & \text{for destimulants} \end{cases} \tag{4}$$

Variable V_j being the j-th column of the V matrix represents gender gaps due to the j-th diagnostic characteristic for all objects (EU-27 member states). Its values indicate the direction of gender imbalance (positive values indicate female disadvantage and negative values indicate male disadvantage) or indicate gender balance (values equal to 0).

Step 2. Transformation of matrix V to form:

$$Z = [z_{ij}]_{27\times 15}, \tag{5}$$

where

$$z_{ij} = \begin{cases} 0 & \text{for } v_{ij} \le 0 \\ v_{ij} & \text{for } v_{ij} > 0 \end{cases}, \tag{6}$$

Rejecting negative values from the matrix V transforms its columns into the form of the stimulants of feminisation of poverty (Z_j).

Step 3. Normalisation of the stimulants of feminisation of poverty:

$$\overset{*}{z}_{ij} = \frac{z_{ij}}{\max_i z_{ij}}, \tag{7}$$

Steps 2 and 3 together are zeroed unitisation with a veto threshold equal to 0 (cf. [104]) denoting gender balance.

Step 4. For each object, the determination of the value of the synthetic variable (taxonomic synthetic measure), which is a weighted arithmetic mean of the form [105]:

$$R_i = \sum_{j=1}^{15} w_j \overset{*}{z}_{ij}, \tag{8}$$

where:

$$\sum_{j=1}^{15} w_j = 1, \tag{9}$$

R—synthetic variable,
R_i—value of the synthetic variable R for the i-th object,
w_j—weight of the j-th normalised stimulant of feminisation of poverty.

The values of the weights depended on the number of diagnostic features related to each of the six SDGs included in the study; they are summarised in Table 2. By setting the weights in this way, each of the six SDGs was equally valid regardless of the number of diagnostic features representing it.

Table 2. Weights for groups of normalised stimulants according to the 2030 Agenda SDGs.

	SDG 1	SDG 3	SDG 4	SDG 5	SDG 6	SDG 8
number of diagnostic features	5	1	4	1	1	3
weight of normalised stimulant	$\frac{1}{30}$	$\frac{1}{6}$	$\frac{1}{24}$	$\frac{1}{6}$	$\frac{1}{6}$	$\frac{1}{18}$

Step 5. Linear ordering of objects according to the decreasing values of the synthetic variable R. A higher place in the ranking (i.e., a lower rank value) indicates a higher level of feminisation of poverty in a given country compared to the other EU-27 member states.

Step 6. Classification of ranked objects into one of four classes of objects similar in terms of the level of feminisation of poverty:

$$\text{class I}: \overline{R} + s(R) < R_i \le 1 \tag{10}$$

$$\text{class II}: \overline{R} < R_i \le \overline{R} + s(R) \tag{11}$$

$$\text{class III}: \overline{R} - s(R) < R_i \le \overline{R} \tag{12}$$

$$\text{class IV}: 0 \le R_i \le \overline{R} - s(R) \tag{13}$$

where:

$\overline{R}$—arithmetic mean of synthetic variable R,
s(R)—standard deviation of synthetic variable R.

EU-27 member states in class I have the highest level of feminisation of poverty, while those in class IV have the lowest.

3. Results

In the first stage of the empirical study carried out, the diagnostic features for men and women presented in Table 1 were transformed based on Formulas (3) and (4), determining the values of gender gaps (V_j). Selected descriptive characteristics of the transformed diagnostic features, i.e., gender gaps, are presented in Table 3.

Table 3. Selected descriptive characteristics of gender gaps.

j	Mean	Median	Min	Max	Skewness	Number of Values of Gender Gap		
						$V_j < 0$	$V_j = 0$	$V_j > 0$
1	1.8	1.5	−0.3	4.7	0.66	2	0	25
2	0.6	0.4	−0.6	2.6	1.12	4	1	22
3	0.4	0.4	−1.6	2.5	0.02	8	1	18
4	−1.6	−1.4	−7.7	1.2	−1.47	23	1	3
5	0.7	0.9	−0.8	3.1	0.59	6	1	20
6	0.7	0.5	−0.3	3.5	1.39	3	3	21
7	−3.2	−3.4	−8.6	1.9	−0.10	25	0	2
8	−7.7	−7.7	−17.8	3.5	0.07	26	0	1
9	−0.1	−0.2	−1.3	1.3	0.48	16	3	8
10	−2.6	−1.9	−13.6	0.4	−1.96	22	1	4
11	20.3	19.9	5.1	37.7	0.10	0	0	27
12	−0.1	0.0	−1.2	0.2	−2.35	11	13	3
13	5.9	4.3	0.3	19.2	1.15	0	0	27
14	9.7	8.3	1.7	19.7	0.64	0	0	27
15	0.2	−0.1	−1.3	5.3	3.54	15	0	12

The values of the V_j variables indicate the direction of the gender imbalance. From the information in Table 3, it can be seen that the three variables V_{11}, V_{13}, V_{14} had only positive values, which means that in all EU-27 member states, in 2020, there was a disadvantage for women due to the diagnostic characteristics: inactive population due to caring responsibilities, young people neither in employment nor in education and training, employment rate—in every country, their values (in percentage) for women were higher than for men. The maximum gender gaps for these three features were the highest among all 15 variables, being respectively: 37.7 pp. (Poland), 19.2 pp. (Czechia), 19.7 pp. (Italy).

The values of the other gender gaps were both positive and negative and, in a few cases, equal to zero. The largest number (as many as 13) of values equal to zero indicating gender balance had variable V_{12}, meaning that in 13 countries, there were equal proportions of women and men having neither a bath, nor a shower, nor indoor flushing toilet in their household. Furthermore, in most of the remaining countries, men were at a disadvantage compared to women due to the previously mentioned feature and only in three countries (Austria, Czechia, Romania) was the situation reversed. The six variables V_4, V_7, V_8, V_9, V_{10}, and V_{15} (i.e., all the characteristics included in the study concerning the quality of education and, in addition, the work, at-risk-of-poverty rate, and the long-term unemployment rate) had a preponderance of negative values (between 16 and 26) indicating that men were generally worse off in terms of these characteristics. It should be stressed that there were no variables representing gender gaps with all negative values, i.e., variables due to which men would always be disadvantaged. The closest to this was variable V_8—only in one country (Germany) was the percentage of female tertiary educational attainment lower than the corresponding percentage of males (by 3.5 pp.).

In the next stage of the empirical research, the transformed diagnostic characteristics (gender gaps V_j) were subjected to unitisation with a veto threshold denoting gender balance (Formulas (5)–(7)), during which stimulants of feminisation of poverty (Z_j) were determined. An aggregate synthetic variable (Formulas (8) and (9)) was then determined, on the basis of which the EU-27 member states were ranked and divided into four classes (Formulas (10)–(13)) according to the level of feminisation of poverty in 2020. The results obtained are presented in Table 4.

Table 4. Values of the taxonomic synthetic measure, ranking and classification of the EU-27 member states.

i	EU-27 Member State	R_i	Rank	Class
1	Austria	0.2012	17	III
2	Belgium	0.1376	24	III
3	Bulgaria	0.2148	14	III
4	Croatia	0.1914	18	III
5	Cyprus	0.2426	9	II
6	Czechia	0.3655	4	I
7	Denmark	0.1004	25	IV
8	Estonia	0.3902	3	I
9	Finland	0.1443	23	III
10	France	0.2028	16	III
11	Germany	0.2680	8	II
12	Greece	0.5561	2	I
13	Hungary	0.2272	13	III
14	Ireland	0.2377	10	III
15	Italy	0.3014	6	II
16	Latvia	0.3193	5	II
17	Lithuania	0.1857	20	III
18	Luxembourg	0.1753	21	III
19	Malta	0.2367	11	III
20	Netherlands	0.0869	26	IV
21	Poland	0.2974	7	II
22	Portugal	0.2101	15	III
23	Romania	0.5598	1	I
24	Slovakia	0.2303	12	III
25	Slovenia	0.1739	22	III
26	Spain	0.1902	19	III
27	Sweden	0.0825	27	IV

The information in Table 4 shows that in class I of the EU-27 countries with the highest level of feminisation of poverty in 2020 in terms of the variables included in the study, there were four countries, namely Romania (1st place in the ranking), Greece (2nd), Estonia (3rd), and Czechia (4th). For these countries, most of the gender gaps and therefore also the stimulants of feminisation of poverty had positive values denoting women's disadvantage, with six of them reaching their maximum. The highest level of feminisation of poverty in Romania was determined by the high values of most of the stimulants (higher than their average values)—as many as 10 of them had positive values indicating the disadvantage of women. In addition, Romania had the highest values of variables Z_7 (1.9 pp.) and Z_{12} (0.2 pp.) among all the countries studied. Greece, which ranked second, had the highest values of the two poverty-inducing variables Z_3 (2.5 pp.) and Z_{15} (5.3 pp.) among the EU-27 countries. Estonia and Czechia had the highest values of the variables Z_6 (3.5 pp.) and Z_{13} (19.2 pp.). In Greece and Czechia, moreover, the largest number (as many as 11) of the stimulants of feminisation of poverty had positive values.

Five countries were in class II with a high middle level of feminisation of poverty: Latvia, Italy, Poland, Germany, Cyprus. For these countries, the six stimulants of feminisation of poverty reached their maximum. Latvia (5th) had the highest value of the Z_4

stimulant (1.2 pp.), Italy (6th)—Z_9 (1.3 pp.) and Z_{14} (19.7 pp.), Poland (7th)—Z_{11} (37.7 pp.), Germany (8th)—Z_8 (3.5 pp.), and Cyprus (9th)—Z_{10} (0.4 pp.).

Class III, with a low middle level of feminisation of poverty, was the most numerous, with as many as fifteen countries in it. Only the last three stimulants of feminisation of poverty reached their maximum there: the highest values of stimulants Z_2 (2.6 pp.) and Z_5 (3.1 pp.) were for Bulgaria (14th place in the ranking) and Z_1 (4.7 pp.) for Lithuania (20th place). The last places in this class (21st–24th) were occupied by Luxembourg, Slovenia, Finland, and Belgium, respectively.

In class IV of the EU-27 countries having the lowest level of feminisation of poverty in 2020, three countries are ranked: Denmark (25th), Netherlands (26th), Sweden (27th). For Netherlands and Sweden, the taxonomic synthetic measures were similar, their lowest values being determined by the very small values of all stimulants, with as many as seven having values equal to 0.

The spatial distribution of the EU-27 country classes is shown in Figure 1.

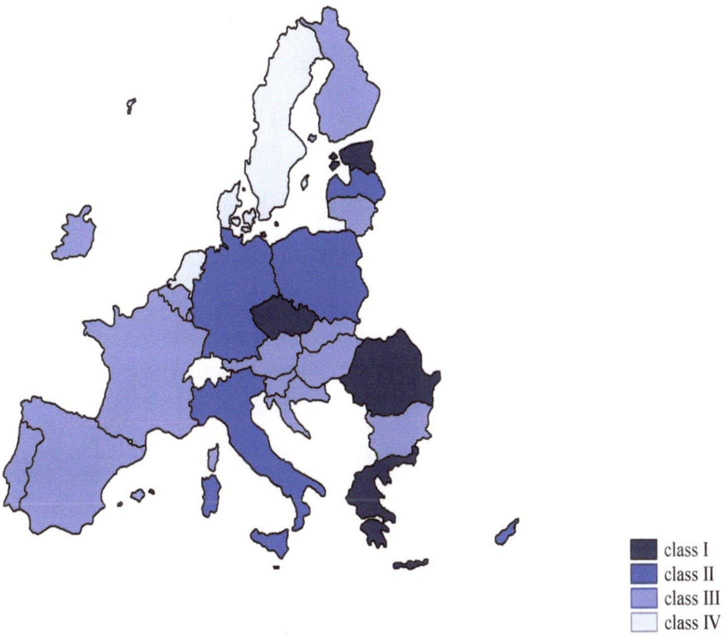

Figure 1. Spatial distribution of the EU-27 member state classes.

The average value of the taxonomic synthetic measure was highest in class I with the highest level of feminisation of poverty at 0.468, in subsequent classes it decreased to reach a value of 0.090 in class IV with the lowest level of feminisation of poverty (Table 4). Similarly, in most cases, the average values of stimulants of feminisation of poverty were highest in class I and in the subsequent classes were increasingly smaller. The exceptions were variables Z_2, Z_8, Z_9, and Z_{12}, but the differences between their mean values in each class were insignificant. Some variables, especially those related to the quality of education, had the fewest positive values (denoting women's disadvantage) of all the variables: variable Z_8 had only one value different from zero (for Germany), Z_7—two (Romania, Czechia), Z_4—three (Czechia, Latvia, Belgium), Z_{12}—three (Romania, Czechia, Austria), Z_{10}—four (Greece, Czechia, Germany, Cyprus). It can be seen that most of the countries mentioned belonged to class I and class II, only Austria and Belgium were from class III. The average values of the selected stimulants of feminisation of poverty in each class are shown in Figure 2.

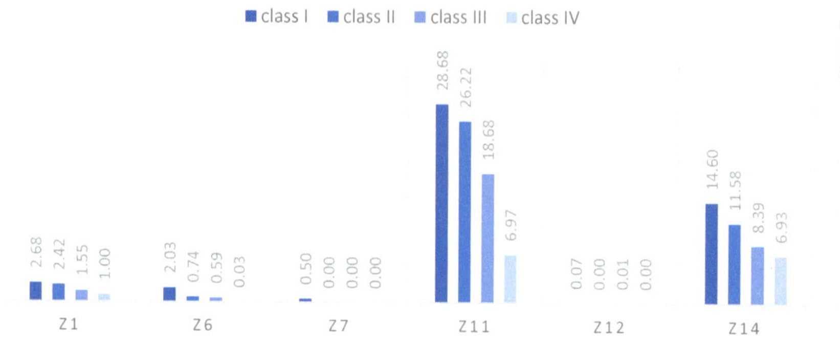

Figure 2. Average values of selected stimulants of feminisation of poverty (in pp.) in classes I–IV.

4. Discussion

The relative poverty indicator AROP used in the empirical study (SDG indicator 1.20 from the Eurostat database) expresses the percentage of people at risk of income poverty with the cut-off point set at 60% of median equivalised income after social transfers. From its preliminary analysis, it can already be seen that in 2020 in the EU-27, women were worse off on average than men—17.4% of women were at risk of monetary poverty, while the corresponding percentage for men was 15.8%. However, the AROP indicator is not the only measure of poverty in Europe; in addition to it, Eurostat's database on SDG 1 contains nine more indicators of poverty reduction (also linked to other SDGs) mainly from the Survey on Income and Living Conditions, EU-SILC [106]. Selected indicators from the Eurostat database on SDG 1 were used, among others, by Piwowar and Dzikuć in their study of poverty and social exclusion in the Visegrad Group Countries [107], Sompolska-Rzechuła and Kurdyś-Kujawska for the assessment of the development of poverty in EU countries [108], Bąk and Perzyńska in their study of poverty and social exclusion in EU-28 [109]. This paper proposes that indicators from the Eurostat database on different goals can be used to assess the level of feminisation of poverty in EU-27 member states.

According to the study, the highest level of feminisation of poverty in 2020 was in Romania, for which as many as 10 of the 15 stimulants of feminisation of poverty had values greater than their mean values, including the highest values among all countries surveyed for the differences between the percentages for women and men early leavers from education and training (1.9 pp.) and having neither a bath, nor a shower, nor indoor flushing toilet in their household (0.2 pp.). Although the latter value is not very large, it should be noted that only in two more countries besides Romania (Greece and Austria) was the proportion of women who did not have access to water and sanitation greater than the corresponding proportion of men—and this diagnostic characteristic is important as one of the causes of period poverty. The analysis of the SDG indicators (i.e., untransformed diagnostic characteristics) by sex further shows that Romania was also the country with the highest proportions of women with no access to water and sanitation (21.3%), early leavers from education and training (16.6%), and experiencing severe material and social deprivation (26.5%) across the European Union. The percentage of women at risk of income poverty after social transfers in Romania was the second highest among the EU-27 countries (24.6%). According to Eurostat data, the reduction in the percentage of the risk of poverty rate, due to social transfers (excluding pensions) in Romania was only 15.8% for women and this was the lowest value in the EU-27.

Other EU-27 member states having the highest level of feminisation of poverty in 2020 are Greece, Estonia, and Czechia. In Greece, representing the southern European model of social policy, unemployment benefits are low and people in need have to rely on family [98]. Meanwhile, the country had the highest gender gap among the EU-27 member states for women and men living in households with very low work intensity

(2.5 pp.) and those in long-term unemployment (5.3 pp.). Analysis of the SDGs by sex further shows that Greece was also the country with the highest number of women of all EU-27 countries living in households with very low work intensity (13%), long-term unemployment (13.5%), and in addition, housing cost overburden (34.2%). At the same time, Greece had the lowest values for two variables that can exacerbate female poverty in the country: the employment rate for Greek women was only 48.7%, and only 71.5% of girls participated in early childhood education. The unfavourable high ranking of Greece and Czechia in terms of feminisation of poverty was determined by the high values of as many as 11 out of 15 poverty stimulants. Czechia additionally had the highest value in the EU-27 of the gap between the percentages for young women and men neither in employment nor in education and training (19.2 pp.). Czechia represents the Central and Eastern European model of social policy, in which social security benefits are the lowest [98]. Despite this, according to an analysis of SDGs by sex, only 11.7% of Czech women were at risk of income poverty after social transfers and this was the lowest value among all EU-27 member states. At the same time, the corresponding percentage for men was also the lowest (7.2%), making the gender gap in this respect one of the largest.

The Benelux and Scandinavian countries with very low values of the poverty feminisation stimulants were ranked last in the 2020 poverty feminisation ranking. Denmark, Netherlands, and Sweden had by far the lowest levels of poverty feminisation. These countries represent a citizen-supportive social-democratic model of social policy, with Sweden being the leading welfare state [109]. An important element of Sweden's policy is the promotion of gender equality, enabling women to fulfil care and family responsibilities while striving for self-reliance and changing gender stereotypical choices [110]. The impact of such a policy is evident in the smallest percentage of women who are inactive due to caring responsibilities (7.5%) and the highest employment rate (77.4%) among the EU-27 countries. Analysis of the SDGs by sex further shows that Sweden and the Netherlands were the only countries among the EU-27 member states where neither women nor men reported having neither a bath, a shower, or an indoor flushing toilet in their household. In contrast, the percentages of women at risk of income poverty after social transfers were not the lowest in these countries, at 17.2% and 13.7% respectively.

Countries with the highest level of feminisation of poverty are mainly from the eastern wall of the EU-27. According to the classification of the International Monetary Fund, Romania is included in the emerging and developing economies, while Greece, Estonia, and Czechia belong to the advanced economies [111]. Countries in class II with a high middle level of feminisation of poverty are mostly advanced economies and even major advanced economies (Germany, Italy). The last places of class III and in class IV were exclusively advanced economies. This ranking of the feminisation of poverty on the basis of gender gaps largely reflects the ranking of the promotion of gender equality determined by the European Institute for Gender Equality on the basis of the Gender Equality Index values. According to the EIGE study on gender equality, in 2020, Sweden (which had the lowest level of feminisation of poverty) was the leader in promoting gender equality, while Greece (which was the second country with the highest level of feminisation of poverty) ranked last in promoting gender equality [69]. It should be noted that, since 2010, Sweden has had the highest and Greece the lowest progress in gender equality in the EU [112]. The distant places in the ranking of the promotion of gender equality in 2020 included Romania (25th), Czechia (22nd), and Estonia (17th), i.e., the countries with the highest level of feminisation of poverty. Germany was ranked only 11th, with a Gender Equality Index approximately equal to its average value for the EU-27 as a whole. The other countries with the lowest levels of feminisation of poverty Denmark (2nd) and Netherlands (5th) were high in the ranking of promoting gender equality in 2020 [69].

It is clear that a country's level of feminisation of poverty is conditional on gender inequality (in this study expressed by gender differences in the values of diagnostic characteristics) and does not necessarily correlate with the level of poverty for women, let alone the level of poverty overall. As an example, Czechia and Sweden belonged to the classes with the highest and lowest feminised poverty levels, respectively, in 2020, while according to Bąk and Perzynska [109] and Sompolska-Rzechuła and Kurdyś-Kujawska [108] in 2018 and 2019 belonged to the classes with the lowest and medium poverty levels, respectively. In 2020, the relative poverty rate of AROP overall was 9.5% and 16.1% in these countries, and 11.7% and 17.2% for women, respectively. The highest AROP values were in Bulgaria (23.8% and 25.8%), which ranked only 14th in the feminisation of poverty ranking. The 8th place in the feminisation of poverty ranking of economically highly developed Germany, which was in the class with the lowest poverty levels in 2019 [108] and should, therefore, not be surprising. In 2020, the proportion of people at risk of income poverty after social transfers overall was 16.8% in Germany and the corresponding proportion of women was 16.1% (i.e., less than in Sweden). Meanwhile, in 2020, Germany was the only EU-27 country in which the proportion of women who had completed tertiary education was lower than the corresponding proportion of men (a difference of 3.5 pp.). Furthermore, in Germany, the percentage of women who were inactive due to caring responsibilities was more than 29 pp. higher than the corresponding percentage of men, and this was one of the highest values among the EU-27 member states. This was probably strongly influenced by the conservative social policy model and the patriarchal family model preferred in Germany, in which women do not work outside the home [109,113]. According to a report by the international humanitarian organisation Oxfam, the social welfare institutions operating in Germany try to reduce inequalities [114]. At the same time, benefits for parents or the unemployed do not always stimulate the poorest to increase their income from work [115].

Reducing poverty and inequality are the challenges of today's world, meanwhile, in late 2019, the COVID-19 pandemic began to spread around the world, leading to increased poverty and worsening inequality. The greatest impact of the pandemic has been on the poorest people. Between 2019 and 2021, the average income of the 40% of the population who are poorest has decreased by 2.2%, while the average income of the 40% who are richest has decreased by only 0.5% [116]. According to the United Nations [117], the pandemic has stalled the unbroken trend of 25 years of poverty reduction, and the combination of its impact with the effects of climate change, the Russian invasion of Ukraine, rising inflation and slowing economic growth means that more and more people will live in extreme poverty [118]. Assuming trends continue, 575 million people will be trapped in extreme poverty in 2030 [119], including more than 340 million women and girls [120].

The COVID-19 pandemic slowed down positive trends towards gender equality. According to Oxfam's 2022 report, women, alongside ethnic minorities and people in developing countries, are among the groups most affected by the increase in inequality as a result of the COVID-19 pandemic—when, with the rising cost of living, the poor get poorer, women get poorer the fastest [121]. In 2020, women were 1.4 times more likely to drop out of the labour market than men, and unpaid domestic work and care work took up three times as much of their time [122]. Compared to pre-pandemic projections, it will take 36 more years to close the gender pay gap, or as many as 136 years [123]. Unfortunately, due to missing data for Greece and Ireland, the gender pay gap value could not be used in this study. Available Eurostat data shows that in the other EU-27 countries in 2020, the percentage difference between the average gross hourly earnings of male and female paid employees ranged from 0.7% in Luxembourg to 22.3% in Latvia, i.e., in all these countries, women were paid less than men.

The European Union is regarded as a world leader in gender equality, but inequalities emerge to varying degrees in its member states [112]. Both women and men experience gender-based inequalities, but it is mostly women who face poverty because of it. The feminisation of poverty is a whole mechanism that makes girls and women more likely to fall into poverty than boys and men [124]. The worse economic situation of women is often a manifestation of the social roles imposed on them by the patriarchal family model but is also sometimes caused by violence or gender discrimination [125]. To quote Tahira Abdullah, "Poverty has a woman's face" [126], and its feminine character is exacerbated by the constant pressure of norms stemming from culture, religion, or tradition. Women are affected by the gender pay gap, unequal intra-household distribution of resources [127], the burden of unpaid domestic work, single motherhood, caring for family members, and menstrual poverty.

Human rights violations are an important aspect of poverty—the marginalisation and social exclusion of poor people are widespread, leaving them feeling a lack of respect and dignity [128]. The repercussions of the feminisation of poverty are not only significant because of the disadvantage to women themselves, the feminisation of poverty has negative consequences that also affect their families. Women, being family managers of poverty, undertake various activities to supplement their meagre budgets [129], at the same time reproducing poverty by sharing it with the children they are raising [124]. The severity of women's poverty should be taken into account and a better situation in the family and society should be ensured. The fight against poverty should be combined with tackling gender inequality in all aspects of life. Meanwhile, as noted by D'Adamo et al., academic interest in SDG 5 topics is almost the lowest of all the SDGs [5]. The United Nations emphasises the need to accelerate action leading to SDG 5 and thus warns that failure to prioritise gender equality could jeopardise the entire Agenda 2030 for Sustainable Development [120].

In 2020, the strength of European women was evident in education, especially in secondary and tertiary education, so they should be further supported in this, but at the same time, men should also be supported in this sphere. Women need support in finding employment if they want to work but are prevented from doing so because of the need to care for relatives. Equality policies should also include the elimination of the gender pay gap. Due to a lack of data, this variable was not included in the empirical study conducted, but the phenomenon of unequal pay for women existed in all EU member states for which data were available. It is therefore necessary to provide access to these data from the remaining countries. Data collection on menstrual poverty should also be developed. In addition, all public statistics characterising the population should be presented by gender, and the way they are measured should be standardised to ensure comparability of information from different countries. To assess and measure the feminisation of poverty, it is useful to use indicators that represent gender differences in many aspects of life, as proposed in this paper.

5. Conclusions

The feminisation of poverty, signifying the disproportionate over-representation of women and girls living in poverty, grew out of gender inequality and, on this basis, should be considered as a multidimensional concept related to different spheres of life. The multidimensionality of the phenomenon of feminisation of poverty resulting from gender inequality requires methods and data that take this aspect into account to assess it. This paper proposes not to directly examine the level of poverty itself; instead, the diagnostic characteristics of female and male poverty were transformed into gender gaps so that they show gender inequalities. When comparing the values of the designated gender gaps, it was noted that all variables included in the empirical study confirmed the existence of a gender imbalance in terms of poverty. Most of the diagnostic features indicated the feminisation of poverty, but in a few aspects, its masculinisation became apparent. In all EU-27 countries in 2020, a significantly higher proportion of women than men were inactive due to caring responsibilities, a lower proportion of women than men aged from 20 to 64 years were in employment, a higher proportion of girls than boys were neither in employment nor in education and training. In contrast, the worse situation for men was revealed in tertiary education—in all countries except Germany, the percentage of men in tertiary educational attainment was lower than the corresponding percentage of women. In most countries, with the exception of Romania and Czechia, the share of male early leavers from education and training was higher.

Undoubtedly, the feminisation of poverty that has grown out of gender inequality is a reality and not a myth. As the empirical study showed, this was the case, albeit to varying degrees, in all EU member states in 2020. High levels of feminisation of poverty were found in both countries whose economies are categorised as emerging market and developing economies (Romania) and as advanced economies (Greece, Estonia, and Czechia), while the class of countries with the lowest feminisation of poverty included only advanced economies (Denmark, Netherlands, Sweden). Regardless, all countries with the highest levels of feminisation of poverty had the lowest values of the Gender Equality Index indicating the greatest gender inequalities, while countries with the lowest levels of feminisation of poverty were leaders in promoting gender equality.

The article's findings are intended to support the achievement of SDG 1 and SDG 5. Due to the cross-cutting nature of the phenomenon of feminisation of poverty, the study also used sustainability indicators from the Eurostat database for other SDGs. Limitations related to the availability of statistical data were encountered during data collection. For several countries, data on the gender pay gap were missing. This important variable was therefore not included in the study, even though an unfavourable gender pay gap for women existed in all EU member states for which data were available. Direct data on period poverty was also not used, as it is not widely collected and published.

Author Contributions: Conceptualization, J.P. and M.K.G.; methodology, J.P.; software, J.P.; validation, J.P.; formal analysis, J.P. and M.K.G.; writing—original draft preparation, J.P. and M.K.G.; writing—review and editing, J.P. and M.K.G.; visualization, J.P. All authors have read and agreed to the published version of the manuscript.

Funding: The publication was partially funded by the Faculty of Economics, West Pomeranian University of Technology in Szczecin: "The International Center for Inclusive Economic Development (IDEAL)".

Institutional Review Board Statement: Not applicable.

Informed Consent Statement: Not applicable.

Data Availability Statement: https://ec.europa.eu/eurostat/web/sdi/database (accessed on 29 July 2024).

Conflicts of Interest: The authors declare no conflicts of interest.

References

1. Codes, A.L.M.D. A Trajetória do Pensamento Científico Sobre Pobreza: Em Direção a Uma Visão Complexa. 2008. Available online: https://repositorio.ipea.gov.br/handle/11058/1489 (accessed on 20 August 2024).
2. Sen, A.; Violence, I. Poverty. *J. Peace Res.* **2008**, *45*, 5–15. [CrossRef]
3. Grusky, D.B.; Kanbur, S.R.; Sen, A.K. *Poverty and Inequality*; Stanford University Press: Stanford, CA, USA, 2006.
4. World Commission on Environment and Development. *Our Common Future*; Oxford University Press: Oxford, UK, 1987.
5. D'Adamo, I.; Di Carlo, C.; Gastaldi, M.; Rossi, E.N.; Uricchio, A.F. Economic Performance, Environmental Protection and Social Progress: A Cluster Analysis Comparison towards Sustainable Development. *Sustainability* **2024**, *16*, 5049. [CrossRef]
6. D'Adamo, I.; Gastaldi, M.; Nallapaneni, M.K. Europe Moves toward Pragmatic Sustainability: A More Human and Fraternal Approach. *Sustainability* **2024**, *16*, 6161. [CrossRef]
7. Transforming Our World: The 2030 Agenda for Sustainable Development. Department of Economic and Social Affairs. United Nations. Available online: https://sdgs.un.org/2030agenda (accessed on 21 July 2022).
8. EC Agenda 2030 European Commission Welcomes New 2030 UN Agenda for Sustainable Development. Available online: http://europa.eu/rapid/press-release_IP-15-5708_en.htm (accessed on 21 July 2022).
9. Bradshaw, S.; Chant, S.; Linneker, B. Gender and poverty: What we know, don't know, and need to know for Agenda 2030. *Gend. Place Cult.* **2017**, *24*, 1667–1688. [CrossRef]
10. Pearce, D. The Feminization of Poverty: Women, Work and Welfare. *Urban Soc. Change Rev.* **1978**, *11*, 28–36.
11. Veeran, V. Feminization of Poverty. In Women's Symposium of International Association of Schools of Social Work in Montreal, Canada. 2000. Available online: https://www.researchgate.net/publication/268506084_FEMINIZATION_OF_POVERTY (accessed on 5 September 2023).
12. Charkiewicz, E. Kobiety i ubóstwo–widzialna ręka neoliberalnego państwa. *Kobiety i ubóstwo w Polsce. Badania i analizy 2008* **2011**, *2010*, 6–24.
13. Dormekpor, E. Poverty and gender inequality in developing countries. *Dev. Ctry. Stud.* **2015**, *5*, 76–102.
14. UNICEF. *The State of the World's Children 2007: Women and Children: The Double Dividend of Gender Equality*; UNICEF: Geneva, Switzerland, 2006; Volume 7.

15. Cook, K. State tactics of welfare benefit minimisation: The power of governing documents. *Crit. Soc. Policy* **2022**, *42*, 241–264. [CrossRef]
16. Blau, F.D.; Kahn, L.M. The gender pay gap. *The Economists' Voice* **2007**, *4*.
17. Bishu, S.G.; Alkadry, M.G. A systematic review of the gender pay gap and factors that predict it. *Adm. Soc.* **2017**, *49*, 65–104. [CrossRef]
18. Gradín, C.; Del Río, C.; Cantó, O. Gender wage discrimination and poverty in the EU. *Fem. Econ.* **2010**, *16*, 73–109. [CrossRef]
19. Amaro, F.; Bastos, A.; Cruz, J.; Proença, I. Mind the Gap: The Effects of Eliminating the Gender Pay Gap on Income and Poverty. *J. Poverty* **2023**, 1–19. [CrossRef]
20. Clancy, J.S.; Skutsch, M.; Batchelor, S. The Gender-Energy-Poverty Nexus: Finding the Energy to Address Gender Concerns in Development. DFID Project CNTR998521. 2003. Available online: https://ris.utwente.nl/ws/portalfiles/portal/5134277/Clancy99gender.pdf (accessed on 5 September 2023).
21. Nguyen, C.P.; Su, T.D. Does energy poverty matter for gender inequality? Global evidence. *Energy Sustain. Dev.* **2021**, *64*, 35–45. [CrossRef]
22. Munien, S.; Ahmed, F. A gendered perspective on energy poverty and livelihoods—Advancing the Millennium Development Goals in developing countries. *Agenda* **2012**, *26*, 112–123. [CrossRef]
23. Nussbaum, M. Promoting women's capabilities. In *Global Tensions: Challenges and Opportunities in the World Economy*; Routledge: London, UK, 2004; p. 241.
24. Chant, S. Exploring the "feminisation of poverty" in relation to women's work and home-based enterprise in slums of the Global South. *Int. J. Gend. Entrep.* **2014**, *6*, 296–316. [CrossRef]
25. Bahar, M.; Hamedanian, F.; Farajiha, M.; Golpaygani, T.S. Women's Access to Family Justice in Iran: Exploring the Main Barriers. *Pertanika J. Soc. Sci. Humanit.* **2018**, *26*, 147–164.
26. Hyde, E.; Greene, M.E.; Darmstadt, G.L. Time poverty: Obstacle to women's human rights, health and sustainable development. *J. Glob. Health* **2020**, *10*, 020313. [CrossRef] [PubMed]
27. Abdourahman, O.I. Time poverty: A contributor to women's poverty. *J. Stat. Afr.* **2010**, *11*, 16–36.
28. Noh, H.; Kim, K.S. Revisiting the 'feminisation of poverty' in Korea: Focused on time use and time poverty. *Asia Pac. J. Soc. Work Dev.* **2015**, *25*, 96–110. [CrossRef]
29. Arora, D. Gender differences in time-poverty in rural Mozambique. *Rev. Soc. Econ.* **2015**, *73*, 196–221. [CrossRef]
30. Gammage, S. Time pressed and time poor: Unpaid household work in Guatemala. *Fem. Econ.* **2010**, *16*, 79–112. [CrossRef]
31. Chant, S. The 'feminisation of poverty' and the 'feminisation' of anti-poverty programmes: Room for revision? *J. Dev. Stud.* **2008**, *44*, 165–197. [CrossRef]
32. Saigaran, N.G.; Karupiah, P.; Gopal, P.S. Gender socialization and capability deprivation on child urban poverty: Experiences of Malaysian Indian women. *Malays. J. Soc. Space* **2018**, *14*, 346–356. [CrossRef]
33. Marcelino, G.C.; Cunha, M.S.D. Multidimensional poverty in Brazil: Evidences for rural and urban areas. *Rev. Econ. E Sociol. Rural* **2023**, *62*, e266430. [CrossRef]
34. World Health Organization; Public Health Agency of Canada. *Preventing Chronic Diseases: A Vital Investment*; World Health Organization: Geneva, Switzerland, 2005.
35. Lukindo, M.; Price, V.; Pike, M. Estimating the impact of menstrual poverty on adolescents in Nova Scotia. *Paediatr Child Health* **2022**, *27*, 421–428. [CrossRef]
36. Menstrual Poverty—Workshop. Policy Department for Citizens' Rights and Constitutional Affairs for the Committee on Women's Rights and Gender Equality. European Parliament. Available online: https://www.europarl.europa.eu/committees/en/menstrual-poverty-workshop-/product-details/20221202WKS04761 (accessed on 22 August 2024).
37. Period Poverty—Why Millions of Girls and Women Cannot Afford Their Periods. UN Women. Available online: https://www.unwomen.org/en/news-stories/explainer/2024/05/period-poverty-why-millions-of-girls-and-women-cannot-afford-their-periods (accessed on 22 August 2024).
38. Rossouw, L.; Ross, H. Understanding Period Poverty: Socio-Economic Inequalities in Menstrual Hygiene Management in Eight Low- and Middle-Income Countries. *Int. J. Environ. Res. Public Health* **2021**, *18*, 2571. [CrossRef]
39. Skalli, L.H. Women and poverty in Morocco: The many faces of social exclusion. *Fem. Rev.* **2001**, *69*, 73–89. [CrossRef]
40. Profeta, P. COVID-19 and Its Economic Impact on Women and Women's Poverty. 2021. Available online: https://www.europarl.europa.eu/RegData/etudes/STUD/2021/693183/IPOL_STU(2021)693183_EN.pdf (accessed on 22 August 2024).
41. Kharas, H.; Dooley, M. Extreme Poverty in the Time of COVID-19. 2021. Available online: https://www.brookings.edu/wp-content/uploads/2021/06/extreme-poverty-during-the-time-of-covid-19.pdf (accessed on 22 August 2024).
42. Gönç Şavran, T.; Suğur, N. The gendered pattern of COVID-19 pandemic. In *Society in the COVID-19 Pandemic: Inequalities, Challenges, and Opportunities*; Pegem Akademi Yayıncılık: Macun, Turkey, 2021; pp. 1–26.
43. Posel, D.; Rogan, M. Women, income and poverty: Gendered access to resources in post-apartheid South Africa. *Agenda* **2009**, *23*, 25–34.
44. Parish, S.L.; Rose, R.A.; Andrews, M.E. Income poverty and material hardship among US women with disabilities. *Soc. Serv. Rev.* **2009**, *83*, 33–52. [CrossRef]
45. Cawthorne, A. *The Straight Facts on Women in Poverty*; Center for American Progress: Washington, DC, USA, 2008; Volume 8, pp. 1–3.

46. Gadalla, T.M. Gender differences in poverty rates after marital dissolution: A longitudinal study. *J. Divorce Remarriage* **2008**, *49*, 225–238. [CrossRef]
47. Smeeding, T.M.; Sandstrom, S. *Poverty and Income Maintenance in Old Age: A Cross-National View of Low Income Older Women*; LIS Working Paper Series; Routledge: London, UK, 2005; p. 398.
48. Kawachi, I.; Kennedy, B.P.; Gupta, V.; Prothrow-Stith, D. Women's status and the health of women and men: A view from the States. *Soc. Sci. Med.* **1999**, *48*, 21–32. [CrossRef] [PubMed]
49. Mukhopadhyay, N.; Sengupta, P.P. (Eds.) *Gini Inequality Index: Methods and Applications*; CRC Press: Boca Raton, FL, USA, 2021.
50. Ramos, M.E.; Gibaja-Romero, D.E.; Ochoa, S.A. Gender inequality and gender-based poverty in Mexico. *Heliyon* **2020**, *6*, e03322. [CrossRef]
51. Rashada, A.S.; Sharaf, M.F. Income inequality and intimate partner violence against women: Evidence from India. In *Frankfurt School-Working Paper Series*; Frankfurt School of Finance & Management, Frankfurt: Frankfurt, Germany, 2016; p. 222.
52. Haataja, A. Unemployment, employment and poverty. *Eur. Soc.* **1999**, *1*, 169–196. [CrossRef]
53. Armstrong, P. Women and unemployment. Atlantis: Critical Studies in Gender. *Cult. Soc. Justice* **1980**, *6*, 1–17.
54. Fryer, D.; Fagan, R. Poverty and unemployment. In *Poverty and Psychology: From Global Perspective to Local Practice*; Springer: Boston, MA, USA, 2003; pp. 87–101.
55. Nisak, S.S.; Sugiharti, L. Gender inequality and women poverty in Indonesia. *Int. J. Innov. Creat. Chang.* **2020**, *11*, 375–387.
56. Lloyd, C.B.; Hewett, P. Educational inequalities in the midst of persistent poverty: Diversity across Africa in educational outcomes. *J. Int. Dev.* **2009**, *21*, 1137–1151. [CrossRef]
57. Edström, J. Indicators for Women's Health in Developing Countries: What They Reveal and Conceal 1. *IDS Bull.* **1992**, *23*, 38–49. [CrossRef]
58. Price, J.; Asgary, R. Women's health disparities in Honduras: Indicators and determinants. *J. Women's Health* **2011**, *20*, 1931–1937. [CrossRef]
59. Abdelaziz, F.B. Women's health and equity indicators. *Int. J. Public Health* **2007**, *52* (Suppl. S1), S1–S2. [CrossRef]
60. Schüler, D. The uses and misuses of the Gender-Related Development Index and Gender Empowerment Measure: A review of the literature. *J. Hum. Dev.* **2006**, *7*, 161–181. [CrossRef]
61. Suharnanik, S.; Sarah, Y.; Putut, L.E.E. Analysis of Gender Development Index to prove its correlation with the decrease of poverty line in Kediri City, Indonesia. *Женщина в Российском Обществе* **2023**, *2*, 37–46.
62. Malik, R. HDI and Gender Development Index: Current Status of Women Development in India. *PRAGATI J. Indian Econ.* **2018**, *5*, 30–43. [CrossRef]
63. Dijkstra, A.G.; Hanmer, L.C. Measuring socio-economic gender inequality: Toward an alternative to the UNDP gender-related development index. *Fem. Econ.* **2000**, *6*, 41–75. [CrossRef]
64. Charmes, J.; Wieringa, S. Measuring women's empowerment: An assessment of the gender-related development index and the gender empowerment measure. *J. Hum. Dev.* **2003**, *4*, 419–435. [CrossRef]
65. Adjei, S.B. Assessing women empowerment in Africa: A critical review of the challenges of the gender empowerment measure of the UNDP. *Psychol. Dev. Soc.* **2015**, *27*, 58–80. [CrossRef]
66. Durbin, E. Towards a gendered human poverty measure. *Fem. Econ.* **1999**, *5*, 105–108. [CrossRef]
67. Muttneja, P. A review of Human Development Index (HDI) and Human Poverty Index (HPI) in the Indian perspective. *Sch. Int. J. Manag. Dev.* **2015**, *2*, 15.
68. Prince, H. Macro-level drivers of multidimensional poverty in sub-Saharan Africa: Explaining change in the Human Poverty Index. *Afr. Eval. J.* **2014**, *2*, 11. [CrossRef]
69. EIGE Report—Gender Equality Index 2020: Digitalisation and the Future of Work, Publications Office of the European Union, Luxembourg. Available online: https://eige.europa.eu/publications/gender-equality-index-2020-digitalisation-and-future-work (accessed on 5 September 2023).
70. Eurostat: Gender Pay Gap Statistics. Available online: https://ec.europa.eu/eurostat/statistics-explained/index.php?title=Gender_pay_gap_statistics (accessed on 5 September 2023).
71. Dijkstra, L.L.; Papadimitriou, E.; Norlén, H. The regional gender equality monitor. In *Measuring Female Disadvantage and Achievement in EU Regions*; Publications Office of the European Union: Luxembourg, 2019. [CrossRef]
72. Schmid, C.B.; Elliot, M. "Why Call It Equality?" Revisited: An Extended Critique of the EIGE Gender Equality Index. *Soc. Indic. Res.* **2023**, *168*, 389–408. [CrossRef]
73. López-Martínez, M.; Riquelme Perea, P.J.; de Maya Matallana, M. Gender equality models in the European Union. *Int. J. Sociol. Soc. Policy* **2022**, *42*, 605–623. [CrossRef]
74. Heidenreich, M. *Territorial and Social Inequalities in Europe: Challenges of European Integration*; Springer Nature: New York, NY, USA, 2022.
75. Horemans, J.; Marx, I. Should we care about part-time work from a poverty perspective? An analysis of the EU15 countries. In *Non-Standard Employment in Europe: Paradigms, Prevalence and Policy Responses*; Palgrave Macmillan UK: London, UK, 2013; pp. 169–189.
76. Horemans, J.; Marx, I.; Nolan, B. Hanging in, but only just: Part-time employment and in-work poverty throughout the crisis. *IZA J. Eur. Labor Stud.* **2016**, *5*, 1–19. [CrossRef]
77. Blagoycheva, H. Employment and the "Working Poor" Phenomenon in the EU. *Int. J. Econ. Bus. Adm.* **2016**, *4*, 3–18. [CrossRef]

78. Gil Ruiz, J.M. Laissez faire, laisser passer: Desigualdad estructural laboral y recortes presupuestarios. *Daimon-Rev. Int. Filosofia.* **2017**, 111–124. [CrossRef]
79. Pérez-Corral, A.L.; Moreno Mínguez, A. Single-parent families, educational gradient, and child deprivation: The cases of Italy and Spain. *Child Indic. Res.* **2022**, *15*, 1821–1846. [CrossRef]
80. Sunikka-Blank, M.; Galvin, R. Single parents in cold homes in Europe: How intersecting personal and national characteristics drive up the numbers of these vulnerable households. *Energy Policy* **2021**, *150*, 112134. [CrossRef]
81. Pantazis, C.; Gordon, D. *Poverty and Social Exclusion in Britain*; Levitas, R., Ed.; Policy Press: Bristol, UK, 2006; pp. 23–24.
82. Anderson, C. The diversity, strength, and challenges of single-parent households. In *Normal Family Processes: Growing Diversity and Complexity*, 3rd ed.; Walsh, F., Ed.; The Guilford Press: New York, NY, USA, 2003; pp. 121–152.
83. Downey, D.B.; Ainsworth-Darnell, J.W.; Dufur, M.J. Sex of parent and children's well-being in single-parent households. *J. Marriage Fam.* **1998**, *60*, 878–893. [CrossRef]
84. Bradshaw, J.; Keung, A.; Chzhen, Y. Cash benefits and poverty in single-parent families. In *The Triple Bind of Single-Parent Families*; Policy Press: Bristol, UK, 2018; pp. 337–358.
85. Artazcoz, L.; Cortès-Franch, I.; Arcas, M.M.; Ollé-Espluga, L.; Pérez, K. Time poverty, health and health-related behaviours in a Southern European city: A gender issue. *J. Epidemiol. Community Health* **2024**, *78*, 284–289. [CrossRef]
86. Lodigiani, R.; Maino, F. Minimum income, active inclusion, and work requirements in Europe: Insights from community service projects introduced by Italian Citizenship Income. *Stato E Mercato* **2022**, *42*, 369–406.
87. Tacoli, C. *Urbanization, Gender and Urban Poverty: Paid Work and Unpaid Carework in the City*; Human Settlements Group, International Institute for Environment and Development: London, UK, 2012; p. 48.
88. Ferrant, G.; Pesando, L.M.; Nowacka, K. *Unpaid Care Work: The Missing Link in the Analysis of Gender Gaps in Labour Outcomes*; OECD Development Center: Boulogne Billancourt, France, 2014.
89. Alarcão, V.; Candeias, P.; Stefanovska-Petkovska, M.; Pintassilgo, S.; Machado, F.L.; Virgolino, A.; Santos, O. Mental Health and Well-Being of Migrant Populations in Portugal Two Years after the COVID-19 Pandemic. *Behav. Sci.* **2023**, *13*, 422. [CrossRef]
90. Benedi Lahuerta, S. ENAR Shadow Report 2010–2011: Racism and Related Discriminatory Practices in Spain. 2012. Available online: https://eprints.soton.ac.uk/385036/1/24._spain.pdf (accessed on 5 September 2023).
91. Sahraoui, N. *Racialised Workers and European Older-Age Care*; Springer International Publishing: New York, NY, USA, 2019.
92. Sánchez-Lopez, C.; de Paz Báñez, M.A. Unqueauty and Poverty in Great recession. differences between EU countries. *Rev. Econ. Mund.* **2016**, *44*, 93–124.
93. McDevitt, S. Social exclusion in the European Union: An organized focus for social policy-making. *Soc. Policy J.* **2003**, *2*, 3–18. [CrossRef]
94. Ostner, I.; Lewis, J. Gender and the Evolution of European Social Policies. In *European Social Policy: Between Fragmentation and Integration*; Leibfried, S., Pierson, P., Eds.; Brookings Publications: Washington, DC, USA, 1995; pp. 159–193.
95. Kazepov, Y. (Ed.) *Rescaling Social Policies: Towards Multilevel Governance in Europe*; Ashgate Publishing, Ltd.: Aldershot, UK, 2010; Volume 38.
96. Leskošek, V. Siromaštvo starijih žena u Sloveniji. *Rev. Za Soc. Polit.* **2019**, *26*, 240.
97. Timonen, V.; Doyle, M. Life-long singlehood: Intersections of the past and the present. *Ageing Soc.* **2014**, *34*, 1749–1770. [CrossRef]
98. Steinhilber, S. *The Gender Impact of Pension Reforms: Case Studies of the Czech Republic, Hungary, and Poland*; Public Pensions; OECD: Paris, France, 2004; p. 243.
99. Das, M.B. The Rising Tide: A New Look at Water and Gender. World Bank, Washington, World Bank. 2017. Available online: https://openknowledge.worldbank.org/handle/10986/27949 (accessed on 5 September 2023).
100. Sara, J.; Grown, C.; Keener, S.; Hatzweld, G.; Kuoh, G. Menstrual Health and Hygiene Empowers Women and Girls: How to Ensure We Get It Right. 2021. Available online: https://blogs.worldbank.org/water/menstrual-health-and-hygiene-empowers-women-and-girls-how-ensure-we-get-it-right (accessed on 12 February 2023).
101. Tarka, D. Własności cech diagnostycznych w badaniach typu taksonomicznego. *Ekon. I Zarządzanie* **2011**, *2*, 194–205.
102. Oesterreich, M.; Perzyńska, J.; Barej-Kaczmarek, E. Application of the TOPSIS procedure for evaluation of socio-economic development of the West Pomeranian Voivodeship districts in years 2004–2017. *Zesz. Nauk. Uniw. Przyr. Humanist. W Siedlcach. Ser. Adm. I Zarządzanie* **2019**, *49*, 79–88. [CrossRef]
103. Bąk, I.; Perzyńska, J. Gender Balance in the European Union Countries as an Important Challenge to the Implementation of The Sustainable Development Goals–Statistical Analysis. In *Innovation Management and information Technology impact on Global Economy in the Era of Pandemic: Proceedings of the 37th International Business Information Management Association Conference (IBIMA), Cordoba, Spain, 30–31 May 2021*; 2021; ISSN 2767-9640. pp. 9007–9015. Available online: https://ibima.org/accepted-paper/gender-balance-in-the-european-union-countries-as-an-important-challenge-to-the-implementation-of-the-sustainable-development-goals-statistical-analysis/ (accessed on 29 July 2024).
104. Szanduła, J. Uwagi do unitaryzacji zmiennych w referencyjnym systemie granicznym. *Przegląd Stat.* **2014**, *61*, 147–168. [CrossRef]
105. Nowak, E. *Metody Taksonomiczne w Klasyfikacji Obiektów Społeczno-Gospodarczych*; PWE: Warszawa, Poland, 1990.

106. Decancq, K.; Goedemé, T.; Van den Bosch, K.; Vanhille, J. The Evolution of Poverty in the European Union: Concepts, Measurement and Data. *ImProvE Working Papers. Methodological Paper* 2013. p. 13/1. Available online: https://www.improve-etn.eu/ (accessed on 2 January 2023).
107. Piwowar, A.; Dzikuć, M. Poverty and Social Exclusion: Is this a Problem in Rural Areas in the Visegrad Group Countries? *Eur. Res. Stud. J.* **2020**, *23*, 45–54. [CrossRef]
108. Sompolska-Rzechuła, A.; Kurdyś-Kujawska, A. Assessment of the Development of Poverty in EU Countries. *Int. J. Environ. Res. Public Health* **2022**, *19*, 3950. [CrossRef]
109. Bąk, I.; Perzyńska, J. Poverty and social exclusion in the context of the implementation of the sustainable development goals. *Eur. Res. Stud. J.* **2021**, *24*, 68–83. [CrossRef]
110. Svensson, E.M.; Gunnarsson, Å. Gender Equality in the Swedish Welfare State. *feminists@law*. 2012, p. 2. Available online: https://journals.kent.ac.uk/index.php/feministsatlaw/article/view/51/160 (accessed on 9 March 2022).
111. International Monetary Fund. World Economic Outlook: Global Manufacturing Downturn, Rising Trade Barriers. Washington, DC. 2019. Available online: https://www.imf.org/en/Publications/WEO/Issues/2019/10/01/world-economic-outlook-october-2019#Statistical%20Appendix (accessed on 8 March 2022).
112. Eurofound and EIGE. *Upward Convergence in Gender Equality: How Close Is the Union of Equality?* Publications Office of the European Union: Luxembourg, 2021. Available online: https://eige.europa.eu/publications/upward-convergence-gender-equality-how-close-union-equality (accessed on 29 July 2024).
113. Zgliczyński, W. Polityka społeczna w Europie—Ewolucja i rozwiązania modelowe. *Stud. BAS* **2017**, *2*, 37–58.
114. OXFAM The Commitment to Reducing Inequality Index 2018. A Global Ranking of Governments Based on What They Are Doing to Tackle the Gap between Rich and Poor. 2018. Available online: https://oxfamilibrary.openrepository.com/bitstream/handle/10546/620553/rr-commitment-reducing-inequality-index-2018-091018-en.pdf (accessed on 10 September 2023).
115. Poverty Watch 2019—Monitoring Ubóstwa Finansowego i Polityki Społecznej Przeciw Ubóstwu w Polsce. EAPN Polska. Available online: http://www.eapn.org.pl/eapn/uploads/2019/10/Poland_Poverty-Watch_2019-ost.pdf (accessed on 8 March 2022).
116. Sánchez Páramo, C.; Hill, R.; Gerszon Mahler, D.; Narayan, A.; Yonzan, N. COVID-19 Leaves a Legacy of Rising Poverty and Widening Inequality. World Bank Blogs. 2021. Available online: https://blogs.worldbank.org/developmenttalk/covid-19-leaves-legacy-rising-poverty-and-widening-inequality (accessed on 5 September 2023).
117. UN DESA 2022 Sustainable Development Outlook 2020. Achieving SDGs in the Wake of COVID-19: Scenarios for Policymakers. Available online: https://sdgs.un.org/sites/default/files/2020-07/SDO2020_Book.pdf (accessed on 25 November 2023).
118. UN the Sustainable Development Goals Report 2022. Available online: https://unstats.un.org/sdgs/report/2022/The-Sustainable-Development-Goals-Report-2022.pdf (accessed on 5 September 2023).
119. UN The Sustainable Development Goals Report 2023: Special Edition. Available online: https://unstats.un.org/sdgs/report/2023/ (accessed on 10 May 2024).
120. The Progress on the Sustainable Development Goals: The Gender Snapshot. 2023. Available online: https://unstats.un.org/sdgs/gender-snapshot/2023/ (accessed on 13 July 2024).
121. Ahmed, N.; Marriott, A.; Dabi, N.; Lowthers, M.; Lawson, M.; Mugehera, L. Inequality Kills. The Unparalleled Action Needed to Combat Unprecedented Inequality in the Wake of COVID-19. 2022. Available online: https://www.oxfam.org/en/research/inequality-kills (accessed on 5 September 2023).
122. Oxfam Media Briefing. Profiting from Pain. The Urgency of Taxing the Rich Amid a Surge in Billionaire Wealth and a Global Cost-of-Living Crisis. 2022. Available online: https://www.oxfam.org/en/research/profiting-pain (accessed on 10 September 2023).
123. Armstrong, M. It Will Take Another 136 Years to Close the Global Gender Gap. World Economic Forum. 2021. Available online: https://www.weforum.org/agenda/2021/04/136-years-is-the-estimated-journey-time-to-gender-equality (accessed on 6 September 2023).
124. Balińska, B.; Chołuj, B.; Desperak, I. *Polityka równości płci., Polska 2007*; Raport; Program Narodów Zjednoczonych ds. Rozwoju (UNDP): Warszawa, Poland, 2007.
125. Malgesini, G.; Cesarini-Sforza, L.; Babović, M.; Leemkuil, S.; Sverrisdóttir, M.; Mareková, S. Gender and Poverty in Europe. EAPN Briefing Note. 2017. Available online: https://www.eapn.eu/wp-content/uploads/2018/03/EAPN-2017-EAPN-Briefing-Gender-and-Poverty-final.pdf (accessed on 6 September 2023).
126. Saeed, H. Poverty Has a Woman's Face. Dawn. Today's Paper. 2013. Available online: https://www.dawn.com/news/802325/poverty-has-a-womans-face (accessed on 1 September 2022).
127. Chant, S. *Dangerous Equations? How Female-Headed Households Became the Poorest of the Poor. Causes, Consequences and Cautions, [w:] A. Cornwall, E. Harrison, A. Whitehead (red.), Feminisms in Development: Contradictions, Contestations and Challenges*; Zed Books: London, UK; New York, NY, USA, 2007.
128. Lister, R. To count for nothing: Poverty beyond the statistics. *J. Br. Acad.* **2015**, *3*, 139–165. [CrossRef]
129. Lister, R. Poverty. In *Key Concepts*; Polity Press: Cambridge, UK, 2004; p. 6.

Disclaimer/Publisher's Note: The statements, opinions and data contained in all publications are solely those of the individual author(s) and contributor(s) and not of MDPI and/or the editor(s). MDPI and/or the editor(s) disclaim responsibility for any injury to people or property resulting from any ideas, methods, instructions or products referred to in the content.

sustainability

Article

Does the Inter-Provincial Floating Population Affect Regional Economic Development in China? An Empirical Analysis

Zhijie Cao [1,2], Ziao Li [1,2,*] and Kexin Zhou [1,2,*]

1 School of Public Administration, Hohai University, Nanjing 211100, China; caozhijie@hhu.edu.cn
2 Center for Chinese Migration, Hohai University, Nanjing 211100, China
* Correspondence: 221314010002@hhu.edu.cn (Z.L.); 230214160004@hhu.edu.cn (K.Z.)

Abstract: In recent decades, significant changes in the urban–rural structure of population mobility have profoundly impacted provincial development, urbanization, and population redistribution in China. Based on China's fifth, sixth, and seventh national population census datasets, this study explores the effects of the inter-provincial floating population on regional economic development through statistical and empirical analysis, identifying both the scale and structural impacts of the floating population on regional economic development. The results found that while the scale of China's floating population has been continuously increasing, the spatial distribution pattern remains relatively unchanged, and the pattern is summed up as low in the middle and high on both sides. The floating population exerts both scale and structural effects on the economic development of both inflow and outflow regions, altering regional populations and production efficiency, and thereby influencing regional economic outcomes. Specifically, this study finds that the inflow population has no significant differential impact on high- and low-density regions. In contrast, the outflow population exhibits a significant differential impact, with the negative impact of the outflow population on low-density regions being more substantial than that on high-density regions. Inter-provincial migration supports achieving sustainable development goals (SDG-8 and 11) by shaping regional economic development. To address these dynamics, the high-density regions of China should transform and upgrade the industrial and population structure by promoting the trend of population return to low-density regions. This can be achieved by transferring low-end industries and low-skilled labor, thereby alleviating the pressure of overcrowding. Meanwhile, low-density regions should seize the opportunities for population return and industrial transfer, implement talent introduction, and accurately undertake industrial transfer. This approach can foster the in-depth development of new urbanization and rural revitalization initiatives, promoting balanced regional growth and sustainability.

Keywords: floating population; economic development; structure effect; population return; urban density

Citation: Cao, Z.; Li, Z.; Zhou, K. Does the Inter-Provincial Floating Population Affect Regional Economic Development in China? An Empirical Analysis. *Sustainability* **2024**, *16*, 7142. https://doi.org/10.3390/su16167142

Academic Editors: Idiano D'Adamo, Massimo Gastaldi and Manoj Kumar Nallapaneni

Received: 14 June 2024
Revised: 11 August 2024
Accepted: 16 August 2024
Published: 20 August 2024

Copyright: © 2024 by the authors. Licensee MDPI, Basel, Switzerland. This article is an open access article distributed under the terms and conditions of the Creative Commons Attribution (CC BY) license (https://creativecommons.org/licenses/by/4.0/).

1. Introduction

Since China's reform and opening up, large-scale cross-regional population migration has not only triggered the reshaping of the spatial distribution of the population but also promoted the reshaping of the regional economic development pattern [1–3]. Population mobility is an activity that optimizes the spatial allocation of labor factors based on the economic development pattern [4]. The floating population pattern and the economic development pattern interact and promote each other [5]. Meanwhile, the spatial agglomeration and dispersion of population mobility are affected by socioeconomic development patterns. Similarly, the floating population pattern also affects the socioeconomic development pattern through the spatial optimal allocation of labor factors [6,7]. According to the National Bureau of Statistics and data from the "China Health and Family Planning Statistical Yearbook", the size of the floating population in China has stabilized at over

200 million people since 2010, reaching a peak of 253 million in 2014, which accounted for 18.5% of the national population. Additionally, data from the seventh national census show that from 2010 to 2020, the floating population grew at an average annual rate of 5.46%, rising to 370 million people by 2020, representing 26.5% of the national population. These figures provide material for examining the issues of population mobility and economic convergence in the new era [8]. This indicates that the population mobility model, in a significant period of transformation, is reshaping the current spatial pattern of population mobility in China and will significantly impact population redistribution, urbanization, and regional development [9,10]. On the one hand, a large population inflow has caused tension in infrastructure services and disrupted social and economic life in the cities to which these people moved [11]. The economic gap between high- and low-density regions has gradually widened [12].

In recent years, along with the restructuring of the global industrial chain and the industrial transformation and upgrading of China's coastal areas, the return of population from coastal areas to the central and western regions has become a new trend, and the spatial pattern of economic development has also shown an inland trend [13]. Under the major trend changes in population development and the new development pattern, the report of the 20th National Congress of the Communist Party of China and the "Population Development Plan" (2016–2023) pointed out that there is an urgent need to optimize the spatial distribution of the population, promote the adaptation of population distribution to the national regional development strategy, and actively guide the population's orderly flow and rational distribution. Straightening out the relationship between China's floating population and economic development is the premise and basis for realizing the coordinated development of the population and regions and promoting the national urbanization and rural revitalization strategy [14]. The movement of people from less developed to more developed regions can help to alleviate poverty [15]. By migrating to areas with better economic opportunities, individuals from poorer provinces can increase their incomes, reducing poverty levels (SDG-1). Empirical analyses have shown that inter-provincial migration can lead to better employment opportunities and higher migrant wages, contributing to poverty reduction [16]. The floating population often fills labor shortages in more economically developed regions, contributing to sustained economic growth (SDG-8). This workforce mobility supports industries in high-demand areas, leading to increased productivity and economic expansion. Studies indicate that the influx of labor from other provinces can boost regional economic growth by enhancing labor market flexibility and efficiency. Regions with a high influx of the floating population tend to invest more in infrastructure and innovation (SDG-9) to support economic activities. This investment can lead to improved transportation networks, communication systems, and technological advancements, fostering overall economic development. The empirical evidence suggests that inter-provincial migration encourages regions to enhance their industrial capacities and infrastructural facilities. The floating population influences urbanization patterns and urban development. Cities that attract large numbers of migrants often need to improve infrastructure, housing, and services to accommodate the growing population [17,18]. This urbanization process can drive sustainable urban development if appropriately managed, leading to more inclusive and resilient cities (SDG-11). Inter-provincial migration can help to reduce economic disparities between regions. Migrants often send remittances back to their home provinces, which can improve living standards and stimulate economic activity in less developed areas. This financial flow helps to balance regional inequalities (SDG-10) and promotes inclusive economic development [19].

So, what is the relationship between the floating population and economic development? Existing studies can be roughly divided into two categories. First, economic factors are the primary factors driving population migration [20]. In classical theory, the main motivation for population mobility decisions is economic motivation; that is, a low level of economic development in the outflow area is conducive to population outflow, while a high level of economic development in the inflow area is conducive to population inflow.

This conclusion has been repeatedly proven. Economic development differences are also considered important factors affecting population mobility [21,22]. The cost–benefit theory analyzes this phenomenon from the perspective of the benefits and costs of labor transfer, and states that the decision-making of population mobility is related to the income gap between regions. Meanwhile, the "push–pull" theory analyzes this issue from the perspective of kinematics and argues that population migration is affected by the economic factors of the place of origin [23]. The dominant push (pull) and the immigration-dominated pull (push) are jointly affected by economic factors [24]. When the push of emigration is greater than the pull, and the pull of emigration is greater than the push simultaneously, population migration will occur [25]. A new economic migration theory was developed based on the "push-pull" theory, which argues that population mobility is affected by the expected income level of the place of emigration. The second category includes the research on the coordinated development between the floating population and economic development. Yang [26] believes that population agglomeration has significant economic growth benefits, and at the same time, economic growth can further exert the population agglomeration effect. As a bridge, the flow of labor factors first acts on the industry and then promotes economic development [27].

The existing literature reveals that researchers have conducted in-depth research on the relationship between the floating population and economic development, laying a solid foundation for further research in this paper. However, still, there are some shortcomings coming into the picture. First, little attention has been paid to quantitative research considering the structural differences in the impact of the migrant population on economic growth. Second, there is a lack of research that completely separates the migrant population from other variables that affect economic growth through theoretical models. This study considers both the shortcomings and the structural effects of the floating population on economic growth through empirical models. At the same time, due to the influence of China's household registration system, the population transfer includes the population flow along with and without the change in household registration. The floating population specifically refers to the second type of temporary population flow when households are separated. This type of population has a strong connection with both (outflow and inflow) places when they enter the system to work and do not completely consume and save in the system decision-making, but transfer capital accumulation back into the economic system from which it originated. This study examines the socio-economic effects of cross-provincial migration, highlighting its significant influence on regional economic development [28]. The findings aim to support the advancement of Sustainable Development Goal 8 (Decent Work and Economic Growth) and Sustainable Development Goal 11 (Sustainable Cities and Communities).

2. Materials and Methods

2.1. Data Collection and Sources

In this study, the 31 provincial-level administrative units in mainland China were targeted, except for Hong Kong, Macao, and Taiwan. This study utilizes the data from the fifth, sixth, and seventh national censuses to subdivide net population flow into population inflow and outflow, incorporating them into a unified analytical framework. Mathematical formulas are used to deduce trends in population flow. We examined the temporal and spatial characteristics of the inter-provincial floating population from 2000 to 2020 and their differential impact on the economic development of different regions in China. Since the inter-provincial outflow population data of the seventh census have not yet been released, this study calculated the inter-provincial outflow population data based on the "China Census Yearbook 2020" and calculated the inter-provincial outflow rate in 2020 [27,28]. The statistical yearbook was compiled by the National Bureau of Statistics and the statistical bureaus of each province and an autonomous region, from which data such as GDP, per capita GDP, regional average wage level, and the proportion of the tertiary industry were obtained. Qiao [29] and others [27,30] proposed that the floating population can be obtained from the difference between the resident population and the registered population. This study adopted a net flow rate (registered population of permanent residents). The method of the registered population was used to obtain the inter-provincial net flow rate [29,30], and the outflow rate, inflow rate, net flow rate, and the inter-provincial population outflow rate in the region in the year 2020 were obtained.

2.2. Theoretical Models

The neoclassical economic growth model states that technological progress, capital, and effective labor are the three main factors that affect economic output. Here, the Douglas production function was extended, and the model introduced the variable of the floating population to study the impact of the national inter-provincial population inflow and outflow on regional economies.

2.2.1. Impact of Inter-Provincial Net Floating Population on Regional Economic Development

Assumption 1. *The Douglas production function is a Haro-neutral production function with constant returns to scale [29].*

$$Y(t) = K(t)^{\alpha}[A(t)L(t)]^{1-\alpha},\ 0 < \alpha < 1 \tag{1}$$

where $Y(t)$ represents output, $K(t)$ represents the capital required for production, $A(t)$ represents technological progress, $L(t)$ represents the effective labor of the system, t represents time, and α is a parameter of the production function.

Assumption 2. *Technological progress is endogenous $A(t)$and capital stock is exogenous $K(t)$, consistent with the hypothesis of Lu Fenggang [31].*

$$A(t) = \lambda K(t)^{\varphi},\ \lambda > 0,\ 0 < \varphi < 1 \tag{2}$$

where λ is the transformation parameter, representing the impact of increased capital on knowledge, and φ is the return-to-scale property of the knowledge production function. $\varphi < 1$ which means that the returns to scale of the knowledge production function is diminishing because with the continuous accumulation of knowledge, the new knowledge generated from the newly added capital will be more and more difficult to obtain.

Assumption 3. *Assume that there is a floating population in the economic system where l is the net mobility rate.*

$$L(t) = (1 + l)P(t) \tag{3}$$

The $P(t)$ represents the total population within the economic system

Substituting Formulas (2) and (3) into Formula (1) to obtain the following formula:

$$\begin{aligned} Y(t) = K(t)^{\alpha}[A(t)L(t)]^{1-\alpha} &= \lambda^{1-\alpha}K(t)^{\alpha+\varphi(1-\alpha)}L(t)^{(1-\alpha)} \\ &= \lambda^{1-\alpha}K(t)^{\alpha+\varphi(1-\alpha)}(1+l)^{1-\alpha}P(t)^{1-\alpha} \end{aligned} \tag{4}$$

Dividing both sides of Formula (5) by the total population $P(t)$ to obtain the Formula (6) corresponding to per capita output $y(t)$:

$$y(t) = \frac{Y(t)}{P(t)} = \lambda^{1-\alpha}K(t)^{\alpha+\varphi(1-\alpha)}(1+l)^{1-\alpha}P(t)^{-\alpha} \tag{5}$$

Take the natural logarithm of both sides of Formulas (4) and (5) at the same time to obtain the equations below:

$$\ln Y(t) = (\alpha + \varphi(1-\alpha))\ln K(t) + (1-\alpha)\ln\lambda + (1-\alpha)\ln(1+l) + (1-\alpha)\ln P(t) \tag{6}$$

$$\ln y(t) = (\alpha + \varphi(1-\alpha)) \ln K(t) + (1-\alpha) \ln \lambda + (1-\alpha) \ln(1+l) - \alpha P(t) \quad (7)$$

In order to more clearly examine the influencing factors of total economic output and per capita economic output, the terms in Formulas (6) and (7) are expanded $\ln(1+l)$ by the third-order Taylor series, and Formulas (8) and (9) are obtained.

$$\ln Y(t) = (\alpha + \varphi(1-\alpha)) \ln K(t) + (1-\alpha) \ln \lambda + (1-\alpha) \ln P(t) + (1-\alpha)(1+l) - \frac{1-\alpha}{2}(1+l)^2 + o(1+l)^3 \quad (8)$$

$$\ln y(t) = (\alpha + \varphi(1-\alpha)) \ln K(t) + (1-\alpha) \ln \lambda - (1-\alpha) P(t) + (1-\alpha)(1+l) - \frac{1-\alpha}{2}(1+l)^2 + o(1+l)^3 \quad (9)$$

By organizing Formulas (8) and (9), expanding and merging the quadratic terms, and removing the infinitesimal quantities, Formulas (10) and (11) can be obtained.

$$\ln Y(t) = (\alpha + \varphi(1-\alpha)) \ln K(t) + (1-\alpha) \ln \lambda + (1-\alpha) \ln P(t) + (1-\alpha) l + (1-\alpha) \quad (10)$$

$$\ln y(t) = (\alpha + \varphi(1-\alpha)) \ln K(t) + (1-\alpha) \ln \lambda - (1-\alpha) P(t) + (1-\alpha) l + (1-\alpha) \quad (11)$$

It can be found that changes in regional total output $K(t)$ and per capita output $Y(t)$ not only depend on asset investment $y(t)$, but also the total population $P(t)$ and net floating population l. From $0 < \alpha < 1$, showing that $1 - \alpha > 0$, it is found that (i) the net inflow of population has a positive impact on regional total output and per capita output; (ii) asset investment has a certain positive impact on regional total output and per capita output; and (iii) the total population has a positive impact on regional total output and a negative impact on per capita output.

2.2.2. Impact of Inter-Provincial Inflows and Outflows on Regional Economic Development

In order to explore the impact of the out-migration population and the in-migration population on the regional economy, this study further sub-divided the net migrant population and used the Douglas production function to understand the impact of the national inter-provincial in-migration population and the out-migration population on the regional economy.

Assumption 4. *The net floating population in this economic system is equal to the incoming population minus the outgoing population, that is,*

$$l = l1 - l2,\ 0 < l1 < 1, 0 < l2 < 1 \quad (12)$$

$$L(t) = (1+l)P(t) = (1 + l1 - l2)P(t) \quad (13)$$

where $P(t)$ represents the total population within the economic system, l is the net mobility rate, $l1$ is the inflow rate, and $l2$ is the outflow rate. The inflow rate $l1$ and the outflow rate $l2$ are, respectively, obtained by using the ratio of the inter-provincial inflow population and the inter-province outflow population to the registered population. Substitute Formula (13)'s values into Formulas (8) and (9) to obtain the following equations:

$$\begin{aligned} \ln Y(t) = (\alpha + \varphi(1-\alpha)) \ln K(t) + (1-\alpha) \ln \lambda + (1-\alpha) \ln P(t) + (1-\alpha)(1 + l1 - l2) \\ - \tfrac{1-\alpha}{2}(1 + l1 - l2)^2 + o(1 + l1 - l2)^3 \end{aligned} \quad (14)$$

$$\ln y(t) = (\alpha + \varphi(1-\alpha)) \ln K(t) + (1-\alpha) \ln \lambda - (1-\alpha) P(t) + (1-\alpha)(1 + l1 - l2) - \frac{1-\alpha}{2}(1 + l1 - l2)^2 + o(1 + l1 - l2)^3 \tag{15}$$

It is found from Formulas (14) and (15) that changes in regional total output and per capita output not only depend on asset investment $K(t)$, but also on total population $P(t)$, inflow rate $l1$, and outflow rate $l2$. Further, by sorting out Formulas (14) and (15), we obtain the following equations.

$$\ln Y(t) = (\alpha + \varphi(1-\alpha)) \ln K(t) + (1-\alpha) \ln \lambda + (1-\alpha) \ln P(t) + (1-\alpha) l1 - (1-\alpha) l2 + (1-\alpha) \tag{16}$$

$$\ln y(t) = (\alpha + \varphi(1-\alpha)) \ln K(t) + (1-\alpha) \ln \lambda - \alpha \ln P(t) + (1-\alpha) l1 - (1-\alpha) l2 + (1-\alpha) \tag{17}$$

In Formulas (16) and (17), if $0 < \alpha < 1$ can be $1 - \alpha > 0$, it can be found that (i) the inflow rate has a positive promoting effect on the regional total output and per capita output, and (ii) the outflow rate has a positive effect on the regional total output and per capita output and further acts as a negative inhibitory effect. This study will verify these influencing relationships from an empirical perspective.

2.3. Empirical Analysis Model

2.3.1. Regression Analysis

According to the above theoretical derivation process, especially Formulas (10) and (11), taking into account the endogeneity problems caused by omitted variables and the inertial effect of economic growth, the first-order lagged explained variable is introduced into the explanatory variables to construct a dynamic panel of data models (Formulas (18) and (19)).

$$\ln Y_{it} = \beta_0 + \beta_1 \ln Y_{it-1} + \beta_2 l_{it} + \beta_3 \ln K_{it} + \beta_4 \ln P_{it} + \varepsilon_{it} \tag{18}$$

$$\ln y_{it} = \gamma_0 + \gamma_1 \ln y_{it-1} + \gamma_2 l_{it} + \gamma_3 \ln K_{it} + \gamma_4 \ln P_{it} + \varepsilon_{it} \tag{19}$$

The empirical model studies the net floating population to identify and measure the scale effect and structural effect of the floating population on regional economic growth. The net mobility parameter β_2 in Formula (18) reflects the scale effect of the floating population on economic development, and the net mobility parameter γ_2 in Formula (19) is used to identify its structural effect on economic development. When $\beta_2 > 0$, it is explained that the net floating population leads to an increase in the regional population and increases the total regional output, as a scale effect. When $\gamma_2 > 0$, it is explained that the net floating population increases the population base, and the per capita output is due to the increase in the production efficiency of the inflowing population, reflected as a structural effect.

According to the above theoretical derivation process, especially Formulas (16) and (17), taking into account the endogeneity problems caused by omitted variables and the inertial effect of economic growth, the first-order lagged explained variable is introduced into the explanatory variables to construct a dynamic panel of data models (Formulas (18) and (19)).

$$\ln Y_{it} = \beta_0 + \beta_1 \ln Y_{it-1} + \beta_2 l1_{it} + \beta_3 l2_{it} + \beta_4 \ln K_{it} + \beta_5 \ln P_{it} + \varepsilon_{it} \tag{20}$$

$$\ln y_{it} = \gamma_0 + \gamma_1 \ln y_{it-1} + \gamma_2 l1_{it} + \gamma_3 l2_{it} + \gamma_4 \ln K_{it} + \gamma_5 \ln P_{it} + \varepsilon_{it} \tag{21}$$

The empirical model studies the variables of the in-migration population and out-migration population to identify and measure the scale effect and structural effect of the floating population on regional economic growth. The inflow rate (β_2) and outflow rate (β_3) in Formula (20) reflect the scale effect of the floating population on economic development. Similarly, the inflow rate (γ_2) and outflow rate (γ_3) in Formula (21) are also used to identify its structural effect on economic development. Taking the inter-provincial population inflow rate as an example, if $\beta_2 > 0$, it is explained that the population inflow leads to an increase in the regional population and a further increase in the total regional output, revealing a scale effect. If $\gamma_2 > 0$, it is explained that the population inflow increases the population base, and the per capita output will be due to the inflow of people. Due to the increase in production efficiency, this reflects a structural effect.

2.3.2. Analysis of Variance

In order to further examine whether the floating population has a differential impact on targeted areas with different levels of urban density, this study introduces the intersectional terms of floating population and urban density for analysis. Urban density refers to the density characteristics and configuration intensity of various urban elements in spatial distribution. To a certain extent, it not only measures the coordination of land resource supply and demand, but is also an important indicator of the level of urbanization to comprehensively evaluate the development trend of a city [32]. This study divided 31 provinces and cities into two types of areas based on measurement indicators such as population density, economic density, and building density, namely high-density urban areas and low-density urban areas [33,34]. At the same time, dummy variables were introduced, meaning that that if the province was a high-density area, the value was 1, and otherwise it was 0. Unbalanced panel data were constructed, and the above empirical models (Formulas (20) and (21)) were expanded to obtain the following four empirical models (Formulas (22)–(25)).

$$\ln Y_{it} = \beta_0 + \beta_1 l1_{it} + \beta_2 l1 \cdot density_{it} + \beta_3 density + \beta_4 \ln K_{it} + \beta_5 \ln P_{it} + \varepsilon_{it} \quad (22)$$

$$\ln Y_{it} = \lambda_0 + \lambda_1 l2_{it} + \lambda_2 l2 \cdot density_{it} + \lambda_3 density + \lambda_4 \ln K_{it} + \lambda_5 \ln P_{it} + \varepsilon_{it} \quad (23)$$

$$\ln y_{it} = \beta_0 + \beta_1 l1_{it} + \beta_2 l1 \cdot density_{it} + \beta_3 density + \beta_4 \ln K_{it} + \beta_5 \ln P_{it} + \varepsilon_{it} \quad (24)$$

$$\ln y_{it} = \lambda_0 + \lambda_1 l2_{it} + \lambda_2 l2 \cdot density_{it} + \lambda_3 density + \lambda_4 \ln K_{it} + \lambda_5 \ln P_{it} + \varepsilon_{it} \quad (25)$$

where $\ln Y_{it}$ represents the total regional output, $\ln y_{it}$ represents the per capita output, K_{it} represents asset investment, P_{it} represents the total regional population, $l1 \cdot density_{it}$ represents the intersection of population outflow rate and urban density, abd $l2 \cdot density_{it}$ represents the intersection of population outflow rate and urban density, among which Formulas (22) and (23) introduce the inflow model and empirical model while Formulas (24) and (25) introduce the outflow rate and the cross term between the outflow rate and the development level.

2.4. Statistics and Tools Used

In this study, Figures 1–3 were produced on the Arc-GIS platform. All data were analyzed using Stata 17 software and Microsoft Office 2010.

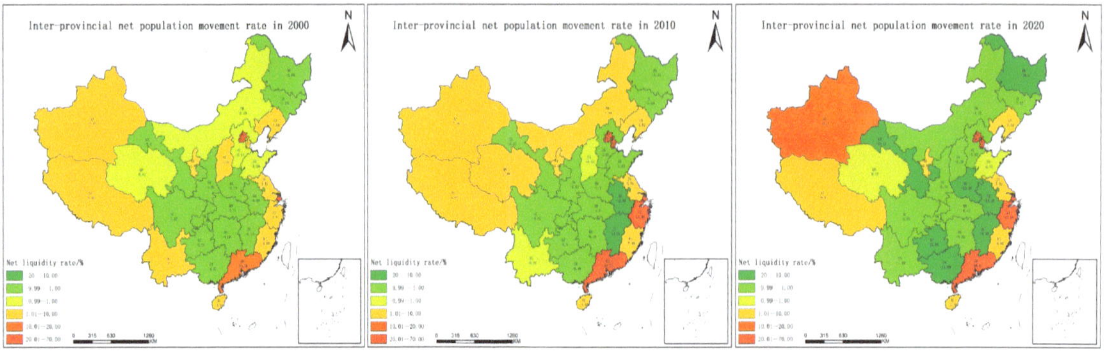

Figure 1. Comparison of inter-provincial net population movement rates (%) from 2000 to 2020.

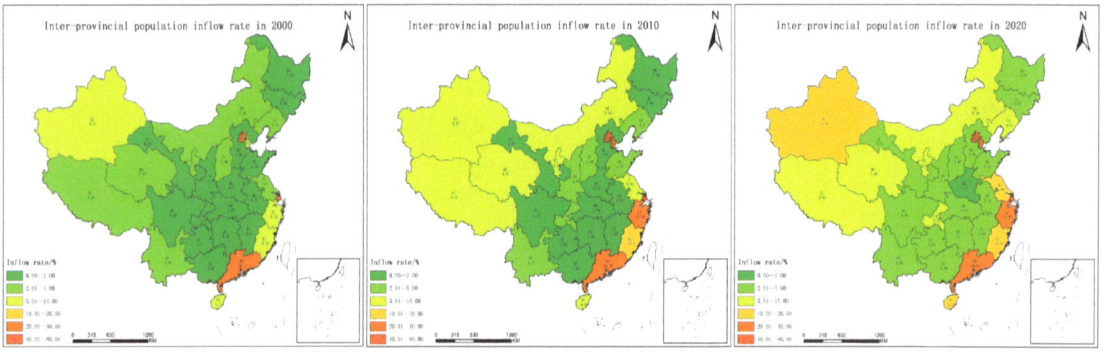

Figure 2. Comparison of cross-provincial population inflow rates (%) from 2000 to 2020.

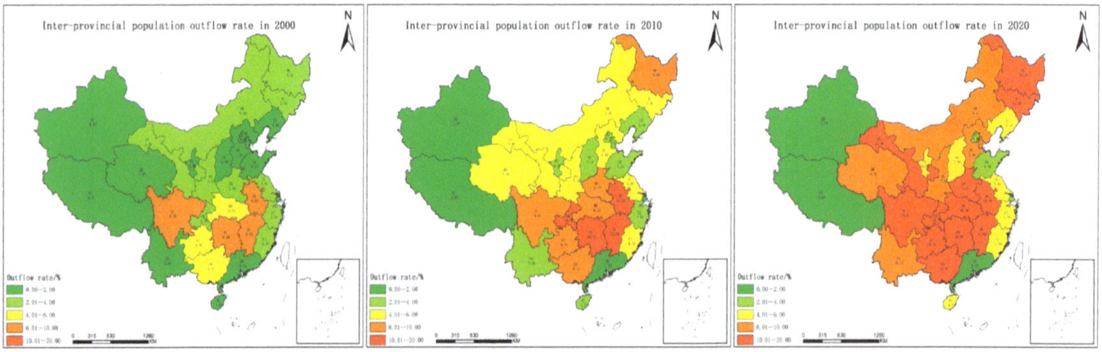

Figure 3. Comparison of inter-provincial population outflow rates (%) from 2000 to 2020.

3. Results and Discussion

3.1. Calculation of Inter-Provincial Floating Population Data

The three-year cross-provincial population net flow, inflow, and outflow distribution maps were drawn using Arc-GIS 10.8 software. In order to facilitate the comparison of the temporal and spatial change characteristics of the floating population between 2000 and 2020, the same scale was used for classification, and the inter-provincial net migration rate, inflow rate, and outflow rate of the three years were classified. The net population flow rate revealed the indicators of population loss and the net increase in the process of population migration in each province. Figure 1 presents the net population flow of each province separately and provides an overview of the spatial distribution of the inter-provincial net population flow rate across the country. The 31 provinces are divided into 6 levels. From 2000 to 2020, China's inter-provincial population flow increased from 42.41 million to 124.83 million, indicating that the scale of China's floating population is constantly increasing [35]. During this period, the spatial pattern of China's population flow did not change significantly, and the overall situation was "low in the middle and high at both ends". The net inflow of inter-provincial floating population is mainly concentrated in two regions: one is the high-density southeast coastal areas, including Shanghai, Beijing, Zhejiang, Guangdong, Tianjin, Jiangsu, and Fujian, because population mobility has obvious cohesion, and the attractiveness of the eastern region to migrant populations still has an absolute advantage [36]; and second is Xinjiang and Tibet, located in the west. With 2008 as the time node, the implementation of the Western development policy accelerated the economic growth of the Western region and triggered a large population inflow [37].

Notably, most of the central regions and provinces show varying degrees of persistent net population outflow losses, such as Anhui, Jiangxi, and Hunan. The main reason for this phenomenon is that since the reform and opening up of China, the economic development gap between regions in China has gradually widened, and the rapid economic development of the eastern coastal areas has created more employment opportunities and higher wages, thereby fostering a large population. For the economically less developed central region, the gap in employment opportunities, wages, education, and healthcare compared to the eastern region is becoming increasingly evident [14]. The phenomenon of population loss is serious, and the spatial distribution presents an obvious "central depression". However, Xinjiang and Tibet in the west have extremely low population outflow rates and high population inflow rates, resulting in the highest inter-provincial net population migration rates in the country. This may be due to the relatively closed traffic environment that caused a low outflow of residents from Tibet and Xinjiang. At the same time, the implementation of China's Western development strategy and the pull of rich natural and mineral resources attracted population inflows, leading to this phenomenon. At the same time, simply using the net flow rate to represent the regional population flow is not comprehensive enough to explore its relationship with economic development. Therefore, this paper divides the net turnover rate into two parts, the inflow rate and the outflow rate, to further explore the relationship between spatial patterns and regional economic development. Figure 2 depicts the inter-provincial population inflow from 2000 to 2020 and divides the 31 provinces into 6 levels.

Figure 2 reveals that Beijing, Tianjin, Zhejiang, Guangdong, and Hainan are the main choices for population inflows. However, changing trends can also be observed in Heilongjiang, Jilin, Liaoning, and other northeastern regions, as well as in Gansu, Sichuan, Guizhou, Hunan, Jiangxi, Anhui, and Shandong. The inflow rate of the population in central and western regions such as Hebei and Hebei has increased significantly, which has played a role in diversion and return. This is reflected in the seventh national population dataset. The data of the seventh national census show that China's inter-provincial population mobility model is still mainly focused on the eastern region, but the proportion has decreased from 84.6% (2005) to 73.54% (2020), a decrease of about 11 percentage points, of which the central region (including the northeast region) and the western region have absorbed 6.5% and 3.5%, respectively [37]. This is mainly because the adjustment of the industrial structure in the eastern region has shifted resource-intensive and labor-intensive industries to the central and western regions, causing this part of the workforce to flow to the Midwest, forming the role of diversion and return, representing new characteristics of the return of labor from the eastern coastal areas to the central and western regions [38]. In addition, the calculation shows that the inter-provincial floating population concentration index dropped from 51.73% in 2000 to 46.60% in 2020, which also proves that the inter-provincial population inflows show a decentralization trend. Figure 3 displays the inter-provincial population outflow from 2000 to 2020 and divides the 31 provinces into 5 levels.

Figure 3 reveals that the population loss in inland and some western provinces is still the most serious and presents a continuous loss situation. Among them, Anhui is the province with the largest population loss, followed by Jiangxi and Guizhou, but the overall distribution pattern of the inter-provincial outflow population only changes slightly. The basic situation of the central and western regions still being the main sources of population outflow has not changed.

In addition, by comparing Figures 2 and 3, it can be found that the ranges of inter-provincial inflow rates from 2000 to 2020 were 23.24%, 62.72%, and 69.91%, and the ranges of outflow rates were 8.17%, 12.99%, and 17.90%, respectively. The extreme difference between the outflow rate and the inflow rate shows a significant upward trend, indicating that the inter-provincial population outflow and inflow gap between different regions in China is increasing year by year. This gap is much larger than the population outflow, which also shows that the change in China's net population flow rate depends to a large extent on the change in the inter-provincial population inflow rate.

3.2. Data Verification

To verify whether the calculation results are in line with the real situation of each province, this study selects the sixth census dataset and uses the above method to calculate the inter-provincial outflow population in 2010, and then calculates the difference between the calculated value and the real value. Through calculation, the total difference between the real value and the estimated value of the outflow population this year is −6.723%, the average error of each province is about −0.216%, and the estimated error range is probably controlled between −8.5% and 5.6%. It can be concluded that the test data error is generally controlled within the effective range, and the measured data are basically consistent with the real data as a whole, so the calculation method is practical and effective.

3.3. Impact of Inter-Provincial Floating Population on Regional Economic Development

3.3.1. Descriptive Statistics of Variables

This study provides descriptive statistics on the variables in the measurement model (Table 1).

Table 1. Descriptive statistical analysis.

Variables	Name	Mean	Standard Deviation	Minimum Value	Maximum Value
Explained variable	Regional total output lnY (CNY)	8.956	1.404	4.766	11.615
	Output per capita lny (CNY)	10.088	1.026	7.944	12.013
Explanatory variables	Fixed asset investment lnK (100 million CNY)	8.369	1.594	4.159	10.989
	Total regional population lnP (number)	17.288	0.893	14.778	18.651
	Inter-provincial inflow rate 11 (%)	7.149	9.089	0.390	42.136
	Inter-provincial outflow rate 12 (%)	5.741	4.855	0.193	21.521
Instrumental variables	Regional average salary lnw (CNY)	10.788	0.506	10.231	12.091
	* Industrial structure lns (%)	3.731	0.238	3.242	4.429

* Industrial structure is obtained by the ratio of the output value of the tertiary industry to the total regional production industry.

3.3.2. Regression Analysis of the Net Floating Population and Economic Development

Comparing the fixed -effect and random-effect methods through the Hausman test, the p value obtained is 0.6284, which indicates that the random effects method is more effective than the fixed effects method. However, the LM test is used to differentiate between mixed regression and random effects. The p-value of the LM test is 0.3409, indicating that the null hypothesis of "there is no individual random effect" cannot be rejected, i.e., mixed effects are considered to be chosen between random effects and mixed effects. Similarly, when comparing fixed effects and mixed regression, the F test p value is 0.447, and the null hypothesis cannot be rejected, i.e., mixed regression is considered to be significantly better than fixed regression. Furthermore, using the clustering robust error LSDV method, we found that the *p*-values of most individual dummy variables are large, so mixed regression should be used.

In order to overcome the endogeneity problem, which leads to biased estimation, the model is further estimated using the generalized moment estimation method (GMM). The premise of using the instrumental variable method is that the instrumental variable is valid. For this purpose, an over-identification test is performed to examine whether the instrumental variable is exogenous—that is, it has nothing to do with the disturbance term. The over-identification test shows that the *p*-value results are 0.6795 and 0.6246, indicating that the instrumental variables are all exogenous. The specific calculation results of Stata software are shown in Table 2.

Table 2. Regression analysis of net floating population and economic development.

Index	$\ln Y$	$\ln y$
l	0.016 ** (12.549)	0.015 ** (10.562)
$\ln K$	0.758 ** (33.987)	0.764 ** (33.198)
$\ln P$	0.383 ** (12.924)	−0.621 ** (−20.622)
$\ln Y_{t-1}$ / $\ln y_{t-1}$	0.068 * (−2.521)	0.068 * (2.520)
c	−4.064 ** (−10.737)	14.096 ** (36.410)
R^2	0.98	0.964

l is the net population mobility rate, $\ln K$ is the fixed asset investment amount, $\ln P$ is the total regional population, c refers to the constant term, $\ln Y_{t-1}$ indicates the GDP, and $\ln y_{t-1}$ represents the GDP per capita. ** and * indicated in Table 2 represent the significant values at 5% ($p < 0.005$) and 10% ($p < 0.01$), respectively, and the t estimator is in the brackets.

Table 2 reveals that the inter-provincial net mobility rates passed the 5% significance test and have a positive impact on regional total output. The results are the same as their theoretical derivation structures. Similarly, the inter-provincial inflow rate also passed the 5% significance test and has a positive impact on per capita output. The result also shows that the impact of fixed asset investment on regional total output and per capita output passed the 5% significance test, and all regression coefficients are significantly positive, which shows that capital investment has a significant impact on the economy. Comparing the regression coefficients of all other variables, this study finds that fixed asset investment has the greatest impact on regional GDP and per capita output. However, fixed investment has a positive effect on regional economic growth [39]. Moreover, the impact of the total regional population on regional total output and per capita output passed the 5% significance test. Likewise, the results are still consistent with the theoretical model. The total regional population has different effects on regional total output and per capita output. Analysis shows that the total regional population has a positive impact on the total regional output, while it has a significant negative impact on per capita output. The first-order lagged regional total output and per capita output both passed the 10% significance test, and the influence coefficients were both positive.

3.3.3. The Impact of Inter-Provincial Inflows and Outflows on Regional Economic Development

Based on the empirical models in Formulas (20) and (21), we employed Stata software to explore the impact of interprovincial population inflow and outflow rates on regional economic development. The specific results of Stata software are shown in Table 3.

Table 3 reveals that the inter-provincial inflow rates passed the 1% significance test, showing a positive impact on the regional total output. The inter-provincial outflow rates passed the 10% significance test, showing a negative impact on the regional total output. The results of the case analysis are consistent with the theoretical derivation structure is the same. Empirical results based on the assumed parameters of the econometric model (Formula (15)) identify the scale effect of the floating population on regional economic growth—that is, population inflow increases the population in the destination, population agglomeration improves the local labor force structure, improves the stock of human capital, and promotes the upgrading of the regional industrial structure, thus increasing the total output of the region [39], while population outflow reduces the population at the destination, causing the region's total output to decline.

Table 3. Regression analysis of floating population and economic development.

Index	ln*Y*	ln*ry*
*I*1	0.0221 *** (0.00361)	0.0113 *** (0.00265)
*I*2	−0.00686 * (0.00474)	−0.0110 * (0.00532)
$\ln K$	0.418 *** (0.0612)	0.401 *** (0.0667)
$\ln P$	0.406 *** (0.0511)	−0.318 *** (0.0622)
$\ln Y_{t-1}$ / $\ln y_{t-1}$	0.263 *** (0.0511)	0.305 *** (0.0562)
c	−3.648 *** (0.432)	9.517 *** (0.955)
R^2	0.9860	0.9505

*I*1 refers to the population inflow rate, *I*2 indicates the population outflow rate, $\ln K$ is the fixed asset investment amount, $\ln P$ is the total regional population, and $\ln Y_{t-1}/\ln y_{t-1}$ indicates the ratio of GDP/GDP per capita. *** and * indicated in Table 3 represent the significant values at 1% ($p < 0.001$) and 10% ($p < 0.01$), respectively, and the t estimator is in the brackets.

Similarly, the inter-provincial inflow rates passed the 1% significance test, showing a positive impact on per capita output. The inter-provincial outflow rates passed the 10% significance test, showing a negative impact on per capita output. The econometric model (Formula (16)) and empirical results show that population inflow and outflow have structural effects, i.e., the inflow and outflow of people change the regional population, causing the total regional output to increase or decrease. At the same time, the structural differences in the floating population are accurately identified by the example model. An increase in the inflow population will increase the production efficiency of the place of origin, resulting in an increase in per capita output, while the outflow population will reduce the population base of the place of departure, but it will not increase the per capita output. The structural difference in high production efficiency causes this part of the outflow population to have a negative impact on the per capita output of the place of departure.

Table 3 further reveals that the impact of fixed asset investment on regional total output and per capita output passed the 1% significance test, and all regression coefficients are significantly positive, which shows that capital investment will have a significant impact on the economy. Comparing the regression coefficients of all other variables, we can find that fixed asset investment has the greatest impact on regional GDP and per capita output. Moreover, the impact of the total regional population on regional total output and per capita output passed the 1% significance test. The results are still consistent with the theoretical model. The total regional population has different effects on regional total output and per capita output. The impact analysis shows that the total regional population has a positive impact on the total regional output, while it has a significant negative impact on per capita output. Furthermore, the first-order lagged regional total output and per capita output both passed the 1% significance test, and the influence coefficients were both positive. This shows that the higher the total output of the previous period, the higher the total output of the current period, which means that the inertial effect of economic growth is significant, and it also means that the more economically developed regions have inherent advantages over the economies of backward regions.

There is a certain two-way effect between regional population mobility and economic development. Population inflow has a significant promotional effect on the economic development of the place of inflow. The large-scale population inflow means the complement and enrichment of human resources in the region, driving regional economic development. At the same time, Ravenstein [21] proposed the law of population migration and mentioned that economic reasons are the main causes of population migration, and economic development brings a large number of job opportunities and makes it easier to attract population inflow. This easily creates a siphon effect where more and more people gather in these high-density urban areas. The lack of human resources caused by population outflow in the outflow area will inhibit the region's economic development. There may be a vicious cycle between continued population outflow and economic development, where the low capital output rate leads to brain drain, further intensifying the vicious cycle [40].

3.3.4. Differential Impact of Inter-Provincial Floating Population on Regional Economic Development

In order to further examine whether the floating population has differential impacts on regions with different levels of economic development [41], this study introduces the cross-term of mobility variables and urban density for analysis. Based on empirical formulas (Formulas (22)–(25)), the calculated results using Stata software are shown in Table 4.

Table 4. Analysis of cross-terms between floating population and urban density.

Index	ln*Y*		ln*y*	
	Equation (1)	**Equation (2)**	**Equation (3)**	**Equation (4)**
ln*K*	0.398 *** (0.0641)	0.398 *** (0.0798)	0.367 *** (0.0622)	0.373 ** (0.0985)
ln*P*	0.347 *** (0.0476)	0.319 *** (0.0376)	−0.327 *** (0.0640)	−0.229 ** (0.0872)
$\ln Y_{t-1}/\ln y_{t-1}$	0.301 *** (0.0497)	0.427 *** (0.0740)	0.328 *** (0.0486)	0.467 *** (0.0738)
*l*1	0.0107 (0.0701)		0.000223 (0.00660)	
*l*2		−0.0337 *** (0.00701)		−0.0334 *** (0.00617)
l1density	0.00715 (0.00681)		0.0113 (0.00609)	
l2density		0.0189 *** (0.00109)		0.0179 *** (0.00458)
c	−2.347 *** (0.607)	−2.071 *** (0.545)	9.654 *** (0.978)	7.692 *** (1.227)
R^2	0.981	0.983	0.939	0.921

ln*K* refers to the fixed asset investment amount, ln*P* is the total regional population, $\ln Y_{t-1}/\ln y_{t-1}$ indicates the ratios of GDP/GDP per capita, *l*1 is the population inflow rate, *l*2 is the population outflow rate, *l1density* is the intersection of inflow population and urban density, and *l2density* is the intersection of outflow population and urban density. *** and ** indicated in Table 4 represent the significant values at 1% ($p < 0.001$) and 5% ($p < 0.05$), respectively, and the t estimator is in the brackets.

Equations (1) and (3) in Table 4 show the cross-term of the inter-provincial inflow rate and urban density and indicate whether it is in the model with regional total output as the explained variable or in the model with per capita output as the explained variable. Table 4 reveals that there is no significant difference in the impact of population inflow on economic development between high-density areas and low-density areas. Further, the results of Equation (2) show that the cross-term between the inter-provincial outflow rate and urban density is significant both in the model with regional total output as the explained variable and in the model with per capita output as the explained variable. However, there is a significant difference in the impact of population outflow on economic development between high-density and low-density areas. Among them, the negative impact of out-migration on low-density areas is greater than that on high-density areas. The results of Equation (4) show that the inter-provincial outflow rate has a negative impact on the per capita output level.

The above discussion shows a weak negative impact on high-density areas, and the impact on low-density areas is greater than that on high-density areas. Analysis of other variables like capital investment demonstrates that they have a positive impact on regional total output and per capita output, and an increase in capital will simultaneously promote regional economic development. The total population has a positive impact on regional total output, and per capita output has a negative effect. The lag period shows a significant positive impact on both regional total output and per capita output. Therefore, the conclusions obtained through empirical analysis in this article are consistent with the results derived from the previous theoretical model.

4. Discussion

The southeastern coastal regions of China, such as Guangdong, Jiangsu, and Zhejiang, are major population inflow areas, whereas central regions like Henan and Anhui experience high population outflow. The southeastern regions possess strong economic capabilities and efficient operations, with economic outputs and per capita GDP leading the nation. These regions have diversified economies dominated by manufacturing and services. The Pearl River Delta and Yangtze River Delta are key hubs for manufacturing and high-tech industries, creating substantial employment opportunities. According to the China Statistical Yearbook 2020, Guangdong's tertiary industry employs 60 million people, accounting for 60% of the province's total employment. These regions also have higher wage levels; for instance, the average salary in Guangdong is CNY 89,000, while in Henan, it is only CNY 55,000 [42]. The significant wage disparity drives the labor force to migrate to higher-paying areas for better income and living conditions. Additionally, high GDP and per capita GDP correlate with higher urbanization rates and better urban infrastructure. According to the "China Statistical Yearbook 2020", Guangdong's urbanization rate is 71.03%, Jiangsu's is 72.63%, while Henan's is only 51.66%. High urbanization rates mean better urban infrastructure and public services, such as quality education and healthcare, which attract large population inflows [43,44].

Conversely, central regions like Henan and Anhui exhibit lower per capita GDP—approximately CNY 41,000 in Henan and CNY 52,000 in Anhui—reflecting slower economic development, lower resident income, and poorer living standards. This economic performance leads to insufficient economic attractiveness, compelling the workforce in these regions to migrate to more economically developed areas in search of better opportunities. The relatively low per capita GDP and lagging economic development in these central provinces result in high population outflow rates [45]. Notably, the southeastern coastal regions of China, with their robust economic indicators, diverse industrial structures, high wage levels, and advanced urbanization, serve as significant population inflow areas. In contrast, due to their relatively low per capita GDP and slower economic growth, central provinces like Henan and Anhui experience high population outflow rates.

Interprovincial migrant populations significantly impact the supply and demand dynamics of the labor market in China [15]. Cross-provincial migration enhances productivity and economic efficiency by increasing the labor supply and labor market participation. The influx of lower-human-capital groups fills gaps in low-skilled and labor-intensive jobs in high-density urban areas, promoting regional economic growth [16]. Meanwhile, high-human-capital groups, unable to obtain high wages in low-density urban areas, proactively migrate to high-density urban areas [17]. The diverse backgrounds and advanced skills of these migrants contribute to regional technological innovation, particularly in science and engineering fields, thereby positively impacting economic growth. With favorable economic opportunities and a well-established legal system, human capital significantly boosts per capita GDP growth [46,47]. However, the influx of large numbers of cross-provincial migrants poses significant challenges to infrastructure and services in high-density urban areas. Increased demand for housing, public transportation, and basic services, coupled with environmental pollution due to high population density, can strain urban resources and lead to management crises [48,49]. The urban challenges induced by the migrant population create a "forcing mechanism" that compels local governments to implement policy interventions. The government leverages think tanks and substantial local finances to conduct forward-looking research and planning for future urban development. Through systematic and comprehensive policy measures, the sustainability of urban development can be enhanced. This includes ensuring basic housing conditions, improving public infrastructure services, promoting social inclusion and community integration, and enhancing environmental risk management. These efforts contribute effectively to the achievement of SDG-11. Given the above research findings, this paper puts forward the following policy recommendations:

(1) For the low-density areas in the inland central and western regions, against the background of the two drivers of urbanization and rural revitalization, the government should fully seize the opportunity window of population return and industrial gradient transfer and accelerate the transformation of the economic growth mode in order to achieve the Sustainable Development Goal (SDG-8) based on improving development quality. At the same time, the government must also use policies as "gravity" to implement active and effective talent introduction policies to activate the talent engine, so that returning talents can settle down and stay stable, and eliminate the vicious cycle between continuous brain drain and economic development.

(2) For high-density areas along the eastern coast, a two-pronged approach of "introduction" and "diversion" should be advocated for in population control, transferring mid- to low-end industries. Further guiding low-skilled labor to flow to low-density areas with higher environmental carrying capacity to balance population distribution patterns in encouraged in order to alleviate the pressure of overpopulation and strained public resources in high-density urban areas. The government should focus on improving the level of public services, improving urban environmental governance, optimizing the urban living environment, and striving to create green and livable modern low-density cities, which will be conducive to realizing SDG-11.

(3) Household registration and social security policies significantly impact population mobility. Strict household registration systems can restrict the free movement of people between different regions, while relaxing these restrictions can facilitate population mobility. For example, China's ongoing reforms to its household registration system are gradually easing settlement conditions in cities, especially in small and medium-sized cities, to promote population mobility and urbanization. This approach helps to encourage the return of migrant populations to low-density areas and improves the service management level for migrant populations in high-density areas. It ensures that migrants enjoy equal rights with local residents regarding employment, education, healthcare, and housing, thereby reducing institutional barriers.

(4) In the process of policy formulation, the government should thoroughly understand and respect the regularity of population mobility, avoiding excessive intervention. By leveraging data analysis and research, the government needs to identify the economic and social factors influencing population movement and develop policies that align with these patterns to support and encourage voluntary migration. Additionally, considering that economic and social environments are dynamic and that the patterns of population mobility will evolve accordingly, policy formulation must possess dynamic adaptability. It is essential to establish a regular policy evaluation mechanism that involves data collection, analysis, and feedback to assess the actual effectiveness of policies. Based on the evaluation results, policies should be flexibly adjusted to

ensure alignment with real-world conditions, thereby enhancing the sustainability of these policies.

To enhance the robustness and comprehensiveness of the analysis, future research should incorporate policy analysis to examine the impact of specific policy measures on regional migration and economic development, including both restrictive and supportive policies. Longitudinal studies are needed to assess the long-term impacts of migration on regional economies, including the integration of migrants and the sustainability of economic growth. Utilizing micro-level data to capture individual migrant experiences and regional variations would provide a more nuanced understanding of migration impacts. Additionally, exploring the long-term effects of the COVID-19 pandemic on regional migration patterns and economic recovery, considering both immediate disruptions and longer-term adaptations, is crucial [50].

5. Conclusions and Limitations

Based on the data of the fifth, sixth, and seventh national censuses, this paper studies the temporal and spatial characteristics of the inter-provincial floating population through theoretical model derivation, a GMM model, and the cross-term method for the 31 provincial-level administrative units in mainland China, except for Hong Kong, Macao, and Taiwan. Based on the empirical results, the following conclusions have been drawn:

(1) The size of the migrant population under study is constantly increasing, but the spatial pattern of population mobility has not changed significantly. The overall pattern is still "low in the middle and high at both ends", although the central region is still the main source of population outflow.
(2) The net inflow of inter-provincial migrants is mainly concentrated in two areas—firstly, the high-density area along the southeastern coast, and secondly, the western regions such as Xinjiang and Tibet—while most central provinces show varying degrees of sustained net population outflow losses.
(3) The net migrant population has a significant promoting effect on the total regional output and per capita output. The inflow population has a significant promoting effect on the total output and per capita output of the place of inflow.
(4) The outflow population has a significant effect on the total output and per capita output of the place of outflow. Migration shows a significant inhibitory effect.
(5) The floating population has scale and structural effects on regional economic development. It affects regional economic development by changing the regional population and affecting regional production efficiency.
(6) There is a certain two-way effect between regional population mobility and economic development. The population inflow has no significant differential impact on high-density and low-density areas, while the out-migration population has a significant differential impact.
(7) The negative impact of the out-migration population on low-density areas is greater than that on high-density areas. Due to the lack of talent attraction in low-density areas, the outflow of talent reduces human capital. For high-density areas, a small portion of the out-migration population will appropriately alleviate urban pressure and reduce local fiscal expenditures, thus causing the out-migration population to have a weak negative impact on high-density areas.

Overall, the inter-provincial floating population in the studied region of China plays a crucial role in regional economic development, directly supporting SDG-8 and 11. By facilitating labor mobility, fostering economic growth, reducing inequalities, promoting sustainable urban development, and driving infrastructural improvements, this demographic trend contributes significantly to achieving the Sustainable Development Goals. This study has several limitations. Firstly, it does not consider the impact of various policy measures on regional migration patterns and economic outcomes. Policies like the Hukou system, urban development strategies, and pandemic-related restrictions can significantly influence migration flows and regional economic dynamics. Secondly, the analysis primarily focuses on the immediate effects of inter-provincial migration on economic growth, without thoroughly examining long-term effects, including the sustainability of growth and the integration of migrants into urban economies. Thirdly, the study period does not fully capture the long-term economic and demographic consequences of the COVID-19 pandemic, which introduced unprecedented disruptions to population mobility and economic activities. Finally, using macroeconomic data may overlook micro-level variations and the individual experiences of migrants, which can provide deeper insights into migration's economic and social implications.

Author Contributions: Z.C. and Z.L. were primarily responsible for writing the manuscript and providing the original data used in this manuscript. K.Z. provided the results of the empirical analysis for the research in this paper. All authors contributed to the article and approved the submitted version. All authors have read and agreed to the published version of the manuscript.

Funding: This work was funded by the Fundamental Research Funds for the Central Universities (B240207112&B240207032).

Institutional Review Board Statement: Not applicable.

Informed Consent Statement: Not applicable.

Data Availability Statement: The data and materials are available from the corresponding author upon reasonable request.

Conflicts of Interest: The authors declare that the research was conducted in the absence of any commercial or financial relationships that could be construed as a potential conflict of interest.

References

1. Chen, S.; Cao, Z. An Analysis of the Concept and Types of Climate Migration. *China Popul. Resour. Environ.* **2012**, *22*, 164–169.
2. Van der Geest, K.; de Sherbinin, A.; Gemenne, F.; Warner, K. Editorial: Climate migration research and policy connections: Progress since the foresight report. *Front. Clim.* **2023**, *5*, 1231679. [CrossRef]
3. Jolivet, D.; Fransen, S.; Adger, W.N.; Fábos, A.; Abu, M.; Allen, C.; Boyd, E.; Carr, E.R.; Codjoe, S.N.; Gavonel, M.F.; et al. COVID-19 responses restricted abilities and aspirations for mobility and migration: Insights from diverse cities in four continents. *Humanit. Soc. Sci. Commun.* **2023**, *10*, 250. [CrossRef] [PubMed]
4. Qing, H.; Kumar, R.; Kumar, A. Climate change and human migration: Perspectives for environmentally sustainable societies. *J. Geochem. Explor.* **2023**, *256*, 107352. [CrossRef]
5. Haas, H.D. Migration and Development: A Theoretical Perspective1. *Int. Migr. Rev.* **2010**, *1*, 227–264. [CrossRef]
6. Yang, C.G.; Zeng, Y.M. Spatial Imbalance, Population Flow and Regional Selection of Foreign Direct Investment—Analysis of Interprovincial Spatial Panel Data in China from 1995 to 2010. *Popul. Res.* **2014**, *6*, 25–39.
7. Mcleman, R. International migration and climate adaptation in an era of hardening borders. *Nat. Clim. Chang.* **2019**, *9*, 911–918. [CrossRef]
8. Jiang, M.; Lu, X. Growth in the Midst of Loss: A Study on Population Mobility and Economic Growth Convergence in Small and Medium-sized Cities. *Mod. Econ. Res.* **2024**, *7*, 17–29. [CrossRef]
9. Duan, C.R.; Xie, D.H.; Lu, L.D. Migration and transformation of China's population. *Popul. Res.* **2019**, *43*, 12–20.
10. Ma, X.H.; Duan, C.R.; Guo, J. Comparative study of four types of floating population. *Chin. Popul. Sci.* **2014**, *5*, 36–46, 126–127.
11. Ma, J.L.; Li, L. Empirical analysis of the relationship between population mobility and regional economic development—Taking Guyuan CNY City, Ningxia as an example. *Northwest Popul.* **2006**, 19–20.
12. Wang, S.B.; Luo, X.L. Population mobility and urbanization effects in underdeveloped areas—Taking Gansu Province as an example. *Urban Probl.* **2022**, *324*, 4–11.
13. Lin, L.Y.; Zhu, Y.; Ke, W.Q. The spatial willingness and policy implications of the return of floating population under the background of coordinated regional development. *Geogr. Res.* **2021**, *40*, 1515–1528.
14. Xiao, J.C.; Hong, H. The evolution trend of my country's inter-provincial population mobility pattern and its urbanization effect. *Urban Probl.* **2020**, *8*, 22–32.
15. Chen, J.; Fan, C.C. China's floating population and its implications for regional development. *Asian Surv.* **2016**, *56*, 529–557.
16. Cai, F. The growth and structural changes of China's employment. *J. Comp. Econ.* **2011**, *39*, 42–57.
17. Chan, K.W. China: Internal Migration. In *The Encyclopedia of Global Human Migration*; Ness, I., Ed.; Wiley-Blackwell: Hoboken, NJ, USA, 2013. [CrossRef]
18. Zhao, Z. Migration and earnings differences: The case of rural China. *Econ. Dev. Cult. Chang.* **1999**, *47*, 767–782. [CrossRef]
19. Francesco, C. Drivers of migration: Why do people move? *J. Travel Med.* **2018**, *25*, tay040. [CrossRef]
20. D'Adamo, I.; Di Carlo, C.; Gastaldi, M.; Rossi, E.N.; Uricchio, A.F. Economic Performance, Environmental Protection and Social Progress: A Cluster Analysis Comparison towards Sustainable Development. *Sustainability* **2024**, *16*, 5049. [CrossRef]

21. Ravenstein, E.G. The laws of migration. *J. Stat. Soc. Lond.* **1885**, *48*, 167–235. [CrossRef]
22. Lee, E.S. A theory of migration. *Demography* **1966**, *3*, 47–57. [CrossRef]
23. Kazlauskiene, A.; Rinkevicius, L. Lithuanian "brain drain" causes: Push and pull factors. *Eng. Econ.* **2006**, *46*, 27–37.
24. Khan, I.; Alharthi, M.; Haque, A.; Illiyan, A. Statistical analysis of push and pull factors of migration: A case study of India. *J. King Saud Univ. Sci.* **2023**, *35*, 102859. [CrossRef]
25. Hear, V.N.; Bakewell, O.; Long, K. Push- Pull Plus: Reconsidering the Drivers of Migration. *J. Ethn. Migr. Stud.* **2018**, *6*, 927–944. [CrossRef]
26. Yang, D.L.; Li, P.G. The Economic Effect of Population Agglomeration: An Empirical Study Based on Instrumental Variables. *Popul. Sci. J.* **2019**, *3*, 28–37.
27. Du, X.M.; Chen, J.B. An Empirical Analysis of the Impact of Population Migration and Flow on the Economy of Various Regions in my country. *Popul. Res.* **2010**, *3*, 77–88.
28. Yang, L.; Feng, R.; Cai, D. Study on the Spatio-Temporal Convergence Mechanism and Effects of Population Mobility on Regional High-Quality Economic Development. *China Soft Sci.* **2024**, *S1*, 172–181.
29. Qiao, X.C.; Huang, Y.H. The Situation of Trans-provincial Floating Population in China—Analysis Based on "Six Census" Data. *Popul. Dev.* **2013**, *1*, 13–28.
30. Sun, J.G.; Shi, T.H.; Xu, Q.Q. Population Migration, Wage Changes and Economic Growth—Spatial Econometric Analysis Based on Interprovincial Panel Data. *Popul. Dev.* **2021**, *4*, 14–23.
31. Lu, F.G. Has Population Loss Affected Economic Growth in Northeast China?—Based on the Estimation Data of Household Population Loss in Northeast China. *Popul. Dev.* **2021**, *5*, 98–110.
32. Wang, H.; Yang, M. Review of high-density urban research and analysis of quantitative indicators. *Shandong For. Sci. Technol.* **2023**, *53*, 116–122.
33. Wang, J.J.; Zhang, M.H.; Wang, N.N. Spatial patterns and influencing factors of the distribution of mobile population in China: A study based on county-level census data. *J. Popul.* **2023**, *45*, 82–96. [CrossRef]
34. Wang, G.X.; Li, M. The Spatial Interaction Between Inter-provincial Migration and Manufacturing Industry Transfer. *Sci. Geogr. Sin.* **2019**, *39*, 183–194. [CrossRef]
35. Dou, X.; Arellano, B.; Roca, J. China's inter-provincial population flow based on the interaction value analysis. *Geogr. Res.* **2018**, *37*, 1848–1861.
36. Hou, Y.F.; Chen, Z.C. Population migration-economic growth convergence puzzle'in China: Based on analysis and testing of neoclassical endogenous economic growth model. *China Popul. Resour. Environ.* **2016**, *26*, 11–19.
37. Zhou, H. Stability and Enlightenment of China's Population Mobility Model—Reflections Based on the Data of the Seventh National Census Bulletin. *China Popul. Sci.* **2021**, 28–41+126–127.
38. He, X.L.; Shi, S.Y. The impact of population mobility on regional economic growth: An empirical analysis based on China's prefecture-level city panel data. *Financ. Econ.* **2021**, *3*, 63–70. [CrossRef]
39. Su, W.Z.; Shen, H.Y. Empirical test of population mobility on economic growth in provincial regions. *Stat. Decis.-Mak.* **2017**, 106–110. [CrossRef]
40. Huang, R.; Liang, Q.J.; Lu, L.C. The relationship between urban population structure and innovation ability—Based on the empirical analysis of Chinese cities. *Urban Dev. Res.* **2014**, *21*, 84–91.
41. Cai, X.; Wang, D. The Sustainability of China's Economic Growth and Labor Contribution. *Econ. Res. J.* **1999**, *10*, 62–68.
42. Wang, S.J.; Wang, Z.C. Research on the spatial consistency of population agglomeration and economic agglomeration in China. *Demogr. J.* **2017**, *39*, 43–50. [CrossRef]
43. Haider, A.; Jabeen, S.; Rankaduwa, W.; Shaheen, F. The Nexus between Employment and Economic Growth: A Cross-Country Analysis. *Sustainability* **2023**, *15*, 11955. [CrossRef]
44. Fleuret, S.; Atkinson, S. *Sustainable Cities, Quality of Life, and Mobility-Related Happiness*; Springer: Cham, Switzerland, 2023. [CrossRef]
45. Borjas, G.J.; Monras, J. Immigration and the Dynamics of Urban Labor Markets. *J. Econ. Geogr.* **2017**, *17*, 503–533. [CrossRef]
46. Rodrik, D. Populism and the economics of globalization. *J. Int. Bus. Policy* **2018**, *1*, 12–33. [CrossRef]
47. Ali, M.; Egbetokun, A.; Memon, M.H. Human Capital, Social Capabilities and Economic Growth. *Economies* **2018**, *6*, 2. [CrossRef]
48. Glaeser, E.L.; Mare, D.C. Cities and Skills. *J. Labor Econ.* **2001**, *19*, 316–342. [CrossRef]
49. Florida, R.; Mellander, C.; Rentfrow, P.J. The Happiness of Cities. *Reg. Stud.* **2011**, *47*, 613–627. [CrossRef]
50. Yang, M.; Xie, Z.Y. Impacts of Fighting COVID-19 on China's Population Flows: An Empirical Study Based on Baidu Migration Big Data. *Popul. Res.* **2020**, *44*, 74–88.

Disclaimer/Publisher's Note: The statements, opinions and data contained in all publications are solely those of the individual author(s) and contributor(s) and not of MDPI and/or the editor(s). MDPI and/or the editor(s) disclaim responsibility for any injury to people or property resulting from any ideas, methods, instructions or products referred to in the content.

sustainability MDPI

Article

Meaningful Youth Engagement in Sustainability Processes in Japan and Finland: A Comparative Assessment

Miki Fukuda * and Eric Zusman

Institute for Global Environmental Strategies, Hayama 240-0115, Japan; zusman@iges.or.jp
* Correspondence: miki.fukuda@iges.or.jp

Abstract: In many countries, young people are justifiably viewed as possessing the energy and ideas required to advance the sustainability agenda. However, the degree to which youth can influence that agenda depends on how meaningfully they engage in Sustainable Development Goals (SDGs) and related processes. Further, the extent to which they can meaningfully engage may vary across countries. Though studies have shed light on the core elements of meaningful youth engagement, those insights have not been used to compare how these elements vary across countries. The paper aims to fill this knowledge gap with a preliminary assessment of youth engagement in sustainability processes in Finland and Japan. The approach used herein draws upon relevant literature to construct an analytical framework comprised of four key elements underpinning meaningful youth engagement: (1) aims and justifications; (2) power-sharing; (3) transparency and accountability; and (4) support. The study then employs text mining, institutional analysis, and key informant interviews to offer a preliminary assessment of how meaningfully youth have engaged in sustainability processes in Finland and Japan. The assessment reveals that youth engagement mechanisms in Finland more clearly explicate aims and justifications, balance power dynamics, and enhance transparency and accountability than in Japan. Both countries could do more to offer support to young people. The article also suggests that additional research is needed on three areas: (1) the effect of underlying socio-cultural differences on youth engagement; (2) the relationship between youth engagement and the performance of SDGs 4, 7, and 13; and (3) the role of education as an upstream enabler of engagement.

Keywords: sustainability; Sustainable Development Goals (SDGs); youth; participation; stakeholder engagement

Citation: Fukuda, M.; Zusman, E. Meaningful Youth Engagement in Sustainability Processes in Japan and Finland: A Comparative Assessment. *Sustainability* **2024**, *16*, 6415. https://doi.org/10.3390/su16156415

Academic Editors: Idiano D'Adamo, Massimo Gastaldi and Manoj Kumar Nallapaneni

Received: 9 January 2024
Revised: 18 July 2024
Accepted: 18 July 2024
Published: 26 July 2024

Copyright: © 2024 by the authors. Licensee MDPI, Basel, Switzerland. This article is an open access article distributed under the terms and conditions of the Creative Commons Attribution (CC BY) license (https://creativecommons.org/licenses/by/4.0/).

1. Introduction

Recent years have made it abundantly clear that the world is not on track to achieve the 17 Sustainable Development Goals (SDGs) that sit at the heart of the 2030 Agenda for Sustainable Development (2030 Agenda). At the midpoint of the 2030 Agenda, there is a growing consensus that the world needs a course change to accelerate progress on the SDGs. The engagement of multiple, diverse stakeholders could arguably inject the energy and ideas needed to trigger this course change. Among different stakeholders that could lead this change, much of the attention has rightly focused on meaningfully engaging the next generation—that is, young people.

Youth participation has been invoked in intergovernmental agreements since the adoption of the UN Convention on the Rights of the Child in 1989 [1]. That foundational agreement underlined that it is children's right to be involved in decisions that affect them and their future. Since that agreement, young people have become increasingly viewed as essential actors and valued participants in advancing the sustainability agenda. Numerous international forums, programs, and resolutions have underlined the growing need for youth engagement—for example, calling for young people's or youth-led organizations' active participation in sustainability processes. Similar calls for the importance of engaging youth can be found in the academic literature on sustainability [2].

At the same time, some additional reflections should inform discussions of youth engagement. Among the reasons for more reflection is that there is no consistently agreed definition of young people; nor are young people a homogenous bloc. On the former point, biological age is often cited as a determinant of youth at both the national and international levels. Yet, even in this case, there exists variation in what constitutes being young—with age cut points ranging from 13 to 35 [3] (p. 17). In addition, young people are very diverse in their backgrounds and identities. Generalization may lead to overlooking and failing to capture the differences in opinions and insights from young people with varying geographical, educational, and socioeconomic backgrounds.

On a related note, there may also be various ways young people participate in sustainability processes. Voting has long been the main means to formally participate in policymaking processes; however, this once widely held view that voting is the chief channel for engagement is becoming less relevant in recent years. It is also less applicable to young people who may be under voting age or more inclined to use other avenues to push for change, especially in progressive policy realms such as sustainability. Instead, there could be multiple forms of participation and varying influences on the decision-making processes in crafting sustainable futures [4,5]. Further, given the diversification of channels, there is also the question of whether the ways that young people engage can have a meaningful effect on sustainability processes. Another pertinent point is that the degree to which young people can meaningfully influence these processes may vary from one country to the next.

The above possibilities suggest a critical question: how do countries compare in terms of meaningful youth engagement in sustainability processes? This paper answers this question with a preliminary assessment of two developed democratic countries: Finland and Japan. This assessment draws on a simple framework comprised of four elements that arguably sit at the core of meaningful youth engagement: (1) aims and justifications; (2) power-sharing; (3) transparency and accountability; and (4) support. It then employs a mixed-method approach to derive some initial insights into how Finland and Japan compare across those elements. This approach entails the text mining of key documents followed by a desk study of official descriptions of key institutions and key informant interviews. The assessment reveals that youth engagement mechanisms in Finland more clearly explicate aims and justifications, balance power dynamics, and enhance transparency and accountability than in Japan. Both countries could do more to offer support to young people.

The remainder of the paper is divided into six sections. The next two sections (Sections 2 and 3) reviews the relevant literature on participation and engagement to develop an analytical framework consisting of four elements: (1) aims and justifications; (2) power-sharing; (3) transparency and accountability; and (4) support. Sections 4 and 5 use this framework to compare Finland and Japan. The final section (Section 6) provides a summary of the paper and recommendations for future work.

2. Literature Review

2.1. Setting the Context

Often, decisions made today affect young people for years to come. In SDG processes, policymakers and practitioners are increasingly acknowledging that these decisions will influence young people into the future. At the same time, there is also a realization that young people are not merely passive supplicants when it comes to those decisions. Rather, they are key stakeholders and future leaders with the potential to shape their future. This growing recognition traces back to the notion of intergenerational equity [6]. That notion was made famous in the 1987 Brundtland Report "Our Common Future" and features in definitions of sustainable development that offer "development that meets the needs of the present without compromising the ability of future generations to meet their own needs".

In the years since the release of the Brundtland Report, landmark agreements would give young people firmer ground to stand upon in sustainability processes. For example, the 1992 Rio Earth Summit adopted Agenda 21, including Chapter 25, titled "Children and Youth in Sustainable Development", which identified young people as pivotal stakeholders in sustainable development. In underlining the central role of the next generation, the chapter clarified that "the involvement of today's youth in environment and development decision-making is critical to the long-term success of Agenda 21". Agenda 21 also led to the creation of nine "Major Groups" that help to organize the different channels for communication and engagement in UN activities related to sustainable development. The Major Group for Children and Youth (MGCY) is one officially recognized group, receiving a mandate from the UN General Assembly and specific entities through agreements and terms of reference.

The 2030 Agenda and its SDGs also recognize the importance of intergenerational equity and acknowledge that the future of humanity and the planet lies in the hands of younger generations. Toward that end, the SDGs developed specific targets relating to youth empowerment, education, capacity building, and (un)employment (e.g., SDG 4; SDG 8.6) and emphasized the importance of participatory and inclusive decision-making and societal development across all ages or other statuses (e.g., SDG 16.7). As implied in the above list of SDGs, young people are not only influenced by the intergenerational disparities that need to be narrowed, they are also expected to shape pathways to a sustainable future or use their unique perspectives to craft creative solutions to thorny problems [7,8]. In addition, young people may hold broader conceptions of sustainability that can help to pull communities and even countries closer together [9]. Further, young people, equipped with sufficient knowledge and skills, can use those broader conceptions to spark positive change in everything from small towns to global arenas [10,11]. As such, there are a growing number of positive examples of youth participation contributing to awareness-raising across generations and community building across regions in sustainability processes [12].

In recent years, the United Nations System has increasingly sought to harness this potential by involving youth in the SDGs and related processes. For instance, the UN has established an office called "the Secretary-General's Envoy on Youth" to foster a collaborative and integrated approach to UN initiatives on youth issues. The UN has supported other similarly focused efforts to expand youth participation, such as "Youth 2030"—an umbrella framework initiated by Secretary-General António Guterres in 2018 to engage young people and solicit ideas on UN activities across three pillars: peace and security, human rights, and sustainable development. In doing so, national governments are urged to make youth participation the norm, rather than the exception, across all decisions, policies, and investments [13]. In response, Member States have been reported to establish a variety of forms to engage young people, including participation in national youth councils, follow-up and review processes, and thematic political consultations. However, there still remains room for improvement when it comes to meaningful engagement [13,14].

2.2. *Meaningful Youth Engagement*

Much of the research on participation comes from political science where there is a long-running emphasis on institutionalized forms of engagement, such as elections. Political science studies often assume that participation entails activities that are intended to directly influence the choice of elected representatives or the organizations behind them [15,16]. However, in recent years, the notion of participation has grown to include forms of engagement beyond the government or formal political institutions while questioning boundaries between political and social spheres. This has led to a broader definition of political participation that stresses activities "indirectly aimed at impacting civil society or attempting to alter systematic patterns of social behavior" [17].

This more expansive view is important when discussing the engagement of young people [18–20]. At the national level, the legal voting age often functions as a threshold for deciding whether young people can participate in activities. Indeed, some young people who are not eligible to vote may need other kinds of institutional mechanisms to participate in public matters. Thus, such a broader view is essential since it is impossible for young people to vote until they reach legal age (though voting rates have also fallen among younger generations). While this declining trend might suggest a disinterest in public affairs, it should not be regarded as prima facie evidence of political apathy [21]. Instead, sociological research on values and behavior changes has revealed that young people opt for alternative and new forms of engagement such as protests, demonstrations, utilization of social networking services (SNSs), consumer actions, and charitable fundraising or volunteer work [22]. Such activities are forms of participation—a form that is neither explicitly social nor political but rather socio-political [23]. If expanding youth engagement is considered a desirable goal, there is also a need to look more at the landscape of means and channels through which young people actively engage [24].

In a similar vein to opening multiple opportunities, young people may need support to enable their involvement in key processes. Such support is warranted because, compared to adult stakeholders, young people may lack confidence and tools or confront barriers to becoming involved. This support could, for instance, prioritize marginalized and vulnerable youth to ensure inclusive decision-making, aiming to "reach the furthest behind first" [25]. Research indicates that active youth are more likely to be from socio-economically advantageous groups, already equipped with the necessary abilities to participate [26,27]. Therefore, empowering youth with knowledge and skills as well as appropriate support is crucial to making their participation meaningful [5,28]. Focusing on the essential elements to make their engagement meaningful is necessary, as surface-level tokenistic participation may actually do more harm than good.

Building upon the above insights, a few studies have looked at what factors affect whether young people meaningfully engage on the SDGs in different contexts. For example, some studies have used survey data from university students in Rome, Italy, to show that taking classes centered on real-world energy projects can motivate greater youth involvement in sustainability initiatives [29]. Other studies also underlined that equipping university students in Malaysia with knowledge of the SDGs increased awareness and action on sustainability [30]. However, not all of the research has been so optimistic. To illustrate, research based on interviews again in Malaysia found that even motivated youth that are working on the SDGs operate in a relatively small circle and have not been able to reach a wider range of international/government organizations or the private sector [31]. Some studies have taken these findings a step further to suggest how to break through these barriers. In this connection, some have offered that one of the keys to meaningful engagement is to enable young people to serve as evaluators of SDG processes, noting that Saharan Africa, the Middle East and North Africa (MENA), and South America have seen progress in this regard [32].

Though the above studies help to illustrate some of the promise and challenges for truly influential or meaningful engagement on the SDGs in particular places, they lack the kinds of insights needed to determine how meaningful participation varies across countries. There is, however, a complementary set of studies that can shed light on some of the core elements of meaningful youth engagement that could be used for a comparative assessment. The next section reflects on some of the key insights from that literature.

2.3. Toward a Comparative Assessment of Meaningful Youth Engagement

The work that can inform a comparative assessment of meaningful engagement has a long history that is not only about youth but engagement more generally. In fact, much of the relevant literature draws upon on the well-known typology known as the "Ladder of Children's Participation [33]". The metaphorical ladder is based on seminal studies from Arnstein (1969) that argued that citizen (adult) participation could be classified as sitting on rungs corresponding to actual levels of influence on a plan and/or program [34]. The lowest three rungs on the more recent "Ladder of Children's Participation", (1) manipulation, (2) decoration, and (3) tokenism, are viewed as forms of "non-participation"; the five rungs above, (4) assigned but informed, (5) consulted and informed, (6) adult-initiated, shared decisions with youth, (7) youth-initiated and directed, and (8) youth-initiated, shared decisions, suggest young people are granted progressively more responsibility and decision-making power. Moving up the rungs implies changing power relations and a more equal distribution of rights between adults and young people [33].

A recent OECD Future of Education and Skills 2030 project offered that the "Sun Model of Co-Agency" has modified this ladder schema yet further. In this useful reformulation, the emphasis is placed on developing fruitful collaborations with adults across degrees of "co-agency" (except the newly added degree of zero or silence where neither young people nor adults believe that young people can contribute and adults lead all activities and make decisions) [35]. This reformulation is particularly helpful because it implies a shift towards joint action and decision-making from young people working with adults, while also recognizing young people can be experts in some areas, especially in youth-related matters [36]. In suggesting the potential for collaborative partnership while also underlining the promise for youth agency and expertise, the Sun Model aligns well with the trends in sustainability processes that began this section. More concretely, it suggests that young people can be active participants and valuable partners in efforts to institutionalize initiatives such as the aforementioned "Youth 2030", as well as to develop specific SDG-related policies where they have strong relevance and interests. Some examples include but are not limited to health and well-being (SDG 3) [37,38], education (SDG 4) [39], employment (SDG 8) [40], and climate change (SDG 13) [41].

Though the Sun Model and some similarly motivated antecedents [42–44] represent breakthroughs for classifying different forms of partnership and participation, one possible drawback is they may not be easy to use for a comparative assessment as the flexibility in the terms used in the model makes comparisons challenging. Simply stated, it can be difficult to employ a co-agency framework for comparatively assessing how meaningfully youth engage in SDG processes. In this connection, several studies offer some insights into the core elements of meaningful youth participation [3,13,37,45,46] or combine several approaches toward that end [7]. Though it is not possible to bring in insights from all of this above work, distilling four elements that can be compared across countries as they encourage and empower young people to participate in mechanisms related to the SDGs: (1) aims and justifications; (2) power-sharing; (3) transparency and accountability; and (4) support.

1. Aims and Justifications: A clear presentation of aims and justification underlying youth engagement is arguably essential for meaningful participation. The need for such a clear presentation stems from the fact that young people have the right to participate and express their views and opinions freely in all matters affecting them (rights justification). In so doing, they can offer suggestions that can help to improve policies and services (efficiency justification). Furthermore, young people can develop knowledge and skills (development justification) and enhance self-esteem through involvement (empowerment justification). It also merits noting that there may be several aims and justifications and these could be combined within a single participatory activity [3,13,47,48].
2. Power-sharing: Another critical dimension of meaningful engagement involves efforts to create balanced power-sharing arrangements. Young people can initiate activities and make decisions on their own. It is also arguably their choice to work together and share decisions with adults—provided that different responsibilities and competencies are clearly stated. It may similarly be important that young people are considered capable and competent and their views are respected. Under these conditions, young people can establish partnerships among other relevant stakeholders, including decision and policy makers. In this context, it merits noting that engagement mechanisms may aim to position young people as equal partners to improve the situation—a sentiment that aligns well with the partnership principle of Agenda 2030 [33,35,42,43,45,46,49,50].

3. Transparency and Accountability: A third important element of meaningful engagement involves whether the processes involving the SDGs are transparent and clear to young people and other stakeholders. With open communication channels, young people can be well-informed on all matters affecting them. For example, they can know what roles they are expected to play along with other stakeholders and what competencies are required to move forward. A possible desirable side effect of transparency is strengthening accountability. That is, transparency around the SDGs will help to allow decision-makers (as well as young people) to be held accountable for their commitments and actions. Similar to the previous elements, there are many references to calls for transparency and accountability in an effort to drive forward implementation [13,14,38,45,46].
4. Support: Last but not least is the issue of support. As noted previously in the article, young people may need support to safely and continuously participate regardless of background, status, or identity. Support may be particularly important for young people from less privileged backgrounds, enabling them to participate with confidence. Such support includes, but is not limited to, financial assistance for travel and accommodations, economic incentives, psychological and mentorship assistance, as well as providing learning and capacity development opportunities. The above needs are embedded in the SDGs No One Left Behind principles, which calls for support to be provided to those who are vulnerable and excluded [7,10,11,13,14,37,45,51].

The above four key elements are selected because they are viewed as core components of meaningful youth participation while aligning with principles underpinning the SDGs. The analytical framework based on these four elements makes it possible to bridging the theoretical question with the empirical analysis [52]. The next part of this paper uses this framework to offer an initial assessment of youth engagement mechanisms or activities related to SDGs' implementation in Finland and Japan. The four aforementioned elements will be helpful in comparing engagement mechanisms between two countries: Japan and Finland.

3. Case Selection and Methods

3.1. Selection of Countries

The four elements described above will be used for a small-n comparison of youth engagement in Japan and Finland. The selection of the two countries follows assumptions in comparative case study research that are frequently used in the social sciences to offer practical policy recommendations. These assumptions highlight the desirability of selecting cases that are similar across most key dimensions while varying in one or more limited sets of areas of interest. The reason for selecting cases with many similarities is to control for the possible confounding effects of issues that may influence a particular policy, program, or process (including engagement mechanisms) but are not the focus of the research. For example, if the two cases selected for this study were from countries of different levels of development and/or had different political systems, one might reasonably argue that inferences drawn about the variation in the engagement mechanisms are really related to the development levels or political systems [53].

As such, Finland and Japan were selected because there are many similarities between the two countries that limit concerns about confounding effects and make them useful for comparative policy-relevant case study research. In fact, there are at least five general similarities. First, both Finland and Japan are developed high-income countries [54]. Second, both countries are parliamentary democracies. Third, both countries have formulated and implemented policies to address environmental challenges, mitigate climate change, and promote sustainable development. Fourth, Finland and Japan are renowned for high-quality education systems that nurture future generations and consistently perform well in international assessments such as the Programme for International Student Assessment (PISA). Fifth, the two countries are recognized for promoting environmental education and Education for Sustainable Development (ESD).

In addition to the five more general similarities, both countries share more specific sets of traits related to engaging young people in implementing the SDGs. For example, Japan established the "Japan Next-Generation Platform" in 2018 under the initiative of the Prime Minister, creating a space for those interested in sustainable development. Similarly, Finland set up "The 2030 Agenda Youth Group" in 2017 under the Finnish National Commission on Sustainable Development, led by the Prime Minister. Both initiatives were referred to as good practices in the Voluntary National Reviews (VNR) shared at the High-level Political Forum (HLPF) to summarize the progress and achievements in implementing SDGs [55,56].

In addition to these general characteristics and specific traits, there are also similarities in the kinds of policy documents that have been developed. Table 1 shows the SDGs' plans and related policy documents that will be used to inform the comparative analysis in this study. In Japan, the SDGs' Implementation Guiding Principles (launched in 2016 and revised in 2019 and 2023) provide an overview of the status of the SDGs as well as its visions, priorities, institutions, and processes in promoting the SDGs. The guiding principles are used to develop an annual SDGs Action Plan in Japan. To some extent paralleling Japan, Finland developed a national 2030 Agenda Roadmap in 2021. That roadmap is a medium-term plan detailing the actions Finland needs to take to achieve the goals of the 2030 Agenda. The most recent version of the roadmap, adopted in 2022, is used in the comparative analysis to follow.

Table 1. SDGs-related documents included in the analysis.

Country	Document Title
Finland	• The 2030 Agenda Roadmap of the Finnish National Commission on Sustainable Development (2022) [57] • Strategy of the National Commission on Sustainable Development 2022–2030 (2022) [58] • Report on the Implementation of the 2030 Agenda for Sustainable Development (2020) [55]
Japan	• SDGs Implementation Guiding Principles Revised Edition (2019) [59] • SDGs Action Plan (2023) * [60] • Report on the implementation of 2030 Agenda ~Toward achieving the SDGs in the post-COVID-19 era~ (2021) [56]

* Only available in Japanese.

As seen above, Finland and Japan demonstrate a shared commitment to engaging young people in the national SDGs' implementation by, for instance, establishing institutionalized mechanisms for these purposes. While there are general similarities, there may be particular differences in how much they facilitate or support meaningful engagement. Further, although a rigorous check of causal claims goes beyond the scope of this article, those differences may also have implications for performance on the SDGs (see also the conclusion of the article). On this point, it should be noted that Japan ranks lower (21st in 2023) than Finland (1st in 2023) in SDG performance (based on the Sustainable Development Report from the Sustainable Development Solutions Network (SDSN)) [61].

The next subsection describes how to draw out those differences and answer the aforementioned research questions: How do countries compare in terms of meaningful youth engagement in sustainability processes?

3.2. Three Approaches

The paper relies on three different but complementary approaches to compare how meaningful youth engagement is in sustainability processes in Finland and Japan (see Figure 1). It merits noting that the approaches build off each other, starting with a rather general text mining of key words that is then complemented by descriptions of formal institutions and key informant interviews organized around the four categories in the literature review. In addition, as is also discussed in the conclusion, these methods can offer a "preliminary" assessment of engagement and could be further complemented by a larger sample of survey data or an expanded number of country case studies (see also Section 6, the Conclusions).

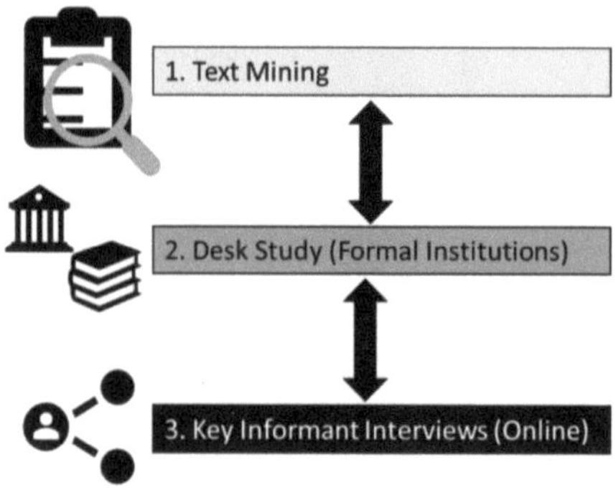

Figure 1. Three research methods.

It is also critical to highlight that, with the exception of the text mining, the data used for this research are qualitative. The use of qualitative as opposed to quantitative data is commonly practiced in comparative institutional analysis as a subfield of social science research. Though not perfect, qualitative data allow for insights into the design of policies and institutions that might not be visible from a quantitative analysis of key features of youth engagement mechanisms. It is also arguably more appropriate to use a qualitative lens for a two-country comparison as it can offer a richer and more nuanced understanding of the design of policies and institutions.

The first approach used for this comparative assessment involved the text mining of key words related to youth engagement in the documents in Table 1. The text mining aimed to gain a first-glance indication of how much youth engage in SDGs and other sustainability processes. Text mining can offer a useful perspective of whether and to what extent the formal mechanisms for engagement are meaningful as it can help to illuminate patterns in language that might not be discernible from a description of institutions. To obtain these initial insights, the authors went through an iterative process of re-reading key SDG documents to determine what words might be associated with youth engagement in the case studies' countries. After repeated re-readings (and the realization that the coverage of youth-related issues was a small part of the key documents), it was decided that the key word list should include a select number of the following admittedly general key words:

- Finland: youth; young people; adolescent(s); education
- Japan: youth; young people; adolescent(s); education; next-generation; wakamono (youth, adolescent); jisedai (next generation); kyoiku (education)

The key words were then counted based on a simple automated keyword search procedure with manual coding to check whether the word was used in a way that was relevant to the scope and substance of the article. This was important because there were some instances where key terms applied to statements about youth and initiatives taken outside the countries (e.g., international aid). In most cases, it was possible to conduct the text mining in English. For Japan's Action Plan, this was not feasible, and translation from Japanese to English was needed. Following the text analysis, the article turned to the two additional methods to obtain an additional perspective on the meaningfulness of that participation that is more closely aligned with core elements in the literature: institutional analysis and key informant interviews.

The second approach therefore involved a careful desk study of formal institutions and policies. Much of the material for this review came from official government and youth organization websites and/or other publication channels. During the desk study, the authors aimed to assess how the engagement mechanisms performed on the four analytical categories detailed in the previous section.

The third and final approach aimed to address concerns that looking only at formal institutions and policies may miss how they actually work in practice. To address this concern, key informant interviews were conducted with young people who are members of the Finnish Agenda 2030 Youth Group and Japan Youth Platform for Sustainability (JYPS), one of Japan's Steering Committee Organizations of the Next-Generation Platform. The interviews sought to obtain perceptions about youth engagement mechanisms and validate or refine some of the inferences drawn from the desk study. To accommodate the needs of the interviewees, the discussions consisted of one group interview and three one-to-one interviews.

The interviews were semi-structured, with questions covering themes including the motivation to participate in the respective platforms and suggestions for better youth engagement mechanisms. Each interview took around one hour. The interviews were conducted mostly online, and every interview was recorded with the interviewee's consent. Rather than being presented as a standalone section, the results of the interviews are woven into the desk study description of the formal institutions and policies (see the following Results Section). Though each of the above approaches has flaws, the combination of different approaches offers a useful way to obtain some preliminary comparative insights into the meaningfulness of engagement.

4. Results

As noted in the previous section, the study used three different techniques to determine whether youth could meaningfully engage in sustainability processes in Finland and Japan. The first part of this subsection brings in text mining and frequency counts of keywords. The latter two parts of this subsection draw upon the desk study and complementary interviews for Finland and Japan, respectively.

4.1. Results of the Text Mining

The text analysis provides an initial indication on how much youth engage in SDG processes. As illustrated in Table 2, Finland's key documents generally have more references to the keywords highlighted in Section 3. This contrast is most apparent when comparing Japan's SDG principles and Finland's roadmap. The former includes 25 references to youth-related terms, while the latter includes 188 such references. It nonetheless merits noting that much of the difference is due to references to "education". In addition, the differences in the number of references are smaller in the other two sets of analyzed documents, including both country's VNRs. To look more closely at whether these first-glance inferences are supported by other evidence, the study reviews the formal institutions and policies for each country.

Table 2. Frequency counts of key words in the reviewed documents.

Japan	Related Terms	Finland	Related Terms
SDGs Principles (2019)	youth (1), young people (2), education (14), next-generation (8)	Roadmap (2022)	youth (9), young people (5), education (168), adolescent (6; all come along with "children")
Action Plan (2023)	Wakamono (11), jisedai (9) *, kyoiku (104)	Strategy (2022)	youth (8), young people (5), education (165), adolescent (6, all come along "children")
VNR (2021)	youth (13), young people (20), next generation (15), education (127), adolescent (1)	VNR (2020)	youth (35), young people (28), education (118)

* There were 24 references, but 9 were only used in the way this research intends.

4.2. Desk Study and Interviews on Finland's Youth Participation in Sustainability Processes

To understand how youth engagement works in Finland, it is important to provide background on sustainable development institutions. That background begins with the National Commission on Sustainable Development and Development Policy Committee (hereafter: National Commission). The National Commission has been chaired by the Prime Minister since 1993 and actively encourages participation. For instance, the National Commission has encouraged a participatory approach to follow-up and review in Finland that is held in high regard for strengthening stakeholder engagement and policy coherence [62]. The national follow-up and review system also includes innovative participatory elements, such as a Citizen Panel that assesses the state of sustainable development and provides information and citizen views to policymakers. Importantly, the citizen panel also reaches out to and integrates inputs from young people.

There are also other measures taken to expand the involvement of young people in SDGs' implementation in Finland. These measures can be divided into two main areas. One involves promoting forms of education that extend beyond formal schooling to lifelong learning in an effort to empower young people to contribute to sustainable development. Initiatives promoting education utilize existing mechanisms to support young people, such as youth work [58] (p. 30 and p. 70).

While there are several mechanisms that promote engagement generally, the article focuses chiefly on the Agenda 2030 Youth Group as a mechanism that enables youth engagement specifically. The Agenda 2030 Youth Group is an institutionalized mechanism for young representatives to participate in the decision-making process on sustainable development. The Agenda 2030 Youth Group was established at the initiative of the then Vice-Chair of the National Commission (Figure 2). It is composed of 12 people and 2 equal-status chairs aged 15 to 28 years from diverse backgrounds across Finland. It is intended to expand the involvement of a wide cross-section of young people. The decision to include multiple youth representatives was based on the belief that appointing a single youth delegate for sustainable development was not enough to capture the diverse range of sentiments from young people in different places and stations in life [55] (p. 26). Table 3 shows the summarized references on the group within the reviewed documents categorized into four analytical categories.

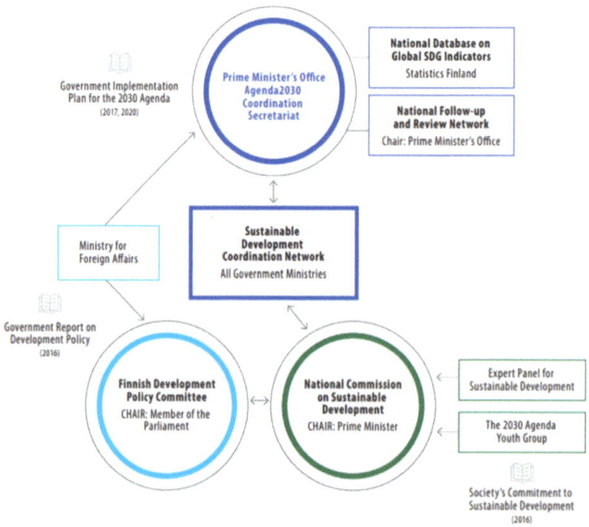

Figure 2. Illustration of National 2030 Agenda structure in Finland [42] (p. 91).

Table 3. References to areas in the analytical categories for the 2030 Agenda Youth Group.

Analytical Categories	Relevant References (Summarized)
Aims and Justifications	• Aim to expand young people's involvement in SD policy to address the issues (VNR, p. 14) • Two expected roles: (1) to spur the Finnish National Commission and bring young people's voices to SD policy processes and public debate; (2) to inform other young people of SDGs in their networks, schools, and leisure activities (VNR, p. 27)
Power-sharing	• Sufficient autonomy on what they will do and what matters they will promote (VNR, p. 27) • Members' recognition of a mechanism that brought them to the center of SD decision-making processes, placing them on equal footing with other parties as contributors to SD (VNR, p. 27) • Members participated in the discussion on roadmap development, and the individual section was developed within (Roadmap, p. 83)
Transparency and Accountability	• The group is an institutionalized mechanism in the national SDGs implementation process (VNR, p. 14) • The member selection is conducted with the support of Finnish Youth Cooperation, Allianssi (VNR, p. 27) • The expected roles to play are clear (VNR, p. 27), and opportunities for participation (both mandatory and voluntary) are listed (VNR, p. 27/Roadmap, p. 83) • The members' evaluating comments on SDG processes are integrated into VNR as part of follow-up and review (VNR, p. 27, 97)
Support	• Finish Youth Cooperation supports the engagement processes, trying to make better representation within the group (VNR, p. 27)

Regarding the aims and justifications, Finland explicitly aims to establish the Agenda 2030 Youth Group to expand opportunities for young people to express their views in the sustainable development policy process. The main justifications for creating this mechanism are that it provides a guarantee on the fundamental right for young people to participate and thereby improve sustainable development-related policies and services. Having this justification and their expected roles made explicit is helpful because it firmly established that young people can contribute substantively to national SDG processes (Group Interview, 2024).

In addition to having a well-defined role, the activities in which young people engage are based on balanced power-sharing. This balance is achieved in several ways. For example, it is mandated that the Agenda 2030 Youth Group regularly attends the National Commission's meetings to bring messages from different youth perspectives. They are also invited to participate in critical SDG processes, such as the central government's policy development, budgeting, and follow-up and review. While the desk study highlighted these more formal channels, interviewees further indicated that there are even more opportunities called by the national government than they attend (Group Interview, 2024).

At the same time, there are guarantees of their autonomy or freedom from political interference in SDG processes, thereby encouraging participation in public discussions and dialogue around the SDGs. Some of these signs of balanced power-sharing are evident in the active participation of young people as panelists or organizers of the stakeholders' events, posting on social media, and media interviews. Young people also organize events by themselves to reach a wider range of young people (Group Interview, 2024). Although it is difficult to pinpoint precisely how an input influences a decision, there are several signs that young people's views are respected in critical decision-making processes. There is also evidence that the effort to share power helps to energize young people to engage in public discussions on sustainable development in partnership with other stakeholders.

The review of publicly available information suggests that youth engagement in Finland is supported by relatively high levels of transparency and accountability. For instance, an open application process facilitated by the Finish Youth Cooperation called Allianssi is used to determine who can join the 2030 Agenda Youth Group. The members are selected to represent diverse backgrounds across Finland, including their interests in the SDGs. This representativeness, when combined with the ability to participate in the key processes mentioned above, may also help to increase transparency and accountability. That is, policymakers may feel the need to share more information with different segments of the youth population and feel more beholden to the interests of this diverse mix of young people. While the documents imply this potential on paper, some members "do not feel that they represent Finnish youth as a whole [55] (p. 28)". It was also indicated that the information about this mechanism was easily reachable to those who are in the "circle", who were already active and interested, but not "regular youth" (Group Interview, 2024).

The one area where Finland appears to perform less well compared to the other elements in the framework is support. For this dimension, the allocated government budget covers the travel, accommodation, and food expenses for the members, although their work is on a voluntary basis; in other words, they do not receive financial incentives for activities (Group Interview, 2024). Based on a review of the available documents, most of the non-material support comes from the aforementioned group, Allianssi. Allianssi serves as an advocacy organization for the youth sector with over 140 national youth and education sector organizations. Funded by the Ministry of Education and Culture, Allianssi endeavors to prevent the exclusion of young people and help young people to develop into responsible members of society and participate in decision-making [63]. According to interviews, Allianssi plays a secretariat function in this mechanism, supporting annual recruitment processes and providing some technical support and advice to their own initiated activities. Though Allianssi helps to a certain degree, some youth raised the necessity of capacity development, especially technical and professional knowledge of sustainability issues. Although potential youth representatives indicate their areas of interest upon their selection, they still need to be knowledgeable about the field to make a substantive contribution (Group Interview, 2024). There might be such a space for improvement to enable their participation with confidence. There are nonetheless limited details of the type of support provided for youth participation in SDG processes—though one might presume that this assistance comes from other mechanisms reviewed herein.

In sum, Finland has created relatively well-established mechanisms aimed at enabling meaningful youth participation in SDG processes. These mechanisms have clear aims and justification, balanced power-sharing, and relatively high transparency and accountability (though there may be scope for work in this area). The amount of support provided may need to be increased.

4.3. Desk Study and Interviews on Japan's Youth Participation in Sustainability Processes

Since its establishment in 2016, the SDGs Promotion Headquarters (hereafter Headquarters), composed of all Cabinet Ministers and led by the Prime Minister, has coordinated SDGs' implementation in Japan. As illustrated in Figure 3, the Headquarters promotes stakeholder consultations through "SDGs Promotion Roundtable Meetings", including representatives from related government agencies, civil society, academia, the private sector, and international organizations. This mechanism also reviews SDG progress, provides inputs into the SDGs Implementation Guiding Principles (hereafter Principles), and determines recipients of the "Japan SDGs Award" to recognize best practices. The Headquarters is responsible for monitoring the progress of measures taken in line with the Principles, as well as raising awareness of the SDGs to increase public understanding and support for engagement with the SDGs [64]. To help achieve the SDGs, the Principles set out three pillars underpinning Japan's "SDGs Model": (1) "Promotion of Society 5.0 that corresponds to the SDGs", (2) "Regional Revitalization driven by the SDGs", and (3) "Empowerment of the next generation and women" [59] (p. 2). Japan has created two main channels to enable young people with various backgrounds to contribute to achieving the SDGs: enhancement of education policies and systems and the "Next-Generation Platform".

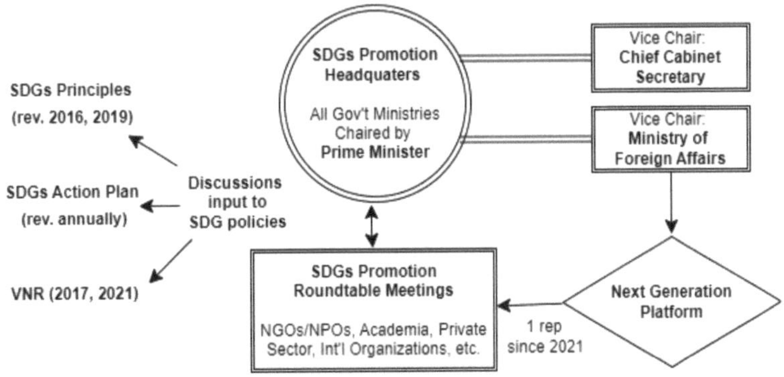

Figure 3. Illustration of the SDG implementation structure in Japan (Source: Authors).

Given the thematic scope of the article, most of the focus is placed on the "Next-Generation Platform". The Next-Generation Platform was established in 2018 under the initiatives of the then Prime Minister. The Platform comprises seven Steering Committee Organizations: AISEC Japan, ETIC, G7/G20 Youth Japan, Junior Chamber International Japan (JCI), Japan Disability Forum (JDF), Japan Model United Nations (JMUN), and Japan Youth Platform for Sustainability (JYPS). These organizations have varied aims and activities and members. Among them, JYPS is the only organization led by young people under 30 years old to advocate youth engagement in the sustainability field [65]. The Platform aims to "deepen the involvement of the next generation, who will play a leading role in promoting the SDGs after 2030 [56] (p. 36)". Table 4 summarizes the Platform's performance based on the four key elements in the analytical framework and the reviewed documents.

Table 4. References to areas in the analytical categories in Japan's Next-Generation Platform.

Analytical Categories	Relevant References (Summarized)
Aims and Justifications	• "Empowerment of the next generation and women" as one of the three pillars of Japan's "SDGs Model" (Principles, p. 2) • Work to accelerate the proactive promotion of SDGs by the next generation and disseminate Japan's SDGs Model regarding SDGs' promotion by the next generation (Principles, p. 2)
Power-sharing	• One youth from the Platform became a member of the "SDGs Promotion Roundtable" (VNR, p. 5)
Transparency and Accountability	• The platform is the institutionalized framework for the promotion of the SDGs (Principles, p. 11) • The activities conducted are showcased, also embedded in the national governmental mechanism related to the SDGs process (VNR, p. 14) • There was a discussion opportunity with the young generation in the VNR preparation (VNR p. 5)
Support	• No description was found related to the support given to the Platform

Regarding the aims and justification, the empowerment of the next generation and women is listed as one of the key elements to support Japan's SDGs Model and to achieve the SDGs. While the importance of young people is recognized, the launch of Japan's Next-Generation Platform is offered as a good example of the government's effort to concretize that vision. However, there is no apparent justification for this initiative. According to the Principles, the Platform is intended to let young people "think about how to promote the SDGs and how to transform society when they become the main players [59] (p. 10)". Despite the stated goal of empowering younger generations, the Platform does not seem to offer an equally explicit justification for engagement—such as guaranteeing rights, utilizing ideas for better policies and services, and developing knowledge, skills, and self-esteem.

For some of the same reasons, it is challenging to assess aims and justifications, and it was also difficult to evaluate power-sharing. The Platform members are likely working on the activities of each representative organization, but there is no reference to the activities initiated by the Platform itself. As mentioned previously, one youth representative from the Platform became a member of the roundtable in December 2021. Their involvement in the roundtable started after the submission of the latest VNR, to discuss issues related to the subsequent follow-up and review processes with other stakeholders. According to the interviews, youth representatives are invited to consultation meetings twice a year to discuss revising the Principles and Action Plan and are given the same mandate as other adult members. However, youth stakeholders also find it difficult to influence the decision-making process. This is because the meetings primarily aim to exchange views between the Headquarters (government) and the non-government stakeholders, with decision-making to be made by the government (Individual Interview, 2023). While they enjoy the same mandate as other adult members, several interviewees suggested that they felt as if they were less of a priority during consultations with lower-ranked government personnel compared to other adult stakeholders. Despite the efforts to hear stakeholders' opinions, including young people, no information was available about the activities initiated by the Platform or the extent of its influence on decision-making processes.

Similarly, there are almost no references to transparency and accountability. Although all the meeting notes of the roundtable discussion are publicly available online, there is no publicly available information about the Platform's mechanisms or activities. The lack of such references is apparent in the limited details on the member selection process: simply stated, the reasons why some organizations were selected as the Steering Committee are not laid out clearly, and references to which organizations act as members of the Platform are also murky. The appointment of one youth representative mentioned may also provide a relatively narrow base for holding decision-makers accountable to different groups of young people. Combined with the limited clarity under the power-sharing category, there may be a need for greater transparency and broader accountability.

Lastly, there is almost no description of the Platform initiatives in terms of support. According to the one related reference found in the SDGs Action Plan, there is no specific budget allocation for operating this initiative for the years 2022 and 2023 [60] (p. 94, 95). The interviews underlined the same point: there is a lack of financial compensation for their activities. The youth representatives invited to the roundtable meetings receive compensation with a daily allowance; however, they do not receive any financial support for their activities. Furthermore, the meetings are often held on weekdays in Tokyo in an in-person format only, which is preferable for business persons; on the other hand, it makes it difficult for youth representatives, especially if they live far from the capital. Of course, financial support is not only a way of enabling young people's participation but it also strongly influences their continuous engagement. Another difficulty in sustaining their activities is member recruitment and capacity development. The vast majority of the members of the Steering Committee Organizations are college students, who mostly stay active for two to three years only. For a similar reason, many young people leave their activities to pursue studies and career development or give up for economic reasons—and this could prevent the involvement of diverse young people. Informally, one of the steering committee organizations, the Junior Chamber, consisting of young working professionals below 40 years old, offered to serve as a quasi-secretariat of the Platform to assist activities (Individual Interview, 2023). However, more formal and structural support, including financial assistance or incentives, is likely needed. Otherwise, it may limit the Platform's effectiveness, potentially excluding less socially and economically privileged youth.

Though the above review suggests there exists scope for improvement, it merits highlighting that it was a significant step for Japan to create a mechanism to engage young people in SDGs' implementation. Importantly, the review suggests areas in which Japan may want to take additional steps. These include making clearer links between aims and justifications and concrete activities. This would entail going beyond simply "think[ing] about how to promote SDGs and how to transform society when they become the main players" to have a right and role to influencing activities now—and into the future. In a similar vein, though awareness-raising is important, it is arguably just as vital to have more balance in power dynamics so that people's views have a discernible impact in the matters affecting them. A clearer understanding of young people's competencies and responsibilities would further lead to more equitable power-sharing while also boosting transparency and accountability. Last but not least, support, particularly clear budget lines, and continuous assistance, including secretariat functions and providing capacity development opportunities, could increase the involvement of young people from more diverse backgrounds.

4.4. *Preliminary Comparative Assessment*

Though the picture painted in this section is based chiefly on text mining and descriptions of institutions and complementary interviews, they nonetheless shed some useful initial light on how meaningful youth engagement is in Japan and Finland. More concretely, the results suggest that there are differences in how much Finland and Japan support youth engagement in sustainability processes. Those differences are succinctly summarized in Table 5 that breaks down the comparison by the analytical categories distilled from the literature review. Table 5 shows that, by and large, Finland's aims and justifications, power-sharing, and transparency and accountability tend be clearer and stronger than Japan's. Meanwhile, both countries could do better in offering support for young people. The next section discusses these findings and their broader implications.

Table 5. Comparison of Finland and Japan.

Analytical Categories	Finland	Japan
Aims and Justifications	There are clear aims and justifications for establishment. They aim to expand opportunities for young people to express their views in SD policy processes, guarantee their fundamental right to participate, and improve SD-related policies and services.	Despite its intention to empower the next generation, the Platform lacks a clear justification. The platform encourages young people to think about promoting SDGs but does not provide justifications such as guaranteeing their rights or utilizing their ideas for policy improvement.
Power-sharing	Group members actively participate in national SDG processes, attend government meetings, and engage in policy development and budgeting. Autonomy in their activities is guaranteed, and they engage in public dialogue, social media, and media interviews to inform young people about SDGs.	The activities conducted by the Platform are not well-described in the documents, making it difficult to assess shared power. Although the representatives sent to the roundtable are given the same mandate as other adult stakeholders, structurally, they have no power in the decision-making.
Transparency and Accountability	Finland's member selection process for group members is relatively open, although there is room for improvement in extending representation. Youth engagement contributes to government accountability, especially in follow-up and review processes.	Japan's reviewed documents lack transparency and accountability, especially in the selection process and representation of the Platform's members. In addition, the Platform's activities are unclear, as well as how influential it is in decision-making processes.
Support	Finland provides some material and non-material support for youth activities in SDG processes through the Youth Council. Their necessary expenses are well-covered, but not for their work.	Little assistance is provided with Platform activities. Some non-material support is provided on a voluntary basis; the lack of financial support, especially, may limit the effectiveness of Platform activities.

5. Discussion

Young people's involvement in SDG processes is important, as it is gaining recognition at the highest levels in Finland and Japan. This realization has paid some dividends; there has been progress in involving youth in the SDGs and related processes through the creation of the 2030 Agenda Youth Group and the Next-Generation Platform in Japan in Finland. However, there is arguably a need to dig deeper into the actual operations of these mechanisms and determine whether they truly facilitate meaningful youth engagement.

In considering ways to dig deeper, it should be noted that analyzing stakeholder engagement in SDG-related documents such as VNRs is an imperfect science; these documents do not always fully articulate their goals or outline relevant activities [66]. These limitations notwithstanding, Finland's approach appears to cover more of the criteria associated with meaningful engagement than Japan's. While recognizing the four analytical categories are interrelated and difficult to disentangle, the following discussion divides the comparison into those categories for ease of presentation.

1. Aims and Justifications: For engagement to be truly meaningful, defining the relevant aims and supporting justifications for participation is essential. In Japan, the aims and justification for including youth are short on details and tend to rest chiefly on the need for information-sharing and awareness—a finding that is not unique to Japan as many countries tend to focus on information-sharing and SDG promotion. In the case of Finland, there tends to be not only a more varied set of aims and activities but also more clearly explicated underlining justifications. The clearer aims and justification would help to identify the appropriate youth to hear their opinions, especially to make progress on youth-related SDGs. In doing so, there would be a more deliberate effort to engage youth from different locations and social segments so as not to simply cater to, for example, socially advantaged groups [44,67,68].
2. Power-sharing: Similar to the first element, Finland appears to have made greater efforts to balance power dynamics. To illustrate, young people are regularly invited to participate in high-level sustainable development meetings while enjoying the autonomy to conduct pertinent activities. Further, this balance has arguably helped to build the confidence needed to launch related activities, such as the formation of a climate-focused sub-group within the Agenda 2030 Youth Group in Finland in 2023 [69]. In Japan, the youth representatives are equally responsible with the adult stakeholders in the roundtable; they are structurally limited to influencing and shaping the trajectory of the decisions. Despite the existence of the Platform, Japan has tended to limit the power and autonomy offered to young people, keeping the scope of those activities circumscribed within their respective organizations.
3. Transparency and Accountability: As noted previously, transparency and accountability are also critical elements to ensuring that young people are meaningfully engaging in the SDG process. In this case, Finland also tends to demonstrate a clearer commitment to these principles and related practices. This is evident in both the relatively open selection of multiple youth representatives as well as the role in important processes that can boost accountability, such as the follow-up and review. In the case of Japan, there is more scope for improvement in explaining how youth representatives are selected and how the mechanism operates, moving from informal agreements to transparent long-term contracts and arrangements [38].
4. Support: Perhaps more so than other forms of engagement, youth engagement necessitates support. Both countries could improve their engagement mechanisms in this regard, ensuring varied participation channels to communicate with a wider youth segment and providing adequate compensation, capacity building, and empowerment, especially for underrepresented and vulnerable youth. In Finland, the continuous backing of youth-led initiatives through official platforms, like Alliansi or National Youth Councils, may offer some forms of support. On the contrary, Japan should take specific measures to support its activities; otherwise, this could eventually engage the limited segment of young people who are already capable and socio-economically privileged. This might inadvertently keep young people who are vulnerable as recipients of "care" rather than as participants to be empowered.

6. Conclusions

At the most general level, participation requires a well-designed set of institutions—that is, a set of rules and structures that continually encourage diverse voices to engage in the public sphere. Creating these institutions to foster the meaningful participation of young people in sustainability processes is increasingly viewed as integral to accelerating progress on the SDGs. Not only can it help to build intergenerational trust and social cohesion, but it also ensures fresh and forward-looking ideas that inform the policies where youth have strong interests, including education, clean energy, and climate change.

Many governments have therefore made commendable efforts to engage young people in these decisions. Some research has developed heuristics to determine the degree to which engagement matters or is meaningful. There have nonetheless been few efforts to use insights from that research to compare youth engagement across countries.

This article employs a novel approach to address this gap in understanding and answer a critical question: how do countries compare in terms of meaningful youth engagement in sustainability processes? That approach uses a framework based on four analytical categories that have been argued to underpin meaningful youth engagement: aims and justification, power-sharing, transparency and accountability, and support. It then employs the framework to assess how Japan and Finland perform in each of these areas. The comparison suggests that Finland's effort to engage youth exhibits greater attention to the areas highlighted in this framework. This is particularly apparent in the first three elements of the framework, where there is a clearer presentation of aims and justifications, more balanced power-sharing, and greater transparency and accountability. Both countries, the article suggests, may need to do more to extend support to young people.

Expanding upon the conclusion, it warrants highlighting that the article also opens channels for future inquiry. As previously mentioned, comparative analysis of relevant policy documents may offer an incomplete picture of how formal institutions actually operate and how young people actually engage. The article addresses this issue to some degree by noting that the conclusions drawn herein are still preliminary. In addition, it also brings in key informant interviews to supplement inferences; however, a clear way forward is expanding the group of interviewees to include not only youth stakeholders but also other relevant stakeholders, such as policy and decision-makers, researchers, and other civil society members.

Another avenue for future exploration lies in examining the influence of demographic differences on participation. For instance, Finland's population is approximately 5.5 million, significantly smaller than Japan's 122 million. Given these demographic disparities, it may be beneficial to compare mechanisms across smaller countries like Finland. It may also be possible to examine differences and similarities with local-level mechanisms in bigger countries like Japan to discern variations in youth engagement mechanisms/strategies and their effectiveness. In addition, it may be useful to examine the influence of the regional level. Not limited to UN initiatives, the Finnish case indicates there might also be some influences from regional frameworks and cooperation, such as the European Union, the Council of Europe, and the Nordic Councils.

Another important consideration that has thus far been underemphasized is socio-cultural differences. Differences in underlying traditions and belief systems may play a significant role in participation and power-sharing. For instance, the Finnish documents highlight a strong tradition of participation, whereas social science studies on values indicate varying preferences for political participation among people across cultures. Comparing the approaches of various countries in involving youth can help to shed light on the reasons for the diverse pathways and effective strategies to enhance youth participation within different cultural contexts.

Further, it may be necessary to revisit the framework employed in this analysis. This research has suggested that Finland may place greater emphasis on downstream participation opportunities. Instead, Japan's emphasis appears to lean towards strengthening upstream formal school education for sustainable development, which was not featured in the framework. The framework's key elements were originally designed to incorporate various ways of contributing to youth engagement in SDG processes, but it admittedly downplays the crucial role of education.

Finally, there may be scope to use the analyses offered herein to better understand how more meaningful youth engagement translates into performance on key SDGs. In this connection, additional work could look at whether there is a correlation between youth engagement on SDGs where there are clear benefits for young people such as SDG 4 (education) and areas where there are clear opportunities to trigger real-world changes such as SDGs 7 (clean energy) and 13 (climate change). In carrying out this research, one might hypothesize that more meaningful engagement is correlated with, for instance, quality education (SDG 4) and a willingness to take concrete actions that conserve energy (SDG 7) and address the climate crisis (SDG 13). Conducting this kind of study would ideally require expanding the number of cases and related assessments of youth engagement mechanisms. Importantly, this study offers a feasible approach for making that comparison.

In sum, there is ample room to delve deeper into the dynamics of youth engagement, accounting for reporting gaps, demographic influences, and cultural nuances, ultimately advancing understandings of how to foster effective youth participation in the SDGs in various contexts. It is also clear that looking more closely at the links between youth engagement and SDG performance would be mutually beneficial for young people and the sustainability of the planet.

Author Contributions: Conceptualization, M.F. and E.Z.; methodology, M.F. and E.Z.; investigation, M.F. and E.Z.; resources, M.F.; writing—original draft preparation, M.F.; writing—review and editing, M.F. and E.Z. All authors have read and agreed to the published version of the manuscript.

Funding: This research was funded by the Environment Research and Technology Development Fund (JPMEERF20221M03) of the Environmental Restoration and Conservation Agency provided by the Ministry of the Environment of Japan.

Institutional Review Board Statement: Relevant ethical considerations were considered and addressed. Informed consent was obtained from all subjects involved in the study.

Informed Consent Statement: Informed consent was obtained from all subjects involved in the study.

Data Availability Statement: The data presented in this study are available on request from the corresponding author. The data are not publicly available due to ethical restrictions.

Conflicts of Interest: The authors declare no conflicts of interest.

References

1. UN General Assembly. *Convention on the Rights of the Child*; Treaty Series; United Nation: Geneva, Switzerland, 1989; Volume 1577, p. 3.
2. Ali, S.M.; Appolloni, A.; Cavallaro, F.; D'Adamo, I.; Di Vaio, A.; Ferella, F.; Gastaldi, M.; Ikram, M.; Kumar, N.M.; Martin, M.A.; et al. Development Goals towards Sustainability. *Sustainability* **2023**, *15*, 9443. [CrossRef]
3. Bárta, O.; Boldt, G.; Lavizzari, A. *Meaningful Youth Political Participation in Europe: Concepts, Patterns and Policy Implications—Research Study*; Council of Europe: Strasbourg, France, 2021.
4. Willems, H.; Heinen, A.; Meyers, C. *Between Endangered Integration and Political Disillusion: The Situation of Young People in Europe*; University of Luxembourg: Luxembourg, 2012.
5. Hwang, S.; Kim, J. *UN SDGs and United Nations and Sustainable Development Goals A Handbook for Youth*; UNESCAP East and North-East Asia: Incheon, Republic of Korea, 2017.
6. El Zoghbi, M.B. Conferences as Learning Spaces on Climate Change and Sustainability. In *Promoting Climate Change Awareness through Environmental Education*; IGI Global: Hershey, PA, USA, 2015; pp. 37–59.
7. Singh, A.; Faiqoh; Zuhra; Putu Natalya, N.; Talata Farida, C.; Nugroho, S.; Mujoko, A.; Alfian Novianto, D.; Tabrani Al-Ikhlas, I.; van Reeuwijk, M. *Operations Research on Meaningful Youth Participation in Indonesia*; Rutgers: Utrecht, The Netherlands, 2016.
8. Fielding, M. Students as Radical Agents of Change. *J. Educ. Change* **2001**, *2*, 123–141. [CrossRef]
9. D'Adamo, I.; Di Carlo, C.; Gastaldi, M.; Rossi, E.N.; Uricchio, A.F. Economic Performance, Environmental Protection and Social Progress: A Cluster Analysis Comparison towards Sustainable Development. *Sustainability* **2024**, *16*, 5049. [CrossRef]
10. Arnold, H.E.; Cohen, F.G.; Warner, A. Youth and Environmental Action: Perspectives of Young Environmental Leaders on Their Formative Influences. *J. Environ. Educ.* **2009**, *40*, 27–36. [CrossRef]

11. Browne, L.P.; Garst, B.A.; Deborah Bialeschki, M. Engaging Youth in Environmental Sustainability: Impact of the Camp 2 Grow Program. *J. Park Recreat. Adm. Fall* **2011**, *29*, 70–85.
12. Cumiskey, L.; Hoang, T.; Suzuki, S.; Pettigrew, C.; Herrgård, M.M. Youth Participation at the Third UN World Conference on Disaster Risk Reduction. *Int. J. Disaster Risk Sci.* **2015**, *6*, 150–163. [CrossRef]
13. United Nations. *Our Common Agenda Policy Brief 3 Meaningful Youth Engagement in Policy and Decision-Making Processes*; United Nations: New York, NY, USA, 2023.
14. United Nations Office of the Secretary-General's Envoy on Youth. *Believe in Better: From Policy to Practice Strengthening Accountability Relationships with and for Young People*; United Nations Office of the Secretary-General's Envoy on Youth: New York, NY, USA, 2021.
15. Verba, S.; Nie, N.H. *Participation in America: Political Democracy and Social Equality*; University of Chicago Press: Chicago, IL, USA, 1987; ISBN 9780226852966.
16. Milner, H. *The Internet Generation: Engaged Citizens or Political Dropouts. The Internet Generation: Engaged Citizens or Political Dropouts*; Tufts University Press: Medford, MA, USA, 2010.
17. Norris, P. *Democratic Phoenix*; Cambridge University Press: Cambridge, UK, 2002; ISBN 9780521811774.
18. Inglehart, R. *Modernization and Postmodernization: Cultural, Economic, and Political Change in 43 Societies*; Princeton University Press: Princeton, NJ, USA, 1997.
19. Dalton, R.J. Citizenship Norms and the Expansion of Political Participation. *Political Stud.* **2008**, *56*, 76–98. [CrossRef]
20. Pickard, S. *Politics, Protest and Young People*; Palgrave Macmillan: London, UK, 2019; ISBN 978-1-137-57787-0.
21. London School of Economics and Political Science. *Youth Participation in Democratic Life*; London School of Economics and Political Science: London, UK, 2013.
22. Barrett, M. Young People's Civic and Political Engagement and Global Citizenship. Available online: https://www.un.org/en/chronicle/article/young-peoples-civic-and-political-engagement-and-global-citizenship (accessed on 28 December 2023).
23. Mannarini, T.; Legittimo, M.; Talò, C. Determinants of Social and Political Participation among Youth. A Preliminary Study. *Psicol. Política* **2008**, *36*, 95–117.
24. Norris, P. Young People & Political Activism: From the Politics of Loyalties to the Politics of Choice? In Proceedings of the Civic Engagement in the 21st Century: Toward a Scholarly and Practical Agenda, University of Southern California, Los Angeles, CA, USA, 1 October 2004.
25. United Nations Sustainable Development Group. *Operationalizing Leaving No One Behind: Good Practice Note for UN Country Teams*; United Nations Sustainable Development Group: New York, NY, USA, 2022.
26. Dezelan, T.; Moxon, D. *Influencing and Understanding Political Participation: The European Perspective*; Publication Office of the European Union: Luxembourg, 2021.
27. UNICEF. *Engaged and Heard! Guidelines on Adolescent Participation and Civic Engagement*; UNICEF: New York, NY, USA, 2020.
28. Moxon, D.; Barta, O.; Pasic, L.; Vanhee, J. *Evaluation of Participant Inclusion Levels within the EU Youth Dialogue*; Publication Office of the European Union: Luxembourg, 2023.
29. Biancardi, A.; Colasante, A.; D'Adamo, I.; Daraio, C.; Gastaldi, M.; Uricchio, A.F. Strategies for Developing Sustainable Communities in Higher Education Institutions. *Sci. Rep.* **2023**, *13*, 20596. [CrossRef]
30. Tuan Ismail, T.N.; Mohd Yusof, M.I.; Ab Rahman, F.A.; Dwi, H. Youth and Their Knowledge on the Sustainable Development Goals (SDGs). *Environ. Behav. Proc. J.* **2022**, *7*, 329–335. [CrossRef]
31. Mohd Yusof, M.I.; Ho Nyuk Onn@Ariffin, M.; Dwi, H. Stakeholder Engagement in Implementation of Youth-Led SDG-Related Programmes in Malaysia. *Environ. Behav. Proc. J.* **2022**, *7*, 323–328. [CrossRef]
32. Montrosse-Moorhead, B.; Bitar, K.; Arévalo, J.; Rishko-Porcescu, A. Revolution in the Making: Evaluation "Done Well" in the Era of the SDGs with a Youth Participatory Approach. In *Evaluation for Transformational Change: Opportunities and challenges for the Sustainable Development Goals*; van den Berg, R.D., Magro, C., Mulder, S.S., Eds.; International Development Evaluation Association: Exeter, UK, 2019; pp. 33–50.
33. Hart, R.A. *Children's Participation*; Routledge: London, UK, 2013; ISBN 9781134172221.
34. Arnstein, S.R. A Ladder Of Citizen Participation. *J. Am. Inst. Plann.* **1969**, *35*, 216–224. [CrossRef]
35. OECD. *Future of Education and Skills 2030 Conceptual Learning Framework: Student Agency for 2030*; OECD: Paris, France, 2018.
36. Brasof, M. *Student Voice and School Governance*; Routledge: London, UK, 2015; ISBN 9781315724041.
37. Ndayala, P.; Kuya, J. *Exploring the Factors That Influence Meaningful Youth Involvement in the Health Care System Management in Western Kenya Final Report Empowering Young People for Access to Quality SRH Service in Kenya View Project Sexual and Reproductive Health and Rights among Adolescents and Youth View Project*; Rutgers & GLUK: Kisumu, Kenya, 2016. [CrossRef]
38. World Health Organization. *Status of Meaningful Adolescent and Youth Engagement (MAYE): Summary Report of the Results of an Accountability Survey Submitted by Signatories of the Global Consensus Statement on MAYE*; World Health Organization: Geneva, Switzerland, 2021.
39. Anja, H.; Sergeant, L. *Planning Education with and for Youth*; UNESCO: Paris, France, 2015.
40. International Labour Organization. *Global Employment Trends for Youth 2022: Investing in Transforming Futures for Young People Executive Summary*; International Labour Organization: Geneva, Switzerland, 2022.
41. Ingaruca, M.; Richard, N.; Carman, R.; Savarala, S.; Jacovella, G.; Baumgartner, L.; Kurukulasuriya, P.; Lister, S.; Flynn, C. *Aiming Higher: Elevating Meaningful Youth Engagement for Climate Action*; United Nations Development Program: New York, NY, USA, 2022.

42. Treseder, P. *Empowering Children & Young People: Training Manual: Promoting Involvement in Decision-Making*; Smith, P.G., Ed.; Save the Children Fund (Great Britain): London, UK, 1997; ISBN 9781899120475.
43. Checkoway, B. What Is Youth Participation? *Child Youth Serv. Rev.* **2011**, *33*, 340–345. [CrossRef]
44. Mackinnon, M.P.; Pitre, S.; Watling, J. *Lost in Translation: (Mis)Understanding Youth Engagement Synthesis Report Charting the Course for Youth Civic and Political Participation*; SoundOut: Reading, UK, 2007.
45. Fletcher, A. *Guide to Students as Partners in School Change*; CIDA: Hull, QC, Canada, 2005.
46. Kellett, M. Three Children and Young People's Participation. In *Children and Young People's Worlds: Developing Frameworks for Integrated Practice*; Montgomery, H., Kellett, M., Eds.; Bristol University Press: Bristol, UK, 2009; pp. 43–60. ISBN 9781847429520.
47. Farthing, R. Why Youth Participation? Some Justifications and Critiques of Youth Participation Using New Labour's Youth Policies as a Case Study. *Youth Policy* **2015**, *109*, 71–97.
48. Lansdown, G. *Conceptual Framework for Measuring Outcomes of Adolescent Participation*; UNICEF: New York, NY, USA, 2019.
49. Cockburn, G. *Meaningful Youth Participation in International Conferences: A Case Study of the International Conference on War-Affected Children, Winnipeg, Canada, September 2000*; CIDA: Hull, QC, Canada, 2001.
50. Shier, H. Pathways to Participation: Openings, Opportunities and Obligations. *Child. Soc.* **2001**, *15*, 107–117. [CrossRef]
51. Van Reeuwijk, M.; Singh, A. Meaningful Youth Participation as a Way to Achieving Success. *Can. J. Child. Rights Rev. Can. Des Droits Des Enfants* **2018**, *5*, 200–222. [CrossRef]
52. Coral, C.; Bokelmann, W. The Role of Analytical Frameworks for Systemic Research Design, Explained in the Analysis of Drivers and Dynamics of Historic Land-Use Changes. *Systems* **2017**, *5*, 20. [CrossRef]
53. Plümper, T.; Troeger, V.E.; Neumayer, E. Case Selection and Causal Inferences in Qualitative Comparative Research. *PLoS ONE* **2019**, *14*, e0219727. [CrossRef]
54. World Bank. World Bank Country and Lending Groups. Available online: https://datahelpdesk.worldbank.org/knowledgebase/articles/906519-world-bank-country-and-lending-groups (accessed on 26 December 2023).
55. Finland Prime Minister's Office. *Finland Report on the Implementation of the 2030 Agenda for Sustainable Development*; Finland Prime Minister's Office: Helsinki, Finland, 2020.
56. Government of Japan. *Report on the Implementation of 2030 Agenda ~Toward Achieving the SDGs in the Post-COVID19 Era~*; Government of Japan: Tokyo, Japan, 2021.
57. Finnish National Commission on Sustainable Development. *The 2030 Agenda Roadmap of the Finnish National Commission on Sustainable Development*; Finnish National Commission on Sustainable Development: Helsinki, Finland, 2022.
58. Finnish National Commission on Sustainable Development. *Strategy of the National Commission on Sustainable Development 2022–2030: A Prosperous and Globally Responsible Finland That Protects the Carrying Capacity of Nature*; Finnish National Commission on Sustainable Development: Helsinki, Finland, 2022.
59. Government of Japan. *SDGs Implementation Guiding Principles Revised Edition*; Government of Japan: Tokyo, Japan, 2019.
60. Government of Japan. *SDGs Action Plan 2023*; Government of Japan: Tokyo, Japan, 2023.
61. SDSN Sustainable Development Report Rankings. Available online: https://dashboards.sdgindex.org/rankings (accessed on 26 December 2023).
62. OECD. Policy Coherence for Sustainable Development Country Profiles (Finland). Available online: https://web-archive.oecd.org/2019-03-15/488087-Country%20Profile%20Finland.pdf (accessed on 26 December 2023).
63. The Finnish National Youth Council Allianssi Allianssi Promotes the Well-Being of Young People. Available online: https://nuorisoala.fi/en/the-finnish-national-youth-council-allianssi/allianssi-promotes-the-well-being-of-young-people/ (accessed on 26 December 2023).
64. OECD. *Policy Coherence for Sustainable Development Country Profiles*; OECD: Paris, France, 2019.
65. Japan Youth Platform for Sustainability About JYPS. Available online: https://en.jyps.website/about-1 (accessed on 13 May 2024).
66. UNDESA. *Multi-Stakeholder Engagement in 2030 Agenda Implementation: A Review of Voluntary National Review Reports (2016–2019)*; UNDESA: New York, NY, USA, 2019.
67. Feldmann-Wojtachnia, E.; Gretschel, A.; Helmisaari, V.; Kiilakoski, T.; Matthies, A.-L.; Meinhold-Henschel, S.; Roth, R.; Tasanko, P. *Youth Participation in Finland and in Germany Status Analysis and Data Based Recommendations*; Nuorisotutkimusseura ry: Helsinki, Finland, 2010.
68. Lochocki, T. *Trends, Causes and Patterns of Young People's Civic Engagement in Western Democracies: A Review of Literature*; Institute for Social Research: Oslo, Norway, 2010.
69. Government Communications Department, Ministry of the Environment. *Youth Participation in Decision-Making on Climate and Nature to Be Strengthened by New Young People's Climate Change and Nature Group*; Government Communications Department, Ministry of the Environment: Helsinki, Finland, 2023.

Disclaimer/Publisher's Note: The statements, opinions and data contained in all publications are solely those of the individual author(s) and contributor(s) and not of MDPI and/or the editor(s). MDPI and/or the editor(s) disclaim responsibility for any injury to people or property resulting from any ideas, methods, instructions or products referred to in the content.

sustainability

MDPI

Article

Methodology for Stakeholder Prioritization in the Context of Digital Transformation and Society 5.0

Ana M. Osorio [1], Luisa F. Úsuga [1], Jaime A. Restrepo-Carmona [2], Isabel Rendón [1], Julián Sierra-Pérez [2,3] and Rafael E. Vásquez [2,3,*]

1 Sustainability Program, Universidad Pontificia Bolivariana, Medellín 050031, Colombia; ana.osoriof@upb.edu.co (A.M.O.); luisafernanda.usuga@upb.edu.co (L.F.Ú.); isabel.rendon@upb.edu.co (I.R.)
2 Corporación Rotorr, Universidad Nacional de Colombia, Cr. 45 26-85, Bogotá 111311, Colombia; jarestrepoca@unal.edu.co (J.A.R.-C.); julian.sierra@upb.edu.co (J.S.-P.)
3 School of Engineering, Universidad Pontificia Bolivariana, Medellín 050031, Colombia
* Correspondence: rafael.vasquez@upb.edu.co; Tel.: +57-4-448-8388

Abstract: This paper addresses a pragmatic and well-articulated qualitative methodology for the identification, prioritization, and consultation of stakeholder groups for a higher education institution as a key element for the organization in the context of digital transformation and Industry 5.0. First, the identification phase required technological surveillance and competitive intelligence, which allowed for defining the organization's stakeholders and their characteristics. Then, the prioritization phase was performed to determine the stakeholders that potentially will have the greatest impact on achieving the institution's strategic objectives to the targets of the Sustainable Development Goals prioritized by the institution, and those who will be most affected (positively or negatively) by the HEI activities. Finally, different methods and technological tools were used for consulting internal and external stakeholders, according to the type of relationship with each group, which allowed the understanding of the perceptions of different stakeholder groups on issues such as gender equity, mental health, regenerative economy, and diversity training. The results are then presented in terms of organizational context, where the concept of stakeholder group was defined by the dynamics of the selected HEI; the prioritized stakeholders include students, employees, academic and research sector, public sector, business sector, social sector, community, archdiocese and diocese, alumni, donors, and benefactors. This approach enabled the identification of issues that became a priority in the university's actions towards the future. Although the presented methodology is mainly qualitative, which can represent a high degree of subjectivity, the stakeholder prioritization exercise provides organizations with inputs for decision making aligned with their needs and expectations. Using such a methodology can help the organization to experience structural changes reflected in improved strategic alignment, understanding, and satisfaction of stakeholders' expectations and needs, enhancement of reputation, risk and conflict mitigation, and the consolidation of long-term healthy and trustworthy relationships, in the context of Society 5.0, where human-centered solutions are expected.

Keywords: stakeholders prioritization; sustainable corporate governance; corporate strategy; stakeholder responsibility; Industry 5.0; Society 5.0

Citation: Osorio, A.M.; Úsuga, L.F.; Restrepo-Carmona, J.A.; Rendón, I.; Sierra-Pérez, J.; Vásquez, R.E. Methodology for Stakeholder Prioritization in the Context of Digital Transformation and Society 5.0. *Sustainability* **2024**, *16*, 5317. https://doi.org/10.3390/su16135317

Academic Editors: Manoj Kumar Nallapaneni, Idiano D'Adamo and Massimo Gastaldi

Received: 24 April 2024
Revised: 25 May 2024
Accepted: 4 June 2024
Published: 21 June 2024

Copyright: © 2024 by the authors. Licensee MDPI, Basel, Switzerland. This article is an open access article distributed under the terms and conditions of the Creative Commons Attribution (CC BY) license (https://creativecommons.org/licenses/by/4.0/).

1. Introduction

In 2015, all the member states of the United Nations (UN) approved the 2030 Agenda for Sustainable Development [1]. The 2030 Agenda aims to end poverty, protect the planet, and ensure prosperity and peace for all human beings [2,3]. The 2030 Agenda, as a global purpose, has required the construction and modeling of different mechanisms that link scenarios of environmental and social order to be more precise in fulfilling its objectives and set goals. The main backbone of the UN Sustainable Development Program is constituted by 17 main Sustainable Development Goals (SDGs), which aim to address current global issues in the areas of economic, social, environmental, and political security [4,5]. In these guidelines generated by the UN through the 2030 Agenda for transforming the world, there is a specific call for the education sector, mainly higher education institutions (HEIs), as they play a key role in increasing students' knowledge about sustainability, transforming their attitudes, and motivating them to promote or participate in sustainable behaviors [6].

The accelerated pace of the world's growth due to business models, digital transformation (DT), process innovation, and other deployed forms of social, economic, and environmental dimensions of sustainability, has generated a temporal and functional categorization of each stage. The term "Industry 5.0" has been included in such sustainable development dialogue as a concept complementing the well-known term "Industry 4.0", since this concept has recently emerged to portray the vision of a future that uses modern technologies for the benefit of society [7,8]. Advocates of Industry 5.0 believe that Industry 4.0 is not the appropriate framework for achieving sustainable development [9], and the European Commission has presented a vision for the future of the European industry in its policy report, named Industry 5.0, which has three central concepts: human-centricity, sustainability, and resilience. The purpose of such central concepts is to integrate a complementary vision between the digital industry and society. Hence, the concept of Society 5.0 also emerges, which aims to balance economic advancement with the resolution of social problems [10,11]. The goal of Society 5.0 is to contribute to the development of an intelligent and human-centric society, allowing all citizens to access a high-quality life, full of comfort and vitality, providing the necessary goods and services for people through the fusion of cyberspace and the physical world.

The integration of humanity into different sectoral environments has required transformations that directly impact the conditions of the economy, markets, digital scenarios, and other structures of social development [12–14]. From here, precise needs arise regarding the effective linkage of new transformation pathways, focusing on the benefits and possibilities of improving human and natural environments [15]. To achieve this, it is necessary to transcend the mere adaptation of organizations to present changes and to generate ecosystems based on a culture of promoting innovation, dynamics of leveraging the digitization of things, and disruptive leadership focused on human capital. As Khairy et al. [16] point out, "In today's accelerated and constantly changing business environment, characterized by rapid technological advancements, unpredictable market conditions, and changing customer expectations, organizational agility has become a critical factor for long-term success".

The transformation from the Fourth Industrial Revolution (Industry 4.0) to 5.0 is currently an ongoing process and is characterized by the integration of advanced technologies such as artificial intelligence, virtual and augmented reality, and advanced robotics, among others [17]. Industry 5.0 opens a scenario of inclusion in issues that directly contribute to environmental sustainability, such as efficient waste management, the use of alternative energies, and the optimization of resource use [18]. Furthermore, it focuses on promoting the formulation of industrial systems that directly impact improving human conditions in their individual and collective forms, such as social well-being, the inclusion of governance principles such as transparency and participation, social appropriation of knowledge, and the creation of standards for social and environmental development [19]. Hence, Industry 5.0 represents a historical moment where it is essential to ensure that technology is used for social welfare purposes [20].

Consequently, skills encompassing emotional intelligence, resilience, empathy, creativity, and critical thinking will emerge as crucial competencies to cultivate a productive and skilled workforce, prepared to meet future demands that include human–machine interactions [21]. Humans excel in areas such as interpersonal interactions, intuition, and complex decision making. Nonetheless, although humans and machines can arrive at the same answer, their different thinking processes result in different paths to conclusions. Building on the aforementioned, it is important to identify those stakeholders who have direct or indirect links to the actions undertaken in line with the integration between industry, its innovations, and developments with the fulfillment of the 2030 Agenda [22]. Therefore, identifying expectations and demands of stakeholders, and integrating them into the strategy of organizations, is a powerful tool for the success of socially responsible positioning [23–25].

As mentioned by Colle [26], stakeholders are all those identifiable groups or individuals on whom an organization depends for its survival, sometimes referred to as primary stakeholders: shareholders, employees, customers, suppliers, and key government agencies. However, on a broader level, a stakeholder is any identifiable group or individual who can affect or be affected by the organization's performance in terms of products, policies, and work [27]. Colle [26] argues that stakeholder marketing is slowly merging with the broader thinking that has emerged in stakeholder management and ethics literature over the past quarter century. However, the prevailing view of stakeholders advocated by many marketing specialists remains primarily pragmatic and company-centered. The position advocated here is that stronger forms of stakeholder marketing are needed to reflect more normative, macro/social, and network-centered orientations [28].

The process of stakeholder prioritization holds pivotal importance in sustainable development, particularly within the context of digital transformation and the advent of Society 5.0. Some recent works show how this prioritization process has been conducted within different contexts, including environmental management [29], circular bioeconomy [30], textile industry [31], electric vehicle industry, [32], and research and innovation projects [33], among others. To understand the needs and expectations of stakeholders in institutions, it is necessary to establish, according to the dynamics of each organization, the concept of stakeholder groups and to compile a list of those who are impacted directly or indirectly by the institution. In the case of HEIs, universities have several stakeholders, and it is impossible to manage them all in the same way, making stakeholder management a complex issue to apply in practice [34,35]. After identifying the stakeholder groups, they should be classified to define consultation strategies and actions.

In this way, this work presents a methodology to show the direct relationship that exists between Industry 5.0 and the strategies of identification, prioritization, consultation, and dynamism of stakeholder groups to contribute to the fulfillment of the objectives and goals proposed in the 2030 Agenda. The proposed methodology has the following contributions: (i) it allows organizations to prioritize their stakeholders about the fulfillment of the 2030 Agenda for Sustainable Development, (ii) it recognizes organizational strategic goals according to their stakeholders, (iii) it is adaptable and flexible to various needs and expectations in different spatial and temporal contexts, and (iv) it is a consultative methodology that is easy to adopt for stakeholders, incorporating Industry 5.0 into decision making. The hypothesis of this work is that organizations that prioritize stakeholder groups using a pragmatic and well-articulated methodology experience structural changes reflected in improved strategic alignment, understanding, and satisfaction of stakeholders' expectations and needs, enhancement of reputation, risk and conflict mitigation, and the consolidation of long-term healthy and trustworthy relationships, in the context of Society 5.0, where human-centered solutions are expected. The organization of the paper is as follows: Section 2 contains the process for identifying, prioritizing, and consulting stakeholders. Section 3 shows the results for a higher education institution. Then, Section 4 contains the discussion, and finally, conclusions are presented in Section 6.

2. Materials and Methods

The process of identifying, prioritizing, and consulting stakeholders of an institution involves a series of steps that can be consolidated into three phases.

2.1. Phase I. Identification

The identification phase comprises three main components, which are described as follows.

- Technological Surveillance and Competitive Intelligence Study. The aim of this study is to analyze quantitative and qualitative valorization methodologies and identify the best models for stakeholder identification [36]. Databases, scientific articles, and company experiences serve as inputs for the first phase.
- Definition of Stakeholders. Based on the results of the technological surveillance and competitive intelligence study, the institution defines the concept of stakeholders and describes their characteristics and purpose. The study provides tools and context for the institution, but the institution formulates the actual concept to describe its stakeholder relationships [37].
- Stakeholder List. The final outcome of this phase is to create the preliminary stakeholder list. This list should contain stakeholders impacted by the institution [37]. It is important that this list includes the institution's mission and strategic plan, especially the strategic lines, programs, and plans.

2.2. Phase II. Prioritization

The prioritization of stakeholders is performed by identifying those who potentially will have the greatest impact on achieving the institution's strategic objectives and those who will be most affected—positively or negatively—by its activities.

Understanding stakeholders as counterparts with whom there must be assertive relationships in favor of a dynamic exchange of ideas and issues to address at the institution, an analysis exercise must be carried out for each of the prioritized stakeholder groups, based on two key variables for the identification and validation of their participation in the institution: contribution and impact. Contribution is the criterion that allows for measuring how much a stakeholder group influences the strategic lines [38], the relationship, and the contribution to achieving the prioritized Sustainable Development Goals' (SDGs) targets [39]; on the other hand, the second variable, impact, allows for identifying the benefit or the impact that stakeholders obtain in the same proposed scope [40].

The stakeholder groups described in the first phase are evaluated based on the variables: contribution and impact, applied to the strategic lines, the relationship, and the prioritization of SDGs and their targets. The rating will range from 0 to 1, with 0 being the value assigned when the criteria have no relationship with the evaluated variables, and the number 1 is assigned to the stakeholder groups that do have some relationship with these variables.

The scoring that allows for obtaining the final result of the stakeholder analysis is defined based on the following percentages:

- Contribution 50%: this percentage corresponds to 20% in strategic lines, 15% in effectiveness in strategic communication, and 15% in contributions to the targets of the prioritized SDGs.
- Impact 50%: this percentage corresponds to 20% in strategic lines, 15% in consultation and relationship, and 15% in incidence of the goals of the prioritized SDGs on stakeholder groups.

The results of the prioritization of stakeholder groups are visualized through a scatter plot, which allows for identifying the degree of contribution and the impact of each one by quadrant. Those groups located in the upper-right quadrant will be selected.

2.3. Phase III. Consultation

To promote the participation of stakeholders, different strategies for internal and external consultation must be designed. Such strategies include workshops, surveys, focus groups, meetings, among others; the use of social networks as a communication channel needs to be promoted to achieve a broader interaction where people can express their ideas and contributions more dynamically [41].

The definition of stakeholders is a joint construction, in which it is important to establish a common definition route, where different areas of the institution participate in all stages of prioritization, from identification and consultation tools to appropriate mechanisms for collecting and analyzing results [37]. In line with this, a consultation exercise needs to be conducted with units that lead relationship matters and that have a direct impact on stakeholders. In the case of an HEI, for instance, these internal stakeholders can include Alumni, Institutional Welfare, Extension, Planning, Teaching, Communications, Marketing, Advanced Training, Research, and Internationalization, among others [42,43]; their opinion on existing stakeholder groups and the consultation methods should be consulted in this process.

3. Results

The methodology for the identification, prioritization, and consultation of stakeholder groups was applied by conducting a case study for the Universidad Pontificia Bolivariana (UPB), which is a private non-profit multicampus HEI, with headquarters located at Medellín, Colombia [44]. Such institution has been promoting the sustainability culture, by creating the Sustainability Office in 2017, and became the first carbon-neutral university in Latin America by 2018 [45].

3.1. Phase I. Identification

3.1.1. Technological Surveillance and Competitive Intelligence Study

The surveillance study carried out by the UPB's Analytics and Context Studies Office was developed in two stages: in the first one, the methods and criteria for prioritizing stakeholder groups were addressed, including the qualitative and quantitative assessment of the criteria and some general guidelines for the analysis of the resulting matrices. The second stage focused on the identification of processes for prioritizing stakeholder groups in national and international HEIs in the context of sustainability strategy; the methodologies of four HEIs were consulted: University of Antioquia [46] (local context), Simón Bolívar University [47] and EAN University [48] (national context), and Royal Melbourne Institute of Technology University [49] (international context).

The identification study yielded the following points for the first stage:

(i) Five sets of criteria for prioritizing stakeholder groups were identified: Power/Interest, Opportunity/Interest, Affectation/Influence, Power/Legitimacy/Urgency (Prominence Matrix), and a criterion integration method.
(ii) The Affectation/Influence matrix was implemented as in [50].
(iii) The criterion integration methodology was applied to the prioritization of stakeholder groups for environmental management in Sharpe's study [29], considering 10 criteria: Level of Interest, Level of Influence, Magnitude of Impact, Probability of Impact, Urgency/Temporal Immediacy, Proximity, Economic Interest, Rights, Equity, and Underrepresented/Underserved Populations.
(iv) The quantitative assessment of prioritization factors/criteria varies depending on the source.
(v) The resulting classification from the assessment process will indicate the level of appropriate treatment for the stakeholder or stakeholder group, and in general, the points of their treatment are addressed in this report.

These results represent an outcome aligned with the central proposal of Industry 5.0, where the inclusion of the human component is prioritized through the identification and prioritization of stakeholder groups, contributing to the theme of human centrality, which is one of the three main concepts of Industry 5.0. The human focus, a central value of Industry 5.0, places humans at the center of production, leading to the prioritization of human needs, ranging from health and safety to self-realization and personal growth [51].

The second stage of the identification phase yielded the following:

(i) The University of Antioquia [46] applies the Affectation/Influence matrix in its stakeholder group prioritization process in the sustainability strategy in the way reported in [50].

(ii) Simón Bolívar University [47], with the support of ARCO Consultores (a consulting company), carried out the identification and prioritization of the university's stakeholder groups in the context of materiality analysis. The work methodology included the following steps: (a) identification of university stakeholder groups, (b) validation of stakeholder groups, and (c) stakeholder group prioritization process.
(iii) The matrix proposed by EAN University [48] considers five criteria: decision making, income generation, business operation, organizational strategy, and reputation.
(iv) The Royal Melbourne Institute of Technology [49] includes the following criteria in its annual sustainability report: representation, dependency, responsibility, and influence.

The attributes related to the effect and influence of stakeholder groups are drivers of an inclusion strategy that takes into account all the components identified in the environment. In this way, resilience patterns were identified, one of the three central concepts of Industry 5.0, thus integrating the human component and the impacts of the industry and the various economic sectors. The main argument of the stakeholder theory is that an organization has relationships with various constituent groups (both internal and external) and can generate and maintain their support by considering and balancing their relevant interests [52].

3.1.2. Sectoral Analysis

An analysis of management and sustainability reports was conducted on various companies and universities to identify the most representative stakeholder groups and the most used prioritization methodologies, supplementing the results of the technological surveillance study. The selected companies are recognized for their sustainability reputation and their national leadership in stakeholders' engagement. The chosen universities were the top 5 of the QS "Universities for Environmental and Social Sustainability" ranking for 2023 [53]. The analyzed organizations and universities include Argos, Bavaria, Enka, AES, Cerrejón, Nutresa, ISA, Postobón, XM, and Uniban, University of California, University of Toronto, University of British Columbia, University of Edinburgh, and University of New South Wales (UNSW). The details of this review and analysis exercise are presented as follows.

The analysis of companies allows for the identification of various strategies they apply to characterize their stakeholder groups. These strategies include the incorporation of Environmental, Social, and Governance (ESG) criteria; the establishment of internal committees; and the creation of strategic maps from sustainable value perspectives, including shareholder value, social and environmental impact, and corporate validity. These companies periodically produce sustainability reports and validate the stakeholder groups every 3 years through materiality analysis.

- Argos: it uses specialized software to monitor external risks and opportunities related to ESG issues across different industries and peers in the construction sector. Through this process, priority stakeholder groups were identified [54].
- AES Colombia: in 2013, AES Colombia identified the most relevant stakeholder groups for the organization through a strategic work session with the AES Colombia Management Committee. During this session, each stakeholder was analyzed, focusing on the impact of the company's operations on them and their impact on the company's operations. This prioritization of stakeholder groups is validated every 3 years by conducting a materiality analysis, during which it is assessed whether there is a need to add or remove any specific group due to the impact of the company's operations and the impact these groups have on the operations [55].
- Cerrejón: it does not specify how stakeholder groups are prioritized. They have identified the following groups: employees and contractors, union members, suppliers, shareholders, customers, governmental and regulatory entities, communities, traditional authorities, non-governmental organizations, trade associations, opinion leaders, civil society, and the media. Annually, they conduct a corporate reputation study [56].

- ISA: for ISA companies, stakeholder groups consist of individuals, organizations, and institutions with whom they build and share common interests and who could be affected by the company's services or activities. The relationship between these groups and the companies is clear, visible, and legitimate [57].
- XM: their direction is based on the company's strategic map, which outlines the organization's objectives from the perspectives of sustainable value: shareholder value, social and environmental impact, and corporate validity. This is achieved through the pillars of its VIDA strategy, which stands (in Spanish) for Green, Innovation, Development, and Articulation. The impacted stakeholder groups include customers, employees, the state, suppliers, and society [58].
- Bavaria: in 2020, the parent company AB InBev conducted a materiality analysis in which they identified key social, environmental, and economic themes prioritized by their stakeholder groups. These were classified in a materiality matrix according to the degree of interest and the potential impact on the business. Civil organizations, non-governmental organizations (NGOs), buyers (retailers), and consumers, regulators, suppliers, investors, and partners are among the external actors considered in the global-level analysis. Bavaria recognizes the importance of working hand in hand with its internal and external stakeholder groups to create high-impact and sustainable projects over time. Currently, they are developing a specific materiality analysis for the operation in Colombia, which will precisely understand the interests and expectations of the stakeholder groups [59].
- Enka: the last update of the Materiality Matrix was carried out in 2021, when the most important issues for our stakeholder groups and the company were defined, which remain valid. It will be reviewed again in 2023, with the aim of including not only the issues that impact the company in its current environment but also those that could affect its future financial performance, in accordance with the new requirements of the Financial Superintendence of Colombia [60].
- Nutresa: the Corporate Governance, Risk Management, and Compliance models of Grupo Nutresa are constantly challenged by the social, political, and economic dynamics of the countries where the organization operates. Therefore, it is fundamental to have an appropriate definition and implementation of controls for the identified risks that, given the environment, could materialize and result in ethical breaches or inappropriate behaviors by third parties and the organization's personnel. This can impact the reputation, the relationship with stakeholder groups, market loss, and legal sanctions [61].
- Postobón: it understands stakeholder groups as representative sets of individuals, groups, or organizations with whom it maintains relationships of trust and credibility, in addition to commitments, in order to achieve its strategic goal of being a sustainable company [62].
- Uniban: for the consultation process, they took a sample of 509 contacts from different stakeholders and received 445 responses, representing an overall participation rate of 87%. Twenty-four percent of the surveys were conducted via email, and 76% by telephone. In these latter cases, some stakeholder groups, such as the community and banana workers, appreciated that Uniban included them in these processes [63].

The analysis of the universities reveals the positioning of the sustainability offices and the integrated work with teaching and research, and highlights the real need to produce reports using international methodologies or, alternatively, prioritize goals with a short- and medium-term scope, in line with global objectives.

- The University of California, Berkeley (UCB), has a sustainability office. The report is available on the website (it is not GRI standard) and does not present a methodology for the identification, consultation, and prioritization of stakeholder groups [64].
- The University of Toronto has a sustainability office and a 10-year strategic plan. It does not produce sustainability reports and lacks a methodology for prioritizing stakeholder groups [65].

- The University of British Columbia has a sustainability strategy based on academic offerings and research. There is a dedicated sustainability team, but they do not produce reports [66].
- The University of Edinburgh has a strategy up to 2030 and a plan that includes 32 commitments to impact the surrounding communities. They do not have a sustainability report or a differentiated strategy for stakeholder groups [67].
- The University of New South Wales (UNSW) has an environmental plan for the period 2022–2024 focused on climate change, physical infrastructure, and efficient use of resources. They have prioritized 9 of the 17 Sustainable Development Goals (SDGs). There is no information related to stakeholder groups provided [68].

3.1.3. Definition of Stakeholders

In accordance with the results of the technological surveillance and sector analysis study, the UPB consolidated the definition of its stakeholder groups; such definition was provided by the Sustainability Office, as follows: "For the University, stakeholders are the actors in the ecosystem of which the UPB is part, with whom it must relate and attend in a strategic and differentiated way through the development and transfer of its value offer, aiming to contribute to transformation, sustainability, and the achievement of everyone's purposes". This redefinition allows for considering individuals and organizations with which the UPB has a direct relationship, in accordance with its mission axes and macroprocesses outlined in the Institutional Development Plan.

3.1.4. Stakeholder´s List

The university has been developing for 87 years with the work of those who have made their contributions from different and relevant roles. For this reason, the identification of UPB's stakeholders has been an exercise valued from the methodology and recognized by its role in the university. Stakeholders have been categorized as internal or external (Figure 1). In this way, a bidirectional relationship of participation has been generated, and different alternatives of communication and consultation methods have been used to allow a continuous, effective, and close dialogue.

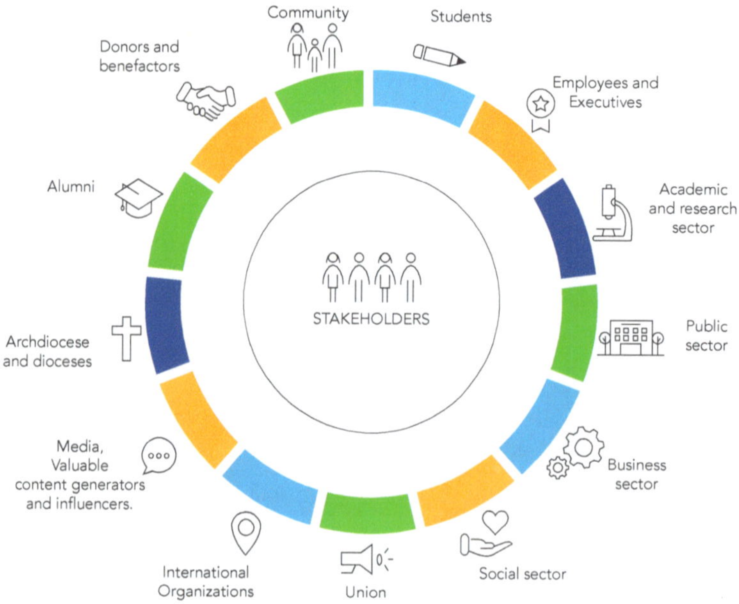

Figure 1. University stakeholders.

3.2. Phase II. Prioritization

As mentioned in the previous chapter, to prioritize stakeholder groups, evaluation criteria were specified. First, the concept of contribution and impact was defined. Following this, the university's four strategic lines were identified, and finally, the goals contributing to the sustainable SDGs prioritized by the UPB were established.

3.2.1. Contribution and Impact

Contribution is the criterion that allows for measuring how much a stakeholder group influences the strategic lines, the relationship, and the contribution to achieving the goals of the prioritized Sustainable Development Goals. On the other hand, the second variable, impact, identifies the benefit or detriment that stakeholder groups receive within the same proposed scopes.

The concept of sustainable development has been adopted by governments worldwide and has deeply rooted itself in the public psyche since its introduction in the report "Our Common Future", proposed by the United Nations World Commission on Environment and Development, often referred to as the Brundtland Commission, over 30 years ago [69].

The incorporation of criteria for achieving the SDGs' targets, based on the identification of stakeholder impacts and contributions, is a strategic mechanism that boosts sustainability as one of the central concepts of Industry 5.0. This is achieved through the actual analysis of contexts in the development of innovation and technological growth, pillars of human centrality and resilience. Although sustainable development is a global cause and society is becoming increasingly sensitive to it, there are many complex challenges, such as the need to align the expectations of various stakeholders, the growing need for innovations in sustainability, implementing social responsibility strategies, and defining and assessing sustainability performance. Furthermore, sustainable development requires collaboration in the form of integrative thinking and action [70,71].

3.2.2. Strategic Lines

The UPB has four strategic lines that were evaluated to identify how they impact and are impacted by different stakeholder groups. First, Institutional Identity strengthens the identity by developing human capabilities and competencies that inspire and guide its actions towards its purpose of social and human transformation. Second, Integral Formation achieves human development based on the values and principles of Christian humanism, covering life's dimensions, ethics, and aesthetics; it promotes scientific training with autonomous, reflective, and critical thinking for integrating care with the world. Third, Creation of Value and Social Impact generates, applies, transfers, and appropriates scientific and social knowledge in a co-creative manner with various actors to effectively contribute to solving contextual problems from local to global levels. Lastly, Sustainability ensures institutional performance based on governance and the value expectations of stakeholder groups in economic, social, and environmental fields, securing the institution's permanence through management founded on the principles of an intelligent and original organization.

3.2.3. SDG Prioritization

As it has been defined by the United Nations [72], "the Sustainable Development Goals (SDGs) represent the universally agreed roadmap to overcome economic and geopolitical divisions, restore trust, and rebuild solidarity. Many proposals are aimed at facilitating the achievement of these goals, such as reforming the international architecture, developing parameters different from Gross Domestic Product (GDP), strengthening digital cooperation, encouraging youth participation in decision-making, transforming education, creating an emergency platform, and promoting a new peace agenda".

To fulfill this roadmap, the UPB carried out a prioritization exercise in 2019 that precisely identified which of the 17 Sustainable Development Goals it impacts directly, in line with its status as a HEI (Figure 2). In this exercise, SDGs 3, 4, 6, 7, 9, 13, 16, and 17 were prioritized. For each SDG prioritized by the university, an exhaustive evaluation of the goals to which it directly contributes was carried out. For example, it was identified that, for SDG 3, the university directly contributes to 6 out of the 13 targets declared in the 2030 Agenda, contributing 46 % towards its fulfillment. Following this prioritization, in 2022, an approach was made to the targets of the prioritized SDGs to determine the institution's contributions to the fulfillment of the 2030 Agenda for Sustainable Development.

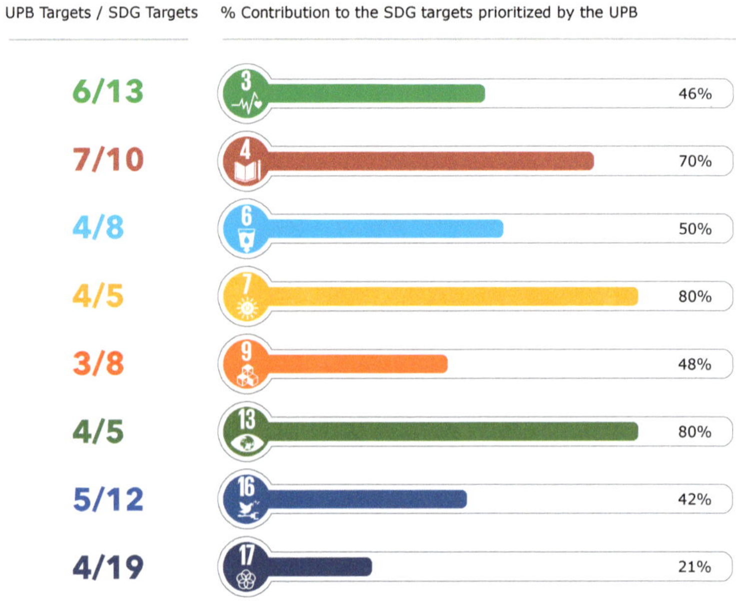

Figure 2. SDG targets prioritized by the UPB.

3.2.4. Internal Consultation

The definition of stakeholder groups is a joint construction in which it is important to establish a common route of definition, where different areas of the university participate in all the prioritization stages, from identification, through consultation tools, to the appropriate mechanisms for the collection and analysis of results. In line with this, a consultation exercise was conducted with the leading units in relationship matters and with a direct impact on stakeholder groups, such as Alumni, Institutional Well-Being, Extension, Planning, Teaching, Communications, Marketing, Advanced Training, Center for the Development of Research and Innovation (CIDI), and Internationalization Office, with the goal of knowing their opinion about the existing stakeholder groups and the consultation methods implemented with them. From this consultation exercise, six responses were obtained to the question, would you add any other stakeholder group? The consulted leaders unanimously responded no. For the other two questions about removing any stakeholder group, responses are shown in Figure 3, and responses on proposals for consultation methods to implement are shown in Figure 4. This last figure shows that the most appropriate consultation mechanisms for the stakeholders are consultation surveys and workshops, followed by a relevance validation exercise, with digital media and focus groups being the last option.

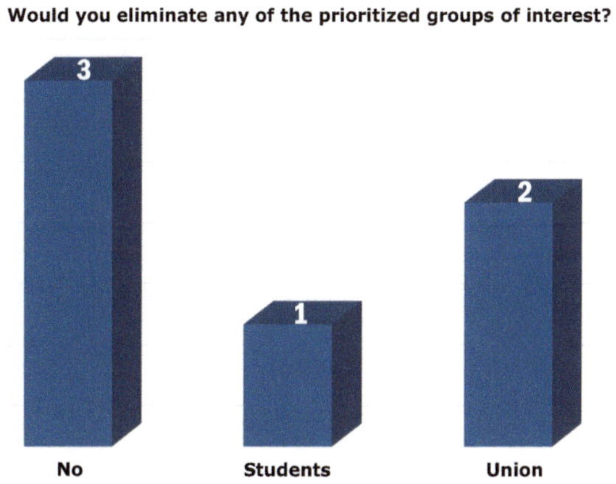

Figure 3. Results of proposal to eliminate interest groups.

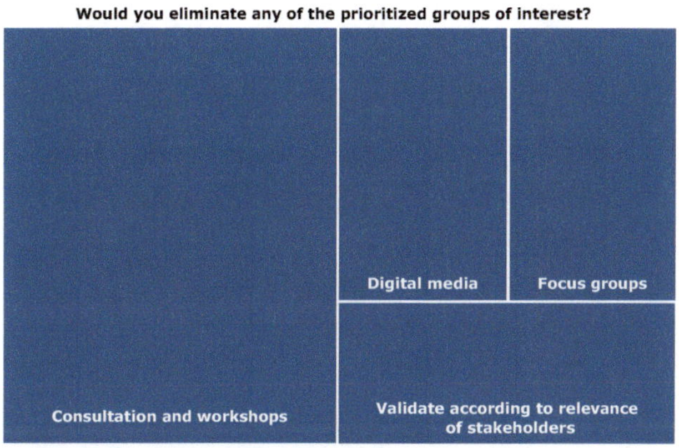

Figure 4. Results with different consultation mechanisms.

3.2.5. Scatter Chart and Interest Groups

With the results obtained from the contribution and impact analysis, using the percentages defined in the methodology, and the questions asked to the units described in the previous section, a scatter plot was generated (Figure 5). The stakeholder groups that were positioned in the upper-right quadrant were selected as priorities. Likewise, a final grouping process was carried out, resulting in the definitive list of stakeholder groups for the UPB.

- Students (undergraduate, graduate, initial education, basic and secondary education, continuing education, and technical and technological training);
- Employees (faculty, administrative staff, and associations);
- Academic and research sector (HEIs, universities, international institutes and academies, research centers and agencies, academic and scientific associations, innovation centers, and public sector);
- Public sector (international governments, national government, and departmental and local governments);
- Business sector (business groups and associations, multilatinas, business groups, companies and international companies, clusters, suppliers, and concessionaires);
- Social sector (NGOs, community organizations, charitable organizations, and media);
- Community (parents of students and applicants, professionals, and professional associations);
- Archdiocese and diocese;

- Alumni;
- Donors and benefactors.

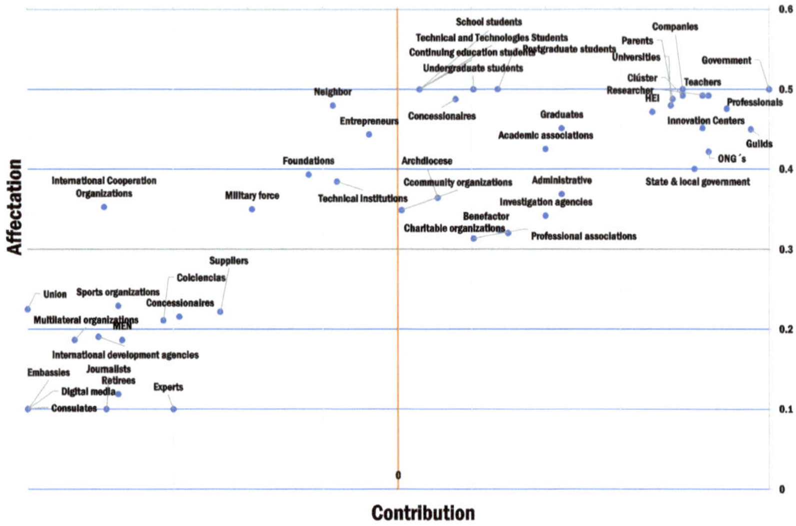

Figure 5. Groups of interest for the UPB.

3.3. *Phase III. Consulting*

Once the stakeholder groups were identified, different methods and technological tools were defined for consulting them, according to the type of relationship with each group. To establish a bidirectional relationship with the stakeholder groups identified and prioritized by the university, the following roadmap was created:

1. Definition of questions related to the Sustainable Development Goals (SDGs) prioritized by the university;
2. Identification of communication channels, methods, and technological tools for consultation;
3. Systematization of the results.

For the student stakeholder group, the consultation was carried out through the university's banner system during the pre-registration process. For employees, a Microsoft Forms questionnaire was created and distributed through all institutional channels. For the other stakeholder groups, mass emails were sent, and targeted approaches were made using Microsoft Forms questionnaires and QR codes.

The result of the systematization provided a broad overview of issues that impact the dimensions of sustainability and become actions of the university. Issues related to gender equity, mental health, regenerative economy, social impact, infrastructure, and safety are highly relevant for stakeholder groups.

4. Discussion

Regarding the fulfillment of the 2030 Agenda, the results evidenced the key role of educational institutions in achieving the SGD's targets, not only in terms of the global challenges in the utilization, demand, and use of natural resources, but also in the consolidation of a critical mass that reflects on its role in society. This supports the idea reported by Filho [73], who stated that universities have a great opportunity to "support the implementation of the SDGs agenda" through teaching, research, and transfer. In his study, Filho points out that universities have revised their curricula and research topics, among other reasons, due to the "increasing demand from students to research and learn about sustainability", making higher education "a key player in promoting engagement on the

SDGs across different sectors and in training students to practice sustainability in their personal and professional lives" [73].

The results obtained with the methodology for the identification, prioritization, and consultation of stakeholder groups for a higher education institution have impacts on sustainable education toward sustainable community development [14,21] as opening a permanent communication channel not only makes them feel relevant and important for the organization's decision making but also ensures that they demand appropriate training in these topics. This allows their positions and initiatives to be considered, creating a long-term critical mass based on global sustainability challenges. For example, students, who are consulted annually, request training through the United Nations Educational, Scientific and Cultural Organization (UNESCO) Chair of Sustainability to have solid arguments and participate in all conversation scenarios. This allows the methodology to serve as an example of how the adoption of new communication forms ensures the success of its implementation.

We conducted a benchmarking exercise on the most representative universities in sustainability worldwide, as per the QS Sustainability Ranking. However, we noticed that these universities do not have any stakeholder definition methodology in their communication channels. This presents an opportunity for our presented methodology to establish long-term strategies that align with stakeholders' needs and expectations. Our approach aligns with the concept of collective intelligence, as defined by Longoria et al. [74], which emphasizes "collaboration and competition among individuals to gather community insights and stimulate designers' creativity. The combination of collective human brain power and modern information technology can be a useful tool for developing sustainable projects".

The presented work emphasizes the importance of establishing reference models with tactical and strategic criteria that allow for the delimitation and consolidation of the definition of stakeholder groups and their long-term impact on the sustainable development of higher education institutions. As mentioned by Srivastava et al. [75], "to maintain sustainability, organizations need to integrate the efforts of all the stakeholders and drive them towards one direction to ensure effective implementation of practices and policies. Since in academic institutions, teachers are key players, the authorities need to focus on their understanding of brand values and commitment towards organizational sustainability".

Defining and prioritizing stakeholder groups is crucial for the development of the idea of human-centricity, which has become a focus in the transition from Industry 4.0 to Industry/Society 5.0. This concept should take into account the organizational, regional, and strategic contexts, and should promote a model where stakeholder groups are central to decision making. In this way, the concept of Industry 5.0 can be integrated into organizational decisions, enhancing transparency and participation. This approach is in line with Srivastava's statement [75]: "Proper training and leadership opportunities can enable employees to participate in decision-making. This can provide more clarity to employees regarding organizational goals and the background behind decisions. Additionally, it can direct their activities towards a common purpose".

Finally, the results obtained by using the stakeholder's prioritization methodology are aligned with the findings the reported by Ghobakhloo et al. [7], which stated, "Scholars believe that the newly introduced Industry 5.0 has the potential to move beyond the profit-centered productivity of Industry 4.0 and to promote sustainable development goals such as human-centricity, socioenvironmental sustainability, and resilience. However, little has been done to understand how this ill-defined phenomenon may deliver its indented sustainability values despite these speculative promises".

5. Limitations and Future Research

Among the most relevant limitations presented by the methodology, it is noteworthy that the results are mainly qualitative and the importance rating has a high degree of subjectivity, depending on the evaluator. This represents a challenge that involves conducting a consultation with hyper-segmented results to a representative sample of the organization, along with a quantitative analysis to facilitate decision making.

The stakeholder prioritization exercise provides organizations with inputs for decision making aligned with their needs and expectations. However, the methodology does not address historical or underlying issues related to information access. To design relevant offers that ensure the closing of gaps, this exercise is consolidated as a starting point for the collection and analysis of information in the digital era.

The methodology has limitations in terms of context and temporality, which requires organizations to update it based on new goals and strategies. Likewise, the prioritization exercise must be accompanied by an analysis of the environment, trends, strengths, and weaknesses, as organizations learn, act, and co-evolve through interaction with their stakeholders.

Although several advantages were identified when prioritizing stakeholders at the UPB, risks must also be considered in future studies, since stakeholders are heterogeneous and changing, which forces the organization to establish expeditious and structured communication routes and to make prioritizations in short periods. Furthermore, it is feasible that, in the long term, value judgments about the importance of one group or another will be involved; hence, quantitative measures are required in future studies. If prioritization does not align with the organization's objectives and strategies, there may be a mismatch between what is prioritized and what is needed to achieve the objectives. Assessing the impact and importance of each stakeholder quantitatively can be complicated, making it difficult to make informed decisions.

6. Conclusions

The prioritization of stakeholders in organizations, and in particular in universities, is a strategic approach that recognizes the importance of identifying and addressing the needs and expectations of the various stakeholders involved in the academic institution. These stakeholders include, but are not limited to, students, faculty, administrative staff, alumni, donors, the local community, governmental bodies, and the industry.

Gathering information and data at universities can be a fundamental task, but it can also present a series of challenges. Some of the most common difficulties organizations face when collecting information and data are related to the lack of access to adequate data sources, as well as the availability and accessibility of the necessary data sources. In many cases, higher education institutions (HEIs) may have difficulty accessing external or internal data, making it challenging to collect verifiable and measurable information. Additionally, the typology of stakeholders determines the difficulty in accessing information, as it is evident that, with some of them, a sensitization and cultural effort is necessary regarding the collection of relevant data and information. For example, efficient consultation mechanisms exist for students, thanks to permanent interaction tools. However, with a stakeholder group like suppliers, the mechanisms are specific and inefficient, often limited to a commercial relationship in some cases.

The relationship between stakeholders and the 2030 Agenda mainly concerns how various interested parties, such as governments, companies, non-governmental organizations (NGOs), civil society, and individual citizens, contribute to the implementation of the Sustainable Development Goals (SDGs) set out in the 2030 Agenda. In this regard, there are several ways in which stakeholders are linked to this agenda, such as participation and collaboration in defining those material issues relevant to HEIs. Stakeholders play a fundamental role in the implementation of the 2030 Agenda by collaborating in the formulation of policies, programs, and projects that promote sustainable development. Governments, for example, can involve businesses, NGOs, and other actors in the planning and execution of initiatives related to the SDGs.

With the exercise of prioritizing and consulting stakeholders, a direct contribution is made to SDG 17 "Partnerships to achieve the SDG", recognizing the importance of collaborative work, dialogue, and participation; the consolidation of partnerships; shared responsibility; joint learning; and the fulfillment of the central and transformative promise of the 2030 Agenda: "Leave No One Behind" (LNOB).

The humanization of stakeholders, in the context of Industry 5.0, refers to the trend of treating the interested parties in the business and technological environment in a more people-centered manner, taking into account not only economic and technical aspects but also social, ethical, and human ones. Industry 5.0, as an evolution of Industry 4.0, is characterized by the integration of cutting-edge technologies such as artificial intelligence, advanced robotics, the Internet of Things, and cybersecurity into industrial processes. The humanization of stakeholders in Industry 5.0 involves aspects such as ethics and social responsibility, from which organizations and companies strive to adopt ethical and socially responsible practices in their decision making and operations. They consider not only the economic benefit but also the social and environmental impact of their actions.

The commitments that must be undertaken with stakeholders are related to open and transparent communication, providing relevant information regularly, and ensuring that the interested parties are informed about important decisions and developments. The most relevant aspects of the relationship between the institution and stakeholders, in light of what is proposed by Industry 5.0, are related to listening to and understanding needs, including them in decision making, fulfilling the commitments made, and being very precise in accountability and transparency in the material matters presented to the interest groups.

Prioritizing interest groups within organizations can lead to significant long-term advantages. This involves focusing efforts on building strong relationships, practicing assertive communication, and allocating resources more effectively. Moreover, understanding the influence or dependence of these groups enables us to anticipate and proactively manage potential risks associated with them. This can help mitigate potential conflicts, address concerns promptly, and gain valuable insights into market trends, customer needs, and innovation opportunities.

Author Contributions: Conceptualization, A.M.O., L.F.Ú., J.A.R.-C., I.R., J.S.-P. and R.E.V.; methodology, A.M.O., L.F.Ú. and I.R.; validation, J.A.R.-C., J.S.-P. and R.E.V.; formal analysis, A.M.O., L.F.Ú., J.A.R.-C., I.R., J.S.-P. and R.E.V.; investigation, A.M.O., L.F.Ú., J.A.R.-C., I.R., J.S.-P. and R.E.V.; writing—original draft preparation, A.M.O., L.F.Ú. and I.R.; writing—review and editing, J.A.R.-C., J.S.-P. and R.E.V.; supervision, A.M.O., J.A.R.-C., J.S.-P. and R.E.V. All authors have read and agreed to the published version of the manuscript.

Funding: This work was developed with the funding of the Universidad Pontificia Bolivariana (UPB) and with the support of the Universidad Nacional de Colombia in association with the Comptroller General of the Republic in the frame of Contract CGR-373-2023.

Institutional Review Board Statement: Not applicable.

Informed Consent Statement: Not applicable.

Data Availability Statement: Data are available on request from the authors.

Conflicts of Interest: The authors declare no conflicts of interest.

Abbreviations

The following abbreviations are used in this manuscript:

DT	digital transformation
ESG	Environmental, Social, and Governance
HEIs	higher education institutions
IoT	Internet of Things (IoT)
NGOs	non-governmental organizations
SDGs	Sustainable Development Goals
UN	United Nations
UNESCO	United Nations Educational, Scientific and Cultural Organization

References

1. United Nations. Transforming Our World: The 2030 Agenda for Sustainable Development. 2023. Available online: https://sdgs.un.org/2030agenda (accessed on 31 October 2023).
2. Graf, A.; Konou, A.A.; Meier, L.; Brattig, N.W.; Utzinger, J. More than seven decades of Acta Tropica: Partnership to advance the 2030 Agenda for Sustainable Development. *Acta Trop.* **2022**, *225*, 106175. [CrossRef] [PubMed]
3. Ali, S.M.; Appolloni, A.; Cavallaro, F.; D'Adamo, I.; Di Vaio, A.; Ferella, F.; Gastaldi, M.; Ikram, M.; Kumar, N.M.; Martin, M.A.; et al. Development Goals towards Sustainability. *Sustainability* **2023**, *15*, 9443 . [CrossRef]
4. Lukin, E.; Krajnović, A.; Bosna, J. Sustainability Strategies and Achieving SDGs: A Comparative Analysis of Leading Companies in the Automotive Industry. *Sustainability* **2022**, *14*, 4000. [CrossRef]
5. Perevoznic, F.M.; Dragomir, V.D. Achieving the 2030 Agenda: Mapping the Landscape of Corporate Sustainability Goals and Policies in the European Union. *Sustainability* **2024**, *16*, 2971. [CrossRef]
6. Leiva-Brondo, M.; Lajara-Camilleri, N.; Vidal-Meló, A.; Atarés, A.; Lull, C. Spanish University Students' Awareness and Perception of Sustainable Development Goals and Sustainability Literacy. *Sustainability* **2022**, *14*, 4552. [CrossRef]
7. Ghobakhloo, M.; Iranmanesh, M.; Mubarak, M.F.; Mubarik, M.; Rejeb, A.; Nilashi, M. Identifying industry 5.0 contributions to sustainable development: A strategy roadmap for delivering sustainability values. *Sustain. Prod. Consum.* **2022**, *33*, 716–737. [CrossRef]
8. Ghobakhloo, M.; Iranmanesh, M.; Foroughi, B.; Rejeb, A.; Nikbin, D.; Tseng, M.L. A practical guide on strategic roadmapping for information and operations technology management: A case study on industry 5.0 transformation. *J. Ind. Prod. Eng.* **2024**, *41*, 397–421. [CrossRef]
9. Leng, J.; Zhong, Y.; Lin, Z.; Xu, K.; Mourtzis, D.; Zhou, X.; Zheng, P.; Liu, Q.; Zhao, J.L.; Shen, W. Towards resilience in Industry 5.0: A decentralized autonomous manufacturing paradigm. *J. Manuf. Syst.* **2023**, *71*, 95–114. [CrossRef]
10. Huang, S.; Wang, B.; Li, X.; Zheng, P.; Mourtzis, D.; Wang, L. Industry 5.0 and Society 5.0—Comparison, complementation and co-evolution. *J. Manuf. Syst.* **2022**, *64*, 424–428. [CrossRef]
11. Slavic, D.; Marjanovic, U.; Medic, N.; Simeunovic, N.; Rakic, S. The Evaluation of Industry 5.0 Concepts: Social Network Analysis Approach. *Appl. Sci.* **2024**, *14*, 1291. [CrossRef]
12. Kraus, S.; Jones, P.; Kailer, N.; Weinmann, A.; Chaparro-Banegas, N.; Roig-Tierno, N. Digital Transformation: An Overview of the Current State of the Art of Research. *Sage Open* **2021**, *11*, 21582440211047576. [CrossRef]
13. Sun, Z.; Wang, W.; Wang, W.; Sun, X. How does digital transformation affect corporate social responsibility performance? From the dual perspective of internal drive and external governance. *Corp. Soc. Responsib. Environ. Manag.* **2024**, *31*, 1156–1176. [CrossRef]
14. Leal-Filho, W.; Salvia, A.L.; Beynaghi, A.; Fritzen, B.; Ulisses, A.; Avila, L.V.; Shulla, K.; Vasconcelos, C.R.; Moggi, S.; Mifsud, M.; et al. Digital transformation and sustainable development in higher education in a post-pandemic world. *Int. J. Sustain. Dev. World Ecol.* **2024**, *31*, 108–123. [CrossRef]
15. Ordonez-Ponce, E. Exploring the Impact of the Sustainable Development Goals on Sustainability Trends. *Sustainability* **2023**, *15*, 16647. [CrossRef]
16. Khairy, H.A.; Baquero, A.; Al-Romeedy, B.S. The Effect of Transactional Leadership on Organizational Agility in Tourism and Hospitality Businesses: The Mediating Roles of Organizational Trust and Ambidexterity. *Sustainability* **2023**, *15*, 14337. [CrossRef]
17. Musarat, M.A.; Irfan, M.; Alaloul, W.S.; Maqsoom, A.; Ghufran, M. A Review on the Way Forward in Construction through Industrial Revolution 5.0. *Sustainability* **2023**, *15*, 13862. [CrossRef]
18. van Erp, T.; Carvalho, N.G.P.; Gerolamo, M.C.; Gonçalves, R.; Rytter, N.G.M.; Gladysz, B. Industry 5.0: A new strategy framework for sustainability management and beyond. *J. Clean. Prod.* **2024**, *461*, 142271. [CrossRef]
19. Lu, Y.; Zheng, H.; Chand, S.; Xia, W.; Liu, Z.; Xu, X.; Wang, L.; Qin, Z.; Bao, J. Outlook on human-centric manufacturing towards Industry 5.0. *J. Manuf. Syst.* **2022**, *62*, 612–627. [CrossRef]
20. Poláková, M.; Suleimanová, J.H.; Madzík, P.; Copuš, L.; Molnárová, I.; Polednová, J. Soft skills and their importance in the labour market under the conditions of Industry 5.0. *Heliyon* **2023**, *9*, e18670. [CrossRef]
21. Thornhill-Miller, B.; Camarda, A.; Mercier, M.; Burkhardt, J.M.; Morisseau, T.; Bourgeois-Bougrine, S.; Vinchon, F.; El Hayek, S.; Augereau-Landais, M.; Mourey, F.; et al. Creativity, Critical Thinking, Communication, and Collaboration: Assessment, Certification, and Promotion of 21st Century Skills for the Future of Work and Education. *J. Intell.* **2023**, *11*, 54. [CrossRef]
22. Pactwa, K.; Woźniak, J.; Jach, K.; Brdulak, A. Including the social responsibility of universities and sustainable development goals in the strategic plans of universities in Europe. *Sustain. Dev.* **2024** . [CrossRef]
23. Castaño-Quintero, C.A.; Díaz-Cáceres, N.; Lozano-Correa, J. *Manual para la Gestión del Relacionamiento con los Grupos de Interés*; Universidad EAN: Bogotá, Colombia, 2012.
24. Taghian, M.; D'Souza, C.; Polonsky, M. A stakeholder approach to corporate social responsibility, reputation and business performance. *Soc. Responsib. J.* **2015**, *11*, 340–363. [CrossRef]
25. ElAlfy, A.; Palaschuk, N.; El-Bassiouny, D.; Wilson, J.; Weber, O. Scoping the Evolution of Corporate Social Responsibility (CSR) Research in the Sustainable Development Goals (SDGs) Era. *Sustainability* **2020**, *12*, 5544. [CrossRef]
26. Colle, S.D. A stakeholder management model for ethical decision making. *Int. J. Manag. Decis. Mak.* **2005**, *6*, 299. [CrossRef]

27. Mu, H.L.; Xu, J.; Chen, S. The impact of corporate social responsibility types on happiness management: A stakeholder theory perspective. *Manag. Decis.* **2024**, *62*, 591–613. [CrossRef]
28. Laczniak, G.R.; Murphy, P.E. Stakeholder Theory and Marketing: Moving from a Firm-Centric to a Societal Perspective. *J. Public Policy Mark.* **2012**, *31*, 284–292. [CrossRef]
29. Sharpe, L.M.; Harwell, M.C.; Jackson, C.A. Integrated stakeholder prioritization criteria for environmental management. *J. Environ. Manag.* **2021**, *282*, 111719. [CrossRef] [PubMed]
30. Santos, J.M.; Fernandes, G. Prioritizing stakeholders to boost collaborative R&I projects benefits: An analytic network process approach. *Procedia Comput. Sci.* **2023**, *219*, 1660–1669 [CrossRef]
31. Singhal, N. Stakeholders sustainable development goals (SDGs) prioritization. *Bus. Strategy Dev.* **2023**, *6*, 986–990. [CrossRef]
32. van der Koogh, M.; Chappin, E.; Heller, R.; Lukszo, Z. Stakeholder prioritizations for electric vehicle charging across time periods. *Transp. Policy* **2023**, *142*, 173–189. [CrossRef]
33. Santos, J.M.R.C.A.; Fernandes, G. Prioritizing Stakeholders in Collaborative Research and Innovation Projects Toward Sustainability. *Proj. Manag. J.* **2024**, 87569728241231266. [CrossRef]
34. Aerts, G.; Cauwelier, K.; de Pape, S.; Jacobs, S.; Vanhondeghem, S. An inside-out perspective on stakeholder management in university technology transfer offices. *Technol. Forecast. Soc. Change* **2022**, *175*, 121291. [CrossRef]
35. Brdulak, A.; Stec, B. Concept of sustainable development at Wrocław University of Science and Technology based on the perspective of selected stakeholder groups. *Oper. Res. Decis.* **2024**, *34*, 61–89. [CrossRef]
36. López-Robles, J.R.; Otegi-Olaso, J.R.; Porto-Gomez, I.; Gamboa-Rosales, H.; Gamboa-Rosales, N.K. Understanding the intellectual structure and evolution of Competitive Intelligence: A bibliometric analysis from 1984 to 2017. *Technol. Anal. Strateg. Manag.* **2020**, *32*, 604–619. [CrossRef]
37. Dmytriyev, S.D.; Freeman, R.E. (Eds.) *R. Edward Freeman's Selected Works on Stakeholder Theory and Business Ethics*; Springer Nature: Berlin/Heidelberg, Germany, 2023. [CrossRef]
38. Falqueto, J.M.Z.; Hoffmann, V.E.; Gomes, R.C.; Onoyama Mori, S.S. Strategic planning in higher education institutions: What are the stakeholders' roles in the process? *High. Educ.* **2020**, *79*, 1039–1056. [CrossRef]
39. Silva, S. Corporate contributions to the Sustainable Development Goals: An empirical analysis informed by legitimacy theory. *J. Clean. Prod.* **2021**, *292*, 125962. [CrossRef]
40. Pedro, E.d.M.; Leitão, J.; Alves, H. Stakeholders' perceptions of sustainable development of higher education institutions: An intellectual capital approach. *Int. J. Sustain. High. Educ.* **2020**, *21*, 911–942. [CrossRef]
41. Talbot, D.; Raineri, N.; Daou, A. Implementation of sustainability management tools: The contribution of awareness, external pressures, and stakeholder consultation. *Corp. Soc. Responsib. Environ. Manag.* **2021**, *28*, 71–81. [CrossRef]
42. Caeiro, S.; Sandoval Hamón, L.A.; Martins, R.; Bayas Aldaz, C.E. Sustainability Assessment and Benchmarking in Higher Education Institutions—A Critical Reflection. *Sustainability* **2020**, *12*, 543. [CrossRef]
43. Núnez Chicharro, M.; Mangena, M.; Alonso Carrillo, M.I.; Priego De La Cruz, A.M. The effects of stakeholder power, strategic posture and slack financial resources on sustainability performance in UK higher education institutions. *Sustain. Accounting, Manag. Policy J.* **2024**, *15*, 171–206. [CrossRef]
44. Osorio, A.M.; Úsuga, L.F.; Vásquez, R.E.; Nieto-Londoño, C.; Rinaudo, M.E.; Martínez, J.A.; Leal Filho, W. Towards Carbon Neutrality in Higher Education Institutions: Case of Two Private Universities in Colombia. *Sustainability* **2022**, *14*, 1774. [CrossRef]
45. Universidad Pontificia Bolivariana. Sostenibilidad UPB. 2024. Available online: https://www.upb.edu.co/es/sostenibilidad (accessed on 30 April 2024).
46. Universidad de Antioquia. UdeA Sostenible. 2023. Available online: https://www.udea.edu.co/wps/portal/udea/web/inicio/institucional/udea-sostenible (accessed on 15 December 2023).
47. Universidad Simón Bolívar. Planeación. 2023. Available online: https://www.unisimon.edu.co/servicios/planeacion/848/3185 (accessed on 15 December 2023).
48. Universidad EAN. EAN Sostenibilidad. 2023. Available online: https://universidadean.edu.co/politica-de-sostenibilidad-y-emprendimiento-sostenible (accessed on 15 December 2023).
49. Royal Melbourne Institute of Technology. Sustainability RMIT. 2023. Available online: https://www.rmit.edu.au/about/our-values/sustainability (accessed on 15 December 2023).
50. Espinal-Ospina, L. Orientaciones Hacia la Formulación de la Estrategia de Sostenibilidad de la Empresa INTEINSA: Gobernanza y Grupos de Interés. Master's Thesis, Master's Program in Sustainability, Universidad Pontificia Bolivariana, Medellín, Colombia, 2023.
51. Wang, B.; Zhou, H.; Li, X.; Yang, G.; Zheng, P.; Song, C.; Yuan, Y.; Wuest, T.; Yang, H.; Wang, L. Human Digital Twin in the context of Industry 5.0. *Robot. Comput.-Integr. Manuf.* **2024**, *85*, 102626. [CrossRef]
52. Morkan, B.; Bertels, H.M.; Sheth, A.; Holahan, P.J. Building megaproject resilience with stakeholders: The roles of citizenship behavior and critical transition mechanisms. *Int. J. Proj. Manag.* **2023**, *41*, 102485. [CrossRef]
53. QS Top Universities. Top Universities for Environmental and Social Sustainability 2023. 2023. Available online: https://www.topuniversities.com/university-rankings/qs-sustainability-ranking/top-universities-environmental-social-sustainability-2023 (accessed on 30 April 2024).
54. Argos. Reporte Integrado 2022. Cementos Argos. 2023. Available online: https://argos.co/reporte-integrado/ (accessed on 30 October 2023).

55. AES. Informe de Sostenibilidad 2022. 2023. Available online: https://www.aescol.com/es/sostenibilidad (accessed on 30 September 2023).
56. Cerrejón. Informe de Sostenibilidad 2022. Cerrejón 2023. Available online: https://www.cerrejon.com/sostenibilidad/informes-de-sostenibilidad (accessed on 30 March 2023).
57. ISA. Reporte Integrado de Gestión 2022. 2023. Available online: https://www.isa.co/es/grupo-isa/reporte-integrado-de-gestion-isa-2022/ (accessed on 30 October 2023).
58. XM. Reporte Integral de Sostenibilidad, Operación y Mercado 2021. 2022. Available online: https://informeanual.xm.com.co/informe/pages/xm/05-grupos-de-interes-impactados.html (accessed on 30 September 2023).
59. Bavaria. Informe Desarrollo Sostenible 2022. 2023. Available online: https://www.bavaria.co/desarrollo-sostenible/informes-de-desarrollo-sostenible-bavaria (accessed on 30 October 2023).
60. ENKA. Informe de Sostenibilidad 2021. 2022. Available online: https://www.enka.com.co/informe-de-sostenibilidad/ (accessed on 30 October 2023).
61. Nutresa. Informe Integrado 2022. 2023. Available online: https://gruponutresa.com/sostenibilidad/nuestra-gestion-en-sostenibilidad/informes-de-sostenibilidad/ (accessed on 30 October 2023).
62. Postobon. Informe de Sostenibilidad 2022. Available online: https://informe2022.postobon.com/ (accessed on 30 October 2023).
63. Uniban. Informe de Sostenibilidad 2021. 2022. Available online: https://uniban.com/wp-content/uploads/Uniban_informe_GRI_2021.pdf (accessed on 30 October 2023).
64. University of California, Berkeley. Berkeley Annual Sustainability Report. 2022. Available online: https://sustainability.berkeley.edu/plans-reports/sustainability-reports (accessed on 30 September 2023).
65. University of Toronto. Sustainability Strategic Plan. 2021. Available online: https://www.utm.utoronto.ca/green/sustainability-progress-report (accessed on 30 September 2023).
66. University of British Columbia. Climate & Sustainability Report. 2021. Available online: https://sustain.ok.ubc.ca/reports/ (accessed on 30 September 2023).
67. University of Edinburgh. Community Plan 2020–2025. 2020. Available online: https://www.ed.ac.uk/local/our-community-plan (accessed on 30 September 2023).
68. University of New South Wales. Environmental Sustainability Plan 2022–2024. 2022. Available online: https://www.sustainability.unsw.edu.au/our-plan (accessed on 30 September 2023).
69. Huang, R. SDG-oriented sustainability assessment for Central and Eastern European countries. *Environ. Sustain. Indic.* **2023**, *19*, 100268. [CrossRef]
70. d Keeys, L.A.; Huemann, M. Project benefits co-creation: Shaping sustainable development benefits. *Int. J. Proj. Manag.* **2017**, *35*, 1196–1212. [CrossRef]
71. D'Adamo, I.; Gastaldi, M. Monitoring the Performance of Sustainable Development Goals in the Italian Regions. *Sustainability* **2023**, *15*, 14094. [CrossRef]
72. United Nations. *The Sustainable Development Goals Report 2023*; Department of Economic and Social Affairs: New York, NY, USA, 2023. Available online: https://unstats.un.org/sdgs/report/2023/ (accessed on 15 December 2023).
73. Filho, W.L.; Salvia, A.L.; Eustachio, J.H.P.P. An overview of the engagement of higher education institutions in the implementation of the UN Sustainable Development Goals. *J. Clean. Prod.* **2023**, *386*, 135694. [CrossRef]
74. Longoria, L.C.; López-Forniés, I.; Sáenz, D.C.; Sierra-Pérez, J. Promoting sustainable consumption in Higher Education Institutions through integrative co-creative processes involving relevant stakeholders. *Sustain. Prod. Consum.* **2021**, *28*, 445–458. [CrossRef]
75. Srivastava, A.P.; Mani, V.; Yadav, M. Evaluating the implications of STAKEHOLDER'S role towards sustainability of higher education. *J. Clean. Prod.* **2019**, *240*, 118270. [CrossRef]

Disclaimer/Publisher's Note: The statements, opinions and data contained in all publications are solely those of the individual author(s) and contributor(s) and not of MDPI and/or the editor(s). MDPI and/or the editor(s) disclaim responsibility for any injury to people or property resulting from any ideas, methods, instructions or products referred to in the content.

sustainability

MDPI

Article

Achieving the 2030 Agenda: Mapping the Landscape of Corporate Sustainability Goals and Policies in the European Union

Florentina Madalina Perevoznic and Voicu D. Dragomir *

Faculty of Accounting and Management Information Systems, Bucharest University of Economic Studies, 010374 Bucharest, Romania; perevoznicflorentina17@stud.ase.ro
* Correspondence: voicu.dragomir@cig.ase.ro

Abstract: The United Nations Sustainable Development Goals (UN SDGs) were introduced in 2015 to advance the 2030 Agenda of sustainable development in all supporting countries. The SDGs are applicable to countries, non-governmental organizations, industries, and companies. In this article, we focus on the contribution of listed companies headquartered in the European Union (EU) to the SDGs. The EU intends to be the front-runner in the race for sustainable development and has adopted comprehensive strategies that mirror the UN SDGs. For this reason, we collected relevant data points from the Refinitiv Eikon database for 1156 companies headquartered in EU countries for the financial year 2022. The data collected refer to contributions to each SDG and the adoption of corporate sustainability policies. Data were statistically analyzed per country and sector to generate a comprehensive image of industry contributions to the SDGs in the EU. By applying a comparative analysis of country-level achievements and policies, the results point to four EU countries that are significant contributors to the SDGs through their economic activities. At the same time, other EU countries are still facing significant challenges in this domain. The socioeconomic considerations for these cases are laid out in the Discussion section. The present article offers a snapshot of corporate contributions to the SDGs as climate and geopolitical challenges become more prominent.

Keywords: sustainable development goals; European Union; corporate policy; Agenda 2030; sustainability goals

Citation: Perevoznic, F.M.; Dragomir, V.D. Achieving the 2030 Agenda: Mapping the Landscape of Corporate Sustainability Goals and Policies in the European Union. *Sustainability* **2024**, *16*, 2971. https://doi.org/10.3390/su16072971

Academic Editors: Idiano D'Adamo, Massimo Gastaldi and Manoj Kumar Nallapaneni

Received: 4 February 2024
Revised: 17 March 2024
Accepted: 25 March 2024
Published: 2 April 2024

Copyright: © 2024 by the authors. Licensee MDPI, Basel, Switzerland. This article is an open access article distributed under the terms and conditions of the Creative Commons Attribution (CC BY) license (https://creativecommons.org/licenses/by/4.0/).

1. Introduction

The 2030 Agenda for Sustainable Development was adopted by the United Nations (UN) in late September 2015 and became effective in January 2016 [1]. However, the Sustainable Development Goals (SDGs) had been proposed in 2012. The adoption year represents a milestone as 13% of policies issued by 71 countries as of 2016 were linked to the SDGs. By 2021, the percentage increased to 52% in 84 countries [2]. The commitment to the 2030 Agenda was undertaken by 193 UN member countries dedicated to transforming and addressing global challenges for a more sustainable and equitable future [2,3]. The 2030 Agenda encompasses 17 SDGs and 169 specific targets providing guidance across environmental, social, and governance (ESG) dimensions for nations seeking a more sustainable development.

The SDGs represent a universal call to action, recognizing five concepts referred to as the 5Ps: people, planet, prosperity, peace, and partnerships [4]. This call requires collective efforts to eradicate poverty and hunger, promote prosperity for all, ensure good health and well-being, provide quality education, achieve gender equality, and protect the planet through climate action. The SDGs were created as a result of negotiations between stakeholders and governmental entities [5], and there may be overlaps or potential contradictions between some of the targets associated with the SDGs. However, mappings between the SDGs and economic, social and environmental (ESG) factors demonstrate the relevance and urgency of the SDGs [6]. Organizations are important actors in the implementation of sustainable business practices to create value for investors, but also stakeholders, with the aim of limiting the effects of climate change and facilitating the transition to a greener economy [7].

The European Union (EU) has sought the position of the frontrunner in the sustainability race. The European Commission and the European Parliament adopted a series of strategies in several domains of sustainable development. Some of these strategies include the European Green Deal [8,9], aiming to make the EU climate-neutral by 2050 through policies across industries that support the Farm to Fork strategy [10]; the Circular Economy Action Plan [11,12] that focuses on promoting sustainable practices in waste management; and the Biodiversity Strategy for 2030 [13], outlining actions to protect nature and restore biodiversity. These strategies are closely coordinated with the corresponding SDGs and are significant for various industries.

The most recent literature (2021–2023) comprises various topics such as sustainability during the COVID-19 pandemic and how this affected the pursuit of the SDGs [14,15]; the analysis of a limited number of SDGs in and outside the European Union [16,17]; and the sustainability-related impacts of specific industries [18]. Our article responds to the call [19] for research to clarify the responsibilities of large multinationals in the domain of energy transition, decarbonization, and environmental protection, as well as in the social realm, in areas such as gender equality and education. We have identified a research gap in the specificity of SDG adoption in each industrial sector in relation to country-specific conditions. This research gap represents a challenge, prompting us to explore the dynamics and corporate behaviors associated with SDG adoption.

While there is a growing recognition of the importance of achieving the SDGs [20,21], a more granular analysis is required to discern the specific factors that influence the successful implementation of sustainable practices. Sector-specific research facilitates the understanding of different development strategies and policies linked to the characteristics of each industry, taking into consideration factors such as available resources and the complexity of the industry in connection to the regulatory frameworks at the national level.

The present research analyzes the applicability of the SDGs and sustainability policies among companies headquartered in each EU member country. The article uses a dual framework that encompasses both country- and industry-based perspectives. The investigation aims to determine to what extent the SDGs align with the various socioeconomic contexts within the EU states. The paper provides insights into national and sectoral dynamics, observing the challenges and opportunities associated with the integration of SDGs and sustainability policies. This granularity allows for the analysis of the different roles of industries in supporting sustainable development. By identifying the countries that significantly contribute to the SDGs, we envisage a wider dissemination of knowledge and an adaptation of effective business models, leading to progress among all nations. The present article enriches academic understanding of SDG implementation by providing an analysis of the national and industry contexts within the European Union.

The structure of the present article is as follows. The literature review is divided between studies approaching SDG implementation at the country level and those focusing on implementation at the industry level, also identifying general research questions. The Materials and Methods section presents the process of collecting data from the Refinitiv Eikon database representative of the 2022 financial year for 1156 companies. The binary data collected represent the status of adherence of EU companies to each of the 17 SDGs, as well as 24 sustainability policies. The analytical questions are discussed in the Results section of the article. Furthermore, in the Discussion section, we adopt a comparative perspective and highlight the best practices of the top countries contributing to the SDGs, versus the situation of the least-contributing countries. The final section identifies the limitations of the research and proposes avenues for future investigation.

2. Literature Review and Research Questions

2.1. SDGs at the Country Level

The UN's 2030 Agenda serves as a common language for all the participating nations and presents a roadmap for companies and non-governmental organizations [22]. To monitor the progress on sustainability aspects, the UN Department of Economic and Social Affairs (UNDESA) proposed sustainable development (SD) indicators based on the 17 SDGs [23]: no poverty; zero hunger; good health and well-being; quality education; gender equality; clean water and sanitation; affordable and clean energy; decent work and economic growth; industry, innovation, and infrastructure; reduced inequalities; sustainable cities and communities; responsible consumption and production; climate action; life underwater and life on land; peace, justice, and strong institutions; and partnership for the goals.

Most countries have already begun the implementation of the SDGs and are monitored through an SDG progress index grouped by regional dimension [24]. In this regard, social inclusion represents a global challenge. Based on this premise, social inclusion extends beyond interpersonal relations, having implications at the macro level in the form of decreased poverty and positive ecological impacts. As observed by Mosse [25], adversities in the field of sustainable development are not isolated events but rather outcomes of strategic decisions. Within the framework of the SDGs, addressing these challenges is critical for sustainable and inclusive development. Poverty, insufficient education, and ecological concerns underscore the pressing need to adjust corporate strategies to achieve the SDGs.

Initial reviews showed slow progress at the national level in setting targets and evaluating policies as there was limited experience of national governments in this domain [26]. The positive impact lies in the integration of the SDGs into national policies that align with one of the 169 specified targets [27]. The previous literature [28] revealed that of the 169 targets, 49 targets (29%) were well-developed, 91 targets (54%) could benefit from increased specificity, and 29 targets (17%) required substantial improvement. The main weaknesses included the inadequate alignment of targets and goals with existing international agreements and political processes, ineffective implementation, and potential conflicts between goals and targets.

In Europe, North America, and Asia, the SDG scores trended toward high and extremely high levels [6]. Before the implementation of the 2030 Agenda, in 2017, the European Commission created a reference framework to monitor the applicability of the SDGs in the European context. The SDG indicator set was created in alignment with the UN's list of global sustainability indicators [23]. Based on the SDG index [24], Finland, Sweden, Denmark, Germany, Austria, France, Norway, Czechia, Poland, and Estonia are the ten most developed countries. Due to the danger posed by climate change [29], Finland set ambitious climate targets, with significant progress [30]. Sweden has a rich history of sustainable development, such as being the frontrunner of Agenda 21 [31], and acted as a role model for the Arctic countries, including Norway, in order to reach the UN goals by 2030 together [32]. Czechia also has a vast knowledge of sustainable development, being a model for other Central and Eastern European countries in achieving the SDGs [33].

Previous results point to the fact that the level of SDG implementation varies widely: higher sustainability is observed in the Nordic and Baltic states, the Netherlands, and Austria, and lower goal achievement is present in southern and eastern countries, especially Romania, Bulgaria, and Greece [34,35]. Researchers have completed country-level sustainability assessments for Romania [36], the Czech Republic, Hungary, Poland, Slovakia, and other countries within the European Union [37,38]. The Czech Republic, Hungary, Poland, and Slovakia, referred to as the V4 (Visegrád Group) countries, have implemented sustainable development strategies and are significantly improving their domestic environmental policies—for example, the Strategy for Responsible Development was implemented by Poland as a response to the 2030 Agenda [39]. For Central and Eastern Europe, an aggregate sustainability index was created as a comparison tool, using multiple methods and various indexes that measure sustainability performance in several areas [40].

2.2. SDGs at the Company Level

Sustainability reporting involves the public disclosure of companies' economic, environmental, and social impacts, along with their adherence to the SDGs and related policies [41]. The Non-financial Reporting Directive (2014/95/EU) [42] required the preparation of non-financial statements in conjunction with the annual reports of EU companies. The non-financial statements must present companies' policies on environmental and social impacts, alongside information on the respect for human rights, diversity on the company's board, anti-corruption, and bribery prevention. The updated Corporate Sustainability Reporting Directive (2022/2464) [43] extended the scope of existing legislation and made reference to the UN SDGs. The newer directive introduces stronger disclosure requirements, with quantitative indicators on progress towards sustainability targets and clear policies on environmental, social, and governance (ESG) aspects. On the other hand, companies are aware that the reporting of sustainability policies, programs, and projects can generate negative feedback from the community due to increased transparency and comparability [44]. For this reason, sustainability reporting is often accused of greenwashing [45] or pinkwashing [46].

Companies that contribute to the SDGs consider that this process is supported by good governance practices such as the presence of independent directors and experts in non-financial reporting [47]. Organizations align with SDG targets that are directly related to their activities in the value chain and tend to focus on those that minimize negative impacts [48,49]. For greater coverage of the SDGs, it is important that researchers observe them beyond the industry-specific objectives [50]. The potential of sustainable supply chain management (SSCM) to contribute to the adoption of the SDGs is realized through optimized processes, the reduction of environmental impacts, the lowering of operational costs, and the promotion of responsible consumption and production.

Previous research on factors that influence non-financial disclosure refers to voluntary disclosure theory, legitimacy theory, and stakeholder theory to elucidate the voluntary non-financial reporting behavior of companies [51]. SDG indicators were identified to monitor existing policies and forecast trends until 2030 based on a data set available for the period 2007–2018 [52]. For the period 2015–2018, the SDG disclosure increased to 58% in Europe for a population of 1732 companies. A sample of 652 enterprises provided information on the SDGs, with a great proportion represented by France and the United Kingdom (UK) [53]. The UK was excluded in the present research as it has not been part of the European Union since 2020. The goals most often prioritized were SDGs 3, 8, 9, 12, and 13 [53]. At the European level, there is no country on track to achieve all the targets and SDGs related to well-being and health, but achieving half of the proposed targets within the 2030 Agenda is feasible if the current level of engagement is maintained [52].

2.3. SDGs at the Industry Level

In 2018, it was difficult for companies to identify the SDGs to which they contributed directly and indirectly. Most companies in the utilities (58%), automotive (58%), retail (57%), and technology (56%) sectors were more likely to report on the SDGs, with the healthcare industry (47%) in the top five contributing industries. In contrast, the financial services (37%), industrials (30%), and oil and gas (28%) industries were less likely to report on the SDGs [54]. The significance of SDG reporting in environmentally sensitive industries (e.g., oil and gas, heavy metals, paper, chemicals, and utilities) has become more pronounced as these industries face increased stakeholder and social pressure and might use SDG reporting as a strategy to repair their reputation [55].

Regarding the metals and mining industry, the most pursued goals are SDGs 3, 8, and 12, which are directly linked to the activities of the companies [56]. As an energy-intensive sector, it is important to understand how companies in this industry reduce costs and increase energy efficiency. In addition to the 2030 Agenda, companies in this sector should consider that the global steel demand will increase by 2050, together with greenhouse gas (GHG) emissions [57]. The researchers found that only 26% of companies in the steel industry based in China made the connection between the company's activity and SDG compliance [56].

Another important actor is the automotive industry, which can easily reach certain goals but at the same time is responsible for severe environmental issues such as air pollution and greenhouse gas emissions, mainly downstream in the value chain [58]. Automotive manufacturers face increasing challenges as the European Parliament approved a strategy to phase out combustion engines in new passenger vehicles starting in 2035 [59]. This approach demands technological advancement in battery production for electric vehicles to make them competitive in the market.

The construction industry is of great importance for the achievement of the SDGs in the EU [60] because it generates large amounts of greenhouse gas emissions from the processing of raw materials [61]. Regarding the construction industry, specific policies have been developed to promote sustainable business models in this sector, namely the European Commission's Construction Products Regulation (CPR) [62]. As indicated in the literature, the construction sector can contribute to SDGs 6, 8, 9, 11, and 12 [60]. The procurement process for non-renewable resources during construction projects has an impact on the achievement of the SDGs, especially SDG 12 [63,64].

SDGs 1, 2, 5, 8, 13, and 15 are extremely relevant to the food industry, especially "eradicating poverty" and "zero hunger" [65,66]. Researchers have observed that, for the social dimension, there is high employment for women in the agricultural industry, especially in regions with low food security. On the contrary, in regions that have sufficient food security, women can work outside the agriculture industry and increase their income. On the other hand, in terms of sustainability, food security does not pose a threat to biodiversity [65]. Together with the food industry, the agriculture sector aims to satisfy the demand for food in a sustainable way, to enhance resource allocation and promote equal access [67]. Sustainable agriculture constructs the base for SDGs 3 and 13 [68] but also for the most relevant SDGs identified by the Food and Agriculture Organization of the UN (FAO): SDGs 2, 6, 8, 12, 14, and 15 [67].

The financial sector has focused on SDG 3 through partnerships with companies in the healthcare industry; SDG 7 through investments related to environmental challenges; SDG 8 through creating space for career and skill development; SDG 9 through the digitalization of internal processes and the development of sustainable services; and SDG 13 by supporting climate research and clients willing to change their business models towards sustainable and eco-friendly activities [69].

As the previous literature has not focused sufficiently on the overall adoption of the SDGs in specific industries, we propose the following research questions, which will be operationalized in more detail:

Q1. How many companies headquartered in the EU apply each SDG and related sustainability policies?

Q2. How many companies in each EU country follow the SDGs and implement sustainability policies?

Q3. Which SDGs and corporate sustainability policies are applied in each industry and EU country?

Q4. What is the relationship between the SDGs and the corresponding policies as applied by the sample companies?

3. Materials and Methods

The research sample for the study comprises publicly traded companies in the 27 European member states. The member states are characterized by the geographic scope and diversity inherent in the European Union, which ensures a comprehensive representation of various economic, cultural, and regulatory environments [70]. The data set was compiled from the Refinitiv Eikon database, which is part of the London Stock Exchange Group (LSEG). Refinitiv is known for its detailed ESG data set, which contains over 450 metrics with a high level of granularity on various dimensions of sustainability [71].

This includes coverage of dimensions such as environmental strategies, resource use efficiency, pollution abatement, waste reduction, environmental innovation, human rights, and product responsibility, among others. Data availability over more than a decade facilitates longitudinal studies and has become an important tool for researchers [71–73]. Refinitiv applies a transparent methodology to collect and verify ESG metrics, with frequent updates that ensure access to the latest and most relevant information. Furthermore, the data set is suitable for conducting cross-country and cross-industry comparative analyses.

The collected data set is cross-sectional for the 2022 financial year, which is the most recent period for analysis at the time of download (end of November 2023). This creates a very recent background for our research. The database comprises 7015 companies, as observed in Table 1, allowing a better understanding of trends and patterns within the European Union business landscape. The results presented in Table 1 are an overview of the availability of data on listed companies in European countries. The sample data complement the series of research papers investigating the role of European companies in advancing SDG adherence within the EU [74,75]. Our study analyzes companies operating within the EU, which, as members of the UN, play an important role in global sustainability efforts. Through our analysis, we provide an enhanced perspective on the contributions made by EU-based companies towards the 2030 Agenda. The focus of data collection was to identify and analyze Boolean data (TRUE or FALSE) representing the adherence or nonadherence of each company to the 17 Sustainable Development Goals (SDGs), as well as information on 24 distinct policies conceptually related to the SDGs.

Table 1. The number and proportion of sample companies with data available in the Refinitiv database, as of November 2023.

Country	Total Number of Listed Companies	Total Number of Companies with Available Data	Percentage of Companies with Available Data
Austria	91	29	31.87%
Belgium	206	44	21.36%
Bulgaria	253	1	0.40%
Croatia	86	0	-
Cyprus	108	0	-
Czech Republic	28	2	7.14%
Denmark	164	61	37.20%
Estonia	32	0	-
Finland	184	69	37.50%
France	950	149	15.68%
Germany	1291	226	17.51%
Greece	154	12	7.79%
Hungary	120	5	4.17%
Ireland	23	12	52.17%
Italy	459	107	23.31%
Latvia	13	0	-
Lithuania	31	0	-
Luxembourg	24	2	8.33%
Malta	33	0	-
Netherlands	149	48	32.21%
Poland	775	26	3.35%
Portugal	53	11	20.75%
Romania	359	5	1.39%
Slovak Republic	43	2	4.65%
Slovenia	116	3	2.59%
Spain	253	45	17.79%
Sweden	1017	297	29.20%
Total	7015	1156	100%

During data collection, we implemented a screening process to safeguard the integrity of data and the validity of our research. Two types of data that were considered not valid for further analysis were: (a) 761 companies lacking industry classification, as this information was absent from the Refinitiv Eikon database, resulting in a sample of 6254 valid records; and (b) "missing" information, i.e., data that were not collected by Refinitiv analysts at the end of November 2023—the timestamp for exporting data for our research.

Consequently, a final sample of 1156 listed companies with valid data was further analyzed. Valid data incorporate companies for which the Boolean data were collected from Refinitiv, fulfilling our objective of comprehensive data collection, including industry classification, total assets of the companies, and data collected for all 17 SDGs and 24 policies relevant to the SDGs. This approach was essential to enhance the reliability and precision of our findings by eliminating potential sources of bias and ensuring that the analyzed data set was representative of companies with available financial and sustainability information.

Additional data collected for analysis incorporate fundamental information such as the names of the companies and their TRBC (Thomson Reuters Business Classification) industry classifications, which were subsequently correlated with NACE [76] industry codes (rev. 2.0) to streamline and enhance the categorization process. This approach aims to create a more compact and coherent framework for analysis. Additionally, the data set includes the total assets reported by the companies, providing the financial context. This set of variables facilitates a detailed exploration of their commitment to sustainability through the lens of both specific SDGs and policies [77,78].

To address research questions Q1–Q4, data transformations and statistical procedures were applied in SPSS. Specifically, we transformed the data downloaded as "TRUE" or "FALSE" into 1 and 0 values to perform different calculations. We created another variable that computes the number of SDGs applied per company and a check variable to mark which company applies at least one SDG. This is also a Boolean value where 1 means the company applies at least one SDG and 0 means the company does not apply any SDG. Subsequently, we created a correspondence table between the SDGs and sustainability policies. To answer the research objectives, we used frequencies and correlations. Frequencies were calculated to observe policy and SDG adherence. Furthermore, descriptive statistics were generated in relation to minimum and maximum values, as well as mean and median values, to provide insights into the applied SDGs and policies. The calculations were carried out considering industry- and country-specific contexts, facilitating the comparative analysis to better understand corporate policies and sustainability goals.

Relevant corporate policies, along with their descriptions, were extracted from the Refinitiv Eikon database. The mapping between the 24 distinct policies and the 17 SDGs is presented in Table A1 (Appendix A). The identification of policies from the Refinitiv database was carried out in response to the research question, based on the assumption that companies are pursuing a specific SDG or the corresponding policy. During the mapping process, key terms and formulations in the policy description were identified and used to create the "word search" column in Table A1. This column served as a tool to identify specific words and concepts within the context of specific targets pertaining to each SDG, thereby establishing an effective mapping process. As a result, the cross-check of policies with the SDGs is based on the identifiers of specific targets. For example, the policies linked to SDG 3 are summarized as p_SDG3 and can be any of the following: p_PER (product environmental responsible use), p_WEF (policy water efficiency), p_EHS (employee health safety policy), and p_SHS (supply chain health and safety).

The mapping approach was introduced to analyze the correlations between the adherence to the SDGs and the policy implementation of the sample companies. This multifaceted data set is a representative instrument of analysis of the intersection between sustainable practices and company self-regulation within the complex setting of the EU economic environment.

To enhance data visualization and pursue a comparative analysis, we created two types of figures. The first type represents the geographical map of the number of companies that apply at least one SDG and one sustainability policy, respectively. The geographical maps were created using the R software environment (version 4.3.2). Specifically, we used the ggplot2 library (version 3.5.0) in conjunction with the Natural Earth database [79]. For the second type of figure, we present heatmaps for the number of companies in each country that apply each SDG and each sustainability policy, respectively. In this regard, we constructed tables in Excel, as presented in the figures, and used the tool named "Conditional Formatting", keeping the main settings for the "Green-Yellow-Red color scale".

4. Results

4.1. Statistical Results on the Application of Each SDG and Sustainability Policy by EU Companies

The present section explores research question Q1 in multiple facets. We start by answering the following detailed questions on the application of SDGs:

- Q1a. How many companies in the EU apply each SDG?
- Q1b. What is the proportion of companies in the EU (in total) that apply each SDG?

In Table 2, we observe that the level of application of the SDGs varies between different goals, with SDGs 3, 7, 8, 12, and 13 having the highest application percentages ranging from 50.3% to 68.3%, with SDG 8 regarding decent work and economic growth at the top of the list. These are the dimensions that are incorporated into most of the companies' operations. The percentages indicate a strong commitment of the companies in our sample to health and safety at work, clean energy, responsible consumption, and climate action. At the opposite pole, SDGs 1, 2, and 14 have the lowest number of companies pursuing these goals, meaning 13.7%, 11.0%, and 14.4%, respectively, of the 1156 companies. The results for SDGs 1 and 2 are similar as there are few companies that pursue these goals. This is because the specific targets allocated to SDG 2—meaning T2.1., T2.2., and T2.3 (related to access to food and nutritional needs for poor and vulnerable people, and the agricultural productivity of small-scale producers)—are interlinked with T1.1., T1.2. and T1.3 (related to reducing poverty, especially for vulnerable people, through social protection systems) [80].

Table 2. The number and proportion of sample companies applying each SDG.

SDG	Companies Applying Each SDG		Companies Not Applying the Respective SDG	
	Count (N)	Proportion	Count (N)	Proportion
SDG 1	158	13.7%	998	86.3%
SDG 2	127	11.0%	1029	89.0%
SDG 3	582	50.3%	574	49.7%
SDG 4	468	40.5%	688	59.5%
SDG 5	649	56.1%	507	43.9%
SDG 6	274	23.7%	882	76.3%
SDG 7	519	44.9%	637	55.1%
SDG 8	789	68.3%	367	31.7%
SDG 9	597	48.4%	559	51.6%
SDG 10	383	33.1%	773	66.9%
SDG 11	407	35.2%	749	64.8%
SDG 12	740	64.0%	416	36.0%
SDG 13	781	67.6%	375	32.4%
SDG 14	166	14.4%	990	85.6%
SDG 15	286	24.7%	870	75.3%
SDG 16	441	38.1%	714	61.9%
SDG 17	390	33.7%	766	66.3%

Note. Total sample N = 1156.

The data in Table 2 highlight the most pursued SDGs [81]. Our article builds on previous results to highlight the various ways EU companies engage with each of the SDGs. Unlike previous studies that evaluated national performance [82], our analysis represents a more granular approach, identifying specific areas of underperformance within countries and empowering policymakers to incentivize the adoption of relevant and effective policies. Furthermore, our study represents an overview of all 17 SDGs, rather than being restricted to industry-specific goals, as observed in earlier literature [16,17,83,84]. There is room for improvement in the pursuit of all goals for the 2030 Agenda, as even the goals with the highest adoption rate do not reach 70% of the application within this sample.

The following two related questions refer to the adoption of relevant sustainability policies by EU-based companies:

- Q1c. How many companies in the EU apply each type of sustainability policy?
- Q1d. What is the proportion of companies in the EU that apply each type of sustainability policy?

Building on the existing literature [85], we identify relevant sustainability policies and elucidate their relationship with the SDGs, thus improving our understanding of the policy landscape in the context of sustainable development. Furthermore, our study goes beyond identifying the presence of these policies by examining the level of commitment and adherence to the SDGs among companies and providing valuable insights on the extent to which policies translate into actions aligned with the SDGs. As shown in Table 3, several policies show high adoption rates, indicating a strong commitment from a significant proportion of companies. Such policies refer to resource reduction (p_ARR, 91.4%), mapped in our research to targets 11.4, 12.2, 14.7, and 15.5; human rights (p_AHR, 90.2%), mapped to targets 1.4, 8.7, and 16.2; energy efficiency (p_EEF, 84.9%). mapped to targets 7.3 and 13.2; and waste reduction initiatives (p_WRI, 83.1%), mapped to targets 6.A, 11.6, 12.4, and 14.1. The high percentages of policy adoption suggest that a substantial number of companies are actively incorporating policies related to energy efficiency, human rights, and resource reduction, linked to SDGs 1, 7, 8, 11, 12, 13, 14, 15, and 16.

Table 3. The number and proportion of sample companies applying each policy.

Policies	Sample Companies Applying Each Policy		Sample Companies Not Applying the Respective Policy	
	Count (N)	Proportion	Count (N)	Proportion
p_ANT (animal testing)	90	7.8%	1066	92.2%
p_ARR (resource reduction)	1057	91.4%	99	8.6%
p_AWB (animal well-being)	65	5.6%	1091	94.4%
p_EEF (energy efficiency)	981	84.9%	175	15.1%
p_EMS (emissions)	994	86.0%	162	14.0%
p_ESC (environmental supply chain)	864	74.7%	292	25.3%
p_FFD (fossil fuel divestment)	29	2.5%	1127	97.5%
p_GBD (green buildings)	283	24.5%	873	75.5%
p_LBW (labeled wood)	54	4.7%	1102	95.3%
p_PER (product environmental responsible use)	727	62.9%	429	37.1%
p_PNS (nuclear safety)	8	0.7%	1148	99.3%
p_SPK (sustainable packaging)	320	27.7%	836	72.3%
p_TRI (takeback and recycling)	151	13.1%	1005	86.9%
p_WEF (water efficiency)	642	55.5%	514	44.5%
p_WRI (waste reduction)	961	83.1%	195	16.9%
p_AHR (human rights)	1043	90.2%	113	9.8%
p_CDL (child labor)	918	79.4%	238	20.6%
p_CDV (career development)	974	84.3%	182	15.7%
p_DOP (diversity and opportunity)	1104	95.5%	52	4.5%
p_EHS (employee health and safety)	1071	92.6%	85	7.4%
p_FAS (freedom of association)	719	62.2%	437	37.8%
p_FLB (forced labor)	900	77.9%	256	22.1%
p_SHS (supply chain health and safety)	610	52.8%	546	47.2%
p_STR (skills training)	987	85.4%	169	14.6%

Note. Total sample N = 1156.

Low application percentages are notable for the fossil fuel divestment policy (p_FFD, 2.5%), nuclear safety policy (p_PNS, 0.7%), and take-back and recycling initiatives (p_TRI, 13.1%) as companies may find these policies less relevant to their operations or there may be challenges in implementing and adhering to these specific types of policies. These policy areas may be relevant to companies that want to improve their commitment to certain goals. In summary, companies prioritize and apply policies differently depending on their industry, size, and strategies.

- Q1e. Is there a relationship between company size and the number of SDGs applied or if a company applies the SDGs?

In our study, we performed a correlation analysis to understand the relationship between adherence to SDGs and company size. This analytical approach can provide valuable insights for stakeholders and policymakers seeking to navigate the interaction between sustainability actions and the economic impact of sample companies. Furthermore, the correlation analysis builds upon previous research [86–88] by offering an examination within the specific context of our sample. The results in Table 4 suggest that there is a positive relationship between company size (total assets) and the number of SDGs applied or if the company applies the SDGs. The correlation coefficient between total assets and the number of SDGs applied is 0.259. This correlation suggests a weak positive relationship between the total assets of a company and the number of SDGs pursued by the sample companies.

Table 4. Correlation between company size (total assets) and the number of applied SDGs or if the company applies any SDGs.

Pearson Correlations	Total Assets	Number of Applied SDGs	Company Applies SDGs
Total Assets	1	0.259 **	0.132 **
Number of Applied SDGs	0.259 **	1	0.730 **
Company Applies SDGs	0.132 **	0.730 **	1

Note. ** $p < 0.01$. N = 997.

The correlation coefficient between total assets and the binary variable indicating whether the SDGs are applied is 0.132. This very weak positive correlation suggests a slight positive relationship between the total assets of a company and whether it pursues any SDGs. Larger companies are slightly more likely to apply some SDGs. Other factors not considered in the analysis can influence the observed relationships. The conclusion is that companies, regardless of size, can choose to apply the SDGs or not based on their sustainable development strategies and business models. While larger companies may be under more peer pressure and stakeholder scrutiny, their public visibility is not enough to convince the board of directors to pursue the relevant SDGs.

- Q1f. Is there a relationship between company size and the number of policies applied, or if the company applies any sustainability policy?

In parallel with our previous correlation analysis that examined the relationship between SDG adherence and company size, we extended our inquiry to explore the correlation between the economic contribution of a company and the application of sustainability policies. This approach allows us to investigate the impact of policy implementation on financial outcomes independently of SDG alignment. Our research builds on the existing literature [89–92] by uncovering patterns or associations between policy implementation and financial outcomes, thus enriching the understanding of the mechanism that drives sustainable development and economic success.

The correlation coefficient of 0.171 between total assets and the number of policies applied indicates a weak positive association (see Table 5). This implies that there is a tendency for companies with higher total assets to have a greater number of applied policies, but the relationship is weak. The coefficient of 0.171 signifies a positive but relatively low correlation (very small statistical effect). As the total assets increase, there is a slight motivation for company boards to adopt multiple policies to deal with higher operational complexity. However, the correlation coefficient of 0.024 suggests that there is no relationship between sustainability commitment and company size. In conclusion, larger companies tend to have more sustainability policies in place, but increased operational complexity is not a sufficient motivation for boards to implement sustainability policies.

Table 5. Correlation between company size (total assets) and the number of applied policies or if the company applies any sustainability policy.

Pearson Correlations	Total Assets	Number of Applied Policies	Sustainability Policies Applied (Binary Variable)
Total Assets	1	0.171 **	0.024
Number of Applied Policies	0.171 **	1	0.292 **
Policies Applied	0.024	0.292 **	1

Note. ** $p < 0.01$. N = 1153.

4.2. Statistical Results on the Application of SDGs and Corporate Sustainability Policies in each EU Country

In this section, we focus on countries of incorporation. We reformulate the analysis in terms of how many companies in each country pursue the SDGs and apply sustainability policies. This analysis addresses research question Q2. We start by investigating the countries in which companies apply at least one SDG. The following question is relevant in this regard:

- Q2a. How many companies in each EU country pursue at least one SDG?

The total number of companies with valid data in each country provides context for the overall size of the business landscape in these nations (see Table 6). Companies incorporated in several countries, such as Austria (93.11%), Czechia (100%), Denmark (90.16%), Ireland (91.67%), the Netherlands (91.67%), Portugal (90.91%), Slovenia (100%), and Spain (91.11%), have high application rates, indicating a significant commitment of most companies to apply at least one SDG. On the other hand, companies from several countries (Bulgaria, Romania, Luxembourg) have a less-pronounced interest in the SDGs. Variations may be influenced by cultural and regulatory differences, even within the European Union, due to distinct national policies and the level of awareness and engagement in sustainability initiatives, influencing companies' propensity to address the SDGs.

A visual representation of the results is provided in Figure 1. The map allows us to understand the patterns of SDG contributions in different geographical regions [93] within the European Union. The observed pattern is defined by a color spectrum from red to green. Figure 1 confirms the results of previous studies on the limited adoption of sustainability initiatives in Eastern Europe [94,95], as evidenced by the prevalence of red or near-red hues on the map, which signal countries where SDG targets are applied by fewer companies. Using color mapping can offer significant decision-making support by fostering a comparative viewpoint across SDG indicators, pinpointing the geographies requiring attention for sustainable development efforts and enabling the prioritization and allocation of resources to specific EU regions in alignment with SDG objectives.

The following question focuses on the application of sustainability-related policies by country:

- Q2b. How many companies in each EU country apply at least one sustainability policy?

Almost all companies in EU countries apply at least one sustainability policy. The uniformity in policy adoption throughout the business landscape observed in Table 6 shows the corporate commitment to responsible business practice. Sweden stands out with a slightly lower application percentage (97.98%) compared to the 100% application rates seen in several other countries. This indicates that a very small proportion of companies in Sweden have not adopted any policy. However, Sweden has the highest proportion of companies in the sample. The highest number of companies that do not apply any policy can be observed again in Sweden, meaning that 6 Swedish companies out of the 297 do not apply any policy.

Table 6. The number and proportion of companies in each country that pursue at least one SDG and adopt at least one sustainability-related policy (N = 1156 companies).

Country	Companies Applying at Least One SDG		Companies Adopting at Least One Sustainability Policy	
	Count	Proportion	Count	Proportion
Austria	27	93.11%	29	100.00%
Belgium	38	86.37%	43	97.73%
Bulgaria	0	0.00%	1	100%
Czechia	2	100.00%	2	100%
Denmark	55	90.16%	60	98.36%
Finland	57	82.61%	69	100%
France	111	74.5%	148	99.33%
Germany	145	64.16%	225	99.56%
Greece	8	66.67%	12	100%
Hungary	3	60.00%	5	100%
Ireland	11	91.67%	12	100%
Italy	92	85.98%	107	100%
Luxembourg	0	0.00%	2	100%
Netherlands	44	91.67%	48	100%
Poland	21	80.77%	26	100%
Portugal	10	90.91%	11	100%
Romania	1	20.00%	5	100%
Slovakia	1	50.00%	2	100%
Slovenia	3	100.00%	3	100%
Spain	41	91.11%	45	100%
Sweden	228	76.77%	291	97.98%
Total	898	77.68%	1146	99.13%

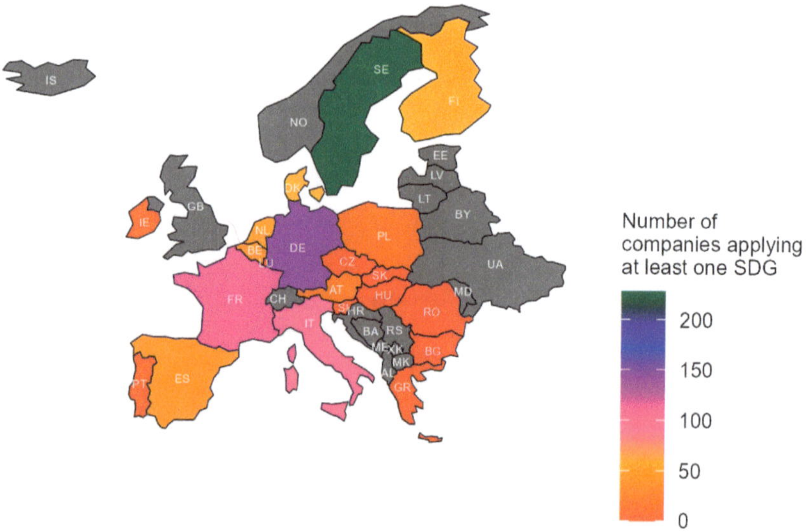

Figure 1. Geographical map of the number of companies in the European Union pursuing at least one SDG. Source: Table 6.

A visual representation of these results is shown in Figure 2, facilitating an understanding of patterns in sustainability policies in diverse geographical regions within the

EU. These policies are crucial to addressing the multifaceted challenges outlined in sustainability agendas. Figure 2 reinforces previous research findings that highlight the varying degrees of implementation of sustainability policies, particularly in certain regions (Eastern Europe), indicated by red or near-red hues, where policy uptake may be lacking [96–98]. Sustainability policies are the internal mechanism by which companies implement the SDGs and other requirements, such as the due diligence process mandated by the Corporate Sustainability Reporting Directive [43]. In this respect, the adoption of sustainability policies is more relevant to the sustainability strategy of EU companies than the self-declared contributions to the SDGs. Policies create a substantive obligation for company managers to prevent harm or facilitate environmental protection and social welfare, as a reflection of EU-level policies [99].

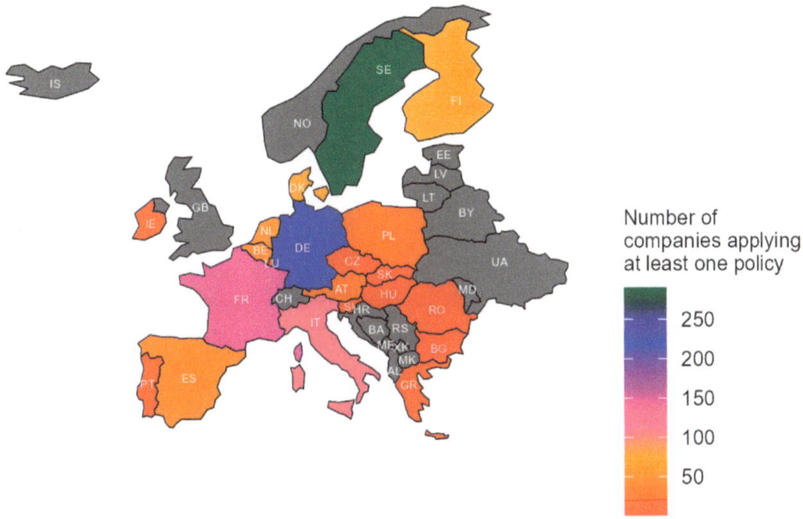

Figure 2. Geographical map of the number of companies in the European Union applying at least one sustainability policy. Source: Table 6.

We continue the exploration of SDGs and policies by country by calculating the descriptive statistics of the extent to which SDGs are applied in each EU country. We will answer the following question:

- Q2c. What are the average and range of the number of SDGs applied per country?

The total number of companies (1156) considered in the analysis provides context for the overall sample size and the breadth of the study (see Table 7). The mean number of SDGs applied per country is calculated as the average number of SDGs adopted by companies within each country. The overall mean across all countries is 6.68, indicating that, on average, companies in EU countries apply approximately six or seven SDGs. There is considerable variation between countries in the mean number of SDGs applied. The means are found in the interval from 0 (e.g., Bulgaria, Luxembourg) to 11.50 (Czechia). In countries with many companies containing valid data (France, Germany, Italy, and Sweden), we can observe that the means lie between 5 and 9. Romania has a relatively low mean (2.60), suggesting that, on average, companies in Romania apply a smaller number of SDGs. This may be due to specific challenges or less emphasis on sustainability practices within the business landscape [100].

Table 7. The mean, median, and range of SDGs applied per country.

Country	SDGs Applied per Country					Total Number of Companies
	Min	Max	Mean	Median	Weighted Mean	
Austria	0.00	17.00	9.17	10.00	0.23	29
Belgium	0.00	17.00	7.61	7.00	0.29	44
Bulgaria	0.00	0.00	0.00	0.00	0.00	1
Czechia	10.00	13.00	11.50	11.50	0.02	2
Denmark	0.00	15.00	6.61	6.00	0.35	61
Finland	0.00	17.00	5.70	6.00	0.34	69
France	0.00	17.00	8.05	9.00	1.04	149
Germany	0.00	17.00	5.52	6.00	1.08	226
Greece	0.00	17.00	8.50	10.50	0.09	12
Hungary	0.00	17.00	7.20	3.00	0.03	5
Ireland	0.00	17.00	7.34	7.00	0.08	12
Italy	0.00	17.00	8.89	9.00	0.82	107
Luxembourg	0.00	0.00	0.00	0.00	0.00	2
Netherlands	0.00	14.00	6.00	6.00	0.25	48
Poland	0.00	17.00	6.85	7.00	0.15	26
Portugal	0.00	15.00	8.91	9.00	0.08	11
Romania	0.00	13.00	2.60	0.00	0.01	5
Slovakia	0.00	7.00	3.50	3.50	0.01	2
Slovenia	5.00	13.00	10.34	13.00	0.03	3
Spain	0.00	17.00	10.20	11.00	0.40	45
Sweden	0.00	17.00	5.39	5.00	1.38	297
Total	0.00	17.00	6.68	7.00	-	1156

All EU countries have companies that do not apply any SDGs, except Czechia (with two listed companies that apply at least ten SDGs) and Slovenia (with three listed companies applying a minimum of five SDGs). The maximum number of SDGs applied is 17, for most of the countries, while companies in Slovakia apply the least SDGs, with a maximum of 7. The median value of the sample (seven) is close to the mean, indicating a relatively symmetric distribution of the number of SDGs applied per country. Some countries, such as Sweden, have means that are close to their respective medians, suggesting a more balanced distribution of the SDG application. Other countries, such as Greece and Spain, have means that are slightly higher than their medians, indicating a positively skewed distribution, with prominent companies adopting a higher number of SDGs.

The weighted mean of each country is calculated by multiplying the mean number of SDGs pursued by the proportion of companies in the total sample. This indicator shows the contribution of the respective countries to the SDGs in the European Union. In our sample, Sweden has the largest weighted mean, followed by France and Germany. Not surprisingly, these countries also have many companies that pursue the SDGs and contribute to sustainable development.

- Q2d. What are the average and range of the number of sustainability policies applied by companies in each country?

The following analysis delves into the application of sustainability policies by companies in each EU country, mirroring the previous analysis concerning the SDGs in each country. The results are presented in Table 8. The overall mean across all countries is 13.45, indicating that, on average, companies in these countries apply approximately 13 to 14 policies out of the 24 analyzed. The mean values range from 7.00 (Bulgaria, Luxembourg) to 15.91 (Spain). This indicates a considerable variation in the average number of policies adopted by companies in different countries. In many cases, the mean values are close to the median values, suggesting a relatively symmetric distribution of policy adoption.

Table 8. The mean, median, and range of the number of sustainability policies applied by sample companies, per country.

Country	Policies Applied by Sample Companies				Weighted Mean	Total Number of Companies
	Min	Max	Mean	Median		
Austria	11.00	18.00	15.00	15.00	0.38	29
Belgium	0.00	19.00	14.23	15.50	0.54	44
Bulgaria	7.00	7.00	7.00	7.00	0.01	1
Czechia	14.00	17.00	15.50	15.50	0.03	2
Denmark	0.00	21.00	14.21	15.00	0.75	61
Finland	2.00	19.00	14.52	15.00	0.87	69
France	0.00	21.00	14.54	16.00	1.87	149
Germany	0.00	20.00	13.71	15.00	2.68	226
Greece	3.00	17.00	13.17	15.50	0.14	12
Hungary	10.00	18.00	13.80	13.00	0.06	5
Ireland	12.00	18.00	14.42	14.00	0.15	12
Italy	3.00	19.00	14.22	15.00	1.32	107
Luxembourg	3.00	11.00	7.00	7.00	0.01	2
Netherlands	5.00	19.00	15.04	16.00	0.62	48
Poland	10.00	20.00	15.08	15.00	0.34	26
Portugal	7.00	18.00	15.18	16.00	0.14	11
Romania	8.00	16.00	11.40	11.00	0.05	5
Slovakia	3.00	17.00	10.00	10.00	0.02	2
Slovenia	10.00	15.00	13.00	14.00	0.03	3
Spain	5.00	21.00	15.91	16.00	0.62	45
Sweden	0.00	20.00	11.01	12.00	2.83	297
Total	0.00	21.00	13.45	15.00	-	1156

The minimum and maximum values represent the range of policy adoption within each country. For example, in Belgium, the minimum is 0, indicating that some companies have not adopted any policy, while the maximum is 19, indicating a high level of policy adoption by some Belgian companies. Some countries, such as Bulgaria and Luxembourg, have relatively lower mean values of 7.00, indicating a lower average level of policy adoption with minimum and maximum ranging between 3 and 11. There are companies in Belgium, Denmark, France, Germany, and Sweden that do not have any policy, according to Refinitiv. However, these countries have several companies that apply a very large number of policies, ranging from 19 to 21. Companies in Denmark, France, and Spain apply a maximum of 21 out of the 24 policies analyzed.

For countries where the mean is lower than the median (e.g., Belgium, Hungary, Romania, Slovakia, and Sweden), there is a potential skewness toward lower policy adoption. Policymakers may explore strategies to encourage wider policy adoption among companies in these countries.

The weighted mean of each country is calculated by multiplying the mean number of policies applied by the proportion of companies in the total sample. This indicator shows the commitment of companies in each country to the implementation of sustainability policies in the EU. In our sample, Sweden has the largest weighted mean, closely followed by Germany. France and Italy occupy the third and fourth positions, respectively. Not surprisingly, these countries also have numerous companies that implement sustainability policies and contribute to sustainable development.

4.3. Statistical Results on the Application of Each SDG and Sustainability Policy per Industry

The following analysis discusses the application of SDGs and sustainability policies per industry (classified using the NACE codes rev. 2.0). This is relevant to this investigation because industries have different impacts, risks, and opportunities related to sustainability [43]. The first point refers to the extent to which the SDGs are applied in each industry. The following question is relevant to this analysis:

- Q3a. What are the average and range of the number of SDGs applied per industry?

There is considerable variation in the mean number of SDGs applied in different industries, ranging from 4.38 in the administrative and support services industry (N) to 12 in the public administration and defense industry and the compulsory social security industry (O). This variation suggests that industries differ in their commitment to and alignment with the SDGs. Table 9 offers more details. Industries such as public administration and defense, compulsory social security (O), and financial and insurance activities (K) have relatively higher mean values of 12.00 and 7.91, respectively. This indicates that companies within these industries, on average, adopt a larger number of SDGs. This may be attributed to regulatory requirements, industry norms, and a strong emphasis on sustainability in these sectors [101]. Industries such as professional, scientific, and technical activities (M) and administrative and support services (N) have lower mean values of 4.50 and 4.38, respectively. In many cases, the mean values are close to the median values, indicating a relatively symmetric distribution of SDG applications within each industry.

Table 9. The mean, median, and range of SDGs applied per industry.

Industry	SDGs Applied per Industry					Total Number of Companies
	Min	Max	Mean	Median	Weighted Mean	
A (Agriculture, Forestry and Fishing)	0.00	12.00	5.46	6.00	0.06	13
B (Mining and Quarrying)	0.00	14.00	6.24	7.00	0.11	21
C (Manufacturing)	0.00	17.00	6.81	6.00	2.49	422
D (Electricity, Gas, Steam and Air Conditioning Supply)	0.00	17.00	8.49	8.00	0.35	47
E (Water Supply; Sewerage, Waste Management and Remediation Activities)	0.00	0.00	0.00	0.00	0.00	0
F (Construction)	0.00	17.00	8.07	7.5	0.31	44
G (Wholesale and Retail Trade; Repair of Motor Vehicles and Motorcycles)	0.00	17.00	5.78	6.00	0.73	146
H (Transportation and Storage)	0.00	17.00	6.90	7.00	0.24	40
I (Accommodation and Food Service Activities)	0.00	17.00	6.19	7.00	0.17	32
J (Information and Communication)	0.00	17.00	6.63	7.00	0.45	79
K (Financial and Insurance Activities)	0.00	17.00	6.97	7.00	0.96	159
L (Real Estate Activities)	0.00	17.00	7.91	7.50	0.30	44
M (Professional, Scientific and Technical Activities)	0.00	17.00	5.09	4.00	0.24	55
N (Administrative and Support Service Activities)	0.00	9.00	4.50	4.50	0.02	6
O (Public Administration and Defense; Compulsory Social Security)	0.00	13.00	4.37	4.50	0.06	16
P (Education)	7.00	17.00	12.00	12.00	0.02	2
Q (Human Health and Social Work Activities)	0.00	16.00	6.10	5.50	0.05	10
R (Arts, Entertainment and Recreation)	0.00	14.00	6.50	6.50	0.11	20
S (Other Service Activities)	0.00	0.00	0.00	0.00	0.00	0
T (Activities of Households as Employers; Undifferentiated Goods and Services Producing Activities of Households for Own Use)	0.00	0.00	0.00	0.00	0.00	0
U (Activities of Extraterritorial Organizations and Bodies)	0.00	0.00	0.00	0.00	0.00	0
Total	0.33	12.62	5.43	5.38	0.32	1156

The maximum number of SDGs applied is 17, which is common in most industries. Companies in the professional, scientific, and technical activities sector (M) apply a maximum of nine SDGs—the least number of SDGs in the present sample. The public administration and defense industry (O) applies a minimum of seven SDGs, compared to all other industries, where the minimum number of applied SDGs is zero.

- Q3b. What are the average and range of the number of sustainability policies applied per industry?

There is considerable variation in the mean number of policies applied across different industries (see Table 10), ranging from 8.50 (education—P) to 14.51 (manufacturing—C). Industries such as manufacturing (C), mining and quarrying (B) and construction (F) have relatively higher mean values of policies applied, indicating a stronger commitment to adopting a larger number of policies on average. Education (P), professional, scientific, and technical activities (M), and administrative and support service activities (N) have lower mean values, suggesting a comparatively weaker adoption of policies within these sectors. Industries such as manufacturing (C), information and communication (J), financial and insurance (K), professional, scientific, and technical activities (M), and arts, entertainment, and recreation (R) include some companies that do not apply any policy. There is no industry with at least 1 company that implemented all 24 policies.

Table 10. The mean, median, and range of policies applied per industry.

Industry	Policies Applied per Industry					Total Number of Companies
	Min	Max	Mean	Median	Weighted Mean	
A (Agriculture, Forestry and Fishing)	10.00	17.00	14.38	15.00	0.1618	13
B (Mining and Quarrying)	3.00	18.00	14.48	15.00	0.2630	21
C (Manufacturing)	0.00	21.00	14.51	15.50	5.2976	422
D (Electricity, Gas, Steam and Air Conditioning Supply)	1.00	18.00	13.43	16.00	0.5458	47
E (Water Supply; Sewerage, Waste Management and Remediation Activities)	0.00	0.00	0.00	0.00	0.0000	0
F (Construction)	3.00	17.00	14.02	15.00	0.5337	44
G (Wholesale and Retail Trade; Repair of Motor Vehicles and Motorcycles)	1.00	21.00	13.12	15.00	1.6566	146
H (Transportation and Storage)	3.00	18.00	13.47	14.00	0.4663	40
I (Accommodation and Food Service Activities)	4.00	20.00	13.06	14.00	0.3616	32
J (Information and Communication)	0.00	19.00	12.63	14.00	0.8633	79
K (Financial and Insurance Activities)	0.00	19.00	12.65	14.00	1.7396	159
L (Real Estate Activities)	8.00	17.00	13.48	14.00	0.5130	44
M (Professional, Scientific and Technical Activities)	0.00	18.00	9.84	11.00	0.4680	55
N (Administrative and Support Service Activities)	5.00	15.00	9.67	10.00	0.0502	6
O (Public Administration and Defense; Compulsory Social Security)	2.00	18.00	13.19	14.50	0.1825	16
P (Education)	8.00	9.00	8.50	8.50	0.0147	2
Q (Human Health and Social Work Activities)	8.00	19.00	13.40	13.50	0.1159	10
R (Arts, Entertainment and Recreation)	0.00	18.00	12.70	13.50	0.2197	20
S (Other Service Activities)	0.00	0.00	0.00	0.00	0.00	0
T (Activities of Households as Employers; Undifferentiated Goods and Services Producing Activities of Households for Own Use)	0.00	0.00	0.00	0.00	0.00	0
U (Activities of Extraterritorial Organizations and Bodies)	0.00	0.00	0.00	0.00	0.00	0
Total	2.67	14.38	10.31	11.07	0.64	1156

In many cases, the mean values are close to the median values, indicating a relatively symmetric distribution of policy adoption within each industry. Across all industries, the overall mean number of policies applied is 12.74, suggesting that, on average, companies across various industries have aligned their practices with an average of 13 policies out of 24 analyzed, with a minimum of 3 policies applied and a maximum of 18 policies applied.

- Q3c. How many companies apply the SDGs in each industry and country (cross-tabulation)?

In Table A2 (Appendix A), the last column on the right provides the overall count of companies applying the SDGs, broken down by industry and EU country. Industries with higher counts indicate a greater prevalence of companies adopting the SDGs. For example, manufacturing (C) and wholesale and retail trade (G) have relatively high counts, suggesting widespread adoption in these sectors. Industries with lower counts have fewer companies actively applying SDGs. For example, mining and quarrying (B) and education (P) have lower counts. The frequencies for each country provide insights into the level of adoption of the SDGs. For instance, Germany (DEU) and Sweden (SWE) have relatively higher counts compared to other countries. Low values for the adoption of SDGs can be observed in Czechia, Hungary, Portugal, Romania, Slovakia, and Slovenia.

- Q3d. How many companies apply sustainability policies in each industry and country (cross-tabulation)?

In Table A3 (Appendix A), each column represents a country, and the numbers indicate the frequency with which companies in that country apply policies in each sector. Industries with higher counts indicate a greater prevalence of companies adopting sustainability policies. For instance, manufacturing (C) and wholesale and retail trade (G) have relatively high counts, suggesting widespread adoption in these sectors. Some industries such as agriculture (A) and mining and quarrying (B) have lower counts, indicating lower adoption and/or fewer companies in these sectors applying policies. The frequencies for each country provide insights into the level of adoption of sustainability policies. For instance, Sweden (SWE) and Germany (DEU) have a relatively higher count compared to other countries, considering that the most advanced industry in terms of sustainability policies is manufacturing (C). The lowest counts on corporate policies between countries are represented by Bulgaria, Czechia, Hungary, Italy, Romania, Slovakia, and Slovenia.

- Q3e. How many companies apply the SDGs in each industry?

In Table A4 (Appendix A), each row corresponds to a specific SDG, and the columns represent different industries. For example, in manufacturing (C), there are 228 companies applying SDG 3, which is related to good health and well-being. Industries such as manufacturing (C), wholesale and retail trade (G), and financial and insurance activities (K) have relatively high counts across multiple SDGs, indicating a comprehensive approach to sustainability. The most pursued SDGs (8, 12, 13, 5, 3, 7, and 9, in descending frequency of application) refer to job creation and economic development, inclusivity and social responsibility, resource efficiency and waste reduction, regulatory compliance and risk mitigation, environmental responsibility, and market demand and consumer preferences. A survey conducted by KPMG in 2022 on 12 countries, territories, and jurisdictions observed that the frequency of prioritized SDGs follows the same pattern; SDG 8 (72%), SDG 12 (58%), SDG 13 (63%), followed by SDG 5 (43%), SDG 3 (49%), SDG 7 (50%) and SDG 9 (50%) [102].

Some industries, like agriculture (A) and mining and quarrying (B), have lower counts for certain SDGs, possibly reflecting the need to focus more on specific sustainability goals within these sectors. Industry-specific challenges and characteristics are linked to resource-intensive practices, economic dependence on the exploitation of natural resources, or the fragmentation of the industry [103].

We analyzed the literature by searching Clarivate Web of Science and Scopus using keywords from the industry's name and "SDGs" (e.g., "Agriculture SDG", "Mining SDGs," etc.). We performed textual analysis to identify the SDGs correlated with the results of the present study. Table A5 (Appendix A) includes the correspondence between the previous literature and the present results. We have identified common contributions to the SDGs for each industry, except water supply, sewerage waste management, and remediation activities (E), as the companies in this industry in our sample did not present adherence to any of the 17 SDGs. Higher SDG matching is observed in agriculture (A), electricity and gas (D), financial (K), education (P), and arts and recreation (R). We identified less SDG matching in wholesale and retail (G), transportation and storage (H), accommodation and food service activities (I), real estate activities (L), professional (M), administrative and support services (N), and human health and social work activities (Q). Table A5 also indicates future research opportunities in the investigation of the SDGs by industry to increase the matching between what companies consider relevant and what researchers consider relevant in terms of corporate contributions to the SDGs.

- Q3f. How many companies apply sustainability policies in each industry?

In Table A6 (Appendix A), each row corresponds to a specific policy, and the columns represent different industries. For example, in manufacturing (C), there are 402 companies applying the policy labeled p_ARR. Policies referring to resource reduction (p_ARR), human rights (p_AHR), diversity and equal opportunity (p_DOP), and employee health and safety (p_EHS) have relatively high frequencies across multiple industries. Similarly, with the pursuit of SDGs across industries, there is a high degree of adoption of policies in manufacturing (C), wholesale and retail trade (G), and financial services (K). In manufacturing (C) and wholesale and retail trade (G), there is a significant focus on the supply chain, leading to a greater adoption of sustainability policies in the process of sourcing raw materials, manufacturing, and distribution.

Integrating sustainable practices can be a competitive advantage in attracting environmentally conscious consumers. On the other hand, financial institutions (K), including banks and investment firms, recognize the financial risks associated with environmental, social, and governance (ESG) factors [71]. As sustainability issues can pose risks to long-term business viability, financial institutions are incorporating sustainability policies to assess and manage these risks in their investment portfolios and help their clients achieve the SDGs [104].

Some policies, like the fossil fuel divestment policy (p_FFD) and labeled wood (p_LBW), are applied by fewer companies, possibly reflecting industry-specific requirements. The lowest value observed is related to nuclear safety policy (p_PNS), which is applied in the electricity supply sector (D) and the construction sector (F) in relation to facility construction or materials for nuclear facilities.

4.4. Statistical Results on the Relationship between SDGs and the Corresponding Policies

The final section of the results proposes an exploration of the relationship between the SDGs and p_SDGs based on the SDG versus policies matrix (Table A1). The correlation is based on mapping the SDGs to corporate policies related to the SDGs. This correlation analysis answers Q4, which was formulated as:

- Q4. What is the relationship between the SDGs and the corresponding policies as applied by the sample companies?

In Table A7 (Appendix A), each cell in the table contains a correlation coefficient between each SDG and its corresponding p_SDG composed of mapped policies (in Table A1). For example, p_SDG1 is based on the occurrence of p_AHR (analytic human rights policy). Based on the mapping in Table A1, correlation coefficients were calculated between the SDG frequencies in columns and the aggregated p_SDG frequencies in rows. Specifically, the correlation coefficient between SDG 1 and its corresponding policy (p_SDG1) is 0.122 (significant at the 0.05 threshold). The positive value (0.122) suggests a very weak positive correlation between SDG 1 and p_SDG1.

The highest correlation can be observed between SDG 13 and the sustainability policies associated with this SDG (p_SDG13). Many of the coefficients are positive and statistically significant, indicating a general positive correlation between the implementation of SDGs and their corresponding policies aggregated by the p_SDGs. However, for most relationships, the statistical effect is either small or medium, but it is not strong. This indicates that the policies adopted by the respective companies are not entirely aligned with the SDGs pursued and reported by the same companies. This is an important result, which will be discussed in the next section.

5. Discussion

5.1. Comparative Perspective on SDG Adoption in the European Union Countries

The data collected from Refinitiv describe not only the parent company from the country of origin but also the application of policies and SDGs to the entire group of companies. Multinational corporations apply their SDGs and sustainability policies throughout the consolidation perimeter, leading to various business behaviors and progress toward the SDGs at various rates [105]. In other words, the sample companies—most of which are multinational enterprises—disseminate their policies in all countries of operation and pursue their underlying SDGs. The following discussion presents the notable contributions of the top performers in our sample, according to the results summarized in the previous tables.

Figure 3 provides a quantitative overview of the engagement of different companies in achieving specific SDGs using the number of companies applying each SDG. Red and orange hues indicate little contribution to the SDGs. Green hues indicate a substantial contribution to the SDGs. Sample companies from Bulgaria and Luxembourg had not reported contributions to any SDG. Companies in Croatia, Cyprus, Estonia, Latvia, Lithuania, and Malta did not have data collected by the Refinitiv analysts (as of November 2023), as observed in Table 1. In this sense, they appear in Figure 3 as lacking participation.

Country/SDG	SDG 1	SDG 2	SDG 3	SDG 4	SDG 5	SDG 6	SDG 7	SDG 8	SDG 9	SDG 10	SDG 11	SDG 12	SDG 13	SDG 14	SDG 15	SDG 16	SDG 17
Austria	7	2	20	18	22	13	19	24	19	11	15	25	26	6	11	18	10
Belgium	6	7	29	23	19	12	24	31	25	14	18	33	36	11	13	14	20
Bulgaria	0	0	0	0	0	0	0	0	0	0	0	0	0	0	0	0	0
Croatia	0	0	0	0	0	0	0	0	0	0	0	0	0	0	0	0	0
Cyprus	0	0	0	0	0	0	0	0	0	0	0	0	0	0	0	0	0
Czech Republic	0	0	1	0	1	2	2	2	2	2	2	2	2	0	1	2	2
Denmark	0	4	32	22	41	12	26	45	19	19	17	51	46	10	11	28	20
Estonia	0	0	0	0	0	0	0	0	0	0	0	0	0	0	0	0	0
Finland	2	3	26	17	33	15	22	48	31	15	17	48	54	4	14	23	21
France	35	25	87	82	93	52	75	101	75	75	62	96	107	33	63	78	61
Germany	22	20	93	93	105	45	84	128	104	61	58	111	135	23	41	55	69
Greece	5	3	7	5	7	3	6	8	7	6	7	8	8	4	6	8	4
Hungary	2	2	2	2	2	2	2	2	3	2	3	3	2	2	2	2	1
Ireland	3	4	7	6	8	1	5	9	6	5	7	8	10	1	4	1	3
Italy	24	22	70	66	74	34	68	89	71	48	60	83	82	25	35	54	46
Latvia	0	0	0	0	0	0	0	0	0	0	0	0	0	0	0	0	0
Lithuania	0	0	0	0	0	0	0	0	0	0	0	0	0	0	0	0	0
Luxembourg	0	0	0	0	0	0	0	0	0	0	0	0	0	0	0	0	0
Malta	0	0	0	0	0	0	0	0	0	0	0	0	0	0	0	0	0
Netherlands	5	4	23	15	22	8	25	35	21	11	15	36	34	4	10	8	12
Poland	6	4	14	16	11	5	12	16	14	9	9	16	19	4	9	5	9
Portugal	2	1	6	6	9	2	9	10	6	4	6	10	10	2	8	3	4
Romania	1	0	1	1	1	0	1	1	1	1	0	1	1	0	1	1	1
Slovak Republic	0	1	0	0	0	1	1	0	1	0	1	0	1	0	1	0	0
Slovenia	1	0	3	2	2	1	3	3	2	2	1	3	3	1	2	1	1
Spain	18	9	30	31	34	17	31	40	34	28	28	33	38	10	13	29	36
Sweden	19	16	131	63	165	49	104	197	118	70	81	173	167	26	41	111	70

Figure 3. Heatmap of the number of companies applying each SDG per country.

Swedish companies have outstanding participation in numerous SDGs; French and German companies demonstrate substantial engagement across most SDGs; and Italian companies engage significantly across multiple SDGs, approaching the level of sustainable development in France and Germany. On the other hand, there is minimal participation of companies in Romania and Bulgaria. More research is needed to understand the reasons behind the lack of participation in these countries.

On the same note, we observe in Figure 4 the heat map of the number of companies that apply each policy per country. Red and orange hues indicate lower adoption of sustainability policies. Green hues indicate a substantial adoption of corporate sustainability policies in the respective countries. Similar to the contributions towards the SDGs, French, German, Italian, and Swedish companies are the ones that engage more with sustainability policies. Policies most frequently applied by the companies are p_PER (product environmental responsible use), p_ARR (analytic resource reduction), p_WEF (water efficiency), p_EEF (energy efficiency), p_ESC (environmental supply chain), p_EHS (employee health), p_STR (skills training), p_CDV (career development), p_DOP (diversity and opportunity), p_FAS (freedom of association), p_CDL (prevention of child labor), p_AHR (human rights policy), and p_FLB (prevention of forced labor).

Country/policy	p_AHR	p_ANT	p_ARR	p_AWB	p_CDL	p_CDV	p_DOP	p_EEF	p_EHS	p_EMS	p_ESC	p_FAS	p_FFD	p_FLB	p_GBD	p_LBW	p_PER	p_PNS	p_SHS	p_SPK	p_STR	p_TRI	p_WEF	p_WRI
Austria	29	0	29	1	28	27	29	29	29	28	24	24	1	28	8	2	26	0	13	7	28	1	16	28
Belgium	39	7	43	2	35	42	43	39	42	42	32	29	3	35	11	2	31	0	22	14	41	7	25	40
Bulgaria	1	0	1	0	0	0	1	0	1	0	1	0	0	0	0	0	0	0	1	0	0	0	0	1
Croatia	0	0	0	0	0	0	0	0	0	0	0	0	0	0	0	0	0	0	0	0	0	0	0	0
Cyprus	0	0	0	0	0	0	0	0	0	0	0	0	0	0	0	0	0	0	0	0	0	0	0	0
Czech Republic	2	0	2	0	1	2	2	2	2	2	2	1	0	1	1	0	2	1	1	0	2	1	2	2
Denmark	58	12	57	3	51	52	60	55	57	56	50	43	2	49	11	2	35	0	43	24	47	13	35	52
Estonia	0	0	0	0	0	0	0	0	0	0	0	0	0	0	0	0	0	0	0	0	0	0	0	0
Finland	67	3	67	7	64	60	69	62	66	65	56	51	0	63	15	6	41	0	44	23	62	8	39	64
France	134	15	141	13	124	133	144	136	139	134	122	106	2	124	57	9	97	3	77	47	137	26	109	138
Germany	206	15	212	14	189	209	214	199	214	200	167	130	4	183	52	7	155	1	117	66	207	35	125	178
Greece	10	0	11	0	10	10	10	11	11	9	10	8	0	10	2	0	7	0	8	2	9	0	9	11
Hungary	5	0	5	0	4	4	5	4	5	3	3	2	0	4	3	0	5	0	3	0	4	3	2	5
Ireland	12	1	12	2	9	12	12	12	12	12	10	4	0	11	3	1	5	0	7	5	12	0	7	12
Italy	99	2	103	6	88	98	104	98	107	93	90	59	1	83	28	1	69	1	66	35	104	12	75	100
Latvia	0	0	0	0	0	0	0	0	0	0	0	0	0	0	0	0	0	0	0	0	0	0	0	0
Lithuania	0	0	0	0	0	0	0	0	0	0	0	0	0	0	0	0	0	0	0	0	0	0	0	0
Luxembourg	1	1	2	0	1	1	1	0	1	0	1	1	0	1	0	0	0	0	0	0	1	0	1	1
Malta	0	0	0	0	0	0	0	0	0	0	0	0	0	0	0	0	0	0	0	0	0	0	0	0
Netherlands	47	5	46	3	42	45	48	45	46	47	36	33	4	42	20	1	35	0	30	18	46	10	29	44
Poland	26	0	26	1	23	26	26	25	26	25	19	18	2	22	12	1	16	0	13	10	26	4	20	25
Portugal	11	0	11	0	11	9	10	10	11	10	11	6	0	9	2	3	8	0	9	4	10	2	10	10
Romania	4	0	5	0	3	5	5	4	5	3	4	2	0	3	0	0	1	0	2	0	5	0	2	4
Slovak Republic	1	0	1	0	1	1	2	1	1	2	1	1	1	1	1	0	2	0	0	0	1	0	1	1
Slovenia	3	0	3	0	2	3	3	3	3	3	2	0	1	2	0	0	2	0	1	0	3	1	2	2
Spain	44	4	44	3	41	44	45	43	45	43	42	41	4	41	22	1	35	2	33	7	45	2	40	45
Sweden	244	25	236	10	191	191	271	203	248	217	181	160	4	188	35	18	155	0	120	58	197	26	93	198

Figure 4. Heatmap of the number of companies applying each policy per country.

Based on our research, SDG 1 ("end poverty") is approached mainly in the manufacturing (C) and financial (K) industries. The contribution of the manufacturing sector to this SDG is characterized by high values due to the strong focus of companies on product development for a low-income population [81]. Observing the heatmap in Figure 3, companies in France, Germany, Italy, and even Spain and Sweden are the ones that approach this goal more. SDG 2 ("zero hunger") has a similar scope in all sectors as it is closely linked to SDG 1, with high values in the manufacturing sector (C). The most significant contribution belongs to manufacturing companies that focus on their value chain. Other industries could adhere to this SDG by supporting organizations such as the World Food Programme.

SDG 3 ("good health and well-being") continues the interlinkage with SDG 2 since hunger can cause diseases and affect the health of the population [81]. The manufacturing sector (C) and the wholesale and retail industry (G) have a high interest in SDG 3, implicitly seeking to attract and retain their personnel. Health should be a subject of interest for all companies by ensuring a safe work environment. Companies investing in education (SDG 4—"quality education") could also target SDG 8 ("decent work") by developing a skilled workforce through their investment in education before hiring and on the job. We observed that companies in sectors such as manufacturing (C), wholesale and retail (G), and financial services (K) apply SDG 4 to a larger extent, but companies in other industries could be more engaged and implement appropriate training programs for their workforce or participate in training partnerships [81].

SDG 5 ("gender equality") creates benefits for the entire economy, as we observe an increase in the country's GDP when the employment gap is reduced [106]. SDG 5 is highly applied by French, German, and Swedish companies, as shown in Figure 3. More companies in various industries should be aware of the gender equality policies.

SDG 6 ("access to water") is mainly applied by the manufacturing (C) industry as water usage levels are higher in cooling and cleaning processes [107]. It is important for companies to track their water consumption and discharge in business activities and in providing sanitation to employees [81].

SDG 7 ("affordable and clean energy") represents a new challenge for all companies as fossil fuels are phased out in the EU [108]. This sustainability goal is relevant not only to the manufacturing sector (C) but also to industries such as electricity and gas (D), construction (F), wholesale and retail (G), information and communication (J), financial services (K), and real estate (L). As in the case of SDG 6, companies should understand their energy usage patterns, especially in the manufacturing industry. In recent years, the hospitality industry (I) has demonstrated greater social and environmental responsibility as it seeks solutions to provide green accommodation practices that could be used as a marketing instrument toward the responsible consumer [109]. Organizations should seek to disseminate their clean energy practices throughout the supply chain when selecting and auditing suppliers.

SDG 8 ("decent work and economic growth") reflects more dimensions, combining SDG 4, SDG 5, and access to finance [81], providing a specific link to SDGs 1 and 3. SDG 8 reflects the challenges related to unemployment and the risk of forced labor. During our research, we observed the applicability of SDG 8 in manufacturing (C) and wholesale and retail (G), as both industries are based on human labor. SDG 9 ("industry, innovation and infrastructure") was approached by the same sectors mentioned above, in addition to information and communication (J), which deals mainly with the innovation side of this specific goal. SDG 10 ("reduced inequalities") is relevant for manufacturing (C), wholesale and retail (G), and financial services (K), mainly as a consequence of their compliance efforts [42,81]. The three mentioned industries also seek continuous innovation, and the diversity of the workforce is a key factor in achieving SDG 10.

As the SDGs are interconnected [77], urban and rural life is supported by agriculture (A), manufacturing (C), and transportation (H). In our research, we observed a high interest in SDG 11 ("sustainable cities and communities") in the manufacturing industry (C). Agriculture (A) and transportation (H) are not significant contributors, although they should diminish their negative effects by pursuing the targets associated with SDG 11. On the other hand, SDG 12 ("responsible consumption and production") is approached by the same industries discussed above, with very high participation from the manufacturing industry (C) as it seeks to become carbon neutral [81]. SDG 13 ("climate action"), based on the Paris Agreement [110], puts pressure on revenue streams and relationships with suppliers and distributors, as the Corporate Sustainability Reporting Directive demands carbon accounting across the value chain [43].

SDG 14 ("life below water") refers to the pollution of the seas directly or indirectly generated by companies in any industry, with a direct impact on the health of the population (SDG 4). High polluters should seek to identify plastic use and toxic substances in their production processes. Sustainable partnerships are needed to implement circular solutions in different value chains [111]. SDG 15 ("life on land") is of great relevance to companies in the manufacturing industry (C) as they recognize "natural capital" as scarce [81]. SDG 15 is considered a challenge for companies as biodiversity is threatened in many parts of the world. SDG 16 ("peace and justice") is linked to corporate transparency that helps policymakers and the public understand firm behavior. Multinational companies should disseminate their codes of conduct in the supply chain to ensure best practices and transparency with respect to their sustainable actions. Finally, SDG 17 ("partnerships for the goals") has no homogenous applicability in the present sample, although it supports partnerships between the private and public sectors to solve complex problems and overcome challenges regarding funding, technology, or transition events.

Looking at the European Union as a whole in relation to the 17 SDGs, and understanding the role of the EU in setting the trend for all member states [52], researchers have concluded that the implementation of the SDGs can be carried through a collective approach, not limited to a particular industry. The following subsections analyze the "leaders" and "laggers" in the sustainability landscape of the European Union, based on the results discussed above. We seek to describe the socioeconomic context of the countries where companies have the highest contribution to the SDGs versus the countries where companies do not yet seem interested in applying the SDGs.

5.2. Contribution of Swedish Companies to the SDGs

In our study, we observed significant participation from Swedish companies; Sweden was the country with the highest share of valid data, as detailed in Table 1. In 2022, the services sector, including wholesale and retail trade (G), hotels and restaurants (I), transport (H), government (N, O), financial services (K), professional services (M), education (P), health care (Q), and real estate (L) contributed 63.5% to the total GDP. Industrial activity contributed 24% to GDP, with manufacturing taking 13% and construction 11%, while agriculture (A) contributed a share of 1.5%. According to the Sustainable Development Report 2023 [112], Sweden scored 85.98%, ranking in the second position.

Swedish companies focus their efforts on achieving SDGs 7, 12, 13, and 14 [113], which are primarily prevalent in the manufacturing (C), financial (K), and wholesale and trade (G) industries. Table A2 indicates that Swedish companies focus their efforts on the SDGs in these sectors. Sweden excels in domains such as the circular economy, eco-friendly agriculture, green infrastructure, an eco-friendly lifestyle, and sustainable living, aiming to become a fossil-fuel-free nation by 2050 through its commitment under the Roadmap 2050 initiative [114].

Currently, Sweden is recognized as an extensively documented innovator and contributor with a high adaptive capacity to address climate change [115]. The ranking comes as a result of environmental protection regulations and policies that are primarily applied in the manufacturing industry (C), which has the highest contribution to SDG 7. Sweden has already achieved the target for 2020 regarding 50% of energy from renewable sources (SDG 7) [116] by investing in hydropower and biomass [117]. Due to its northern geographical location, companies and the population are large consumers of energy. In this regard, Swedish companies have invested more in renewable energy sources than companies in other countries, making a greater step toward clean energy [116,118–120]. Regarding air pollution, the metropolitan population's exposure to PM2.5 is below the WHO threshold [121] and is continuously decreasing (SDG 11). Climate change vulnerabilities, such as surface run-off and marine debris (SDG 14), consisting of industrial pellets, plastic, cigarettes, and paraffin pieces [122], are observed in the Baltic Sea [123] and linked to various industries, mainly manufacturing (C), retail and wholesale (G), and construction (F).

Companies headquartered in Sweden actively promote gender policies [124], benefiting from the most comprehensive legal framework for gender equality (SDG 5) [112]. Women hold 47% of parliamentary seats, as well as more than 40% of management positions, and SDG 5 is generally observed in every industry analyzed, with more focus on manufacturing (C) and financial services (K). Together with Sweden, European developed countries that are part of the G7, including France, Germany, and Italy, maintain a strong position regarding the applicability of SDGs [125].

5.3. Contribution of German Companies to the SDGs

Based on our research, in the list of countries with more than 100 companies with available data, Germany is the second country, as seen in Table 6. According to the Sustainable Development Report [112], Germany holds the fourth position in terms of sustainability performance, achieving an index of 83.36. In Germany, companies in the services sector account for 62.7% of GDP, which includes wholesale and retail trade (G), hotels and restaurants (I), transport (H), government services (N, O), financials (K), professional services (M), education (P), health care (Q), and real estate (L) [126]. In Germany, services also include the information and telecommunication sector (J) [127]. The second largest category is industry (26.9% of the GDP), which is divided between manufacturing (C), with a proportion of 18%, and construction (F), with 8.9%. Agriculture (A) contributes 0.9% to Germany's GDP.

The German labor market has excelled in effectively utilizing the skills of its workforce. It is observed that SDG 8 is tackled in construction (F), wholesale and retail (G), information and telecommunication (J), and agriculture (A)—industries that have a high contribution to the GDP. In this regard, companies in these sectors provide substantial hourly wages, while the unemployment rate is less than half of the OECD average. Although there is room for enhancement, efforts have been made to improve educational results and significantly increase childcare enrollment (SDG 4). There is a high level of proficiency, above the OECD average, in functional skills such as numeracy, literacy, and digital skills (SDG 4) [128]. Engagement in lifelong learning surpasses the OECD average but has the potential for further improvement (SDG 4). The vocational education and training system facilitates the integration of young people into the labor market, with the majority of German youth either employed, pursuing education, or undergoing training (SDG 8) [129]. However, women are still underrepresented in decision-making positions, but there is a positive trend in this regard [128,130]. Women spend more time in unpaid care and domestic work than men, with a gap lower than the OECD average [128].

Germany has established ambitious sustainability goals for 2030 [129]. These goals include a minimum reduction of 65% in GHG emissions and the goal of reaching carbon neutrality by 2045 [129]. Although currently recognized by the OECD as a leader in waste management, Germany is above the average in retail food waste (SDG 2). Manufacturing (C), real estate (L), and agriculture (A) contribute to proficient waste management and support Germany's leading position in recycling (SDGs 11 and 12). However, around 80% of freshwater, terrestrial, mountain, and marine areas, deemed crucial for biodiversity, lack protection (SDGs 14 and 15), although German companies implement policies regarding environmental and biodiversity protection.

5.4. Contribution of French Companies to the SDGs

In our study, France holds the third place as seen in Table 1. France holds the sixth position according to the Sustainable Development Report 2023 with an overall SDG adoption score of 82.05 [112]. It has a high standard of living and quality of life supported by social security systems and access to healthcare, as well as essential goods and services [124]. In France, companies in the services sector account for 70.7% of GDP [131]. This sector includes wholesale and retail trade (G), hotels and restaurants (I), transport (H), government (N, O), financials (K), professional services (M), education (P), health care (Q), and real estate (L). Industry accounts for 16.8% of the total GDP, specifically manufacturing (C) and construction (F). We identified in Table 6 that, from a total of 149 companies that contained valid data, 74.50% apply at least one SDG, especially in the two mentioned categories: services and industry, which represent 82.88% of the total number of companies that apply at least one SDG.

France is committed to implementing education policies within the curriculum (SDG 4). However, certain challenges persist. Based on the OECD, French students' proficiency in reading and mathematics falls below the minimum target levels (SDG 4), registering an indicator of 26.9 [128]. Labor market challenges (SDG 8) faced by France come as a result of the situation in the construction, wholesale and retail, transport, education, and healthcare industries, due to the slow growth in labor productivity and high unemployment rate [124]. The professions in these types of industries are considered difficult by 61% of recruitment companies due to a lack of candidates, inadequate professional preparation of candidates, work conditions, lack of financial motivation, or difficult access to the workplace [132]. Despite this fact, France maintains an income poverty rate below the OECD average (SDG 1).

Gender inequality is also lower than the OECD average, supported by legal frameworks that promote gender equality (SDG 5). However, concerns arise regarding the time women spend on unpaid care and housework compared to men, although the gap is below the OECD average [133]. Additionally, one-third of management positions are occupied by women. SDG 5 is specifically addressed in government services (N, O), with an increase in the indicator related to seats held by women in the national parliament [124].

France's environmental performance is sustained by self-owned nuclear electricity, leading to a decrease in GHG emissions (SDGs 9 and 13). In particular, in France, protected areas cover one-third of terrestrial regions and half of territorial waters, surpassing the 2020 Aichi biodiversity targets [134]. Moreover, beyond 80% of freshwater, terrestrial, and marine areas are considered protected areas (SDGs 14 and 15). However, companies in the manufacturing industry (C) are significant contributors to domestic material use (SDG 12). The manufacturing industry (C) contributes negatively to the SDGs by increasing marine pollution, exceeding the OECD average in terms of marine debris and nutrient pollution (SDG 14), as well as air pollution (SDGs 11 and 12).

5.5. Contribution of Italian Companies to the SDGs

As observed in Table 1, Italy is the fourth among the countries with over 100 companies that had available ESG data in Refinitiv Eikon. Italy ranks 24th in the Sustainable Development Report [112] with an index of 78.80. Since 3 February 2016, Italy has established the Italian Alliance for Sustainable Development (ASviS), comprising 270 member organizations [135]. This alliance plays a vital role in raising awareness and mobilizing stakeholders to achieve the SDGs. Italian companies show positive trends toward achieving SDGs 2, 5, 7, 9, 12, and 17, while SDGs 11, 13, 4, and 1 present worsening trends [136].

Companies in the services sector contribute to 64.3% of the Italian GDP. This sector includes wholesale and retail trade (G), hotels and restaurants (I), transport (H), government services (N, O), financial services (K), professional services (M), education (P), health care (Q), and real estate (L) [126]. Industry accounts for 23.8% of Italy's GDP, of which manufacturing (C) represents 15% and construction (F) represents 8.8%. Agriculture represents 1.8% of GDP.

The growth of GDP and labor productivity (SDG 8) have been low or negative in the last 10 years [137], with a decrease in GDP for the year 2022 compared to 2021. Economic recovery is attributed to the construction industry (F) and the manufacturing industry (C), as well as certain services such as accommodation and food service activities (I) and transport and storage (H), which increased the value added per person employed [136]. The employment rate has increased, approaching the OECD average, but the unemployment rate remains high [112,136]. The proportion of young people (between 15 and 29 years of age) who are not engaged in any level of education or employment (SDG 8) is above the OECD average and exceeds the values reported by Spain and Germany [128].

Italy excels in waste management, witnessing a decline in the share of municipal waste sent to landfills in 2022 [112]. Italian companies lead in the OECD ranking on recycling (SDGs 11 and 12) [128], especially in the manufacturing sector (C), as shown in Table A4 (Appendix A). Italy falls below the OECD average on food waste from households and retailers (G), especially due to the COVID-19 pandemic period, making it comparable to major European countries, but it is close to the OECD average when it comes to restaurants (I) (SDG 12).

Approximately 25% of terrestrial, mountain, and marine areas, along with 15% of freshwater areas crucial to biodiversity, remain unprotected (SDGs 14 and 15), and the IUCN Red List of Threatened Species [138] confirms a decline in biodiversity (SDG 15). While air pollution is decreasing in general, emissions in large cities remain high (SDG 11) due to industries such as transport (H), wholesale and retail (G), manufacturing (C), construction (F), and agriculture (A) in the metropolitan area (Table A2). This poses a health threat to the population (SDG 3) [139] since Italy, similar to France and Germany, registers high values of PM 2.5 pollution (SDG 11) [121,140]. Italy experiences lower mortality from non-communicable diseases compared to the OECD average (SDG 3), reflecting an effective healthcare system [112].

Social challenges in Italy are related to adults not having minimum proficiency in numeracy and literacy (SDG 4), as well as low participation in learning, compared to the OECD average and countries such as Spain and Sweden [128]. Low labor productivity values and high unemployment rates are also linked to the deficiency in digital skills of Italian adults and young people (SDG 4). Gender equality (SDG 5) in Italy is below average [112], and women are underrepresented in the parliament and managerial positions. The gender gap in unpaid work also exceeds the OECD average [112]. In summary, Italy's economic and social environment is characterized by contrasting aspects and uneven contributions to the SDGs.

5.6. The Case of Romanian Companies Regarding the SDGs

Romania and Bulgaria (in the next section) are EU countries with valid data but with very little contribution to the SDGs. They are ranked at the end of the scale of countries whose companies issue sustainability reports (analyzed by Refinitiv) but have little progress towards the SDGs and apply relatively few sustainability policies. According to the Sustainable Development Report 2023, Romania ranks 35th with a percentage of progress of 77.5 [112]. In 2022, the GDP of Romania was shared between services (57.6%) including hotels and restaurants (I), transport (H), government services (N, O), financials (K), professional services (M), education (P), health care (Q), and real estate (L), complemented by industry (28.9%) and agriculture (4.5%) [141].

Romania has the potential to make significant contributions to SDG 1, as agriculture (A) has well-developed supply chains [81]. This would also contribute to the fight against poverty (SDG 1) outside the European Union as the United Nations considers agriculture to be the best opportunity to reduce poverty. Improvements in food accessibility are necessary to reduce poverty [142]. Romania has a high rate of poverty (16.2%) [143], with only one more EU member—Bulgaria—registering a higher rate (16.5%). There is also a high proportion of malnutrition (SDG 2). Companies' indifference to SDG 2 is compensated by the existence of the Romanian Food Bank [144], which helps food chain actors, specifically in the wholesale and retail industry (G), to improve the economic value of different types of products by minimizing their storage and disposal costs.

There is a significant link between agriculture (A) and SDG 2. Our sample includes 13 companies in agriculture (A), but none of them are in Romania. Agriculture (A) in Romania employs approximately 26.7% of the working population (SDG 8)—one of the highest rates in Europe [145]—as Romania focuses on the promotion of sustainable agriculture. Efforts towards SDG 3 are seen in high-polluter industries such as electricity and gas (D), agriculture (A), wholesale and retail trade (G), the food industry (I), transport (H), and construction (F) [146].

The education system ranks well below the EU average (SDG 4) due to school abandonment, poor infrastructure, and lack of investment in sustainable education [147]. Variations depend on the region and whether they are urban or rural [148]. In rural areas, the main reasons for school dropout are linked to the lack of construction or modernization of educational buildings, purchase of IT equipment, free transportation to and from school, and the existence of after-school programs.

Romania has a competitive position as energy-intensive companies benefit from a low dependence on imported energy resources. Romania's geographical position and natural resources facilitate the use of wind and solar power, as well as hydropower [149]. Small and medium enterprises (SMEs) in Romania face challenges in finding partners that share sustainability values and actions.

Despite the existence of legal norms related to social protection expenditure, social services, and poverty prevention, Romania is often found at the bottom of the EU charts, revealing substantial gaps in these areas [150]. While there are favorable results for several key indicators, Romania still has a significant journey before reaching the level of achievement seen in the world's most advanced nations [151]. In the corporate environment, SDG 4 is approached through training sessions for skills related to sustainable development [152].

5.7. The Case of Bulgarian Companies Regarding the SDGs

Bulgaria ranks 44th, with a percentage of progress of 74.6%, according to the Sustainable Development Report 2023 [112]. Bulgaria's GDP in 2022 was divided between services (59.6%) including hotels and restaurants (I), transport (H), government services (N, O), financial services (K), professional services (M), education (P), health care (Q), and real estate (L) [153], followed by industry (25.5%) and agriculture (3.9%). In Bulgaria, the private sector has the resources to contribute significantly to the SDGs through improved access to quality education (SDG 4) [154].

A survey conducted by CSR AdviceBox observed that in a sample of 300 companies, two-thirds of companies are aware of the SDGs, and more than half are promoting the SDGs through corporate events. The most relevant SDGs for these companies were SDGs 4 and 13 [154]. Companies offering goods and services in a B2B transaction identified climate change as a greater risk, while companies offering products to end-users are more accountable with regard to SDGs 1, 2, and 3. As a general matter, Bulgarian companies focus on environmental actions linked to air quality (SDG 3 and 11), water management (SDG 6), marine protection (SDG 14), and the protection of nature and biodiversity (SDG 15). Bulgarian companies observe progress related to SDGs 4, 8, and 10 [155].

Bulgaria stands out in Europe with a prominent presence of women in the information and communication industry (J), reaching 27% [154]. Additionally, 53% of scientists and engineers in Bulgaria are women. Efforts are underway to reduce the gender pay gap (SDG 5), promote equality in decision-making processes, and combat gender-based violence. The challenges related to poverty and inequality, both within and between regions, hinder progress in the implementation of the SDGs [156]. Relatively poor performance in reading, mathematics, and science [157] underscores the need to increase efforts to provide quality education (SDG 4) to everyone [128].

Considering a different survey based on 30 Bulgarian companies, it is observed that most companies are aware of the SDGs but implementation is lacking due to challenges created by bureaucracy and insufficient financial resources in the company's budget [158]. Lack of financial allocation and subsidies for sustainable development could be another reason for Bulgaria and Romania lagging behind the EU cohort [154].

5.8. Comparing the Rankings of European Union Countries on SDG and Policy Adoption

In Table 11 we present a comparison of the rankings of European Union countries regarding their SDGs and policy adoption. Weighted mean rankings are derived from the present research, specifically from Tables 7 and 8. These are compared to the rankings derived from the Sustainable Development Report 2023 [112] for this subsample of EU countries only. In the cited report, the data on which the scoring methodology is based are drawn from international organizations (OECD, UNICEF, WHO, etc.) and civil society organizations and networks. The report data [112] are relevant at the country level and do not directly reflect the contribution of (multinational) companies to the achievement of the SDGs. However, it is relevant to compare these rankings to discover whether companies headquartered in a certain country follow the national trend in pursuing the SDGs.

The impact of EU leader countries on progress toward the SDGs is significant, with Sweden emerging as a consistent frontrunner, often ranking within the top three performers. Table 11 confirms the prominence of Swedish companies in their commitment to the SDGs, reflecting the nation's proactive stance on sustainability initiatives. Furthermore, previous research [159] has highlighted several other leading countries within the EU, such as Denmark, Germany, Ireland, Luxembourg, the Netherlands, Slovenia, Belgium, Austria, and Finland, each demonstrating notable contributions to SDG adherence. Our study contributes new insights to the existing literature by delving deeper into the behavior of companies headquartered in the EU and validating the high rankings of Sweden, Germany, France, Italy, and Spain. However, it should be noted that Luxembourg, while recognized for its overall consistency in SDG adherence, does not rank as highly in our results, with the difference being attributed to the dynamics of Luxembourg's corporate landscape. Despite the country's efforts toward alignment with the SDGs, individual companies do not consistently report or prioritize sustainability initiatives. As such, our study sheds light on the relationship between national policy frameworks and corporate practices, offering various perspectives to policymakers, stakeholders, and researchers.

Table 11. Comparison of the rankings of EU countries regarding their SDGs and policy adoption.

Country	Rank by Weighted Mean on SDG Adoption	Rank by Weighted Mean on Policy Adoption	Rank According to the Sustainable Development Report [112]
Austria	10	10	5
Belgium	8	9	19
Bulgaria	20	20	44
Croatia	-	-	12
Cyprus	-	-	59
Czech Republic	17	17	8
Denmark	6	6	3
Estonia	-	-	10
Finland	7	5	1
France	3	3	6
Germany	2	2	4
Greece	12	13	28
Hungary	15	15	22
Ireland	13	12	17
Italy	4	4	24
Latvia	-	-	14
Lithuania	-	-	37
Luxembourg	21	21	33
Malta	-	-	41
Netherlands	9	7	20
Poland	11	11	9
Portugal	14	14	18
Romania	18	16	35
Slovak Republic	19	19	23
Slovenia	16	18	13
Spain	5	8	16
Sweden	1	1	2

There is a strong correlation between the SDG adoption by companies and the SDG rank of the country of incorporation (Spearman's rho = 0.644, $p < 0.01$, based on the data from Table 11). Similarly, there is a strong correlation between the companies' policy adoption and the SDG rank of the country of incorporation (Spearman's rho = 0.642, $p < 0.01$). The correlations between these ranks are not perfect, but they validate the results of the present article. It is apparent that companies with headquarters in a certain country largely follow government policies in terms of sustainable development and social welfare. However, the European Union is based on the principle of economic integration, so multinational companies disseminate their goals and policies in all countries in which they are active, without restrictions. Therefore, there are positive spillover effects from companies in high-ranking countries (such as Sweden) that disseminate their sustainability policies and practices in other EU countries where they have subsidiaries and perform economic activities. Multinational companies can contribute to the SDGs in any of the countries in which they operate.

6. Conclusions

This article makes several contributions to the scientific literature and EU policy. Our research describes the business landscape of the European Union in relation to the applicability of the SDGs. Relying on high-quality data from the Refinitiv Eikon database, we analyzed the differences between EU countries with respect to the contribution of listed companies to each SDG. The results are very relevant to national and supranational regulators (like the European Commission). As observed in the macroeconomic reports from Eurostat [23], the differences between EU member states are profound and far-reaching. While it is true that multinational companies apply their sustainability practices and pursue their goals in the value chain and multiple countries, it is also true that some countries have an economic landscape that is actively engaged with the SDGs, while other countries are lagging in the EU cohort. For example, Romania has 85 companies listed on the Bucharest Stock Exchange [160] but only 5 are analyzed by Refinitiv (as of November 2023) and only 1 is reported to pursue any SDG. At the other end of the spectrum, Sweden has the largest share of companies with valid ESG data in the Refinitiv database and makes the most significant contribution to the SDGs at the EU level.

The present research has implications for companies but also governments and sectoral organizations. By conducting a comparative analysis of policy commitments and alignment with the SDGs, our research sheds light on the efficacy of existing policies in driving sustainable development outcomes. Our paper creates an avenue for policy frameworks that help companies adhere to the SDGs. By identifying gaps in corporate policy implementation, policymakers, corporate stakeholders, and NGOs can improve regulatory interventions and promote sustainable development objectives. The pragmatic view of sustainability encapsulates the current representation of adherence to the SDGs among companies in the EU and creates an understanding of sustainable business models. Governments can put pressure on companies (specifically large corporations and multinational enterprises) to improve their corporate social responsibility (CSR) and environmental policies. The ESG scores of companies should align with the SDGs to provide a tool for managers and boards to measure their progress and facilitate the adoption of sustainable business models [161]. The pragmatic development of ESG scoring systems promotes a higher level of responsibility and encourages managers to better disclose information on how each company plans to achieve the relevant SDGs.

Multinational corporations must adhere to sustainable development goals and policies, as outlined in the OECD Guidelines, within the nations they operate [162]. These policies emphasize crucial aspects of sustainability: economic, social, and environmental contributions. It is important for companies to uphold human rights, fulfill obligations to those impacted by their operations, foster collaborations with communities by promoting local capacity building and development, advocate for education and skill enhancement through employment, raise awareness among employees, eliminate discriminatory practices, and apply corporate governance principles [163]. The SDGs are explicitly mentioned in the Corporate Sustainability Reporting Directive [43], which encourages large companies to report their contribution to each SDG. This article offers a different perspective of the EU landscape regarding sustainable development in the economic domain.

We observed the trend in the EU, illustrated in this paper in Figures 1 and 2, indicating that companies in Eastern Europe exhibit lesser adherence to SDGs and sustainability policies compared to leading contributors like Sweden, Germany, France, or Italy. To achieve the common sustainability objectives among nations, the European Commission created various tools to "leave no one behind". The EU "Whole-of-Government" [164] refers to legislative acts implemented by member states through their ministries, public administrations, or other agencies to achieve the application of policies through better regulation and engagement with stakeholders. Voluntary National Reviews (VNRs) of SDGs represent how EU countries assess their progress and share their experience and challenges faced in the implementation of SDGs [23]. The European Commission engages numerous stakeholders in the process. In this regard, the European Commission encourages stakeholders to actively participate and submit their opinions, in line with SDG 16, providing feedback to policymakers via an online portal. These reviews are used for VNRs, and for the discussion in the present paper. The other relevant tool to understand the contribution to each SDG is the statistics database published by Eurostat, which offers users access to comprehensible data to monitor the SDGs and compare the evolution throughout the European Union.

Several limitations must be highlighted. First, the analysis is limited to the sample of companies reviewed by Refinitiv. The present results are not representative of other EU companies that may contribute to the SDGs but have not been reviewed by Refinitiv. The analysis captures the contribution of certain very large companies headquartered in each EU country to achieving the SDGs. At the same time, the results do not capture the contribution of large but unlisted companies or SMEs. At the same time, it is known that companies may be supported by NGOs (foundations) and state institutions to contribute to the SDGs, but this is not captured in this analysis. Second, the sample is cross-sectional for 2022, so it does not capture an evolution of the phenomenon; it only presents a snapshot of the contribution of EU companies to the SDGs in 2022. Third, the quantitative analysis describes the status of a socioeconomic phenomenon but does not infer causalities. The causal links between various factors and the contributions to the SDGs require the application of other methods of investigation and longitudinal analysis over a longer period. Finally, the equivalence matrix between the SDGs and sustainability policies does not suggest a perfect conceptual equivalence between the SDGs and the respective policies. Also, it cannot be said that the implementation of these policies within the analyzed companies contributes to achieving the corresponding SDGs.

The present investigation opens several avenues for future studies. Researchers can turn directly to sustainability reports, using automated data collection methods, to extract and synthesize disclosures that relate to companies' contributions to the SDGs. In this way, a much larger sample of listed and unlisted companies compliant with the Corporate Sustainability Reporting Directive can be covered. Using a different methodological approach, a qualitative analysis of the link between sustainability policies and SDGs can be carried out at the level of a sample of large companies. Researchers can investigate whether the contribution to the SDGs is made by explicitly applying sustainability policies so that the mapping of the SDGs to policies is explicit and relevant. The contribution of companies can be determined on each SDG or on certain SDGs grouped thematically by analyzing sustainability policies. This type of analysis should discuss the sectoral specificities and corporate factors of the adoption of certain SDGs.

Country-level or regional analyses can provide insight into socioeconomic and regulatory factors for adopting SDGs and corporate sustainability policies. In this sense, institutional theory [165] can provide a conceptual foundation on which to explain why certain sustainability concerns are more prevalent in some countries and sectors. Stakeholders can monitor progress toward the SDGs using external indexes that would evaluate business models, corporate strategies, and managerial behaviors under national circumstances and within each industry.

Author Contributions: Conceptualization, F.M.P. and V.D.D.; methodology, F.M.P. and V.D.D.; software, F.M.P. and V.D.D.; validation, V.D.D.; formal analysis, F.M.P.; investigation, F.M.P.; resources, F.M.P.; data curation, F.M.P.; writing—original draft preparation, F.M.P. and V.D.D.; writing—review and editing, V.D.D.; visualization, V.D.D.; supervision, V.D.D. All authors have read and agreed to the published version of the manuscript.

Funding: This research received no external funding.

Institutional Review Board Statement: Not applicable.

Informed Consent Statement: Not applicable.

Data Availability Statement: Data available from Refinitiv Eikon.

Acknowledgments: F.M.P. is grateful for the kind support of the Doctoral School of Accounting, Bucharest University of Economic Studies, Romania. The authors would like to thank the Editors and Reviewers for their kind consideration of this article and the numerous comments which significantly improved each version.

Conflicts of Interest: The authors declare no conflicts of interest.

Acronym List

ASviS	Alleanza Italiana per lo Sviluppo Sostenibile
CPR	Construction Products Regulation
CSR	Corporate Social Responsibility
ESG	Environmental, Social and Governance
EU	European Union
FAO	Food and Agriculture Organization of the United Nations
GDP	Gross Domestic Product
GHG(s)	Greenhouse Gas(es)
IEA	International Energy Agency
IUCN	International Union for Conservation of Nature
LSEG	London Stock Exchange Group
MDG	Millennium Development Goals
NACE	Nomenclature statistique des Activités économiques dans la Communauté Européenne (the European Nomenclature of Economic Activities)
NGO	Non-Governmental Organization
OECD	Organisation for Economic Co-operation and Development
SD	Sustainable Development
SDG	Sustainable Development Goals
SMEs	Small and Medium-sized Enterprises
SSCM	Sustainable Supply Chain Management
TRBC	Thomson Reuters Business Classification
UK	United Kingdom
UN	United Nations
UNDESA	United Nations Department of Economic and Social Affairs
V4	Visegrád Group countries
WHO	World Health Organization

Appendix A

Table A1. Matrix of sustainability policies linked to SDGs.

Policy	Index (Refinitiv)	Word Search	SDG and Related Targets																
			1	2	3	4	5	6	7	8	9	10	11	12	13	14	15	16	17
p_ANT	Animal Testing	"animal", "test"		2.5										12.3					
p_ARR	Analytic Resource Red Policy	"natural", "resources"											11.4	12.2		14.7	15.5		
p_AWB	Animal Well-being	"animal"															15.C		
p_EEF	Policy Energy Efficiency	"energy"							7.3										
p_EMS	Policy Emissions	"emission"									9.4								
p_ESC	Policy Env Supply Chain	"impact", "supply", "chain"							7.B					12.3		14.2			
p_FFD	Fossil Fuel Divestment Policy	"fossil", "fuel"							7.A					12.C					
p_GBD	Green Buildings	"green", "office" "building"											11.7						
p_LBW	Labeled Wood	"forest", "wood"						6.6									15.1		
p_PER	Product Environmental Responsible Use	"product", "responsible", "cost", "effective" "emissions", "pollution" "noise"			3.9			6.3			9.4					14.1			
p_PNS	Policy Nuclear Safety	"nuclear", "disaster"											11.5						
p_SPK	Policy Sustainable Packaging	"pack *", "containers" "biodegradable"																	
p_TRI	Takeback Recycling Initiatives	"take back", "recyc *"												12.5					
p_WEF	Policy Water Efficiency	"water"			3.9			6.4						12.4					
p_WRI	Waste Reduction Initiatives	"waste", "recycle", "reduce" "reuse", "substitute", "treat"						6.A					11.6	12.4		14.1			

Table A1. *Cont.*

Policy	Index (Refinitiv)	Word Search	SDG and Related Targets																
			1	2	3	4	5	6	7	8	9	10	11	12	13	14	15	16	17
p_AHR	Analytic Human Rights Policy	"child", "labor", "forced" "vulnerable"	1.4							8.7								16.2	
p_CDL	Policy Child Labor	"child", "labor"								8.7								16.2	
p_CDV	Policy Career Development	"skills", "training", "career"				4.4				8.6									
p_DOP	Policy Diversity Opportunity	"diversity", "equality", "minorities", "race" "ethnicity", "religion"					5.C					10.2							
p_EHS	Employee Health Safety Policy	"health", "safety", "accidents" "employ *"			3.8					8.8				12.4					
p_FAS	Policy Freedom of Association	"freedom", "association"																16.10	
p_FLB	Policy Forced Labor	"forced", "labour"								8.7									
p_SHS	Supply Chain HS Policy	"health", "safety"			3.8					8.8									
p_STR	Policy Skills Training	"skills", "training"				4.4				8.6									

Note. Asterisks (*) are used to format the search phrase to take into account word suffixes.

Table A2. Number of companies that apply the SDGs in each country and industry.

Industry	AUT	BEL	BGR	CZE	DNK	FIN	FRA	DEU	GRC	HUN	IRL	ITA	LUX	NLD	POL	PRT	ROU	SVK	SVN	ESP	SWE	Total
A	0	0	0	0	2	0	1	3	0	0	1	0	0	1	1	0	0	0	0	0	3	12
B	1	0	0	0	1	1	2	2	1	0	0	0	0	2	1	0	0	0	0	1	3	15
C	10	15	0	0	21	29	36	58	2	2	4	36	0	17	3	3	1	0	1	6	89	333
D	1	1	0	1	2	1	4	6	2	0	0	7	0	1	1	3	0	0	0	4	2	36
E	0	0	0	0	0	0	0	0	0	0	0	0	0	0	0	0	0	0	0	0	0	0
F	2	2	0	0	4	3	7	2	0	0	1	3	0	1	0	1	0	0	0	2	12	40
G	0	3	0	0	4	6	14	24	0	0	0	11	0	1	5	1	0	0	0	3	32	104
H	1	2	0	0	7	1	3	6	1	0	0	4	0	3	1	1	0	0	0	1	2	33
I	1	1	0	0	1	1	6	2	0	0	1	1	0	0	0	0	0	0	0	3	7	24
J	2	2	0	0	4	7	5	11	0	1	0	7	0	1	2	1	0	0	1	3	13	60
K	6	12	0	1	6	4	15	15	2	0	4	15	0	11	7	0	0	1	1	9	18	127
L	2	0	0	0	0	2	2	7	0	0	0	0	0	1	0	0	0	0	0	3	23	40
M	0	0	0	0	3	1	8	3	0	0	0	3	0	1	0	0	0	0	0	4	13	36
N	0	0	0	0	0	0	0	0	0	0	0	1	0	2	0	0	0	0	0	0	1	4
O	1	0	0	0	0	0	3	2	0	0	0	1	0	0	0	0	0	0	0	0	3	10
P	0	0	0	0	0	0	0	0	0	0	0	0	0	0	0	0	0	0	0	0	2	2
Q	0	0	0	0	0	1	1	2	0	0	0	1	0	0	0	0	0	0	0	0	3	8
R	0	0	0	0	0	0	4	2	0	0	0	2	0	2	0	0	0	0	0	2	2	14
S	0	0	0	0	0	0	0	0	0	0	0	0	0	0	0	0	0	0	0	0	0	0
T	0	0	0	0	0	0	0	0	0	0	0	0	0	0	0	0	0	0	0	0	0	0
U	0	0	0	0	0	0	0	0	0	0	0	0	0	0	0	0	0	0	0	0	0	0
Total	27	38	0	2	55	57	111	145	8	3	11	92	0	44	21	10	1	1	3	41	228	898

Note. Industries are expressed as NACE codes rev.2.0. Country abbreviations follow the ISO country codes (three letters).

Table A3. Number of companies that adopted sustainability policies in each country and industry.

Industry	AUT	BEL	BGR	CZE	DNK	FIN	FRA	DEU	GRC	HUN	IRL	ITA	LUX	NLD	POL	PRT	ROU	SVK	SVN	ESP	SWE	Total
A	0	0	0	0	2	0	0	2	0	0	1	0	0	1	1	0	0	0	0	0	0	7
B	1	0	0	0	1	1	2	3	0	0	0	0	0	2	2	0	0	0	0	1	7	20
C	10	17	0	0	21	27	49	85	2	1	3	39	0	12	3	1	2	0	1	8	110	391
D	1	1	0	1	2	1	5	9	2	0	0	7	0	1	1	3	1	0	0	4	3	42
E	0	0	0	0	0	0	0	0	0	0	0	0	0	0	0	0	0	0	0	0	0	0
F	2	2	0	0	7	4	9	5	1	1	1	3	0	2	0	1	0	0	0	3	16	57
G	0	2	0	0	1	4	5	8	0	0	0	7	0	0	4	1	0	0	0	2	9	43
H	1	1	0	0	3	0	1	4	1	0	0	0	0	2	0	1	0	0	0	1	2	17
I	1	0	0	0	0	0	3	0	0	0	1	0	0	0	0	0	0	0	0	2	1	8
J	1	2	0	0	5	11	15	28	0	1	0	12	1	1	3	2	1	0	1	3	30	117
K	6	5	1	1	5	3	8	21	3	1	3	14	1	4	9	0	0	2	1	9	25	122
L	3	0	0	0	0	2	2	10	0	0	0	0	0	1	0	0	0	0	0	3	23	44
M	0	2	0	0	3	2	12	7	0	0	0	3	0	1	0	0	0	0	0	4	18	52
N	0	0	0	0	0	0	2	2	0	0	0	1	0	2	0	0	0	0	0	0	4	11
O	0	0	0	0	0	0	0	0	0	0	0	0	0	0	0	0	0	0	0	0	0	0
P	0	0	0	0	0	0	0	0	0	0	0	0	0	0	0	0	0	0	0	0	0	0
Q	0	0	0	0	0	1	1	4	0	0	0	1	0	0	0	0	0	0	0	0	3	10
R	0	0	0	0	1	0	4	3	1	0	0	0	0	2	0	0	0	0	0	1	6	18
S	0	0	0	0	0	0	0	2	0	0	0	0	0	0	0	0	0	0	0	0	2	4
T	0	0	0	0	0	0	0	0	0	0	0	0	0	0	0	0	0	0	0	0	0	0
U	0	0	0	0	0	0	0	0	0	0	0	0	0	0	0	0	0	0	0	0	0	0
Total	29	43	1	2	60	69	148	225	12	5	12	107	2	48	26	11	5	2	3	45	291	1146

Note. Industries are expressed as NACE codes rev.2.0. Country abbreviations follow the ISO country codes (three letters).

Table A4. Number of companies that apply SDGs for each industry.

SDG/ Industry	A	B	C	D	F	G	H	I	J	K	L	M	N	O	P	Q	R	Total
SDG 1	0	1	58	13	7	15	3	3	6	35	5	7	0	1	1	1	2	158
SDG 2	5	0	58	9	6	16	2	5	4	13	2	4	0	1	1	1	0	127
SDG3	6	10	228	18	19	76	20	12	34	79	29	27	2	2	1	8	11	582
SDG 4	1	7	147	19	16	62	19	13	42	83	12	20	4	5	2	6	10	468
SDG 5	5	12	226	26	26	76	21	18	51	97	29	33	4	8	2	4	11	649
SDG 6	2	9	137	17	13	27	5	5	7	25	11	5	0	2	1	2	6	274
SDG 7	5	10	183	36	30	48	19	9	29	85	33	13	1	3	1	3	11	519
SDG 8	10	12	292	35	36	94	30	21	54	110	33	28	4	7	2	8	13	789
SDG 9	5	11	209	32	31	55	21	12	50	70	20	20	1	8	1	5	8	559
SDG 10	0	3	128	18	14	46	18	15	33	52	18	16	3	4	2	2	11	383
SDG 11	1	4	125	27	32	28	11	10	35	75	38	12	1	1	2	1	4	407
SDG 12	10	14	300	32	32	85	26	17	49	92	31	26	1	7	1	5	12	740
SDG 13	9	15	297	36	32	86	30	21	50	116	36	24	2	8	2	6	11	781
SDG 14	0	3	83	15	7	14	8	4	4	15	3	3	1	1	1	1	3	166
SDG 15	6	5	117	25	17	29	5	6	10	39	16	4	0	1	1	1	4	286
SDG 16	3	10	146	19	23	47	16	16	36	67	16	19	1	9	1	5	7	441
SDG 17	3	5	138	22	14	40	22	11	30	56	16	19	2	2	2	2	6	390
TOTAL	71	131	2872	399	355	844	276	198	524	1109	348	280	27	70	24	61	130	-

Note. Industries are expressed as NACE codes rev.2.0.

Table A5. SDGs observed in previous studies versus the current study.

Industry	SDGs Identified in Previous Studies	Most Prominent SDGs Identified in the Current Study	Matching SDGs
A (Agriculture, Forestry and Fishing)	SDG 2 [18,166], 12 [166,167], 13 [167], 14 [166,168], 15 [166,167]	SDG 2, 8, 12, 13, 15	SDG 2, 12, 13, 15
B (Mining and Quarrying)	SDG 3 [169], 8 [169], 13 [169]	SDG 5, 8, 12, 13	SDG 8, 13
C (Manufacturing)	SDG 1 [170], 2 [170], 3 [171], 6 [172], 8 [171–173], 9 [171,172], 10 [172], 11 [172], 12 [171–174]	SDG 3, 5, 8, 12, 13	SDG 3, 8, 12
D (Electricity, Gas, Steam and Air Conditioning Supply)	SDG 1 [175], 3 [175], 4 [175], 6 [175], 7 [175–178], 9 [175,176,178], 11 [176], 12 [175,176,178], 13 [176,177]	SDG 7, 9, 12, 13	SDG 7, 9, 12, 13
E (Water Supply; Sewerage, Waste Management and Remediation Activities)	SDG 6 [179–181], 7 [182], 11 [182], 12 [182,183]	-	SDGs not identified in our study
F (Construction)	SDG 12 [184], 13 [184]	SDG 7, 8, 9, 11, 12, 13	SDG 12, 13
G (Wholesale and Retail Trade; Repair of Motor Vehicles and Motorcycles)	SDG 12 [84,185]	SDG 8, 12, 13	SDG 12
H (Transportation and Storage)	SDG 3 [186], 7 [186], 11 [186], 12 [186,187]	SDG 8, 12, 13, 17	SDG 12
I (Accommodation and Food Service Activities)	SDG 12 [188]	SDG 5, 8, 12, 13	SDG 12
J (Information and Communication)	SDG 1, 2, 3, 4, 5, 8, 9, 10, 11, 16, 17 [189]	SDG 5, 8, 9, 12, 13	SDG 5, 8
K (Financial and Insurance Activities)	SDG 1 [190], 3 [190,191], 4 [190,191], 5 [190], 8 [190,191], 9 [191], 10 [191], 13 [190], 16 [191]	SDG 5, 8, 12, 13	SDG 5, 8, 13
L (Real Estate Activities)	SDG 11 [192]	SDG 7, 8, 11, 12, 13	SDG 11
M (Professional, Scientific and Technical Activities)	SDG 4, 8, 9, 10 [193]	SDG 3, 5, 8, 12	SDG 8
N (Administrative and Support Service Activities)	SDG 5 [194], 6 [194], 10 [194], 11 [195], 13 [194], 16 [195], 17 [195]	SDG 4, 5, 8	SDG 5
O (Public Administration and Defense; Compulsory Social Security)	SDG 5 [194], 6 [194], 10 [194], 11 [196], 13 [194]	SDG 5, 9, 13, 16	SDG 5, 13
P (Education)	SDG 3, 4, 5, 8, 9, 10, 11, 12, 13, 16, 17 [197]	SDG 4, 5, 8, 10, 11, 13, 17	SDG 4, 5, 8, 10, 11, 13, 17
Q (Human Health and Social Work Activities)	SDG 3 [105,198], 10 [199], 16 [200]	SDG 3, 4, 8, 13	SDG 3
R (Arts, Entertainment and Recreation)	SDG 2 [201], 3 [201,202], 4 [201,202], 5 [201,202], 8 [201,202], 10 [201,202], 11 [201,203], 12 [202], 13 [201,202], 16 [201,203], 17 [202]	SDG 3, 5, 7, 8, 10, 12, 13	SDG 3, 5, 8, 10, 12, 13

Table A6. Number of companies that apply sustainability policies for each industry.

Policy/ Industry	A	B	C	D	F	G	H	I	J	K	L	M	N	O	P	Q	R	Total
p_ANT	0	0	49	0	0	11	0	0	0	7	0	20	0	0	0	3	0	90
p_ARR	13	20	402	40	42	128	37	30	70	142	42	42	5	14	2	10	18	1057
p_AWB	1	0	33	1	0	17	0	7	0	3	1	1	0	0	0	0	1	65
p_EEF	13	19	384	36	40	115	35	25	61	133	41	36	4	13	1	9	16	981
p_EMS	12	20	374	40	42	116	37	28	62	139	43	34	5	13	2	10	17	994
p_ESC	11	18	350	37	37	104	32	24	58	97	34	28	1	12	0	8	13	864
p_FFD	0	0	0	0	0	0	0	0	0	28	1	0	0	0	0	0	0	29
p_GBD	1	2	71	8	13	39	13	9	22	71	12	11	2	3	0	2	4	283
p_LBW	0	0	34	0	2	9	1	1	0	3	0	2	0	0	0	0	2	54
p_PER	7	14	294	40	36	85	11	10	48	107	32	17	0	11	0	3	12	727
p_PNS	0	0	0	6	2	0	0	0	0	0	0	0	0	0	0	0	0	8
p_SPK	4	4	181	2	2	67	9	9	19	7	0	2	0	1	0	3	10	320
p_TRI	2	1	77	2	1	26	1	3	26	4	0	3	0	1	1	1	2	151
p_WEF	11	19	286	35	21	58	22	14	30	71	29	19	1	10	0	5	11	642
p_WRI	10	18	372	39	40	114	33	27	64	125	39	39	2	14	1	9	15	961
p_AHR	13	20	399	40	41	133	38	27	69	139	41	36	5	15	2	9	16	1043
p_CDL	13	19	370	36	38	114	35	24	59	112	31	27	3	15	0	6	16	918
p_CDV	11	19	356	40	42	119	38	30	65	136	38	35	6	13	2	9	15	974
p_DOP	13	21	410	46	43	137	39	32	74	150	44	45	6	15	2	9	18	1104
p_EHS	13	21	400	43	44	133	40	31	69	140	41	47	5	15	2	10	17	1071
p_FAS	8	16	299	31	30	80	26	16	40	93	27	21	4	12	0	7	9	719
p_FLB	13	19	366	37	34	111	34	23	58	111	31	26	3	14	0	5	15	900
p_SHS	8	15	260	30	25	71	22	17	41	58	25	14	0	8	0	6	10	610
p_STR	10	19	357	42	42	128	36	31	63	135	41	36	6	12	2	10	17	987
TOTAL	187	304	6124	631	617	1915	539	418	998	2011	593	541	58	211	17	134	254	-

Note. Industries are expressed as NACE codes rev.2.0. Policy abbreviations are included in Table A1.

Table A7. Pearson correlation matrix between the frequencies of SDGs (columns) and p_SDGs (corresponding sustainability policies on rows).

Correl.	SDG 1	SDG 2	SDG 3	SDG 4	SDG 5	SDG 6	SDG 7	SDG 8	SDG 9	SDG 10	SDG 11	SDG 12	SDG 13	SDG 14	SDG 15	SDG 16	SDG 17
p_SDG1	**0.122** **	0.106 **	0.221 **	0.182 **	0.243 **	0.149 **	0.203 **	0.326 **	0.208 **	0.176 **	0.169 **	0.324 **	0.313 **	0.110 **	0.162 **	0.187 **	0.179 **
p_SDG2	0.077 **	**0.151** **	0.120 **	−0.020	0.014	0.145 **	−0.048	−0.022	−0.069 *	0.005	−0.110 **	0.039	0.011	0.135 **	0.092 **	−0.023	0.034
p_SDG3	0.071 *	0.046	**0.152** **	0.156 **	0.177 **	0.117 **	0.164 **	0.235 **	0.169 **	0.121 **	0.137 **	0.245 **	0.258 **	0.086 **	0.111 **	0.139 **	0.150 **
p_SDG4	0.105 **	0.086 **	0.198 **	**0.252** **	0.274 **	0.160 **	0.257 **	0.304 **	0.218 **	0.167 **	0.179 **	0.299 **	0.333 **	0.094 **	0.147 **	0.220 **	0.224 **
p_SDG5	0.086 **	0.076 **	0.177 **	0.170 **	**0.212** **	0.111 **	0.188 **	0.282 **	0.185 **	0.144 **	0.125 **	0.263 **	0.295 **	0.089 **	0.124 **	0.153 **	0.155 **
p_SDG6	0.126 **	0.121 **	0.189 **	0.182 **	0.200 **	**0.182** **	0.234 **	0.266 **	0.189 **	0.150 **	0.165 **	0.303 **	0.335 **	0.116 **	0.188 **	0.154 **	0.171 **
p_SDG7	0.109 **	0.100 **	0.202 **	0.179 **	0.238 **	0.161 **	**0.235** **	0.310 **	0.206 **	0.128 **	0.189 **	0.319 **	0.358 **	0.130 **	0.167 **	0.165 **	0.198 **
p_SDG8	0.075 *	0.067 *	0.153 **	0.156 **	0.195 **	0.094 **	0.161 **	**0.247** **	0.155 **	0.123 **	0.140 **	0.213 **	0.213 **	0.051	0.098 **	0.139 **	0.135 **
p_SDG9	0.122 **	0.106 **	0.192 **	0.236 **	0.249 **	0.177 **	0.274 **	0.339 **	**0.237** **	0.176 **	0.218 **	0.330 **	0.369 **	0.102 **	0.168 **	0.211 **	0.192 **
p_SDG10	0.086 **	0.076 **	0.177 **	0.170 **	0.212 **	0.111 **	0.188 **	0.282 **	0.185 **	**0.144** **	0.125 **	0.263 **	0.295 **	0.089 **	0.124 **	0.153 **	0.155 **
p_SDG11	0.092 **	0.091 **	0.181 **	0.162 **	0.227 **	0.127 **	0.204 **	0.301 **	0.185 **	0.113 **	**0.145** **	0.292 **	0.303 **	0.106 **	0.131 **	0.166 **	0.154 **
p_SDG12	0.063 *	0.067 *	0.165 **	0.149 **	0.208 **	0.085 **	0.154 **	0.271 **	0.167 **	0.125 **	0.132 **	**0.236** **	0.257 **	0.079 **	0.099 **	0.141 **	0.137 **
p_SDG13	0.096 **	0.116 **	0.190 **	0.198 **	0.250 **	0.154 **	0.252 **	0.323 **	0.212 **	0.130 **	0.208 **	0.317 **	**0.362** **	0.116 **	0.173 **	0.170 **	0.161 **
p_SDG14	0.089 **	0.079 **	0.176 **	0.151 **	0.212 **	0.125 **	0.194 **	0.275 **	0.184 **	0.097 **	0.156 **	0.264 **	0.270 **	**0.092** **	0.119 **	0.142 **	0.134 **
p_SDG15	0.104 **	0.108 **	0.197 **	0.177 **	0.228 **	0.149 **	0.220 **	0.303 **	0.197 **	0.130 **	0.180 **	0.312 **	0.336 **	0.125 **	**0.154** **	0.170 **	0.192 **
p_SDG16	0.122 **	0.106 **	0.221 **	0.182 **	0.243 **	0.149 **	0.203 **	0.326 **	0.208 **	0.176 **	0.169 **	0.324 **	0.313 **	0.110 **	0.162 **	**0.187** **	0.179 **

Notes. * $p \leq 0.05$, ** $p < 0.01$. Correlation coefficients in bold are on the diagonal line of this matrix and represent the relationship between each SDG and the corresponding policy, as mapped in Table A1.

References

1. Department of Economic and Social Affairs. Transforming Our World: The 2030 Agenda for Sustainable Development. Available online: https://sdgs.un.org/2030agenda (accessed on 12 February 2023).
2. Chalmers, A.W.; Klingler-Vidra, R.; van de Wijs, P.P.; van der Lugt, C.; Bailey, T. Carrots & Sticks: Beyond Disclosure in ESG and Sustainability Policy. Annual Report. Available online: https://www.carrotsandsticks.net/ (accessed on 12 February 2023).
3. Fukuda-Parr, S.; McNeill, D. Knowledge and Politics in Setting and Measuring the SDG s: Introduction to Special Issue. *Glob. Policy* **2019**, *10*, 5–15. [CrossRef]
4. United Nations Economic and Social Commission for Western Asia. The 5Ps of the Sustainable Development Goals. Available online: https://www.unescwa.org/sites/default/files/inline-files/the_5ps_of_the_sustainable_development_goals.pdf (accessed on 12 February 2023).
5. Tosun, J.; Leininger, J. Governing the Interlinkages between the Sustainable Development Goals: Approaches to Attain Policy Integration. *Glob. Chall.* **2017**, *1*, 1700036. [CrossRef]
6. Bali Swain, R.; Yang-Wallentin, F. Achieving Sustainable Development Goals: Predicaments and Strategies. *Int. J. Sustain. Dev. World Ecol.* **2020**, *27*, 96–106. [CrossRef]
7. Jimenez, D.; Franco, I.B.; Smith, T. A Review of Corporate Purpose: An Approach to Actioning the Sustainable Development Goals (SDGs). *Sustainability* **2021**, *13*, 3899. [CrossRef]
8. European Commission. The European Green Deal. Available online: https://commission.europa.eu/strategy-and-policy/priorities-2019-2024/european-green-deal_en (accessed on 31 January 2024).
9. European Council. European Green Deal. Available online: https://www.consilium.europa.eu/en/policies/green-deal/ (accessed on 31 January 2024).
10. European Commission. Farm to Fork Strategy. Available online: https://food.ec.europa.eu/horizontal-topics/farm-fork-strategy_en (accessed on 31 January 2024).
11. European Commission. Circular Economy Action Plan. Available online: https://environment.ec.europa.eu/strategy/circular-economy-action-plan_en (accessed on 31 January 2024).
12. European Commission. A New Circular Economy Action Plan for a Cleaner and More Competitive Europe. Available online: https://eur-lex.europa.eu/resource.html?uri=cellar:9903b325-6388-11ea-b735-01aa75ed71a1.0017.02/DOC_1&format=PDF (accessed on 31 January 2024).
13. European Commission. Biodiversity Strategy for 2030. Available online: https://environment.ec.europa.eu/strategy/biodiversity-strategy-2030_en (accessed on 3 January 2024).
14. Zhang, D.; Wang, C.; Dong, Y. How Does Firm ESG Performance Impact Financial Constraints? An Experimental Exploration of the COVID-19 Pandemic. *Eur. J. Dev. Res.* **2023**, *35*, 219–239. [CrossRef]
15. Ranjbari, M.; Shams Esfandabadi, Z.; Zanetti, M.C.; Scagnelli, S.D.; Siebers, P.-O.; Aghbashlo, M.; Peng, W.; Quatraro, F.; Tabatabaei, M. Three Pillars of Sustainability in the Wake of COVID-19: A Systematic Review and Future Research Agenda for Sustainable Development. *J. Clean. Prod.* **2021**, *297*, 126660. [CrossRef]
16. Bereczk, A.; Hodine, B.H.; Sasvari, P. Prevalence of the Sustainable Cities and Communities Goal (SDG-11) in the Research Activity of V4 Countries. In Proceedings of the Central and Eastern European eDem and eGov Days 2023, Budapest, Hungary, 14–15 September 2023; ACM: New York, NY, USA; pp. 125–131.
17. Grzebyk, M.; Stec, M.; Hejdukova, P. Implementation of Sustainable Development Goal 8 in European Union Countries—A Measurement Concept and a Multivariate Comparative Analysis. *Sustain. Dev.* **2023**, *31*, 2758–2769. [CrossRef]
18. Craparo, G.; Cano Montero, E.I.; Santos Peñalver, J.F. Trends in the Circular Economy Applied to the Agricultural Sector in the Framework of the SDGs. *Environ. Dev. Sustain.* **2023**. *ahead of print*. [CrossRef]
19. Ali, S.M.; Appolloni, A.; Cavallaro, F.; D'Adamo, I.; Di Vaio, A.; Ferella, F.; Gastaldi, M.; Ikram, M.; Kumar, N.M.; Martin, M.A.; et al. Development Goals towards Sustainability. *Sustainability* **2023**, *15*, 9443. [CrossRef]
20. Kleespies, M.W.; Dierkes, P.W. The Importance of the Sustainable Development Goals to Students of Environmental and Sustainability Studies—A Global Survey in 41 Countries. *Humanit. Soc. Sci. Commun.* **2022**, *9*, 218. [CrossRef]
21. Cernev, T.; Fenner, R. The Importance of Achieving Foundational Sustainable Development Goals in Reducing Global Risk. *Futures* **2020**, *115*, 102492. [CrossRef]
22. Delabre, I.; Alexander, A.; Rodrigues, C. Strategies for Tropical Forest Protection and Sustainable Supply Chains: Challenges and Opportunities for Alignment with the UN Sustainable Development Goals. *Sustain. Sci.* **2020**, *15*, 1637–1651. [CrossRef]
23. Eurostat. *Sustainable Development in the European Union: Monitoring Report on Progress towards the SDGs in an EU Context*, 2023rd ed.; European Union: Luxembourg, 2023.
24. Sachs, J.D.; Lafortune, G.; Fuller, G.; Drumm, E. *Implementing the SDG Stimulus: Sustainable Development Report 2023*; Dublin University Press: Dublin, Ireland, 2023.
25. Mosse, D. A Relational Approach to Durable Poverty, Inequality and Power. *J. Dev. Stud.* **2010**, *46*, 1156–1178. [CrossRef]
26. Allen, C.; Metternicht, G.; Wiedmann, T. Initial Progress in Implementing the Sustainable Development Goals (SDGs): A Review of Evidence from Countries. *Sustain. Sci.* **2018**, *13*, 1453–1467. [CrossRef]
27. Breuer, A.; Janetschek, H.; Malerba, D. Translating Sustainable Development Goal (SDG) Interdependencies into Policy Advice. *Sustainability* **2019**, *11*, 2092. [CrossRef]

28. Hák, T.; Janoušková, S.; Moldan, B. Sustainable Development Goals: A Need for Relevant Indicators. *Ecol. Indic.* **2016**, *60*, 565–573. [CrossRef]
29. Wang, Y.; Adebayo, T.S.; Ai, F.; Quddus, A.; Umar, M.; Shamansurova, Z. Can Finland Serve as a Model for Other Developed Countries? Assessing the Significance of Energy Efficiency, Renewable Energy, and Country Risk. *J. Clean. Prod.* **2023**, *428*, 139306. [CrossRef]
30. International Trade Administration. Finland—Country Commercial Guide. Available online: https://www.trade.gov/country-commercial-guides/finland-energy (accessed on 28 December 2023).
31. Eckerberg, K.; Forsberg, B. Implementing Agenda 21 in Local Government: The Swedish Experience. *Local Environ.* **1998**, *3*, 333–347. [CrossRef]
32. Bie, Q.; Wang, S.; Qiang, W.; Ma, X.; Gu, Z.; Tian, N. Progress toward Sustainable Development Goals and Interlinkages between Them in Arctic Countries. *Heliyon* **2023**, *9*, e13306. [CrossRef]
33. Šebestová, J.; Sroka, W. Sustainable Development Goals and SMEs Decisions: Czech Republic vs. Poland. *J. East. Eur. Cent. Asian Res.* **2020**, *7*, 39–50. [CrossRef]
34. Fura, B.; Wojnar, J.; Kasprzyk, B. Ranking and Classification of EU Countries Regarding Their Levels of Implementation of the Europe 2020 Strategy. *J. Clean. Prod.* **2017**, *165*, 968–979. [CrossRef]
35. Lazăr, D.; Minea, A.; Purcel, A.-A. Pollution and Economic Growth: Evidence from Central and Eastern European Countries. *Energy Econ.* **2019**, *81*, 1121–1131. [CrossRef]
36. Benedek, J.; Ivan, K.; Török, I.; Temerdek, A.; Holobâcă, I. Indicator-based Assessment of Local and Regional Progress toward the Sustainable Development Goals (SDGs): An Integrated Approach from Romania. *Sustain. Dev.* **2021**, *29*, 860–875. [CrossRef]
37. Bompard, E.F.; Corgnati, S.P.; Grosso, D.; Huang, T.; Mietti, G.; Profumo, F. Multidimensional Assessment of the Energy Sustainability and Carbon Pricing Impacts along the Belt and Road Initiative. *Renew. Sustain. Energy Rev.* **2022**, *154*, 111741. [CrossRef]
38. Saladini, F.; Betti, G.; Ferragina, E.; Bouraoui, F.; Cupertino, S.; Canitano, G.; Gigliotti, M.; Autino, A.; Pulselli, F.M.; Riccaboni, A.; et al. Linking the Water-Energy-Food Nexus and Sustainable Development Indicators for the Mediterranean Region. *Ecol. Indic.* **2018**, *91*, 689–697. [CrossRef]
39. Sobczak, E.; Bartniczak, B.; Raszkowski, A. Implementation of the No Poverty Sustainable Development Goal (SDG) in Visegrad Group (V4). *Sustainability* **2021**, *13*, 1030. [CrossRef]
40. Lamichhane, S.; Eğilmez, G.; Gedik, R.; Bhutta, M.K.S.; Erenay, B. Benchmarking OECD Countries' Sustainable Development Performance: A Goal-Specific Principal Component Analysis Approach. *J. Clean. Prod.* **2021**, *287*, 125040. [CrossRef]
41. Rosati, F.; Faria, L.G.D. Addressing the SDGs in Sustainability Reports: The Relationship with Institutional Factors. *J. Clean. Prod.* **2019**, *215*, 1312–1326. [CrossRef]
42. European Parliament and the Council of the European Union. Directive 2014/95/EU of the European Parliament and of the Council of 22 October 2014 Amending Directive 2013/34/EU as Regards Disclosure of Non-Financial and Diversity Information by Certain Large Undertakings and Groups. *Off. J. Eur. Union* **2014**, *L330*, 1–9.
43. European Parliament and the Council of the European Union. Directive (EU) 2022/2464 of the European Parliament and of the Council of 14 December 2022 Amending Regulation (EU) No 537/2014, Directive 2004/109/EC, Directive 2006/43/EC and Directive 2013/34/EU, as Regards Corporate Sustainability Reporting (Text with EEA Relevance). *Off. J. Eur. Union* **2022**, *L322*, 15.
44. Fonseca, L.; Ferro, R. Influence of Firms' Environmental Management and Community Involvement Programs in Their Employees and in the Community. *FME Trans.* **2015**, *43*, 370–376. [CrossRef]
45. Inês, A.; Diniz, A.; Moreira, A.C. A Review of Greenwashing and Supply Chain Management: Challenges ahead. *Clean. Environ. Syst.* **2023**, *11*, 100136. [CrossRef]
46. De Gennaro, D.; Piscopo, G. Pinkwashing and Mansplaining: Individual and Organizational Experiences of Gender Inequality at Work during the COVID-19 Pandemic. *Cult. Organ.* **2023**, *29*, 298–314. [CrossRef]
47. Pizzi, S.; Del Baldo, M.; Caputo, F.; Venturelli, A. Voluntary Disclosure of Sustainable Development Goals in Mandatory Non-financial Reports: The Moderating Role of Cultural Dimension. *J. Int. Financ. Manag. Account.* **2022**, *33*, 83–106. [CrossRef]
48. Van Zanten, J.A.; Van Tulder, R. Multinational Enterprises and the Sustainable Development Goals: An Institutional Approach to Corporate Engagement. *J. Int. Bus. Policy* **2018**, *1*, 208–233. [CrossRef]
49. Nilsson, M.; Chisholm, E.; Griggs, D.; Howden-Chapman, P.; McCollum, D.; Messerli, P.; Neumann, B.; Stevance, A.-S.; Visbeck, M.; Stafford-Smith, M. Mapping Interactions between the Sustainable Development Goals: Lessons Learned and Ways forward. *Sustain. Sci.* **2018**, *13*, 1489–1503. [CrossRef] [PubMed]
50. Leal Filho, W.; Viera Trevisan, L.; Paulino Pires Eustachio, J.H.; Dibbern, T.; Castillo Apraiz, J.; Rampasso, I.; Anholon, R.; Gornati, B.; Morello, M.; Lambrechts, W. Sustainable Supply Chain Management and the UN Sustainable Development Goals: Exploring Synergies towards Sustainable Development. *TQM J.* **2023** *ahead of print*. [CrossRef]
51. Christensen, H.; Hail, L.; Leuz, C. *Mandatory CSR and Sustainability Reporting: Economic Analysis and Literature Review*; National Bureau of Economic Research: Cambridge, MA, USA, 2019; p. w26169.
52. Ionescu, G.H.; Firoiu, D.; Tănasie, A.; Sorin, T.; Pîrvu, R.; Manta, A. Assessing the Achievement of the SDG Targets for Health and Well-Being at EU Level by 2030. *Sustainability* **2020**, *12*, 5829. [CrossRef]
53. Hummel, K.; Szekely, M. Disclosure on the Sustainable Development Goals—Evidence from Europe. *Account. Eur.* **2022**, *19*, 152–189. [CrossRef]

54. KPMG. How to Report on the SDGs. Available online: https://assets.kpmg.com/content/dam/kpmg/xx/pdf/2018/02/how-to-report-on-sdgs.pdf (accessed on 18 December 2023).
55. Emma, G.-M.; Jennifer, M.-F. Is SDG Reporting Substantial or Symbolic? An Examination of Controversial and Environmentally Sensitive Industries. *J. Clean. Prod.* **2021**, *298*, 126781. [CrossRef]
56. Hatayama, H. The Metals Industry and the Sustainable Development Goals: The Relationship Explored Based on SDG Reporting. *Resour. Conserv. Recycl.* **2022**, *178*, 106081. [CrossRef]
57. IEA. Iron and Steel Technology Roadmap. Available online: https://www.iea.org/reports/iron-and-steel-technology-roadmap (accessed on 4 March 2024).
58. Lisowski, S.; Berger, M.; Caspers, J.; Mayr-Rauch, K.; Bäuml, G.; Finkbeiner, M. Criteria-Based Approach to Select Relevant Environmental SDG Indicators for the Automobile Industry. *Sustainability* **2020**, *12*, 8811. [CrossRef]
59. European Parliament Fit for 55: Zero CO2 Emissions for New Cars and Vans in 2035. Available online: https://www.europarl.europa.eu/news/en/press-room/20230210IPR74715/fit-for-55-zero-co2-emissions-for-new-cars-and-vans-in-2035 (accessed on 8 March 2024).
60. Diaz Gonçalves, T.; Saporiti Machado, J. Origins of the Sustainability Concept and Its Application to the Construction Sector in the EU. *Sustainability* **2023**, *15*, 13775. [CrossRef]
61. Alawneh, R.; Mohamed Ghazali, F.E.; Ali, H.; Asif, M. Assessing the Contribution of Water and Energy Efficiency in Green Buildings to Achieve United Nations Sustainable Development Goals in Jordan. *Build. Environ.* **2018**, *146*, 119–132. [CrossRef]
62. European Commission. Construction Products Regulation (CPR). Available online: https://single-market-economy.ec.europa.eu/sectors/construction/construction-products-regulation-cpr_en (accessed on 1 March 2024).
63. Opoku, A.; Deng, J.; Elmualim, A.; Ekung, S.; Hussien, A.A.; Buhashima Abdalla, S. Sustainable Procurement in Construction and the Realisation of the Sustainable Development Goal (SDG) 12. *J. Clean. Prod.* **2022**, *376*, 134294. [CrossRef]
64. Yu, A.T.W.; Yevu, S.K.; Nani, G. Towards an Integration Framework for Promoting Electronic Procurement and Sustainable Procurement in the Construction Industry: A Systematic Literature Review. *J. Clean. Prod.* **2020**, *250*, 119493. [CrossRef]
65. Ghufran, M.; Aldieri, L.; Pyka, A.; Ali, S.; Bimonte, G.; Senatore, L.; Vinci, C.P. Food Security Assessment in the Light of Sustainable Development Goals: A Post-Paris Agreement Era. *Environ. Dev. Sustain.*. [CrossRef]
66. Agovino, M.; Cerciello, M.; Gatto, A. Policy Efficiency in the Field of Food Sustainability. The Adjusted Food Agriculture and Nutrition Index. *J. Environ. Manag.* **2018**, *218*, 220–233. [CrossRef]
67. Pânzaru, R.L.; Firoiu, D.; Ionescu, G.H.; Ciobanu, A.; Medelete, D.M.; Pîrvu, R. Organic Agriculture in the Context of 2030 Agenda Implementation in European Union Countries. *Sustainability* **2023**, *15*, 10582. [CrossRef]
68. Bennetzen, E.H.; Smith, P.; Porter, J.R. Agricultural Production and Greenhouse Gas Emissions from World Regions—The Major Trends over 40 Years. *Glob. Environ. Change* **2016**, *37*, 43–55. [CrossRef]
69. Li, Y.; Rockinger, M. Unfolding the Transitions in Sustainability Reporting. *Sustainability* **2024**, *16*, 809. [CrossRef]
70. Robaina, M.; Villar, J.; Pereira, E.T. The Determinants for a Circular Economy in Europe. *Environ. Sci. Pollut. Res.* **2020**, *27*, 12566–12578. [CrossRef]
71. Bătae, O.M.; Dragomir, V.D.; Feleagă, L. The Relationship between Environmental, Social, and Financial Performance in the Banking Sector: A European Study. *J. Clean. Prod.* **2021**, *290*, 125791. [CrossRef]
72. Pozzoli, M.; Pagani, A.; Paolone, F. The Impact of Audit Committee Characteristics on ESG Performance in the European Union Member States: Empirical Evidence before and during the COVID-19 Pandemic. *J. Clean. Prod.* **2022**, *371*, 133411. [CrossRef]
73. Disli, M.; Yilmaz, M.K.; Mohamed, F.F.M. Board Characteristics and Sustainability Performance: Empirical Evidence from Emerging Markets. *Sustain. Account. Manag. Policy J.* **2022**, *13*, 929–952. [CrossRef]
74. Krasodomska, J.; Zieniuk, P.; Kostrzewska, J. Reporting on Sustainable Development Goals in the European Union: What Drives Companies' Decisions? *Compet. Rev. Int. Bus. J.* **2023**, *33*, 120–146. [CrossRef]
75. Rosati, F.; Rodrigues, V.P.; Cosenz, F.; Li-Ying, J. Business Model Innovation for the Sustainable Development Goals. *Bus. Strategy Environ.* **2023**, *32*, 3752–3765. [CrossRef]
76. Eurostat. NACE Rev. 2—Statistical Classification of Economic Activities. Available online: https://ec.europa.eu/eurostat/web/products-manuals-and-guidelines/-/ks-ra-07-015 (accessed on 1 November 2023).
77. Bennich, T.; Weitz, N.; Carlsen, H. Deciphering the Scientific Literature on SDG Interactions: A Review and Reading Guide. *Sci. Total Environ.* **2020**, *728*, 138405. [CrossRef]
78. Ronzon, T.; Sanjuán, A.I. Friends or Foes? A Compatibility Assessment of Bioeconomy-Related Sustainable Development Goals for European Policy Coherence. *J. Clean. Prod.* **2020**, *254*, 119832. [CrossRef] [PubMed]
79. Natural Earth Data Natural Earth. Free Vector and Raster Map Data at 1:10 m, 1:50 m, and 1:110 m Scales. Available online: https://www.naturalearthdata.com/downloads/ (accessed on 7 March 2024).
80. Mainali, B.; Luukkanen, J.; Silveira, S.; Kaivo-oja, J. Evaluating Synergies and Trade-Offs among Sustainable Development Goals (SDGs): Explorative Analyses of Development Paths in South Asia and Sub-Saharan Africa. *Sustainability* **2018**, *10*, 815. [CrossRef]
81. Deloitte. Sustainable Development Goals A Business Perspective. Available online: https://www2.deloitte.com/content/dam/Deloitte/nl/Documents/risk/deloitte-nl-risk-sdgs-from-a-business-perspective.pdf (accessed on 31 January 2024).
82. Murphy, E.; Walsh, P.P.; Murphy, E. Nation-Based Peer Assessment of Europe's Sustainable Development Goal Performance. *PLoS ONE* **2023**, *18*, e0287771. [CrossRef]

83. Chen, J.; Huang, S.; Kamran, H.W. Empowering Sustainability Practices through Energy Transition for Sustainable Development Goal 7: The Role of Energy Patents and Natural Resources among European Union Economies through Advanced Panel. *Energy Policy* **2023**, *176*, 113499. [CrossRef]
84. Amos, R.; Lydgate, E. Trade, Transboundary Impacts and the Implementation of SDG 12. *Sustain. Sci.* **2020**, *15*, 1699–1710. [CrossRef]
85. Correa-Mejía, D.A.; Correa-García, J.A.; García-Benau, M.A. Analysis of Double Materiality in Early Adopters. Are Companies Walking the Talk? *Sustain. Account. Manag. Policy J.* **2024**, *15*, 299–329. [CrossRef]
86. Al Lawati, H.; Hussainey, K. Does Sustainable Development Goals Disclosure Affect Corporate Financial Performance? *Sustainability* **2022**, *14*, 7815. [CrossRef]
87. Martí-Ballester, C.-P. Analysing the Financial Performance of Sustainable Development Goals-Themed Mutual Funds in China. *Sustain. Prod. Consum.* **2021**, *27*, 858–872. [CrossRef]
88. Khan, P.A.; Johl, S.K.; Akhtar, S. Firm Sustainable Development Goals and Firm Financial Performance through the Lens of Green Innovation Practices and Reporting: A Proactive Approach. *J. Risk Financ. Manag.* **2021**, *14*, 605. [CrossRef]
89. Abughniem, M.S.; Aishat, M.A.H.A.A.A.; Hamdan, A.H. Corporate Sustainability as an Antecedent to the Financial Performance: An Empirical Study. *Pol. J. Manag. Stud.* **2019**, *20*, 35–44. [CrossRef]
90. Siedschlag, I.; Yan, W. Do Green Investments Improve Firm Performance? Empirical Evidence from Ireland. *Technol. Forecast. Soc. Chang.* **2023**, *186*, 122181. [CrossRef]
91. Qalati, S.A.; Barbosa, B.; Iqbal, S. The Effect of Firms' Environmentally Sustainable Practices on Economic Performance. *Econ. Res.-Ekon. Istraživanja* **2023**, *36*, 2199822. [CrossRef]
92. Pham, D.C.; Do, T.N.A.; Doan, T.N.; Nguyen, T.X.H.; Pham, T.K.Y. The Impact of Sustainability Practices on Financial Performance: Empirical Evidence from Sweden. *Cogent Bus. Manag.* **2021**, *8*, 1912526. [CrossRef]
93. Song, L.; Zhan, X.; Zhang, H.; Xu, M.; Liu, J.; Zheng, C. How Much Is Global Business Sectors Contributing to Sustainable Development Goals? *Sustain. Horiz.* **2022**, *1*, 100012. [CrossRef]
94. Fortea, C.; Zlati, M.L.; Lazarescu, I. Analysis of the Sustainable Development of the Eastern European Countries from the Perspective of the Transition to the Green Economy. *Ovidius Univ. Ann. Econ. Sci. Ser.* **2023**, *23*, 914–922.
95. Huang, R. SDG-Oriented Sustainability Assessment for Central and Eastern European Countries. *Environ. Sustain. Indic.* **2023**, *19*, 100268. [CrossRef]
96. Jakovljevic, M.; Cerda, A.A.; Liu, Y.; García, L.; Timofeyev, Y.; Krstic, K.; Fontanesi, J. Sustainability Challenge of Eastern Europe—Historical Legacy, Belt and Road Initiative, Population Aging and Migration. *Sustainability* **2021**, *13*, 11038. [CrossRef]
97. Markowski, Ł.; Kotliński, K.; Ostrowska, A. Sustainable Consumption and Production in the European Union—An Attempt to Assess Changes and Convergence from the Perspective of Central and Eastern European Countries. *Sustainability* **2023**, *15*, 16485. [CrossRef]
98. Bălăcescu, A.; Zaharia, M.; Gogonea, R.-M.; Căruntu, G.A. The Image of Sustainability in European Regions Considering the Social Sustainability Index. *Sustainability* **2022**, *14*, 13433. [CrossRef]
99. Ahlström, H.; Sjåfjell, B. Why Policy Coherence in the European Union Matters for Global Sustainability. *Environ. Policy Gov.* **2023**, *33*, 272–287. [CrossRef]
100. Rădoiu, A. Romania's Challenges and Perspectives in Achieving Sustainable Development Goals. *Int. Conf. Knowl.-Based Organ.* **2020**, *26*, 251–256. [CrossRef]
101. United Nations Global Compact; KPMG SDG Industry Matrix. Available online: https://assets.kpmg.com/content/dam/kpmg/xx/pdf/2017/05/sdg-financial-services.pdf (accessed on 31 January 2024).
102. KPMG. Reporting on the UN Sustainable Development Goals. Available online: https://kpmg.com/xx/en/home/insights/2022/09/survey-of-sustainability-reporting-2022/sdg.html (accessed on 22 January 2024).
103. United Nations. Development Programme Global Trends: Challenges and Opportunities in the Implementation of the SDGs. Available online: https://www.undp.org/publications/global-trends-challenges-and-opportunities-implementation-sdgs (accessed on 2 February 2024).
104. Boffo, R.; Patalano, R. ESG Investing: Practices, Progress and Challenges. Available online: https://www.oecd.org/finance/ESG-Investing-Practices-Progress-Challenges.pdf (accessed on 31 January 2024).
105. Yang, S.; Zhao, W.; Liu, Y.; Cherubini, F.; Fu, B.; Pereira, P. Prioritizing Sustainable Development Goals and Linking Them to Ecosystem Services: A Global Expert's Knowledge Evaluation. *Geogr. Sustain.* **2020**, *1*, 321–330. [CrossRef]
106. The Economist. Closing the Gap. Available online: https://www.economist.com/special-report/2011/11/26/closing-the-gap (accessed on 1 February 2024).
107. Förster, J. Cooling for Electricity Production Dominates Water Use in Industry. Available online: https://ec.europa.eu/eurostat/statistics-explained/index.php?title=Archive:Water_use_in_industry (accessed on 1 February 2024).
108. Covert, T.; Greenstone, M.; Knittel, C.R. Will We Ever Stop Using Fossil Fuels? *J. Econ. Perspect.* **2016**, *30*, 117–138. [CrossRef]
109. Abdou, A.H.; Hassan, T.H.; El Dief, M.M. A Description of Green Hotel Practices and Their Role in Achieving Sustainable Development. *Sustainability* **2020**, *12*, 9624. [CrossRef]
110. United Nations. Paris Agreement. Available online: https://unfccc.int/files/essential_background/convention/application/pdf/english_paris_agreement.pdf (accessed on 3 February 2024).

111. Deloitte. Professionalising the Ocean Cleanup. Available online: https://www2.deloitte.com/nl/nl/pages/over-deloitte/articles/the-ocean-cleanup.html (accessed on 3 February 2024).
112. Sustainable Development Solutions Network. Sustainable Development Report 2023. Available online: https://s3.amazonaws.com/sustainabledevelopment.report/2023/sustainable-development-report-2023.pdf (accessed on 19 January 2024).
113. United Nations. Towards Sustainable Welfare—The 2030 Agenda. Available online: https://www.eea.europa.eu/themes/sustainability-transitions/sustainable-development-goals-and-the/country-profiles/sweden-country-profile-sdgs-and (accessed on 25 January 2024).
114. Government of Sweden. The Global Goals and the 2030 Agenda for Sustainable Development. Available online: https://www.government.se/government-policy/the-global-goals-and-the-2030-Agenda-for-sustainable-development/ (accessed on 27 March 2024).
115. Saeed, S.; Makhdum, M.S.A.; Anwar, S.; Yaseen, M.R. Climate Change Vulnerability, Adaptation, and Feedback Hypothesis: A Comparison of Lower-Middle, Upper-Middle, and High-Income Countries. *Sustainability* **2023**, *15*, 4145. [CrossRef]
116. Sweden Sverige. Energy Use in Sweden. Available online: https://sweden.se/climate/sustainability/energy-use-in-sweden (accessed on 5 January 2024).
117. Insightful Team. How Sweden Continues to Set the Benchmark as One of the World's Most Sustainable Countries. Available online: https://www.insightvacations.com/blog/sweden-worlds-most-sustainable-countries/ (accessed on 5 January 2024).
118. Hexagon. Hexagon's R-Evolution Launches Its Renewable Energy Project Portfolio with a Focus on Digitalising Solar Production. Available online: https://hexagon.com/company/newsroom/press-releases/2021/hexagons-r-evolution-renewable-energy-project (accessed on 1 January 2024).
119. Vattenfall. Sustainable Production. Available online: https://group.vattenfall.com/what-we-do/roadmap-to-fossil-freedom/sustainable-production (accessed on 31 January 2024).
120. Scania. Renewable Fuels. Available online: https://www.scania.com/group/en/home/innovation/technology/renewable-fuels.html (accessed on 31 January 2024).
121. World Health Organization. *WHO Global Air Quality Guidelines: Particulate Matter (PM2.5 and PM10), Ozone, Nitrogen Dioxide, Sulfur Dioxide and Carbon Monoxide*; WHO European Centre for Environment and Health: Bonn, Germany, 2021; ISBN 978-92-4-003422-8.
122. Haseler, M.; Balciunas, A.; Hauk, R.; Sabaliauskaite, V.; Chubarenko, I.; Ershova, A.; Schernewski, G. Marine Litter Pollution in Baltic Sea Beaches—Application of the Sand Rake Method. *Front. Environ. Sci.* **2020**, *8*, 599978. [CrossRef]
123. Neste. The Baltic Sea—Vulnerable and Unique. Available online: https://journeytozerostories.neste.com/sustainability/baltic-sea-vulnerable-and-unique#dc90d05a (accessed on 25 January 2024).
124. OECD. *The Short and Winding Road to 2030: Measuring Distance to the SDG Targets*; OECD: Paris, France, 2022; ISBN 978-92-64-34190-6.
125. Die Bundesregierung G7: The Group of the Seven. Available online: https://www.g7germany.de/g7-en/g7-summit/g7-members (accessed on 19 January 2024).
126. The World Bank. Value Added (% of GDP)—Germany. Available online: https://data.worldbank.org/indicator/NV.SRV.TOTL.ZS?locations=DE (accessed on 22 January 2024).
127. Statista. Germany: Share of Economic Sectors in Gross Domestic Product (GDP) in 2022. Available online: https://www.statista.com/statistics/295519/germany-share-of-economic-sectors-in-gross-domestic-product/ (accessed on 22 January 2024).
128. Eurostat. Sustainable Development Indicators. Available online: https://ec.europa.eu/eurostat/databrowser/explore/all/tb_eu?lang=en&subtheme=sdg&display=list&sort=category (accessed on 19 January 2024).
129. Sustainable Development Goals Knowledge Platform. Germany. Available online: https://sustainabledevelopment.un.org/memberstates/germany (accessed on 19 January 2024).
130. European Institute for Gender Equality Gender Equality Index. Available online: https://eige.europa.eu/gender-equality-index/2023/country (accessed on 27 February 2024).
131. The World Bank. Services, Value Added (% of GDP)—France. Available online: https://data.worldbank.org/indicator/NV.SRV.TOTL.ZS?end=2022&locations=FR&start=2022&view=chart (accessed on 19 January 2024).
132. République Française. Pôle Emploi: BMO 2023. Available online: https://statistiques.pole-emploi.org/bmo (accessed on 19 January 2023).
133. Organisation for Economic Co-Operation and Development Employment: Time Spent in Paid and Unpaid Work, by Sex. Available online: https://stats.oecd.org/index.aspx?queryid=54757 (accessed on 5 January 2024).
134. Convention on Biological Diversity Aichi Biodiversity Targets. Available online: https://www.cbd.int/sp/targets/ (accessed on 13 January 2024).
135. Alleanza Italiana per lo Sviluppo Sostenibile. ASviS—Who We Are. Available online: https://asvis.it/who-we-are/ (accessed on 25 January 2024).
136. Istituto Nazionale di Statistica. 2022 SDGs Report. Available online: https://www.istat.it/it/files//2023/05/2022-SDGS-Report_Inglese.pdf (accessed on 25 January 2024).
137. The World Bank. GDP—Italy. Available online: https://data.worldbank.org/indicator/NY.GDP.MKTP.CD?locations=IT (accessed on 25 January 2024).
138. The IUCN Red List of Threatened Species. The Mediterranean Red List of Species. Available online: https://www.iucnredlist.org/ (accessed on 25 January 2024).

139. European Commission. Air Quality: Commission Urges Italy to Take Action against Small Particulate Matter (PM10) to Safeguard Public Health. Available online: https://ec.europa.eu/commission/presscorner/detail/ET/IP_17_1046 (accessed on 25 January 2024).
140. Jaume, T.; Ripoll, A.; Banyuls, L.; Ortiz, A.G.; Soares, J. *Status Report of Air Quality in Europe for Year 2022, Using Validated and up-to-Date Data (ETC-HE Report 2023/2)*; European Environment Agency: Copenhagen, Denmark, 2023.
141. HitHorizons. Industry Breakdown of Companies in Romania. Available online: https://www.hithorizons.com/eu/analyses/country-statistics/romania (accessed on 26 January 2024).
142. Christiaensen, L.; Demery, L.; Kuhl, J. The (Evolving) Role of Agriculture in Poverty Reduction—An Empirical Perspective. *J. Dev. Econ.* **2011**, *96*, 239–254. [CrossRef]
143. OECD. *Poverty Rate*; OECD: Paris, France, 2021.
144. Federatia Bancilor pentru Alimente din Romania Banca Pentru Alimente Bucuresti. Available online: https://bucuresti.bancapentrualimente.ro/ (accessed on 27 January 2024).
145. Eurostat. Annual National Accounts. Available online: https://ec.europa.eu/eurostat/databrowser/explore/all/economy?lang=en&subtheme=na10.nama10&display=list&sort=category (accessed on 26 January 2024).
146. Climate Trade. The World's Most Polluting Industries. Available online: https://climatetrade.com/the-worlds-most-polluting-industries/ (accessed on 29 January 2024).
147. The Department of Sustainable Development, as Part of the Working Apparatus of the Romanian Government Romania's Sustainable Development Strategy 2030. Available online: https://dezvoltaredurabila.gov.ro/files/public/10000001/Romania-Sustainable-Development-Strategy-2030-en.pdf (accessed on 5 January 2024).
148. Stănescu, I.; Bilan, A. Voluntary Subnational Review 2023: Localising the UN 2030 Agenda Sustainable Development Goals in Romanian Municipalities and Communes. Available online: https://hlpf.un.org/sites/default/files/vnrs/2023/VNR%202023%20Romania%20Report%20Subnational%20Report.pdf (accessed on 27 January 2024).
149. Sustainable Development. Goals Knowledge Platform Romania. Available online: https://sustainabledevelopment.un.org/memberstates/romania (accessed on 29 January 2024).
150. Eurostat. Expenditure on Social Protection (2010–2021). Available online: https://ec.europa.eu/eurostat/databrowser/product/view/spr_exp_sum (accessed on 2 February 2024).
151. Firoiu, D.; Ionescu, G.H.; Băndoi, A.; Florea, N.M.; Jianu, E. Achieving Sustainable Development Goals (SDG): Implementation of the 2030 Agenda in Romania. *Sustainability* **2019**, *11*, 2156. [CrossRef]
152. Radu, O.-M.; Dragomir, V.D.; Ionescu-Feleagă, L. The Link between Corporate ESG Performance and the UN Sustainable Development Goals. *Proc. Int. Conf. Bus. Excell.* **2023**, *17*, 776–790. [CrossRef]
153. World Bank. Bulgaria. Available online: https://data.worldbank.org/indicator/NV.SRV.TOTL.ZS?locations=BG (accessed on 29 January 2024).
154. United Nations. Economic and Social Affairs Voluntary National Review: Sustainable Development Goals—Bulgaria. Available online: https://sustainabledevelopment.un.org/content/documents/26289VNR_2020_Bulgaria_Report.pdf (accessed on 29 January 2024).
155. Batembergska, L.; Dimitrova, N.; Georgiev, Y. *The Link between the Implementation of the UN Sustainable Development Goals in Bulgaria and the Migration*; Bulgarian Platform for International Development: Sofia, Bulgaria, 2020; Available online: https://gcap.global/wp-content/uploads/2021/02/Monitoring-report-2020-BPID.pdf (accessed on 29 January 2024).
156. Ionescu, G.H.; Jianu, E.; Patrichi, I.C.; Ghiocel, F.; Țenea, L.; Iancu, D. Assessment of Sustainable Development Goals (SDG) Implementation in Bulgaria and Future Developments. *Sustainability* **2021**, *13*, 12000. [CrossRef]
157. OECD. PISA 2022 Results. Available online: https://www.oecd.org/publication/pisa-2022-results/ (accessed on 2 February 2024).
158. Dimitrov, I. Challenges for the Bulgarian Companies in the Implementation of Sustainability Practices. *Eur. J. Mark. Econ.* **2023**, *6*, 1–14. [CrossRef]
159. Kuc-Czarnecka, M.; Markowicz, I.; Sompolska-Rzechuła, A. SDGs Implementation, Their Synergies, and Trade-Offs in EU Countries—Sensitivity Analysis-Based Approach. *Ecol. Indic.* **2023**, *146*, 109888. [CrossRef]
160. CEIC Data. Romania Number of Companies: Bucharest Stock Exchange (BVB): Listed. Available online: https://www.ceicdata.com/en/romania/bucharest-stock-exchange-number-of-companies/number-of-companies-bucharest-stock-exchange-bvb-listed (accessed on 4 February 2024).
161. Markopoulos, E.; Barbara Ramonda, M. An ESG-SDGs Alignment and Execution Model Based on the Ocean Strategies Transition in Emerging Markets. In Proceedings of the 13th AHFE International Conference on Creativity, Innovation Entrepreneurship, New York, NY, USA, 24–28 July 2022.
162. OECD. *OECD Guidelines for Multinational Enterprises on Responsible Business Conduct*; OECD: Paris, France, 2023; ISBN 978-92-64-60598-5.
163. Global Reporting Initiative Synergies between the OECD Guidelines for Multinational Enterprises (MNEs) and the GRI 2002 Sustainability Reporting Guidelines. Available online: https://www.oecd.org/corporate/mne/35150230.pdf (accessed on 19 January 2024).
164. European Commission. EU 'Whole-of-Government' Approach. Available online: https://commission.europa.eu/strategy-and-policy/sustainable-development-goals/eu-whole-government-approach_en (accessed on 16 March 2024).

165. Avram, V.; Calu, D.A.; Dumitru, V.F.; Dumitru, M.; Glăvan, M.E.; Jinga, G. The Institutionalization of the Consistency and Comparability Principle in the European Companies. *Energies* **2018**, *11*, 3456. [CrossRef]
166. Miglietta, P.P.; Coluccia, B.; Pacifico, A.M.; Malorgio, G. Tracking on Food and Agriculture-Related SDG Indicators in the Mediterranean Region. *New Medit* **2023**, *4*, 56–71. [CrossRef]
167. Högbom, L.; Abbas, D.; Armolaitis, K.; Baders, E.; Futter, M.; Jansons, A.; Jõgiste, K.; Lazdins, A.; Lukminė, D.; Mustonen, M.; et al. Trilemma of Nordic–Baltic Forestry—How to Implement UN Sustainable Development Goals. *Sustainability* **2021**, *13*, 5643. [CrossRef]
168. Johnson, A.F.; Lidström, S.; Kelling, I.; Williams, C.; Niedermüller, S.; Poulsen, K.V.; Burgess, S.; Kent, R.; Davies, W. The European Union's Fishing Activity Outside of European Waters and the Sustainable Development Goals. *Fish Fish.* **2021**, *22*, 532–545. [CrossRef]
169. Ivic, A.; Saviolidis, N.M.; Johannsdottir, L. Drivers of Sustainability Practices and Contributions to Sustainable Development Evident in Sustainability Reports of European Mining Companies. *Discov. Sustain.* **2021**, *2*, 17. [CrossRef]
170. Marchetti, S.; Secondi, L. The Economic Perspective of Food Poverty and (In)Security: An Analytical Approach to Measuring and Estimation in Italy. *Soc. Indic. Res.* **2022**, *162*, 995–1020. [CrossRef]
171. Mancini, L.; Valente, A.; Barbero Vignola, G.; Sanyé Mengual, E.; Sala, S. Social Footprint of European Food Production and Consumption. *Sustain. Prod. Consum.* **2023**, *35*, 287–299. [CrossRef]
172. Perello-Marin, M.R.; Rodríguez-Rodríguez, R.; Alfaro-Saiz, J.-J. Analysing GRI Reports for the Disclosure of SDG Contribution in European Car Manufacturers. *Technol. Forecast. Soc. Change* **2022**, *181*, 121744. [CrossRef]
173. Malik, A.; Lafortune, G.; Carter, S.; Li, M.; Lenzen, M.; Kroll, C. International Spillover Effects in the EU's Textile Supply Chains: A Global SDG Assessment. *J. Environ. Manag.* **2021**, *295*, 113037. [CrossRef]
174. Caldeira, C.; De Laurentiis, V.; Corrado, S.; Van Holsteijn, F.; Sala, S. Quantification of Food Waste per Product Group along the Food Supply Chain in the European Union: A Mass Flow Analysis. *Resour. Conserv. Recycl.* **2019**, *149*, 479–488. [CrossRef]
175. Vögele, S.; Govorukha, K.; Mayer, P.; Rhoden, I.; Rübbelke, D.; Kuckshinrichs, W. Effects of a Coal Phase-out in Europe on Reaching the UN Sustainable Development Goals. *Environ. Dev. Sustain.* **2023**, *25*, 879–916. [CrossRef]
176. Momete, D.C. Salient Insights on the Performance of EU Member States on the Road towards an Energy-Efficient Future. *Energies* **2023**, *16*, 925. [CrossRef]
177. Zhao, X.; Ramzan, M.; Sengupta, T.; Deep Sharma, G.; Shahzad, U.; Cui, L. Impacts of Bilateral Trade on Energy Affordability and Accessibility across Europe: Does Economic Globalization Reduce Energy Poverty? *Energy Build.* **2022**, *262*, 112023. [CrossRef]
178. Aydin, M.; Erdem, A. Analyzing the Impact of Resource Productivity, Energy Productivity, and Renewable Energy Consumption on Environmental Quality in EU Countries: The Moderating Role of Productivity. *Resour. Policy* **2024**, *89*, 104613. [CrossRef]
179. Cai, J.; Zhao, D.; Varis, O. Match Words with Deeds: Curbing Water Risk with the Sustainable Development Goal 6 Index. *J. Clean. Prod.* **2021**, *318*, 128509. [CrossRef]
180. Dilekli, N.; Cazcarro, I. Testing the SDG Targets on Water and Sanitation Using the World Trade Model with a Waste, Wastewater, and Recycling Framework. *Ecol. Econ.* **2019**, *165*, 106376. [CrossRef]
181. Ezbakhe, F.; Giné-Garriga, R.; Pérez-Foguet, A. Leaving No One behind: Evaluating Access to Water, Sanitation and Hygiene for Vulnerable and Marginalized Groups. *Sci. Total Environ.* **2019**, *683*, 537–546. [CrossRef]
182. Caglar, A.E.; Gökçe, N.; Şahin, F. Sustaining Environment through Municipal Solid Waste: Evidence from European Union Economies. *Environ. Sci. Pollut. Res.* **2023**, *31*, 6040–6053. [CrossRef]
183. Halkos, G.E.; Aslanidis, P.S.C. Promoting Sustainable Waste Management for Regional Economic Development in European Mediterranean Countries. *Euro-Mediterr. J. Environ. Integr.* **2023**, *8*, 767–775. [CrossRef]
184. Poranek, N.; Łaźniewska-Piekarczyk, B.; Lombardi, L.; Czajkowski, A.; Bogacka, M.; Pikoń, K. Green Deal and Circular Economy of Bottom Ash Waste Management in Building Industry—Alkali (NaOH) Pre-Treatment. *Materials* **2022**, *15*, 3487. [CrossRef]
185. Corrado, S.; Rydberg, T.; Oliveira, F.; Cerutti, A.; Sala, S. Out of Sight out of Mind? A Life Cycle-Based Environmental Assessment of Goods Traded by the European Union. *J. Clean. Prod.* **2020**, *246*, 118954. [CrossRef]
186. Onat, N.C.; Abdella, G.M.; Kucukvar, M.; Kutty, A.A.; Al-Nuaimi, M.; Kumbaroğlu, G.; Bulu, M. How Eco-efficient Are Electric Vehicles across Europe? A Regionalized Life Cycle Assessment-based Eco-efficiency Analysis. *Sustain. Dev.* **2021**, *29*, 941–956. [CrossRef]
187. Barona, J.; Ballini, F.; Canepa, M. Circular Developments of Maritime Industrial Ports in Europe: A Semi-Systematic Review of the Current Situation. *J. Shipp. Trade* **2023**, *8*, 25. [CrossRef]
188. Beretta, C.; Hellweg, S. Potential Environmental Benefits from Food Waste Prevention in the Food Service Sector. *Resour. Conserv. Recycl.* **2019**, *147*, 169–178. [CrossRef]
189. Lyulyov, O.; Pimonenko, T.; Saura, J.R.; Barbosa, B. How Do E-Governance and e-Business Drive Sustainable Development Goals? *Technol. Forecast. Soc. Change* **2024**, *199*, 123082. [CrossRef]
190. Gutiérrez-Fernández, M.; Gallego-Sosa, C.; Fernández-Torres, Y. Commitment to Sustainability in Large European Banks and Its Relationship with Board Gender Diversity: A 2030 Agenda Perspective. *J. Manag. Organ.* **2023**, *ahead of print*, 1–27. [CrossRef]
191. Sardianou, E.; Stauropoulou, A.; Evangelinos, K.; Nikolaou, I. A Materiality Analysis Framework to Assess Sustainable Development Goals of Banking Sector through Sustainability Reports. *Sustain. Prod. Consum.* **2021**, *27*, 1775–1793. [CrossRef]
192. Kovalivska, S.; Shcherbyna, A.; Nikolaiev, V. Intensification of Investment in the Renovation of Residential Real Estate in the Context of the Sustainable Development Goals. *Balt. J. Econ. Stud.* **2020**, *6*, 184–195. [CrossRef]

193. Cucco, P.; Maselli, G.; Nesticò, A.; Ribera, F. An Evaluation Model for Adaptive Reuse of Cultural Heritage in Accordance with 2030 SDGs and European Quality Principles. *J. Cult. Herit.* **2023**, *59*, 202–216. [CrossRef]
194. Bisogno, M.; Cuadrado-Ballesteros, B.; Rossi, F.M.; Peña-Miguel, N. Sustainable Development Goals in Public Administrations: Enabling Conditions in Local Governments. *Int. Rev. Adm. Sci.* **2023**, *89*, 1223–1242. [CrossRef]
195. Meuleman, L. Public Administration and Governance for the SDGs: Navigating between Change and Stability. *Sustainability* **2021**, *13*, 5914. [CrossRef]
196. Bielecka, E.; Calka, B. Towards Sustainable Development Exemplified by Monitoring Land Use Efficiency in Europe Using SDG 11.3.1. *Misc. Geogr.* **2022**, *26*, 208–214. [CrossRef]
197. Malešević Perovic, L.; Mihaljević Kosor, M. The Efficiency of Universities in Achieving Sustainable Development Goals. *Amfiteatru Econ.* **2020**, *22*, 516–532. [CrossRef]
198. Pereira, M.A.; Marques, R.C. The 'Sustainable Public Health Index': What If Public Health and Sustainable Development Are Compatible? *World Dev.* **2022**, *149*, 105708. [CrossRef]
199. Peña-Sánchez, A.R.; Ruiz-Chico, J.; Jiménez-García, M. Dynamics of Public Spending on Health and Socio-Economic Development in the European Union: An Analysis from the Perspective of the Sustainable Development Goals. *Healthcare* **2021**, *9*, 353. [CrossRef]
200. Bickler, G.; Morton, S.; Menne, B. Health and Sustainable Development: An Analysis of 20 European Voluntary National Reviews. *Public Health* **2020**, *180*, 180–184. [CrossRef]
201. Dai, J.; Menhas, R. Sustainable Development Goals, Sports and Physical Activity: The Localization of Health-Related Sustainable Development Goals through Sports in China: A Narrative Review. *Risk Manag. Healthc. Policy* **2020**, *13*, 1419–1430. [CrossRef]
202. Baena-Morales, S.; Jerez-Mayorga, D.; Delgado-Floody, P.; Martínez-Martínez, J. Sustainable Development Goals and Physical Education. A Proposal for Practice-Based Models. *Int. J. Environ. Res. Public Health* **2021**, *18*, 2129. [CrossRef]
203. Quinn, B.; Colombo, A.; Lindström, K.; McGillivray, D.; Smith, A. Festivals, Public Space and Cultural Inclusion: Public Policy Insights. *J. Sustain. Tour.* **2021**, *29*, 1875–1893. [CrossRef]

Disclaimer/Publisher's Note: The statements, opinions and data contained in all publications are solely those of the individual author(s) and contributor(s) and not of MDPI and/or the editor(s). MDPI and/or the editor(s) disclaim responsibility for any injury to people or property resulting from any ideas, methods, instructions or products referred to in the content.

sustainability MDPI

Article

Fuzzy Logic Method for Measuring Sustainable Decent Work Levels as a Corporate Social Responsibility Approach

Alma Nataly Abundes-Recilla [1], Diego Seuret-Jiménez [2], Martha Roselia Contreras-Valenzuela [1,*] and José M. Nieto-Jalil [3,*]

1 Faculty of Chemical Sciences and Engineering (FCQeI), Autonomous University of Morelos State (UAEM), Avenida Universidad 1001, Colonia Chamilpa, Cuernavaca, Morelos CP 62209, Mexico; alma.abundesre@uaem.edu.mx
2 Research Centre in Engineering and Applied Sciences, Autonomous University of Morelos State (UAEM), Avenida Universidad 1001, Colonia Chamilpa, Cuernavaca, Morelos CP 62209, Mexico; dseuret@uaem.mx
3 Tecnologico de Monterrey, School of Engineering and Sciences, Reserva Territorial Atlixcáyotl, Puebla CP 72453, Mexico
* Correspondence: marthacv@uaem.mx (M.R.C.-V.); jnietoj@tec.mx (J.M.N.-J.); Tel.: +52-7773297039 (M.R.C.-V.); Tel.: +52-2229090496 (J.M.N.-J.)

Abstract: The purpose of this study was to propose an interactive computer system that utilises the MATLAB Fuzzy Logic Designer to measure the level of implementation of SDG 8, which focuses on sustainable decent work (SDW) and economic growth. This study used policies and laws as parameters to determine the presence or absence of SDW. The fuzzy method was implemented in car windshield manufacturing in the auto parts industry as a case study to define and quantify work conditions and to determine the level of sustainable decent work (SDWL). The study described environmental conditions, such as noise, lighting, and heat stress; ergonomic factors, such as exposure time, the mass of the object manipulated, and lifting frequency; and organisation at work, such as workplace violence, salary, and workday, as linguistic variables. The level of the presence or absence of SDW was defined as their membership functions. The resulting vectors determined the absence of SDW with a score of 1.5 in two linguistic variables: environmental conditions and ergonomic factors. Some features of SDW in the linguistic variable organisation at work had an SDW score of 5. The SDWL vector determined a final score of 1.24, indicating the absence of decent work in production areas. This study found that the workers suffer a lack of long and healthy lives and a bad standard of living without economic growth due to work-related musculoskeletal disorders and work illnesses, increasing their out-of-pocket spending and catastrophic health expenses. As a CSR approach, assessing SDWLs helped managers improve policies and work conditions.

Keywords: corporate social responsibility; fuzzy logic; decent work; ergonomic risk evaluation; work conditions

Citation: Abundes-Recilla, A.N.; Seuret-Jiménez, D.; Contreras-Valenzuela, M.R.; Nieto-Jalil, J.M. Fuzzy Logic Method for Measuring Sustainable Decent Work Levels as a Corporate Social Responsibility Approach. *Sustainability* **2024**, *16*, 1791. https://doi.org/10.3390/su16051791

Academic Editors: Manoj Kumar Nallapaneni, Massimo Gastaldi and Idiano D'Adamo

Received: 11 January 2024
Revised: 5 February 2024
Accepted: 8 February 2024
Published: 22 February 2024

Copyright: © 2024 by the authors. Licensee MDPI, Basel, Switzerland. This article is an open access article distributed under the terms and conditions of the Creative Commons Attribution (CC BY) license (https://creativecommons.org/licenses/by/4.0/).

1. Introduction

Measuring the level of sustainable and decent work (SDW) in industrial processes can be challenging due to several factors. Firstly, the concept of SDW is multifactorial, making it difficult to measure. Secondly, the lack of information on the subject further complicates the measurement. Finally, each industrial process has different work safety parameters, which makes it difficult to establish universal standards for measuring SDW. Addressing these issues, a method for measuring sustainable decent work levels (SDWLs) that utilises fuzzy logic is proposed in this paper. The SDWL interactive computer system was built using MATLAB Fuzzy Logic Designer. It was tested to validate its performance in a case study to determine the level of SDW engaged in producing car windshields in the auto parts industry. Many firms do not currently have formal Corporate Social Responsibility (CSR) policies and programs implemented as a part of their value chain [1] because they contain features that require firms to have and maintain a sustainable production process and that creates firstly external constraints (environmental government policies), which influence their competition in the market trend, and secondly, domestic pressures, such as excessive yields of manufacture and extra-economic support in measuring process and operation improvements [2], thus the firms neglect their responsiveness about the work conditions under which personnel perform their tasks.

In general, work is any activity, whether intellectual or material, that people do to achieve a desired outcome, regardless of the level of technical expertise required for a particular profession or trade [3]. Thus, a worker is an individual who provides personal work to an entity in a subordinate position, and this employment normally is the means by which society allocates and utilises limited resources for a better quality of life, reallocating funds to meet present and future needs [4,5]. However, the concept of SDW does not apply to this study in cases where resources are obtained through unemployment subsidies or social subsidies for people who are unable to work because the subordinate position does not exist. Besides protecting the planet, sustainable development promotes prosperity in education, health, social protection, and job opportunities [6]. Protecting workers' rights and encouraging safe and secure working places and environments is part of economic growth [7]. The United Nations (UN) proposed SGD 8 as part of their 2030 Agenda for Sustainable Development [8], which focuses on "Decent Work and Economic Growth". This goal aims to promote sustainable decent work (SDW) by allowing individuals to engage in productive activities that offer reasonable income, workplace security, and social protection for their families. SDW also encourages personal development and social integration. Employees of any workplace, including men and women, people with or without disabilities, and minors, interact with the physical conditions of their workplace environment. Therefore, measuring the absence or presence of SDW in workplaces must be essential for guaranteeing families' economic growth. Unsafe and insecure working places and environments cause occupational illness, which reduces the economic growth of workers and their families because they cannot work due to the consequences of disease or accident, increasing out-of-pocket spending and catastrophic health expenses like outpatient care, hearing equipment, wheelchairs, therapy for musculoskeletal disorders, and unsubsidised medications, among others [9]. Unfortunately, many organisations and countries fail to consider the impact of work conditions on health and the expenses associated with it as part of SDW. Therefore, companies play a significant role in advancing sustainable development if they adopt voluntary actions to implement the three sustainability dimensions (economic, environmental, and social) in their organisations [10]. Voluntary sustainability standards are crucial to ensure that fundamental human rights, worker health and safety, and environmental impacts are considered. However, a clear governance framework must be in place to enhance the credibility of voluntary efforts and promote coordination and alignment across initiatives. Unfortunately, not all countries have established such a framework [11].

The evaluation of the presence or absence of SDW has been present in some way in the research literature; the most representative works for this article are described as follows: Barford et al. [12] focus on group surveys and statistics in low-income countries to identify empirical and conceptual drivers of youth perspectives on SDW scarcity; however, implications about the work were not considered. A qualitative approach to validate the decent work scale (DWS) in French society was proposed by Vignoli et al. [13], but only some components of SDW (life satisfaction, family–work conflict, and meaningful work) were considered; nonetheless, job safety and health were not included in their study. The links between three components of health (general health, symptoms, and healthy behaviours) concerning work-generated fatigue were examined by Duffy et al. [14], i.e., they determined the presence or absence of SDW through fatigue. One of the dimensions established by Yildirim et al. [15] that adds value to SDW is safe working conditions, which positively impact the motivation of workers; nevertheless, industrial conditions were not defined. Lout et al. [16] establish safety at work as one of the main elements in measuring the absence of SDW. Finally, Yan et al. [17] select a scale to measure the perception of decent work by measuring the well-being of workers in the workplace. The International Labour Organization (ILO) has devised a global measurement of decent work using questionnaires, but it does not include information about work conditions or suggestions for improving them. As a result, the ILO recommended developing an investigation into unsafe industrial work conditions [18]. Few works directly address work conditions related to SDW and CSR. The three most relevant for this investigation include the one developed by Raufflet et al. [19], who acquired and described the concept of "regulatory scripts", defined as the practices proportioned by a group of enterprises in mining and oil in response to international CSR standards; their study includes an analysis of work environmental systems and safety and health conditions from an index control point of view. The other work, by Kwon and Park [20], proposed a quantitative method of realising the responsible development of emerging technologies via text analysis of future-oriented web data and scientific publication data; this technique allows developers of emerging technologies to consider social concerns and human norms alongside more typical engineering ideals. Finally, Hadj [21] analysed the role played by CSR in improving the responsible innovation and the competitiveness of North African small and medium-sized enterprises by focusing on the commitments toward internal and external stakeholders and environmental management and formulating a sustainable development model based on responsible innovation in small and medium-sized enterprises. The works in this section considered the voices of stakeholders, and some considered CSR policies; they proposed decision-making based on user requirements considering SDW with a CSR focus. Analysing the information described above, it was possible to establish that previous investigations have not addressed fuzzy logic methods in combination with labour environmental factors to develop a new SDW evaluation method as a CSR approach. In their work, Ali S. et al. [22] highlight the need to ensure the robustness of results obtained through alternative scenarios. Therefore, this paper proposes new scenarios for SDG 8. Thus, this article sets a new frame of reference in studying sustainable decent work. Considering that, through a solid evaluation of data, fuzzy logic allows the incorporation of a technique that minimises the differences in point of view during decision-making. Using the fuzzy method to determine the level of SDW during the evaluation of work task conditions makes it possible to define the level of a presence or absence of SDW. Consequently, it was possible to determine whether the physical work environment complies with the sustainable decent work concept. This article reports a proposed method for evaluating the presence or absence of SDW inside workplaces by analysing three main groups of conditions: environmental conditions, ergonomic factors, and organisation at work, which directly impact the quality of the working lives and personal lives of occupationally exposed personnel (OEP). The application was developed in the MATLAB Fuzzy Logic Toolbox (FLT) [23] and applied to evaluate six areas in the production of car windshields in the auto parts industry. This paper's contribution and research's novelty lie in:

1. The design of a fuzzy logic method to define and quantify work conditions to determine the level of sustainable decent work as a CSR approach.
2. The definition of the presence or absence of sustainable decent work using policies and laws relating to sustainability and safety and health in work.
3. A proposed CSR approach tool that establishes a multivariable record for measuring and monitoring the level of sustainable decent work in the workplace.

This hypothesis will demonstrate that the fuzzy logic tool is helpful as a CSR approach in the results and discussion sections.

The rest of the paper is organised as follows: The materials and methods are described in detail in Section 2. The results of this study are presented in Section 3. The discussion and conclusion of the investigation are presented in Sections 4 and 5, respectively.

2. Materials and Methods

As was mentioned above, SDW and economic growth promote and allow individuals to engage in productive activities that offer three features: reasonable income, workplace security, and social protection for their families. They also encourage personal development and social integration [8]. Their definitions are extensive and multifactorial, so in this research, the scope will be limited only to evaluations in industries that seek to comply with SDG 8 guidelines. In this context, three variables representative of these features were defined: environmental conditions, ergonomic factors, and organisation at work. To deeply measure workplace security, environmental conditions and ergonomic factors were chosen because they represent the characteristics of the process that directly affect the workers' health and cause work-related illnesses and musculoskeletal disorders impacting their standard of living and well-being, sometimes forever. Organisation at work measures reasonable income and personal development through salary per day; social integration was measured through hours per workday and violence at work. In the case of formal work, all employees have social protection for their families; consequently, this feature was not considered for measurement in this study.

In the language of sustainability, some concepts are imprecise or vague (fuzziness); for example, how do we measure "responsibility" or the quality of being "decent"? Usually, a company can be defined as highly responsible with their employees or not, or a work can be decent or not; in both cases, the meaning depends on who expresses the sentence and the expertise in the topic. Therefore, the main concern of fuzzy logic is representing, manipulating, and drawing inferences from such imprecise statements [24]. For this study, the term decent was defined as fuzzy in the sentence, so it cannot be determined with absolute precision. However, it is used in SDW and CSR decision-making. Representing the sentence "it is a sustainable decent work" was difficult to assert whether it was true or false. Thus, SDW was defined as a function and its components as variables (*Xs*). The relationship between SDW and *Xs* becomes a matter of degree and depends on policies and laws relating to sustainability, safety, and health in work to create different subsets, which can be represented as follows:

$$Subset_{SDW}(x) = \begin{cases} 1 \; if \; x \in SDW \;\; presence \\ 0.5 \; x \in SDW \; some \; presence \\ 0 \; if \; x \notin SDW \; absence \end{cases} \tag{1}$$

The individual mathematical fuzzy models for the presence or absence of SDW were determined by its components X_{SDW} in Equations (2)–(11) during the fuzzification process.

The methodology proposed for the research is divided into two stages: Phase I defines the fuzzy rules based on three groups of fuzzy sets. Phase II involves programming the Fuzzy Logic Toolbox to determine the SDWL.

2.1. Problem Description

Recent research has focused on applying surveys and developing statistics to identify perspectives on SDW and classify the relationship between components of health and poor motivation. However, the relationship between environmental conditions, ergonomic factors, and organisation at work to determine SDW level is often neglected. This study has applied a fuzzy logic method as a socially responsible approach. This has resulted in a significant problem: defining the degree of membership in a set of work conditions to express the degree to which some condition exists (in our case, SDW), despite its vagueness/fuzziness, in decision-making. To address this issue, we have developed a fuzzy logic method that considers the relationship between safety and health risk levels and workstation conditions. We first specify in advance parameters for the fuzzy variables that define risks and hazards for workers in industrial workplaces. Then, we define membership rules as elements of association and segmentation. Finally, an SDWL is introduced by adding a socially responsible point of view. In subsequent sections, we describe this fuzzy process to better understand how it works.

2.2. Case Study

The case study was developed in a company in Mexico; therefore, the applicable salaries and regulations correspond to this country. In manufacturing car windshields, the production process imposes safety conditions and risks that can harm workers' health; therefore, direct labour workers were considered major internal stakeholders for this study. To identify whether tasks were carried out in a socially responsible environment, a fuzzy logic method for measuring sustainable decent work levels was developed and implemented in six production departments (laminating, tempering, post glass, supply chain, and quality control). For developing the evaluation, only a sample of 10% of workers were interviewed due to the process involving a total of 3000 workers divided into three work shifts. Tasks in all departments require manual material handled with a mass of windshields from 11 kg to 26 kg and a quota of 2100 pieces per shift. The daily salary varies from USD 12 to USD 18, depending on the work department. Tasks involve overtime and high stress.

2.3. Phase I: Fuzzification

To apply fuzzy logic (FL) methods, it was necessary to declare the SDW variables and safety and health parameters that should be measured. Next, the parameters for measurement were converted into appropriate fuzzy sets to define their vagueness; this step is called fuzzification [25]. FL is a mathematical formalism used in this investigation to emulate the ability to correctly evaluate work conditions based on linguistic data. Therefore, FL admits information regarding variables (environmental conditions, ergonomic factors, and organisation at work) to build the fuzzy sets. The relationships between sets of variables were combined to determine decisions [26]. Fuzzy degrees are membership percentages in a fuzzy set. Vagueness/fuzziness expresses a degree of some condition that exists and represents the "level" of presence or absence of SDW.

2.3.1. Defining Fuzzy Models and Fuzzy Sets

Levels of Risk in the Fuzzy Model: As fuzzy choices, the membership functions with three levels of presence of SDW were determined in Equation (2) using the following parameters:

- Low level of risk—presence of SDW—environmental conditions, ergonomic factors, and organisation at work in workplaces which generate minimum fatigue or stress and lead to work-related illnesses in the long term—scored between 0 and 3;
- Medium level of risk—some presence of SDW—environmental conditions, ergonomic factors, and organisation at work in workplaces which generate fatigue and stress and lead to work-related illnesses in the medium term—scored between 2 and 7;
- High level of risk—absence of SDW—environmental conditions, ergonomic factors, and organisation at work in workplaces which generate extreme fatigue and stress and lead to work-related illnesses in the short term—scored between 6 and 10.

$$Level\ of\ Risk_{SDW}(x) = \begin{cases} 1\ if\ 6\ points \leq x \leq 10\ points \\ 0.5\ if\ 2\ points \leq x \leq 7\ points \\ 0\ if\ 0\ points \leq x \leq 3\ points \end{cases} \quad (2)$$

To define the parameters for the environmental conditions used as fuzzy sets, international safety and health parameters were considered as follows:

Noise Fuzzy Model: The European, Asian, and Latin American legislation establishes that for an 8 h workday, the levels of noise exposure without protection before damage is produced must be less than 80 dB [27–29]. The ILO states that the levels workers are exposed to in an 8 h workday should not exceed 90 dB and defines that the maximum permissible exposure, in terms of average daily noise levels, can vary, depending on the country, from 80 to 85 or 90 dBA, with accumulation factors of 3, 4, or 5 dBA [30]. In some countries, such as Japan, permissible noise levels are set between 50 and 85 dBA, depending on the type of work performed and considering the physical and mental workload [31]. In the case of pregnant female workers, the protection of the unborn baby's hearing organ must also be included; according to the Spanish Society of Gynaecology and Obstetrics (SEGO), tasks should not be carried out in excessive-noise conditions (more than 80 dB) after the 20th–22nd week of gestation [32]. Therefore, the fuzzy model was determined in Equation (3) with the following parameters:

- Low level of risk—presence of SDW—noise less than 80 dB;
- Medium level of risk—some presence of SDW—noise between 80 dB and 90 dB;
- High level of risk—absence of SDW—noise higher than 90 dB.

$$Noise_{SDW}(x) = \begin{cases} 1 & if\ x < 80\ \text{dB} \\ 0.5\ if & 80\ \text{dB} \le x \le 90\ \text{dB} \\ 0 & if\ 90\ \text{dB} \le x \end{cases} \tag{3}$$

Lighting Fuzzy Model: In some Latin American countries, it is established that a minimum of 200 lux is required to perform a task with a simple vision [33]. The ILO has established a comprehensive reference framework by defining lighting levels depending on the type of task being performed: for example, for tasks with limited visual requirements: from 200 lux to 300 lux; for tasks with normal visual requirements: from 500 lux to 1000 lux; and for special or high-precision work: from 1000 lux to 20,000 lux (for example, surgeries) [34]. In Japan, the illuminance of working conditions was defined per the type of work: for example, 300 lux or more for precision work, 150 lux for ordinary work, and 70 lux for rough work [35]. Therefore, the fuzzy parameters for the lighting of machinery, office spaces, and inspection were set as follows:

- Low visual requirement: low level of risk—presence of SDW—lighting from 70 lux to 300 lux;
- Normal: medium level of risk—some presence of SDW—lighting from 200 lux to 750 lux;
- Demanding: high level of risk—absence of SDW—lighting higher than 500 lux.

$$Lighting_{SDW}(x) = \begin{cases} 1\ if\ 70\ \text{lux} \le x \le 300\ \text{lux} \\ 0.5\ if\ 200\ \text{lux} \le x \le 750\ \text{lux} \\ 0\ if\ 500\ \text{lux} \le x \end{cases} \tag{4}$$

Heat Stress Fuzzy Model: The ILO Encyclopaedia considers a cold work environment to be where the temperature is below 20 °C and a sensation of thermal neutrality to be between 20 and 26 °C [36] in light or sedentary work conditions, wherein the estimation of the thermal stress that a worker is subjected to is made through the WBGT index (wet bulb and black globe) [36,37]. Therefore, the fuzzy parameters for hot or cold environmental conditions measured via the WBGT index indoors at workplaces were:

- Cold: low level of risk—presence of SDW—heat stress from 10 °C and under to 23 °C;
- Thermal neutrality: medium level of risk—some presence of SDW—heat stress between 20 °C and 30 °C;
- Hot: high level of risk—absence of SDW—heat stress from 27 to 45 °C, considering that from values close to 40 °C, heat disorders begin to occur in humans.

$$Heat\ stress_{SDW}(x) = \begin{cases} 1\ if\ 10\ ^{\circ}\text{C} \le x \le 23\ ^{\circ}\text{C} \\ 0.5\ if\ 20\ ^{\circ}\text{C} \le x \le 30\ ^{\circ}\text{C} \\ 0\ if\ 27\ ^{\circ}\text{C} \le x \le 45\ ^{\circ}\text{C} \end{cases} \tag{5}$$

In the case of ergonomic parameters defined according to the standard ISO 11228-1:2021 [38], in a task related to manual material handling, the time of exposition is considered an unfavourable condition when the mass of the handled object is over 25 kg and the frequency of lifting is above 900 movements per shift. A cumulated mass of 6500 kg in a shift of 8 h should not be exceeded. Thus, the fuzzy sets were defined as follows [39]:

Time of Exposition Fuzzy Model:

- Low level of risk—presence of SDW—exposition time of 0 min to 80 min;

- Medium level of risk—some presence of SDW—exposition time between 60 min and 120 min;
- High level of risk—absence of SDW—exposition time of 100 min to 180 min or more.

$$Time\ of\ exposition_{SDW}(x) = \begin{cases} 1\ if\ 0\ \text{min} \leq x \leq 80\ \text{min} \\ 0.5\ if\ 60\ \text{min} \leq x \leq 120\ \text{min} \\ 0\ if\ 100\ \text{min} \leq x \leq 180\ \text{min}\ or\ more \end{cases} \quad (6)$$

Mass of the Handled Object Fuzzy Model

- Low level of risk—presence of SDW—mass of 0 kg to 10 kg;
- Medium level of risk—some presence of SDW—mass between 7 kg and 15 kg;
- High level of risk—absence of SDW—mass of 13 kg to 25 kg or more.

$$Mass\ of\ the\ object_{SDW}(x) = \begin{cases} 1\ if\ 0\ \text{kg} \leq x \leq 10\ \text{kg} \\ 0.5\ if\ 7\ \text{kg} \leq x \leq 15\ \text{kg} \\ 0\ if\ 13\ \text{kg} \leq x \leq 25\ \text{kg} \end{cases} \quad (7)$$

Frequency of Handling Fuzzy Model

- Low level of risk—presence of SDW—handling frequency of 0 movements to 700 movements;
- Medium level of risk—some presence of SDW—handling frequency between 600 movements and 1100 movements;
- High level of risk—absence of SDW—handling frequency of 900 to 1800 movements or more.

$$Frequency\ of\ handling_{SDW}(x) = \begin{cases} 1\ if\ 6\ \text{movements} \leq x \leq 10\ \text{movements} \\ 0.5\ if\ 2\ \text{movements} \leq x \leq 7\ \text{movements} \\ 0\ if\ 900\ \text{movements} \leq x \leq 1800\ \text{movements}\ or\ more \end{cases} \quad (8)$$

To define the organisation at work, the parameters contained in the Mexican standard NOM 035 STPS:2018 were used as a score from 0 to 13 points for evaluating psychosocial factors concerning workplace violence, salary, and the workday [40].

Workplace Violence Fuzzy Model

- Low level of risk—presence of SDW—workplace violence score of 0 points to 8 points;
- Medium level of risk—some presence of SDW—workplace violence score between 9 points and 12 points;
- High level of risk—absence of SDW—workplace violence score of more than 13 points.

$$Workplace\ violence_{SDW}(x) = \begin{cases} 1\ if\ 0\ \text{points} \leq x \leq 8\ \text{points} \\ 0.5\ if\ 9\ \text{points} \leq x \leq 12\ \text{points} \\ 0\ if\ 13\ \text{points} \leq x \end{cases} \quad (9)$$

Salary (in México) Fuzzy Model

- Low level of risk—presence of SDW—salary of more than USD 30 per day;
- Medium level of risk—some presence of SDW—salary between USD 12 and USD 30 per day;
- High level of risk—absence of SDW—salary of USD 6 to USD 12 per day.

$$Salary_{SDW}(x) = \begin{cases} 1\ if\ 30\ \text{USD} \leq x \\ 0.5\ if\ 12\ \text{USD} \leq x \leq 30\ \text{USD} \\ 0\ if\ 6\ \text{USD} \leq x \leq 12\ \text{USD} \end{cases} \quad (10)$$

Workday Fuzzy Model

- Low level of risk—presence of SDW—workday between 4 h and 8 h;
- Medium level of risk—some presence of SDW—workday between 7 h and 10 h;
- High level of risk—absence of SDW—workday between 9 h and 12 h or more.

$$Workday_{SDW}(x) = \begin{cases} 1\, if\, 4\,\mathrm{h} \leq x \leq 8\,\mathrm{h} \\ 0.5\, if\, 7\,\mathrm{h} \leq x \leq 10\,\mathrm{h} \\ 0\, if\, 9\,\mathrm{h} \leq x \leq 12\,\mathrm{h}\, or\, more \end{cases} \tag{11}$$

2.3.2. Defining Linguistic Variables and Fuzzy Rules

Our objective is to predict the different levels of SDW from the following three variables: environmental conditions, ergonomic factors, and organisation at work. To build the linguistic model, it was necessary to define fuzzy IF–THEN rules, and input variables and consequent sentences were defined as linguistic variables as follows:

- Variable → Environmental conditions. Linguistic variables → Noise, Lighting, and Heat_Stress;
- Variable → Ergonomic factors. Linguistic variables → Exposition_Time, Mass_Object, Lifting_Frequency;
- Variable → Organisation at work. Linguistic variables → Workplace_Violence, Salary, Workday;
- Membership function → absence of SDW;
- Membership function → some features of SDW;
- Membership function → presence of SDW.

Because FL is an intuitive model used for determining the level of the presence of sustainable decent work, two ranges of evaluation were needed. First, an assessment of risk in SDW with a score from 1 to 10 points was defined, considering 1 as the absence of SDW and 10 as the presence of SDW. The second was for the SDWL, with a score given from 1 to 10, where 1 defines the absence of SDW, and 10 establishes the presence of SDW. The fuzzification of variables to conform to the universe of discourse and linguistic variables, fuzzy choices, and consequent sentences to define the SDWL is presented in Figure 1 and was used to determine the fuzzy rules.

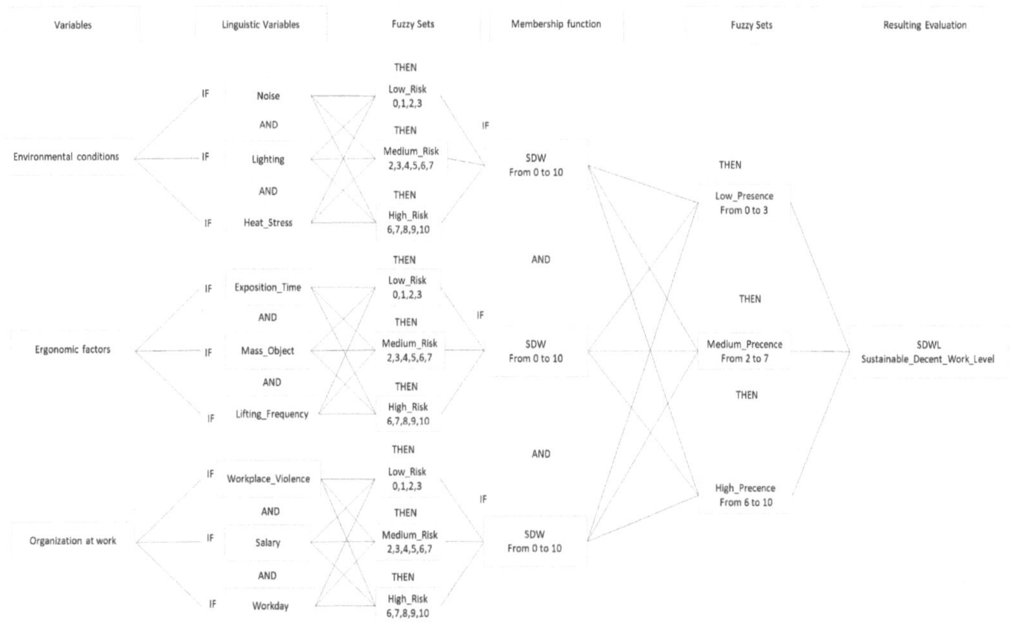

Figure 1. Universe of discourse between linguistic variables, fuzzy choices, and membership functions to define SDWLs.

Considering the rule that if N is the number of linguistic variables and S is the number of linguistic values, then the possible number of fuzzy rules will be S raised to N as indicated by Equation (12).

$$Number\ of\ FR = S^N \tag{12}$$

Therefore, 108 rules were determined, 27 for each variable, including the resulting evaluation, as is shown in Table 1. All combinations of the linguistic variables through the IF, AND, and THEN represent the inference stage to define the level of SDW for each work condition.

2.4. Phase II: Defuzzification

Programming the Fuzzy Rules in the Fuzzy Logic Designer

Mathematically, a fuzzy set is one in which the contra-domain is the interval (0, 1), and the domain is the universe [39]. Thus, if the degree of membership is closer to 1, the more included the element will be in each set, i.e., there is a level of presence of SDW. If the degree of membership is closer to 0, fewer elements will be included in each set [41], and there is the absence of SDW. Therefore, FL determines the inferential mechanism needed to reach the output value related to the SDW level, numerically determining its presence or absence. The fuzzy sets define the universe of discourse (universe of possible actions). They can be represented graphically as a function when the universe of discourse X (or underlying domain) is continuous (not discrete), as is shown in Figure 2. Once the fuzzy set was defined, the groups of 27 rules defined in Table 1 for each linguistic variable were entered into the Mamdani-type inference in the Fuzzy Logic Toolbox; an example is shown in Figure 3, and the rest of the rules are shown in the figures of Appendix A.

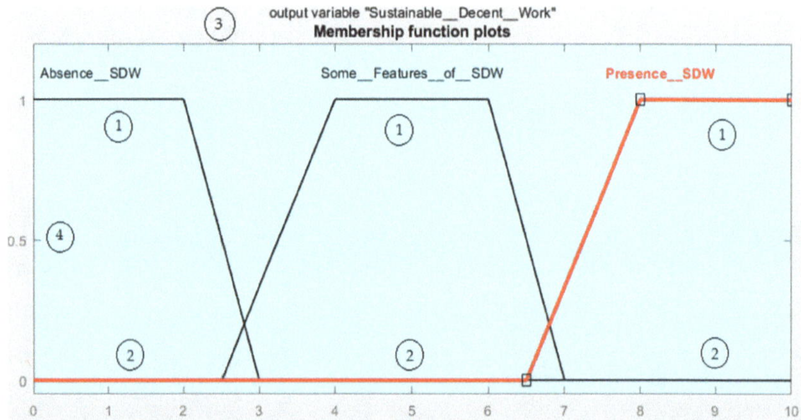

Figure 2. Example of a graphic representation of a membership function built by the Fuzzy Logic Designer: (1) fuzzy sets for each risk level of decision, (2) domain for each fuzzy set, (3) linguistic variable, and (4) membership score. The red line identifies the selected fuzzy set, and the black boxes are positioned in the parameter limits of the function. The black lines in the membership function plot indicate unselected fuzzy sets.

Finally, the defuzzification consists of converting the fuzzy sets into a crisp single value that is the precise representation of the fuzzy sets [25]. The defuzzified values represent the level of sustained decent work in any workplace, ranging from 1 to 10.

Table 1. Fuzzy rules. We have made a change to the letter style by using italics for variables.

	Environmental Conditions				Ergonomic Factors				Organisation at Work			
	IF	IF	IF		IF	IF	IF		IF	IF	IF	
		AND	AND	THEN		AND	AND	THEN		AND	AND	THEN
Rule No.	**Noise**	**Lighting**	**Heat_Stress**	**SWD Risk**	**Exposition_Time**	**Mass_Object**	**Lifting_Frequency**	**SDW Risk**	**Workplace_Violence**	**Salary**	**Workday**	**SDW Risk**
1	Low	Low	Low	Absence	High	High	High	Absence	Low	Low	Low	Some
2	Medium	Low	Low	Absence	High	High	Medium	Absence	Low	Low	Medium	Some
3	High	Low	Low	Absence	High	High	Low	Absence	Low	Low	High	Absence
4	Low	Low	Medium	Some	High	Medium	High	Absence	Low	Medium	Low	Presence
5	Medium	Low	Medium	Some	High	Medium	Medium	Absence	Low	Medium	Medium	Presence
6	High	Low	Medium	Absence	High	Medium	Low	Absence	Low	Medium	High	Some
7	Low	Low	High	Absence	High	Low	High	Absence	Low	High	Low	Presence
8	Medium	Low	High	Absence	High	Low	Medium	Some	Low	High	Medium	Presence
9	High	Low	High	Absence	High	Low	Low	Presence	Low	High	High	Presence
10	Low	Medium	Low	Absence	Medium	High	High	Absence	Medium	Low	Low	Absence
11	Medium	Medium	Low	Absence	Medium	High	Medium	Absence	Medium	Low	Medium	Absence
12	High	Medium	Low	Absence	Medium	High	Low	Some	Medium	Low	High	Absence
13	Low	Medium	Medium	Presence	Medium	Medium	High	Some	Medium	Medium	Low	Absence
14	Medium	Medium	Medium	Presence	Medium	Medium	Medium	Some	Medium	Medium	Medium	Absence
15	High	Medium	Medium	Some	Medium	Medium	Low	Some	Medium	Medium	High	Absence
16	Low	Medium	High	Absence	Medium	Low	High	Some	Medium	High	Low	Absence
17	Medium	Medium	High	Absence	Medium	Low	Medium	Some	Medium	High	Medium	Absence
18	High	Medium	High	Absence	Medium	Low	Low	Presence	Medium	High	High	Absence
19	Low	High	Low	Absence	Low	High	High	Some	High	Low	Low	Absence
20	Medium	High	Low	Absence	Low	High	Medium	Some	High	Low	Medium	Absence
21	High	High	Low	Absence	Low	High	Low	Presence	High	Low	High	Absence
22	Low	High	Medium	Presence	Low	Medium	High	Some	High	Medium	Low	Absence
23	Medium	High	Medium	Presence	Low	Medium	Medium	Some	High	Medium	Medium	Absence
24	High	High	Medium	Some	Low	Medium	Low	Presence	High	Medium	High	Absence
25	Low	High	High	Absence	Low	Low	High	Presence	High	High	Low	Absence
26	Medium	High	High	Absence	Low	Low	Medium	Presence	High	High	Medium	Absence
27	High	High	High	Absence	Low	Low	Low	Presence	High	High	High	Absence

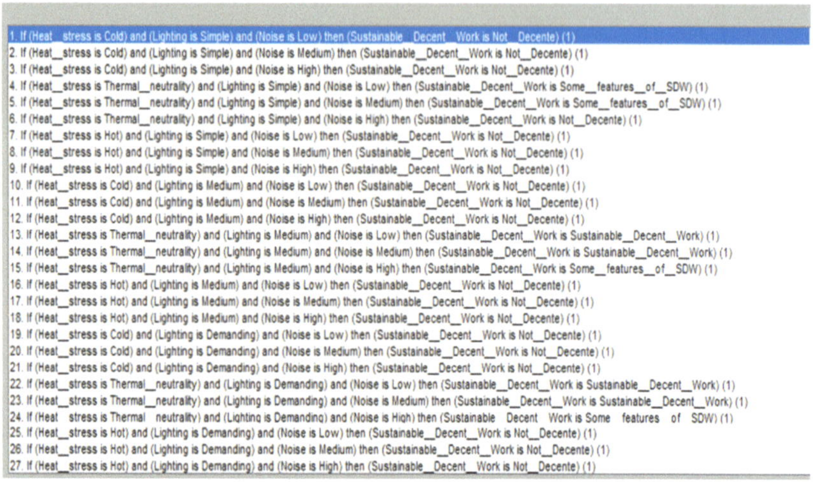

Figure 3. Fuzzy rules for the linguistic variable SDWL in the module Rule Editor of the Fuzzy Logic Toolbox Designer.

3. Results

This section presents the results of this fuzzy logic method for measuring sustainable decent work levels (SDWLs) applied to a case study in the auto parts industry. It considers three essential assumptions: (1) The evaluation is restricted to three variables: environmental conditions, ergonomic factors, and organisation at work. (2) Each variable was assessed by three membership functions that represent, respectively, the absence of SDW, some features of SDW, or the presence of SDW. (3) The evaluation result represents the level that defines the absence or presence of SDW. To define out-of-pocket spending, a questionnaire was applied only to 330 direct labour workers (66 workers from each department) but not to middle managers. To identify whether there is a presence or absence of SDW in their environmental conditions, five production departments (laminating, tempering, post glass, supply chain, and quality control) were measured. The frequency of conditions found is presented in Table 2.

Table 2. Frequency of environmental conditions organised by membership function.

			Departments					
Linguistic Variables	Membership Functions	Ranks	Laminating	Tempering	Post Glass	Supply Chain	Quality Control	Total
Noise	Low_Risk	<80 db	2	10	0	1		13
	Medium_Risk	80–90 db	39	35	16	10		100
	High_Risk	>90 db	6	2	0	1		9
Lighting	Low_Risk	70–300 lux	43	55	42	0	1	42
	Medium_Risk	200–750 lux	92	91	7	49	23	262
	High_Risk	>600 lux	0	0	0	0		0
Heat_Stress	Low_Risk	10–23 °C	0	0	0	10		10
	Medium_Risk	20–30 °C	7	10	0	40		57
	High_Risk	27–45 °C	15	8	50	3	25	101
			Total					594

To define the level of SDW for the set environmental condition, the three highest frequencies were evaluated with the following parameters: Noise 90 db, Lighting 500 lux, and Heat_Stress 32 °C. This fuzzy evaluation is shown in Figure 4a. The columns in yellow are the membership vectors, which represent graphically the rules for each membership function, and the column in blue represents the evaluation result. For example, in the Heat_Stress membership vector equal to 32 °C, the rules 7, 8, 9, 16, 17, 18, 25, 26, and 27 were activated in the high-risk zone; in the Lighting membership vector equal to 500 lux, the rules 10, 11, 12, 13, 14, 15, 16, 17, and 18 were activated in the low-risk zone; and in the Noise membership vector equal to 90 db, the rules 3, 6, 9, 12, 15, 18, 21, 24, and 27 were activated in the high-risk zone. Considering that the solution does not represent a direct relationship between columns and rows but evaluates memberships between rules is important. The resultant vector establishes that degree of membership. Therefore, the combination of all memberships activates rule 18 with a score of 1.5 in Absence_SDW, as shown in the membership function plots at the top of Figure 4a. To evaluate the ergonomic factors, the highest mass manipulated was considered to be 26 kg (windshield mass), with a frequency of 2100 liftings and an exposition time of 300 min per shift. The evaluation results are shown in Figure 4b and indicate the absence of SDW with a score of 1.5.

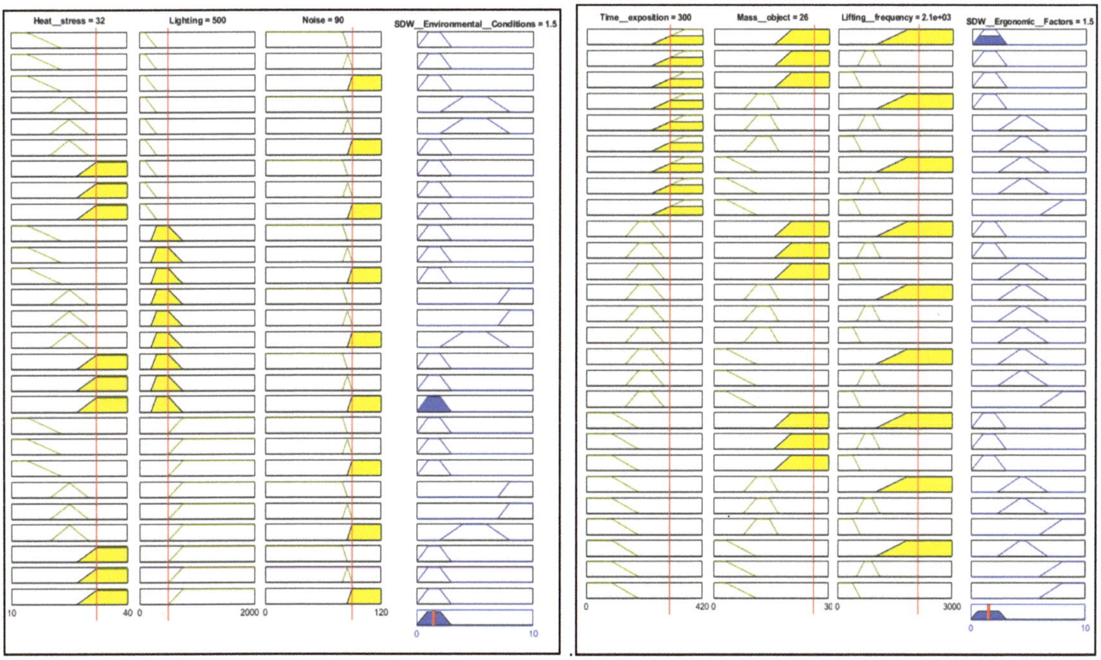

(**a**) (**b**)

Figure 4. Individual evaluation by a set of linguistic variables: (**a**) ruler viewer for environmental conditions, with an SDW resultant score of 1.5; (**b**) ruler viewer for ergonomic factors, with an SDW resultant score of 1.5. Both cases indicate the absence of SDW. The yellow vectors indicate the rules that have been activated for each linguistic set, while the blue vector represents the SDW level (the resulting evaluation). The red line on the vectors corresponds to the value inputted for each linguistic variable.

A salary of USD 12 per day for direct labour workers was considered to evaluate the set organisation at work. The result from the evaluation of violence at work was positive with 10 points; twenty-one per cent of workers reported having suffered violence at work (see Table 3). Finally, the weekday was 12 h each third day, totalling 48 h per week.

The evaluation results are shown in Figure 5a and indicate some features of SDW with a resulting score of 5. Finally, the individual scores from each set were evaluated to define a final SDWL (see Figure 5b). The resulting score for the SDWL was 1.24, indicating that the production process of car windshields is carried out under conditions where there is an absence of sustained decent work.

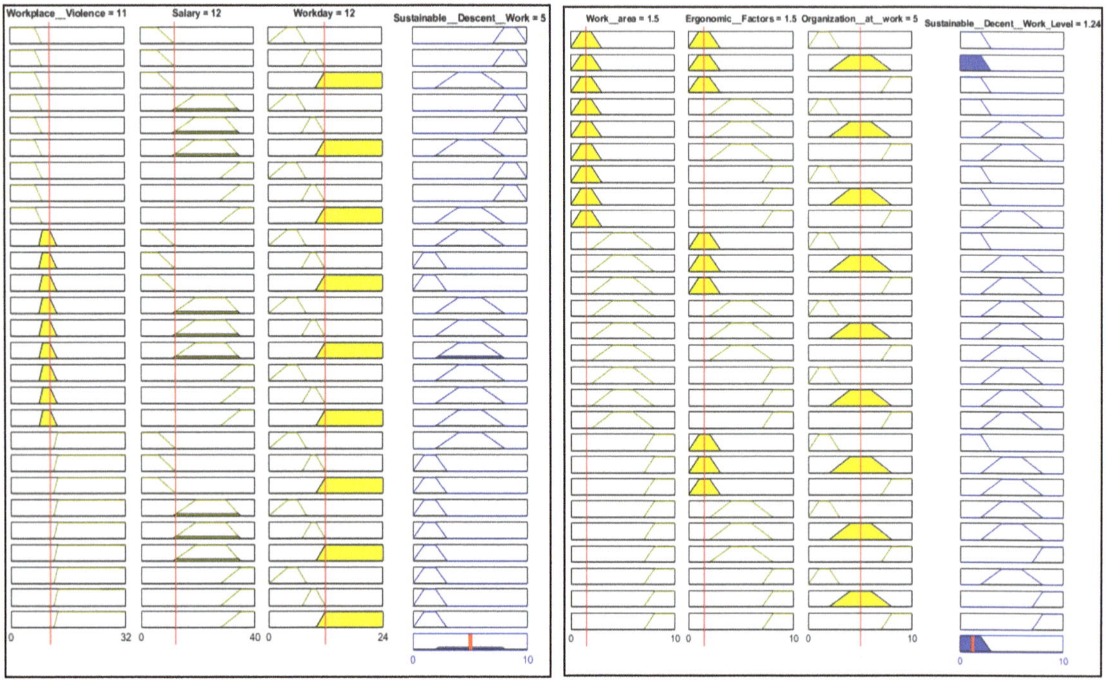

(**a**) (**b**)

Figure 5. Individual evaluation by a set of linguistic variables: (**a**) ruler viewer for the organisation at work with an SDW resultant score of 5, indicating some features of SDW; (**b**) ruler viewer for the final SDWL, with a resultant score of 1.24, indicating the absence of sustainable decent work. The yellow vectors indicate the rules that have been activated for each linguistic set, while the blue vector represents the SDW level (the resulting evaluation). The red line on the vectors corresponds to the value inputted for each linguistic variable.

Consequently, the company is required to modify its internal socially responsible policies. An example of an evaluation for the departments is presented in Appendix B. The results of 33 individual evaluations of workstations are shown in Table 4, wherein in 14 cases, there was an absence of SDW; in the other 14 cases, there were some features of SDW, and only 4 cases showed the presence of SDW.

The resulting score of 1.24 for SDWL negatively impacts each worker's family economy, as was observed in the results of the out-of-pocket spending and catastrophic health expense interviews for 330 workers (see Table 3). For example, at some point in their life, 70% of the interviewed workers had paid between USD 1 and USD 50 for health expenses out of pocket due to work at least once per month; 31% had paid between USD 101 and USD 500 for health expenses out of pocket due to work at least once per month; and almost 3% had paid more than USD 500 for health expenses out of pocket due to work at least once per month. In the case of violence at work, only 18 workers stated that they had suffered violence at work; however, 45 workers refused to answer the questionnaire for fear of repression from their immediate boss: both cases were considered as violence, totalling 21%.

Table 3. Responses from 330 workers interviewed across 33 workstations and three shifts regarding their out-of-pocket spending, catastrophic health expenses, and violence at work.

Question	Questions Answered with "Yes"	Questions Answered with "No"	Unanswered Question for Fear of Reprisals
1. Have you ever developed any work-related illnesses like back pain, loss of hearing, or a respiratory illness?	78	188	45
2. Have you experienced any work-related accident that temporarily prevents work?	86	180	45
3. Have you experienced any work-related pain or injuries that temporarily prevent work?	169	97	45
4. Have you been treated for occupational diseases by public health services (free of charge)?	185	81	45
5. Have you received treatment for work-related injuries or illnesses from a private healthcare provider in addition to the treatment you had already received?	150	116	45
6. Have you ever paid for out-of-pocket health expenses due to work at least once per month, between USD 1 and USD 50?	223	43	45
7. Have you ever paid for out-of-pocket health expenses due to work at least once per month, between USD 51 and USD 100?	43	233	45
8. Have you ever paid for out-of-pocket health expenses due to work at least once per month, between USD 101 and USD 500?	105	161	45
9. Have you ever paid for out-of-pocket health expenses due to work at least once per month for more than USD 500?	9	246	45
10. Have you ever suffered any type of abuse or threats from a superior that you consider to be violence at work?	18	267	45

Table 4. Results of 33 evaluations of individual workstations.

Workstation	Environmental Conditions	Ergonomics Factors	Organisation at Work	SDWL	Results
1 Laminating	5	5	5	5.84	Some features of SDW
2 Laminating	3	6	7	5.8	Some features of SDW
3 Laminating	7	4	5	5.81	Some features of SDW
4 Laminating	8	6	8	1.78	Absence of SDW
5 Laminating	7	4	7.5	3.23	Absence of SDW
6 Laminating	7	6	5	5.8	Some features of SDW
7 Laminating	4	4	4	5.84	Some features of SDW
8 Quality Control	8	3	8	1.78	Absence of SDW
9 Quality Control	8	3	3	5.8	Some features of SDW
10 Quality Control	8	2	2	8.84	Presence of SDW
11 Quality Control	10	1	1	8.9	Presence of SDW
12 Quality Control	7	3	3	5.8	Some features of SDW
13 Quality Control	7	2	2	5.81	Some features of SDW
14 Supply Chain	1.5	7.69	1.5	1.59	Absence of SDW
15 Supply Chain	4.53	3.79	4.03	5.83	Some features of SDW
16 Supply Chain	4	1.34	5	5.84	Some features of SDW
17 Supply Chain	4.04	2.38	5.67	5.79	Some features of SDW
18 Tempering	2.35	1.06	3.38	2.69	Absence of SDW
19 Tempering	1.5	3.85	7.32	1.81	Absence of SDW
20 Tempering	5	7.12	8.5	1.71	Absence of SDW
21 Tempering	8.69	8.22	1.5	1.53	Absence of SDW
22 Tempering	3.45	4.5	1.5	5.83	Some features of SDW
23 Tempering	2.35	1.24	5	2.69	Absence of SDW
24 Tempering	8.74	6.63	3.5	3.2	Absence of SDW
25 Post Glass	3.48	1.4	4	5.83	Some features of SDW
26 Post Glass	8.69	1.24	6.5	8.87	Presence of SDW
27 Post Glass	1.5	3.88	7	1.74	Absence of SDW
28 Post Glass	5	1.06	8.5	1.54	Absence of SDW
29 Post Glass	8	2.35	1.5	8.82	Presence of SDW
30 Post Glass	7	8.47	3	1.74	Absence of SDW
31 Post Glass	3.5	1.07	1.5	5.84	Some features of SDW
32 Post Glass	2.3	8.23	5	1.65	Absence of SDW
33 Post Glass	1.5	3.85	2.32	1.81	Absence of SDW

4. Discussion

This section discusses the results obtained from the fuzzy logic method for measuring sustainable decent work levels (SDWLs) applied to a case study in the production of car windshields as a validation. First, the theoretical implications are discussed to analyse the impact of the SDWLS and compare them with the current situation to define this study's contributions to the existing literature and how it will be helpful for managers as a socially responsible approach, especially in the auto parts industry. Then, practical implications were defined to establish its contribution to real industrial situations.

The SDWL can classify information from each workstation and organise memberships between linguistic variables and risk levels to define the presence or absence of sustainable decent work. Comparing the linguistic variables of environmental conditions, ergonomics, and organisation at work with the variables defined in [10]—quality of life, equal opportunity, and workers' rights—the proposed SDWL added an alternative way to analyse the social component of sustainability. Regarding the current situation in fuzzy logic models, Parra-Dominguez et al. [42] found that there are voices in the scientific community considering whether sustainable decent growth is suitable for measuring the progress of the 2030 Agenda and providing models for the pursuit of a long and healthy life, access to knowledge, and a good standard of living; in this context, the SDWL defines new critical criteria establishing 10 fuzzy models based on measurable international parameters, which impact important human aspects like violence at work and work-related illnesses. However, comparing results from vectors was not possible because the variables and their fuzzy models were not the same. In the case of the auto parts industry, the study proposed in [43] included ergonomic factors and organisation at work as variables to analyse Volvo's production system. Compared with that study, the SDWL incorporates new parameters for work conditions and organisation at work with a sustainable emphasis rather than a production focus. Therefore, SDWLs can be used by the industry as a tool to meet voluntary sustainability standards like the proposals in [11].

The Equations (2) to (11) generate the resulting vectors which determine the absence of SDW with a score of 1.5 in two linguistic variables (see Figure 4a,b): environmental conditions and ergonomic factors; although these results vary marginally, in general, the models solves adequately. Having the same value in these results was unexpected, given that these linguistic variables include evaluation parameters that are very different. This allows us to assume that the sensitivity of the method requires improvement. On the other hand, the result of some features of SDW in the organisation at work sector having an SDW score of 5 was expected because the salary range is suitable for a company located in Mexico despite the work conditions. However, the resulting vector determines that the production of car windshields with a score of 1.24 has a poor SDWL. In the current case study, the effects of the poor SDW were supported by the results of the out-of-pocket spending and catastrophic health expenses interviews applied to 330 workers (see Table 4). The economic impact on workers occurs because public health is deficient in rehabilitation after accidents or musculoskeletal disorders. Therefore, workers must go to private means to achieve adequate rehabilitation and return to work. Considering both the workers' salary and the sample size of interviewed workers, we can infer that these workers suffer a lack of a long and healthy life and a bad standard of living without economic growth due to a poor SDW (the cause of work-related musculoskeletal disorders and work illnesses), increasing their out-of-pocket spending and catastrophic health expenses.

Therefore, these findings contribute to defining the most suitable task conditions and their parameters using policies and laws relating to sustainability to explain correctly the presence or absence of sustainable decent work. As a practical implication, the SDWL is helpful for managers in real-life situations, particularly in the case of producing car windshields in the auto parts industry (as shown in the case study). For example, from the economic implications, it was possible to identify safety and health parameters and safety costs by focusing on the linguistic rules highlighted in the resultant vectors. For example, in the noise vector, ergonomic factors vector, and workday vector, the rules over the safety conditions were activated; this allows for identifying improvements for the prevention program. This promotes establishing better CSR policies for places with identical risk conditions, saving time and implementation costs.

Finally, it is crucial to note that the outcomes of the SDWL evaluation are not definitive. Thus, further tests by supplementing more linguistic variables will be necessary to define new fuzzy rules and membership functions to determine the mathematical sensibility and method behaviour. In future research, the case study results will be classified according to gender to define if there are differences in SDW regarding work conditions, ergonomic factors, and organisation at work between males and females who work in the auto parts industry. Moreover, a new study will be implemented in cardboard manufacturing to define the SDWL's usability as a CSR tool to meet voluntary sustainability standards.

5. Conclusions

In conclusion, a fuzzy logic method built in MATLAB Fuzzy Logic Designer was proposed in this work. This interactive computer system considers the relationship between safety and health risk levels and workstation conditions to identify the level of sustainable decent work in the production of car windshields in the auto parts industry. The method included environmental conditions, ergonomic factors, and organisation at work as linguistic variables related to three membership functions: the absence of SDW, some features of SDW, and the presence of SDW. The resulting vector determined there to be an absence of SDWL in the production of car windshields with a score of 1.24. Our evaluation indicates that workers in the production of car windshields suffer a lack of a long and healthy life and a bad standard of living without economic growth. Therefore, the SDWL can help managers implement voluntary sustainability standards and improve CSR policies, as was established in SDG 8. However, the outcomes of the SDWL are not conclusive. Moreover, further tests could be limited if more linguistic variables are added due to that fact that exponentially increasing the rules can be challenging to program, affecting the mathematical sensibility and method behaviour.

Author Contributions: Conceptualisation, methodology, writing—original draft preparation, and writing—review and editing: M.R.C.-V.; Conceptualisation, methodology, and Software: D.S.-J.; software and formal analysis and investigation: A.N.A.-R.; methodology, writing—original draft preparation, and writing—review and editing: J.M.N.-J. All authors have read and agreed to the published version of the manuscript.

Funding: This research received no external funding.

Institutional Review Board Statement: Not applicable.

Informed Consent Statement: Not applicable.

Data Availability Statement: Section A include the dataset generated and analysed during the present research.

Acknowledgments: The authors of this work thank Instituto Tecnológico de Monterrey (ITESM) and Universidad Autónoma del Estado de Morelos (UAEM) for supporting the development of this article.

Conflicts of Interest: The authors declare no conflicts of interest.

Appendix A. Fuzzy Rules

```
1. If (Heat__stress is Cold) and (Lighting is Simple) and (Noise is Low) then (SDW__Environmental__Conditions is Absence___SDW) (1)
2. If (Heat__stress is Cold) and (Lighting is Simple) and (Noise is Medium) then (SDW__Environmental__Conditions is Absence___SDW) (1)
3. If (Heat__stress is Cold) and (Lighting is Simple) and (Noise is High) then (SDW__Environmental__Conditions is Absence___SDW) (1)
4. If (Heat__stress is Thermal__neutrality) and (Lighting is Simple) and (Noise is Low) then (SDW__Environmental__Conditions is Some__features__of__SDW) (1)
5. If (Heat__stress is Thermal__neutrality) and (Lighting is Simple) and (Noise is Medium) then (SDW__Environmental__Conditions is Some__features__of__SDW) (1)
6. If (Heat__stress is Thermal__neutrality) and (Lighting is Simple) and (Noise is High) then (SDW__Environmental__Conditions is Absence___SDW) (1)
7. If (Heat__stress is Hot) and (Lighting is Simple) and (Noise is Low) then (SDW__Environmental__Conditions is Absence___SDW) (1)
8. If (Heat__stress is Hot) and (Lighting is Simple) and (Noise is Medium) then (SDW__Environmental__Conditions is Absence___SDW) (1)
9. If (Heat__stress is Hot) and (Lighting is Simple) and (Noise is High) then (SDW__Environmental__Conditions is Absence___SDW) (1)
10. If (Heat__stress is Cold) and (Lighting is Medium) and (Noise is Low) then (SDW__Environmental__Conditions is Absence___SDW) (1)
11. If (Heat__stress is Cold) and (Lighting is Medium) and (Noise is Medium) then (SDW__Environmental__Conditions is Absence___SDW) (1)
12. If (Heat__stress is Cold) and (Lighting is Medium) and (Noise is High) then (SDW__Environmental__Conditions is Absence___SDW) (1)
13. If (Heat__stress is Thermal__neutrality) and (Lighting is Medium) and (Noise is Low) then (SDW__Environmental__Conditions is Presence__SDW) (1)
14. If (Heat__stress is Thermal__neutrality) and (Lighting is Medium) and (Noise is Medium) then (SDW__Environmental__Conditions is Presence__SDW) (1)
15. If (Heat__stress is Thermal__neutrality) and (Lighting is Medium) and (Noise is High) then (SDW__Environmental__Conditions is Some__features__of__SDW) (1)
16. If (Heat__stress is Hot) and (Lighting is Medium) and (Noise is Low) then (SDW__Environmental__Conditions is Absence___SDW) (1)
17. If (Heat__stress is Hot) and (Lighting is Medium) and (Noise is Medium) then (SDW__Environmental__Conditions is Absence___SDW) (1)
18. If (Heat__stress is Hot) and (Lighting is Medium) and (Noise is High) then (SDW__Environmental__Conditions is Absence___SDW) (1)
19. If (Heat__stress is Cold) and (Lighting is Demanding) and (Noise is Low) then (SDW__Environmental__Conditions is Absence___SDW) (1)
20. If (Heat__stress is Cold) and (Lighting is Demanding) and (Noise is Medium) then (SDW__Environmental__Conditions is Absence___SDW) (1)
21. If (Heat__stress is Cold) and (Lighting is Demanding) and (Noise is High) then (SDW__Environmental__Conditions is Absence___SDW) (1)
22. If (Heat__stress is Thermal__neutrality) and (Lighting is Demanding) and (Noise is Low) then (SDW__Environmental__Conditions is Presence__SDW) (1)
23. If (Heat__stress is Thermal__neutrality) and (Lighting is Demanding) and (Noise is Medium) then (SDW__Environmental__Conditions is Presence__SDW) (1)
24. If (Heat__stress is Thermal__neutrality) and (Lighting is Demanding) and (Noise is High) then (SDW__Environmental__Conditions is Some__features__of__SDW) (1)
25. If (Heat__stress is Hot) and (Lighting is Demanding) and (Noise is Low) then (SDW__Environmental__Conditions is Absence___SDW) (1)
26. If (Heat__stress is Hot) and (Lighting is Demanding) and (Noise is Medium) then (SDW__Environmental__Conditions is Absence___SDW) (1)
27. If (Heat__stress is Hot) and (Lighting is Demanding) and (Noise is High) then (SDW__Environmental__Conditions is Absence___SDW) (1)
```

Figure A1. Fuzzy rules for the linguistic variable of environmental conditions.

1. If (Time_exposition is High_risk) and (Mass_object is High_risk) and (Lifting_frequency is High_risk) then (SDW_Ergonomic_Factors is Absence_SDW) (1)
2. If (Time_exposition is High_risk) and (Mass_object is High_risk) and (Lifting_frequency is Medium_risk) then (SDW_Ergonomic_Factors is Absence_SDW) (1)
3. If (Time_exposition is High_risk) and (Mass_object is High_risk) and (Lifting_frequency is Low_risk) then (SDW_Ergonomic_Factors is Absence_SDW) (1)
4. If (Time_exposition is High_risk) and (Mass_object is Medium_risk) and (Lifting_frequency is High_risk) then (SDW_Ergonomic_Factors is Absence_SDW) (1)
5. If (Time_exposition is High_risk) and (Mass_object is Medium_risk) and (Lifting_frequency is Medium_risk) then (SDW_Ergonomic_Factors is Some_features_of_SDW) (1)
6. If (Time_exposition is High_risk) and (Mass_object is Medium_risk) and (Lifting_frequency is Low_risk) then (SDW_Ergonomic_Factors is Some_features_of_SDW) (1)
7. If (Time_exposition is High_risk) and (Mass_object is Low_risk) and (Lifting_frequency is High_risk) then (SDW_Ergonomic_Factors is Some_features_of_SDW) (1)
8. If (Time_exposition is High_risk) and (Mass_object is Low_risk) and (Lifting_frequency is Medium_risk) then (SDW_Ergonomic_Factors is Some_features_of_SDW) (1)
9. If (Time_exposition is High_risk) and (Mass_object is Low_risk) and (Lifting_frequency is Low_risk) then (SDW_Ergonomic_Factors is Presence_SDW) (1)
10. If (Time_exposition is Medium_risk) and (Mass_object is High_risk) and (Lifting_frequency is High_risk) then (SDW_Ergonomic_Factors is Absence_SDW) (1)
11. If (Time_exposition is Medium_risk) and (Mass_object is High_risk) and (Lifting_frequency is Medium_risk) then (SDW_Ergonomic_Factors is Absence_SDW) (1)
12. If (Time_exposition is Medium_risk) and (Mass_object is High_risk) and (Lifting_frequency is Low_risk) then (SDW_Ergonomic_Factors is Some_features_of_SDW) (1)
13. If (Time_exposition is Medium_risk) and (Mass_object is Medium_risk) and (Lifting_frequency is High_risk) then (SDW_Ergonomic_Factors is Some_features_of_SDW) (1)
14. If (Time_exposition is Medium_risk) and (Mass_object is Medium_risk) and (Lifting_frequency is Medium_risk) then (SDW_Ergonomic_Factors is Some_features_of_SDW) (1)
15. If (Time_exposition is Medium_risk) and (Mass_object is Medium_risk) and (Lifting_frequency is Low_risk) then (SDW_Ergonomic_Factors is Some_features_of_SDW) (1)
16. If (Time_exposition is Medium_risk) and (Mass_object is Low_risk) and (Lifting_frequency is High_risk) then (SDW_Ergonomic_Factors is Some_features_of_SDW) (1)
17. If (Time_exposition is Medium_risk) and (Mass_object is Low_risk) and (Lifting_frequency is Medium_risk) then (SDW_Ergonomic_Factors is Some_features_of_SDW) (1)
18. If (Time_exposition is Medium_risk) and (Mass_object is Low_risk) and (Lifting_frequency is Low_risk) then (SDW_Ergonomic_Factors is Presence_SDW) (1)
19. If (Time_exposition is Low_risk) and (Mass_object is High_risk) and (Lifting_frequency is High_risk) then (SDW_Ergonomic_Factors is Absence_SDW) (1)
20. If (Time_exposition is Low_risk) and (Mass_object is High_risk) and (Lifting_frequency is Medium_risk) then (SDW_Ergonomic_Factors is Absence_SDW) (1)
21. If (Time_exposition is Low_risk) and (Mass_object is High_risk) and (Lifting_frequency is Low_risk) then (SDW_Ergonomic_Factors is Absence_SDW) (1)
22. If (Time_exposition is Low_risk) and (Mass_object is Medium_risk) and (Lifting_frequency is High_risk) then (SDW_Ergonomic_Factors is Some_features_of_SDW) (1)
23. If (Time_exposition is Low_risk) and (Mass_object is Medium_risk) and (Lifting_frequency is Medium_risk) then (SDW_Ergonomic_Factors is Some_features_of_SDW) (1)
24. If (Time_exposition is Low_risk) and (Mass_object is Medium_risk) and (Lifting_frequency is Low_risk) then (SDW_Ergonomic_Factors is Presence_SDW) (1)
25. If (Time_exposition is Low_risk) and (Mass_object is Low_risk) and (Lifting_frequency is High_risk) then (SDW_Ergonomic_Factors is Some_features_of_SDW) (1)
26. If (Time_exposition is Low_risk) and (Mass_object is Low_risk) and (Lifting_frequency is Medium_risk) then (SDW_Ergonomic_Factors is Presence_SDW) (1)
27. If (Time_exposition is Low_risk) and (Mass_object is Low_risk) and (Lifting_frequency is Low_risk) then (SDW_Ergonomic_Factors is Presence_SDW) (1)

Figure A2. Fuzzy rules for the linguistic variable of ergonomic factors.

1. If (Workplace_Violence is Low) and (Salary is Low) and (Workday is Low) then (SDW_Organization_at_Work is Presence_SDW) (1)
2. If (Workplace_Violence is Low) and (Salary is Low) and (Workday is Medium) then (SDW_Organization_at_Work is Presence_SDW) (1)
3. If (Workplace_Violence is Low) and (Salary is Low) and (Workday is High) then (SDW_Organization_at_Work is Some_features_of_SDW) (1)
4. If (Workplace_Violence is Low) and (Salary is Medium) and (Workday is Low) then (SDW_Organization_at_Work is Presence_SDW) (1)
5. If (Workplace_Violence is Low) and (Salary is Medium) and (Workday is Medium) then (SDW_Organization_at_Work is Presence_SDW) (1)
6. If (Workplace_Violence is Low) and (Salary is Medium) and (Workday is High) then (SDW_Organization_at_Work is Some_features_of_SDW) (1)
7. If (Workplace_Violence is Low) and (Salary is High) and (Workday is Low) then (SDW_Organization_at_Work is Presence_SDW) (1)
8. If (Workplace_Violence is Low) and (Salary is High) and (Workday is Medium) then (SDW_Organization_at_Work is Presence_SDW) (1)
9. If (Workplace_Violence is Low) and (Salary is High) and (Workday is High) then (SDW_Organization_at_Work is Some_features_of_SDW) (1)
10. If (Workplace_Violence is Medium) and (Salary is Low) and (Workday is Low) then (SDW_Organization_at_Work is Some_features_of_SDW) (1)
11. If (Workplace_Violence is Medium) and (Salary is Low) and (Workday is Medium) then (SDW_Organization_at_Work is Absence_SDW) (1)
12. If (Workplace_Violence is Medium) and (Salary is Low) and (Workday is High) then (SDW_Organization_at_Work is Absence_SDW) (1)
13. If (Workplace_Violence is Medium) and (Salary is Medium) and (Workday is Low) then (SDW_Organization_at_Work is Some_features_of_SDW) (1)
14. If (Workplace_Violence is Medium) and (Salary is Medium) and (Workday is Medium) then (SDW_Organization_at_Work is Some_features_of_SDW) (1)
15. If (Workplace_Violence is Medium) and (Salary is Medium) and (Workday is High) then (SDW_Organization_at_Work is Some_features_of_SDW) (1)
16. If (Workplace_Violence is Medium) and (Salary is High) and (Workday is Low) then (SDW_Organization_at_Work is Some_features_of_SDW) (1)
17. If (Workplace_Violence is Medium) and (Salary is High) and (Workday is Medium) then (SDW_Organization_at_Work is Some_features_of_SDW) (1)
18. If (Workplace_Violence is Medium) and (Salary is High) and (Workday is High) then (SDW_Organization_at_Work is Some_features_of_SDW) (1)
19. If (Workplace_Violence is High) and (Salary is Low) and (Workday is Low) then (SDW_Organization_at_Work is Some_features_of_SDW) (1)
20. If (Workplace_Violence is High) and (Salary is Low) and (Workday is Medium) then (SDW_Organization_at_Work is Absence_SDW) (1)
21. If (Workplace_Violence is High) and (Salary is Low) and (Workday is High) then (SDW_Organization_at_Work is Absence_SDW) (1)
22. If (Workplace_Violence is High) and (Salary is Medium) and (Workday is Low) then (SDW_Organization_at_Work is Absence_SDW) (1)
23. If (Workplace_Violence is High) and (Salary is Medium) and (Workday is Medium) then (SDW_Organization_at_Work is Absence_SDW) (1)
24. If (Workplace_Violence is High) and (Salary is Medium) and (Workday is High) then (SDW_Organization_at_Work is Absence_SDW) (1)
25. If (Workplace_Violence is High) and (Salary is High) and (Workday is Low) then (SDW_Organization_at_Work is Absence_SDW) (1)
26. If (Workplace_Violence is High) and (Salary is High) and (Workday is Medium) then (SDW_Organization_at_Work is Absence_SDW) (1)
27. If (Workplace_Violence is High) and (Salary is High) and (Workday is High) then (SDW_Organization_at_Work is Absence_SDW) (1)

Figure A3. Fuzzy rules for the linguistic variable of organisation at work.

Appendix B. Examples of Evaluations by Department

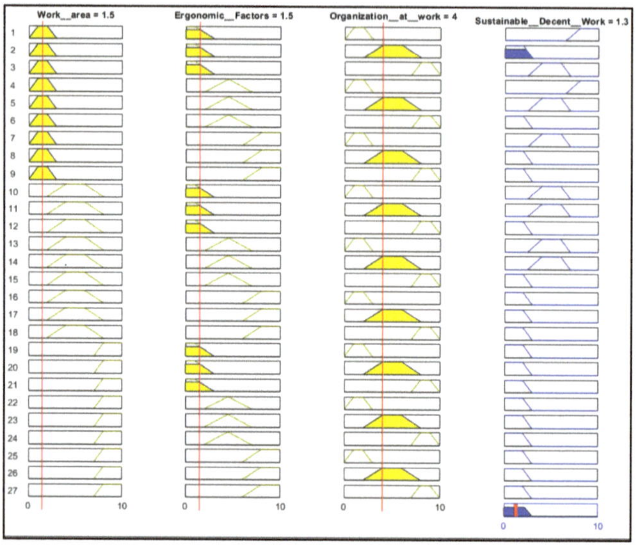

Figure A4. Individual evaluation for the laminating department: environmental conditions with a score of 1.5, ergonomic factors with a score of 1.5. The organisation at work has a score of 4. The final SDWL, with a score of 1.3, indicates the absence of sustainable decent work. The yellow vectors indicate the rules that have been activated for each linguistic set, while the blue vector represents the SDW level (the resulting evaluation). The red line on the vectors corresponds to the value inputted for each linguistic variable.

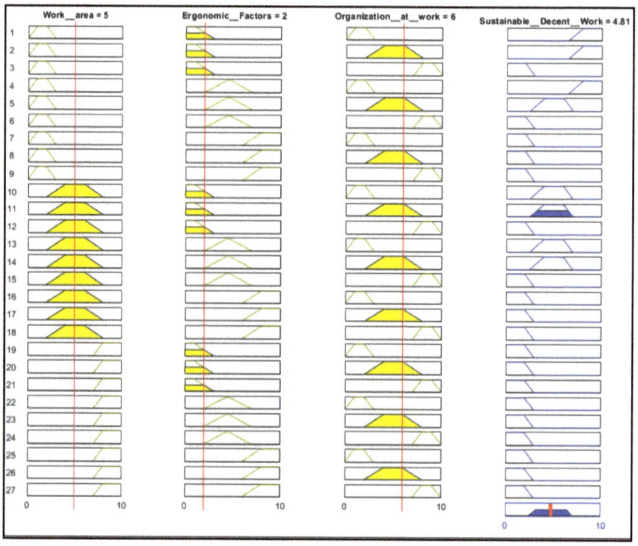

Figure A5. Individual evaluation for the quality control department: environmental conditions with a score of 5, ergonomic factors with a score of 2. The organisation at work has a score of 6. The final SDWL, with a score of 4.81, indicates some presence of sustainable decent work. The yellow vectors indicate the rules that have been activated for each linguistic set, while the blue vector represents the SDW level (the resulting evaluation). The red line on the vectors corresponds to the value inputted for each linguistic variable.

References

1. Zadjafar, M.A.; Gholamian, M.R. A Sustainable Inventory Model by Considering Environmental Ergonomics and Environmental Pollution, Case Study: Pulp and Paper Mills. *J. Clean. Prod.* **2018**, *199*, 444–458. [CrossRef]
2. Sánchez-Infante Hernández, J.P.; Yañez-Araque, B.; Moreno-García, J. Moderating Effect of Firm Size on the Influence of Corporate Social Responsibility in the Economic Performance of Micro-, Small- and Medium-Sized Enterprises. *Technol. Forecast. Soc. Chang.* **2020**, *151*, 119774. [CrossRef]
3. Congress of the United Mexican States. *Federal Labor Law of the United Mexican States*; Diario Oficial de la Federación, CDMX: Mexico City, Mexico, 2024.
4. Larrouyet, M.C. Desarrollo Sustentable: Origen, Evolución y Su Implementación Para el Cuidado del Planeta. Master's Thesis, Universidad Nacional de Quilmes, Bernal, Argentina, 2015.
5. Plata De Plata, D.; Plata Díaz, O. Ambiente, Economía, Tecnología y Sociedad: Componentes Clave Para El Desarrollo Sostenible. *Multiciencias* **2009**, *9*, 7–12.
6. The United Nations. 17 Goals to Transform Our World. Available online: https://www.un.org/sustainabledevelopment/ (accessed on 24 November 2023).
7. United Nations. Decent Work and Economic Growth: Why It Matters. Available online: https://www.un.org/sustainabledevelopment/wp-content/uploads/2018/09/Goal-8.pdf (accessed on 26 July 2023).
8. United Nations. Goal 8 Decent Work and Economic Growth. *Promote Inclusive and Sustainable Growth, Employment and Decent Work for All.* Available online: https://www.un.org/sustainabledevelopment/economic-growth/ (accessed on 27 July 2023).
9. Rodríguez, M. Gasto de Bolsillo y Gastos Catastróficos En Salud En Hogares Mexicanos. *Carta Económica Reg.* **2021**, *0*, 59–83. [CrossRef]
10. Mensah, J. Sustainable Development: Meaning, History, Principles, Pillars, and Implications for Human Action: Literature Review. *Cogent Soc. Sci.* **2019**, *5*, 1653531. [CrossRef]
11. Balderas, A.; Luna, E.; Voora, V.; Larrea, C. *Voluntary Standards and Initiatives for Carbon Management: Navigating the Landscape*; IISD: Winnipeg, MB, Canada, 2024.
12. Barford, A.; Coombe, R.; Proefke, R. Against the Odds: Young People's High Aspirations and Societal Contributions amid a Decent Work Shortage. *Geoforum* **2021**, *121*, 162–172. [CrossRef]
13. Vignoli, E.; Prudhomme, N.; Terriot, K.; Cohen-Scali, V.; Arnoux-Nicolas, C.; Bernaud, J.L.; Lallemand, N. Decent Work in France: Context, Conceptualization, and Assessment. *J. Vocat. Behav.* **2020**, *116*, 103345. [CrossRef]
14. Duffy, R.D.; Prieto, C.G.; Kim, H.J.; Raque-Bogdan, T.L.; Duffy, N.O. Decent Work and Physical Health: A Multi-Wave Investigation. *J. Vocat. Behav.* **2021**, *127*, 103544. [CrossRef]
15. Yildirim, U.; Toygar, A.; Çolakoğlu, C. Compensation Effect of Wages on Decent Work: A Study on Seafarers Attitudes. *Mar. Policy* **2022**, *143*, 105155. [CrossRef]
16. Lout, G.; Fitzpatrick, J.; Garcia Lozano, A.J.; Finkbeiner, E. Decent Work in a Seascape of Livelihoods: Regional Evaluation of the Shrimp and Groundfish Fishery of the Guianas-Brazil Shelf. *Mar. Policy* **2022**, *144*, 105231. [CrossRef]
17. Yan, Y.; Geng, Y.; Gao, J. Measuring the Decent Work of Knowledge Workers: Constructing and Validating a New Scale. *Heliyon* **2023**, *9*, e17945. [CrossRef]
18. International Labour Organizatio. *Measurement of Decent Work: Discussion Paper for the Tripartite Meeting of Experts on the Measurement of Decent Work*; International Labour Organizatio: Geneva, Switzerland, 2008.
19. Raufflet, E.; Barin Cruz, L.; Bres, L. An Assessment of Corporate Social Responsibility Practices in the Mining and Oil and Gas Industries. *J. Clean. Prod.* **2014**, *84*, 256–270. [CrossRef]
20. Kwon, H.; Park, Y. Proactive Development of Emerging Technology in a Socially Responsible Manner: Data-Driven Problem Solving Process Using Latent Semantic Analysis. *J. Eng. Technol. Manag.* **2018**, *50*, 45–60. [CrossRef]
21. Hadj, T.B. Effects of Corporate Social Responsibility towards Stakeholders and Environmental Management on Responsible Innovation and Competitiveness. *J. Clean. Prod.* **2020**, *250*, 119490. [CrossRef]
22. Ali, S.M.; Appolloni, A.; Cavallaro, F.; D'Adamo, I.; Di Vaio, A.; Ferella, F.; Gastaldi, M.; Ikram, M.; Kumar, N.M.; Martin, M.A.; et al. Development Goals towards Sustainability. *Sustainability* **2023**, *15*, 9443. [CrossRef]
23. The MathWorks Inc. Fuzzy Logic Toolbox Diseñe y Simule Sistemas de Lógica Difusa. Available online: https://la.mathworks.com/products/fuzzy-logic.html (accessed on 19 July 2023).
24. Nguyen, H.T.; Walker, C.L.; Walker, E.A. *A First Curse in Fuzzy Logic*, 4th ed.; CRC Press: Boca Raton, FL, USA; Taylor & Francis Group: New York, NY, USA, 2019.
25. Klir, G.J.; Yuan, B. *Fuzzy Sets and Fuzzy Logic. Theory and Applications*, 1st ed.; Prentice Hall PTR: Hoboken, NJ, USA, 1995.
26. Díaz-Contreras, C.A.; Aguilera-Rojas, A.; Guillén-Barrientos, N. Lógica Difusa vs. Modelo de Regresión Múltiple Para La Selección de Personal Fuzzy Logic vs. Multiple Regression for Selection Personnel. *Ingeniare Rev. Chil. Ing.* **2014**, *22*, 547–559. [CrossRef]
27. European Agency for Safety and Health at Work: An Introduction to Noise at Work. Factsheet 56—CE_2004_2169_EN.Indd. Available online: https://osha.europa.eu/sites/default/files/Factsheets_56_-_Informationen_zu_Larm_bei_der_Arbeit.pdf (accessed on 10 January 2024).
28. Official Journal of the European Union. *On the Minimum Health and Safety Requirements Regarding the Exposure of Workers to the Risks Arising from Physical Agents (Noise)*; Official Journal of the European Union: Brussels, Belgium, 2003; pp. 4238–4244.
29. Secretaria del Trabajo y Previsión Social. *NORMA Oficial Mexicana NOM-011-STPS-2001, Condiciones de Seguridad e Higiene En Los Centros de Trabajo Donde Se Genere Ruido*; STPS (Secretarïa del Trabajo y Previsión Social): Ciudad de México, México, 2002.

30. Suter, A.H. Noise. Standards and Regulations. In *Encyclopaedia of Occupational Health and Safety—Part VI. General Hazards;* International Labour Organization (ILO): Geneva, Switzerland, 2011; p. 47.
31. Ministry of the Environment Government of Japan. Cabinet Order for Implementation of the Noise Regulation Law. Appendix I. Available online: https://www.env.go.jp/en/laws/air/noise/ap.html (accessed on 26 July 2023).
32. Näf Cortés, R.R. *Guía Práctica Para El Análisis y La Gestión Del Ruido Industrial,* 1st ed.; FREMAP Government of Spain, Ed.; FREMAP Government of Spain: Madrid, Spain, 2013.
33. Secretaría del Trabajo y Previsión Social. *NORMA Oficial Mexicana NOM-025-STPS-2008, Condiciones de Iluminación En Los Centros de Trabajo;* Diario Oficial de la Federación: Ciudad de México, México, 2008.
34. Hernández Calleja, A.; Ramos Pérez, F. Lighting, Conditions Required for Visual Comfort. In *Encyclopaedia of Occupational Health and Safety—Part VI. General Hazards;* International Labour Organization (ILO): Geneva, Switzerland, 2011; p. VI.
35. Ministry of Health, Labour and Welfare. *Ordinance on Industrial Safety and Health;* Ministry of Health, Labour and Welfare: Tokyo, Japan, 2006. Available online: https://www.japaneselawtranslation.go.jp/en/laws/view/3878/en#je_pt3ch5 (accessed on 10 January 2024).
36. Parsons, K.C. Heat and Cold. Assessment of Heat Stress and Heat Stress Indices. In *Encyclopaedia of Occupational Health and Safety—Part VI. General Hazards;* International Labour Organization (ILO): Geneva, Switzerland, 2011; p. 42.
37. Secretaría del Trabajo y Previsión Social. *NORMA Oficial Mexicana NOM-015-STPS-2001, Condiciones Térmicas Elevadas o Abatidas-Condiciones de Seguridad e Higiene;* Diario Oficial de la Federación: Ciudad de México, México, 2001.
38. *ISO 11228-1:2021(E);* Ergonomics-Manual Handling—Part 1: Lifting, Lowering and Carrying. International Standard Organization—ISO: Geneva, Switzerland, 2021.
39. Contreras-Valenzuela, M.R.; Seuret-Jiménez, D.; Hdz-Jasso, A.M.; León Hernández, V.A.; Abundes-Recilla, A.N.; Trutié-Carrero, E. Design of a Fuzzy Logic Evaluation to Determine the Ergonomic Risk Level of Manual Material Handling Tasks. *Int. J. Environ. Res. Public Health* **2022**, *19*, 6511. [CrossRef]
40. Secretaría del Trabajo y Previsión Social. *NORMA Oficial Mexicana NOM-035-STPS-2018, Factores de Riesgo Psicosocial en el Trabajo-Identificación, Análisis y Prevención;* STPS: Mexico City, Mexico, 2018.
41. Ghasemi, F.; Mahdavi, N. A New Scoring System for the Rapid Entire Body Assessment (REBA) Based on Fuzzy Sets and Bayesian Networks. *Int. J. Ind. Ergon.* **2020**, *80*, 103058. [CrossRef]
42. Parra-Domínguez, J.; Alonso-García, M.; Corchado, J.M. Fuzzy Logic to Measure the Degree of Compliance with a Target in an SDG—The Case of SDG 11. *Mathematics* **2023**, *11*, 2967. [CrossRef]
43. Neumann, W.P.; Ekman, M.; Winkel, J. Integrating Ergonomics into Production System Development—The Volvo Powertrain Case. *Appl. Ergon.* **2009**, *40*, 527–537. [CrossRef] [PubMed]

Disclaimer/Publisher's Note: The statements, opinions and data contained in all publications are solely those of the individual author(s) and contributor(s) and not of MDPI and/or the editor(s). MDPI and/or the editor(s) disclaim responsibility for any injury to people or property resulting from any ideas, methods, instructions or products referred to in the content.

sustainability

MDPI

Article

A Business Case for Climate Neutrality in Pasture-Based Dairy Production Systems in Ireland: Evidence from Farm Zero C

Theresa Rubhara [1,*], James Gaffey [1,2,*], Gavin Hunt [2], Fionnuala Murphy [2,3], Kevin O'Connor [2,4], Enda Buckley [5] and Luis Alejandro Vergara [2,3]

1 Circular Bioeconomy Research Group, Munster Technological University, V92 HD4V Tralee, Ireland
2 BiOrbic Bioeconomy SFI Research Centre, University College Dublin, Belfield, D04 V1W8 Dublin, Ireland; gavin.hunt@biorbic.com (G.H.); fionnuala.murphy@ucd.ie (F.M.); kevin.oconnor@ucd.ie (K.O.); luis.vergara@ucdconnect.ie (L.A.V.)
3 School of Biosystems and Food Engineering, University College Dublin, Belfield, D04 V1W8 Dublin, Ireland
4 School of Biomolecular and Biomedical Science, University College Dublin, Belfield, D04 V1W8 Dublin, Ireland
5 Carbery Group, Ballineen Co., P47 YW77 Cork, Ireland
* Correspondence: theresarubhara@gmail.com (T.R.); james.gaffey@mtu.ie (J.G.)

Abstract: Agriculture in Ireland is responsible for producing and exporting healthy, nutritional food pivotal for meeting the Sustainable Development Goals (SDGs) such as global food security, economic development and sustainable communities. However, the agricultural sector, dominated by a large bovine population, faces the challenge of reducing greenhouse gas (GHG) emissions to reach climate neutrality by 2050. The objective of the current study was to model the environmental and economic impact of simultaneously applying farm-level climate change mitigation strategies for a conventional grass-based dairy farm in Ireland. An average farm of 52 ha with a spring-calving herd of 93 was used as a reference scenario to create a business case. Partial budgeting was used to calculate the annual net benefit. A cradle-to-grave life cycle assessment (LCA) was used to model the reduction in GHG emissions, which was expressed as kg of carbon dioxide equivalent per kilogram of fat- and protein-corrected milk (kg CO_2-eq/kg FPCM). The baseline for average emissions was 0.960 kg CO_2-eq/kg FPCM. An average farm would reduce its annual emissions by 12% to 0.847 kg CO_2-eq/kg FPCM in Scenario 1, where climate change mitigation strategies were applied on a minimal scale. For Scenario 2, the emissions are reduced by 36% to 0.614 kg CO_2-eq/kg FPCM. In terms of annual savings on cash income, an increase of EUR 6634 and EUR 18,045 in net savings for the farm are realised in Scenarios 1 and 2, respectively. The business case provides evidence that farms can move towards climate neutrality while still remaining economically sustainable.

Keywords: climate-neutral agriculture; greenhouse gases; net benefit; global warming

Citation: Rubhara, T.; Gaffey, J.; Hunt, G.; Murphy, F.; O'Connor, K.; Buckley, E.; Vergara, L.A. A Business Case for Climate Neutrality in Pasture-Based Dairy Production Systems in Ireland: Evidence from Farm Zero C. *Sustainability* **2024**, *16*, 1028. https://doi.org/10.3390/su16031028

Academic Editors: Manoj Kumar Nallapaneni, Idiano D'Adamo and Massimo Gastaldi

Received: 28 November 2023
Revised: 15 January 2024
Accepted: 16 January 2024
Published: 25 January 2024

Copyright: © 2024 by the authors. Licensee MDPI, Basel, Switzerland. This article is an open access article distributed under the terms and conditions of the Creative Commons Attribution (CC BY) license (https://creativecommons.org/licenses/by/4.0/).

1. Introduction

Irish agriculture has the potential to become a global leader in sustainable food systems through the production, marketing and management of low-carbon food. According to Ireland's Department of Agriculture, Food and the Marine [1], the agricultural sector contributed 9.5% of Irish merchandising exports and approximately EUR 18.7 billion to the value of agri-food exports in 2022. The sector underpins much of rural Ireland, with over 170,400 (7.1% of total employment) people employed in the agri-food sector [1]. Ireland exports 90% of its food products to 160 countries worldwide, contributing directly to the Sustainable Development Goal (SDG) of global food security (SDG2) and economic development [1]. Despite such a positive economic contribution, the agricultural sector, is, however, associated with negative environmental impacts, such as greenhouse gas (GHG) emissions, loss of natural habitats and diversity due to intensive agriculture and monoculture, a decline in air and water quality and deforestation. This poses a threat to the achievement of the SDG13, SDG14 and SDG15 targets of ensuring environmental sustainability. As consumers become more aware of the various production systems and how their consumer choices can have an impact on the environment [2], the demand for products that are sustainably produced will continue to increase. Farmers could integrate greater innovation in the production of food products to ensure that food is produced in an environmentally sustainable manner. Currently, Ireland produces approximately 1.1 million tonnes of food waste each year, which results in a carbon footprint of 3.6 Mt CO_2eq [3]. As such, increasing production to feed the world, is not, in and of itself, sustainable. It should be noted that the SDGs towards sustainability, previously listed, are all interlinked and not always antagonistic. For instance, with responsible production and consumption (SDG12), food losses can be minimized, natural resources less depleted and greater food security can be achieved [4].

Unlike other European countries where transport and energy industries are the major GHG emitters, in Ireland, agriculture accounted for about 37.5% of the nation's total GHG emissions in 2021 [5], approximately double the emissions from the energy industry. The fact that agriculture continues to contribute the largest sectoral percentage of GHG emissions to the national inventory is a major cause for concern for such an important industry. Ireland has regularly fallen short of its climate change emission targets and currently faces a very challenging target of reducing overall emissions by 51% by 2030 and achieving net zero emissions by 2050 [6,7]. The largest share of GHG emissions in Irish agriculture relates to ruminant production and is predominantly a result of rumen methane and nitrous oxide from soils. The 2021 Irish Farm Sustainability Report [8] indicated that the amount of GHG emissions from an average Irish farm rose in 2020, largely due to an increase in herd size, in addition to a 3.3% and 6.2% increase in fertiliser use and liming, respectively [5].

With over 80% of the agricultural land in Ireland being grassland [9], the grass-based nature of livestock production in the country offers positive environmental opportunities in terms of manure recycling, the integration of livestock and crops for feed, low feed–food competition, biodiversity, soil quality and organic carbon content [10]. Grass, a relatively cheap but abundant feed source, also gives Irish dairy farmers a competitive edge in terms of lower costs and higher profits [11]. Results from the National Farm Survey (NFS) [12] show that dairy remains the most economically and socially sustainable farming system in Ireland; however, the continuous expansion of the bovine population has resulted in higher agricultural GHG at the national level due to higher methane emissions. Ireland aims to achieve a climate-neutral food system by 2050 [13]. Climate-neutral agriculture is defined as net zero emissions of agricultural GHG emissions, implying that the total GHGs (expressed in the carbon dioxide equivalent) released into the atmosphere by sources are equal to or less than the carbon absorbed by carbon sinks [14,15]. In grass-fed systems like Ireland, such as Australia [16–18] and New Zealand [19], much of the research towards climate-neutral agriculture has put a major emphasis on carbon sequestration modelling or strategies that require major land use changes and capital investment. To reduce emissions in agriculture, robust but practical measures are required to be implemented at the farm level. In previous research in Ireland, strategies such as clover, multispecies swards, slurry management and the use of protected urea have shown positive environmental impacts [20–22]. However, such research and analyses of the impact of the mitigation strategies have largely been conducted in isolation. In contrast, the Farm Zero C (FZC) initiative, which is the basis for the current paper, combines at least 15 strategies at once. FZC uses a holistic and pragmatic approach to transform a conventional farm into a more sustainable farm, with the overall aim of achieving a climate-neutral dairy farm. To achieve this, FZC undertakes an interdisciplinary program of work to reduce emissions, targeting several areas:

I. Soil and grassland management: measuring and increasing soil carbon organic stocks through soil and grassland management practices such as incorporating clover and growing multispecies swards;

II. Animal diet and breeding: trialling different types of diets and anti-methane additives that can alter animal digestion, reducing the amount of methane emitted by cows;
III. Renewable energy: producing and using renewable energy on the farm where possible to reduce the farm's reliance on carbon-emitting fossil fuels.

The research completed to date has demonstrated that the combined strategies can significantly reduce the emissions at Farm Zero C as the life cycle assessment (LCA) modelling of GHG emissions has shown a decrease in emissions from 0.86 kg CO_2-eq/kg FPCM in 2018 to 0.66 kg CO_2-eq/kg FPCM in 2022. Previous studies have shown that adoption decisions for climate change mitigation practices among farmers are not based solely on the environmental impact of the strategies [23]. Farmers are likely to adopt innovations which they perceive to have economic returns [24]; for instance, those arising from increased efficiencies, economies of scale and financial incentives [23]. Various economic-oriented studies have been published focusing on the costs of climate change mitigation in Ireland; however, most of these either used input-output models or policy analysis to provide a broad perspective on the economic impacts of climate change mitigation at the national or regional level [25,26]. For instance, The Irish Marginal Abatement Cost Curve (MACC) study provides a detailed cost analysis of the climate mitigation strategies across all farming systems with the absolute emission reduction pathways at the national level [26]. This current paper, on the other hand, builds on data arising from Farm Zero C combined with available information on climate change mitigation costs to provide a case study on how a combination of strategies can be applied for economic and environmental sustainability at the farm level. A business case, based on the implementation of a selection of the Farm Zero C climate-neutral strategies, is modelled under specified assumptions to determine the economic and environmental impact of a set of mitigation measures applied at different levels. The objective of this paper is to provide evidence that climate neutrality at the farm level produces opportunities for cost reduction and revenue growth, thus contributing positively to SDG13's targets of combating climate change.

2. Materials and Methods

2.1. Description of the Farm

A case study dairy farm in Ireland which closely resembles an average Irish dairy farm adopted from the National Farm survey data and agricultural factsheet [1,12] was used for formulating the Holistic FZC business case scenarios. The business case assumes that the case study farm reduces its emissions to a certain level year on year until it reaches climate neutrality by 2050. The physical farm components are important for evaluating the economic and environmental analysis and these are summarized in Table 1 below. An average dairy farm of 52 ha located in the southern part of Ireland stocked at 2.2 LU/ha, which currently uses none of the FZC climate mitigation strategies is presented as the baseline scenario. This represents a typical Irish pasture-based, spring-calving dairy farm where cows spend an average of 241 days on grass.

Table 1. Values used to describe case study farm.

Variable	Description
Farm Size	52 ha
Soil drainage	Average
Herd size	93 dairy cows
Replacement rate	22%
Productivity	5700 L/cow/yr @4.21% fat and 3.57% protein KG MS/Cow—455

Table 1. *Cont.*

Variable	Description
Chemical nitrogen fertiliser use	220 kg/ha (50% Urea & 50% CAN)
Concentrate	1100 kg/cow/yr
Grazing management	241 days per annum
Animals culled	20 Mature heads @ 550 kg live weight
Slurry spreading method	Splash plate
Slurry spreading season	50% in Summer 50% in Spring
Manure storage	Pit storage for the mature herd and heifers, solid storage for calves

2.2. Mitigation Measure Selection

Since June 2021, the FZC holistic climate-neutral strategies have been tested and demonstrated on a commercial dairy farm, Shinagh Farm, at Bandon, Co., Cork, Ireland. The selected mitigation strategies, for the current business case, were based on the findings from these research, demonstration and analysis activities. Table 2 presents a summary of the assumptions made, and economic and environmental impacts based on evidence from the FZC trials and other research outcomes. It should be noted that Shinagh farm is not representative of the conventional farm in Ireland, as it is highly resource efficient with a relatively low carbon footprint of 0.66 kg CO_2-eq/kg FPCM as compared to 0.96 kg CO_2-eq/kg FPCM for an average farm in 2022.

The information in Table 2 includes all trials and strategies implemented at FZC except for green biorefinery and soil carbon sequestration. Holistic livestock management strategies which have been shown to increase technical efficiencies including maximizing utilisation of grass, improving grassland management by incorporating clover and MSS, optimising slurry for organic nitrogen, and improving the economic breeding index were implemented based on previous trials, and assumptions were made based on this work [22,27,28]. Innovative technologies such as feed and slurry additives were trialled at FZC and the results were used to model the conventional farm. The strategies were chosen on the basis that they are both practical to implement in the short run and effective in reducing GHG emissions. Whilst the use of the green biorefinery for grass has demonstrated a high potential for reducing emissions from imported feed, it may be more challenging to implement in the short term, as proper planning is required to address issues relating to the initial investment cost and the ownership of equipment [29]. One option would be for farmers to come together as a cooperative and purchase a biorefinery where they would process their own grass. Similarly, without evidence-based measurement, reporting and verification (MRV) of carbon sequestration it is difficult to model the level of soil organic carbon and the associated costs [30]. The FZC is taking steps to implement MRV, but more data are required to include it in the model.

The mitigation measures included were aimed at reducing emissions per unit of output (CO_2-eq/kg FPCM), rather than absolute farm emissions. This approach allows the inclusion of strategies which increase production and sometimes the farm's emissions such as the economic breeding index (EBI) and extending the grazing season but significantly reduce emissions per kg of product [31]. Where available, costs of inputs such as fertiliser were adopted from the central statistics office; however, for novel technologies such as feed and slurry additives which are not readily available on the market, the FZC prices were used, which may be more or less the same as commercial prices.

Table 2. Summary of mitigation strategies and assumptions.

Strategy	Target Emission Source	Environmental Impacts	Economic Impacts	Assumptions
Reduce chemical N use through White clover, Red clover and Multispecies swards	Fertiliser use	Reduces nitrous oxide emissions and nitrate losses to water Reduces the upstream impacts associated with fertiliser production	Reduction of fertiliser costs Incremental reseeding costs	Nitrogen fertiliser reduction to 150 kg/ha [22] No changes to Dry Matter (DM) yield [21,32]
Grazing management	Manure management	Manure left on pasture which has lower methane emissions than stored [33]	Savings from less silage and less concentrate feed Higher milk solids	Farmer either reduces concentrate or increases productivity An extra week on grass reduces total GHG by 1% [31]
Protected Urea	Fertiliser use	Reduces N_2O and NH_3 losses [20]	Protected urea is cheaper per kg N than calcium ammonium nitrate (CAN) though slightly more expensive than urea	Cost is based on nitrogen value only; phosphorous (P) and potassium (k) costs remain constant
Slurry management through: Spreading all slurry in Spring Use of Low Emission Slurry Spreading (LESS) Chemically amend slurry	Manure management	Spreading slurry in Spring ensures less N is lost as NH_3 Reduces N losses through NH_3 Reduces ammonia and methane emissions during slurry storage slurry	Approximately 0.4 kgN/m^3 more is saved in Spring than in Summer thus reducing total fertiliser costs [34] Reduces demand for chemical fertiliser thus reducing N_2O losses [34] The extra cost of the chemical amendment	The value of N retained only is considered, P&K values remain constant Extra spreading cost EUR 20/h when LESS is used instead of Splash plate, assuming splash plate spreads @ 34 m^3/h and trailing shoe @ 28 m^3/h [27,35] Chemical amendment cost was estimated at EUR 2/m^3 slurry (estimates from FZC trials)
Use native feeds	Feed production	Reduces GHG emissions associated with imported soya and grain	The cost of native ingredients is higher than conventional feed	Native feeds cost EUR 25 more per tonne than conventional feeds (estimates from FZC)
Anti-methane additives (Bovaer)	Animal digestion	Reduces CH_4 emissions	The extra cost of the dietary additive	Dietary additives cost approximately EUR 75/cow/yr (estimates from FZC) Milk yield remains constant
Reduce replacement rate	All hotspots		Costs are reduced as the farmer has less young stock to rear	Rearing a heifer from the calf for 24 months costs approximately EUR 1500 [36]
Use renewable sources to reduce energy inputs	Farm Energy	Reduces CO_2 emissions	Investment costs for the renewable energy equipment	Potential savings or costs were not included in the analysis as they represent an investment cost which differs across different technologies
Increase productivity by 5% (EBI and management)	All hotspots	For every EUR 10 increase in EBI, GHG emissions decline by 1% per unit of product [28]	Increases farm revenue from sales of extra milk solids	The total farm, emissions do not decrease but as productivity increases the quantity of GHG per kg of FPCM reduces

2.3. Modelling Different Scenarios

Two scenarios were used to model the likely impacts of different levels of application for several mitigation strategies when applied simultaneously. Scenario 1 (S1) represents the minimal implementation case option which includes strategies that can be implemented in the short term at a low scale without major costs (low-hanging fruit) usually adopted by the risk-averse farmers, as shown in Table 3. These include reducing fertiliser use, incorporating clover and MSS in grasslands, slurry management, and reducing feed concentrate and replacement rate. Reducing chemical fertiliser use and switching to protected urea has been estimated to reduce nitrous oxide emissions by 5.4% for the Irish agricultural sector in the MACC [35]. However, reducing the quantity of chemical fertiliser on its own can result in low productivity and creates a risk of the grassland being less self-sufficient to feed the animals [37]. White clover incorporation in grassland has been shown to reduce chemical fertiliser requirement to 150 kg/ha [22,38,39] and using red clover in silage can completely replace chemical fertiliser requirements [32]. The modelling for the FZC business case considered the area required for grass-clover swards and MSS to maintain sward productivity, which would result in a 16% and 46% reduction in fertiliser use for S1 and S2, respectively, as shown in Table 3. Scenario 2 (S2), on the other hand, involves a larger scaling of immediate technologies as well as the adoption of innovations such as the use of additives, representing a model that would be more likely to be adopted by a risk-tolerant farmer. The Bovaer (3NOP) additive has been trialled at FZC for methane reduction, and based on these results and the results of previous studies, a 28% reduction was estimated during housing with a 10% reduction throughout the grazing period [40–42]. The environmental impact is limited to GHG emissions (carbon footprint) only. The description of the strategies employed in the baseline and the two scenarios (S1 and S2) is summarised in Table 3.

Table 3. Scenarios under consideration.

Strategy	Baseline	S1	S2
Reduce Chemical N	No change (220 kgN/ha)	To 185 kg N/ha (16%) -Include white clover on 25% of pasture area -Include red clover on 10% of silage area -Include MSS on 10% of pasture area	To 150 kg N/ha (45%) -Include white clover on 50% of pasture area -Include red clover on 25% of silage area -Include MSS on 20% of pasture area
Grazing management	235 Days grazing Season	Extend grazing season by 7 days	Extend grazing season by 14 days
Protected urea	0% of chemical N	50% of chemical N	100% of chemical N
Slurry spreading season	50% Summer, 50% Spring	50% in Summer, 50% in Spring	80% in Spring, 20% in Summer
Slurry spreading method	Splash plate	LESS	LESS
Chemically amend slurry	0% Slurry	0% of slurry	100% of slurry
Native feeds	0% of feed	50% of the diet is native	100% of the diet is native
Reduce feed concentrate	No change	By 5%	By 10%
Anti-methanogenic feeds (Bovaer)	No change	No change	Throughout the year—housing + grazing (28% reduction during housing and 10% when grazing).
Reduce replacement rate	No change (22%)	To 20%	To 18%
Use renewable sources to reduce energy inputs	No change	By 25%	By 50%
Increase production of milk solids (EBI—management practices)	No change	No changes	By 5%

2.4. Economic Impact Analysis

The scenarios were modelled to determine the changes in net profit under different levels of mitigation. Firstly, a whole farm budget was prepared for a typical average dairy farm using farm-level data and the 2022 National Farm Survey (NFS) results to create a baseline [12]. The basic components of the dairy budget were adopted from the Teagasc Profit Monitor tool, which is a digital tool used by farmers to assess their profit/loss over a period of time [43]. Components of the data include the revenue, variable costs, fixed costs and net profit as presented in Equation (1).

$$\text{Profit or loss} = \text{Total Revenue} - \text{Total Cost} \quad (1)$$

where profit or loss is the amount remaining after removing total costs from the gross total revenue [44].

Farm gross revenue was calculated by combining milk sales receipts, cow sales, and the average value of calves sold. Market values for revenue items were adopted from [45] and [46] as follows: farm gate milk price of EUR 0.41/L of milk, EUR 1300/culled cow and EUR 169/calf. Costs were split into variable costs and fixed costs. Variable costs are those costs that vary per scale of production [47,48]. The major variable costs were fertiliser, concentrates, reseeding, replacement rearing, and contractor costs. Fixed costs are costs which do not vary with the level of production ha [43]. Fixed costs include machinery running and lease costs, hired labour, repairs and maintenance and overheads. Expenses such as depreciation and loan repayment were not included in order to simplify the model.

Partial Budget Analysis

A partial budget was then used to check the overall changes in net profit under the two scenarios. The changes resulting from adopting and implementing the mitigation strategies only affect a part of the business and mostly the direct costs, as such, a partial budget was applicable for the analysis of the impact of such changes [49]. Using a partial budget, one can evaluate whether a change in management will increase or decrease profit [50]. The method does not determine profit; rather, it checks the changes in net profit which is recorded as net benefit:

$$\text{Net benefit} = \text{Total benefit change} - \text{Total cost change} \tag{2}$$

$$\text{Total benefit Change} = \text{Extra revenue} + \text{Cost saved} \tag{3}$$

$$\text{Total cost change} = \text{Total cost increased} + \text{Revenue forgone} \tag{4}$$

where total benefit change is the summation of extra revenue increased plus cost saved, and total cost change is the total cost increased plus the revenue forgone [49].

The partial budget economic analysis involves understanding the changes in costs and/or revenues associated with various climate change mitigation strategies demonstrated through FZC. The net benefit, expressed in EUR/Farm/year, can also be referred to as net annual profit, which is the total of the marginal benefits accruing from the net savings on each strategy. A positive margin implies an increase in net profits, whereas a negative figure implies that the introduction of the mitigation strategies reduces net farm profit.

2.5. Environmental Impact Analysis

To quantify the environmental footprint of a conventional farm, a LCA model was initially developed for the FZC farm using 2018 and 2022 data, and subsequently, data were adapted for the average farm. The LCA methodology, guided by the International Standardization Organization's (ISO) framework of goal and scope definition, inventory analysis, impact assessment, and interpretation (ISO 14040:2006) was used to calculate the global warming potential [51]. The cradle-to-farm gate system boundary was used, and the functional unit was the kg carbon dioxide equivalent per kilogram of fat- and protein-corrected milk (kg CO_2-eq/kg FPCM). The methodology and functional unit used measured the GHG intensity rather than absolute emissions. In essence, a reduction in kg CO_2-eq/kg FPCM means the kg of milk on the dairy farm was produced with a lower carbon footprint [5]. The life cycle inventory analysis was carried out using the LCA model developed at FZC (yet to be published) and average figures from the 2022 National Farm Survey data (highlighted in Table 1) were used to simulate the baseline farm. The model was populated using data on animal performance, fertiliser application, manure management, forage production and energy consumption. The calculations are based on the Intergovernmental Panel on Climate Change (IPCC) tier 2 and tier 3 by using country-specific emissions factors from different sources including Ireland's National Inventory Report 2022 [52], Ireland's Informative Inventory Report 2018 [53], the European

Environment Agency's 2019 Air Pollutants Report [54], and the IPCC's 2019 updates to its 2006 publication [55,56].

3. Results

3.1. Baseline

Using the variables described in Table 1 the baseline scenario was first modelled to find the net profit and GHG emissions of the farm before any climate change mitigation measure was applied. Table 4 provides a major summary of the costs, revenue and profit from the case study farm. Total revenue from the sale of milk, culled cows and calves was EUR 255,678. Total variable costs were calculated as EUR 116,087 with the high costs of fertiliser and concentrate the major factor in the high variable costs. The net profit was estimated to be EUR 84,265 for the baseline. Similarly, the net profit for S1 and S2 was estimated as EUR 90,900 and EUR 102,311, respectively.

Table 4. Dairy enterprise budget.

	Baseline	S1	S2
Annual concentrates fed (kg/cow)	1100 kg	1045 kg	990 kg
Milk yield (L/cow)	530,100 L	530,100 L	556,605 L
Milk sales (EUR)	12,337 EUR	12,506 EUR	12,844 EUR
Meat sales (EUR)	38,337 EUR	38,506 EUR	38,844 EUR
Total Sales(EUR)	255,678 EUR	255,847 EUR	267,052 EUR
Variable costs			
Concentrates (EUR)	43,682 EUR	41,491 EUR	39,145 EUR
Fertiliser (EUR)	18,533 EUR	14,574 EUR	10,998 EUR
Reseeding (EUR)	1633 EUR	2163 EUR	2799 EUR
Additives (EUR)	-	-	8079 EUR
Replacements rearing (EUR)	21,769 EUR	20,269 EUR	17,269 EUR
Contractor costs (EUR)	15,874 EUR	16,529 EUR	16,529 EUR
Veterinary and breeding (EUR)	14,596 EUR	14,596 EUR	14,596 EUR
Total variable costs (EUR)	116,087 EUR	109,622 EUR	109,416 EUR
Gross margin (EUR)	139,591 EUR	146,225 EUR	157,636 EUR
Fixed costs			
Total fixed costs (EUR)	55,325 EUR	55,325 EUR	55,325 EUR
Net Income(cash) (EUR)	84,266 EUR	90,900 EUR	102,311 EUR
Net Savings(cash) (EUR)	-	6634 EUR	18,045 EUR

For the environmental metrics, only the GHG emissions were considered and the absolute farm emissions for the baseline was 534 tonnes CO_2. The emission intensity expressed per unit of product using LCA was modelled to be 0.96 $kgCO_2eq/kg$ FPCM, before any form of intervention as shown in Figure 1. The main emission sources were animal digestion (0.505 kg), manure management (0.130 kg) and fertiliser use (0.155 kg). Animal digestion constitutes 52% of the farm emissions.

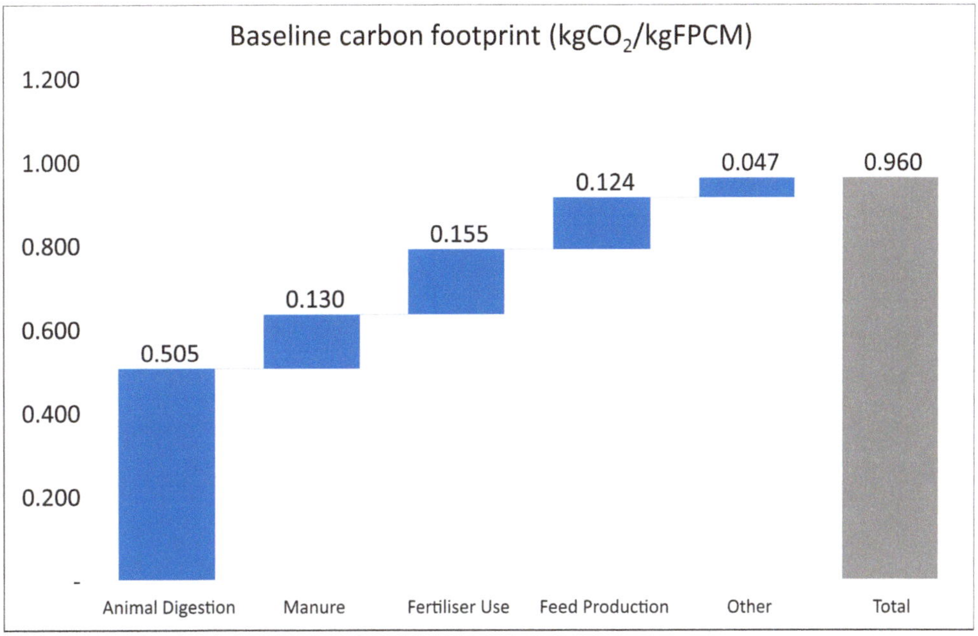

Figure 1. Baseline GHG emissions.

3.2. *Changes in Net Profit under Different Scenarios*

The partial budget results are shown in Table 5. Savings were realised from fertiliser savings, savings in concentrate and the reduced cost of rearing heifers for replacement. Incremental costs were also realised for reseeding and contractors' costs for both scenarios. The cost of slurry and anti-methanogenic additives were the major costs of the mitigation strategies employed. Extra revenue was realised from extra milk sales, replacement sales and fertiliser savings.

Table 5. Net benefit from different scenarios.

Cost/Benefit (EUR)	S1	S2
Cost saved		
Fertiliser savings (EUR)	3959 EUR	7535 EUR
Concentrate (EUR)	2191 EUR	4537 EUR
Replacement rate (EUR)	1500 EUR	4500 EUR
Extra revenue(EUR)		
Extra sales (EUR)	169 EUR	11,374 EUR
Extra costs incurred		
Reseeding (EUR)	−530 EUR	−1166 EUR
Additives (EUR)	0 EUR	−8079 EUR
Contractor costs (EUR)	−655 EUR	−655 EUR
Net Benefit	6634 EUR	18,045 EUR

All costs and benefits are expressed in euros (EUR). The negative sign represents a loss or cost.

3.2.1. Costs Saved

Major costs forgone were the reduced fertiliser, concentrate and replacement heifer rearing costs. Fertiliser reduction was due to the use of clover and MSS swards to reduce the need for chemical N fertiliser. The switching from calcium ammonium nitrate (CAN) to protected urea also resulted in large savings because 1 kg of protected urea would cost EUR 0.18 less than CAN. Generally, large savings of EUR 3959 and EUR 7535 were realised in S1 and S2, respectively. This could be attributed to higher fertiliser prices experienced in 2022 and 2023. The reduction in concentrate resulted in net savings of EUR 3649 and EUR 4536, respectively, in S1 and S2. Extra savings were also realised from the reduced cost of rearing replacement heifers.

3.2.2. Extra Revenue

For Scenario 2, extra sales were recognized from increased productivity, as annual milk yield per cow increased from 530,100 L to 556,605 L due to a higher EBI and improved management. This results in an increase of EUR 11,374 in revenue under S2.

3.2.3. Extra Costs

Under Scenario 2, feed and slurry additives were the major costs amounting to EUR 8079.

The reseeding costs were the common costs in both S1 and S2, as clover and MSS would need more frequent reseeding than the grass swards. However, these reseeding costs were offset by the large fertiliser savings.

3.2.4. Net Profit/Loss

The economic modelling showed an increase in net farm profit in both Scenarios 1 and 2. Under S1, where the farmer applied minimum measures for climate change mitigation, a net farm profit increase of EUR 6634 was achieved, and a larger profit (EUR 18,045) was realised for S2. Though high costs were realised for the use of additives, these costs were neutralised by savings from reduced fertiliser use and increased productivity. A positive figure shows that the overall impact of the intervention provides a profit rather than a loss. Farmers may be more willing to adopt the interventions modelled in both scenarios as they have a positive net economic benefit.

3.3. *Environmental Impact*

Following LCA modelling of climate change mitigation strategies for both scenarios, changes were noted across the main categories of emission sources as shown in Figure 2. The GHG emissions were reduced to 0.847 kg of CO_2-eq/kgFPCM and 0.614 kg of CO_2-eq/kgFPCM, in S1 and S2, respectively.

3.3.1. Animal Digestion

Methane emissions from enteric fermentation were the dominant source of GHG emissions, contributing 56% (0.458 kg of CO_2-eq) of the total GHG emissions under S1 and 62% (0.409 kg of CO_2-eq) per kilogram of FPCM under S2. A 28% methane emission reduction during the housing period and a 10% reduction during the grazing season were assumed for S2, where anti-methane additives were used. This resulted in a 12% emission reduction from animal digestion in S2.

3.3.2. Manure Management

As compared to the baseline, manure accounted for a greater percentage of total emissions than fertiliser, after climate change mitigation strategies were applied. By extending the grazing season and applying manure management of chemically amending slurry, applying slurry during favourable weather conditions and using LESS for spreading slurry, the emissions from manure decreased from 0.13 to 0.122 and 0.073 kg CO_2-eq/kgFPCM, in S1 and S2, respectively.

3.3.3. Fertiliser Use

Nitrogen fertiliser use is responsible for nitrous oxide emissions during the fertiliser application and production processes. The use of protected urea and a reduction in the quantity of nitrogen fertiliser applied saw the fertiliser emissions drop by 3.3% and 7.1% for S1 and S2, respectively.

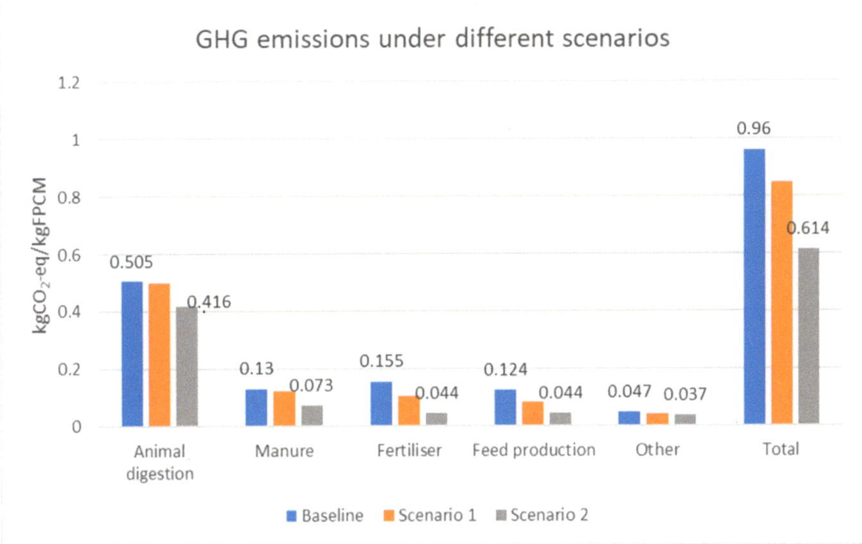

Figure 2. Changes in GHG emissions under different scenarios.

3.3.4. Feed Production

Ireland mainly depends on feed imports such as soya from Brazil and grain from America for concentrate ingredients, which have higher GHG emissions from land use changes and transport than local ingredients. A reduction in feed concentrates and sourcing of EU-produced feed ingredients had a significant reduction in emissions from feed production, with a decrease from the baseline of 0.124 to 0.083 and 0.44 kg CO_2-eq/kgFPCM, in S1 and S2, respectively.

3.3.5. Other

The category "other" represented all emissions from the farm which may not fall into the broader categories, for example, farm energy use. Energy demand reductions for farm applications, such as manure spreading and fertiliser applications, would result in lower on-farm CO_2 emissions.

3.3.6. Net Environmental Impact

The overall results show a net reduction in GHG emissions of 12% from 0.96 kg of CO_2-eq/kgFPCM to 0.847 under S1, and in S2 where the farm-level strategies are employed at a larger scale there was a 36% reduction in emissions the final footprint was 0.614 kg CO_2-eq/kgFPCM.

4. Discussion

Irish farmers need to move towards climate neutrality by adopting robust but practical technologies which ensure both the environmental and economic sustainability of dairy farms. The national emission reduction target of 25% is based on absolute farm emissions using the IPCC methodology [5]. In the FZC case study, the LCA methodology, which considers emissions from upstream processes such as fertiliser and feed production, was used. The results show the reduction in emissions per kilogram of milk, which is an important measure at the farm level for farmers to be able to understand how their environmental decisions affect the farm profit.

4.1. Economic Impact

The baseline represents an average commercial dairy farm which is economically sustainable, as evidenced by a net profit of EUR 84,265. The higher revenues, as compared to the previous seasons, can be attributed to higher farmgate milk prices [12]. According to the NFS results, fertiliser and concentrate feeds are the major variable costs, with fertiliser contributing almost 20% to the total variable costs [12]. The results are in line with [11], which indicates that more grass in the diet increases the profitability of the farming system, and increasing more concentrate feeds is not always the most economically rational choice. The reduction in the quantity of concentrate feeds is not expected to reduce the productivity of the farm, as the assumption is that any dry matter shortfall is compensated for by grassland management practices such as extending the grazing season [44]. Farmers are expected to reduce their concentrate feed not only due to the environmental benefits but also the realisation of cost reduction opportunities. This is in line with evidence from an online survey of 396 Irish farmers, which showed 73% support for the paradigm of maximizing milk from forage and minimizing concentrate use [57] for both environmental and economic gains. Major incremental gains were also realised from increasing the EBI of the herd. The EBI is an efficiency tool which aims to improve the genetic merit of animals for increased profitability. In line with [24], the increase in dairy productivity would result in reduced emission intensity but absolute farm emissions may not drop.

Though large methane emission reductions were realised from the anti-methane additives in S2, it was determined from this analysis that the additives were the major incremental costs in the dairy system under S2. The costs of anti-methane additives represent a major challenge to the adoption of the strategy, as previous research has shown that costly mitigation strategies are a huge disincentive [23].

Other Opportunities for Revenue Generation

The partial budget results in Section 3 indicate the likely increase in net profit by adopting the current farm-level FZC mitigation opportunities; however, by following specific MRV standards, the products can be certified as "carbon reduced" or "carbon free". Where an MRV procedure can be established and the emission abatement can be attributed to specific strategies, the farmers can potentially obtain money from carbon credits or market premiums. In the EU, a tonne of CO_2 is expected to cost EUR 140 by 2030 [58]; hence, under the S2 scenario, farmers have the potential to earn up to EUR 20,580 from abating 147 tonnes of CO_2. A recent Irish consumer survey of 1500 adults showed that 72% of respondents were willing to pay more for dairy products, provided they see the evidence that the increase is going to embed the latest environmental initiatives in production [59]. This view is reiterated by [60]'s findings on consumers' perceptions of carbon footprint labels for dairy products in Italy, which shows that consumers would be willing to pay extra when they are fully aware of the products and claims made about the carbon footprint.

4.2. Environmental Impact

In contrast to indoor systems where manure is the largest contributor to GHG emissions, in grass-based systems, total emissions consist mainly of methane emissions from enteric fermentation [33]. Scenario 2 resonates with target resource use efficiency systems simulated by [61,62] whose results concur with the current study. Both studies found that methane emissions in target farms would contribute a larger percentage to overall GHG than the baseline current farm with lower efficiency. The results highlight the need to reduce methane emissions in grass-based systems. The additive Bovaer (3NOP) has shown consistency in methane reduction with an average of 30% reduction when administered in feed for dairy cows [42,63,64]. However, there are practical issues around administration of the additives in grass-based systems. The adoption rate of additives in general is expected to be low, as there are issues regarding social acceptance and the cost of the additives [65]. Previous studies have suggested the use of slow-release bolus to incorporate the additives in pasture-based systems where 95% of the animals' diet is from grazed forage [41,63].

The reduction in fertiliser emissions in both S1 and S2 is a result of nitrous oxide reduction. Nitrous oxide emissions account for 25% of the agricultural sector's emissions in Ireland [35]. In line with [22,37], the FZC case study highlights the importance of incorporating clover and minimising chemical fertiliser for grassland productivity. This also concurs with [33], who showed that nitrogen surplus from chemical nitrogen per hectare was positively correlated to the GHG emission intensity of milk. According to [66], animal excreta and urine are the biggest sources of N_2O per year in grasslands, followed by manure applications. Manure acts as an emission source for both methane and nitrous oxide, and the quantity emitted is linked to environmental conditions, type of management and composition of the manure [67]. By extending the grazing season and applying manure management strategies, fewer emissions are released from the storage and application of manure. The overall reduction in emissions from manure management in both S1 and S2 results shows that though extending the grazing season results in more N emissions from excreta deposited on grassland, the reduction in emissions from stored manure will be higher than the marginal increase from manure deposits [33]. In contrast to confinement systems, in grass-based systems, the total GHG emissions associated with feed production are predominantly from grass [61].

The results show that to achieve climate neutrality in dairy systems by 2050, a holistic approach which combines different mitigation strategies at significant but reasonable levels (e.g., S2) of application is required. Farmers are still able to achieve a notable emission reduction of 12% under S1 without incorporating new technologies such as anti-methane additives. The results concur with studies by [62], in Ireland and [19] in New Zealand, which showed that combining climate mitigation strategies that increase production efficiencies, resulted in substantial emission reduction for grass-based systems. The business case underlines the significance of using multiple measures in reducing climate change as there are no quick fixes to achieving net zero emissions [68]. It is important to highlight that efficient use of resources can offer additional benefits, other than reducing costs and carbon dioxide emissions. A study by [62] has already shown that moving to a target-efficient system would reduce freshwater eutrophication, acidification, and non-renewable energy depletion in Irish dairy. As more consumers become more responsible for their purchasing behaviour by purchasing environmentally friendly products [59,60], and reducing food waste and losses, this case study shows that there are potential positive ripple effects of holistic sustainable production [69]. Such solutions are important to achieve all three dimensions of environmental, economic and social sustainability [69].

Opportunities for Further Emission Reduction

While work remains to further reduce the emission footprint of the farm towards net zero, Shinagh Farm has several planned activities which can help to improve these scenarios. The project plans to implement a grass biorefinery and anaerobic digestion plant in 2024. The benefits of grass biorefinery to improve the use of grassland on dairy farms have previously been highlighted in Ireland through projects such as Biorefinery Glas and FZC. In this approach, fresh grass can be converted into multiple protein sources, including a press cake which is suitable for feeding ruminants and a leaf protein concentrate (LPC) which is suitable for feeding monogastric animals, such as pigs and poultry. Previous work from [70] has highlighted the potential for press cake to replace silage in dairy cow diets, achieving comparable milk yields, while offering a reduction in nutrient (nitrogen and phosphorus) excrement losses and delivering a higher nitrogen use efficiency. Work by [71,72] has shown that the extracted protein LPC can serve as a suitable replacement for imported soya bean meal in the diets of pigs. By creating "off-farm" products, the biorefinery approach can help the farm to achieve further environmental benefits by enabling a redistribution of the environmental impacts associated with grassland production. The inclusion of anaerobic digestion to produce biogas using farm residual streams is also expected to add further improvements to the current scenario. For example, ref. [73] has previously shown that small-scale anaerobic digestion of cattle slurry, co-digested with some grass from

Irish farms, can meet the farm's energy needs with surplus energy exported, representing between 73% and 79% of the total energy generated, with all scenarios investigated offering a net CO_2 emission reduction of approximately 173,237 kg CO_2-eq.yr^{-1}. In addition to the slurry, the residual streams or by-products such as grass whey and press cake from the grass biorefinery can also be utilised as a feedstock for biogas production, helping to further improve the sustainability and circularity of the farm model [72]. It is anticipated that a further 0.3 to 1.1 tonne CO_2/ha/yr reduction can be obtained from carbon sequestration from grassland management practices already employed at FZC as well as hedgerows.

5. Conclusions

To meet the GHG targets at the national and EU level without jeopardizing the economic viability of the sector, Irish agriculture needs to adopt practical climate mitigation strategies. Using partial budget analysis and LCA assessment to measure the change in farm profit and GHG emissions under different scenarios, the business case for an average Irish dairy farm was formed based on the FZC holistic approach. The FZC approach reiterates the importance of adopting win–win approaches also highlighted in the Teagasc MACC curve, such as the inclusion of clover, protected urea, slurry management and reducing feed concentrates immediately, as they result in lower operational costs. Evidence from S1 shows that by implementing these win–win solutions even at a small scale, a 16% reduction in GHG emissions can be achieved. Incremental costs are realised especially from methane additives, slurry amendments and the use of native feeds. Biogenic methane is the major GHG in grass-based systems; therefore, the use of methane additives for emission reduction should be considered a priority. As highlighted by the business case, anti-methane additives are costly. Subsidies or other financial policy incentives should be considered to foster the uptake of additives, especially during the period when the animals are housed as the additives are most effective.

While significant sustainability improvements can be achieved by implementing the current farm-level mitigation strategies at a higher scale (S2), these steps alone may not be sufficient to achieve climate neutrality, as shown by the reduction to 0.614 kg CO_2-eq/kg FPCM in S2. This means that there is a need for more research into additional climate mitigation measures in order to reach net zero emissions on the farm. Targeting net zero ensures that the environmental sustainability goals are achieved without compromising food security. More research should be invested towards the MRV of soil carbon sequestration potential of grasslands and hedgerows so that the contribution of soil organic carbon could be incorporated in future business cases. Other ways to further reduce the emissions include anaerobic biodigesters for renewable energy, and the implementation of biorefineries to improve the efficiency of grassland use. Consumers will also be crucial in driving the demand for climate-neutral agriculture; therefore, consumer-side policies should be aimed at increasing awareness of the climate change challenge. In addition, multi-actor partnerships would be crucial in the dissemination of information on climate change mitigation across the agricultural sector. Stakeholders like producer associations, dairy companies, cooperatives and advisory organisations should continue to advise farmers on low-carbon farming. The holistic approach to sustainable agricultural production can be instrumental in achieving other SDGs including food security, responsible production and consumption, and life in water and on the ground.

Author Contributions: Conceptualization, K.O., J.G., E.B. and F.M.; methodology, L.A.V., F.M., T.R. and G.H.; validation, K.O., J.G., F.M., E.B. and G.H.; formal analysis, T.R. and L.A.V.; investigation, data curation, G.H., T.R. and L.A.V.; writing—original draft preparation, T.R., L.A.V. and G.H.; writing—review and editing, K.O., J.G., F.M. and E.B.; visualization, L.A.V.; supervision, J.G.; project administration, G.H. and J.G.; funding acquisition, K.O., F.M., J.G. and E.B. All authors have read and agreed to the published version of the manuscript.

Funding: This research was funded through the Farm Zero C project by Science Foundation Ire-land's Zero Emissions Challenge, grant number 19/FIP/ZE/7558. The authors would like to acknowledge this funding.

Institutional Review Board Statement: Not applicable.

Informed Consent Statement: Not applicable.

Data Availability Statement: Data is available upon a formal request.

Conflicts of Interest: Author E.B. is employed by Carbery group and the company was a partner in the FZC project.The authors declare no conflicts of interest.

References

1. Department of Agriculture Food and the Marine. Fact Sheet on Irish Agriculture. 2023. Available online: https://assets.gov.ie/246024/593efa4b-d404-41c2-8-08-eb459e1326cc.pdf (accessed on 11 September 2023).
2. Gaffey, J.; McMahon, H.; Marsh, E.; Vehmas, K.; Kymäläinen, T.; Vos, J. Understanding Consumer Perspectives of Bio-based Products—A Comparative Case Study from Ireland and the Netherlands. *Sustainability* **2021**, *13*, 6062. [CrossRef]
3. Department of the Environment, Climate and Communication. A Waste Action Plan for a Circular Economy. 2020. Available online: https://assets.gov.ie/8647/dcf554a4-0fb7-4d9c-9714-0b1fbe7dbc1a_3.pdf (accessed on 15 January 2024).
4. Ali, S.M.; Appolloni, A.; Cavallaro, F.; D'Adamo, I.; Di Vaio, A.; Ferella, F.; Gastaldi, M.; Ikram, M.; Kumar, N.M.; Martin, M.A.; et al. Development Goals towards Sustainability. *Sustainability* **2023**, *15*, 9943. [CrossRef]
5. Duffy, P.; Black, D.; Fahey, B.; Hyde, A.; Kehoe, E.; Kent, T.; MacFarlane, B.; Monoghan, J.; Murphy, J.; Ponzi, J.; et al. Ireland's National Inventory Report 2023. 2023. Available online: https://www.epa.ie/publications/monitoring--assessment/climate-change/air-emissions/NIR-2023-Final_v3.pdf (accessed on 21 October 2023).
6. Government of Ireland. CLIMATE ACTION PLAN 2021 Securing Our Future. 2021. Available online: https://assets.gov.ie/203546/a183a324-40ed-49c9-b630-bab0fbdd2ce2.pdf (accessed on 4 April 2023).
7. Department of Environment, Climate and Communication. Ireland's Long-Term Strategy on Greenhouse Gas Emissions Reduction. 2023. Available online: https://assets.gov.ie/255743/35b2ae1b-effe-48af-aaf3-156dc5b01ee6.pdf (accessed on 12 October 2023).
8. Buckley, C.; Donnellan, T.; Moran, B.; Lennon, J.; Brennan, J.; Colgan, J.; Curley, A.; Deane, L.; Doyle, T.; Harnett, P.; et al. *Teagasc National Farm Survey 2021 Sustainability Report*; Teagasc, Athenry, Co.: Galway, Ireland, 2022.
9. Central Statitics Office. Environmental Indicators Ireland: Irish Land Use Categories 1990–2020. Available online: https://www.cso.ie/en/releasesandpublications/ep/p-eii/environmentalindicatorsireland2022/landuse/ (accessed on 2 July 2023).
10. O'Mara, F.; Richards, K.G.; Shalloo, L.; Donnellan, T.; Finn, J.A.; Lanigan, G. Sustainability of Ruminant Livestock Production in Ireland. *Anim. Front.* **2021**, *11*, 32–43. [CrossRef] [PubMed]
11. Ramsbottom, G.; Horan, B.; Pierce, K.M.; Berry, D.P.; Roche, J.R. A Case Study of Longitudinal Trends in Biophysical and Financial Performance of Spring-Calving Pasture-Based Dairy Farms. *Int. J. Agric. Manag.* **2020**, *9*, 33–44.
12. Dillon, E.; Donnellan, T.; Moran, B.; Lennon, J. Teagasc National Farm Survey 2022. 2023. Available online: https://www.teagasc.ie/publications/2023/teagasc-national-farm-survey-2022.php (accessed on 3 October 2023).
13. Department of Agriculture, Food and the Marine. Fact Sheet on Irish Agriculture Key Indicators for Agri-Food Sectors. 2020. Available online: https://assets.gov.ie/88632/eff46189-8124-4072-9526-c49f995833b9.pdf (accessed on 5 November 2023).
14. Chen, R.; Zhang, R.; Han, H. Climate Neutral in Agricultural Production System: A Regional Case from China. *Environ. Sci. Pollut. Res.* **2021**, *28*, 33682–33697. [CrossRef] [PubMed]
15. Searchinger, T.; Zionts, J.; Wirsenius, S.; Peng, L.; Beringer, T.; Dumas, P.; Rahbek, C.; Kvist Johannsen, V.; Jacobesen, B.; Bredahl, J.; et al. A Pathway to Carbon Neutral Agriculture in Denmark. 2021. Available online: https://d30mzt1bxg5llt.cloudfront.net/public/uploads/PDFs/carbon-neutral-agriculture-denmark.pdf (accessed on 3 August 2023).
16. Kingwell, R. Making agriculture Carbon Neutral amid a Changing Climate: The Case of South-Western Australia. *Land* **2021**, *10*, 1259. [CrossRef]
17. Sanderman, J. Can Management Induced Changes in the Carbonate System Drive Soil Carbon Sequestration? A Review with Particular Focus on Australia. *Agric. Ecosyst. Environ.* **2012**, *155*, 70–77. [CrossRef]
18. Kragt, M.E.; Pannell, D.J.; Robertson, M.J.; Thamo, T. Assessing Costs of Soil Carbon Sequestration by Crop-Livestock Farmers in Western Australia. *Agric. Syst.* **2012**, *112*, 27–37. [CrossRef]
19. Yang, W.; Rennie, G.; Ledgard, S.; Mercer, G.; Lucci, G. Impact of Delivering 'Green' Dairy Products on Farm in New Zealand. *Agric. Syst.* **2020**, *178*, 102747. [CrossRef]
20. Forrestal, P.; Somers, C.; Plunkett, M.; Wall, D.; O'dwyer, T. Protected Urea: What Is It, Does It Work, and Is It Cost Effective? 2019. Available online: https://www.teagasc.ie/media/website/environment/climate-change/Andy-Boland--Patrick-Forrestal-Protected-Urea-April-2019-resized.pdf (accessed on 2 April 2023).
21. Mccarthy, K.; Mcaloon, C.; Lynch, B.; Pierce, K.; Mulligan, F.; Totty, V.; Greenwood, S.; Bryant, R.; Edwards, G. The effect of a zero-grazed perennial ryegrass, perennial rye-grass and white clover, or multispecies sward on the dry matter intake and milk production of dairy cows. *Anim. Sci. Proc.* **2021**, *12*, 96. [CrossRef]

22. Egan, M.; Galvin, N.; Hennessy, D. Incorporating White Clover (*Trifolium Repens* L.) into Perennial Ryegrass (*Lolium Perenne* L.) Swards Receiving Varying Levels of Nitrogen Fertilizer: Effects on Milk and Herbage Production. *J. Dairy Sci.* **2018**, *101*, 3412–3427. [CrossRef]
23. Farstad, M.; Melås, A.M.; Klerkx, L. Climate Considerations aside: What Really Matters for Farmers in Their Implementation of Climate Mitigation Measures. *J. Rural. Stud.* **2022**, *96*, 259–269. [CrossRef]
24. Läpple, D. Information about Climate Change Mitigation: What Do Farmers Think? *EuroChoices* **2023**, *22*, 74–80. [CrossRef]
25. Geoghegan, C.; O'donoghue, C.; Loughrey, J. The Local Economic Impact of Climate Change Mitigation in Agriculture. *Bio-Based Appl. Econ.* **2022**, *11*, 323–337. [CrossRef]
26. Lanigan, G.J.; Hanrahan, K.; Richards, K.G.; Lanigan, G.; Black, K.; Donnellan, T.; Crosson, P.; Beausang, C.; Buckley, C.; Lahart, B.; et al. *MACC 2023: An Updated Analysis of the Greenhouse Gas Abatement Potential of the Irish Agriculture and Land-Use Sectors between 2021 and 2030 Prepared by Teagasc Climate Centre*; Teagasc: Oak Park, Carlow, Ireland, 2023.
27. Lalor, S.T.J.; Schröder, J.J.; Lantinga, E.A.; Oenema, O.; Kirwan, L.; Schulte, R.P.O. Nitrogen Fertilizer Replacement Value of Cattle Slurry in Grassland as Affected by Method and Timing of Application. *J. Environ. Qual.* **2011**, *40*, 362–373. [CrossRef] [PubMed]
28. Lahart, B.; Shalloo, L.; Herron, J.; O'Brien, D.; Fitzgerald, R.; Boland, T.M.; Buckley, F. Greenhouse Gas Emissions and Nitrogen Efficiency of Dairy Cows of Divergent Economic Breeding Index under Seasonal Pasture-Based Management. *J. Dairy Sci.* **2021**, *104*, 8039–8049. [CrossRef]
29. Biorefinery Glas. Grass Biorefinery. 2020. Available online: https://biorefineryglas.eu/wp-content/uploads/2020/02/factsheet-1.pdf (accessed on 27 November 2023).
30. Visser, S.; Lhermite, E.; Keesstra, S. The Innovation Potential for the Irish Agri-Food Sector 2. Carbon Farming. 2023. Available online: https://www.climate-kic.org/wp-content/uploads/2023/03/EIT-Climate-KIC_Report_Dealing-with-climate-change-and-sustainability-targets.pdf (accessed on 30 June 2023).
31. Cahill, L.; Patton, D.; Reilly, B.; Pierce, K.M.; Horan, B. Grazing Season Length and Stocking Rate Affect Milk Production and Supplementary Feed Requirements of Spring-Calving Dairy Cows on Marginal Soils. *J. Dairy Sci.* **2023**, *106*, 1051–1064. [CrossRef] [PubMed]
32. Johnston, D.J.; Laidlaw, A.S.; Theodoridou, K.; Ferris, C.P. Performance and Nutrient Utilisation of Dairy Cows Offered Silages Produced from Three Successive Harvests of Either a Red Clover–Perennial Ryegrass Sward or a Perennial Ryegrass Sward. *Ir. J. Agric. Food Res.* **2020**, *59*, 42–55. [CrossRef]
33. O'Brien, D.; Hennessy, T.; Moran, B.; Shalloo, L. Relating the Carbon Footprint of Milk from Irish Dairy Farms to Economic Performance. *J. Dairy Sci.* **2015**, *98*, 7394–7407. [CrossRef]
34. Forrestal, P. Manure and It's Management-Focus on Cattle Slurry. In Proceedings of the Soil Fertility Conference: Optimising Soil and Fertiliser for Sustainable Grassland Management, Kilkeny, Ireland, 17 October 2018.
35. Lanigan, G.J.; Donnellan, T.; Lanigan, G.; Hanrahan, K.; Paul, C.; Shalloo, L.; Krol, D.; Forrestal, P.; Farrelly, N.; O'brien, D.; et al. *An Analysis of Abatement Potential of Greenhouse Gas Emissions in Irish Agriculture 2021–2030 Prepared by the Teagasc Greenhouse Gas Working Group Authors*; Teagasc: Oak Park, Carlow, Ireland, 2019.
36. Agritech.ie. Prepare for Calving 2023. Available online: https://agritech.ie/preparing-for-calving-2023/ (accessed on 3 August 2023).
37. Ruelle, E.; Delaby, L.; Shalloo, L.; O'Donovan, M.; Hennessy, D.; Egan, M.; Horan, B.; Dillon, P. Modelling the Effects of Stocking Rate, Soil Type, Agroclimate Location and Nitrogen Input on the Grass DM Yield and Forage Self-Sufficiency of Irish Grass-Based Dairy Production Systems. *J. Agric. Sci.* **2022**, *160*, 235–249. [CrossRef]
38. McClearn, B.; Shalloo, L.; Gilliland, T.J.; Coughlan, F.; McCarthy, B. An Economic Comparison of Pasture-Based Production Systems Differing in Sward Type and Cow Genotype. *J. Dairy Sci.* **2020**, *103*, 4455–4465. [CrossRef]
39. Yan, M.J.; Humphreys, J.; Holden, N.M. The Carbon Footprint of Pasture-Based Milk Production: Can White Clover Make a Difference? *J. Dairy Sci.* **2013**, *96*, 857–865. [CrossRef]
40. Martinez-Fernandez, G.; Duval, S.; Kindermann, M.; Schirra, H.J.; Denman, S.E.; McSweeney, C.S. 3-NOP vs. Halogenated Compound: Methane Production, Ruminal Fermentation and Microbial Community Response in Forage Fed Cattle. *Front. Microbiol.* **2018**, *9*, 1582. [CrossRef]
41. Cummins, S.; Lanigan, G.J.; Richards, K.G.; Boland, T.M.; Kirwan, S.F.; Smith, P.E.; Waters, S.M. Solutions to Enteric Methane Abatement in Ireland. *IJAFR* **2022**, *61*, 353–371. [CrossRef]
42. Melgar, A.; Lage, C.F.A.; Nedelkov, K.; Räisänen, S.E.; Stefenoni, H.; Fetter, M.E.; Chen, X.; Oh, J.; Duval, S.; Kindermann, M.; et al. Enteric Methane Emission, Milk Production, and Composition of Dairy Cows Fed 3-Nitrooxypropanol. *J. Dairy Sci.* **2021**, *104*, 357–366. [CrossRef]
43. Hanrahan, L.; McHugh, N.; Hennessy, T.; Moran, B.; Kearney, R.; Wallace, M.; Shalloo, L. Factors Associated with Profitability in Pasture-Based Systems of Milk Production. *J. Dairy Sci.* **2018**, *101*, 5474–5485. [CrossRef] [PubMed]
44. Ramsbottom, G.; Horan, B.; Berry, D.P.; Roche, J.R. Factors Associated with the Financial Performance of Spring-Calving, Pasture-Based Dairy Farms. *J. Dairy Sci.* **2015**, *98*, 3526–3540. [CrossRef] [PubMed]
45. Board Bia. Export Performance and Prospects Report 2022–2023. 2023. Available online: https://www.bordbia.ie/globalassets/bordbia.ie/industry/2022---2023-export-performance--prospects-final.pdf (accessed on 21 June 2023).
46. National Dairy Council. Dairy in a Healthy and Sustainable Irish and European Food System. *Dairy Sustainability Newsletter. Issue No 003*. 2022. Available online: https://ndc.ie/wp-content/uploads/2022/09/Dairy-Newsletter_Issue-2_Web.pdf (accessed on 17 August 2023).

47. Dillon, E.; Donnellan, T.; Moran, B.; Lennon, J. *Teagasc National Farm Survey 2020 Results Rural Economy Development Programme*; Teagasc: Athenry, Galway, Ireland, 2021.
48. Ramsbottom, G.; Cromie, A.R.; Horan, B.; Berry, D.P. Relationship between Dairy Cow Genetic Merit and Profit on Commercial Spring Calving Dairy Farms. *Animal* **2012**, *6*, 1031–1039. [CrossRef] [PubMed]
49. Soha, M.E.D. The Partial Budget Analysis for Sorghum Farm in Sinai Peninsula, Egypt. *Ann. Agric. Sci.* **2014**, *59*, 77–81. [CrossRef]
50. Jerlström, J.; Huang, W.; Ehlorsson, C.J.; Eriksson, I.; Reneby, A.; Comin, A. Stochastic Partial Budget Analysis of Strategies to Reduce the Prevalence of Lung Lesions in Finishing Pigs at Slaughter. *Front. Vet. Sci.* **2022**, *9*, 957975. [CrossRef] [PubMed]
51. *ISO 14040:2006*; Environmental Management-Life Cycle Assessment-Principles and Framework. Environmental Management System Requirements. International Organization for Standardization: Geneva, Switzerland, 2004.
52. Duffy, P.; Black, K.; Fahey, D.; Hyde, B.; Kehoe, A.; Monoghan, S.; Murphy, J.; Ryan, A.M.; Ponzi, J. *Ireland's National Inventory Report 2022*; The Environemntal Protection Agency; Johnstown Castle, Co.: Wexford. Ireland, 2022; Available online: https://www.epa.ie/publications/monitoring--assessment/climate-change/air-emissions/Ireland-NIR-2022_Merge_v2..pdf (accessed on 12 April 2023).
53. Duffy, P.; Hyde, B.; Ryan, A.M.; Alam, M.S. *Ireland's Informative Inventrory Report 2018*; The Environemntal Protection Agency; Johnstown Castle, Co.: Wexford, Ireland, 2018; Available online: https://www.epa.ie/publications/monitoring--assessment/climate-change/air-emissions/Ireland-IIR-2018_Final.pdf (accessed on 12 April 2023).
54. European Environment Agency. *EMEP/EEA Air Pollutant Emission Inventory Guidebook 2019: Technical Guidance to Prepare National Emission Inventories*; Publications Office of the European Union: Copenhagen, Denmark, 2019; Available online: https://www.eea.europa.eu/publications/emep-eea-guidebook-2019/download (accessed on 6 April 2023).
55. Intergovernmental Panel on Climate Change. Refinement to the 2006 IPCC Guidelines for National Greenhouse Gas Inventories. *Chapter 11 N2O Emissions from Managed Soils and CO2 Emissions from Lime and Urea Application*. 2019. Available online: https://www.ipcc-nggip.iges.or.jp/public/2006gl/pdf/4_Volume4/V4_11_Ch11_N2O&CO2.pdf (accessed on 13 April 2023).
56. Intergovernmental Panel on Climate Change. Refinement to the 2006 IPCC Guidelines for National Inventories, Chapter 10 Emissions from Livestock and Manure Management 2019. 2019. Available online: https://www.ipcc-nggip.iges.or.jp/public/2019rf/pdf/4_Volume4/19R_V4_Ch10_Livestock.pdf (accessed on 13 April 2023).
57. Shortall, O.K.; Lorenzo-Arribas, A. Dairy Farmer Practices and Attitud es Relating to Grass-Based, High-Feed-Input, and Indoor Production Systems in Ireland. *J. Dairy Sci.* **2022**, *105*, 375–388. [CrossRef] [PubMed]
58. KPMG. Agri-Food 2030: How the Sector Can Turn Disruption into Opportunity Your Partner for What's Next. 2022. Available online: https://www.epa.ie/our-services/monitoring--assessment/climate-change/ghg/agriculture/ (accessed on 4 September 2023).
59. Board Bia. Buyer Guide: Grass Fed Dairy from Ireland Buyer Guide: Grass Fed Dairy from Ireland 2. 2022. Available online: https://www.bordbia.ie/globalassets/bordbia2020/industry/sector-profile-blocks--images/dairy-brochures/buyer-guide---grass-fed-irish-dairy.pdf (accessed on 4 September 2023).
60. Canavari, M.; Coderoni, S. Consumer Stated Preferences for Dairy Products with Carbon Footprint Labels in Italy. *Agric. Food Econ.* **2020**, *8*, 1–16. [CrossRef]
61. O'Brien, D.; Shalloo, L.; Patton, J.; Buckley, F.; Grainger, C.; Wallace, M. A Life Cycle Assessment of Seasonal Grass-Based and Confinement Dairy Farms. *Agric. Syst.* **2012**, *107*, 33–46. [CrossRef]
62. Herron, J.; O'Brien, D.; Shalloo, L. Life Cycle Assessment of Pasture-Based Dairy Production Systems: Current and Future Performance. *J. Dairy Sci.* **2022**, *105*, 5849–5869. [CrossRef]
63. Van Wesemael, D.; Vandaele, L.; Ampe, B.; Cattrysse, H.; Duval, S.; Kindermann, M.; Fievez, V.; De Campeneere, S.; Peiren, N. Reducing Enteric Methane Emissions from Dairy Cattle: Two Ways to Supplement 3-Nitrooxypropanol. *J. Dairy Sci.* **2019**, *102*, 1780–1787. [CrossRef]
64. Yu, G.; Beauchemin, K.A.; Dong, R. A Review of 3-Nitrooxypropanol for Enteric Methane Mitigation from Ruminant Livestock. *Animals* **2021**, *11*, 3540. [CrossRef]
65. Beauchemin, K.A.; Ungerfeld, E.M.; Eckard, R.J.; Wang, M. Review: Fifty Years of Research on Rumen Methanogenesis: Lessons Learned and Future Challenges for Mitigation. In *Animal*; Cambridge University Press: Cambridge, UK, 2020; Volume 14, pp. S2–S16. [CrossRef]
66. Rivera, J.E.; Chará, J. CH_4 and N_2O Emissions From Cattle Excreta: A Review of Main Drivers and Mitigation Strategies in Grazing Systems. *Front. Sustain. Food Syst.* **2021**, *5*, 657936. [CrossRef]
67. Misselbrook, T.; Hunt, J.; Perazzolo, F.; Provolo, G. Greenhouse Gas and Ammonia Emissions from Slurry Storage: Impacts of Temperature and Potential Mitigation through Covering (Pig Slurry) or Acidification (Cattle Slurry). *J. Environ. Qual.* **2016**, *45*, 1520–1530. [CrossRef]
68. Emmet-Booth, J.P.; Dekker, S.; O'brien, P. Climate Change Mitigation and the Irish Agriculture and Land Use Sector. 2019. Available online: https://www.climatecouncil.ie/councilpublications/councilworkingpaperseries/Working%20Paper%20No.%205.pdf (accessed on 6 February 2023).
69. D'Adamo, I.; Desideri, S.; Gastaldi, M.; Tsagarakis, K.P. Sustainable Food Waste Management in Supermarkets. *Sustain. Prod. Consum.* **2023**, *43*, 204–216. [CrossRef]
70. Serra, E.; Lynch, M.B.; Gaffey, J.; M Sanders, J.P.; Koopmans, S.; Markiewicz-Keszycka, M.; McKay, Z.C.; Pierce, K.M. Biorefined Press Cake Silage as Feed Source for Dairy Cows: Effect on Milk Production and Composition, Rumen Fermentation, Nitrogen and Phosphorus Excretion and in Vitro Methane production. *Livest. Sci.* **2022**, *267*, 105135. [CrossRef]

71. Gaffey, J.; O'Donovan, C.; Murphy, D.; O'Connor, T.; Walsh, D.; Vergara, L.A.; Donkor, K.; Gottumukkala, L.; Koopmans, S.; Buckley, E.; et al. Synergetic Benefits for a Pig Farm and Local Bioeconomy Development from Extended Green Biorefinery Value Chains. *Sustainability* **2023**, *15*, 8692. [CrossRef]
72. Ravindran, R.; Koopmans, S.; Sanders, J.P.M.; McMahon, H.; Gaffey, J. Production of Green Biorefinery Protein Concentrate Derived from Perennial Ryegrass as an Alternative Feed for Pigs. *Clean Technol.* **2021**, *3*, 656–669. [CrossRef]
73. O'Connor, S.; Ehimen, E.; Pillai, S.C.; Lyons, G.; Bartlett, J. Economic and Environmental Analysis of Small-Scale Anaerobic Digestion Plants on Irish Dairy Farms. *Energies* **2020**, *13*, 637. [CrossRef]

Disclaimer/Publisher's Note: The statements, opinions and data contained in all publications are solely those of the individual author(s) and contributor(s) and not of MDPI and/or the editor(s). MDPI and/or the editor(s) disclaim responsibility for any injury to people or property resulting from any ideas, methods, instructions or products referred to in the content.

sustainability

MDPI

Article

Exploring the Impact of the Sustainable Development Goals on Sustainability Trends

Eduardo Ordonez-Ponce

Faculty of Business, Athabasca University, Athabasca, AB T9S 3A3, Canada; eduardo.ordonez@athabascau.ca

Abstract: The SDGs have made a significant contribution to the sustainability movement, being used by many organisations from across sectors all over the world as their sustainability framework. However, have they impacted the previous trend of sustainability challenges just because of their existence? This article aims to contribute to answering this question by statistically comparing the trends in the sustainability performance of the SDGs before and after they were launched in 2015. Data were collected for every SDG and their trends were quantitatively assessed using non-parametric tests, finding that most of the SDGs have not significantly improved and that most of the sustainability indicators are still performing poorly in developing countries. While this research is exploratory and does not assess the direct impact of the SDGs on sustainability, it suggests that for the most part, the SDGs have not significantly changed sustainability trends since they were launched in 2015, which is a concerning finding. This article should serve as a wake-up call to design more suitable sustainability frameworks as the SDGs expire in 2030, and for those using them to be more critical of their reach rather than being satisfied with a framework that although helping will not achieve its main goal.

Keywords: SDGs; sustainability trends; SDG progress

Citation: Ordonez-Ponce, E. Exploring the Impact of the Sustainable Development Goals on Sustainability Trends. *Sustainability* **2023**, *15*, 16647. https://doi.org/10.3390/su152416647

Academic Editors: Idiano D'Adamo, Massimo Gastaldi and Manoj Kumar Nallapaneni

Received: 10 November 2023
Revised: 30 November 2023
Accepted: 5 December 2023
Published: 7 December 2023

Copyright: © 2023 by the author. Licensee MDPI, Basel, Switzerland. This article is an open access article distributed under the terms and conditions of the Creative Commons Attribution (CC BY) license (https://creativecommons.org/licenses/by/4.0/).

1. Introduction

The Sustainable Development Goals (SDGs), launched in 2015 by the United Nations in agreement with 195 countries, have been classified as the greatest agreement ever achieved [1], and many academics, universities, corporations, governments and NGOs are following their lead under the seeming assumption that by doing so they are contributing to local and global sustainability. For some, there has been a shift in the sustainability movement since the SDGs were launched [2,3], as an important milestone in the sustainability journey. However, even though a responsibility cannot be directly assigned to them, it is important to assess how successful they have been in contributing to addressing the sustainability challenges they have identified as being crucial to be tackled. How much has really changed since the SDGs were launched? How have the previous trends in poverty reduction, economic output or biodiversity loss, to name a few, been impacted by the implementation of the SDGs? While the direct relationship between the existence of the SDGs and any progress or deterioration in sustainability challenges cannot be established, finding that things have not really improved would stain the impact and importance that many of us have assigned to the SDGs. Certainly, any sustainability challenge cannot be overcome just because the SDGs were launched, have been promoted and are widely used as a sustainability framework. But also, since they are ubiquitous nowadays, it is fair to ask if sustainable development has really improved thanks, although in part, to their existence. Answering this question is the objective of this article. By not knowing this, there is a serious risk of using their name as a way to greenwash our actions [4,5], and as stated by Ali and colleagues, "sustainable washing does not solve problems" but integrated solutions are what is required [6] (p. 3). This is one of the reasons why answering the simple question of whether they have shifted the trend in sustainable development is necessary. This research aims to explore this question and assess whether there have been significant changes in sustainability challenges thanks to the existence of the SDGs and their implementation. Furthermore, this article is a humble invitation to carefully assess the real weight of the SDGs in the sustainability journey, which is subject to their value and limitations. It is an invitation to those of us conducting research, developing policies, designing strategies or undertaking simple but honest actions to realise that their existence is not enough to contribute to sustainable development and that we must stay away from any potential "SDG washing" if we really want to make deep contributions to our development. This can be considered the immediate implication of this research, but perhaps more importantly, the aim is to use these findings and the discussion we hope to create with them to think ahead of time about the future sustainability framework or strategy we will aim to implement starting in 2030, when the SDGs expire, leaving, most likely, a lot of work still to do.

While the question proposed by this article may seem an obvious one to address, academic work has been mostly focused on the positive value of the SDGs trying to understand how helpful they and every one of them has been in achieving their aim, how they have been used, where and by whom. All certainly contribute to sustainability science and knowledge [6]. However, although it is hard to justify the limited existence of other studies like this, which take a broader and more critical approach, it is crucial to question our own approaches and the frameworks we use so we do not become unconsciously trapped in our own beliefs and short-term agendas, but look to the future and contribute to real change.

2. Literature

The SDGs are about halfway through their span, so it is a good moment to assess whether there have been any changes in sustainability trends since their launch in September 2015. Indeed, assessing their impact on improving the current state of the identified goals can only be made indirectly, since these are such complex challenges that no organisation would be able to tackle alone successfully [7,8]. However, they have been identified as crucial [1] and as causing a shift in the sustainability journey [2], but it cannot be argued that just by their existence that things have changed, just as it is hard to directly link what organisations have done, or declare doing, in terms of contributing to these achievements, with any potential progress [9]. Despite that, partly due to the gap in the literature it is necessary to explore what has happened with the identified sustainability challenges once the SDGs were launched versus what was happening with them before 2015. As their deadline approaches and indicators show that they will not be reached by 2030 [10,11], assessing the success of their existence would help us and local and global policymakers to prepare for 2030 and replace them with a more appropriate and realistic approach.

2.1. SDG Research and Practice

The literature shows that the SDGs have been studied widely in specific and general terms, with a constantly growing number of articles being published on the topic over the years [6]. Across continents, types of organisations and their relevance, there is a global agreement on and a desire to achieve sustainable development [6]. Although some have already argued that they will hardly be achieved by 2030 and provide recommendations to shift route [10], which was probably expected due to the major endeavour they still represent, certainly some improvements at least are expected. But not only that, despite some initial optimism [10] and all the good parts of the SDGs [3,12], including how they could allow corporations to shift their strategies towards specific goals [13], many have highlighted their complexities, limitations and contradictions, e.g., [14–18]. Among them, scholars highlight that they are not legally binding without presenting any obligations, are voluntary so self-interest may be a driver leading to their selection and engagement and that there are no mechanisms for any organisation to innovate in their sustainability agenda [19,20]. Perhaps more challenging are what some have called their inherent contradictions between achieving economic growth (SDG#8) and good health and well-being

(SDG#3), climate action (SDG#13) or protecting life on land (SDG#14), if we continue following the same socio-economic paradigm [21]. On top of these, we cannot forget the impact that the COVID-19 pandemic had on their progress [22,23], along with other crises we are still facing, e.g., the climate crisis, immigration, wars and political polarisation [11]. These have made things even more challenging, as the same challenges we are trying to address hinder the achievement of the SDGs.

Despite the impression that the SDGs have been adopted by many organisations across the world, it is hard to assess the extent of their contributions and the impact of their actions. For example, it is relatively easy to find multinational corporations referring to the SDGs as a strategic framework to structure their sustainability actions [9,24,25], governments at different levels informing how they have incorporated the SDGs into their agendas, e.g., [26–28], collaborative academic initiatives integrating the SDGs into their teaching and research programs, e.g., [29,30] and several other organisations who have adjusted their work in line with the SDGs, e.g., [31–33]. While these are presented as independent initiatives, they depend on each other to ensure some kind of progress and impact on sustainability [34,35]. While the SDGs seem to be everywhere nowadays, the real impact of organisations declaring them as a priority can hardly be measured.

2.2. SDG Impact and Progress

As mentioned, the literature is limited with respect to measuring and monitoring not only the impact of the SDGs on sustainability but the SDGs' general impact, which is still a necessity [36]. When addressing these areas, what is found is mostly at the initial stage of the process and not measuring concrete impact, but rather proposing tools, frameworks or ideas for organisations to operationalise their potential impact on the SDGs, which again speaks to their complexity. For example, B Lab created an SDG Action Manager to help organisations measure their progress [35], although according to Heras-Saizarbitoria et al. [37], it has been limited in its use; SDGs-IAE was developed to qualitatively assess SDG targets on energy projects [38]; van Zanten and van Tulder [39] proposed a nexus approach for companies to manage and assess their contributions to the SDGs in an integrated manner and not in silos; and the World Business Council for Sustainable Development [33] developed the initiative SDG Sector Roadmaps to support businesses in optimising their contributions to the SDGs. Research that has focused on measuring the performance of the SDGs has done so, for example, at the country level, highlighting how their relative position against others is subject to the selected methods and indicators [40], in particular sectors such as manufacturing [41], with respect to firms' ESG performance indicators [42], versus multinationals' GRI disclosure standards [43] or relative to their organisational profile according to whether they focus on environmental or social goals [44].

As stated in the literature, if we aim to contribute to the design of better instruments or frameworks to contribute to sustainable development, measuring and monitoring not only the progress of the SDGs but their progress towards sustainability are crucial [45,46]. However, this type of work has been limited to our knowledge. One of the few articles that tries to understand whether firms changed their practices' breadth and depth due to the SDGs is that of Whittingham et al. [9], who found a "clear but nuanced yes" to their question. However, this is based on an analysis of corporate sustainability reports without any verification of what they found in them. Another recent work is that of D'Adamo and Gastaldi [47] who used a methodological approach to evaluate the SDGs focusing on monitoring their outcomes. While this presents a useful methodological design, its scope is the performance of Italian regions, with relevant contributions to their particular sustainability challenges.

Not surprisingly then, the sustainability challenges highlighted by the SDGs have poorly progressed since they were launched in 2015. According to the latest Sustainable Development Report, not a single SDG is on track to be achieved by 2030, there has been limited progress on environmental and biodiversity goals, health-related goals are off-track, as well as those concerning housing and institutions [11]. Some progress has been made though with respect to infrastructure, including water and energy; however, this varies largely across countries [11]. According to the UN, less than 20% of the SDG targets have been achieved, with about 15% in reversal progress and the rest having limited or no progress at all. Hence, what has been the real value of the SDGs other than being a common framework we use to position and talk about sustainability?

2.3. SDG Progress and Data Selection

Identifying which targets to address to assess progress on the SDGs is certainly challenging, especially when we aim to measure their performance globally, whereas countries are prioritising actions towards certain SDGs according to their own challenges [12]. According to scholars, another important challenge is how to consider the existing range of approaches and tools, including indicator-based assessments, benchmarking, target mapping or systems analysis techniques [48–50], to support a coherent analysis towards evidence-based decision making when they inform different stages of policy again based on country-level goals [51]. The challenge of which targets to prioritise under the statement that the SDGs and their targets are interconnected has been addressed by some scholars but, as expected, this is always context-specific through analyses in a particular country, not globally [51–54]. Furthermore, while for countries there is a risk in their priorisation techniques due to potentially arbitrary or politically motivated decisions [51], this article does not aim to provide hard conclusions from its findings but to shed some light into sustainability progress, accepting that assessing progress on the SDGs is subject to the selection of targets and assessment methods [52]. Finally, there is the practical component since the SDG targets are 169, which makes it an extensive number to manage and for many of them there is just none or not enough data from reliable sources to determine whether their trends have significantly changed from before the SDGs were launched to now [55], which is the aim of this article.

3. Methods

To achieve the purpose of this article and answer the proposed research question, the SDGs were considered as a framework and, based on their targets, specific variables were identified and assessed to determine progress or failures on the goals, similar to what D'Adamo and Gastaldi carried out in Italy [47]. Following this rationale, secondary data were collected from various reliable sources that presented information from the year 2000 to 2022 as a way to assess the periods pre- and post-launch of the SDGs. Additionally, since the SDGs are measured at the country level [16], the data had to be organised per country and then clustered into geographic regions to make numbers more manageable. Then, two groups of data were prepared for analysis, the pre-SDG (2000–2015) and post-SDG datasets (2016–2022), similar to what Whittingham and colleagues did [9]. For example, to assess the progress of SDG#1—No Poverty, poverty rate per country was selected as the variable to be assessed. SDG#17—Partnerships for the Goals was not included in the assessment since this is not a sustainability goal per se but an approach to achieve the other sixteen goals [47,56].

The general hypothesis is that significant statistical changes have not occurred since the SDGs were launched with respect to the trends those goals followed before. Since tests showed that the data were not normally distributed, non-parametric paired tests (Wilcoxon Signed Ranked Tests) were used to determine whether statistically significant differences existed between the two proposed periods [57]. Table 1 shows a summary of the dataset and the selected variables for analysis. As stated above, results from this and any other analysis on SDGs' progress depend on the selected targets [52], so the targets selected in this case were the ones that to the best of our knowledge best represent the respective SDGs.

Table 1. The SDGs, assessed variables, and sources of data.

SDG	Assessed Variables—Associated SDG Target	Source
1—No Poverty	Poverty rate (%)—1.1	[58]
2—Zero Hunger	Prevalence of undernourishment rate (% of population)—2.1.1	[59]
3—Good Health and Well-Being	Maternal mortality rate per 100,000 live births—3.1.1	[60]
4—Quality Education	Out-of-school children rate—4.1	[61]
5—Gender Quality	Female education (%) Higher Level—5.1	[62]
6—Clean Water and Sanitation	Drinking water safely managed service (%)—6.1	[63]
7—Affordable and Clean Energy	Share of modern renewables in total final energy consumption (%)—7.2.1	[64]
8—Decent Work and Economic Growth	GDP annual growth per capita (USD)—8.1.1	[58]
9—Industry, Innovation and Infrastructure	Research and Development expenditure as a percentage of GDP—9.5.1	[65]
10—Reduced Inequalities	Gini index—10.3	[58]
11—Sustainable Cities and Communities	Population living in slums (% of urban population)—11.1.1	[66]
12—Responsible Consumption and Production	Fossil fuels consumption (tons)—12.2.2	[67]
13—Climate Action	CO_2 emissions (metric tons per capita)—13.2	[68]
14—Life Below Water	Mean percentage of each marine key biodiversity area covered by protected area—14.5	[69]
15—Life On Land	Tree cover loss (ha)—15.2	[70]
16—Peace, Justice and Strong Institutions	Rule of law index—16.3	[71]

4. Results

As seen from Table 2 and Figure 1, the results suggest that most of the sustainability goals (10/16) have not significantly improved after the SDGs were launched (2000–2015 vs. 2016–2022), indicating that there has not been significant progress on most of the goals since the SDGs were launched in 2015. While six sustainability goals were found to have statistically significant improvements ($p < 0.05$), namely SDG#2—Zero Hunger, SDG#3—Good Health and Well-Being, SDG#6—Clean Water and Sanitation, SDG#7—Affordable and Clean Energy, SDG#11—Sustainable Cities and Communities, and SDG#14—Life Below Water; eight goals did not change significantly ($p > 0.05$), despite most of them having improved; and two goals showed a statistically significant decline (SDG#8—Decent Work and Economic Growth, and SDG#15—Life on Land), which is very concerning. When crossing these findings with how the SDGs are organised in the wedding cake proposed by Rockström and Sukhdev [56] so we can understand them better, it can be seen that while economic sustainability has progressed (the top layer of the cake), this has not been statistically significant (industry and inequalities) and decent work and economic growth has significantly decreased; social sustainability (middle layer) shows half of its SDGs with significant progress (hunger, health, energy and cities), whereas the other 50% were found to have improved but not significantly (poverty, education, gender equality and peace); and environmental sustainability (bottom layer of the cake) shows half of those SDGs significantly improving (water and oceans), but land biodiversity decreased significantly and climate did not improve. Figure 1 shows the SDGs and their progress, stagnation or decline. Details of test results can be found in Appendix A.

Table 2. Results on SDG progress.

SDG	Assessed Variable	Improvement (Global Level)	Exception (Regions)	Sig Dif ($p < 0.05$)
1—Poverty	Poverty rate	Yes	Middle East and Northern Africa	No
2—Hunger	Undernourishment rate	Yes	Middle East and Northern Africa	Yes
3—Health	Maternal mortality rate	Yes	-	Yes
4—Education	Out-of-school children rate	Yes	East Asia and the Pacific, Eastern Europe and Central Asia, Latin America and the Caribbean	No
5—Gender	Female education rate (tertiary education)	Yes	Latin America and the Caribbean	No
6—Water	Drinking water safely managed	Yes	-	Yes
7—Energy	Modern renewable energy consumption share	Yes	Middle, Northern and Western Africa, and Western Asia	Yes
8—Economy	GDP per capita annual growth	No	Sub-Saharan Africa	Yes
9—Innovation	Research and development expenditure (% of GDP)	Yes	East Asia and the Pacific, Europe and Central Asia, European Union, and South Asia	No
10—Inequality	Gini index	Yes	South Asia	No
11—Cities	Population living in slums (% of urban population)	Yes	-	Yes
12—Consumption and Production	Fossil fuels use	No	Europe	No
13—Climate	CO_2 emissions per capita	No	Central Europe and the Baltics, Europe and Central Asia, European Union, North America	No
14—Oceans	Protected marine areas	Yes	-	Yes
15—Land Biodiversity	Tree cover loss	No	North America	Yes
16—Justice	Rule of law index	Yes	East Asia and the Pacific, Eastern Europe and Central Asia	No

With respect to where the SDGs are performing better or worse, in general terms, the developing world is struggling, which is where sustainability challenges are most severe [16]. When assessing the SDGs that have significantly improved (SDG#2—Hunger, SDG#3—Health, SDG#6—Water, SDG#7—Energy, SDG#11—Cities, and SDG#14—Oceans), results suggest some exceptions in the Middle East and Northern and Western Africa, although these are significantly better on water, cities and oceans. Similarly, countries from East Asia and the Pacific, and Eastern Europe and Central Asia do not show significant improvements in quality of education (SDG#4), industry and innovation (SDG#9) or justice (SDG#16); and countries from Latin America and the Caribbean have not significantly improved their education (SDG#4) nor their gender equality levels (SDG#5), just like South Asian nations on inequality (SDG#10). On the contrary, Europe has improved, although not significantly, its fossil fuel use (SDG#12) and its CO_2 emissions per capita (SDG#13) as North American countries have, which have also reduced their tree cover loss (SDG#15). Details are in Appendix A.

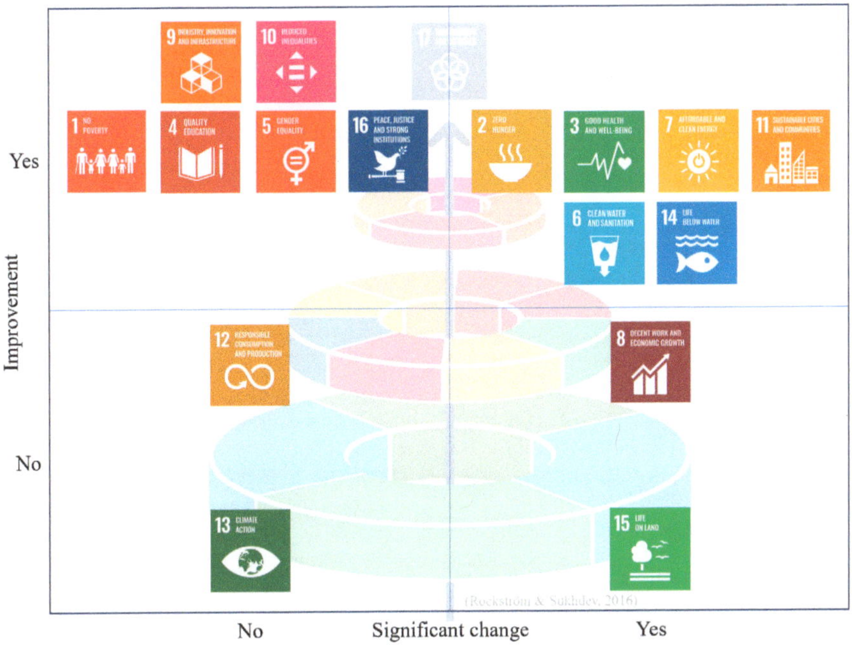

Figure 1. SDGs organised according to their significance of progress (based on Rockström and Sukhdev (2016) SDG wedding cake [56]).

5. Discussion

The purpose of this article is to assess whether the trend that sustainability challenges followed before the SDGs were launched in 2015 has significantly changed over the years. According to some, the SDGs have changed the discourse around sustainability, but has that been enough to modify the trend of development for the better? Indeed, since the SDGs were launched, we have seen companies of all sizes, governments, universities and civil society organisations using the SDGs as the framework to consider when referring to their sustainability policies and practices. However, it is still not clear whether the trends followed by the sustainability challenges have significantly changed due to the very existence of the SDGs, or if these have been used just as a way to organise what organisations do, present it more clearly, and somehow continue doing what they were doing without any emphasis nor accepting the urgency that Agenda 2030 and the SDGs posess.

In general terms, results from this exploratory analysis suggest that most of the sustainability challenges have not significantly progressed since the SDGs were launched, confirming what others have stated in terms of their limited progress and low probability of success [10,11], which is an alarming finding. Following the rationale proposed by Rockström and Sukhdev [56], we see that only half of the SDGs for the supporting system, i.e., the biosphere, have improved significantly (SDG#6—Clean Water and Sanitation, and SDG#14—Life Below Water), while the other two have deteriorated—life on land significantly so, putting at risk not only the accomplishment of the societal and economic SDGs that these are supporting but also the sustainability of the planet and our ways of life. In this respect, we found that the trend of tree cover loss has significantly increased, along with all the impact that it has on terrestrial biodiversity, and that CO_2 emissions per capita have increased despite global agreements such as the Paris Accord, although the latter not significantly. These results are aligned with the latest version of the planetary boundaries, which shows that by now, we have passed not three, like in 2009, nor four, as in 2015, but six of the nine identified boundaries to continue living on a safe planet [72].

Results also suggest that while half of the SDGs focused on the economy have improved, none of them have changed significantly. Conversely, SDG#8—Decent Work and Economic Growth and SDG#12—Responsible Consumption and Production have not improved as measured through GDP annual growth per capita and fossil fuel use, respectively. The former has actually decreased significantly globally. These findings question why the improvements made by industry, innovation and infrastructure (SDG#9) have not led to an increase in decent work and economic growth (SDG#8) as well as the development of more responsible consumption and production practices (SDG#12). The lack of statistical significance of the improvements made by the industry, innovation and infrastructure (SDG#9) is a way to explain the poor performance of SDG#8 on decent work and economic growth and SDG#12 on responsible consumption and production. It seems that the increase made in R&D expenditure has not yet translated into economic growth, nor a reduction or shift from fossil fuel consumption. Perhaps more time is needed for these to further complement and affect each other. However, due to the complexity of these challenges, if positive results are reached, they may come well after 2030 when the SDGs were supposed to be accomplished. This is a task left for the new framework.

Finally, hopeful news is that four out of the eight SDGs that are society-focused have significantly improved, namely a reduction in the undernourishment rate (SDG#2) and the maternal mortality rate (SDG#3), a higher share of modern renewable energy consumption (SDG#7) and a reduction of slums (SDG#11). However, these improvements are not aligned with significant advancements in reducing poverty (SDG#1), improving the quality of education (SDG#4), progress in gender equality (SDG#5) or the betterment of justice across the world (SDG#16). Again, the limited and challenging timeline set to achieve the SDGs is not helping, as we hope to see, for example, how better nutrition and living conditions should positively impact poverty and education. These are results we hope to see within the timeline of the next sustainability framework, but we at least hoped to find positive trends. These are examples of how the three dimensions of sustainability are not independently progressing, which should be directly related. But perhaps more important and challenging is how all the SDGs are coherently integrated into a successful process that allows them all to flourish, which is one of the criticisms they have faced as they aim to accomplish goals that seem contradictory [21], which is an even tougher challenge and something also confirmed in these results. How to tackle the contradictions presented by some SDGs is one of the biggest tests to address.

Undoubtedly, all the SDGs are complex challenges that not only need the collective participation of all actors to be accomplished, but they also need time, and 15 years does not seem to be enough for them to be reached, let alone half that time. However, whether we achieve them or not by 2030, positive trends were expected to be found due to the urgency of addressing them for the betterment of humanity. All the actions performed by diverse organisations across the world on the SDGs can be hardly monitored or measured, other than referring to what they say they do towards the SDGs, so this research aims to shed some light on that and indirectly determine whether we as humanity have been doing more than what we were doing before the SDGs were launched in 2015 to reach our sustainability challenges, or we have just followed the same approach or even declined in our efforts. However, despite that challenge, measuring and monitoring our performance on sustainability goals remains a need [36], which, although subject to selected methods and indicators [40,52], will be a key consideration [45,46] as we start thinking about the new sustainability framework to take place when the SDGs expire in 2030.

From this analysis, the findings and discussion a few practical implications can be presented for policymakers designing new sustainability frameworks, practitioners working on organisational strategies and plans to tackle sustainability challenges, and academics aiming not only to understand how this evolves but also to contribute to that goal. First, it is not enough to have a communicationally successful framework, but concrete actions and progress must be achieved, otherwise we all risk contributing to SDG washing more than to sustainable development. Second, while the dimensions of sustainable development are hardly progressing coherently and consistently, it is even harder to accomplish an integrated sustainability approach as the SDG contradictions highlighted by many [14–18] seem to be playing a crucial role in the lack of progress we have made so far. This needs to be well-considered in the design of new sustainability frameworks. That said, this does not suggest avoiding conflicting challenges such as economic development and the protection of nature but finding smart, innovative and feasible ways to accomplish both and make them work together. Third, despite the bias that selecting some assessment methods or indicators may bring to a determined analysis [52], measuring and monitoring as well as reporting, as we mostly see nowadays, are crucial to understanding where we are at, how we are doing, what to fix and whether we will reach our goals. These three points must be considered when thinking about any framework or strategy to achieve sustainability.

6. Conclusions

The SDGs are a useful framework to promote and position sustainability that has become enormously relevant as society faces extreme social, economic and environmental challenges. Sustainability has gained from the SDGs and organisations from all sectors using them have placed the topic at the top of local and international agendas. However, understanding whether sustainability has progressed thanks to the existence of the SDGs is still an open and relevant question to be answered. This is the first aim of this article.

Findings show that only some social (SDGs#2, 3, 7 and 11) and environmental (SDGs#6 and 14) goals have significantly improved since the SDGs were launched, suggesting that their existence does not imply sustainability progress nor success, which is important to keep in mind as we move forward in the sustainability journey and, as the SDGs expire in a few years, start thinking about a new framework. Furthermore, as highlighted by many, the contradictions that seem to be integral to their design play a role in the lack of progress they have achieved. While this is an intrinsic limitation of what the SDGs actually are and this article does not tackle that challenge directly, it is important to understand the type of effect they have had, although indirectly, so that all of us who use the SDGs as a sustainability framework are well aware of their strengths and limitations. Moreover, while the communications component of the SDGs has been extremely successful not only for the sustainability movement but for all those positioning their actions under the SDG framework, it is crucial to keep an eye on using them, or any other sustainability framework, as a way to consciously or unconsciously greenwash our actions without deeply contributing to the final overarching goal. Certainly, this risk of SDG washing must be contemplated when considering the contradicting features of some sustainable development goals. Finally, as a local and global goal, measuring and monitoring mechanisms must be put in place to objectively assess our performance. This not only should show the path we are following but also keep us away from greenwashing our actions. All of this should be particularly relevant for practitioners and policymakers involved in the design of sustainability strategies and frameworks.

Along with the practical implications discussed above, this article also has some theoretical implications for researchers. First, most of the SDG research has focused on what organisations are doing and whether they are contributing or not to some or all of the SDGs. This article presents an overall perspective on SDGs' progress with a critical analysis on the impact the SDGs have had on sustainable development. While this does not deny that there have been some improvements in some areas, it shows that, versus the trends of sustainable development before the SDGs were launched in 2015, in general things have not significantly improved, which is something that has been intentionally or involuntarily assumed by many of us assigning that power to the SDGs. Second, frameworks such as the

SDGs must be carefully considered as a powerful tool to achieve goals. Although they are probably the greatest agreement ever achieved, they have mainly been useful to highlight what we do to make us look like we are doing a lot, but they have not been able to help us accomplish what they were supposed to. Their contradictions certainly do not help, but also our own responsibility is at play when aiming to achieve them independently and in an integrated manner. We refer to them without going further and assessing their impact critically. This article is an invitation to follow that important path. Third, this article highlights the importance of assessing these instruments through monitoring and measuring mechanisms, not only reporting, so that as academics we can contribute to more appropriate and tough decisions to be discussed and made.

Certainly, this research has its limitations. As can be seen, it is exploratory and it attempts to assess whether the trends of improvement or deterioration around sustainability challenges have changed since the SDGs were launched in 2015. To make that assessment, specific targets associated with every SDG were selected, one per SDG, and based on their performance conclusions were made. Indeed, this is a limited approach, since other targets associated with the SDGs may have performed differently. However, the selected variables are among the key dimensions to determine the improvement or deterioration of the SDGs, so it can be argued that if any of them fails the SDGs are not performing as required by the sustainability agenda. However, we understand that assessing progress on the SDGs and their targets is subject to the selection of those targets and the used methods. Another limitation that is surely associated with the SDGs is the short period of assessment, i.e., just 6 to 7 years or even less in some cases since the SDGs were launched depending on the available data, which is a very short period to see significant changes. However, absolute changes were not sought but variations in the respective trends. These limitations are also invitations for other researchers to continue understanding this phenomenon.

Funding: This research received no external funding.

Institutional Review Board Statement: Not applicable.

Informed Consent Statement: Not applicable.

Data Availability Statement: Data is contained within the article.

Conflicts of Interest: The author declares no conflict of interest.

Appendix A

Table A1. SDG#1—No Poverty.

Region	Before	After	Reduced Poverty Rate? *
East Asia and Pacific	19.00	1.74	Yes
Europe and Central Asia	5.29	2.47	Yes
Latin America and Caribbean	8.40	4.33	Yes
Middle East and North Africa	2.84	7.85	No
South Asia	28.24	11.78	Yes
Sub-Saharan Africa	46.17	35.98	Yes
World	19.39	9.43	Yes

Test Statistic: 3.000; $p > 0.05$; * Assessed variable: poverty rate (%); Source: The World Bank; Range: 2000–2021.

Table A2. SDG#2—Zero Hunger.

Region	Before	After	Reduced Undernourishment? *
East Asia and Pacific	7.59	4.08	Yes
Europe and Central Asia	2.98	2.59	Yes
High-income countries	2.69	2.65	Yes
Latin America and Caribbean	8.01	6.99	Yes
Middle East and North Africa	8.02	8.72	No
South Asia	17.27	13.92	Yes

Table A2. *Cont.*

Region	Before	After	Reduced Undernourishment? *
Sub-Saharan Africa	20.78	18.98	Yes
World	10.35	8.08	Yes

Test Statistic: 3.000; $p < 0.05$; * Assessed variable: prevalence of undernourishment (% of the population); Source: The World Bank; Range: 2001–2020.

Table A3. SDG#3—Good Health and Well-being.

Region	Before	After	Reduced Mortality? *
Africa	674.75	554.56	Yes
Americas	62.47	61.59	Yes
Eastern Mediterranean	266.88	186.98	Yes
Europe	18.10	11.96	Yes
South-East Asia	247.73	131.48	Yes
Western Pacific	56.50	40.69	Yes
World	1326.43	987.26	Yes

Test Statistic: 0.000; $p < 0.05$; * Assessed variable: maternal mortality rate per 100,000 live births; Source: WHO; Range: 2000–2020.

Table A4. SDG#4—Quality Education.

Region	Before	After	Fewer Children out of School? *
East Asia and Pacific	3.94	5.83	No
Eastern Europe and Central Asia	1.99	2.32	No
Eastern and Southern Africa	21.25	9.98	Yes
Latin America and Caribbean	1.98	2.23	No
Middle East and North Africa	9.50	3.07	Yes
South Asia	22.45	8.50	Yes
West and Central Africa	31.88	22.02	Yes
Europe and Central Asia	n/a	1.80	n/a
World	15.01	7.17	Yes

Test Statistic: 6.000; $p > 0.05$; * Assessed variable: out-of-school children rate; Source: UNICEF; Range: 2010–2020.

Table A5. SDG#5—Gender Equality.

Region	Before	After	Increased Female Education? *
East Asia and Pacific	15.58	17.33	Yes
Europe and Central Asia	21.96	37.7	Yes
Latin America and Caribbean	17.71	7.3	No
Middle East and North Africa	17.88	35.8	Yes
South Asia	5.33	10.92	Yes
Sub-Saharan Africa	3.71	4.87	Yes
World	11.97	12.02	Yes

Test Statistic: 23.000; $p > 0.05$; * Assessed variable: female education (%) higher level; Source: WHO; Range: 2000–2019.

Table A6. SDG#6—Clean Water and Sanitation.

Region	Before	After	More Drinking Water Safely Managed? *
Central and Southern Asia	51.91	60.60	Yes
Europe and Northern America	93.03	95.43	Yes
Latin America and the Caribbean	73.80	75.32	Yes
Northern Africa and Western Asia	72.64	78.19	Yes
Sub-Saharan Africa	21.32	28.96	Yes
World	62.54	67.70	Yes

Test Statistic: 21.000; $p < 0.05$; * Assessed variable: drinking water safely managed service (%); Source: United Nations; Range: 2000–2020.

Table A7. SDG#7—Affordable and Clean Energy.

Region	Before	After	Increased Renewable Energy? *
Australia and New Zealand	11.01	12.63	Yes
Central Asia	3.19	3.39	Yes
Eastern Africa	11.46	11.74	Yes
Eastern Asia	4.43	8.87	Yes
Europe	9.74	13.70	Yes
Latin America and the Caribbean	23.56	25.36	Yes
Middle Africa	17.34	14.23	No
North America	8.70	11.51	Yes
Northern Africa	4.35	3.86	No
Oceania (exc. Australia and New Zealand)	10.42	10.74	Yes
Southern Africa	4.39	5.71	Yes
Southern Asia	9.13	10.88	Yes
South-eastern Asia	8.70	11.92	Yes
Western Asia	4.71	4.01	No
Western Africa	6.57	6.01	No
World	8.33	10.90	Yes

Test Statistic: 108.000; $p < 0.05$; * Assessed variable: share of modern renewables in total final energy consumption (%); Source: IEA; Range: 2000–2019.

Table A8. SDG#8—Decent Work and Economic Growth.

Region	Before	After	Larger GDP Annual Growth? *
Africa Eastern and Southern	4.03	1.86	No
Africa Western and Central	7.95	-2.18	No
Central Europe and the Baltics	5.85	1.94	No
East Asia and Pacific	7.19	−1.54	No
Europe and Central Asia	4.48	3.04	No
Latin America and Caribbean	1.71	0.90	No
Middle East and North Africa	4.17	1.18	No
North America	4.89	1.28	No
Pacific Island small states	2.72	−0.57	No
South Asia	6.94	3.17	No
Sub-Saharan Africa	−0.75	−0.55	Yes
World	1.98	0.91	No

Test Statistic: 1.000; $p < 0.05$; * Assessed variable: GDP annual growth per capita; Source: The World Bank; Range: 2000–2021.

Table A9. SDG#9—Industry, Innovation and Infrastructure.

Region	Before	After	Increase in R&D? *
Central Europe and the Baltics	0.86	1.44	Yes
East Asia and Pacific	2.34	1.65	No
Europe and Central Asia	1.75	1.50	No
European Union	1.92	1.45	No
Latin America and Caribbean	0.64	0.67	Yes
Middle East and North Africa	0.91	1.57	Yes
North America	2.59	2.97	Yes
South Asia	0.72	0.63	No
World	2.02	2.27	Yes

Test Statistic: 24.500; $p > 0.05$; * Assessed variable: research and development expenditure (% of GDP); Source: The World Bank; Range: 2000–2020.

Table A10. SDG#10—Reduced Inequality.

Region	Before	After	Reduced Inequality? *
East Asia and Pacific	3.39	2.84	Yes

Table A10. *Cont.*

Region	Before	After	Reduced Inequality? *
Europe and Central Asia	8.40	6.85	Yes
Latin America and Caribbean	7.27	6.07	Yes
Middle East and North Africa	0.94	0.76	Yes
South Asia	0.85	1.64	No
Sub-Saharan Africa	3.25	2.51	Yes
World	4.53	3.66	Yes

Test Statistic: 4.000; $p > 0.05$; * Assessed variable: Gini Index; Source: The World Bank; Range: 2000–2021.

Table A11. SDG#11—Sustainable Cities and Communities.

Region	Before	After	Reduction in People Living in Slums? *
Africa Eastern and Southern	57.99	54.80	Yes
Africa Western and Central	62.67	48.69	Yes
Latin America and Caribbean	25.00	16.38	Yes
South Asia	55.02	50.91	Yes
Sub-Saharan Africa	60.28	51.90	Yes
World	260.97	211.77	Yes

Test Statistic: 0.000; $p < 0.05$; * Assessed variable: people living in slums (% of urban population); Source: The World Bank; Range: 2000–2020.

Table A12. SDG#12—Responsible Consumption and Production.

Region	Before	After	Less Fossil Fuels? *
Africa	383,991,855	396,224,729	No
Asia and the Pacific	3,306,638,570	3,789,079,737	No
Europe	1,409,854,499	1,335,227,496	Yes
Latin America and the Caribbean	454,981,983	455,914,025	No
North America	1,766,377,903	1,871,667,759	No
World	8,553,931,287	10,062,117,427	No

Test Statistic: 31.000; $p > 0.05$; * Assessed variable: fossil fuels consumption (tons); Source: UNEP; Range: 2000–2019.

Table A13. SDG#13—Climate Action.

Region	Before	After	Reduced Fossil Fuels? *
Central Europe and the Baltics	6.70	6.45	Yes
East Asia and Pacific	4.68	6.05	No
Europe and Central Asia	7.47	4.33	Yes
European Union	7.47	6.39	Yes
Latin America and Caribbean	2.59	2.62	No
Middle East and North Africa	4.98	5.44	No
North America	17.91	5.82	Yes
South Asia	1.02	1.46	No
Sub-Saharan Africa	0.78	3.63	No
World	4.40	4.59	No

Test Statistic: 27.000; $p > 0.05$; * Assessed variable: CO_2 emissions (metric tons per capita); Source: The World Bank; Range: 2000–2019.

Table A14. SDG#14—Life Below Water.

Region	Before	After	More Protected Areas? *
Oceania (exc. Australia and New Zealand)	17.82	20.65	Yes
Northern Africa and Western Asia	21.97	25.02	Yes
Eastern and South-Eastern Asia	28.59	32.61	Yes
Central and Southern Asia	33.98	34.28	Yes
Sub-Saharan Africa	38.06	41.12	Yes

Table A14. *Cont.*

Region	Before	After	More Protected Areas? *
Latin America and the Caribbean	37.74	41.48	Yes
Australia and New Zealand	51.48	54.48	Yes
Europe and Northern America	61.80	64.36	Yes
World	43.17	46.03	Yes

Test Statistic: 45.000; $p < 0.05$; * Assessed variable: mean percentage of each marine key biodiversity area covered by protected areas; Source: United Nations; Range: 2010–2019.

Table A15. SDG#15—Life on Land.

Region	Before	After	Less Tree Cover Loss? *
East Asia and Pacific	3,430,732	4,737,269	No
Europe and Central Asia	3,942,659	6,926,284	No
Latin America and the Caribbean	4,706,182	6,274,758	No
Middle East and North Africa	10,846	23,717	No
North America	4,362,358	4,255,521	Yes
South Asia	96,642	186,419	No
Sub-Saharan Africa	1,954,293	4,102,907	No
World	18,503,712	26,506,875	No

Test Statistic: 33.000; $p < 0.05$; * Assessed variable: tree cover loss (hectares); Source: Global Forest Watch; Range: 2001–2022.

Table A16. SDG#16—Peace, Justice and Strong Institutions.

Region	Before	After	Improved Justice? *
East Asia and Pacific	9.046	8.997	No
Eastern Europe and Central Asia	8.280	6.798	No
EU + EFTA + North America	15.741	19.459	Yes
Latin America and Caribbean	8.541	16.312	Yes
Middle East and North Africa	3.706	3.828	Yes
South Asia	2.513	2.684	Yes
Sub-Saharan Africa	8.527	12.787	Yes
World	56.353	70.866	Yes

Test Statistic: 31.000; $p > 0.05$; * Assessed variable: Rule of Law Index; Source: World Justice Project; Range: 2014–2018.

References

1. George, G.; Howard-Grenville, J.; Joshi, A.; Tihanyi, L. Understanding and Tackling Societal Grand Challenges through Management Research. *Acad. Manag. J.* **2016**, *59*, 1880–1895. [CrossRef]
2. Biermann, F.; Kanie, N.; Kim, R.E. Global Governance by Goal-Setting: The Novel Approach of the UN Sustainable Development Goals. *Curr. Opin. Environ. Sustain.* **2017**, *26–27*, 26–31. [CrossRef]
3. Caiado, R.G.G.; Filho, W.L.; Quelhas, O.L.G.; de Mattos Nascimento, D.L. A Literature-based Review on Potentials and Constraints in the Implementation of the Sustainable Development Goals. *J. Clean. Prod.* **2018**, *198*, 1276–1288. [CrossRef]
4. Munro, V. The Universal Sustainable Development Goals for Purpose and Change. In *CSR for Purpose, Shared Value and Deep Transformation. The New Responsibility*; Munro, V., Ed.; Emerald Publishing Limited: Bingley, UK, 2020; pp. 85–117. [CrossRef]
5. Van Zanten, J.A.; van Tulder, R. Multinational Enterprises and the Sustainable Development Goals: An Institutional Approach to Corporate Engagement. *J. Int. Bus. Policy* **2018**, *1*, 208–233. [CrossRef]
6. Ali, S.M.; Appolloni, A.; Cavallaro, F.; D'Adamo, I.; Di Vaio, A.; Ferella, F.; Gastaldi, M.; Ikram, M.; Kumar, N.M.; Martin, M.A.; et al. Development Goals towards Sustainability. *Sustainability* **2023**, *15*, 9443. [CrossRef]
7. Bryson, J.M.; Crosby, B.C.; Stone, M.M. The Design and Implementation of Cross-Sector Collaborations: Propositions from the Literature. *Public Adm. Rev.* **2006**, *66*, 44–55. [CrossRef]
8. Selsky, J.W.; Parker, B. Cross-Sector Partnerships to Address Social Issues: Challenges to Theory and Practice. *J. Manag.* **2005**, *31*, 849–873. [CrossRef]
9. Whittingham, K.L.; Earle, A.G.; Leyva-De la Hiz, D.I.; Argiola, A. The Impact of the United Nations Sustainable Development Goals on Corporate Sustainability Reporting. *Bus. Res. Q.* **2023**, *26*, 45–61. [CrossRef]
10. Leal Filho, W.; Trevisan, L.V.; Rampasso, I.S.; Anholon, R.; Dinis, M.A.P.; Brandli, L.L.; Sierra, J.; Salvia, A.L.; Pretorius, R.; Nicolau, M.; et al. When the alarm bells ring: Why the UN sustainable development goals may not be achieved by 2030. *J. Clean. Prod.* **2023**, *407*, 137108. [CrossRef]

11. Sachs, J.D.; Lafortune, G.; Fuller, G.; Drumm, E. *Sustainable Development Report 2023*; Dublin University Press: Dublin, Ireland, 2023. [CrossRef]
12. Allen, C.; Metternicht, G.; Wiedmann, T. Initial Progress in Implementing the Sustainable Development Goals (SDGs): A Review of Evidence from Countries. *Sustain. Sci.* **2018**, *13*, 1453–1467. [CrossRef]
13. Sullivan, K.; Thomas, S.; Rosano, M. Using Industrial Ecology and Strategic Management Concepts to Pursue the Sustainable Development Goals. *J. Clean. Prod.* **2018**, *174*, 237–246. [CrossRef]
14. Hák, T.; Janoušková, S.; Moldan, B. Sustainable Development Goals: A Need for Relevant Indicators. *Ecol. Indic.* **2016**, *60*, 565–573. [CrossRef]
15. Hickel, J. The Contradiction of the Sustainable Development Goals: Growth versus Ecology on a Finite Planet. *Sustain. Dev.* **2019**, *27*, 873–884. [CrossRef]
16. Salvia, A.L.; Filho, W.L.; Brandli, L.L.; Griebeler, J.S. Assessing Research Trends Related to Sustainable Development Goals: Local and Global Issues. *J. Clean. Prod.* **2019**, *208*, 841–849. [CrossRef]
17. Spaiser, V.; Ranganathan, S.; Swain, R.B.; Sumpter, D. The Sustainable Development Oxymoron: Quantifying and Modelling the Incompatibility of Sustainable Development Goals. *Int. J. Sustain. Dev. World Ecol.* **2017**, *24*, 457–470. [CrossRef]
18. Stafford-Smith, M.; Griggs, D.; Gaffney, O.; Ullah, F.; Reyers, B.; Kanie, N.; Stigson, B.; Shrivastava, P.; Leach, M.; O'Connell, D. Integration: The Key to Implementing the Sustainable Development Goals. *Sustain. Sci.* **2017**, *12*, 911–919. [CrossRef]
19. Jones, P.; Hillier, D.; Comfort, D. The Sustainable Development Goals and Business. *Int. J. Sales Retail. Mark.* **2016**, *5*, 38–48.
20. Spangenberg, J.H. Hot Air or Comprehensive Progress? A Critical Assessment of the SDGs. *Sustain. Dev.* **2017**, *25*, 311–321. [CrossRef]
21. Rockström, J. 5 Transformational Policies for a Prosperous and Sustainable World. 7 November 2018. Available online: https://www.youtube.com/watch?v=Rv-tDrv__mc (accessed on 18 July 2023).
22. Nundy, S.; Ghosh, A.; Mesloub, A.; Albaqawy, G.A.; Alnaim, M.M. Impact of COVID-19 Pandemic on Socio-Economic, Energy-Environment and Transport Sector Globally and Sustainable Development Goal (SDG). *J. Clean. Prod.* **2021**, *312*, 127705. [CrossRef]
23. Shulla, K.; Voigt, B.-F.; Cibian, S.; Scandone, G.; Martinez, E.; Nelkovski, F.; Salehi, P. Effects of COVID-19 on the Sustainable Development Goals (SDGs). *Discov. Sustain.* **2021**, *2*, 15. [CrossRef]
24. Ordonez-Ponce, E.; Talbot, D. Multinational Enterprises' Sustainability Practices and Focus on Developing Countries: Contributions and Unexpected Results of SDG Implementation. *J. Int. Dev.* **2023**, *35*, 201–232. [CrossRef]
25. Van Tulder, R.; Rodrigues, S.B.; Mirza, H.; Sexsmith, K. The UN's Sustainable Development Goals: Can Multinational Enterprises Lead the Decade of Action? *J. Int. Bus. Policy* **2021**, *4*, 1–21. [CrossRef]
26. Brisbane City Council. United Nations Sustainable Development Goals. Vision and Strategy. Available online: https://www.brisbane.qld.gov.au/about-council/governance-and-strategy/vision-and-strategy/united-nations-sustainable-development-goals (accessed on 1 July 2023).
27. Gobierno de Perú. Objetivos de Desarrollo Sostenible (ODS) [Sustainable Development Goals (SDG)]. Available online: https://www.gob.pe/34191-objetivos-de-desarrollo-sostenible-ods (accessed on 1 July 2023).
28. Gouvernement du Québec. Sustainable Development. Policies and Orientations. Available online: https://www.quebec.ca/en/government/policies-orientations/sustainable-development (accessed on 1 July 2023).
29. PRME. Management Education and the Sustainable Development Goals: Transforming Education to Act Responsibly and Find Opportunities. PRME Secretariat. 2016. Available online: https://d1ngk2wj7yt6d4.cloudfront.net/public/uploads/PDFs/SDGBrochurePrint.pdf (accessed on 1 July 2023).
30. SDSN. SDSN Canada. Networks. Available online: https://www.unsdsn.org/canada (accessed on 1 July 2023).
31. GRI. Integrating SDGs into Sustainability Reporting. Available online: https://www.globalreporting.org/public-policy-partnerships/sustainable-development/integrating-sdgs-into-sustainability-reporting/ (accessed on 1 July 2023).
32. OECD. OECD and the Sustainable Development Goals: Delivering on Universal Goals and Targets. Available online: https://www.oecd.org/dac/sustainable-development-goals.htm (accessed on 1 July 2023).
33. WBCSD. SDG Sector Roadmap. Available online: https://www.wbcsd.org/Programs/People-and-Society/Sustainable-Development-Goals/SDG-Sector-Roadmaps (accessed on 1 July 2023).
34. Clarke, A.; MacDonald, A. Outcomes to Partners in Multi-Stakeholder Cross-Sector Partnerships: A Resource-Based View. *Bus. Soc.* **2019**, *58*, 298–332. [CrossRef]
35. Park, K.; Grimes, M.G.; Gehman, J. Becoming a Generalized Specialist: A Strategic Model for Increasing your Organizations SDG Impact while Minimizing Externalities. In *Handbook on the Business of Sustainability*; George, G., Haas, M., Joshi, H., McGahan, A., Tracey, P., Eds.; Elgar: Cheltenham, UK, 2022. [CrossRef]
36. Liu, Y.; Huang, B.; Guo, H.; Liu, J. A Big Data Approach to Assess Progress Towards Sustainable Development Goals for Cities of Varying Sizes. *Commun. Earth Environ.* **2023**, *4*, 66. [CrossRef]
37. Heras-Saizarbitoria, I.; Urbieta, L.; Boiral, O. Organizations' Engagement with Sustainable Development Goals: From Cherry-Picking to SDG-Washing? *Corp. Soc. Responsib. Environ. Manag.* **2022**, *29*, 316–328. [CrossRef]
38. Castor, J.; Bacha, K.; Nerini, F.F. SDGs in Action: A Novel Framework for Assessing Energy Projects against the Sustainable Development Goals. *Energy Res. Soc. Sci.* **2020**, *68*, 101556. [CrossRef]

39. Van Zanten, J.A.; van Tulder, R. Improving Companies' Impacts on Sustainable Development: A Nexus Approach to the SDGS. *Bus. Strategy Environ.* **2021**, *30*, 3703–3720. [CrossRef]
40. Apollonia, M.; Schiltz, F. Measuring Sustainable Development Goals Performance: How to Monitor Policy Action in the 2030 Agenda Implementation? *Ecol. Econ.* **2019**, *164*, 106373. [CrossRef]
41. Chand, M. Strategic Assessment and Mitigation of Risks in Sustainable Manufacturing Systems. *Sustain. Oper. Comput.* **2021**, *2*, 206–213. [CrossRef]
42. Khaled, R.; Ali, H.; Mohamed, E.K.A. The Sustainable Development Goals and Corporate Sustainability Performance: Mapping, Extent and Determinants. *J. Clean. Prod.* **2021**, *311*, 127599. [CrossRef]
43. Ordonez-Ponce, E.; Khare, A. GRI 300 as a Measurement Tool for the United Nations Sustainable Development Goals: Assessing the Impact of Car Makers on Sustainability. *J. Environ. Plan. Manag.* **2021**, *64*, 47–75. [CrossRef]
44. Vílchez, V.F.; Carrasco, P.O.; Bernardo, F.A.S. SDGwashing: A Critical View of the Pursuit of SDGs and its Relationship with Environmental Performance. *J. Environ. Plan. Manag.* **2022**, *65*, 1001–1023. [CrossRef]
45. Huan, Y.; Liang, T.; Li, H.; Zhang, C. A Systematic Method for Assessing Progress of Achieving Sustainable Development Goals: A Case Study of 15 Countries. *Sci. Total Environ.* **2021**, *752*, 141875. [CrossRef] [PubMed]
46. De Neve, J.-E.; Sachs, J.D. The SDGs and Human Well-Being: A Global Analysis of Synergies, Trade-Offs, and Regional Differences. *Sci. Rep.* **2020**, *10*, 15113. [CrossRef] [PubMed]
47. D'Adamo, I.; Gastaldi, M. Monitoring the Performance of Sustainable Development Goals in the Italian Regions. *Sustainability* **2023**, *15*, 14094. [CrossRef]
48. International Council for Science (ICSU). *A Guide to SDG Interactions: From Science to Implementation*; International Council for Science: Paris, France, 2017.
49. Sustainable Development Solutions Network. Getting Started with the Sustainable Development Goals: A Guide for Stakeholders. Sustainable Development Solutions Network. 2015. Available online: https://irp-cdn.multiscreensite.com/be6d1d56/files/uploaded/getting-started-guide-FINAL-PDF-.pdf (accessed on 2 November 2023).
50. United Nations Development Group. Mainstreaming the 2030 Agenda—Reference Guide to United Nations Country Teams. United Nations Development Group. 2016. Available online: https://unsdg.un.org/resources/mainstreaming-2030-agenda-reference-guide-united-nations-country-teams (accessed on 2 November 2023).
51. Allen, C.; Metternicht, G.; Wiedmann, T. Prioritising SDG Targets: Assessing Baselines, Gaps and Interlinkages. *Sustain. Sci.* **2019**, *14*, 421–438. [CrossRef]
52. Allen, C.; Reid, M.; Thwaites, J.; Glover, R.; Kestin, T. Assessing National Progress and Priorities for the Sustainable Development Goals (SDGs): Experience from Australia. *Sustain. Sci.* **2020**, *15*, 521–538. [CrossRef]
53. Bandari, R.; Moallemi, E.A.; Lester, R.E.; Downie, D.; Bryan, B.A. Prioritising Sustainable Development Goals, Characterising Interactions, and Identifying Solutions for Local Sustainability. *Environ. Sci. Policy* **2022**, *127*, 325–336. [CrossRef]
54. Weitz, N.; Carlsen, H.; Nilsson, M.; Skånberg, K. Towards Systemic and Contextual Priority Setting for Implementing the 2030 Agenda. *Sustain. Sci.* **2018**, *13*, 531–548. [CrossRef]
55. Nilashi, M.; Keng Boon, O.; Tan, G.; Lin, B.; Abumalloh, R. Critical Data Challenges in Measuring the Performance of Sustainable Development Goals: Solutions and the Role of Big-Data Analytics. *Harv. Data Sci. Rev.* **2023**, *5*. [CrossRef]
56. Rockström, J.; Sukhdev, P. The SDGs Wedding Cake. Stockholm Resilience Centre. Available online: https://www.stockholmresilience.org/research/research-news/2016-06-14-the-sdgs-wedding-cake.html (accessed on 3 February 2022).
57. Bailey, C. *Quantitative Analysis in Exercise and Sport Science*; University of North Texas Libraries: Denton, TX, USA, 2021; Available online: https://dx.doi.org/10.12794/sps.ot-quantitative-analysis-exss (accessed on 3 February 2022). [CrossRef]
58. The World Bank. Poverty and Inequality Platform. Available online: https://pip.worldbank.org/home (accessed on 1 July 2023).
59. The World Bank. Prevalence of Undernourishment (% of Population). Data. Available online: https://data.worldbank.org/indicator/SN.ITK.DEFC.ZS (accessed on 1 July 2023).
60. WHO. Maternal Mortality Ratio (Per 100 000 Live Births). The Global Health Observatory. Available online: https://www.who.int/data/gho/data/indicators/indicator-details/GHO/maternal-mortality-ratio-(per-100-000-live-births) (accessed on 1 July 2023).
61. UNICEF. Education Overview. Available online: https://data.unicef.org/topic/education/overview/ (accessed on 1 July 2023).
62. WHO. Integrated WHO Nutrition Global Databases. NLiS Data Search. Available online: https://www.who.int/data/nutrition/nlis/data-search (accessed on 1 July 2023).
63. United Nations. UN Water. Available online: https://sdg6data.org/en/tables (accessed on 1 July 2023).
64. EIA. World Energy Balances. Available online: https://www.iea.org/reports/sdg7-data-and-projections/modern-renewables (accessed on 10 November 2023).
65. The World Bank. Research and Development Expenditure (% of GDP). Data. Available online: https://data.worldbank.org/indicator/GB.XPD.RSDV.GD.ZS (accessed on 1 November 2023).
66. The World Bank. Population Living in Slums (% of Urban Population). Data. Available online: https://data.worldbank.org/indicator/EN.POP.SLUM.UR.ZS (accessed on 1 November 2023).
67. UNEP. Global Material Flows Database. International Resource Panel. Available online: https://www.resourcepanel.org/global-material-flows-database (accessed on 7 July 2023).

68. The World Bank. CO_2 Emissions (Metric Tons per Capita). Data. Available online: https://data.worldbank.org/indicator/EN.ATM.CO2E.PC (accessed on 7 July 2023).
69. United Nations Statistics Division. UN Data. Available online: http://data.un.org/Explorer.aspx?d=EDATA (accessed on 1 July 2023).
70. Global Forest Watch. Forest Monitoring Designed for Action. Available online: https://www.globalforestwatch.org/map/?mapMenu=eyJtZW51U2VjdGlvbiI6InNlYXJjaCJ9 (accessed on 7 July 2023).
71. World Justice Project. Rule of Law Index 2022. World Justice Project. 2022. Available online: https://worldjusticeproject.org/rule-of-law-index/downloads/WJPIndex2022.pdf (accessed on 7 July 2023).
72. Richardson, K.; Steffen, W.; Lucht, W.; Bendtsen, J.; Cornell, S.E.; Donges, J.F.; Drüke, M.; Fetzer, I.; Bala, G.; von Bloh, W.; et al. Earth Beyond Six of Nine Planetary Boundaries. *Sci. Adv.* **2023**, *9*, eadh2458. [CrossRef] [PubMed]

Disclaimer/Publisher's Note: The statements, opinions and data contained in all publications are solely those of the individual author(s) and contributor(s) and not of MDPI and/or the editor(s). MDPI and/or the editor(s) disclaim responsibility for any injury to people or property resulting from any ideas, methods, instructions or products referred to in the content.

sustainability

MDPI

Article

Advancing Biodiesel Production System from Mixed Vegetable Oil Waste: A Life Cycle Assessment of Environmental and Economic Outcomes

Farayi Musharavati [1], Khadija Sajid [2], Izza Anwer [3], Abdul-Sattar Nizami [2,*], Muhammad Hassan Javed [2], Anees Ahmad [2] and Muhammad Naqvi [4]

1 Department of Mechanical and Industrial Engineering, Qatar University, Doha 2713, Qatar; farayi@qu.edu.qa
2 Sustainable Development Study Centre, Government College University, Lahore 54000, Pakistan; khadijahsajid1998@gmail.com (K.S.); hassanjavedduggal@gmail.com (M.H.J.); aneesjamal180@gmail.com (A.A.)
3 Department of Transportation Engineering and Management, University of Engineering and Technology, Lahore 54890, Pakistan; engr-izza@uet.edu.pk
4 College of Engineering and Technology, American University of the Middle East, Egaila 54200, Kuwait; sayed.naqvi@aum.edu.kw
* Correspondence: asnizami@gcu.edu.pk

Abstract: This study aims to evaluate the environmental and economic performance of biodiesel production from mixed vegetable oil waste using the life cycle assessment (LCA) model. Due to its huge potential, Pakistan is taken as a case study. It produces 468,842 tons of vegetable oil waste annually. As no biodiesel production plant exists to process it, the environmental performance of biodiesel prototypes has not been investigated. Therefore, the current study is conducted to support the design of a plant to produce biodiesel from mixed oil waste. An attributional LCA was conducted using ReCiPe (H) and found that 400 kg of biodiesel can be produced from 1 t of mixed oil waste. The results, based on a functional unit of 1 ton, showed that biodiesel production from mixed vegetable oil waste is more eco-friendly than the existing landfilling practices with a global warming potential of 1.36×10^{-4} kg CO_2 eq, human toxicity of 5.31 kg 1.4 DB eq, ozone depletion potential of 0.00271 kg CFC-11 eq, eutrophication potential of 0.0118 kg P eq, acidification potential of 123 kg SO_2 eq, and photochemical ozone formation of 51.4 kg NO_x eq. Scenario modelling was conducted using electricity from photovoltaic solar cells, which decrease fine particulate matter formation from 44.5 to 0.725 kg $PM_{2.5}$ eq, instead of using electricity from a grid to the plant. Hotspot identification was carried out to highlight the effects of individual impact categories. An economic analysis showed that 638,839 USD/year revenue would be generated. Generating energy from discarded vegetable oils through biodiesel production presents a sustainable and economically viable approach. This process benefits the environment and contributes to cost savings by reducing waste disposal in landfills. Furthermore, it aligns with the principles of a circular economy, in which resources are reused and recycled. It also supports the pursuit of the United Nations' Sustainable Development Goals (SDGs), particularly SDG-7, which focuses on affordable and clean energy, and SDG-12, which emphasizes responsible consumption and production.

Keywords: vegetable oil; biodiesel production; waste-to-energy (WtE); life cycle assessment (LCA); sustainable development goals (SDGs)

Citation: Musharavati, F.; Sajid, K.; Anwer, I.; Nizami, A.-S.; Javed, M.H.; Ahmad, A.; Naqvi, M. Advancing Biodiesel Production System from Mixed Vegetable Oil Waste: A Life Cycle Assessment of Environmental and Economic Outcomes. *Sustainability* **2023**, *15*, 16550. https://doi.org/10.3390/su152416550

Academic Editors: Marc A. Rosen, Manoj Kumar Nallapaneni, Idiano D'Adamo and Massimo Gastaldi

Received: 8 November 2023
Revised: 28 November 2023
Accepted: 1 December 2023
Published: 5 December 2023

Copyright: © 2023 by the authors. Licensee MDPI, Basel, Switzerland. This article is an open access article distributed under the terms and conditions of the Creative Commons Attribution (CC BY) license (https://creativecommons.org/licenses/by/4.0/).

1. Introduction

Pakistan generates 49.6 million tons of solid waste annually, with a yearly increase of 2.4% [1]. The government of Pakistan estimates that over 16,500 tons of municipal waste is generated every day, resulting in a weekly total of 87,000 tons of solid waste [2]. Additionally, the current solid waste management system is facing significant challenges due to inadequate equipment, low public awareness, and a lack of urban planning. Thus, the lack of sound waste management practices creates serious environmental issues threatening the population's welfare and health. A total of 60–70% of the country's solid waste is collected, dumped, buried, or burned on vacant lots [3]. Reusing mixed vegetable oil waste instead of disposing it can create more mixed oil waste. This increase in the generation of such waste poses several challenges to its efficient management. Moreover, kitchens and food industries produce about 16.5 million tons of vegetable oil waste annually. This waste is usually disposed of in landfills and municipal solid waste or discharged into sewage systems [4].

At the start of the Industrial Revolution (the 20th century), energy utilization rapidly increased due to the increasing population and better living standards. In 2030, a 53% increase in global energy consumption is expected [5]. Currently, energy is primarily derived from natural gas (24%), coal (30%), and crude oil (33%), which are all fossil fuels [6,7]. The excessive use of non-renewable fossil fuels puts the energy security of people with limited access to these resources at risk, leading to climate change. Therefore, there is a need to find alternative energy sources to fossil fuels to guarantee energy security and tackle climate change [8].

In terms of economic development, energy is essential as it provides some necessary services to maintain the quality of human life and economic activity [9]. Pakistan's crude oil production in 2019 was 4.3 million metric tons, satisfying only 20% of the country's total petroleum needs. Alternatives to petroleum-based crude oil for diesel fuel are a major consideration. Hence, biodiesel production is gaining more attention as a direct replacement for crude oil petroleum as a blended component that is 100% renewable and biodegradable, as well as produces lower exhaust emissions compared to conventional diesel fuels [10]. In 2018–2019, Pakistan imported fossil fuels and imported 17.20 million tons of crude oil [3]. In this country, the electricity and transport sectors are the key users of fossil fuels. About 50% more energy is required for the transport and electricity sectors [11]. Moreover, Pakistan needs 10% blended biodiesel in fossil diesel by 2025 [12]. Therefore, lab-scale research on biodiesel production has frequently been conducted by organizations and universities in the country [1]. According to [13], various organizations in Pakistan have been developing biodiesel prototypes from diverse biomass sources. Moreover, many universities of the country have prepared biodiesel prototypes mainly from non-edible waste oils such as those from Jatropha seed oil.

Solid waste, including 4000 tons/day and 32.6 Mt/year of municipal solid waste (MSW), is appropriate for transformation into various waste-to-energy (WtE) forms [14]. Waste-to-energy is a process that reduces greenhouse gas emissions, recovers metals, and generates clean energy from waste materials. One form of material and energy recovery from waste is biodiesel production, mainly from mixed vegetable oil waste that reduces the burden on landfills and helps in energy recovery. Therefore, it is a secondary fuel that can manage various forms of urban and municipal waste, thereby improving waste handling [15]. In addition, waste is passed through a series of processes in which all the non-combustibles are removed for its production as follows: oil extraction, pretreatment, esterification, transesterification, and biodiesel refining, which can be applied after the process of purification to obtain biodiesel with a purity of 98% [16].

An optimized approach is used to check the sustainability of the process by performing a life cycle assessment (LCA), which is one of the most common sustainability assessment decision-making tools for assessing the impacts of different products or processes and environmental performance [17]. A product's life cycle starts with raw material extraction, then it is produced, transported, used, consumed, and disposed of, and finally its emissions and waste management are considered [18,19]. The LCA is a suitable tool for identifying a current project's environmental benefits and drawbacks and comparing them with those of conventional systems. Therefore, it helps policy and decision makers implement the process with minimal environmental impacts [20]. Environmental applications of LCA have increased worldwide over the last few years because it assesses the environmental impacts of a current project throughout its life cycle. Although all processes result in resource consumption, emissions, and environmental impacts, an LCA looks at the process of the environment as a sink and a source and assesses the impact of different environmental impact categories such as human toxicity (HTP), global warming potential (GWP), ozone depletion potential (ODP), eutrophication potential (EP), and acidification potential (AP) [21]. In addition, it is the key factor for developing bioenergy support policies, including GHG savings, energy savings, and environmental and social acceptability [22].

This study examines the current generation and composition of mixed vegetable oil waste along with management practices and presents a design for the line production of biodiesel that can convert mixed oil waste into the formation of biodiesel. However, biodiesel production shows the potential and feasibility of biofuel as a substitute energy source and replacement for crude oil in the transportation sector. The goal is to reduce 10% of landfilled waste and 65% of MSW to be recycled by 2030. This study supports two UN Sustainable Development Goals (SDGs). Firstly, it aligns with SDG 7 for affordable and clean energy by showing how biodiesel from vegetable oil waste can be a practical and sustainable alternative to conventional energy. Secondly, our research supports SDG 12 for responsible consumption and production by recycling waste into energy, thus reducing environmental impacts and advancing sustainable practices. These two SDGs highlight the importance of our work, as emphasized in a recent editorial on sustainable energy research [23]. This study aims to evaluate a biodiesel production system's environmental and economic outcomes, utilizing mixed vegetable oil waste through a comprehensive LCA. This study thoroughly analyses material and energy flows and soil, water, and air emissions. The sustainability of the entire process was evaluated through an LCA using Gabi software (Version 10.0.0.71). A techno-economic analysis was also conducted to determine the economic feasibility of producing biodiesel for use as a fuel.

2. Materials and Methods

Section 2, 'Materials and Methods', is divided into four subsections: 'Section 2.1. Waste Characterization', 'Section 2.2. Study Design', 'Section 2.3. Life Cycle Assessment (LCA)', and 'Section 2.4. Economic Assessment'. These subsections provide information about the composition of mixed vegetable oil waste, the methodology of biodiesel production, the LCA's objectives and boundaries, and the financial aspects of biodiesel production, respectively.

The oil extraction phase includes different steps such as seed decortications, the filtering of oil, and the expulsion of oil from seeds as shown in Figure 1.

Moreover, the processes of transesterification and esterification, which need numerous inputs, such as catalysts in the form of the acid H_2SO_4, the alkali NaOH/KOH, alcohol in the form of methanol, and electricity.

2.1. Waste Characterization

In Pakistan, biodiesel from non-edible vegetable oil yields 0.09 thousand barrels daily. The physical characterization of the major components of mixed oil waste includes rapeseed oil (44.5%), castor oil (23.0%), waste cooking oil (11.5%), and Jatropha oil (21.0%). Moreover, out of every 100 tons of mixed oils processed, a portion is used to produce crude glycerol while the remainder can be used to make biodiesel.

2.2. Study Design

This study consists of a design for the line production of an extensive-scale biodiesel production plant from mixed vegetable oil waste as shown in Figure 2.

Soybean, palm, and peanut oils are vitally utilized in food industries and planted crops in Pakistan. These edible oils have enriched sources and a strong potential to produce biodiesel in large amounts. Thus, 1.25 of palm oil produces 1 L of biodiesel, and 1 L of biodiesel is produced from 1.3 L of soybean oil. Jatropha oil is the main source of biodiesel from Jatropha seeds in Africa and Asia. Jatropha seeds contain 30–35 wt.% oil which can be converted into biodiesel. Waste cooking oil also has the highest capacity to produce biodiesel, and 89% of biodiesel is produced from WCO in the UK, similarly to many other countries.

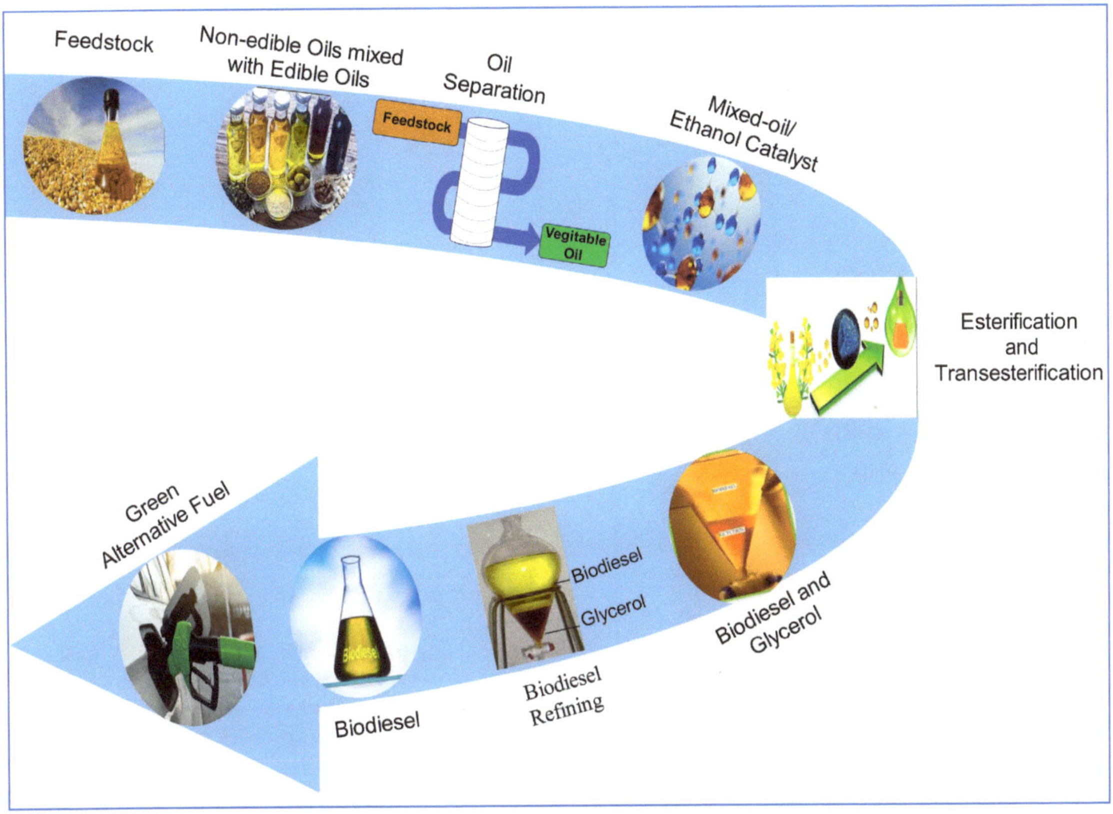

Figure 1. Schematic diagram of biodiesel production line.

Biodiesel production includes the following processes. The first step to produce biodiesel is the extraction of oil, in which the oil filtration, seed pressing, and decortication steps are carried out. The next step is the pretreatment, which is required to process feedstocks prior to their alteration into biodiesel. This step minimizes negative impacts on biodiesel production, such as suspended particles, polymers, FFAs, water, and gums. The pretreatment of oil has been shown to avoid soap formation during transesterification, eventually leading to an increased biodiesel yield. The third process involves two-step transesterification. Most non-edible oils have a higher content of FFAs from the pretreatment step; this amount must be reduced to 0.5–1% using an esterification reaction with an acid catalyst. The acid catalyst H_2SO_4 with methanol reduces the amount of FFAs. Hence, the transesterification process is carried out with the alkali catalysts NaOH or KOH with methanol. In this process, KOH is mostly preferred because of its low price, great productivity, and moderate yield. Therefore, KOH also decreases the oil's tendency to form soap. When KOH is used as a catalyst, it produces crude glycerol, and it is easier to separate this from the produced biodiesel using NaOH. Base transesterification produces a 98% biodiesel yield. Crude glycerol is generated as a byproduct that accounts for 10–12% of biodiesel. It is used as a processed industrial raw material that plays an important role in biodiesel chain sustainability and is the major bottleneck in producing biodiesel chains.

Therefore, crude biodiesel is subjected to wet washing to remove further impurities, such as catalysts, glycerol, soap, and residual alcohol, to obtain a purified biodiesel.

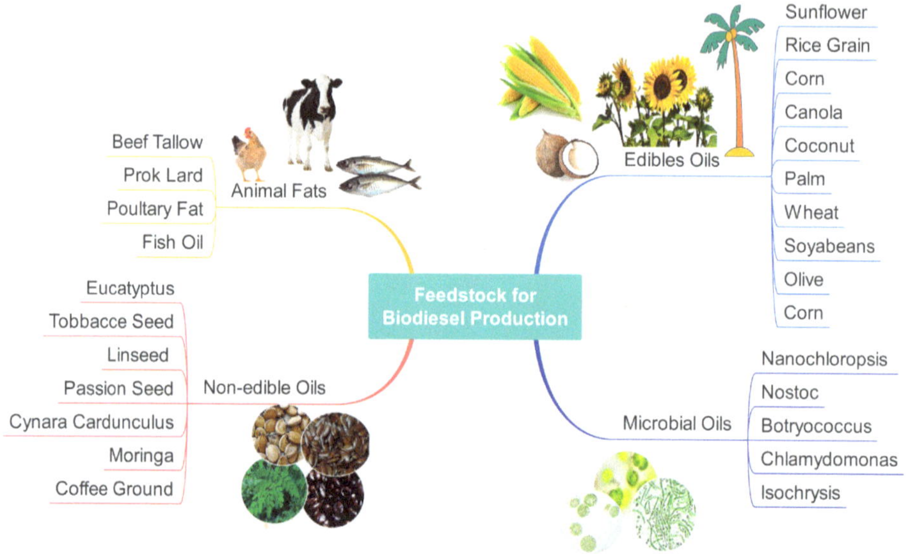

Figure 2. Various feedstock used to produce biodiesel.

2.3. *Life Cycle Assessment (LCA)*

An LCA is a method used to evaluate the sustainability of a product, process, or service by assessing its environmental impacts and benefits throughout its entire life cycle. Examining biodiesel production and consumption helps us to understand their ecological consequences. This study follows a standardized LCA approach in line with ISO 14040 guidelines and uses the Gabi LCA software (Version 10.0.0.71) as the primary tool for our analysis. The LCA process encompasses four fundamental stages: the goal and scope, life cycle inventory, life cycle impact assessment, and interpretation of the results, as outlined by [18,19].

2.3.1. Goal and Scope

Our research aims to evaluate how producing biodiesel from various vegetable oils affects the environment, and this evaluation is conducted through an LCA. From an environmental standpoint, this study aims to determine the feasibility of producing biodiesel using various vegetable oils, following the methodological framework outlined in ISO as in [18,19]. One of the primary goals of the LCA process is to assess the environmental impacts associated with biodiesel production from diverse vegetable oil sources. However, this study evaluates the environmental assessment using an attributional LCA approach, which focuses on quantifying the hotspots and key environmental issues at various stages of biodiesel production. Moreover, system boundaries and functional units are both part of the goal and scope. Figure 3 shows the process of an LCA of biodiesel production.

System Boundaries

The scope of this study is gate-to-gate, along with system boundaries that consist of the following:

- The zero-burden assumption is selected, indicating that the biodiesel plant sources its inputs from various vegetable oils, including palm, soybean, castor, and waste cooking oils, which are all considered to have no environmental impact.

- The biodiesel production process considers both direct emissions generated on site and indirect emissions resulting from the use of electricity and fuel.
- The system boundary encompasses the collection of various vegetable oils from multiple sites and their transportation to the central facility. It extends through the biochemical treatment processes within a biodiesel production facility.

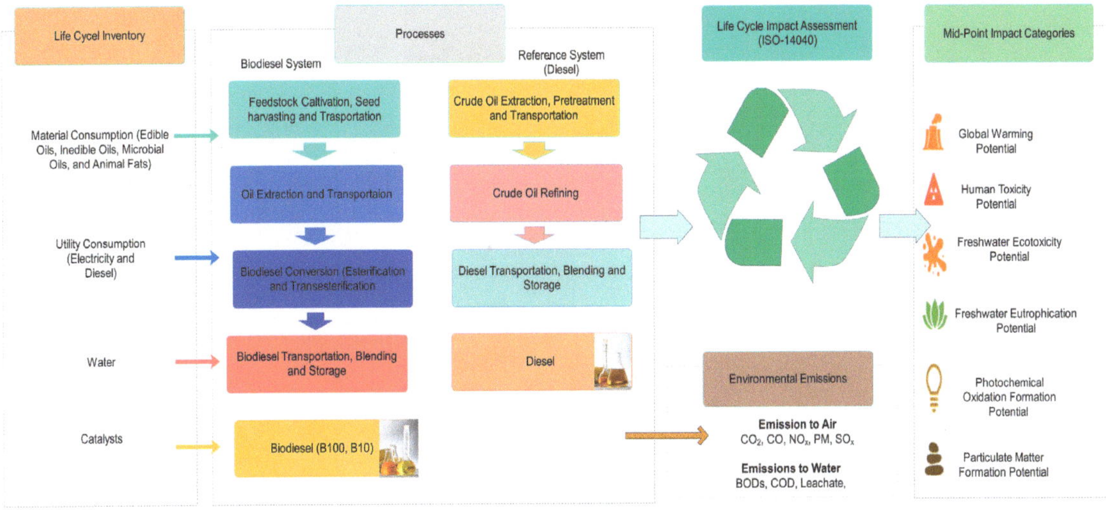

Figure 3. The life cycle assessment (LCA) of biodiesel production.

Functional Unit

The functional unit (FU) is a vital foundational standard for calculations and comparisons in LCA. One of the distinguishing features of LCA within environmental assessment methodologies is the selection of an FU. FU is an essential element in LCA analyses, allowing the comparison of results across different studies. It measures the system's function under study and establishes a reference point to normalize all inputs and outputs.

In the context of waste management systems, the choice of FU is closely linked to the system's inputs, goal and scope, and system boundaries, particularly how waste quantities are managed and processed. When assessing the environmental impact of biodiesel production from mixed vegetable oil waste sources, different FUs may be considered depending on specific factors like mass balance, transport distances, and energy considerations. In this study, the selected FU is 1 ton (1000 kg) of vegetable oil waste utilized in biodiesel production. All inputs and outputs in the analysis are standardized to this 1 ton functional unit, ensuring consistency and comparability across the study.

2.3.2. Life Cycle Inventory (LCI)

Life cycle inventory (LCI) is the second step in the analysis of LCA, which consists of measuring field data for which all the inputs and outputs of the system are considered and calculated. The data needed for biodiesel production include the production and composition of mixed vegetable oil, emission factors, fuel needs, and electricity requirements. In addition, the data obtained from the Gabi software serve as a basis for building the system model of the process. The system boundary, waste flow with mass balance, electricity consumption, and fuel requirements are shown in Figure 4.

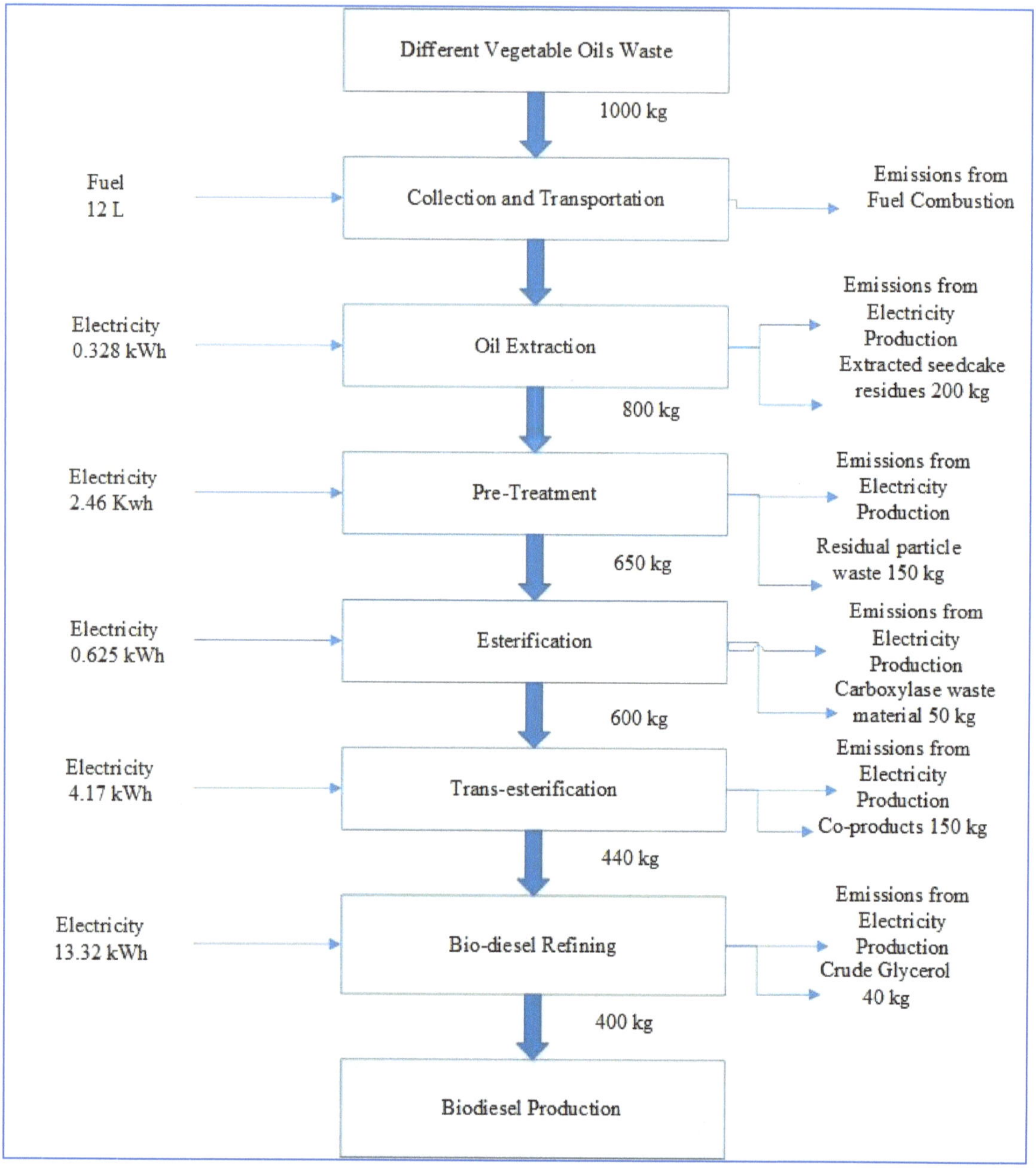

Figure 4. System boundary, vegetable oil waste stream flow with mass balance, electricity, and fuel consumption.

Table 1 presents a mass balance for the biodiesel production process. It lists the inputs, such as different oil wastes, amounting to 1 ton and energy consumption, which is measured in kWh per ton for electricity and litres for fuel. The final products include 400 kg of biodiesel, by-products such as metals and steam, and losses due to material losses and recycled waste. Additionally, the table attributes emissions into air, including NO_x and CO_2, and emissions into water, such as NO_3, as a result of the production process.

Table 1. Mass balance for the biodiesel production process.

Category	Quantities	Units	Amount
Inputs			
Material	Different Oils' Waste	ton	1
Energy Consumption	Electricity	kWh/ton	20.278
	Fuel	L	12
Outputs			
Main products	Biodiesel	kg	400
Other products	Metals	kg	0.08
	Steam	kg	0.009
	Material Losses	kg	2.2
Landfill	Recycled Waste	kg	45.3
Emissions into air	NO_x	kg	0.003405
	N_2O	kg	2.94×10^{-5}
	CH_3	kg	3.25×10^{-7}
	CO_2	kg	0.00895
	NMVOC	kg	0.01
Emissions into water	NO_3	kg	3.00×10^{-5}

2.3.3. Production and Composition of Mixed Vegetable Oil Waste

In Pakistan, 80% of the total consumption of edible vegetable oil stems from imports, while the remaining is produced locally. In 2022, the total consumption of edible vegetable oil, including palm, soybean, and peanut oils, was predicted to be approximately 3.9 million tons annually. However, domestic edible oil production was predicted to increase to 1 million tons in 2022. Palm oil accounts for a major portion of total domestic consumption (71%). Soybean oil accounts for 24% of total edible oil consumption.

LESCO (Lahore Electric Supply Company) is an electricity distributor in Lahore, Kasur, Okra, and Sheikhupura regions. The company's main objective is to ensure that the people in these areas can access a consistent and reliable electricity supply. To meet growing demand, LESCO uses a diverse mix of energy from various sources to supply electricity to the plant. According to [14], the breakdown of electricity sources is as follows: 29% hydro, 24% LNG, 14% furnace oil, 10% imported coal, 11% natural gas, 6% renewables, 4% nuclear, and 2% local. The electricity bills from WAPDA (Water and Power Development Authority) were assessed to determine the plant's power requirements for 2022. WAPDA has been assigned the duties of planning, executing, and investigating projects and schemes for generation, transmission, and distribution of power, as well as the water supply and prevention of water logging. This analysis calculated the electricity needs for each specific process within the plant. The plant operates for a total of 250 days a year, 5 days a week with 8 h of operation each day. On average, the plant consumes 20.27 kWh per ton of production.

2.3.4. Life Cycle Impact Assessment (LCIA)

Life cycle impact assessment (LCIA) is the most important stage of any LCA study. Many impact indicators, like inflows, outflows, mass, and energy consumption, are merged into LCIA to produce a single number for environmental performance. Human toxicity (carcinogenic and non-carcinogenic), photochemical ozone formation, resource depletion, acidification, ozone depletion, eutrophication, and global warming are some environmental impact categories. As a result, the four steps of the LCIA standard approach are normalization, weighting, characterization, and classification. However, ReCiPe (H) is the methodology utilized in the current study to evaluate the effects of LCIA.

2.4. Economic Assessment

The economic evaluation of this study takes operational costs and the returns on capital investment into account. Therefore, byproduct glycerol, material recovery, a reduction in landfill costs, and scrap metals are all advantages of biodiesel production. Tangible and intangible benefits include preventing infections and odors from landfills and leachate poisoning of groundwater. Additionally, the annual cost is subtracted from the benefits received to determine the net benefit annually obtained from the method. Equation (1) is used to calculate the net benefits. In the meantime, Equations (2) and (3) compute the overall costs and benefits.

Economic indicators such as payback period (PB) and net present value (NPV) are used to determine a process's economic viability [24]. The difference between cash inflows and outflows of PV over a specific period of time is referred to as NPV. Moreover, it is a method used to determine a project's economic feasibility in terms of capital costs as well as the suitability of capital budgeting and investments.

Equations (1)–(3) are as follows.

$$\text{Net Benefits} = \text{Total Benefits} - \text{Total Cost} \tag{1}$$

$$\text{Total Benefits} = \text{Biodiesel} + \text{Metals} + \text{MR} \tag{2}$$

$$\text{Total Cost} = \text{LC} + \text{UC} + \text{EC} + \text{MC} \tag{3}$$

In this equation, MR stands for material recovery, LC for labor cost, UC for utility cost, EC for energy consumption cost, and MC for maintenance cost. However, as capital costs are one-time expenses for a project, they are not considered in Equation (3). Therefore, Equation (4) can be used to calculate NPV.

$$\text{NPV} = \sum\nolimits_{t=1} n \frac{\text{Ct}}{(1+\text{k})\text{t}} - \text{Co} \tag{4}$$

where, Ct = Net cash inflow
Co = Capital cost
n = project lifespan
k = discount
t = time

The PP includes the net cash inflows generated from the initial investment because it is the expected time for the initial investment recovery. It is calculated via Equation (5).

$$\text{PP} = \frac{\text{Capital investment or initial investment cost}}{\text{Net Cash inflows(per year)}} \tag{5}$$

2.4.1. Life Cycle Costing (LCC)

Internal and external costs are considered in life cycle costing (LCC). Internal costs include the price of biodiesel l production, while external costs are the emissions during a process's life cycle stages [25]. Equation (6) is used to calculate LCC.

$$\text{Life cycle costing (LCC)} = \text{Internal cost (IC)} + \text{External cost (EC)} \tag{6}$$

2.4.2. Internal and External Costs

Operational and capital expenses are included in internal costs [26]. Maintenance, electricity, utility, and labor expenditures are all included in the operational costs of a plant. On the other hand, capital costs are associated with the plant's building, installation, equipment, and shipping. All associated costs of the current plants with similar facilities in other countries were also researched to obtain a reliable capital cost estimate. Equation (7) is used to determine the IC.

$$\text{IC} = \text{C}_\text{l} + \text{Cu} + \text{Cr} + \text{Cm} + \text{C}_\text{M} \tag{7}$$

where C_l, C_u, C_r, C_m, and C_M stand for labor, utility, raw material, maintenance, and management costs, respectively.

External cost is the damage cost that is linked with environmental emissions. It can be calculated via Equation (8).

$$EC = \sum_{k=1}^{7} C_k \times E_{k,lc} \quad (8)$$

where Ek, lc represents the emissions as determined via Gabi software. The coefficient values for Ck, which includes CO, CO_2, NO_x, SO_2, CH_4, PM, and NMVOC emissions, are obtained from the literature [21]. Table 2 shows the emissions values.

Table 2. Life cycle emissions and external cost coefficient of the biodiesel production process.

Pollutants	Coefficient [a]	Emissions [b]
CO_2	44	5.85×10^{-12}
CH_4	305	18.1
CO	828	10.3
SO_2	7485	104
NO_x	4712	49.7
NMVOC	2352	19.5
PM	8574	0.01209

[a] USD/ton; [b] Gabi (kg).

3. Results and Discussions

This section has four subsections, covering biodiesel production from mixed vegetable oil waste (Section 3.1), an environmental assessment (Section 3.2) of the impacts of biodiesel production through a life cycle approach, scenario modelling and assumptions (Section 3.3) considering different scenarios, and an in-depth economic evaluation (Section 3.4) of biodiesel production.

3.1. Biodiesel Yield and Properties

The present study's first step is identifying the percentage recovery of biodiesel based on the country's sources of mixed vegetable oil waste. Considering the quantity and composition of different types of vegetable oil production waste, a biodiesel production line for a large-scale plant is designed. The characterization and composition of different types of vegetable oil waste were taken from secondary data. The composition of mixed vegetable oil waste is mainly complex, and biodiesel is formed by a chain of hydrocarbons formed with two oxygen atoms, making it biologically active.

Biodiesel is an alternative fuel that can be derived from vegetable oils. Other alternative fuels include vegetable oil micro-emulsions, pyrolysis products of vegetable oils, and vegetable oils mixed with diesel oil. Methyl and ethyl esters can also be produced from vegetable oil or animal fat. In addition, biodiesel can also be produced from mixed vegetable oil, such as waste cooking oil, palm oil, Jatropha seed oil, castor oil, peanut oil, soybean oil, and algal oil [27]. Moreover, biodiesel can be used pure, as B100, or blended with diesel fuel with the blend denoted as BXX, in which XX is the biodiesel percentage in the blend. Hence, the most common ratio is B20, which is 80% diesel and 20% biodiesel [21].

One of the major benefits of biodiesel production is the low content of sulphur. In its chemical composition, oxygen is present, so its combustion is complete and reduces carbon monoxide, unburnt hydrocarbons, and particulate emissions, among with other contaminants [28]. Meanwhile, biodiesel can be used in any diesel engine without modifications and blended with fossil diesel in any proportion since they share similar properties [29]. Compared to fossil diesel, biodiesel has a lower calorific value of about 10% and performs worse at low temperatures. It also tends to solidify in extremely cold conditions, requiring specific additives. In addition, a byproduct, glycerin, is obtained during its production process, which can be used in cosmetic and pharmaceutical industries after purification [30].

Biodiesel from different vegetable oil feedstocks is produced in many countries, including the USA, Brazil, Argentina, Thailand, Malaysia, Singapore, China, Indonesia, and India [28]. The USA produces 1.6 billion gallons of biodiesel, mainly from soybean (40%), canola (20%), palm (20%), and tallow (20%) oils. Brazil produces 6.8 million cubic meters of biodiesel from soybeans (80%), tallow (10%), and other vegetable oils (10%) [31]. China produces 2.43 billion liters of biodiesel from cooking vegetable waste (100%). India produces 185 million liters of biodiesel from cooking oil waste. The Philippines produces 203 million liters of biodiesel from coconut oil (100%) [6].

Some studies have shown that the calorific values of mixed vegetable oil seeds meet the energy demands to produce biodiesel. Non-edible vegetable oil results in a higher production of biodiesel. Jatropha seed oil has a calorific value of 37.27 KJ. Castor oil and cooking oil waste have calorific values of 35.50 KJ and 35.7 KJ, respectively [26]. On the other hand, edible vegetable oils, including palm, coconut, and jojoba oils, show calorific values of 37.30 KJ, 38.10 KJ, 39.86 KJ, respectively, and peanut and soybean oils have higher energy contents [32]. However, several methods exist in the literature for theoretically calculating biodiesel yields [33]. When calculating the theoretical amount of biodiesel that can be produced, it is assumed that one mole of mixed vegetable oil waste will yield three moles of biodiesel, with a 100% yield. However, the actual amount of biodiesel produced is determined using the density, volume, and molar weight. The density of mixed oils is calculated based on the weight and volume as follows:

$$\rho_{oil} = \frac{m_{oil}}{v_{oil}} \tag{9}$$

ρ is the density, m is the mass, and v is the volume of oil.

Hence, the real amount produced by biodiesel is calculated by measuring the produced volume of oil, density of the biodiesel, and mass of biodiesel.

$$\frac{\rho_{biodiesel} \times V_{biodiesel}}{M_{biodiesel}} = \text{real amount of produced biodiesel} \tag{10}$$

Therefore, to obtain a 100% yield of biodiesel, the equation is as follows:

$$n = \frac{\text{real amount of produced biodiesel}}{\text{therotical amount of biodiesel}} \tag{11}$$

The benefits of the production of biodiesel include a decrease in the environmental impacts of MSW, its stable thermal and energy properties, a reduction in the share of landfilled waste, lower leachate production, the conservation of fossil fuels, decreased GHG emissions, and lower pollutant content [34]. However, despite its benefits, its challenges include high capital costs, landfilling options, unstable market conditions, and the availability of industries for co-combustion [1]. Meanwhile, another application of biodiesel production is in the transportation sector because it is a clean, renewable, and biodegradable alternative to conventional fossil diesel. Biodiesel produces fewer pollutants and a lower carbon output than other petroleum products. Compared to petroleum diesel, biodiesel produces less sulfur dioxide, fewer soot particles, and fewer unburnt hydrocarbons. By using biodiesel, people can realize health benefits because they need to spend less on healthcare products. Moreover, biodiesel can also be used to generate energy and electricity and provide heat. Nowadays, the concept of bioheat has continued to grow and depend on biofuels such as biodiesel being used as a source of energy [21]. In addition, a comparison of different studies for the line production of biodiesel is given in Table 3. Through different assessment models, the life cycle of biodiesel production is assessed by using different system inputs and outputs with different system boundaries and functional units. It is as follows.

Table 3. Different studies on the environmental impact assessment of biodiesel production.

System Boundaries	System Outputs	Functional Unit	Assessment Method	Highlights	Reference
Gate-to-gate, waste pretreatment, oil extraction, esterification, transesterification, and commercial biodiesel transportation	Waste (solid and liquid), heat, electricity biodiesel, and glycerol	1 ton	ReCiPe 2016	Biodiesel production is promising, leading to lower levels of carbon dioxide	Present study
Cradle-to-gate, oil waste collection, and oil waste pretreatment	Potassium sulfate, distillation residue, and wastewater	1 ton	ReCiPe 2008	Emissions from combustion and certain chemicals are major environmental issues in biodiesel production.	[35]
Gate-to-gate, pretreatment, transesterification, biodiesel washing, biodiesel dehydration, crude glycerol neutralization, and methanol recovery	Biodiesel, glycerol, electricity, heat, phosphate, free fatty acids, and dipotassium	1 MJ	IMPACT 2002+	Sensitivity analysis showed neutralizing crude oil, consumption of electricity, and methanol impact human health and climate change	[36]
Cradle-to-gate, waste collection, and transportation	Biodiesel	1 ton	CML	Transesterification by alkali catalyst contributes to environmental burdens.	[37,38]
Cradle-to-grave, fresh oil production, pretreatment, collection, transportation, combustion, and transesterification	Biodiesel, methanol, and glycerol	1 kg	-	Due to sulfur compounds in WCO and contaminants released during the cooking process, the resulting biodiesel produces more sulfur dioxide emissions than diesel.	[39]
Cradle-to-cradle, pretreatment, collection, transportation, oil, and transesterification process	Gas emissions (waste), heat, wastewater, glycerol, and biodiesel	1 ton	CML with all categories and Eco-indicator 99	Transesterification process had a significant environmental impact due to increased electricity consumption.	[40]
Oil waste collection and transportation, waste esterification, transesterification, and pure biodiesel	Waste disposal, glycerol, and biodiesel	1 kg	Eco-indicator 99	When a territory is small, centralized production is more eco-friendly, but as the territory increases, decentralization becomes more advisable.	[41,42]

3.2. Environmental Assessment Using LCA

In this context, the study's system boundary is gate-to-gate, and the FU is set at 1 ton of mixed vegetable oil waste. Our analysis of the results employs the ReCiPe Midpoint (H) LCIA method. The data on the quantity, materials, fuel, and energy consumption of the mixed vegetable oil waste were collected from the published literature on the country's overall production of biodiesel [43]. However, the other data (the electricity emissions) were taken from the database. Hence, the LCIA results are converted into impact categories (classification). The classified results are collected for each category indicator (characterization). The ReCiPe methodology used in this study is the follow-up of CML 2002 and Eco-indicator 99 methodologies. The indicator scores are measured in the same way as the Eco-indicator 99 methodology, and this approach uses both midpoint and endpoint modelling [11]. In addition, it is a coordinated LCIA method at a midpoint level [23]. It covers 18 midpoint impact categories, including ozone formation (terrestrial ecosystems, human health), ozone depletion, marine ecotoxicity, freshwater ecotoxicity, human toxicity (cancer, non-cancer), terrestrial acidification, terrestrial ecotoxicity, land use and water consumption, fossil depletion, climate change, ionizing radiation, resource depletion, fine particulate matter formation, and marine eutrophication. The endpoint indicators are mainly grouped into three main categories: ecosystems, resources, and human health.

3.2.1. Midpoint Assessment

The environmental categories in which the substances are released into the environment are determined by the changes made to the natural environment and are included in the midpoint assessment. These are also known as impact categories. Table 4 summarizes the midpoint results for biodiesel production. Overall, our findings showed that biodiesel production has huge potential to enhance the environment in terms of all effect categories. The highest reduction is found in global warming potential (1.36×10^{-4} kg CO_2 eq.), fossil depletion (3.29×10^{-3} kg oil eq.), ozone depletion (0.00271 kg CFC-11 eq.), and all ecotoxicity impacts (freshwater: 0.647 kg 1.4 DB eq., freshwater eutrophication: 0.0118 kg P eq., marine eutrophication: 0.134 kg N eq., and marine ecotoxicity: 9.07 kg 1.4 DB eq.). The following categories are covered under the LCIA: photochemical ozone formation, ozone depletion, human toxicity, ecotoxicity (terrestrial, freshwater, and marine), the depletion of fossils, acidification, the potential for global warming, eutrophication (marine and freshwater), ionizing radiation, resource depletion, and particle formation. The findings of the biodiesel production's midpoint assessment are shown in Table 4.

Table 4. Midpoint assessment of biodiesel production.

Impact Categories	Unit	Values
Climate change, default, excl. biogenic carbon	kg CO_2 eq.	1.36×10^4
Fine particulate matter formation	kg PM2.5 eq.	44.5
Fossil depletion	kg oil eq.	3.29×10^3
Freshwater consumption	m^3	326
Freshwater ecotoxicity	kg 1.4 DB eq.	0.647
Freshwater eutrophication	kg P eq.	0.0118
Human toxicity, cancer	kg 1.4-DB eq.	5.31
Human toxicity, non-cancer	kg 1.4-DB eq.	1.29×10^3
Ionizing radiation	kBq Co-60 eq. to air	27.4
Land use	Annual crop eq. per year	287
Marine ecotoxicity	kg 1.4-DB eq.	9.07
Marine eutrophication	kg N eq.	0.134
Metal depletion	kg Cu eq.	2.8
Photochemical ozone formation, ecosystems	kg NO_x eq.	51.5
Photochemical ozone formation, human health	kg NO_x eq.	51.4
Stratospheric ozone depletion	kg CFC-11 eq.	0.00271
Terrestrial acidification	kg SO_2 eq.	123
Terrestrial ecotoxicity	kg 1.4-DB eq.	1.2×10^4

Biodiesel production contributes to a decrease in pollutant emissions without causing an increase in greenhouse gas (GHG) emissions. Consequently, several studies have indicated that burning can elevate the concentration of CO_2 in the atmosphere. However, carbon absorption throughout a plant's life cycle can offset this increase in emissions [33]. The GWP of the biodiesel process is 136 × 10^{-4} kg CO_2 eq. Eutrophication (freshwater or marine) is aquatic nutrient enrichment brought on by H_3PO_4 and PO_4^{3-}; it causes environmental deterioration. The NP is measured in kg P or N eq. and primarily from landfill or diesel emissions. As a result, open burning emits pollutants into the atmosphere, which eventually settle with rain [44]. Emissions from sedimentation raise the productivity and nutrient levels in water bodies. Thus, algae absorb nutrients that are needed by other aquatic organisms [45]. Eventually, they are decomposed by bacteria and all die. This situation leads to a decrease in the level of DO as the amount of oxygen available to living aquatic organisms decreases [8]. The marine eutrophication of the process is 0.134 kg N eq., and the freshwater eutrophication is 0.0118 kg P eq.

The production of biodiesel is a feasible option for reducing the potential for terrestrial acidification because it results in fewer emissions of NO_x and NH_3 during the processing stage. It is expressed in kg SO_2 eq. as the unit of terrestrial acidification. The decrease in emissions harms plant and animal life and causes the acidity of soil or aquatic ecosystems to decrease. Thus, one effective method for reducing the potential effects of terrestrial acidification is the manufacture of biodiesel. In addition, the terrestrial acidification potential of biodiesel was 123 kg SO_2 eq. Human toxicity is classified into effects that cause cancer and effects that do not cause cancer, and it is related to the maximum daily intake for human toxicity. It is mainly caused by heavy metals, hydrogen sulfide, nitrogen oxides, and formaldehyde and is measured in kg 1.4 DB eq [46]. In the current study, the human toxicity potential (non-cancer) was 1.29 × 10^{-3} kg 1.4 DB eq and the HTP (cancer) was 5.31 kg 1.4 DB eq.

Ozone layer depletion causes damage to human health and ecosystems. However, more Ultraviolet B (UVB) radiation is now at the Earth's surface, which is bad for ecosystems and human health. Natural elements, including methane, water, nitrogen dioxide, and halogenated components, are the main contributors to ozone depletion [47]. Ozone depletion has significantly impacted the ecosystem. Thus, the industrial use of very stable halocarbon gases has led to the formation of halogen compounds in the stratosphere. These gases are found in landfills, and their presence poses a sepulcher environmental threat [48]. The ozone depletion potential of biodiesel was calculated as 0.00271 kg CFC-11 eq.

Compounds that are reactive in the atmosphere and the photochemical ozone formation process can harm human health and the environment. Moreover, various volatile organic compounds (VOCs) produced by activities; the use solvents and motor vehicles are significant sources of this type of pollution. The main contributors to its creation are NO_x, NMVOC, and CH_4, while NO_x is generated during transportation. The value of photochemical ozone formation for human health is 51.4 kg NO_x eq. The particles released into the atmosphere are referred to as the particulate matter formation. PM10 is the term for any organic and inorganic compounds with a diameter of less than 10 m, such as SO_x, NO_x, NH_3, and VOCs [49]. It negatively affects health, leading to respiratory disorders. Therefore, because it necessitates more energy-intensive waste collection and treatment processes, the landfill is the least preferable alternative in terms of air pollution [50]. The PMF in this study was 44.5 kg $PM_{2.5}$ eq. Ecotoxicity refers to the effect of toxic substances on wetland ecosystems and forests. The effect on oceans is known as marine ecotoxicity, while the effect of toxic substances on freshwater bodies like rivers and lakes is known as freshwater ecotoxicity [51]. According to the current study, the marine and freshwater ecotoxicity values were about 9.07 and 0.647 kg 1.4 DB eq, respectively.

Ionization is the environmental release of radioactive elements that cause a higher radiation potential. A significant amount can result in immediate fatalities, severe radiation burns, or acute consequences. This is a result of the radioactive elements found in rocks and soils of landfills [7]. The ionizing radiation potential of the current model was 27.4 kBq Co-60 eq. to air. Resource depletion is the consumption of natural resources. The electricity consumption for 1 t biodiesel production is 21.75 kWh. Studies indicate a decrease in fossil fuels that are mostly used in the power sector [52]. The generation of biodiesel was

3.29×10^{-3} oil eq., which is essentially no fossil depletion. The electricity generated from biodiesel can balance out the electricity used to produce biodiesel [32]. The metal depletion potential was 2.8 kg Cu eq.

While LCAs have been used in South Africa, India, Russia, and Brazil over the past 15 years, they have also been extensively used in several European nations [53]. The production line for mixed vegetable oil waste biodiesel, its composition, and its percentage recovery affects the fuel's economic and environmental advantages. Biodiesel has many uses, and its manufacturing characteristics vary depending on location [17]. Furthermore, the direct emissions of WtE facilities and the LC performance of biodiesel are the waste composition, recovery efficiency, and type of biological treatment. Since biodiesel may be used as a substitute for petroleum diesel, it boosts energy security, improves the environment and air quality, and uses less energy during production than conventional fossil fuels [54]. Therefore, decreased eutrophication and acidification are brought on by the reduction in NO_x emissions [55]. The current study carried out an LCA of the generation of biodiesel from mixed vegetable oil waste.

3.2.2. Normalized Results

The environmental category units are different for each category. Hence, they cannot be compared. The results are normalized, in which the category indicators are divided by a reference value. Moreover, normalized results signify the average environmental impact that a single statistical person exerts, and they are expressed in person equivalent (PE) units [17]. The normalized results are shown in Table 5. The current research paper utilizes the reCiPe 2016 V1.1 (H), global (PE) eliminating biogenic carbon, midpoint normalization built-in Gabi program. Human toxicity (non-cancer), ozone formation, and particulate matter formation have proportionately bigger contributions to the production of biodiesel, while ecotoxicity and climate change have moderate effects. The impacts of land use on eutrophication are minimal.

Table 5. Normalized LCIA results of the biodiesel production process.

Categories	Unit	Values
Ecosystems		
Climate change freshwater ecosystems	species. yr	4.17×10^{-7}
Climate change terrestrial ecosystems	species. yr	0.0153
Freshwater consumption, freshwater ecosystems	species. yr	2.27×10^{-7}
Freshwater consumption, terrestrial ecosystems	specie. yr	0.0016
Freshwater ecotoxicity	species. yr	1.8×10^{-7}
Freshwater eutrophication	species. yr	3.17×10^{-6}
Land use	species. yr	0.00102
Marine ecotoxicity	species. yr	3.81×10^{-7}
Marine eutrophication	species. yr	8.84×10^{-8}
Photochemical ozone formation, ecosystems	species. yr	0.00266
Terrestrial acidification	species. yr	0.0104
Terrestrial ecotoxicity	species. yr	5.49×10^{-5}
Human Health		
Climate change human health	DALY	3.8
Fine particulate matter formation	DALY	8.39
Freshwater consumption, human health	DALY	0.184
Human toxicity, cancer	DALY	0.00529

Table 5. *Cont.*

Categories	Unit	Values
Human Health		
Human toxicity, non-cancer	DALY	0.0886
Ionizing radiation	DALY	6.98×10^{-5}
Photochemical ozone formation, human health	DALY	0.014
Stratospheric ozone depletion	DALY	0.000431
Resources		
Metal depletion	$	420
Fossil depletion	$	7.4×10^{-4}

3.2.3. Hotspot Identification

To regulate the main contributors' stages in the life cycle of biodiesel production, a comparison of the contributions of the individual processes of biodiesel production is shown in Figure 5. This shows a comparison between the involvement of individual processes and landfills. The impact of oil extraction on the overall impact categories is minimal and hence considered negligible. The impact of the pretreatment stage on all the impact categories is very low and almost the same for each. However, the impacts of the esterification and transesterification processes on freshwater, terrestrial, and marine ecotoxicity are almost the same and account for <20%. The stage of biodiesel refinery has the highest contribution, primarily on the global warming potential, human toxicity potential, eutrophication, and acidification potential. In addition, along with the refining process, landfills account for 45% of human toxicity and marine ecotoxicity. The major reason for their higher impacts is the consumption of electricity and heat. Therefore, these two processes are more fuel-intensive than those in the biodiesel production system. The modelling results highly stimulate the data on electricity consumption. Electricity is a cleaner technology, but the emissions from it are at the time of production. The electricity supplied to a plant by LESCO is from the grid mix, and it mostly involves non-renewables. Our overall results show that the impacts are the same across the categories considered. Hence, the system boundary of the present study is from gate to gate; only biodiesel-obtaining processes were considered. Thus, for a plant, only collection and transportation are considered. Another study also shows the usage of electricity and transportation as the dominant stages [4].

3.2.4. Endpoint Assessment

The term "endpoint assessment" (also known as "damage categories") refers to how much of a material is released into the environment before it causes harm. These categories also cover the environment. Endpoint indicators combined all effect subcategories into three major categories: ecosystems, human health, and resources. The results of the biodiesel production of our endpoint assessment are shown in Table 6.

3.3. Scenario Modeling and Assumptions

The midpoint results of the LCA of the current model (Figure 6) show that the electricity supplied to the plant by LESCO is from a grid mix, in which the major contribution is from non-renewables. Therefore, all the secondary emissions of electricity are considered in the LCA. However, in scenario modelling, the electricity supply is assumed to be from photovoltaic solar cells instead of the grid mix. A comparison of scenario modelling and the current model is shown in Figure 6.

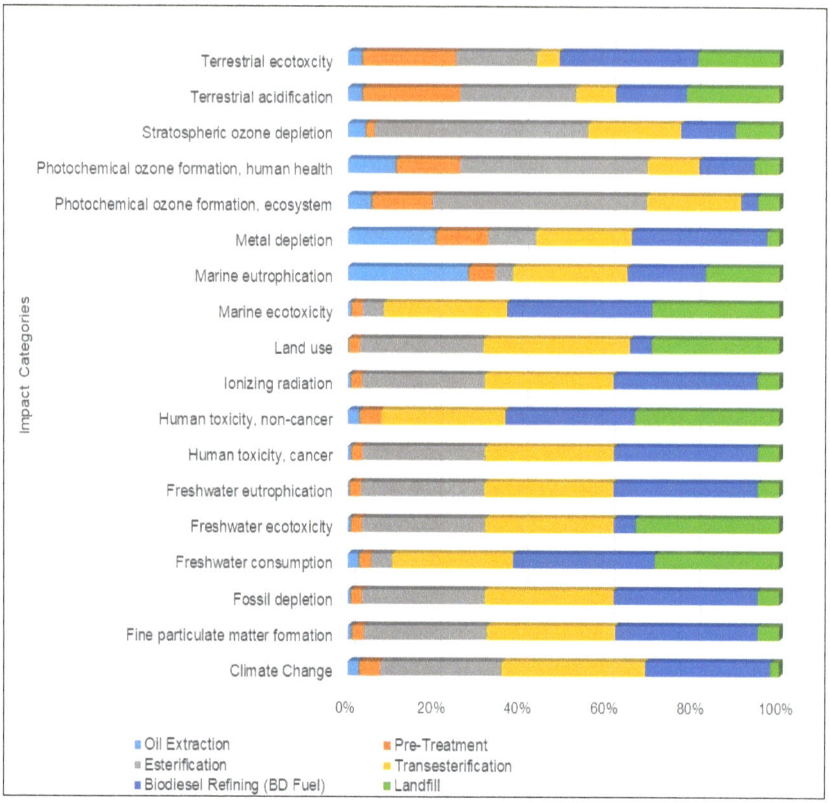

Figure 5. Role of individual processes in the overall environmental impact of the biodiesel production process to midpoint categories linked to landfill emissions.

Table 6. Endpoint results of life cycle characterization of the biodiesel production line.

Categories	Unit	Values
	Ecosystems	
Climate change terrestrial ecosystems	species. yr	3.82×10^{-5}
Climate change freshwater ecosystems	species. yr	1.04×10^{-9}
Photochemical ozone formation, ecosystems	species. yr	6.64×10^{-6}
Freshwater consumption, freshwater ecosystems	species. yr	5.68×10^{-10}
Freshwater consumption, terrestrial ecosystems	species. yr	4.01×10^{-6}
Land use	species. yr	2.55×10^{-6}
Marine ecotoxicity	species. yr	9.52×10^{-10}
Marine eutrophication	species. yr	2.21×10^{-10}
Freshwater ecotoxicity	species. yr	4.5×10^{-10}
Freshwater eutrophication	species. yr	7.93×10^{-9}
Terrestrial acidification	species. yr	2.61×10^{-5}
Terrestrial ecotoxicity	species. yr	1.37×10^{-7}

Table 6. *Cont.*

Categories	Unit	Values
Human health		
Climate change, human health	DALY	0.0127
Human toxicity, cancer	DALY	1.76×10^{-5}
Human toxicity, non-cancer	DALY	0.000295
Fine particulate matter formation	DALY	0.028
Freshwater consumption, human health	DALY	0.000614
Ionizing radiation	DALY	2.33×10^{-7}
Stratospheric ozone depletion	DALY	1.44×10^{-6}
Photochemical ozone formation, human health	DALY	4.67×10^{-5}
Resources		
Fossil depletion	$	247
Metal depletion	$	1.4

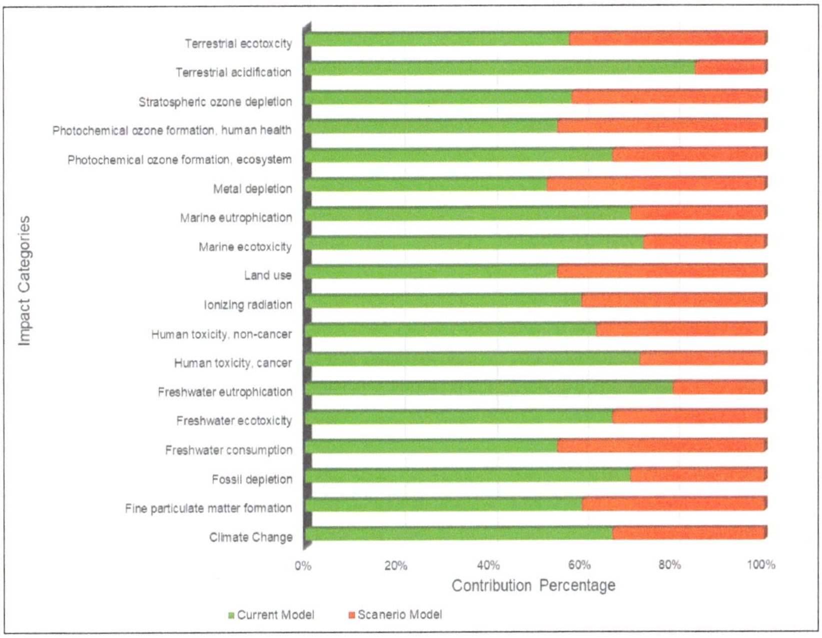

Figure 6. Comparison of current model versus scenario model of biodiesel production process. (In the current model, electricity is supplied from the grid mix, while in the scenario model, it is supplied from photovoltaic solar cells).

Thus, it is apparent that the fine particle matter formation decreases from 44.5 to 0.725 kg $PM_{2.5}$ eq., and the fossil depletion increases from 3.29×10^{-3} to 196 kg oil eq. The effect on freshwater consumption in the current model is 326 m^3, while it slightly decreased to 279 m^3 in the scenario model. Similarly, the terrestrial acidification decreases from 123 to 2.15 kg SO_2 eq., and human toxicity and cancer decrease from 5.31 to 0.405 kg 1.4 DB eq., respectively. In addition, the effect on metal depletion increases from 2.8 to 12.6 kg Cu eq. The midpoint results for the biodiesel production of the scenario model are in Table 7.

Table 7. Midpoint results of life cycle characterization of biodiesel production line of the scenario model.

Impact Categories	Unit	Values
Climate change, default, excl. biogenic carbon	kg CO_2 eq.	654
Fine particulate matter formation	kg PM2.5 eq.	0.725
Fossil depletion	kg oil eq.	196
Freshwater consumption	m^3	279
Freshwater ecotoxicity	kg 1.4 DB eq.	0.136
Freshwater eutrophication	kg P eq.	0.00111
Human toxicity, cancer	kg 1.4-DB eq.	0.405
Human toxicity, non-cancer	kg 1.4-DB eq.	194
Ionizing radiation	kBq Co-60 eq. to air	8.05
Land use	Annual crop eq. per year	30.2
Marine ecotoxicity	kg 1.4-DB eq.	4.33
Marine eutrophication	kg N eq.	0.00765
Metal depletion	kg Cu eq.	12.6
Photochemical ozone formation, ecosystems	kg NO_x eq.	2.68
Photochemical ozone formation, human health	kg NO_x eq.	2.62
Stratospheric ozone depletion	kg CFC-11 eq.	0.000159
Terrestrial acidification	kg SO_2 eq.	2.15
Terrestrial ecotoxicity	kg 1.4-DB eq.	1.33×10^4

Biodiesel production from mixed vegetable oil waste reduces the problems of the waste disposal/handling of waste, reduces emissions by avoiding them, and provides economic benefits. There are some weaknesses in current SWM practices, such as waste handling and the local fleet industry for handling waste. The lack of coordination among stakeholders, including municipalities and the informal sector, treatment technology and management, and initial capital investments are the main difficulties in implementing a sound SWM model [56]. Moreover, there is a dire need to improve the SW sector through proper stakeholder management and coordination. In addition, economic incentives should be given to level up applicable enterprises and implement a sound solid waste management model [57].

Biodiesel produces a clean-burning, renewable alternative fuel to conventional petroleum-based fuels [58]. It improves energy balance and security. Thus, locally manufactured biodiesel can be directly substituted for conventional diesel fuels. Biodiesel produced from soybeans has a positive energy balance that shows a higher yield in a higher amount of energy for every unit of fossil fuel consumed [59]. Moreover, biodiesel production also reduces emissions and improves air quality because of its lower life cycle rating and overall lower emissions which improve air quality. Other biodiesel applications include its use as fuel filters, in oil spill cleanups, as heating oil, and in biodiesel electricity generators.

In Pakistan, vegetable oil is mainly used to treat biodiesel, either with ethanol or methanol to synthesize it. The basic reason for using methanol worldwide is its lower price; coal is the main production source. Around 180 billion tons of coal reserves are in Pakistan, the fifth largest in the world. However, in Pakistan, ethanol production is also higher because its 76 operational sugar mills produce 300,000 tons of cane daily. Some distillery units have a capacity to produce 2 million tons of molasses to form 400,000 tons of ethanol. Excess ethanol can either be used for gasohol or to produce biodiesel. The production capacity of these units is approximately 400,000 tons. The country needs to export up to 80,200 tons, after which about 318,000 tons of ethanol would remain and could be used for biodiesel synthesis. Therefore, this stock is necessary to increase biodiesel production in Pakistan. In 2021, biodiesel production in the country was 0.09 thousand barrels per day; still, the country has a high feasibility of producing biodiesel in large amounts [4].

According to the Alternative Renewable Energy Policy of 2019, by 2025, Pakistan will generate 20% of its energy from renewables, and by 2030, it will generate 30%, promoting the use of alternative energy resources. In recent years, thermal energy has comprised 63%

of the energy mix, while renewable energy has suitably contributed 1.1% [60]. Contrary to the above policy, the current scenario does not contribute to either meeting the target by 2025 or reducing emissions. Furthermore, Pakistan can reach this goal sustainably by exploiting the potential of renewable energy. As a result, the biodiesel production model will support this policy, as well as waste management firms and municipalities, while taking financial limits into account. The government should also provide incentives in the form of subsidies to encourage stakeholders to participate in the execution of the program.

3.4. *Life Cycle Cost and Economic Assessment*

The economic assessment results show the viability of biodiesel production from mixed vegetable oil waste. The benefits of biodiesel production include biodiesel, material (glycerol), and metal recovery, as well as the conservation of land in terms of landfilling. Table 8 shows the revenue generation from a biodiesel production plant. Biodiesel is traded at 0.83 USD/kg, generating 2460.67 USD/day in revenue. Recovered materials and metals are traded at 0.755 USD/bag and 0.672 USD/kg, generating 224.380 USD/day and 199.721 USD/day in revenue, respectively. The total income generation by the biodiesel production plant is 1821.46 USD/day for 1 ton of processed mixed oil waste. Moreover, the biodiesel production plant produces 22 kg each month. Per day, the production cost is 2135.460 USD, and the monthly income generated by the plant is 57,796.617 USD. The yearly income generation by the plant is 638,839.631 USD. The income generated by the biodiesel plant is shown in Table 8.

Table 8. The production, working days, total waste processed, and income generated by the biodiesel plant.

Product Type	Mixed Oil Waste	Material Recovery	Metal	Total
Total waste (kg)	1000	-	-	1000
Working Days	22	22	22	
Percentage in waste	66.25	33.67	0.08	100
Per-day production (kg)	2534	2800	75	5229
Per-day cost ($)	2001.681	98.44258	35.33681	2135.460
Per-month income ($)	55,045.43	1995.776	755.41101	57,796.617
Per-year income ($)	596,443.87	33,428.841	8966.920	638,839.631

Biodiesel = 0.83 USD/kg, Material Recovery = 0.755 USD/bag, Metals= 0.672 USD/kg.

The 20 USD/ton operational cost is considered excellent, and the 3–4 year payback period is economically feasible. The operational cost of the current study is 20 USD/ton, significantly closer than that, and the PP of the initial capital investment is 4 years. As mentioned in Section 2.4, the LCC includes external and internal costs. Equation (7) was used to calculate the internal cost of biodiesel, which was calculated to be 24.33 USD/ton. The external cost was calculated using Equation (8), estimated at 3558.16 USD/ton. Therefore, the LCC of biodiesel calculated via Equation (6) was 3634.9 USD/ton. Table 8 shows the economic assessment results. Moreover, the overall economic assessment results are shown in Table 9.

Table 9. Results of economic assessment.

Costs	USD/Year
Capital Costs	
Capital cost	878,665.35
Installation cost	25,065.85
Operation and Maintenance	
Maintenance cost	3387.61
Utility cost	30,037.31
Labor cost	50,827.92
Electricity cost	19,981.72
Total cost	104,234.56
Benefits	
Biodiesel	596,443.87
Recovery	33,428.84
Metals	8966.920
Total benefits	638,839.631
LCC (USD/ton)	3634.9
NPV	4,648,132.82
PB	4 Years

3.5. Energy Resource for Achieving Sustainable Production

A comprehensive analysis of the environmental impacts of gasoline, diesel, and biodiesel using the LCA reveals that biodiesel significantly reduces greenhouse gas emissions compared to traditional fuels. However, it also increases particulate matter (PM10) emissions, nitrous oxide, nitrogen oxides (NOx), and nutrients that contribute to eutrophication [61]. This balanced view is essential for planning a sustainable transportation system, considering both the environmental benefits and the challenges of biodiesel. Transportation companies in Malaysia need help for adopting biodiesel [62]. A differentiation strategy could help policymakers promote biodiesel usage more effectively by addressing identified barriers [63].

Consumer attitudes towards cellulosic ethanol, another renewable energy source, were explored in the United States. The survey data analysis revealed strong public support and willingness to pay more for cellulosic ethanol. This highlights the significance of consumer perceptions in the adoption of sustainable fuels [64]. The findings indicate a significant interest in alternative fuels, with the environment, energy consumption, climate change concerns, and gasoline prices being key determinants of one's willingness to pay [65,66]. These regional insights reveal the complex interplay between environmental impacts, policy challenges, and consumer attitudes in the context of sustainable biodiesel and renewable resource mobility initiatives [67].

The current study's findings provide key information about the environmental and economic aspects of biodiesel production from mixed vegetable oil waste. The conversion of mixed and different vegetable (edible and non-edible) oils in the production of biodiesel leads to huge benefits in terms of energy generated, reductions in emissions, and reductions in the amount of waste sent to landfills [68]. Biodiesel can be produced locally from a variety of feedstock, reducing our dependence on imported fossil fuels. This can enhance energy security and promote local economic development [69]. However, it also helps in achieving a circular economy and our sustainability goals.

4. Conclusions and Future Directions

Pakistan's energy needs can be met, and indigenous renewable energy sources in Pakistan are highly significant. Furthermore, additional research and development on renewable energies are needed to improve the effects of consumption. Considering a country's economic and environmental conditions, this study was designed to investigate the feasibility of biodiesel production from mixed vegetable oil waste. A medium-scale 1 t (1000 kg) biodiesel plant was designed, and from 1000 kg of mixed vegetable oil waste, 400 kg of biodiesel can be produced. Pakistan can address its energy supply disparities by effectively implementing biodiesel in energy production. This would require supplying energy for household cooking, powering vehicles, and supporting industrial processes, including electricity generation. Nonetheless, a more comprehensive and thoughtful approach to research is needed to promote renewable energy technologies and establish clear biodiesel policies for the government. This should not be marginal but rather a deliberate focus on strengthening initial local research initiatives. Our research highlights the significance of aligning with the United Nations' SDGs. In particular, our work contributes to the progress of SDG-7, which stresses the importance of accessible and clean energy, and SDG-12, which promotes responsible consumption and production.

An LCA was performed to estimate the current project's environmental impacts. The functional unit was 1 t. Thus, three steps (classification, characterization, and normalization) were performed. In addition, midpoint and endpoint assessments were also conducted. The calculated midpoint impacts were CC: 1.36×10^{-5} kg CO_2 eq, HT: 5.31 kg 1.4 DB eq, OD: 0.00271 kg CFC-11 eq, AP: 123 kg SO_2 eq, and POF: 51.4 kg NO_x eq. To determine the main contributors' stages in the life cycle of biodiesel, the relative contribution by individual biodiesel type was calculated. The percentage share of ecotoxicity is greater and has an impact, particularly on marine ecotoxicity and human toxicity. Thus, its collection and transportation at plants show that usage and transportation are leading stages. This process is more fuel-intensive than other processes. To further alleviate the impacts, scenario modelling was conducted, in which the electricity supply was from photovoltaic solar cells. As a result, the global warming potential increases from 1.36×10^{-5} to 2.91×10^{-5} kg CO_2 eq., and the fine particle matter formation and freshwater ecotoxicity also decrease from 44.5 to 0.725 kg $PM_{2.5}$ eq. and 0.647 to 0.136 kg 1.4 DB eq., respectively. Furthermore, the effect on freshwater consumption in the current model is 326 m^3, which decreased to 279 m^3 in the scenario model. Likewise, human toxicity (cancer) and marine eutrophication decreased from 5.31 to 0.405 kg 1.4 DB eq. and 0.134 to 0.00765 kg N eq., respectively. The economic analysis showed that biodiesel is traded at 0.83 USD/kg, generating 2460.67 USD/day (753,132.84 PKR/day). Recovered materials and metals are traded at 0.755 USD/bag and 0.672 USD/kg, generating 224.380 USD/day and 199.721 USD/day in revenue, respectively. Furthermore, the total income generated by the biodiesel plant is 1821.46 USD per day (0.6 million PKR/day) for 100 tons of processed mixed vegetable oil waste. The yearly income generated by the plant is 0.6 million USD (195 million PKR). The payback period of the initial capital investment is four years.

Author Contributions: Conceptualization, F.M.; methodology, K.S.; data curation, I.A.; review and editing, A.-S.N.; software, M.H.J. and A.A.; final editing, M.N. All authors have read and agreed to the published version of the manuscript.

Funding: This research received no external funding.

Institutional Review Board Statement: Not applicable.

Informed Consent Statement: Not applicable.

Data Availability Statement: No new data were created or analyzed in this study.

Conflicts of Interest: The authors declare no conflict of interest.

References

1. International Trade Administration (2022). Pakistan—Country Commercial Guide. Available online: https://www.trade.gov/country-commercial-guides/pakistan-waste-management (accessed on 15 October 2023).
2. Irfan, M.; Zhao, Z.Y.; Panjwani, M.K.; Mangi, F.H.; Li, H.; Jan, A.; Ahmad, M.; Rehman, A. Assessing the energy dynamics of Pakistan: Prospects of biomass energy. *Energy Rep.* **2020**, *6*, 80–93. [CrossRef]

3. Anwar, J. Analysis of energy security, environmental emission and fuel import costs under energy import reduction targets: A case of Pakistan. *Renew. Sustain. Energy Rev.* **2016**, *65*, 1065–1078. [CrossRef]
4. Kumar, B.; Verma, P. Life cycle assessment: Blazing a trail for bio resources management. *Energy Convers. Manag.* **2021**, *10*, 100063.
5. IEA CEM. Global energy outlook: International Energy Agency. *Sustain. Energy Rev.* **2017**, *13*, 903–909.
6. Mofijur, M.; Masjuki, H.H.; Kalam, M.A.; Atabani, A.E.; Fattah, I.R.; Mobarak, H.M. Comparative evaluation of performance and emission characteristics of Moringa oleifera and Palm oil-based biodiesel in a diesel engine. *Ind. Crops Prod.* **2014**, *53*, 78–84. [CrossRef]
7. Nejat, P.; Jomehzadeh, F.; Taheri, M.M.; Gohari, M.; Majid, M.Z.A. A global review of energy consumption, CO_2 emissions and policy in the residential sector (with an overview of the top ten CO_2 emitting countries). *Renew. Sustain. Energy Rev.* **2015**, *43*, 843–862. [CrossRef]
8. Yusuf, N.N.; Kamarudin, S.K.; Yaakub, Z. Overview on the current trends in biodiesel production. *Energy Convers. Manag.* **2011**, *52*, 2741–2751. [CrossRef]
9. Mahesh, S.E.; Ramanathan, A.; Begum, K.M.S.; Narayanan, A. Biodiesel production from waste cooking oil using KBr impregnated CaO as catalyst. *Energy Convers. Manag.* **2015**, *91*, 442–450. [CrossRef]
10. Khan, N.A.; Dessouky, H. Prospect of biodiesel in Pakistan. *Renew. Sustain. Energy Rev.* **2009**, *13*, 1576–1583. [CrossRef]
11. Rafique, M.M.; Rehman, S. National energy scenario of Pakistan–Current status, future alternatives, and institutional infrastructure: An overview. *Renew. Sustain. Energy Rev.* **2017**, *69*, 156–167. [CrossRef]
12. Chakrabarti, M.H.; Ali, M.; Baroutian, S.; Saleem, M. Techno economic comparison between B10 of Eruca Sativa L. and other indigenous seed oils in Pakistan. *Process Saf. Environ. Prot.* **2010**, *89*, 165–171. [CrossRef]
13. Chakrabarti, M.H.; Usmani, J.N.; Ali, M.A. Techno-economic evaluation of two non-edible vegetable oil based biodiesel in Pakistan. *NED Univ. J. Res.* **2010**, *7*, 43–54.
14. Prasoulas, G.; Gentikis, A.; Konti, A.; Kalantzi, S.; Kekos, D.; Mamma, D. Biodiesel Production from Food Waste Applying the Multienzyme System Produced On-Site by Fusarium oxysporum F3 and Mixed Microbial Cultures. *Fermentation* **2020**, *6*, 39. [CrossRef]
15. Rupani, P.F. Current status and future perspectives of solid waste management in Iran: A critical overview of Iranian metropolitan cities. *Environ. Sci. Pollut. Res.* **2019**, *26*, 32777–32789. [CrossRef] [PubMed]
16. Kaisan, F.O.; Anafi, J.; Nuszkowski, D.M.; Kulla, D.M. Calorific value, flash point and cetane number of biodiesel from cotton, Jatropha and neem binary and multi-blends with diesel. *Biofuels* **2017**, *11*, 321–327. [CrossRef]
17. Dastjerdi, B.; Strezov, V.; Rajaeifar, M.A.; Kumar, R.; Behnia, M.A. systematic review on life cycle assessment of different waste to energy valorization technologies. *J. Clean. Prod.* **2021**, *290*, 125747. [CrossRef]
18. *ISO 14040:1997*; Life Cycle Assessment-Principle and Guidelines. International Organization for Standardization: Geneva, Switzerland, 1997.
19. *ISO 14040:2006*; Environmental Management-Life Cycle Assessment-Principles and Framework. International Organization for Standardization: Geneva, Switzerland, 2006.
20. Hosseinzadeh-Bandbafha, H.; Tabatabaei, M.; Aghbashlo, M.; Rehan, M.; Nizami, A.S. Determining key issues in life-cycle assessment of waste biorefineries. In *Waste Biorefinery*; Elsevier: Amsterdam, The Netherlands, 2020; pp. 515–555.
21. Ahmad, M.; Khan, M.; Sultana, S. Systematic studies of Oil yielding Plants and their applications as a Bio-diesel resource in Pakistan. In Proceedings of the CSF-HEC Workshop on Promoting Industry Academia Linkages for Developing the Knowledge Based Enterprise Sector in Pakistan, Islamabad, Pakistan, 31 January 2007.
22. Bali, J.S.; Sankanna, C. Performance and emission characteristics of waste cooking oil as biodiesel in CI Engine. *Int. J. Curr. Eng. Technol.* **2017**, *4*, 38–42.
23. Ali, S.M.; Appolloni, A.; Cavallaro, F.; D'Adamo, I.; Di Vaio, A.; Ferella, F.; Gastaldi, M.; Ikram, M.; Kumar, N.M.; Martin, M.A.; et al. Development Goals towards Sustainability. *Sustainability* **2023**, *15*, 9443. [CrossRef]
24. Bozbas, K. Biodiesel as an alternative motor fuel: Production and policies in the European Union. *Renew. Sustain. Energy Rev.* **2008**, *12*, 542–552. [CrossRef]
25. Yang, J.; Xu, M.; Zhang, X.; Hu, Q.; Sommerfeld, M.; Chen, Y. Life-cycle analysis on biodiesel production from microalgae, water footprint and nutrients balance. *Bioresour. Technol.* **2010**, *102*, 159–165. [CrossRef]
26. Zhang, Z.; Liu, H.; Tang, Y.; Xiang, Y.; Gao, W. Design of a comprehensive experiment of the synthesis of biodiesel catalyzed by CaO. *In IOP Conference Series: Earth Environ. Sci.* **2020**, *450*, 012055. [CrossRef]
27. Ibrahim, H.; Ahmed, A.S.; Bugaje, I.M.; Ugwumma, C.D. Optimized Biodiesel Production from Waste Cooking Oil (WCO) using Calcium Oxide (CaO) Nano-catalyst and its Efficacy for Biodiesel Production. *Sci. Rep.* **2019**, *3*, 90–93.
28. Musa, I.A. The effects of alcohol to oil molar ratios and the type of alcohol on biodiesel production using transesterification process. *Egypt. J. Pet.* **2016**, *25*, 21–31. [CrossRef]
29. Marchetti, J.M.; Errazu, A.F. Technoeconomic study of supercritical biodiesel production plant. *Energy Convers. Manag.* **2008**, *49*, 2160–2164. [CrossRef]
30. Hossain, T. Production of Biodiesel Using Alkaline Based Catalysts from Waste Cooking Oil: A Case Study. *Procedia Eng.* **2015**, *105*, 638–645.
31. Meng, X.; Chen, G.; Wang, Y. Biodiesel production from waste cooking oil via alkali catalyst and its engine test. *Fuel Process Technol.* **2008**, *89*, 851–857. [CrossRef]

32. Wendi, W. Effect of Reaction Temperature and Catalyst Concentration. In Proceedings of the 5th Sriwijaya International Seminar on Energy-Environmental Science and Technology, Palembang, Indonesia, 10–11 September 2014; Volume 1, pp. 32–37.
33. Verma, P.; Sharma, M.P.; Dwivedi, G. Impact of alcohol on biodiesel production and properties. *Renew. Sustain. Energy Rev.* **2015**, *56*, 319–333. [CrossRef]
34. Steenblik, R. Liberalization of Trade in Renewable Energy and Associated Technologies: Biodiesel, Solar Thermal and Geothermal Energy. In *OECD Trade and Environment Working Paper*; OECD: Paris, France, 2004.
35. Foteinis, S.; Chatzisymeon, E.; Litinas, A.; Tsoutsos, T. Used-cooking-oil biodiesel: Life cycle assessment and comparison with first-and third-generation biofuel. *Renew. Energy* **2020**, *153*, 588–600. [CrossRef]
36. Aghbashlo, M.; Almasi, F.; Jafari, A.; Nadian, M.H.; Soltanian, S.; Lam, S.S.; Tabatabaei, M. Describing biomass pyrolysis kinetics using a generic hybrid intelligent model: A critical stage in sustainable waste-oriented biorefineries. *Renew. Energy* **2021**, *170*, 81–91. [CrossRef]
37. Liang, S.; Ming, X.; Tianzhu, Z. Life cycle assessment of biodiesel production in China. *Bioresour. Technol.* **2013**, *129*, 72–77. [CrossRef]
38. Bradley, T.; Rajaeifar, M.A.; Kenny, A.; Hainsworth, C.; Del Pino, V.; Del Valle Inclán, Y.; Heidrich, O. Life cycle assessment of microalgae-derived biodiesel. *Int. J. Life Cycle Assess.* **2023**, *28*, 590–609. [CrossRef]
39. Siregar, K. Study of life cycle assessment in biodiesel production from crude palm oil and its benefits for the sustainability of oil palm industry in Aceh province Indonesia. *IOP Conf. Ser. Earth Environ. Sci.* **2021**, *644*, 012017. [CrossRef]
40. Gholami, A.; Fathollah, P.; Akbar, M. Recent advances of biodiesel production using ionic liquids supported on nanoporous materials as catalysts: A review. *Front. Energy Res.* **2020**, *8*, 144. [CrossRef]
41. Akram, F.; Ul Haq, I.; Raja, S.I.; Mir, A.S.; Qureshi, S.S.; Aqeel, A.; Shah, F.I. Current trends in biodiesel production technologies and future progressions: A possible displacement of the petro-diesel. *J. Clean. Prod.* **2022**, *370*, 133479. [CrossRef]
42. Maheshwari, P. A review on latest trends in cleaner biodiesel production: Role of feedstock, production methods, and catalysts. *J. Clean. Prod.* **2022**, *355*, 131588. [CrossRef]
43. Wang, Z.; Wang, Q.; Jia, C.; Bai, J. Thermal evolution of chemical structure and mechanism of oil sands bitumen. *Energy* **2022**, *244*, 1. [CrossRef]
44. Tshizanga, N.; Funmilayo, E. Optimization of biodiesel production from waste vegetable oil and eggshell ash. *S. Afr. J. Chem. Eng.* **2017**, *23*, 145–156.
45. Yano, J.; Aoki, T.; Nakamura, K.; Yamada, K.; Sakai, S.-I. Life cycle assessment of hydrogenated biodiesel production from waste cooking oil using the catalytic cracking and hydrogenation method. *Waste Manag.* **2015**, *38*, 409–423. [CrossRef]
46. Silitonga, A.S.; Masjuki, H.H.; Mahlia, T.M.I.; Ong, H.C.; Atabani, A.E.; Chong, W.T. A global comparative review of biodiesel production from jatropha curcas using different homogeneous acid and alkaline catalysts: Study of physical and chemical properties. *Renew. Sustain. Energy Rev.* **2020**, *24*, 514–533. [CrossRef]
47. Cheng, Z.; Guo, Z.; Fu, P.; Yang, J.; Wang, Q. New insights into the effects of methane and oxygen on heat/mass transfer in reactive porous media. *Int. Commun. Heat Mass Transf.* **2021**, *129*, 105652. [CrossRef]
48. Li, H.; Li, G.; Li, L. Comparative investigation on combustion characteristics of ADN-based liquid propellants in inert gas and oxidizing gas atmospheres with resistive ignition method. *Fuel* **2023**, *334*, 126742. [CrossRef]
49. Liu, L.; Wu, Y.; Wang, Y.; Wu, J.; Fu, S. Exploration of environmentally friendly marine power technology -ammonia/diesel stratified injection. *J. Clean. Prod.* **2022**, *380*, 135014. [CrossRef]
50. Yang, X.; Liu, K.; Han, X.; Xu, J.; Bian, M.; Zheng, D.; Xie, H.; Zhang, Y.; Yang, X. Transformation of waste battery cathode material $LiMn_2O_4$ into efficient ultra-low temperature NH_3-SCR catalyst: Proton exchange synergistic vanadium modification. *J. Hazard. Mater.* **2023**, *459*, 132209. [CrossRef] [PubMed]
51. Colombo, K.; Ender, L. The study of biodiesel production using CaO as a heterogeneous catalytic reaction. *Egypt. J. Pet.* **2017**, *26*, 341–349. [CrossRef]
52. Liu, Z.; Tang, P.; Hou, K.; Zhu, L.; Zhao, J.; Jia, H.; Pei, W. A Lagrange-multiplier-based Reliability Assessment for Power Systems Considering Topology and Injection Uncertainties. *IEEE Trans. Power Syst.* **2023**, 1–11. [CrossRef]
53. Sajid, Z.; Khan, F.; Zhang, Y. Process simulation and life cycle analysis of biodiesel production. *Renew. Energy* **2016**, *85*, 945–952. [CrossRef]
54. Kawentar, W.A.; Budiman, A. Synthesis of Biodiesel from Second-Used Cooking Oil. *Energy Procedia* **2013**, *32*, 190–199. [CrossRef]
55. Karolline, J.; Cruzeira, S.; Freitas, C. Organic, conventional and sustainable palm oil (RSPO): Formation of 2- and 3-MCPD esters and glycidyl esters and influence of aqueous washing on their reduction. *Food Res. Int.* **2021**, *102*, 1407–1414.
56. Wang, Y.; Wang, Q.; Li, Y.; Wang, H.; Gao, Y.; Sun, Y.; Wang, B.; Bian, R.; Li, W.; Zhan, M. Impact of incineration slag co-disposed with municipal solid waste on methane production and methanogens ecology in landfills. *Bioresour. Technol.* **2023**, *377*, 128978. [CrossRef]
57. Wang, B.; Gupta, R.; Bei, L.; Wan, Q.; Sun, L. A review on gasification of municipal solid waste (MSW): Syngas production, tar formation, mineral transformation and industrial challenges. *Int. J. Hydrogen Energy* **2023**, *48*, 26676–26706. [CrossRef]
58. Liu, L.; Tang, Y.; Liu, D. Investigation of future low-carbon and zero-carbon fuels for marine engines from the view of thermal efficiency. *Energy Rep.* **2022**, *8*, 6150–6160. [CrossRef]
59. Kan, Y.; Li, J.; Zhang, S.; Gao, Z. Novel bridge assistance strategy for tailoring crosslinking networks within soybean-meal-based biocomposites to balance mechanical and biodegradation properties. *Chem. Eng. J.* **2023**, *472*, 144858. [CrossRef]

60. Tan, Y.; Abdullah, M.O.; Hipolito, C.N. Comparison of Biodiesel Production between Homogeneous and Heterogeneous Base Catalysts. *Appl. Mech. Mater.* **2016**, *833*, 71–77. [CrossRef]
61. Koroneos, C. Comparative LCA of the use of biodiesel, diesel and gasoline for transportation. *J. Clean. Prod.* **2012**, *20*, 14–19.
62. Zailani, S.; Iranmanesh, M.; Sean Hyun, S.; Ali, M.H. Barriers of Biodiesel Adoption by Transportation Companies: A Case of Malaysian Transportation Industry. *Sustainability* **2019**, *11*, 931. [CrossRef]
63. Dias, L.C.; Passeira, C.; Malça, J.; Freire, F. Integrating life-cycle assessment and multi-criteria decision analysis to compare alternative biodiesel chains. *Ann. Oper. Res.* **2016**, *312*, 1359–1374. [CrossRef]
64. Johnson, D.M.; Halvorsen, K.E.; Solomen, B.D. Upper Midwestern US consumers and ethanol: Knowledge, beliefs and consumption. *Biomass Bioenergy* **2011**, *35*, 1454–1464. [CrossRef]
65. Ma, Z.; Zhang, C.; Chen, C. Analyzing the factors that influence Chinese consumers' adoption of the biodiésel: The private vehicles owner's investigating in Beijing. *Renew. Sustain. Energy Rev.* **2014**, *37*, 199–206. [CrossRef]
66. Demirbas, A. Importance of biodiesel as transportation fuel. *Energy Policy* **2007**, *35*, 4661–4670. [CrossRef]
67. D'Adamo, I.; Gastaldi, M.; Piccioni, J.; Rosa, P. The role of automotive flexibility in supporting the diffusion of sustainable mobility initiatives: A stakeholder attitudes assessment. *Glob. J. Flex. Syst. Manag.* **2023**, *24*, 459–481. [CrossRef]
68. Hosseinzadeh-Bandbafha, H.; Nizami, A.S.; Kalogirou, S.A.; Gupta, V.K.; Park, Y.K.; Fallahi, A.; Sulaiman, A.; Ranjbari, M.; Rahnama, H.; Aghbashlo, M.; et al. Environmental life cycle assessment of biodiesel production from waste cooking oil: A systematic review. *Renew. Sustain. Energy Rev.* **2022**, *161*, 112411. [CrossRef]
69. Carneiro, M.L.N.; Pradelle, F.; Braga, S.L.; Gomes, M.S.P.; Martins, A.R.F.; Turkovics, F.; Pradelle, R.N. Potential of biofuels from algae: Comparison with fossil fuels, ethanol and biodiesel in Europe and Brazil through life cycle assessment (LCA). *Renew. Sustain. Energy Rev.* **2017**, *73*, 632–653. [CrossRef]

Disclaimer/Publisher's Note: The statements, opinions and data contained in all publications are solely those of the individual author(s) and contributor(s) and not of MDPI and/or the editor(s). MDPI and/or the editor(s) disclaim responsibility for any injury to people or property resulting from any ideas, methods, instructions or products referred to in the content.

sustainability

MDPI

Article

A Bibliometric Analysis of Circular Economies through Sustainable Smart Cities

Ernesto D. R. Santibanez Gonzalez [1,*], Vinay Kandpal [2], Marcio Machado [3,4], Mauro Luiz Martens [3] and Sushobhan Majumdar [5]

1 CES4.0 and Industrial Engineering Department, Faculty of Engineering, Universidad de Talca, Los Niches s/n, Edificio I+D, Room 13, Curicó 3340000, Chile
2 Department of Management Studies, Graphic Era Deemed to be University, Dehradun 248002, India; vinayaimca@gmail.com
3 Graduate Program in Business Administration, Paulista University—UNIP, São Paulo 04026-002, Brazil; marciocmachado@uol.com.br or marcio.machado@docente.unip.br (M.M.); mauro.martens@gmail.com (M.L.M.)
4 Department of Business Administration—Pontifícia, Universidade Católica de, São Paulo 05014-901, Brazil
5 Former Research Fellow, Department of Geography, Jadavpur University, Kolkata 700054, India; sushobhan91@gmail.com
* Correspondence: santibanez.ernesto@gmail.com

Abstract: The rapid pace of urbanization has presented cities worldwide with a range of complex issues pertaining to the management of resources, reduction of waste, and promotion of sustainable practices. The concepts of circular economies and sustainable smart cities have arisen as viable solutions, converging to establish a revolutionary framework for the future of urban living. This study conducts a bibliometric analysis using literature focusing on the past ten years (2013–2022) of research on the circular economy and smart cities using VOSviewer. The most frequently used Scopus database was used to extract bibliometric data. 163 articles were considered for the analysis. This study utilizes co-authorship, co-occurrence, citation analysis and bibliographic coupling of author keywords while grap0hically mapping the bibliographic material using VOS viewer software Version 1.6.19. The bibliographic analysis reveals that the significant themes published in journals revolve around "circular economy", "Sustainable development", "sustainability", "smart city", "waste management", "recycling", "Sustainability", "climate change", "smart technologies", "municipal solid waste", "renewable energy", and "planning". The results would provide a robust base for more research in this area. The research work paves the way for future research in the related areas and issues of the domain, as it is an emerging issue in research, and many problems are untapped.

Keywords: bibliometric analysis; circular economy; smart cities; urbanization; smart technologies; VOSviewer

Citation: Santibanez Gonzalez, E.D.R.; Kandpal, V.; Machado, M.; Martens, M.L.; Majumdar, S. A Bibliometric Analysis of Circular Economies through Sustainable Smart Cities. *Sustainability* **2023**, *15*, 15892. https://doi.org/10.3390/su152215892

Academic Editors: Manoj Kumar Nallapaneni, Idiano D'Adamo, Massimo Gastaldi and Claudio Sassanelli

Received: 29 August 2023
Revised: 1 November 2023
Accepted: 3 November 2023
Published: 13 November 2023

Copyright: © 2023 by the authors. Licensee MDPI, Basel, Switzerland. This article is an open access article distributed under the terms and conditions of the Creative Commons Attribution (CC BY) license (https://creativecommons.org/licenses/by/4.0/).

1. Introduction

Recently, the issues of smart cities, climate change, and the circular economy have attracted extensive scientific research on a global scale. According to predictions from the United Nations, 68% of the world's population will reside in cities by the year 2050, up from around 55%. More than 70% of the world's CO_2 emissions are produced by industrial and motorized transportation systems, which use fossil fuels and are dependent on distant infrastructure made of carbon-intensive materials. Cities and metropolitan areas account for approximately 70% of the world's Gross Domestic Product [1–4]. This urbanization process creates Environmental, Social, and Governance (ESG) issues, stressing the significance of developing environmentally sound methods to reduce resource consumption, such as urban mining. Smart cities use technology and data to optimize resource allocation and reduce waste and energy consumption. For example, smart cities can use sensors and analytics to monitor and manage energy consumption in buildings, optimize traffic flow to reduce fuel consumption and emissions, and use data to identify areas where waste can be reduced and recycled. The concept of the "circular city" is a recent addition to a series of urban sustainability concepts that advocate for fundamental alterations in urban planning, construction, and city development. Nevertheless, this aspect frequently receives criticism due to its inherent ambiguity. Experimentation is a commonly employed approach to urban government to achieve revolutionary goals amid significant uncertainty and ambiguity. However, it is important to note that experimentation is susceptible to manipulation by many stakeholders with different and sometimes self-interested agendas [5]. On the other hand, the goal of a circular economy is to move away from the traditional "take-make-dispose" linear paradigm and towards a more sustainable one where resources are used for as long as feasible while waste and pollution are reduced. This involves designing products for reusability, recyclability, and repair ability and creating closed-loop systems where waste from one process becomes a resource for another. The adoption of sustainable practices has emerged as a strategic approach to gain a competitive edge in the marketplace. The effect of this phenomenon has resulted in significant changes to organizations' behavior in the market, internal organizational structures, interactions with suppliers and customers, as well as the innovation of their product and service offerings [6]. Considering the ongoing process of global urbanization, it becomes imperative to discern and use novel urban development paradigms and tactics to effectively address the complexities associated with sustainable development. The increasing complexity of urban environments necessitates the development of a comprehensive framework for assessing the circular economy in cities, as they grapple with ongoing obstacles in achieving full circularity [7,8].

The Sustainable Development Goals (SDGs) serve as a collective agenda for nations and serve as a focal point for progressive policy discussions. The analysis and monitoring of the Sustainable Development Goals (SDGs) are crucial measures in assessing potential remedial actions [9]. The primary goal of the Sustainable Development Goals (SDGs) is to foster the attainment of a society that is characterized by inclusivity, resilience, safety, and sustainability. Policy makers, entrepreneurs, and residents are tasked together with a significant endeavor to maximize the efficiency of land utilization. The pursuit of sustainability in urban development poses a multifaceted challenge due to its intricate nature, characterized by a network of interdependent linkages. Determining the best equilibrium point within this system presents practical obstacles [10].

The concepts of "Smart Cities and Circular Economies" can be complementary by facilitating more effective resource use, lowering waste and emissions, and enhancing the tracking and management of resources across the metropolitan system. Smart city technology can also assist in the transition towards circular economy practices. For example, smart city technologies can help track and trace materials throughout the supply chain and create more effective waste management systems that reduce the amount of waste sent to landfills. Similarly, circular economy principles can help guide the development of smart city technologies by emphasizing the need for sustainable, long-lasting, and resource-efficient solutions. A "Smart City" is a well-developed urban area that uses cutting-edge, integrated infrastructure, sensors, electronics, and networks interfaced with computerized systems made up of databases, tracking, and decision-making algorithms to be efficient, safe, and environmentally friendly. All structures are developed, constructed, and maintained, whether they are used for power, water, transportation, energy efficiency, etc. Smart city policies encourage innovative ways of planning, organizing, and administering cities and their flows on the one hand while also imposing a new moral order on the city by establishing technical criteria to discriminate between "good" and "bad" cities. The implementation of sustainable urban logistics is crucial to achieve urban sustainability objectives. Within the European context, the European Union (EU) established a notably ambitious objective in 2011, aiming to achieve urban logistics that are essentially devoid of carbon dioxide (CO_2) emissions by the year 2030. Nevertheless, the extent to which these European Union (EU) aims have influenced the development of policies at the national and metropolitan levels remains uncertain [11].

A smart city is one where information and communications technology (ICT) and traditional infrastructure are coordinated and integrated using new and innovative digital technologies [12]. The circular economy (C.E.) may benefit from new digital technologies such as Big Data, Machine Learning, Artificial Intelligence (A.I.), Internet of things (I.O.T.) and Blockchain. Several global issues are believed to have solutions in these digital technologies and sustainable business models, especially those related to the circular economy's transition [13]. A circular economy is a crucial sustainability strategy to fight against climate change. Several governments at various levels have been establishing a long-term circular economy vision. The growth of circular economy capabilities depends on digital progress. Artificial intelligence fueled by big data analytics has recently gained importance [14]. The adoption of a circular economy (CE) is becoming more widely acknowledged as a viable approach to address urgent sustainability issues at the urban level. Indicator-based frameworks, also known as integrated systems of indicators, are widely recognized as valuable instruments for monitoring the process of this shift. It is noteworthy that the majority of frameworks primarily comprise environmental indicators, with only three frameworks incorporating indicators that encompass the various dimensions of sustainable development, namely environmental, social, economic, and governance aspects [15]. The management of electronic waste, commonly referred to as e-waste, has emerged as a pressing concern in the contemporary era characterized by rapid technological advancements. Despite the identification and development of several methods aimed at enhancing e-waste recycling efficiency, the improper disposal of outdated items by end users remains a prevalent issue, as observed in different studies [16].

Cities can become less resilient due to flooding, water pollution, adverse health consequences, inadequate repair and maintenance of water and wastewater systems, rapid urbanization, climate change, poor solid waste management, and water scarcity and pollution [17]. A revolutionary idea called "smart cities" is quickly gaining acceptance since it offers solutions to severe urban problems such as traffic, pollution, energy use, and waste management. "Digital cities", "green cities", and "knowledge cities" are a few examples of older urban planning ideas that have been combined to create "smart city" ideas. Thus, a smart city is a progressive, long-term vision of an enhanced metropolitan area that seeks to reduce its implications on the environment and develop the quality of life for its citizens. It is critical to take the necessary steps to protect our world and modify our wasteful approach to natural resources since the world is continually being pushed to adapt to global climate change and new foreign and internal dangers. Many municipalities worldwide have already started moving in this direction, offering everything from smart parking places to smart benches for solar-powered charging of portable electronics.

The potential of smart cities to address environmental challenges and waste management represents a serious issue that necessitates more profound academic research and policy makers [18]. Modern cities aim to become smarter, yet one of the biggest obstacles is how to process trash effectively. Citizens must be encouraged to interact with modern technology and utilize it daily, especially in emerging economies. The growth of IoT technology has enhanced the need for designing and implementing waste management systems that engage and include the public in the waste management process [19]. The construction industry has a detrimental impact on the environment because of the resources it uses, the energy it consumes, and the waste it produces. The "Circular Economy" (C.E.) is a new paradigm that has the potential to dramatically improve the sustainability of this sector [20].

The amount of Green House Gases (GHGs) discharged by each organization or activity is measured by carbon footprints (CFs). A starting step towards adopting sustainable educational practices could be to report the number of CFs in CO_2 from educational campuses [21]. Smart technology can significantly solve today's major population problems and lay the groundwork for a sustainable future. Today's key challenges are ensuring balanced economic development of society and reducing the effects of global warming. Much research should be performed on topics such as efficient energy conversion technologies, integrating renewable energy systems, enabling the circular economy framework, integrating processes effectively, and other concerns crucial to the public [22]. Ref. [23] demonstrated a statistically significant association that was the contrary. According to their findings, the analysis sample's plastic recycling rate decreases as educational attainment rises. The authors attribute the result to higher opportunity costs for households with higher levels of education. The concept of smart cities has become a prominent subject of scholarly investigation, with a significant focus on technological aspects in the generated knowledge. In this context, the absence of social intelligence, cultural artefacts, and environmental qualities necessary for ICT-related urban innovation is highlighted by the research being advocated [24].

The delivery of public services and the transition from a "take-make-dispose" to a "circular economy" are two areas where current socio demographic expansion poses new challenges for Czech cities. We must introduce new policies to increase city residents' participation, awareness of the issue, and support for the reforms. It is fascinating that younger individuals make more plastic waste than older people do. Two possibilities are possible: either these groups consume differently, with younger consumers buying more plastic packaging products, or younger consumers are more eager to sort and recycle plastic garbage [25]. Since only 42% of post-consumer plastic packaging waste is recycled in Europe, European Regulation 2018/852 set the crucial target of a 55% plastic packaging waste recycling rate by 2030. Plastic Circle was developed as a project supported by the European Union's Horizon 2020 research and innovation as an initiative to promote packaging recycling, improve all stages of garbage pickup, and encourage responsible consumption [26]. Stakeholder Value Creation (SVC) is a fundamental theoretical concept under Stakeholder Theory, as stakeholder-oriented management is primarily concerned with fulfilling the needs of stakeholders [27,28]. The concept of Stakeholder Value Creation (SVC) has the potential to contribute to the promotion of urban sustainability. The Sustainable Development Goals (SDGs) serve as effective instruments for examining and evaluating sustainable development processes. The relationship between social value creation (SVC) and urban sustainability is evident, but the specific impact of SVC on the achievement of the Sustainable Development Goals (SDGs) remains unclear. Social value creation (SVC) plays a significant role in advancing several Sustainable Development Goals (SDGs) inside urban areas, particularly SDGs 11, 17, 9, and 8. The contribution of smart sustainable cities (SVC) to the Sustainable Development Goals (SDGs) is primarily centered around consensus building, as well as the establishment of innovative ecosystems. These key elements play a crucial role in advancing the objectives of the SDGs [29]. The integration of social, economic, and institutional dimensions within the framework of urban sustainability is comprehensive. However, there is a need for improved integration of the Environmental Dimension. Hence, the present dyadic phenomenon might be categorized as either unsustainable or characterized by weak sustainability [30].

The recycling of end-of-life vehicles (ELVs) and the associated methods for assessing their quality in Malaysia are subject to limitations and face numerous significant obstacles. The obstacles encompass the absence of suitable recycling procedures that fully optimize material recovery, as well as concerns regarding the quality and dependability of components utilized in the implementation of circular economy principles [31]. Promoting recycling and reuse methods at the household level can also significantly impact waste creation. Paper, plastic, glass, metal, textile, kitchen, and garden garbage are just a few of the nine material types of waste that have had waste practices related to generation, reuse, and recycling documented and examined [32]. The literature on a crucial area such as smart cities contributing to the circular economy is relatively limited. From the researcher's point of view, much work needs to be conducted in this area. There is a necessity for a study abroad that reviews the available literature and organizes the knowledge and conclusions from past studies. Thus, this research will provide significant awareness of the concept of smart cities and circular economies and cover how smart cities contribute to circular economies. It will also include other vital areas addressed in the literature and suggest untouched areas for future research. Readers will gain an up-to-date perspective on the circular economy by reading this paper, which will also identify the most prominent authors, publications, and theme structure output relevant to the circular economy. The establishment of a robust collaboration between individuals with social and technological expertise poses a novel and significant undertaking for academics across various disciplines. The management of product end-of-life has garnered significant attention, with a focus on developing technologies that effectively handle this waste. This approach aims to assess the economic and environmental advantages via the lens of the circular economy idea [33].

The purpose of this bibliometric analysis is to comprehensively examine the scholarly terrain concerning the incorporation of circular economies in sustainable smart cities. This analysis aims to provide stakeholders with a structured overview of this intricate and dynamic field, facilitating their understanding and navigation within it. This study covers data of 10 years. The remaining part of this study addresses various sections. The following section addresses the research methodology, and the next section addresses the data analysis; the next section addresses discussions and findings leading to the scope for further research directions, implications of the study, and conclusions.

2. Materials and Methods

The current study uses bibliometric analysis to assess the literature's effectiveness and intellectual and social structure on sustainable cities and the circular economy. Bibliometric analysis is a scientific method for analyzing literature in which publication and citation data are analyzed using quantitative methods [34–36]. The current study explicitly follows [35] a four-step method for bibliometric analysis, which entails defining the objectives and scope of the study, selecting the appropriate analytical techniques, gathering the necessary data, conducting the research, and summarizing the results. The technique's ability to manage enormous amounts of bibliographic information is its vital benefit [37]. Similarly, it helps in analyzing a significant amount of data for decision-making that researchers may have ignored. It helps in the exploration and analysis of past data related to a research topic, with the help of which investigators can identify concealed patterns in their studies [38]. Appendix A shows Top Cited 15 Articles Included for Bibliometric Analysis.

The research performs a bibliometric analysis to analyze the dispersed work in the smart city area and its impact on creating a circular economy and assess the significant trends in its theoretical and intellectual association. The work attempts to answer the subsequent research questions, which translate the scope of the research work.

(RQ1) How has the research publication productivity in smart cities and circular economies evolved?

(RQ2) Which journals and researchers are the top performers in the field of smart cities and circular economies?

(RQ3) What are the collaborative networks in smart cities and circular economies?

(RQ4) What are the most searchable topics and themes on smart cities in becoming circular economies?

The dataset for the analysis was extracted using "Smart city" and "Circular Economy" as keywords. After fetching the results, the following filters were applied to refine the results: information related to these documents in the title, authors, abstracts, and keywords was extracted after using the above filters. After the extraction, the file was exported in plain text format, and then search result extraction was performed using VOSviewer. The file was imported in the same format for further processing.

A total of 163 articles were found in the Scopus database accessed on 21 December 2022. The search was limited to document type (research article and review), source type (journal), language (English), and year of publication (2013 to 2022). Full-text articles were considered for analysis. The data for the bibliometric analysis were collected from the Scopus database and exported in a ".csv" (Microsoft Excel) format for use [39]. Scopus was selected as the database for extracting the articles for this study because it is the largest database of intellectual papers and is considered the most acceptable choice among the various multidisciplinary databases [35,40,41].

3. Bibliometric Analysis and Findings

According to the research questions, we conducted an in-depth analysis of the publication trend over ten years related to Circular Economy and Smart City with the help VOSviewer. In particular, results about influential articles, publication productivity, prominent themes and keywords, and promising application areas are related to the first, second, third, and fourth research questions.

The scientific output of this study is described in this section, with the results broken down into co-occurrence mapping, co-authorship mapping, bibliographic coupling, citation, and co-citation analysis. The findings from the bibliometric analysis are presented based on the research questions they address.

3.1. Publication Output

The bibliometric data frame, comprising the period from 2013 to 2022, as no time filter was used while fetching the database. This shows that this area lacked research work in the last ten years with only 163 articles during the above period, and improvement is required in the amount of collaboration among the authors.

The study of smart cities and circular economies has garnered considerable attention in recent times because of the worldwide focus on sustainability, the complexities of urbanization, and improvements in technology. To offer a comprehensive analysis of the progression of research publishing production in these respective domains, it is possible to categorize it both chronologically and thematically. The origins of smart cities and circular economies can be attributed to the discourse surrounding sustainability and urban development throughout the latter part of the 20th century.

Although the specific terminology of "smart cities" and "circular economies" may not have been widely used, scholarly investigations commenced to prioritize sustainable urban design, effective resource allocation, and the utilization of technology to augment urban living conditions. The concepts of "smart cities" and "circular economies" were formally introduced and distinguished in the academic literature throughout the early 2000s.

The proliferation of the Internet and digital technologies has had a significant impact on the development of cities that utilize technology to enhance governance, infrastructure, and services for their citizens. Simultaneously, there emerged apprehensions regarding the depletion of resources, management of waste, and damage of the environment, prompting the formulation of the circular economy framework. This framework places emphasis on the principles of recycling, reusing, and waste reduction. During the 2010s, there was a notable increase in scholarly investigations centered on smart city solutions and circular economy concepts, driven by the rapid growth of urban populations and the escalating effects of climate change. The fields under consideration experienced a significant surge in publication output during the latter half of the 2010s, which can be attributed to the heightened worldwide concern and fascination around these domains. Research has explored the potential of data analytics, the Internet of Things (IoT), and smart governance to facilitate circularity within urban contexts. The research productivity in this convergence field had significant growth, as evidenced by several publications that concentrated on case studies, best practices, and technology breakthroughs aimed at advancing smart and sustainable urban settings.

The emergence of technologies such as artificial intelligence (AI), blockchain, and 5G is expected to contribute to the ongoing expansion of research productivity, enabling the exploration of novel aspects pertaining to intelligent, sustainable, and circular urban environments.

Annual Scientific Production and Growth Trend:

The number of publications on Smart City contribution towards Circular Economy is shown in Figure 1. The annual scientific output in the research domain shows an upward trend from 2013–2022. We have seen a good number of publications in this domain between

2018 and 2021, reaching a level of 48 in 2021. It reflects a growing interest among researchers globally, and it's gradually emerging as an upcoming field of interest.

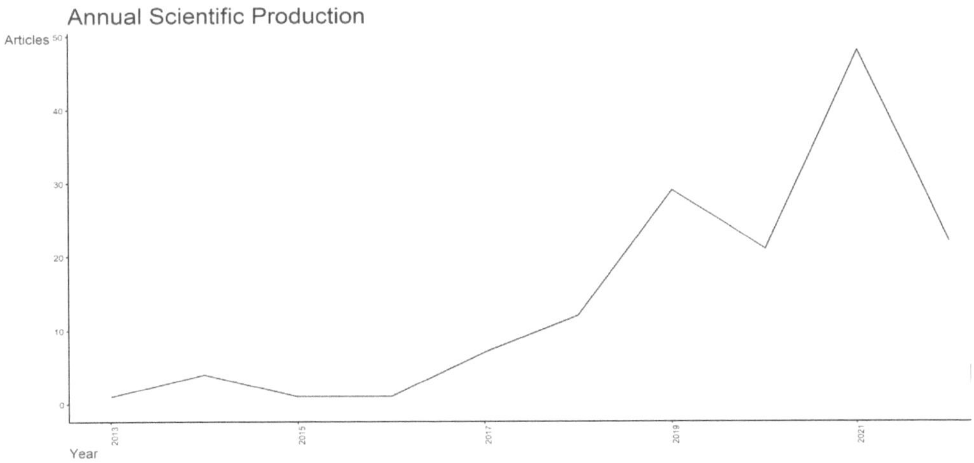

Figure 1. Scientific output.

3.2. *Co-Occurrence Mapping*

3.2.1. Analysis Based on All Keywords

With the help of the full counting method, all keywords were considered as the analysis unit in the co-occurrence mapping. This study also imposed some restrictions on the field of investigation. For instance, a limiting factor was established as a minimum of five (5) instances of a keyword. Therefore, only 47 keywords out of 1401 from 163 articles met the criterion. The links refer to the occurrence of two items together (for example, two keywords). The total number of cited references between any two items is represented by the overall link strength [42]. The occurrences represent the number of articles in which the keyword was found [4].

Figure 2 illustrates the keywords most frequently used by the authors and the most critical topics in this research area. The keywords that appeared most were "Circular Economy" (total link strength 343), which had the highest frequency of appearance, followed by "Sustainable Development (total link strength 253), "Smart cities" (total link strength 244), "Waste Management" (total link strength 132), "recycling" (total link strength 131) and "Sustainability" (total link strength 131) and "climate change" (total link strength 84) as shown in Table 1.

Network visualization was also used to show how often the terms occurred together [43]. Figure 2 illustrates how 1401 keywords were able to group into 5 clusters and 560 links: cluster 1 (red) had 14 items, cluster 2 (green) had 13, cluster 3 (blue) had 11, cluster 4 (yellow) had 7, and cluster 5 (purple) had 2. The size of the circles and texts in each cluster indicates how frequently they occur alongside other keywords. The lines and item distances simultaneously display the relatedness and connections between the keywords. Three distinct visualizations—network visualization, overlay visualization and density visualization—can be used by VOSviewer to depict bibliometric maps.

The overlay visualization in smart cities and circular economies is shown in Figure 3. The update for each phrase is displayed in the visualization overlay. The uniqueness of each phrase is indicated by its color. The area's level of impact increases with color brightness. Circular economies and smart cities are the current research hot topics. Consequently, this could be a crucial subject for future research to identify various themes associated with smart cities and the circular economy.

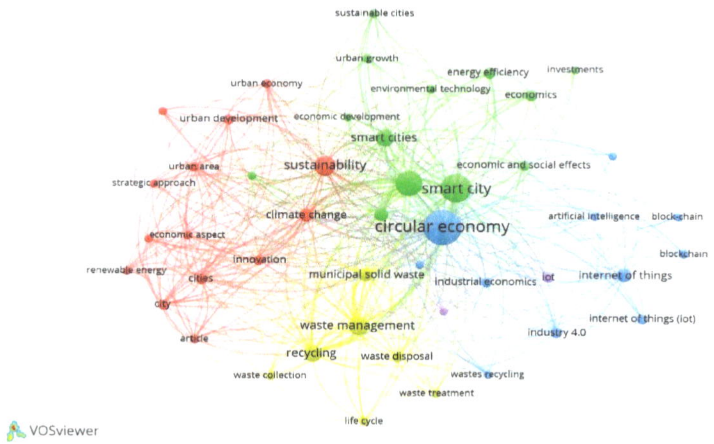

Figure 2. Network analysis with keywords (VOSviewer software).

Table 1. Top Keywords.

Item	Total Link Strength
Circular Economy	343
Sustainable Development	253
Smart City	244
Waste Management	132
Recycling	131
Sustainability	131
Climate Change	84
Municipal Solid Waste	69
Internet of Things	66
Industry 4.0	54

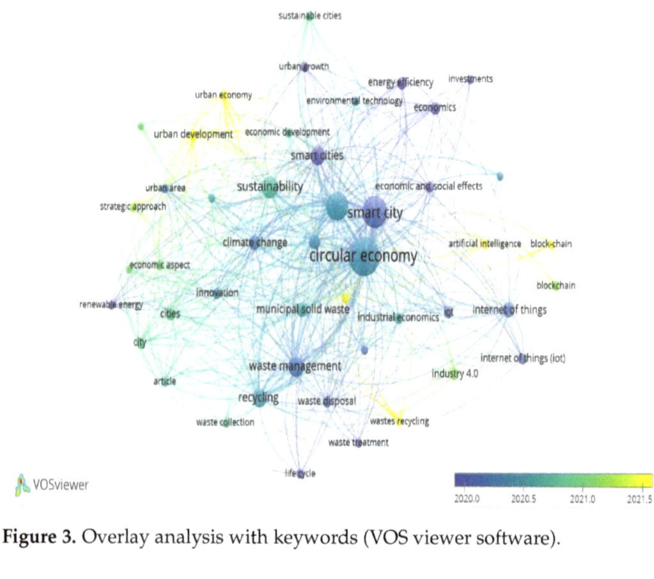

Figure 3. Overlay analysis with keywords (VOS viewer software).

According to the density visualization shown in Figure 4, the more frequently a term appears, the darker or brighter the yellow hue is, and the larger the diameter of the term's circle. This indicates that there are more studies being conducted on relevant topics. There will be fewer studies on a phrase as its color ages and becomes more similar to the background color. According to Figure 4, many studies have been conducted on smart cities, the circular economy, sustainability, waste management, recycling, climate change, artificial intelligence, blockchain, Internet of Things, and their economic and social implications. According to the results of the data mapping of the articles gathered, the keywords that most frequently appeared are the circular economy, smart city, recycling, waste management, Internet of Things, blockchain, and sustainability. We can use this information to search for smart cities and circular economy studies.

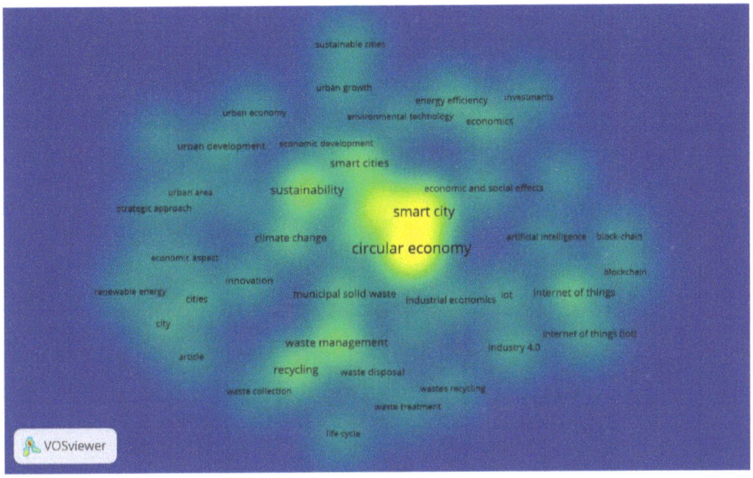

Figure 4. Density Visualization of Keywords.

3.2.2. Co-Occurrence Network of Most Frequently Used Author Keywords

A total of 550 keywords from 163 research publications were given a threshold of 2, resulting in 78 distinct keywords that were used more than twice as shown in Figure 5. The number of highly referenced articles overall is represented by the bubble size, while link strength and clustering are indicated by the line thickness and color, respectively.

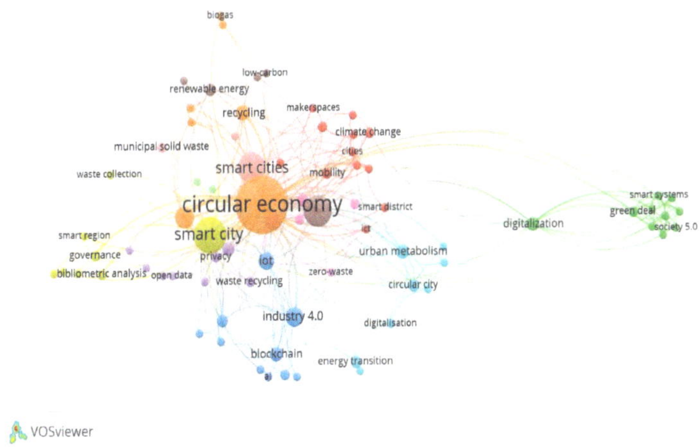

Figure 5. Co-occurrence network of most frequently used author keywords.

3.3. Co-Authorship Visualization Analysis

The authors, connected institutions, and countries that publish on circular economies and smart cities were examined using the function module of VOS viewer's coauthorship visualization. The threshold value for the minimum number of documents was set at two, and the minimum citation was left at one to make it simple to identify the well-known authors who had made contributions in the form of publications in this field. Some of the 551 authors, however, were not linked to the other writers in the network. Only 26 authors met the criteria as shown in Figure 6.

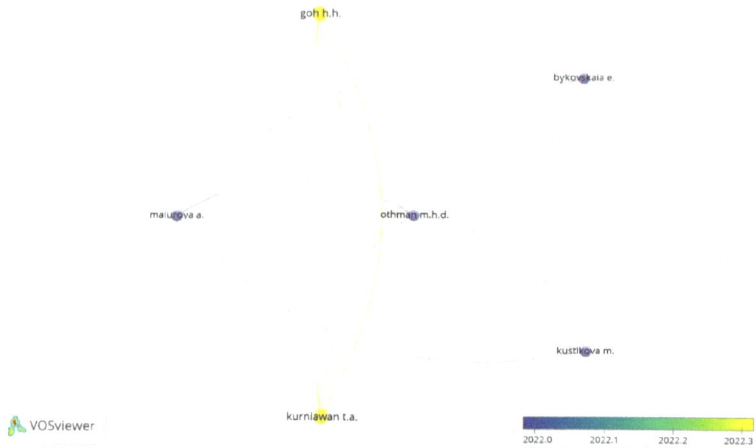

Figure 6. Co-Authorship Analysis of Collaboration of Author (Overlay Visualization).

Based on the bibliographic data collected from the Scopus database, a country co-authorship network visualization map was created (Figure 7) with VOSviewer. In the process of mapping Figure, the minimum document threshold of a nation was set at two, and the minimum number of citations of a country was set at one. Thirty-nine countries out of fifty-seven countries met the thresholds. Figure 7 shows seven clusters between countries involved in smart cities and circular economy research. Researchers from Italy, United Kingdom, China, Romania, and the US are prominently networking countries working in this area.

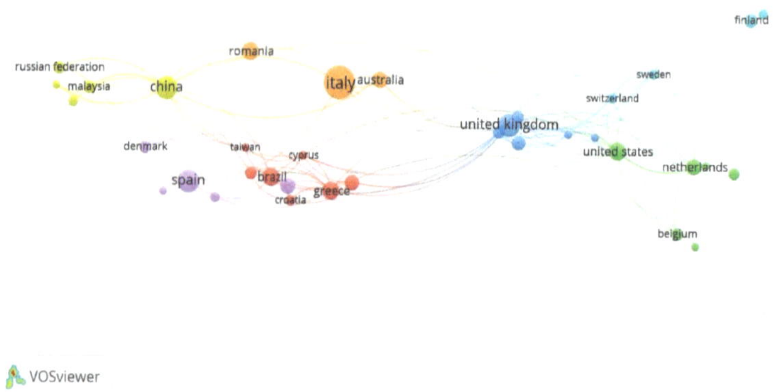

Figure 7. Coauthor network visualization analysis of countries/regions.

The northern regions of Italy exhibit superior performance, with the province of Trento ranking at the forefront. This is followed by the region of Valle d'Aosta and the province of Bolzano, so reaffirming the triumvirate that had already emerged in the preceding year.

An intriguing observation pertains to the expansion of central regions, which exhibit a tendency to approach a value similar to that of the northern regions. Notably, the regions of Toscana, Marche, and Lazio exhibit commendable performance. Furthermore, it has been verified that the southern areas consistently occupy the lowest positions in the ranking, with the sole exception of Abruzzo. Italy presents itself as a strategic focal point for the promotion and advancement of the Sustainable Development Goals (SDGs) to transform the Mediterranean region into a pivotal hub within the future global economy [9].

Regarding the obstacles impeding the implementation of Circular Economy (CE) practices, three distinct categories have been identified: financial, bureaucratic, and regulatory. Companies believe that besides enhancing their brand name and practices, this approach has resulted in advantageous outcomes, including decreasing carbon dioxide emissions and regenerating components within a perpetual cycle. The transition towards a long-term perspective and mindset involves considering not just the economic impact but also the overall performance of the organisation and the consequences of each action. It is important to acquire a circular mindset that can effectively guide the transition to a Circular Economy (CE) or the creation of a Circular Supply Chain (CSC) rather than solely focusing on individual circular initiatives [44].

Italy has been a key contributor to the advancement and advocacy of the circular economy, both domestically and within the European context. The ability of small and medium-sized enterprises (SMEs) to adopt a circular business model and achieve success is contingent upon the level of support provided by various factors. These include the establishment of a company culture that embraces environmental sustainability, with both staff and managers demonstrating a "green" mindset. Additionally, the presence of a local or regional network comprising other SMEs and supporting entities plays a crucial role in facilitating information sharing and raising awareness. Lastly, the advantages derived from cultivating a "green" image and being acknowledged as a supplier committed to sustainability by customers contribute significantly to the overall success of SMEs in transitioning to a circular business model [45]. The nation actively encourages innovation by providing financial resources and assistance to startups and businesses that prioritize circular economy ideas and solutions [46].

3.4. Citation Analysis

3.4.1. Most Cited Authors

Citation is the most frequent method for assessing the influence of an author, journal, and country since it allows for the quick identification of significant works in the chosen area. By taking 1 document and 1 citation as a minimum number of papers and citations of an author as a criterion for citation analysis, we have 444 authors meeting the thresholds out of 551 authors of 163 articles as shown in Table 2. A research paper written by [17] has 372 citations, followed by researchers such as [18,20,22,34,47–50].

Table 2. Most Cited Authors.

Name of Author/s	Documents	Citations
Koop et al.	1	372
Esmaelian et al.	1	232
Nizetic et al	1	234
Hens et al.	1	259
Pencarelli	1	171
Fatimah et al.	1	243
Chauhan et al.	1	146
Sodiq et al.	1	148
Norouzi et al.	1	118

3.4.2. Most Cited Countries

Figure 8 and Table 3 shows that Italy leads the number of publications and United States lead in the number of citations. United Kingdom leads the total link strength. Countries such as India, China, Romania, Colombia, Cuba, Chile, and Belgium are connected, as reflected in Cluster 1. Cluster 2 includes Australia, Iran, Italy, Jordan, Qatar, South Africa, and Spain. The third cluster comprises the Czech Republic, Denmark, France, Indonesia, Slovakia, and Ukraine. The fourth cluster includes Brazil, Germany, Malaysia, Singapore, and the United States. Canada, Croatia, Greece, Russia, and the UK are significant collaborators in Cluster Five. Researchers from India has 13 documents with 187 citations and Researchers from Spain has 11 documents with 171 citations.

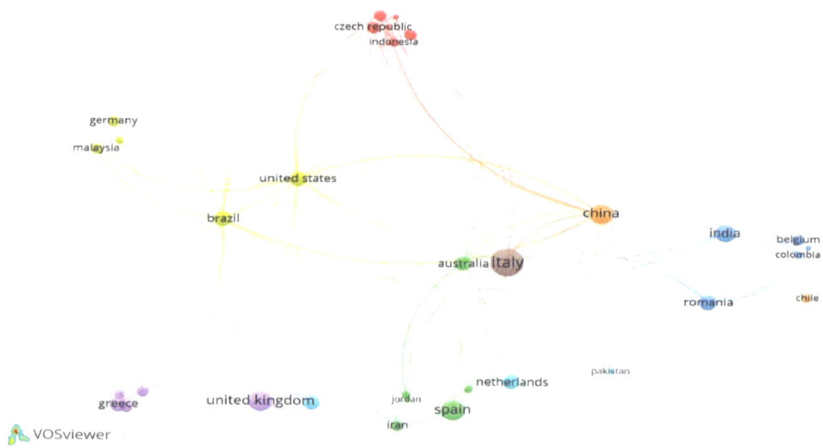

Figure 8. Citation Analysis of Most Cited Countries.

Table 3. Country-wise Production.

Country	Documents	Citations	Total Link Strength
United Kingdom	15	119	22
China	11	215	16
Brazil	10	504	13
Greece	7	249	13
Italy	23	241	13
Malaysia	5	110	11
United States	7	530	10
Cyprus	4	20	9
France	3	55	9
Netherlands	6	260	9

Taking 1 document and 1 citation as a minimum number of papers and citations of an author as a criterion for citation analysis, we have 77 journals meeting the threshold out of 117 journals. Figure 9 reflects the top cited journals. The Journal of Cleaner Production, Sustainability, Waste Management, Environment, Development and Sustainability tops the list of most cited sources as shown in Table 4.

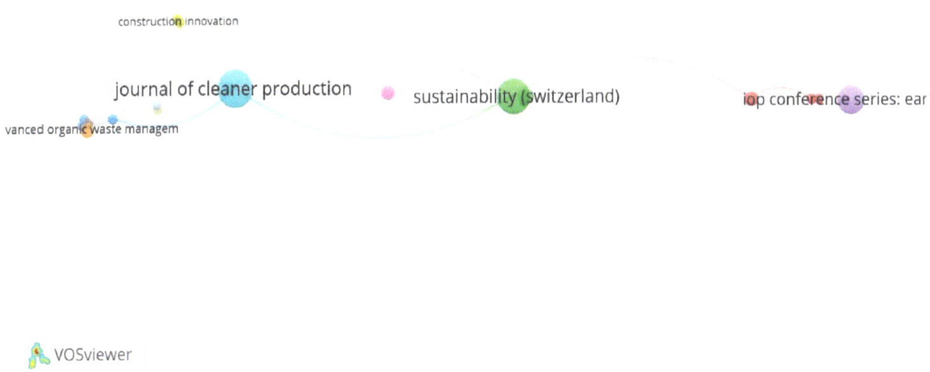

Figure 9. Analysis of Most Cited Sources/Journals.

Table 4. Most Cited Sources/Journals.

Source	Documents	Citations
Journal of Cleaner Production	13	851
Sustainability	9	137
IOP Conference Series: Earth and Environmental Science	7	9
Circular Economy and Sustainability	2	2
Environment, Development and Sustainability	2	372
Journal of Urban Regeneration and Renewal	2	17

3.5. Bibliometric Analysis of Bibliographic Coupling of Authors, Institutions and Countries

Bibliographic coupling is something that is observed when two authors make references to a shared collection of prior works, so suggesting a certain degree of thematic or topical affinity in their respective research endeavors. Bibliographic coupling pertaining to older works may suggest that the writers have established a distinct expertise within a specific field of research. This particular area of expertise, which is grounded in extensive historical or foundational knowledge, enables individuals to conduct more thorough investigations into subjects and generate more profound understandings. When two documents cite the same source, this is known as "bibliographic coupling" [51–53]. The bibliographic coupling map of authors is shown in Figure 10. Citations are used in bibliographic coupling to describe the similarities between two texts, authors, institutions, or nations. This method is grounded on the notion that two papers citing a third paper are highly connected and should be grouped in the visualization map's cluster solution. Using a criterion of a minimum of 1 citation of 163 documents, a total of 108 met the threshold. Eleven clusters were obtained from the analysis. Cluster 1 includes 13 items, and the research area is the circular economy (shown in red). The Red Cluster was anchored by [54], whose research focuses on how automated smartphone recycling could be supported by Artificial Intelligence (A.I.) and could also act as an enabler for Circular Smart Cities (C.S.C.). Research work conducted by researchers [55] from the same cluster considers the circular economy from the perspective of sustainable development, which is one of the main goals of contemporary societies. The main characteristic of a circular economy is the requirement to increase resource efficiency through waste reduction and recycling.

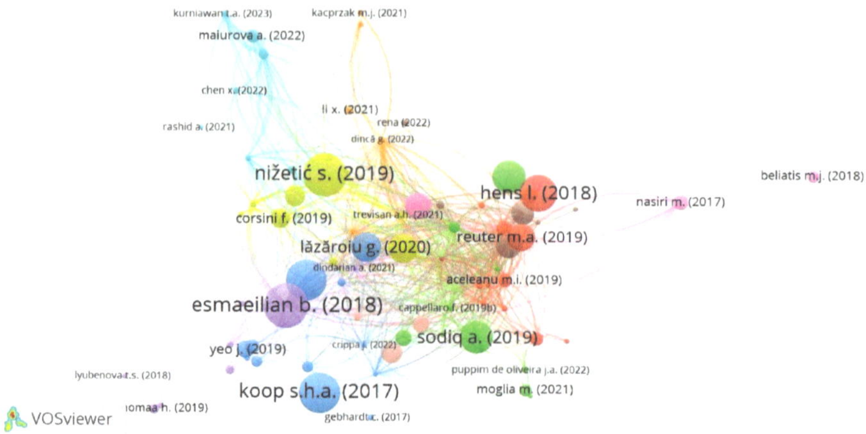

Figure 10. Bibliometric Coupling of Authors (Authors Own Work).

The other clusters are anchored by Arias (green) with 12 items, anchored by Benltoufa (blue) with 11 items, and anchored by Andrade (yellow) with 11 items. The green cluster mainly focuses on research work related to green infrastructure [56] and the environment, highlighting nature and biodiversity [57,58] circular strategies and urban regeneration [59] and the Internet of Things [60]. The blue cluster focuses on smart cities, waste management [48,61], the circular economy, urban branding [48,62], smart technology, and sustainable development [63]. The yellow cluster focuses on smart cities and the circular economy [64], sustainable consumption and the sharing economy [65], municipal solid waste and energy recovery [66], and sustainable business [67].

The analysis of bibliographic coupling among universities with respect to older publications demonstrates a common focus on fundamental knowledge and research approaches. The aforementioned concentration of efforts, together with the advantages derived from extensive immersion in well-established knowledge, can elucidate the exceptional performance of these establishments within their respective fields. Researchers can gain a more comprehensive perspective by actively engaging with non-recent scholarly works. The use of a holistic perspective enables researchers to establish linkages between historical and contemporary discoveries, so cultivating a more comprehensive comprehension of their own discipline. A comprehensive understanding of existing research and scholarship is essential to prevent institutions from duplicating efforts in areas that have already undergone extensive investigation. Alternatively, individuals have the capacity to expand upon preexisting knowledge, so challenging conventional limits while maintaining a foundation in established ideas. Bibliographic coupling of institutions occurs when publications from two institutions reference publications from a third common institution. Figure 11 shows the bibliographic coupling of the institutions with network visualization. The minimum number of publications for an organization was one, and the minimum number of citations was one. Of the 360 organizations, 280 met the thresholds. For each of the 280 organizations, the total strength of the bibliographic coupling links with other organizations was calculated. The organizations with the highest total link strength were selected. Key Laboratory of Poyang Lake Environment and Resource.

Utilization, Ministry of Education, China, was at the top of this list with two publications, 17 citations, and a network strength of 2289. We can observe some other institutions that are contributing to the field and dominate the coupling and anchor its most significant clusters: Academy of Romanian Scientist, Bucharest, Romania), University of Buffalo (New York, NY, USA), Western New England University (Springfield, MA, USA), Georgia Institute of Technology (Atlanta, GA, USA), Massachusetts Institute of Technology (Cambridge, MA, USA), and Pontifical Catholic University of Paraná, Curitiba, (Paraná, Brazil).

Figure 11. Bibliographic coupling of institutions (network visualization).

Figure 12 presents the bibliographic coupling of countries. Bibliographic coupling of countries occurs when publications from two countries reference publications from a third country. According to this graph, scholars from Brazil make a substantial contribution to this field of study, along with those from Canada, Croatia, Cyprus, Poland, and Greece. Figure 12 also shows how frequently coupling occurs among other nations, including Australia, Finland, Germany, India, Mexico, and Romania.

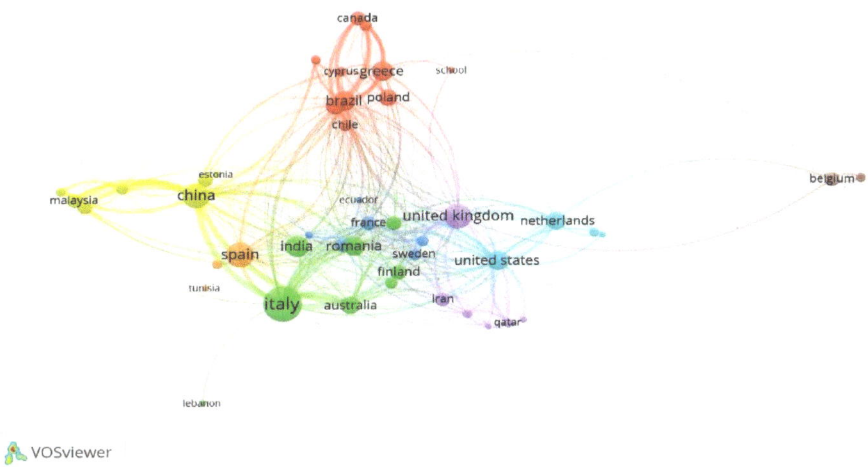

Figure 12. Bibliographic coupling of countries.

4. Conclusions

This study highlights the scope of smart cities and circular economy research using bibliometric analysis, collecting datasets from 2003 to 2022. It further portrays the research topic's theoretical, intellectual, and community structure. The present study tries to integrate the fragmented literature on the topic with the help of VOSviewer. The dataset for this study was extracted from the Scopus database due to its extensive coverage and quality. The dataset shows that the research domain shows an upward trend. Even though it is still in its initial stage, the circular economy is a crucial sustainability strategy that global and corporate leaders use in the battle against climate change. By incorporating relevant elements of the economy, technology, mobility, quality of life, and other areas that contribute to the well-being of its citizens, a smart city is highly developed, inventive, and environmentally friendly [68]. Rapid environmental degradation, a growing number of cars on the road, and a shortage of resources are issues that circular cities and a circular economy can resolve. Our cities are developing into more livable and sustainable places. It is essential to emphasize the value of smart cities through pilot projects that will serve as the testing grounds for the later expansion of CE concepts on a broader, more global scale.

The phenomena of urbanization and climate change are compelling cities to navigate unexplored trajectories to achieve sustainable outcomes. Numerous urban areas are embracing the captivating notion of the 'Circular Economy' (CE) as a guiding principle for this shift. The concept of the Circular Economy (CE) proposes a novel approach to the management of resource flows within economies, aiming to create a closed-loop system [69]. The circular economy aims to shift away from a linear economy to one that is more circular and turns waste into resources. All interested parties, including the public sector, business sector, and public, must work together and be willing to do this. Many European Union member states currently lack the requisite waste-handling infrastructure. It is crucial to establish precise long-term policy goals to direct actions and expenditures, create systems and methods for waste treatment, and prohibit recyclable materials from being used at the bottom of the waste hierarchy [70]. A circular economy in cities will require unprecedented levels of cross-sector and public–private cooperation in the twenty-first century. The time has come to make the most of the numerous opportunities circular cities offer and establish a system that will benefit the environment, society, and economy over the long run. The circular economy has emerged as one of the most widely used theories to address environmental problems. However, research on the circularity subject is still developing, particularly at the firm level [71]. Cities and metropolitan areas can lead circular economies while supporting the use of renewable energy, energy efficiency, sustainable production and consumption, sustainable transportation, resource conservation, and sustainable waste management. Goal 11 of the Sustainable Development Goals 2030 Agenda was to "Make cities and human settlements inclusive, safe, resilient, and sustainable" to achieve these objectives. Research is required to determine how sustainable finance may improve the circular economy and investment opportunities. The circular economy exhibits considerable potential in facilitating the attainment of various Sustainable Development Goals (SDGs), encompassing SDG 7 pertaining to Affordable and Clean Energy, SDG 8 concerning decent work and economic growth, SDG 11 addressing sustainable cities and consumption, SDG 12 focusing on responsible consumption and production, SDG 13 targeting climate action, SDG 14 emphasizing life below water, and SDG 15 centering on life on land [6].

The term "smart city" was first introduced in the 1990s, but research started gaining momentum in the early 2010s. The circular economy as a concept has been around for decades, but research on the topic has increased in recent years, particularly since the publication of the [72] report "Towards the Circular Economy" in 2012. Technology is a crucial part of the global endeavor to reach net zero, but as we move from our current behaviors to a more climate-friendly society, its adoption necessitates practical sacrifices [73].

There are several journals and researchers that are top performers in the field of Smart Cities and Circular Economy. Journal of Cleaner Production, Waste Management, Sustainability, Environment, Development and Sustainability are the journals with the maximum contribution to smart cities and circular economies. Overall, these journals and researchers represent some of the top performers in the field of smart cities and circular economy and have contributed significantly to the knowledge and understanding of these topics. Research publication productivity in both smart cities and the circular economy has experienced significant growth in recent years, reflecting the increasing interest and importance of these fields in addressing environmental and social challenges. The study found that there is a growing body of research on smart and circular cities, with a steady increase in research publications and citations over the past decade. The study also found that research on smart cities and the circular economy is becoming increasingly interconnected, with more studies exploring the potential synergies between the two concepts. The United States, Brazil, China, and Italy have the most citations and research publications. Countries such as India, China, Romania, Colombia, Cuba, Chile, and Belgium have a good collaboration network. Taking a circular approach can also tackle many other socio-economic problems afflicting cities, for example, providing access to affordable accommodation, expanding and diversifying the economic base, and building more engaged and collaborative communities in cities [74]. Cleaner Production (CP) entails the reduction of energy and materials consumption, as well as the replacement of environmentally and health-hazardous products with less damaging alternatives [47]. In order to gain a more comprehensive understanding of the dynamics, difficulties, and facilitators of sustainable consumer behaviour within the circular economy, it is imperative to conduct longitudinal studies [75].

The analysis revealed that the research on smart cities and the circular economy is multidisciplinary, involving researchers from various fields, including urban planning, engineering, economics, and environmental science. The most common research topics were "Circular Economy", "Sustainable Development", "Smart cities", "Waste Management", "recycling", "Sustainability" and "climate change". Refs. [18,22,34,48–50,76,77] were the top cited researchers. The study suggests that there is a growing recognition of the potential for smart cities and the circular economy to work together to create more sustainable and livable urban environments. The research in this area is becoming more integrated and sophisticated over time. The intersection of smart cities and the circular economy has generated interest and study in recent years, and several topics and themes have emerged as the most searchable. Here are some examples:

- Smart and circular infrastructure: This involves using smart technologies to optimize the use of infrastructure and resources in a circular economy context. Examples include the use of sensor networks and data analytics to improve energy efficiency, waste management, and mobility.
- Circular business models: This focuses on the development of business models that enable circularity in the city, such as product-as-a-service, sharing economy, and closed-loop systems. Smart technologies such as blockchain and IoT can facilitate implementing and scaling these models.
- Circular design and manufacturing: This involves the integration of circular principles into the design and manufacturing of products and materials, such as using recycled or renewable materials and designing for disassembly and repair. Smart technologies can support the implementation of circular design principles, such as digital fabrication and 3D printing.
- Circular supply chain management: This topic involves using smart technologies to optimize the circular supply chain, such as real-time tracking of materials and products and using data analytics to improve resource efficiency and reduce waste.
- Circular and smart governance: This topic focuses on the role of governance in enabling and supporting the transition to a circular and smart city. Examples include policies and regulations promoting circular business models and smart technologies and using open data and citizen engagement to support circular and smart city initiatives.

4.1. Potential Areas for Future Research

The bibliometric analysis shows that smart cities and circular economies are emerging areas for policymakers and researchers. In the last ten years, less research has been conducted, and keyword analysis shows that the research primarily focused on smart cities, circular economies, waste management, climate change, technology, and sustainability. The research study also depicts numerous studies on the adoption and diffusion of mobile banking. One should consider including alternative metrics, sometimes referred to as altmetrics, in addition to standard citation counts. Altmetrics encompass a broader range of indicators, such as online mentions, social media shares, and other forms of digital engagement.

The integration of bibliometric analysis with qualitative research methodologies allows for a more comprehensive exploration of the content within publications, enabling the examination of nuanced conversations, case studies, and contextual factors.

In light of potential geographical variations in the adoption of circular economies within smart cities, it is advisable to conduct distinct bibliometric analysis for different areas or nations. This approach can yield more context-specific insights.

To enhance the comprehensiveness of data sources, it is recommended to include gray literature, conference proceedings, and non-academic publications. By incorporating these sources, a broader spectrum of discussions can be captured.

4.2. Implications of the Study

The current study will help marketers' practitioners frame and implement various plans and policies related to smart cities and build a circular economy covering issues such as climate change, energy efficiency, climate change, and waste management. By identifying the most frequently cited authors, publications, and keywords, this analysis can help researchers to understand the research landscape and to identify gaps and opportunities for further research. This can help to identify potential collaborators and opportunities for joint research initiatives and highlight gaps in the collaboration that need to be addressed. By identifying the most highly cited publications and the most frequently used keywords, policymakers and practitioners can better understand the key concepts and approaches in this field and use this information to inform their work. Overall, a bibliometric analysis of smart cities and the circular economy can provide valuable insights for researchers, policymakers, and practitioners, helping to identify trends and gaps in research, highlighting opportunities for collaboration, and informing policy and practice in this critical and rapidly evolving field.

As urban areas embark on the process of enhancing their services to tackle urgent sustainability concerns, policymakers and urban planners encounter the task of comprehending intricate, multifaceted systems and assessing the potential impact of proposed investments or policies on these systems [78]. Digital technology has the potential to decrease the quantity of Municipal Solid Waste (MSW) that is not recycled, while also preserving natural resources and lowering both operational expenses and Green House Gas (GHG) emissions. The process of digitalization enhances the resilience of cities by reinforcing local waste management practices in order to effectively address the global COVID-19 pandemic [79]. According to the bibliometric analysis, policymakers and scholars are increasingly interested in circular economies and smart cities. Little research has been conducted in the last ten years. A keyword analysis reveals that this study mainly concentrated on smart cities, circular economies, waste management, climate change, technology, and sustainability. It is nevertheless easy to overlook significant articles despite efforts to mark at a base as complete as possible, particularly those that are not peer-reviewed or published in less accessible places, such as conference proceedings. Circular economies and smart cities are still relatively fresh and developing ideas. Because of this, there can be variations in how authors define and use these terms, which could cause problems with comparison and interpretation. This research paves the way for future research in the related areas and issues of the domain, as it is an emerging issue in research, and many problems are untapped. Additionally, it highlights the emerging themes of the field on which future research can be conducted. Pragmatic sustainability presents a practical and flexible strategy for attaining a sustainable future, whereas sustainable education equips individuals with the requisite knowledge and resources to actively contribute to this overarching goal. The simultaneous integration of these principles has the potential to guide society towards a state of enhanced equilibrium and a more mutually beneficial cohabitation with our planet.

Author Contributions: Conceptualization, E.D.R.S.G., V.K., M.M., M.L.M. and S.M.; Methodology, E.D.R.S.G., V.K. and S.M.; Validation, V.K.; Formal analysis, V.K. and S.M.; Writing—original draft, V.K. and S.M.; Writing—review & editing, E.D.R.S.G., V.K., M.M., M.L.M. and S.M. All authors have read and agreed to the published version of the manuscript.

Funding: This research received no external funding.

Institutional Review Board Statement: Not applicable.

Informed Consent Statement: Not applicable.

Data Availability Statement: Data sharing is not applicable—no new data are generated, or the article describes entirely theoretical research.

Conflicts of Interest: The authors declare that there is no conflict of interest.

Appendix A

Table A1. Top Cited 15 Articles Included for Bibliometric Analysis.

S. No	Title	Journal	Citations	Reference No
1	The challenges of water, waste and climate change in cities	Environment, Development and Sustainability	372	2017, 19(2), 385–418
2	On the evolution of "Cleaner Production" as a concept and a practice	Journal of Cleaner Production	259	2018, Volume 172, 3323–3333
3	Industry 4.0-based sustainable circular economy approach for smart waste management system to achieve sustainable development goals: A case study of Indonesia	Journal of Cleaner Production	364	2020, Volume 269, 122263
4	The future of waste management in smart and sustainable cities: A review and concept paper	Waste Management	232	2018, 81, 177–195
5	Smart technologies for promotion of energy efficiency, utilization of sustainable resources and waste management	Journal of Cleaner Production	234	2019, Volume 231, 565–591
6	The digital revolution in the travel and tourism industry	Information Technology & Tourism	171	2020, 22(3), 455–476
7	The interplay of circular economy with industry 4.0 enabled smart city drivers of healthcare waste disposal	Journal of Cleaner Production	146	2021, Volume 279, 123854
8	Towards modern sustainable cities: Review of sustainability principles and trends	Journal of Cleaner Production	148	2019, Volume 227, 972–1001
9	Circular economy in the building and construction sector: A scientific evolution analysis	Journal of Building Engineering	118	2021, 44, 102704
10	Promoting digital transformation in waste collection service and waste recycling in Moscow (Russia): Applying a circular economy paradigm to mitigate climate change impacts on the environment	Journal of Cleaner Production	57	2022, Volume 354, 131604
11	The First Two Decades of Smart-City Research: A Bibliometric Analysis	Journal of Urban Technology	607	2017, 24(2), 3–27
12	Circular economy business models: The state of research and avenues ahead	Business Strategy and the Environment	195	2020, 29, 3006–3024.
13	Linking circular economy and digitalization technologies: A systematic literature review of past achievements and future promises	Technological Forecasting and Social Change	185	2022, 177, 121508
14	Circular cities	Urban Studies	134	2019, 56(13), 2746–2762
15	Smart cities of the future	The European Physical Journal Special Topics	2806	2012, 214, 481–518

References

1. Dasgupta, S.; Lall, S.; Wheeler, D. Cutting Global Carbon Emissions: Where Do Cities Stand? Available online: https://blogs.worldbank.org/sustainablecities/cutting-global-carbon-emissions-where-do-cities-stand (accessed on 17 September 2022).
2. United Nations 68% of the World Population Projected to Live in Urban Areas by 2050, Says UN. Available online: https://www.un.org/development/desa/en/news/population/2018-revision-of-world-urbanization-prospects.html (accessed on 21 July 2022).
3. Farhadi, E.; Pourahmad, A.; Ziari, K.; Faraji Sabokbar, H.; Tondelli, S. Indicators Affecting the Urban Resilience with a Scenario Approach in Tehran Metropolis. *Sustainability* **2022**, *14*, 12756. [CrossRef]
4. Wang, J.; Chen, L.; Chen, L.; Zhao, X.; Wang, M.; Ju, Y.; Xin, L. City-Level Features of Energy Footprints and Carbon Dioxide Emissions in Sichuan Province of China. *Energies* **2019**, *12*, 2025. [CrossRef]
5. Winslow, J.; Coenen, L. Sustainability Transitions to Circular Cities: Experimentation between Urban Vitalism and Mechanism. *Cities* **2023**, *142*, 104531. [CrossRef]
6. Ali, S.M.; Appolloni, A.; Cavallaro, F.; D'Adamo, I.; Di Vaio, A.; Ferella, F.; Gastaldi, M.; Ikram, M.; Kumar, N.M.; Martin, M.A.; et al. Development Goals towards Sustainability. *Sustainability* **2023**, *15*, 9443. [CrossRef]
7. Birgovan, A.-L.; Lakatos, E.; Cioca, L.-I.; Pacurariu, R.; Ciobanu, G.; Rada, E. How Should We Measure? A Review of Circular Cities Indicators. *Int. J. Environ. Res. Public Health* **2022**, *19*, 5177. [CrossRef]
8. Kurniawan, T.A.; Maiurova, A.; Kustikova, M.; Bykovskaia, E.; Othman, M.H.D.; Goh, H.H. Accelerating Sustainability Transition in St. Petersburg (Russia) through Digitalization-Based Circular Economy in Waste Recycling Industry: A Strategy to Promote Carbon Neutrality in Era of Industry 4.0. *J. Clean. Prod.* **2022**, *363*, 132452. [CrossRef]
9. D'Adamo, I.; Gastaldi, M. Monitoring the Performance of Sustainable Development Goals in the Italian Regions. *Sustainability* **2023**, *15*, 14094. [CrossRef]
10. D'Adamo, I.; Gastaldi, M.; Ioppolo, G.; Morone, P. An Analysis of Sustainable Development Goals in Italian Cities: Performance Measurements and Policy Implications. *Land Use Policy* **2022**, *120*, 106278. [CrossRef]
11. Shrestha, S.; Haarstad, H. Do EU Goals Matter? Assessing the Localization of Sustainable Urban Logistics Governance Goals in Norwegian Cities. *Cities* **2023**, *137*, 104317. [CrossRef]
12. Batty, M.; Axhausen, K.W.; Giannotti, F.; Pozdnoukhov, A.; Bazzani, A.; Wachowicz, M.; Ouzounis, G.; Portugali, Y. Smart Cities of the Future. *Eur. Phys. J. Spec. Top.* **2012**, *214*, 481–518. [CrossRef]
13. Chauhan, C.; Parida, V.; Dhir, A. Linking Circular Economy and Digitalisation Technologies: A Systematic Literature Review of Past Achievements and Future Promises. *Technol. Forecast. Soc. Change* **2022**, *177*, 121508. [CrossRef]
14. Bag, S.; Pretorius, J.H.C.; Gupta, S.; Dwivedi, Y.K. Role of Institutional Pressures and Resources in the Adoption of Big Data Analytics Powered Artificial Intelligence, Sustainable Manufacturing Practices and Circular Economy Capabilities. *Technol. Forecast. Soc. Change* **2021**, *163*, 120420. [CrossRef]
15. Papageorgiou, A.; Henrysson, M.; Nuur, C.; Sinha, R.; Sundberg, C.; Vanhuyse, F. Mapping and Assessing Indicator-Based Frameworks for Monitoring Circular Economy Development at the City-Level. *Sustain. Cities Soc.* **2021**, *75*, 103378. [CrossRef]
16. Jabbour, C.J.C.; Colasante, A.; D'Adamo, I.; Rosa, P.; Sassanelli, C. Comprehending E-Waste Limited Collection and Recycling Issues in Europe: A Comparison of Causes. *J. Clean. Prod.* **2023**, *427*, 139257. [CrossRef]
17. Koop, S.H.A.; van Leeuwen, C.J. The Challenges of Water, Waste and Climate Change in Cities. *Environ. Dev. Sustain.* **2017**, *19*, 385–418. [CrossRef]
18. Esmaeilian, B.; Wang, B.; Lewis, K.; Duarte, F.; Ratti, C.; Behdad, S. The Future of Waste Management in Smart and Sustainable Cities: A Review and Concept Paper. *Waste Manag.* **2018**, *81*, 177–195. [CrossRef] [PubMed]
19. Al-Jabi, M.; Diab, M. IoT-Enabled Citizen Attractive Waste Management System. In Proceedings of the 2017 2nd International Conference on the Applications of Information Technology in Developing Renewable Energy Processes & Systems (IT-DREPS), Amman, Jordan, 6–7 December 2017; pp. 1–5.
20. Norouzi, M.; Chàfer, M.; Cabeza, L.F.; Jiménez, L.; Boer, D. Circular Economy in the Building and Construction Sector: A Scientific Evolution Analysis. *J. Build. Eng.* **2021**, *44*, 102704. [CrossRef]
21. Kulkarni, S.D. A Bottom up Approach to Evaluate the Carbon Footprints of a Higher Educational Institute in India for Sustainable Existence. *J. Clean. Prod.* **2019**, *231*, 633–641. [CrossRef]
22. Nižetić, S.; Djilali, N.; Papadopoulos, A.; Rodrigues, J.J.P.C. Smart Technologies for Promotion of Energy Efficiency, Utilization of Sustainable Resources and Waste Management. *J. Clean. Prod.* **2019**, *231*, 565–591. [CrossRef]
23. Hage, O.; Söderholm, P. An Econometric Analysis of Regional Differences in Household Waste Collection: The Case of Plastic Packaging Waste in Sweden. *Waste Manag.* **2007**, *28*, 1720–1731. [CrossRef]
24. Mora, L.; Bolici, R.; Deakin, M. The First Two Decades of Smart-City Research: A Bibliometric Analysis. *J. Urban Technol.* **2017**, *24*, 3–27. [CrossRef]
25. Rybová, K.; Slavík, J. Ageing Population of Cities—Implications for Circular Economy in the Czech Republic. In Proceedings of the 2017 Smart City Symposium Prague (SCSP), Prague, Czech Republic, 25–26 May 2017; pp. 1–5.
26. Roche Cerasi, I.; Sánchez, F.V.; Gallardo, I.; Górriz, M.; Torrijos, P.; Aliaga, C.; Franco, J. Household Plastic Waste Habits and Attitudes: A Pilot Study in the City of Valencia. *Waste Manag. Res.* **2021**, *39*, 679–689. [CrossRef] [PubMed]
27. Bridoux, F.; Stoelhorst, J.W. Microfoundations for Stakeholder Theory: Managing Stakeholders with Heterogeneous Motives. *Strateg. Manag. J.* **2014**, *35*, 107–125. [CrossRef]

28. Tantalo, C.; Priem, R.L. Value Creation through Stakeholder Synergy. *Strateg. Manag. J.* **2016**, *37*, 314–329. [CrossRef]
29. Beck, D.; Ferasso, M.; Storopoli, J.; Vigoda-Gadot, E. Achieving the Sustainable Development Goals through Stakeholder Value Creation: Building up Smart Sustainable Cities and Communities. *J. Clean. Prod.* **2023**, *399*, 136501. [CrossRef]
30. Beck, D.; Ferasso, M. Bridging 'Stakeholder Value Creation' and 'Urban Sustainability': The Need for Better Integrating the Environmental Dimension. *Sustain. Cities Soc.* **2023**, *89*, 104316. [CrossRef]
31. Molla, A.H.; Moghtaderi, S.H.; Harun, Z.; Jedi, A.; Manoj Kumar, N. Insights into End-of-Life Vehicle Recycling and Its Quality Assessment Systems in Malaysia Reveals the Need for a New Stakeholder-Centric Approach for Vehicle Waste Management. *Prod. Manuf. Res.* **2023**, *11*, 2236676. [CrossRef]
32. Pandey, R.U.; Surjan, A.; Kapshe, M. Exploring Linkages between Sustainable Consumption and Prevailing Green Practices in Reuse and Recycling of Household Waste: Case of Bhopal City in India. *J. Clean. Prod.* **2018**, *173*, 49–59. [CrossRef]
33. D'Adamo, I. Adopting a Circular Economy: Current Practices and Future Perspectives. *Soc. Sci.* **2019**, *8*, 328. [CrossRef]
34. Pencarelli, T. The Digital Revolution in the Travel and Tourism Industry. *Inf. Technol. Tour.* **2020**, *22*, 455–476. [CrossRef]
35. Donthu, N.; Kumar, S.; Mukherjee, D.; Pandey, N.; Lim, W.M. How to Conduct a Bibliometric Analysis: An Overview and Guidelines. *J. Bus. Res.* **2021**, *133*, 285–296. [CrossRef]
36. Mukherjee, D.; Kumar, S.; Donthu, N.; Pandey, N. *Research Published in Management International Review from 2006 to 2020: A Bibliometric Analysis and Future Directions*; Springer: Berlin/Heidelberg, Germany, 2021; Volume 61, ISBN 0123456789.
37. Ramos-Rodrígue, A.R.; Ruíz-Navarro, J. Changes in the Intellectual Structure of Strategic Management Research: A Bibliometric Study of the *Strategic Management Journal*, 1980–2000. *Strateg. Manag. J.* **2004**, *25*, 981–1004. [CrossRef]
38. Daim, T.U.; Rueda, G.; Martin, H.; Gerdsri, P. Forecasting Emerging Technologies: Use of Bibliometrics and Patent Analysis. *Technol. Forecast. Soc. Change* **2006**, *73*, 981–1012. [CrossRef]
39. Nobanee, H.; Al Hamadi, F.Y.; Abdulaziz, F.A.; Abukarsh, L.S.; Alqahtani, A.F.; Alsubaey, S.K.; Alqahtani, S.M.; Almansoori, H.A. A Bibliometric Analysis of Sustainability and Risk Management. *Sustainability* **2021**, *13*, 3277. [CrossRef]
40. Bartol, T.; Budimir, G.; Dekleva-Smrekar, D.; Pusnik, M.; Juznic, P. Assessment of Research Fields in Scopus and Web of Science in the View of National Research Evaluation in Slovenia. *Scientometrics* **2014**, *98*, 1491–1504. [CrossRef]
41. Norris, M.; Oppenheim, C. Comparing Alternatives to the Web of Science for Coverage of the Social Sciences' Literature. *J. Informetr.* **2007**, *1*, 161–169. [CrossRef]
42. Guo, Y.M.; Huang, Z.L.; Guo, J.; Li, H.; Guo, X.R.; Nkeli, M.J. Bibliometric Analysis on Smart Cities Research. *Sustainability* **2019**, *11*, 3606. [CrossRef]
43. Hossain, S.; Batcha, M.S.; Atoum, I.; Ahmad, N.; Al-Shehri, A. Bibliometric Analysis of the Scientific Research on Sustainability in the Impact of Social Media on Higher Education during the COVID-19 Pandemic. *Sustainability* **2022**, *14*, 16388. [CrossRef]
44. Carissimi, M.C.; Creazza, A.; Fontanella Pisa, M.; Urbinati, A. Circular Economy Practices Enabling Circular Supply Chains: An Empirical Analysis of 100 SMEs in Italy. *Resour. Conserv. Recycl.* **2023**, *198*, 107126. [CrossRef]
45. Rizos, V.; Behrens, A.; Van der Gaast, W.; Hofman, E.; Ioannou, A.; Kafyeke, T.; Flamos, A.; Rinaldi, R.; Papadelis, S.; Hirschnitz-Garbers, M.; et al. Implementation of Circular Economy Business Models by Small and Medium-Sized Enterprises (SMEs): Barriers and Enablers. *Sustainability* **2016**, *8*, 1212. [CrossRef]
46. Colombi, C.; D'Itria, E. Fashion Digital Transformation: Innovating Business Models toward Circular Economy and Sustainability. *Sustainability* **2023**, *15*, 4942. [CrossRef]
47. Hens, L.; Block, C.; Cabello-Eras, J.J.; Sagastume-Gutierez, A.; Garcia-Lorenzo, D.; Chamorro, C.; Herrera Mendoza, K.; Haeseldonckx, D.; Vandecasteele, C. On the Evolution of "Cleaner Production" as a Concept and a Practice. *J. Clean. Prod.* **2018**, *172*, 3323–3333. [CrossRef]
48. Fatimah, Y.A.; Govindan, K.; Murniningsih, R.; Setiawan, A. Industry 4.0 Based Sustainable Circular Economy Approach for Smart Waste Management System to Achieve Sustainable Development Goals: A Case Study of Indonesia. *J. Clean. Prod.* **2020**, *269*, 122263. [CrossRef]
49. Chauhan, A.; Jakhar, S.K.; Chauhan, C. The Interplay of Circular Economy with Industry 4.0 Enabled Smart City Drivers of Healthcare Waste Disposal. *J. Clean. Prod.* **2021**, *279*, 123854. [CrossRef] [PubMed]
50. Sodiq, A.; Baloch, A.A.B.; Khan, S.A.; Sezer, N.; Mahmoud, S.; Jama, M.; Abdelaal, A. Towards Modern Sustainable Cities: Review of Sustainability Principles and Trends. *J. Clean. Prod.* **2019**, *227*, 972–1001. [CrossRef]
51. Fujita, K.; Kajikawa, Y.; Mori, J.; Sakata, I. Detecting Research Fronts Using Different Types of Weighted Citation Networks. *J. Eng. Technol. Manag.* **2014**, *32*, 129–146. [CrossRef]
52. Kessler, M.M. Bibliographic Coupling between Scientific Papers. *Am. Doc.* **1963**, *14*, 10–25. [CrossRef]
53. Shibata, N.; Kajikawa, Y.; Takeda, Y.; Matsushima, K. Comparative Study on Methods of Detecting Research Fronts Using Different Types of Citation. *J. Am. Soc. Inf. Sci. Technol.* **2009**, *60*, 571–580. [CrossRef]
54. Abou Baker, N.; Szabo-Müller, P.; Handmann, U. A Feature-Fusion Transfer Learning Method as a Basis to Support Automated Smartphone Recycling in a Circular Smart City. In Proceedings of the Lecture Notes of the Institute for Computer Sciences, Social Informatics and Telecommunications Engineering, Virtual, 18–19 December 2021; Paiva, S., Lopes, S.I., Zitouni, R., Gupta, N., Lopes, S.F., Yonezawa, T., Eds.; Springer International Publishing: Cham, Switzerland, 2021; pp. 422–441.
55. Aceleanu, M.; Serban, A.C.; Suciu, M.-C.; Biţoiu, T. The Management of Municipal Waste through Circular Economy in the Context of Smart Cities Development. *IEEE Access* **2019**, *7*, 133602–133614. [CrossRef]

56. Puppim de Oliveira, J.A.; Bellezoni, R.A.; Shih, W.; Bayulken, B. Innovations in Urban Green and Blue Infrastructure: Tackling Local and Global Challenges in Cities. *J. Clean. Prod.* **2022**, *362*, 132355. [CrossRef]
57. Arias, A.; Otamendi-Irizar, I.; Grijalba, O.; Oregi, X.; Hernandez-Minguillon, R.J. Surveillance and Foresight Process of the Sustainable City Context: Innovation Potential Niches and Trends at the European Level. *Sustainability* **2022**, *14*, 8795. [CrossRef]
58. Fortes, S.; Hidalgo-Triana, N.; Sánchez-la-Chica, J.-M.; García-Ceballos, M.-L.; Cantizani-Estepa, J.; Pérez-Latorre, A.-V.; Baena, E.; Pineda, A.; Barrios-Corpa, J.; García-Marín, A. Smart Tree: An Architectural, Greening and ICT Multidisciplinary Approach to Smart Campus Environments. *Sensors* **2021**, *21*, 7202. [CrossRef] [PubMed]
59. Cappellaro, F.; Cutaia, L.; Innella, C.; Meloni, C.; Pentassuglia, R.; Porretto, V. Investigating Circular Economy Urban Practices in Centocelle, Rome District. *Environ. Eng. Manag. J.* **2019**, *18*, 2145–2153.
60. Damianou, A.; Khan, M.A.; Marios Angelopoulos, C.; Katos, V. Threat Modelling of IoT Systems Using Distributed Ledger Technologies and IOTA. In Proceedings of the DCOSS 2021: The 17th Annual International Conference on Distributed Computing in Sensor Systems, Virtual, 14–16 July 2021; pp. 404–413. [CrossRef]
61. Benltoufa, A.N.H.S.; Jaafar, F.; Maraoui, M.; Said, L.; Zili, M.; Hedfi, H.; Labidi, M.; Bouzidi, A.; Jrad, B.B.; Belhadj Salah, H. From Smart Campus to Smart City: Monastir Living Lab. In Proceedings of the 2017 International Conference on Engineering and Technology (ICET 2017), Antalya, Turkey, 21–23 August 2017; pp. 1–6. [CrossRef]
62. Crippa, J.; Silva, M.G.; Ribeiro, N.D.; Ruschel, R. Urban Branding and Circular Economy: A Bibliometric Analysis. *Environ. Dev. Sustain.* **2023**, *25*, 2173–2200. [CrossRef]
63. Gebhardt, C. Humans in the Loop: The Clash of Concepts in Digital Sustainability in Smart Cities. In *Sustainability in a Digital World: New Opportunities through New Technologies*; Osburg, T., Lohrmann, C., Eds.; Springer International Publishing: Cham, Switzerland, 2017; pp. 85–93. ISBN 978-3-319-54603-2.
64. Andrade, R.O.; Yoo, S.G. A Comprehensive Study of the Use of LoRa in the Development of Smart Cities. *Appl. Sci.* **2019**, *9*, 4753. [CrossRef]
65. Corsini, F.; Laurenti, R.; Meinherz, F.; Appio, F.P.; Mora, L. The Advent of Practice Theories in Research on Sustainable Consumption: Past, Current and Future Directions of the Field. *Sustainability* **2019**, *11*, 341. [CrossRef]
66. Dashti, A.; Noushabadi, A.S.; Asadi, J.; Raji, M.; Chofreh, A.G.; Klemeš, J.J.; Mohammadi, A.H. Review of Higher Heating Value of Municipal Solid Waste Based on Analysis and Smart Modelling. *Renew. Sustain. Energy Rev.* **2021**, *151*, 111591. [CrossRef]
67. Khan, I.S.; Ahmad, M.O.; Majava, J. Industry 4.0 and Sustainable Development: A Systematic Mapping of Triple Bottom Line, Circular Economy and Sustainable Business Models Perspectives. *J. Clean. Prod.* **2021**, *297*, 126655. [CrossRef]
68. Tahir, Z.; Malek, J.A. Main Criteria in the Development of Smart Cities Determined Using Analytical Method. *Plan. Malays.* **2016**, *14*, 59373340. [CrossRef]
69. Prendeville, S.; Cherim, E.; Bocken, N. Circular Cities: Mapping Six Cities in Transition. *Environ. Innov. Soc. Transit.* **2018**, *26*, 171–194. [CrossRef]
70. Dincă, G.; Milan, A.A.; Andronic, M.L.; Pasztori, A.M.; Dincă, D. Does Circular Economy Contribute to Smart Cities' Sustainable Development? *Int. J. Environ. Res. Public Health* **2022**, *19*, 7627. [CrossRef] [PubMed]
71. Afum, E.; Agyabeng-Mensah, Y.; Baah, C.; Agyapong, G.K.Q.; Lascano Armas, J.A.; Al Farooque, O. Prioritizing Zero-Waste Performance and Green Differentiation Advantage through the Prism of Circular Principles Adoption: A Mediated Approach. *J. Clean. Prod.* **2022**, *361*, 132182. [CrossRef]
72. Ellen MacArthur Foundation. Transitioning to a Circular Economy. 2022. Available online: https://ieg.worldbankgroup.org/evaluations/transitioning-circular-economy (accessed on 21 July 2023).
73. Dwivedi, Y.K.; Hughes, L.; Kar, A.K.; Baabdullah, A.M.; Grover, P.; Abbas, R.; Andreini, D.; Abumoghli, I.; Barlette, Y.; Bunker, D.; et al. Climate Change and COP26: Are Digital Technologies and Information Management Part of the Problem or the Solution? An Editorial Reflection and Call to Action. *Int. J. Inf. Manag.* **2022**, *63*, 102456. [CrossRef]
74. Williams, J. Circular Cities. *Urban Stud.* **2019**, *56*, 2746–2762. [CrossRef]
75. Ferasso, M.; Beliaeva, T.; Kraus, S.; Clauss, T.; Ribeiro-Soriano, D. Circular Economy Business Models: The State of Research and Avenues Ahead. *Bus. Strategy Environ.* **2020**, *29*, 3006–3024. [CrossRef]
76. Shojaei, A.; Ketabi, R.; Razkenari, M.; Hakim, H.; Wang, J. Enabling a Circular Economy in the Built Environment Sector through Blockchain Technology. *J. Clean. Prod.* **2021**, *294*, 126352. [CrossRef]
77. Lakatos, E.S.; Yong, G.; Szilagyi, A.; Clinci, D.S.; Georgescu, L.; Iticescu, C.; Cioca, L.I. Conceptualizing Core Aspects on Circular Economy in Cities. *Sustainability* **2021**, *13*, 7549. [CrossRef]
78. Damianou, A.; Vayona, A.; Demetriou, G.; Katos, V. An Actionable Maturity Planning Model for Smart, Circular Cities. *Cities* **2023**, *140*, 104403. [CrossRef]
79. Maiurova, A.; Kurniawan, T.A.; Kustikova, M.; Bykovskaia, E.; Othman, M.H.D.; Singh, D.; Goh, H.H. Promoting Digital Transformation in Waste Collection Service and Waste Recycling in Moscow (Russia): Applying a Circular Economy Paradigm to Mitigate Climate Change Impacts on the Environment. *J. Clean. Prod.* **2022**, *354*, 131604. [CrossRef]

Disclaimer/Publisher's Note: The statements, opinions and data contained in all publications are solely those of the individual author(s) and contributor(s) and not of MDPI and/or the editor(s). MDPI and/or the editor(s) disclaim responsibility for any injury to people or property resulting from any ideas, methods, instructions or products referred to in the content.

sustainability

MDPI

Article

How to Shape Local Public Acceptance of Not-in-My-Backyard Infrastructures? A Social Cognitive Theory Perspective

Wenling Bao [1], Yu Chen [1], Caiyun Cui [2], Bo Xia [3], Yongjian Ke [4], Martin Skitmore [5] and Yong Liu [1,*]

1 School of Civil Engineering and Architecture, Zhejiang Sci-Tech University, Hangzhou 310018, China; blh20190421@163.com (W.B.); chenyuu021@163.com (Y.C.)
2 School of Civil Engineering and Architecture, North China Institute of Science and Technology, Langfang 065201, China; cuicaiyun@163.com
3 School of Architecture and Built Environment, Queensland University of Technology (QUT), Brisbane, QLD 4001, Australia; paul.xia@qut.edu.au
4 School of Built Environment, University of Technology Sydney, Ultimo, NSW 2007, Australia; yongjian.ke@uts.edu.au
5 Faculty of Society and Design, Bond University, Robina, QLD 4229, Australia; rm.skitmore@qut.edu.au
* Correspondence: jhly1007@zstu.edu.cn

Abstract: Acceptance by the local public is the key determinant for the successful implementation of NIMBY (Not-In-My-Backyard) infrastructures and may be shaped in different ways among different infrastructure types. Based on social cognitive theory (SCT), this study clarifies the specific mechanism shaping local public acceptance of NIMBY facilities with two types of hazardous effects (i.e., pollution and psychological exclusion) using a large-scale questionnaire survey and structural equation model. The results reveal that, firstly, SCT provides a solid theoretical basis for exploring the mechanism under the joint action of environmental and personal factors. Secondly, it is verified that self-efficacy indirectly predicts local public acceptance by influencing perceived risk. The effect of the positive affect tag is mediated by perceived risk in shaping acceptance of polluting facilities but not of psychologically excluded facilities. In general, people tend to have a lower perceived risk, higher perceived benefit, stronger sense of self-efficacy, and more positive attitude when faced with the siting of psychologically excluded NIMBY facilities over polluting ones. These findings are helpful for planning and decision-making of NIMBY facilities with different types of hazardous impacts, reducing NIMBY conflicts and promoting the construction of NIMBY infrastructures. Furthermore, it contributes to the achievement of Sustainable Development Goal (SDG) 16 (promoting peaceful and inclusive societies for sustainable development) and (SDG) 11 (building inclusive, safe, resilient and sustainable cities and human settlements).

Keywords: NIMBY facilities; public acceptance; social cognitive theory; polluting; psychologically excluded

Citation: Bao, W.; Chen, Y.; Cui, C.; Xia, B.; Ke, Y.; Skitmore, M.; Liu, Y. How to Shape Local Public Acceptance of Not-in-My-Backyard Infrastructures? A Social Cognitive Theory Perspective. *Sustainability* **2023**, *15*, 15835. https://doi.org/10.3390/su152215835

Academic Editors: Manoj Kumar Nallapaneni, Idiano D'Adamo and Massimo Gastaldi

Received: 7 October 2023
Revised: 7 November 2023
Accepted: 8 November 2023
Published: 10 November 2023

Copyright: © 2023 by the authors. Licensee MDPI, Basel, Switzerland. This article is an open access article distributed under the terms and conditions of the Creative Commons Attribution (CC BY) license (https://creativecommons.org/licenses/by/4.0/).

1. Introduction

The editorial in the Sustainable Development Goals states that "Inclusive and participatory governance is a cornerstone of sustainable development, ensuring that decision-making processes are transparent, accountable, and responsive to the needs of all stakeholders" [1]. The siting, construction, and operation of NIMBY facilities are a complex process involving multiple stakeholders, including the government, the local public and relevant experts. Therefore, based on the specific mechanism in shaping local residents' acceptance towards NIMBY facilities, the corresponding governance methods are discussed to ensure a transparent decision-making process in the site selection, construction, and operation of NIMBY facilities, enhance public participation and responsive to the needs of all stakeholders. This is a concrete response to the idea of achieving sustainability presented in the editorial.

Local residents are worried that the construction of some high-pollution and high-energy projects will bring great pressure on the local environment, leading to the deterioration of air, water, soil, and other environmental qualities, as well as potential health risks, so they take a highly emotional and collective opposition or even resistance behavior, which is called the NIMBY phenomenon [2]. As SDG 16 refers to "promoting peaceful and inclusive societies for sustainable development" and SDG 11 refers to "building inclusive, safe, resilient and sustainable cities and human settlements", NIMBY conflicts can lead to social discord and hinder the construction of NIMBY facilities, which is not conducive to the achievement of the SDGs. Therefore, exploring the specific mechanism in shaping local residents' acceptance towards NIMBY infrastructures, enhancing public acceptance, and reducing NIMBY conflicts will help achieve SDG 16 and SDG 11.

NIMBY (Not-In-My-Backyard) facilities (e.g., nuclear power plants, waste-to-energy facilities, etc.) face considerable and strong opposition from the local communities in which they are situated [3,4] because they have a potentially negative impact (e.g., smell and illness) on the local environment and health of the public, even though their development can enhance the comprehensive carrying capacity of cities. NIMBY infrastructures can be divided into polluting facilities and psychologically excluded facilities, the former impacting human physiological health (e.g., waste incineration power plants and sewage treatment plants). Polluting facilities will produce a lot of harmful gases during operation, affecting carbon emissions. According to existing research, the increase in carbon emissions is a threat to human health and safety [5]. And the latter concerns human mental health (e.g., funeral homes and prisons) [6,7]. NIMBY facilities with different types of hazardous effects have different public acceptability [8,9].

Previous studies show that the mechanism shaping local public acceptance of NIMBY facilities (hereinafter termed "the mechanism") is related to many factors, including perceived risk, perceived income, social trust, fairness/justice, and transparency in decision-making [10,11]. For example, [12,13] show public acceptance of nuclear energy to be positively correlated with perceived benefits and [14] finds the public acceptance of waste-to-energy (WTE) projects depends on the perceived potential danger. Other studies show public acceptance of nuclear power plants to be significantly influenced by ecological problems, geographical location, and the perceived benefits involved [3], while Li et al., (2019) indicate that the three main aspects affecting public acceptance of dangerous chemical factories to be perceived risk, distrust of the government, and the "positive affect tag" of social cognitive theory (SCT) [15]. More recently, other researchers have also established an extended SCT model based on the benefit-risk perception trade-off of public acceptance of electric power NIMBY facilities [16]. Emotional dependence (e.g., local dependence, place identification, and place dependence) on the local area also seriously affects public acceptance [17,18]. The detailed influencing factors and related conclusions are shown in Table 1.

Table 1. Brief summary of factors affecting public acceptance towards NIMBY facilities.

Factors	Findings	Literature
Perceived risk	Higher level of perceived risk leads to lower level of acceptance	Liu et al., 2018 [4] Wang et al., 2021a [16]
Perceived benefit	Higher level of perceived benefit leads to higher level of acceptance	Chung and Kim, 2009 [2] Ong et al., 2022 [19]
Perceived fairness	Higher level of perceived fairness leads to higher level of acceptance	Wolsink et al., 2010 [11] Liu et al., 2018 [4]
Public trust	Higher level of public trust leads to higher level of acceptance	Zhou et al., 2022 [20] Chung and Kim, 2009 [2]
Emotional state	More positive emotions lead to higher level of acceptance	Li et al., 2019 [15] Wang et al., 2021a [16]
Demographic characteristics	Local residents who are male, elder, or highly educated hold higher level of acceptance	Ren et al., 2016 [21] Finucane et al., 2013 [22]

From social, psychological, and cultural perspectives, different theories have been applied for a long time to seek the determinants of the public's willingness to accept NIMBY facilities and explore the mechanism involved. For example, people with different demographic characteristics have different subjective conscious judgments and acceptance [21,22].

However, research into this issue is currently insufficient. First, many theories have been applied to describe the mechanism; for example, research based on stakeholder theory shows that effective public participation can reduce NIMBY conflicts, but does not take into account the social impact [19,23,24]. Studies based on affective heuristic theory show that personal perception and emotion can effectively control public acceptance, but ignore the role of external factors [15]. Some scholars combined the theory of planned behavior and the theory of protective motivation to prove that public acceptance is controlled by an individual's perceived behavior [19,23]; in addition, under the perspective of values theory, previous studies have explored the influence of personal values, beliefs, and emotions on public acceptance of nuclear power and nuclear energy, only considering the influence of personal factors [25–27]. Nevertheless, such theories often only consider the influence of personal factors on the mechanism, including personal emotion, personal perception, subjective norms, perceived behavior control, etc., but rarely focus on the mechanism under the joint action of personal and environmental factors.

Second, previous studies have pointed out that people have different degrees of acceptance of NIMBY facilities with different types of hazardous effects [28]. However, no detailed analysis has been made of the mechanism.

Third, studies normally focus on single facilities, such as nuclear power plants and WTE incineration facilities [29,30], with little attention paid to multi-case studies.

To bridge these research gaps, the present study is conducted within a viable and classic analytical framework from the perspective of SCT for understanding the mechanism under different types of hazardous effects involving the joint action of personal and environmental factors. Social cognitive theory is a powerful theory of human intent and behavior that does a good job of extending internal determinants to the outside and is recognized as the established theory for exploring patterns of behavior [28]. It points out that people's behavior intention or behavior pattern is controlled and shaped by personal factors and environmental factors [30–32]. From the perspective of social cognitive theory, public acceptance is a decision jointly affected by personal cognition and social environment, which can provide a solid theoretical basis for studying the mechanism in shaping local residents' public acceptance of NIMBY infrastructures under the joint action of personal factors and environmental factors. In contrast with previous work, multi-case studies are also undertaken to enable more general and persuasive research conclusions to be made.

In the following chapters, Section 2 presents the relevant literature review and the corresponding research hypothesis. Section 3 reports on the research design, including research framework, questionnaire design and sample data collection and analysis. Section 4 describes the research results, including descriptive analysis, *t*-test analysis, and structural equation analysis. Section 5 provides a discussion of the research results. Section 6 summarizes the overall research content, practical significance and shortcomings of the study.

2. Literature Review and Research Hypotheses

2.1. NIMBY Facilities

O'hare put forward the concept of NIMBY facilities in 1977, which refers to those having a positive effect on society and generating benefits for the public to share, while their costs tend to be localized and have certain negative externalities [33]. Previous studies have shown that the urban resilience index system consists of four dimensions, including infrastructure resilience [34], and the development of infrastructure can promote the acceleration of urbanization. Although NIMBY facilities can increase a city's capacity, the negative impacts on the environment, reputation, and property values often generate strong opposition, with people questioning why they should be burdened by problems caused by others [27,35]. The negative externalities associated with NIMBY facilities have triggered many well-known conflicts in China, such as the Yuhang Jiufeng waste incineration plant incident in Zhejiang Province [10], the Maoming PX project incident in Guangdong Province [10], and the Lianyungang nuclear cycle project incident in Jiangsu Province [36].

NIMBY facilities include not only risk-gathering amenities (nuclear power plants, substations, etc.) and polluting facilities (garbage incineration power plants, sewage treatment plants, etc.), but also those that create a sense of unhappiness, such as funeral homes, drug rehabilitation centers, prisons, etc. [9,37]. At the same time, risk-gathering NIMBY facilities are characterized by having a high risk, large scale, and being small in number, which have an obvious industrial nature.

2.2. Formation of Public Acceptance

Public acceptance plays a pivotal role in the construction and operation of large-scale urban infrastructures and affects whether or not NIMBY facilities can be smoothly built and put into operation [15]. The mechanism is also complex [3] and has been the subject of many long-term studies, revealing the involvement of such factors as trust, public perception, and self-efficacy [3].

In WTE facilities, the researchers demonstrated that public acceptance of NIMBY facilities is not only related to how well the public knows about the facility [38], but also is related to perceived physical and psychological distance [20]. Junjun et al. (2021) propose a model called cognition and emotion coupling to explain how individual behavior is influenced and, by using a structural equation model (SEM) to measure the interdependence and causal relationships between different factors, establish the mechanism for shaping local public acceptance [39]. For NIMBY risk perception, Kraft and Clary (1991) introduce a basic model that indicates there to be a positive correlation between public attitudes towards NIMBY facilities and the perceived risks associated with their proximity [40]. For nuclear power generation projects, Park et al., (2014) point out that the importance of trust in the inspectorate is emphasized as a factor influencing the public's decision to oppose or reluctantly accept such projects [41]. Additionally, a comparison between the health risks and benefits of nuclear power reveals that developing nuclear energy within the framework of risk-benefit analysis is feasible [42]. In a survey of public perceptions of biomass energy projects in the UK, the key determinant of public acceptance is identified as public trust [43]. Furthermore, for nuclear power, the public's acceptance of nuclear energy is positively correlated with the perceived benefit [16].

2.3. Social Cognitive Theory

In the late 1970s, SCT was introduced by the American psychologist Bandura as a classic theory in pedagogy and social psychology. Building upon social learning theory (SLT), Bandura emphasizes the concept of self-efficacy and proposes that individual beliefs in their own abilities are a crucial factor influencing their motivational behavior [44]. SCT encompasses several key components, including ternary reciprocal determinism, observational learning, and self-efficacy, among which ternary reciprocal determinism is the central focus [31]. Bandura extensively researched the dynamic and mutually influential relationship between individuals, environment, and behaviors. He conceptualized personal factors, environmental factors, and behavioral factors as theoretical entities that are both independent and interconnected, leading to a mutual determination [45], as shown in Figure 1. Of these, personal factors include self-efficacy, self-control, and result expectation; environmental factors include social environment, social fairness, and social trust; and behavioral factors include individual acceptance, rejection, and choice [32]. Research into SCT is currently more in-depth and extensive, which has been widely applied to knowledge-sharing management, enterprise management, behavioral willingness to participate, information system acceptance behavior, innovation and entrepreneurship behavior, education, human resources, and other fields [46,47]. Bandura proposes a five-factor structure model in his subsequent studies of SCT, pointing out that self-efficacy can act on outcome expectations and thus interfere with the occurrence of behavioral patterns [48]. Outcome expectation, a common psychological determinant, usually represents an individual's anticipation of certain behavioral consequences, including perceived risks, benefits, rewards, and sanctions [49].

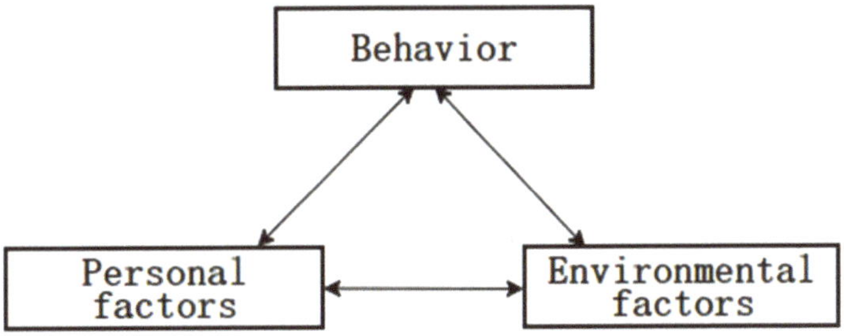

Figure 1. Three-way interaction diagram of social cognitive theory.

2.4. Perceived Risk, Perceived Benefit, and Public Acceptance

As a crucial psychological factor in behavioral research, perceived risk is considered to be a reaction to psychological activities and a subjective judgment of the negative impacts of potentially dangerous things or behaviors [50]. Due to the lack of knowledge, the public's risk cognitive judgment of some unfamiliar or potentially dangerous risk sources is mostly based on emotion and lacks objective understanding [51]. Previous studies have shown that a high-risk perception of NIMBY facilities will lead to lower public acceptance [2,52].

Perceived benefit refers to the perceived possibility that the measures taken positively impact the results [53]. Specifically, when people act in ways they think will benefit them, the result is expected to remain positive for a long time [31]. Although NIMBY facilities pose threats to the physical and mental health, living environment, and social reputation of the local public, their development creates society management benefits, social benefits, economic benefits, etc. [15]. The perceived benefit is regarded as a critical factor in determining the public acceptance of NIMBY facilities [54,55]. In previous studies of the public's attitude towards potentially dangerous facilities or things (e.g., nuclear energy, nuclear power facilities), it has been confirmed that perceived benefits are significantly positively correlated with public acceptance [23,55,56], prompting these hypotheses:

H1. *Perceived risk has a negative impact on the public acceptance of NIMBY facilities.*

H2. *Perceived benefit has a positive impact on the public acceptance of NIMBY facilities.*

2.5. Positive Affect Tag

SCT provides a profound theoretical perspective for understanding the role of emotional experience. As a sign or information for judging one's own ability, emotional state is one of the four key sources of self-efficacy [57]. Studies have shown that people can measure their self-confidence by their emotional state when participating in an action [58]. When individuals experience positive emotional states, such as happiness, they are more inclined to anticipate success than when they experience negative emotional states, such as anxiety or stress. This positive emotional state tends to result in higher levels of self-efficacy [59,60]. Affect heuristic theory holds that emotion can be understood as a feeling state [61]. During the site selection, construction, and operation of NIMBY facilities, the public primarily relies on its subjective feelings and perceptions. It has been demonstrated that there is a higher perception of benefit and a lower perception of risk when people have positive emotions [62,63], prompting these hypotheses:

H3a. *A positive affect tag has a negative impact on perceived risk.*

H3b. *A positive affect tag has a positive impact on perceived benefit.*

H3c. *A positive affect tag has a positive impact on self-efficacy.*

2.6. Self-Efficacy

Self-efficacy refers to an individual's belief in their ability to perform a certain behavior in a certain situation, which is a key internal force driving the occurrence of behavior patterns [64,65]. It is defined as the local public's self-assurance to improve the efficiency of the construction and operation of the NIMBY facilities and to avoid the risks they create [16].

Studies indicate that people are more likely inclined to refuse to perform a certain behavior when they lack confidence in themselves to perform it; in contrast, they are often willing to accept such behavior when they can guarantee their ability to avoid the risks and enjoy the benefits [66]. In other words, self-efficacy can directly affect behavior patterns and guide the occurrence of behaviors [67]. On the other hand, the influence of self-efficacy on behavior patterns is mediated by outcome expectations (e.g., perceived risk, perceived income) [68]. Self-efficacy can significantly affect perceived risks and perceived benefits [53]. Research into the public acceptance of NIMBY facilities has verified that those with high self-efficacy can see a greater perceived benefit [68], while the perceived risks of people with low self-efficacy are amplified, leading to lower acceptance of NIMBY facilities [68], prompting these hypotheses:

H4a. *Self-efficacy has a negative impact on perceived risk.*

H4b. *Self-efficacy has a positive impact on the public acceptance of NIMBY facilities.*

H4c. *Self-efficacy has a positive impact on perceived benefits.*

2.7. Social Environment

According to SCT, behavior is influenced by personal factors and restricted by external environmental factors. Individuals in different situations have different behavior patterns [69,70]. As an inseparable part of SCT, the social environment potentially impacts individual perception and behavior patterns [71]. Within the SCT framework, on the one hand, the social environment directly affects behavior and guides people's behavioral intentions or choices [31,68]. On the other hand, environmental reactions cause changes in individual subjective emotions and cognition (self-efficacy), which further affect behavioral intention. In other words, the behavioral pattern results from the interaction of individual cognition (self-efficacy) and the social environment [64]. In addition, the latter can also interfere with behavioral intention, attitudes, and choices by adjusting the expectation of results, such as perceived risk and perceived benefit [28,41], prompting these hypotheses:

H5a. *The social environment has a positive impact on self-efficacy.*

H5b. *The social environment has a negative impact on perceived risk.*

H5c. *The social environment has a positive impact on perceived benefits.*

H5d. *The social environment has a positive impact on the public acceptance of NIMBY facilities.*

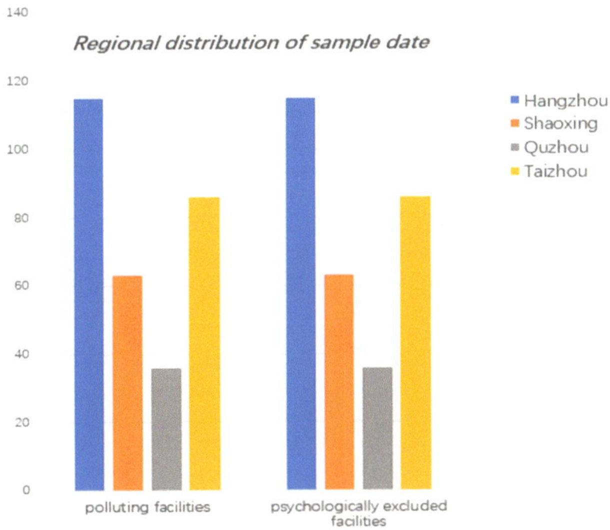

Figure 3. Regional distribution of sample data.

Table 2. Demographic characteristics of the respondents.

Profile	Category	Frequency (%)		
		Type A	Type B	Zhejiang
Gender	Male	122 (49.6%)	130 (48.7%)	33.7 million (52.1%)
	Female	124 (50.4%)	137 (51.3%)	30.9 million (47.8%)
Age	18–34	56 (22.8%)	66 (24.7%)	9.8 million (23.2%)
	35–59	118 (47.9%)	131 (49.1%)	20.4 million (48.3%)
	≥60	72 (29.3%)	70 (26.2%)	12.1 million (28.5%)
Education Level	≤Junior High School	132 (53.6%)	124 (46.4%)	
	Senior High School	51 (20.7%)	60 (22.5%)	
	Junior College	34 (13.8%)	35 (13.1%)	
	Undergraduate	27 (11.0%)	32 (12.0%)	
	≥Graduate	2 (0.8%)	16 (6.0%)	

Note: Types A and B denote polluting and psychologically excluded NIMBY facilities, respectively.

3.4. Data Analysis

The data analysis process is divided into three steps. Firstly, the respondents' attitudes towards polluting and psychologically excluded facilities are assessed using descriptive statistical analysis. Secondly, SEM is used as it is a widely utilized statistical method in academic and professional domains and is employed in numerous research studies, including those examining contractor's construction waste management behaviors [72], as well as individual's food waste behaviors [73], which is well suited to empirically test the hypothesized relationships and compare the mechanism with different types of hazardous effects. Finally, the *t*-test is used to compare the differences in local public opinions of facilities with different types of negative effects because it measures the significance of the differences in mean values between the two data groups.

4. Results

4.1. Descriptive Statistics

Table 3 summarizes the results of the descriptive statistical analysis of the survey data of polluting and psychologically excluded facilities. For polluting facilities, the mean scores of perceived risk and perceived benefit are 3.21~3.50 and 1.74~2.20, respectively, indicating the local residents' perceived risk degree to polluting facilities is relatively high, yet the perceived benefit is very low. The average score range of self-efficacy, positive affect tag, social environment, and public acceptance is between one and three, which means that people hold a quite negative attitude toward polluting facilities. For psychologically excluded facilities, the perceived risk and perceived benefit scores range from 2.26 to 2.69 and 2.23 to 2.42, respectively, revealing that local residents believe such facilities may not bring them much risk and benefit. Moreover, the mean scores of the other four control variables, including self-efficacy, positive affect tag, social environment, and public acceptance, are higher than three, which suggests the local residents have a relatively greater positive attitude to psychologically excluded facilities, with stronger inclusiveness and higher acceptance.

Table 3. Statistical results of the descriptive variables.

Variables	Item	Type A		Type B	
		Mean	Std. Dev	Mean	Std. Dev
Perceived risk	Q1	3.50	1.038	2.38	1.144
	Q2	3.21	1.052	2.46	1.153
	Q3	3.44	1.069	2.69	1.292
	Q4	3.23	1.045	2.26	1.020
Perceived benefit	Q5	1.95	0.848	2.23	0.985
	Q6	2.20	0.986	2.42	1.094
	Q7	1.96	0.880	2.33	1.077
	Q8	2.00	0.872	2.33	0.996
	Q9	1.74	0.765	2.25	1.041
Self-efficacy	Q10	2.29	0.977	3.09	0.930
	Q11	2.15	0.960	3.13	1.001
	Q12	1.93	0.915	2.74	0.920
Positive affect tag	Q13	2.12	1.004	2.90	0.871
	Q14	2.12	0.972	2.96	0.858
	Q15	2.40	1.080	3.35	0.830
Social environment	Q16	2.42	1.124	3.12	1.081
	Q17	2.26	0.958	3.10	1.014
	Q18	2.51	0.996	3.19	0.923
Public acceptance	Q19	2.49	1.023	3.21	0.968
	Q20	2.12	0.969	3.01	0.964
	Q21	1.97	0.956	2.97	1.023

4.2. *t*-Test Analysis

Table 4 shows the *t*-test results, indicating that there are certain differences in the mechanism with different types of hazardous effects. From the perspective of perceived risk, perceived benefit, self-efficacy, positive affect tag, social environment, and public acceptance, the p values are all less than 0.001, which indicates that there are significant differences in the mechanism shaping the acceptance of facilities with different types of hazardous effects: polluting and psychologically excluded facilities. At the same time, the standard deviations are all between zero and two, and most of them are between zero and one, which indicates that the data are small in dispersion, close in aggregation, and close to the true value. In general, the acceptance of polluting facilities is much lower than that of psychologically excluded facilities.

4.3. Structural Equation Analysis

Table 5 shows, through SEM analysis, the model fitting indices for the total sample and two kinds of sub-samples for polluting and psychologically excluded facilities. These show that the indices of all samples have reached the recommended values except for the psychologically excluded facilities sample, in which the GFI is slightly less than 0.9, which indicates that the survey data fit the model sufficiently well.

Table 4. Comparison of public acceptance mechanism of different types of NIMBY facilities.

Item	Type	Average Value	Standard Deviation	t-Value	Sig
PR	type A	3.35	0.889	10.626	***
	type B	2.45	1.022		
PB	type A	1.97	0.675	−4.865	***
	type B	2.31	0.906		
SE	type A	2.12	0.844	−11.653	***
	type B	2.98	0.831		
PAT	type A	2.21	0.913	−11.835	***
	type B	3.07	0.723		
SEN	type A	2.40	0.896	−9.209	***
	type B	3.14	0.923		
PA	type A	2.19	0.840	−11.484	***
	type B	3.06	0.873		

Note: *** $p < 0.001$.

Table 5. Recommended value of fit indices and actual value.

Fitness Index	χ^2/df	RMR	RMSEA	GFI	NFI	IFI	TLI	CFI
Recommended value	<3	<0.08	<0.08	≥0.9	≥0.9	≥0.9	≥0.9	≥0.9
Model A	1.761	0.040	0.054	0.902	0.910	0.959	0.951	0.959
Model B	1.889	0.055	0.059	0.892	0.911	0.956	0.947	0.955
Total	2.120	0.041	0.047	0.936	0.953	0.956	0.947	0.955

Note: See Table 2 note.

It can be seen from the above analysis that the SCT provides a solid theoretical basis for exploring the mechanism under the joint action of personal factors and environmental factors, with the latter having a significant influence.

As Figure 4 shows, all the hypothetical relationships are well supported except for H2 and H3b in Model A (polluting facilities) and H2, H3a, H3b, and H4a in Model B (psychologically excluded facilities). In addition, perceived risk is significantly positively correlated with self-efficacy and significantly negatively correlated with the positive affect tag for polluting facilities but not significantly so for psychologically excluded facilities.

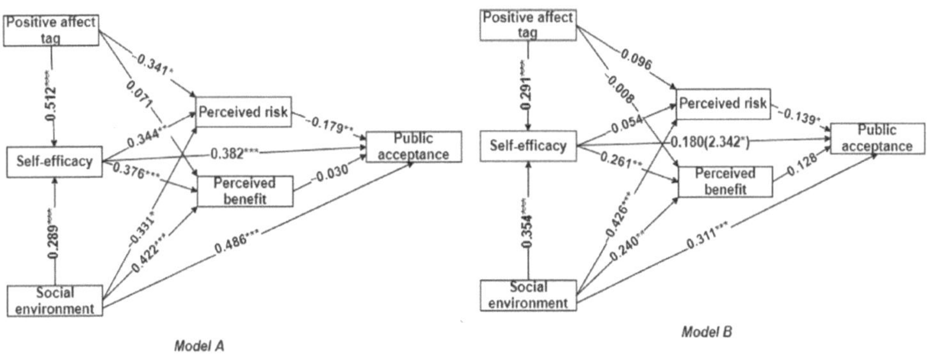

Figure 4. Standardized path coefficients. *Notes:* *** $p < 0.001$, ** $p < 0.01$, * $p < 0.05$.

5. Discussion

The present study shows that, firstly, SCT provides a solid theoretical basis for exploring the specific mechanism shaping local public acceptance of NIMBY facilities under the joint action of personal and environmental factors. Secondly, there is a significant difference in the mechanism with different types of hazardous effects. Compared with polluting facilities, the local public has a more positive attitude towards psychologically excluded facilities. Although nimbyism is usually regarded as negative, it is interesting that the public seems more accepting of psychologically excluded nimbyism than of polluting nimbyism.

According to the SEM analysis, three important findings are obtained. First, a perceived benefit is not an important determinant of local public acceptance of NIMBY facilities, which is quite different from most previous research results [56,63,74]. On the one hand, NIMBY facilities have created a series of hazards to the local public. They are so unbearable and have expressed strong concerns about the negative external effects and pay more attention to the risk. In particular, it will be difficult for people to change their attitude towards NIMBY facilities when they have already formed a fixed perception that they threaten them [75]. On the other hand, studies have confirmed that it is only when local people are experiencing serious economic difficulties that they can change their attitudes [76]. However, in the context of rapid economic development, the benefits that NIMBY facilities can create are negligible. Consequently, public acceptance is mainly controlled by perceived risk rather than perceived benefit.

Second, the positive affect tag indirectly predicts public acceptance through perceived risk for polluting facilities, which corresponds with existing studies [15,77]. However, strangely, no similar results have been found in terms of psychologically excluded facilities, and the role of the positive affect tag on perceived risk is not significant. Studies have shown that emotion plays a key role in determining the public's attitude toward NIMBY facilities [78]. Due to the lack of relevant professional knowledge or information, emotion can become the main source of the public's risk perceptions. In our research, local public risk perception of polluting NIMBY facilities mainly depends on emotional judgment because of the lack of relevant professional knowledge. As for the psychologically excluded facilities, most of the surrounding NIMBY staff residents have more professional knowledge about the facilities and the degree of threat involved, which leads to many local residents tending not to rely entirely on emotional judgment of the risks involved.

Finally, the influence of self-efficacy on the perceived risk is confirmed in terms of polluting NIMBY facilities, but such an effect is not significant in terms of psychologically excluded facilities. A previous study has shown that self-efficacy may not be consistently associated with perceived risk: even if people believe they can manage to avoid the risks associated with the facility, they can still be convinced that the facility poses a significant risk to their mental health, living environment, etc. [79]. In addition, perceived risk is usually influenced by factors other than personal self-efficacy, such as community participation, trust in facility operators, and the perceived benefit of facilities [4,41], which may even affect people with a high sense of self-efficacy. Another explanation is that the relationship between self-efficacy and perceived risk may be related to the specific characteristics of NIMBY facilities. The project's success depends on the perceived benefits involved and a comprehensive understanding of their characteristics [80]. In other words, psychologically excluded NIMBY facilities themselves may not be seriously harmful, leading to the public's low level of perceived risk despite a low sense of self-efficacy.

The *t*-test results raise three main issues. First, the local public's perceived risk of polluting NIMBY facilities is much greater than that of the psychologically excluded facilities. In contrast, the perceived benefit is much weaker. People tend to think more about physical health than mental health when considering the potential benefits and risks associated with each facility. Polluting NIMBY facilities create more obvious and real risks to the local public than psychological exclusion. For example, odors and pollutants discharged from landfill sites or waste treatment plants will affect air quality and harm human health [4,10]. But drug rehabilitation centers and prisons may not have any direct physical impact on the surrounding environment. Moreover, the operation of polluting NIMBY facilities is usually more difficult for laypeople to understand. Studies have shown that people are often afraid of what they do not understand, and for unfamiliar or potentially threatening technologies or facilities, a lack of expertise usually leads to a higher perception of risk [41,81]. In addition, environmental issues have become more prominent over time, while mental health has only recently begun to receive more attention from society [82]. People may be more aware of the risks related to pollution because they have been exposed to pollution for a long time through news reports and social media platforms.

Second, from the two dimensions of self-efficacy and positive affect tag, the public has lower self-efficacy and more negative emotions towards polluting NIMBY facilities. It has been shown that people tend to experience lower self-efficacy when they think that threats are difficult to manage and uncontrollable [83]. The risks created by polluting NIMBY facilities are tangible and uncontrollable, for example, the pollutants such as sulfur dioxide, nitrogen oxides, and particulate matter produced by garbage incineration. These pollutants lead to respiratory diseases, including asthma and bronchitis [10]. Moreover, people are usually faced with unknown and ambiguous problems owing to the complicated operations and management procedures of polluting NIMBY facilities, often leading to lower self-efficacy and increased negative emotions [84].

Thirdly, when the public considers psychologically excluded NIMBY facilities, the social environment is perceived to be less disrupted. Previous studies have shown that a credible and fair social environment can improve the public's judgment of the value of NIMBY facilities and enhance public acceptance [11,28]. Polluting NIMBY facilities are usually opposed by local communities because most of their costs tend to be localized [4]. Therefore, the local government is more inclined to avoid opposition through opaque and unfair decision-making procedures (e.g., the decision-announcement-defense (DAD) model) [85]. As a part of the social environment, distributive justice also causes different public views of the social environment and affects the acceptance of NIMBY facilities [16]. Compared with psychologically excluded NIMBY facilities, polluting facilities create a stronger sense of unfair cost-benefit distribution to the local public. It may be an important reason for the public to show the otherness in the social environment dimension when they faced NIMBY facilities with different types of hazardous effects.

In nimbyism, the construction of NIMBY facilities will not only be resisted by the local public, but also by local administrators or politicians [86]. "NIMBY being accompanied by NIMTO" is a common phenomenon, especially in urban planning and development. The NIMBY (Not-In-My-Backyard) term is generally given a negative connotation regarding all types of local opposition led by opportunistic behavior of residents, while NIMTO (Not-In-My-Term-Of-Office) is the opposition of local administrators or politicians, they are reluctant to approve the construction of these facilities during their term of office because they fear it will affect their political future [87,88]. Additionally, nimbyism may, in some cases, be used by powerful groups to preserve the status quo, prevent social and economic change, and use their wealth and influence on lobby governments against changes to the status quo, even if those changes are socially and environmentally beneficial [89].

These research findings not only have profound theoretical significance but also have certain practical application value. The research results are conducive to the planning and decision-making of NIMBY facilities with different types of hazardous effects and the formulation of related policies and provide certain practical references for relevant government departments and decision-makers in the siting, construction, and operation of NIMBY facilities, so as to better deal with potential environmental or health risks, improve public satisfaction, and maintain social order and stability, and to promote the high-quality and sustainable development of NIMBY infrastructure projects in China.

6. Conclusions

Given the key role of public acceptance in the successful incorporation of NIMBY facilities into a community, the present study conducts a viable and classic analytical framework for understanding the specific mechanism involved in shaping this acceptance under the joint action of personal and environmental factors from the perspective of SCT and further analyzes the differences in the mechanism with different types of negative effects by SEM and *t*-tests. The results reveal that, firstly, SCT does provide a solid theoretical basis for exploring the mechanism under the joint action of personal and environmental factors.

Secondly, self-efficacy indirectly predicts public acceptance by influencing perceived risk. The effect of the positive affect tag on public acceptance is mediated by the perceived risk in shaping the local public's acceptance of polluting NIMBY facilities, but not for psychologically excluded facilities. Moreover, the public acceptance of NIMBY facilities with different types of hazardous effects is mainly affected by the perceived risk but not controlled by a perceived benefit. In general, the public tends to possess lower perceived risk, higher perceived benefit, a stronger sense of self-efficacy, and a more positive attitude than polluting NIMBY facilities when faced with the siting of psychologically excluded facilities.

Based on these findings, this study has the following practical significance and application value. The research results are helpful for decision-makers, relevant governments, and enterprises to reasonably introduce relevant policies and measures, provide practical references, and better deal with potential risks in the location, construction, and operation of NIMBY infrastructures. In addition, it can help contain NIMBY conflicts, maintain social stability and harmony, and contribute to the achievement of Sustainable Development Goal (SDG) 16 (promoting peaceful and inclusive societies for sustainable development) and (SDG) 11 (Building inclusive, safe, resilient and sustainable cities and human settlements). In practical applications, we can formulate different plans for different types of NIMBY facilities, take different benefit compensation measures, risk response measures, etc., to promote the sustainable and high-quality development of different types of NIMBY facilities.

Although this study counters the deficiency of previous research into the public acceptance of NIMBY infrastructures by exploring the differences in the mechanism shaping the local public's acceptance of NIMBY facilities with different types of hazardous effects, it has certain practical significance to promote the high-quality development of NIMBY infrastructures in China; there are also some limitations. The research data are restricted to Zhejiang province and the influence of regional differences is not considered. Moreover, the applicability of these research conclusions in other countries has not been verified, and extrapolating the research conclusions to other countries and contexts requires us to have a deep understanding of the cultural and social background of other countries, consider cultural differences, and conduct localized research to ensure that our research conclusions can be effectively applied to other countries. On the one hand, specific regions or populations that future research could focus on, for example, the central region, the western region, and even in the whole country, can turn this limitation into an opportunity for further study. On the other hand, further work is needed to explore how cross-culturalism affects the shaping of local public acceptance of NIMBY infrastructures and to establish the extent to which the findings generalize to other parts of China and similarly placed regions worldwide.

Author Contributions: All authors contributed to this study's conception and design. Material preparation, data collection and analysis were performed by W.B. and Y.C. Conceptualization, methodology, and project administration were performed by C.C. and Y.K. Critical revision of the content of the paper is performed by B.X., Y.K., Y.L. and M.S.; funding acquisition was performed by Y.L. The first draft of the manuscript was written by W.B. and all authors commented on previous versions of the manuscript. All authors have read and agreed to the published version of the manuscript.

Funding: This work was supported by the National Natural Science Foundation of China (NSFC) (Grant Nos. 72072165 and 72001079), and the Hangzhou Philosophy and Social Science Planning Project (Grant No. M22JC105).

Institutional Review Board Statement: Not applicable, no human or animal concerned.

Informed Consent Statement: Informed consent was obtained from all subjects involved in the study.

Data Availability Statement: Data available on request due to restrictions, e.g., privacy or ethical, the data presented in this study are available on request from the corresponding author.

Acknowledgments: The authors would like to thank the experts for their time and the frank discussions.

Conflicts of Interest: The authors declare no conflict of interest.

References

1. Ali, S.M.; Appolloni, A.; Cavallaro, F.; Adamo, I.D.; Di Vaio, A.; Ferella, F.; Gastaldi, M.; Ikram, M.; Kumar, N.M.; Martin, M.A.; et al. Development Goals towards Sustainability. *Sustainability* **2023**, *15*, 9443. [CrossRef]
2. Chung, J.B.; Kim, H. Competition, economic benefits, trust, and risk perception in siting a potentially hazardous facility. *Landsc. Urban Plan.* **2009**, *91*, 8–16. [CrossRef]
3. Jenkins Smith, H.C.; Silva, C.L.; Nowlin, M.C.; DeLozier, G. Reversing nuclear opposition: Evolving public acceptance of a permanent nuclear waste disposal facility. *Risk Anal.* **2011**, *31*, 629–644. [CrossRef]
4. Liu, Y.; Sun, C.; Xia, B.; Cui, C.; Coffey, V. Impact of community engagement on public acceptance towards waste-to-energy incineration projects: Empirical evidence from China. *Waste Manag.* **2018**, *76*, 431–442. [CrossRef] [PubMed]
5. Wang, X.; Shen, Y.; Su, C. Spatial—Temporal evolution and driving factors of carbon emission efficiency of cities in the Yellow River Basin. *Energy Rep.* **2023**, *9*, 1065–1070. [CrossRef]
6. Simsek, C.; Elci, A.; Gunduz, O.; Taskin, N. An improved landfill site screening procedure under NIMBY syndrome constraints. *Landsc. Urban Plan.* **2014**, *132*, 1–15. [CrossRef]
7. Vrijheid, M. Health effects of residence near hazardous waste landfill sites: A review of epidemiologic literature. *Environ. Health Perspect.* **2000**, *108*, 101–112.
8. Chen, Y.; Yazdani, M.; Mojtahedi, M.; Newton, S. The impact on neighbourhood residential property valuations of a newly proposed public transport project: The Sydney Northwest Metro case study. *Transp. Res. Interdiscip. Perspect.* **2019**, *3*, 100070. [CrossRef]
9. Shan, S.; Duan, X.; Ji, W.; Zhang, T.; Li, H. Evolutionary game analysis of stakeholder behavior strategies in 'Not in My Backyard' conflicts: Effect of the intervention by environmental Non-Governmental Organizations. *Sustain. Prod. Consum.* **2021**, *28*, 829–847. [CrossRef]
10. Liu, Y.; Ge, Y.; Xia, B.; Cui, C.; Jiang, X.; Skitmore, M. Enhancing public acceptance towards waste-to-energy incineration projects: Lessons learned from a case study in China. *Sustain. Cities Soc.* **2019**, *48*, 101582. [CrossRef]
11. Wolsink, M. Contested environmental policy infrastructure: Socio-political acceptance of renewable energy, water, and waste facilities. *Environ. Impact Assess. Rev.* **2010**, *30*, 302–311. [CrossRef]
12. Wang, Y.; Li, H.; Zuo, J.; Wang, Z. Evolution of online public opinions on social impact induced by NIMBY facility. *Environ. Impact Assess. Rev.* **2019**, *78*, 106290. [CrossRef]
13. Chung, J.; Kim, E. Public perception of energy transition in Korea: Nuclear power, climate change, and party preference. *Energy Policy* **2018**, *116*, 137–144. [CrossRef]
14. Davies, A.R. Environmental justice as subtext or omission: Examining discourses of anti-incineration campaigning in Ireland. *Geoforum* **2006**, *37*, 708–724. [CrossRef]
15. Li, W.; Zhong, H.; Jing, N.; Fan, L. Research on the impact factors of public acceptance towards NIMBY facilities in China-A case study on hazardous chemicals factory. *Habitat Int.* **2019**, *83*, 11–19. [CrossRef]
16. Wang, Y.; Shen, C.; Bartsch, K.; Zuo, J. Exploring the trade-off between benefit and risk perception of NIMBY facility: A social cognitive theory model. *Environ. Impact Assess. Rev.* **2021**, *87*, 106555. [CrossRef]
17. Ho, S.S.; Oshita, T.; Looi, J.; Leong, A.D.; Chuah, A.S. Exploring public perceptions of benefits and risks, trust, and acceptance of nuclear energy in Thailand and Vietnam: A qualitative approach. *Energy Policy* **2019**, *127*, 259–268. [CrossRef]
18. Scannell, L.; Gifford, R. Defining place attachment: A tripartite organizing framework. *J. Environ. Psychol.* **2010**, *30*, 1–10. [CrossRef]
19. Ong, A.K.S.; Prasetyo, Y.T.; Salazar, J.M.L.D.; Erfe, J.J.C.; Abella, A.A.; Young, M.N.; Chuenyindee, T.; Nadlifatin, R.; Redi, A.A.N.P. Investigating the acceptance of the reopening bataan nuclear power plant: Integrating protection motivation theory and extended theory of planned behavior. *Nucl. Eng. Technol.* **2022**, *54*, 1115–1125. [CrossRef]
20. Zhou, Q.; Xu, M.; Liu, Y.; Cui, C.; Xia, B.; Ke, Y.; Skitmore, M. Exploring the effects of spatial distance on public perception of waste-to-energy incineration projects. *Waste Manag.* **2022**, *143*, 168–176. [CrossRef]
21. Ren, X.; Che, Y.; Yang, K.; Tao, Y. Risk perception and public acceptance toward a highly protested Waste-to-Energy facility. *Waste Manag.* **2016**, *48*, 528–539. [CrossRef]
22. Finucane, M.L.; Slovic, P.; Mertz, C.K.; Flynn, J.; Satterfield, T. Gender, Race and Perceived Risk: The 'White-Male' Effect. In *The Feeling of Risk*; Routledge: London, UK, 2013; pp. 125–139.
23. Zhang, T.; Shen, D.; Zheng, S.; Liu, Z.; Qu, X.; Tao, D. Predicting unsafe behaviors at nuclear power plants: An integration of Theory of Planned Behavior and Technology Acceptance Model. *Int. J. Ind. Ergon.* **2020**, *80*, 103047. [CrossRef]
24. Sun, L.; Yung, E.H.; Chan, E.H.; Zhu, D. Issues of NIMBY conflict management from the perspective of stakeholders: A case study in Shanghai. *Habitat Int.* **2016**, *53*, 133–141. [CrossRef]

25. Bidwell, D. The role of values in public beliefs and attitudes towards commercial wind energy. *Energy Policy* **2013**, *58*, 189–199. [CrossRef]
26. Whitfield, S.C.; Rosa, E.A.; Dan, A.; Dietz, T. The Future of Nuclear Power: Value Orientations and Risk Perception. *Risk Anal.* **2009**, *29*, 425–437. [CrossRef]
27. Xia, D.; Li, Y.; He, Y.; Zhang, T.; Wang, Y.; Gu, J. Exploring the role of cultural individualism and collectivism on public acceptance of nuclear energy. *Energy Policy* **2019**, *132*, 208–215. [CrossRef]
28. Wang, Y.; Shen, C.; Zuo, J.; Rameezdeen, R. Same tune, different songs? Understanding public acceptance of mega construction projects: A comparative case study. *Habitat Int.* **2021**, *118*, 102461. [CrossRef]
29. Hou, G.; Chen, T.; Ma, K.; Liao, Z.; Xia, H.; Yao, T. Improving social acceptance of waste-to-energy incinerators in China: Role of place attachment, trust, and fairness. *Sustainability* **2019**, *11*, 1727. [CrossRef]
30. Huang, L.; Ban, J.; Sun, K.; Han, Y.; Yuan, Z.; Bi, J. The influence of public perception on risk acceptance of the chemical industry and the assistance for risk communication. *Saf. Sci.* **2013**, *51*, 232–240. [CrossRef]
31. Schunk, D.H.; DiBenedetto, M.K. Motivation and social cognitive theory. *Contemp. Educ. Psychol.* **2020**, *60*, 101832. [CrossRef]
32. Beauchamp, M.R.; Crawford, K.L.; Jackson, B. Social cognitive theory and physical activity: Mechanisms of behavior change, critique, and legacy. *Psychol. Sport Exerc.* **2019**, *42*, 110–117. [CrossRef]
33. O'Hare, M. *"Not on My Block You Don't"—Facilities Siting and the Strategic Importance of Compensation*; Massachusetts Institute of Technology Laboratory of Architecture and Planning: Cambridge, MA, USA, 1977.
34. Shi, Y.; Zhang, T.; Jiang, Y. Digital Economy, Technological Innovation and Urban Resilience. *Sustainability* **2023**, *15*, 9250. [CrossRef]
35. Wu, Y.; Wang, J.; Hu, Y.; Ke, Y.; Li, L. An extended TODIM-PROMETHEE method for waste-to-energy plant site selection based on sustainability perspective. *Energy* **2018**, *156*, 1–16. [CrossRef]
36. Huang, L.; He, R.; Yang, Q.; Chen, J.; Zhou, Y.; Hammitt, J.K.; Lu, X.; Bi, J.; Liu, Y. The changing risk perception towards nuclear power in China after the Fukushima nuclear accident in Japan. *Energy Policy* **2018**, *120*, 294–301. [CrossRef]
37. Zhao, H.; Ge, Y.; Zhang, J. Evaluation on the implementation effect of public participation in the decision-making of NIMBY facilities. *PLoS ONE* **2022**, *17*, e263842. [CrossRef]
38. Schively, C. Understanding the NIMBY and LULU phenomena: Reassessing our knowledge base and informing future research. *J. Plan. Lit.* **2007**, *21*, 255–266. [CrossRef]
39. Junjun, Z.; Liukai, Y.; Gang, M.; Huixin, M.; Yangyang, J. Residents' acceptance towards waste-to-energy facilities: Formation, diffusion and policy implications. *J. Clean. Prod.* **2021**, *287*, 125560.
40. Kraft, M.E. Citizen Participation and the Nimby Syndrome: Public Response to Radioactive Waste Disposal. *West. Political Q.* **1991**, *44*, 299–328. [CrossRef]
41. Park, E.; Ohm, J.Y. Factors influencing the public intention to use renewable energy technologies in South Korea: Effects of the Fukushima nuclear accident. *Energy Policy* **2014**, *65*, 198–211. [CrossRef]
42. Dai, J.; Li, S.; Bi, J.; Ma, Z. The health risk-benefit feasibility of nuclear power development. *J. Clean. Prod.* **2019**, *224*, 198–206. [CrossRef]
43. Upreti, B.R.; Horst, D.V.D. National renewable energy policy and local opposition in the UK: The failed development of a biomass electricity plant. *Biomass Bioenergy* **2004**, *26*, 61–69. [CrossRef]
44. Bandura, A. Social cognitive theory: An agentic perspective. *Annu. Rev. Psychol.* **2001**, *52*, 21–41. [CrossRef] [PubMed]
45. Bandura, A. Social cognitive theory of self-regulation. *Organ. Behav. Hum. Decis. Process.* **1991**, *50*, 248–287. [CrossRef]
46. Lent, R.W.; Brown, S.D. Social cognitive career theory at 25: Empirical status of the interest, choice, and performance models. *J. Vocat. Behav.* **2019**, *115*, 103316. [CrossRef]
47. Eccles, J.S.; Wigfield, A. From expectancy-value theory to situated expectancy-value theory: A developmental, social cognitive, and sociocultural perspective on motivation. *Contemp. Educ. Psychol.* **2020**, *61*, 101859. [CrossRef]
48. Bandura, A. Health promotion by social cognitive means. *Health Educ. Behav.* **2004**, *31*, 143–164. [CrossRef]
49. Lowry, P.B.; Zhang, J.; Wu, T. Nature or nurture? A meta-analysis of the factors that maximize the prediction of digital piracy by using social cognitive theory as a framework. *Comput. Hum. Behav.* **2017**, *68*, 104–120. [CrossRef]
50. Moyer, R.M.; Song, G. Cultural predispositions, specific affective feelings, and benefit—Risk perceptions: Explicating local policy elites' perceived utility of high voltage power line installations. *J. Risk Res.* **2019**, *22*, 416–431. [CrossRef]
51. Slovic, P. Perception of risk. *Science* **1987**, *236*, 280–285. [CrossRef]
52. Ross, V.L.; Fielding, K.S.; Louis, W.R. Social trust, risk perceptions and public acceptance of recycled water: Testing a social-psychological model. *J. Environ. Manag.* **2014**, *137*, 61–68. [CrossRef]
53. Tumlison, C.; Song, G. Cultural values, trust, and benefit-risk perceptions of hydraulic fracturing: A comparative analysis of policy elites and the general public. *Risk Anal.* **2019**, *39*, 511–534. [CrossRef]
54. Liu, T.; Yau, Y. Institutional inadequacies and successful contentions: A case study of the LULU siting process in Hong Kong. *Habitat Int.* **2014**, *44*, 22–30. [CrossRef]
55. Jang, Y.; Park, E. Social acceptance of nuclear power plants in Korea: The role of public perceptions following the Fukushima accident. *Renew. Sustain. Energy Rev.* **2020**, *128*, 109894. [CrossRef]
56. Guo, Y.; Ru, P.; Su, J.; Anadon, L.D. Not in my backyard, but not far away from me: Local acceptance of wind power in China. *Energy* **2015**, *82*, 722–733. [CrossRef]

57. Bandura, A. *Self-Efficacy: The Exercise of Control*; Freedom and Company: New York, NY, USA, 1997.
58. Britner, S.L.; Pajares, F. Sources of science self-efficacy beliefs of middle school students. *J. Res. Sci. Teach.* **2006**, *43*, 485–499. [CrossRef]
59. Medrano, L.A.; Flores-Kanter, E.; Moretti, L.; Pereno, G.L. Effects of induction of positive and negative emotional states on academic self-efficacy beliefs in college students. *Psicol. Educ.* **2016**, *22*, 135–141. [CrossRef]
60. He, W.; Wong, W. Affective state contributes to creative self-efficacy: Evidence from an experimental study of emotion induction. *Think. Ski. Creat.* **2022**, *45*, 101061. [CrossRef]
61. Finucane, M.L.; Alhakami, A.; Slovic, P.; Johnson, S.M. The affect heuristic in judgments of risks and benefits. *J. Behav. Decis. Mak.* **2000**, *13*, 1–17. [CrossRef]
62. Ascher, T.J.; Wilson, R.S.; Toman, E. The importance of affect, perceived risk and perceived benefit in understanding support for fuels management among wildland—Urban interface residents. *Int. J. Wildland Fire* **2012**, *22*, 267–276. [CrossRef]
63. Visschers, V.H.; Keller, C.; Siegrist, M. Climate change benefits and energy supply benefits as determinants of acceptance of nuclear power stations: Investigating an explanatory model. *Energy Policy* **2011**, *39*, 3621–3629. [CrossRef]
64. Lin, H.; Chang, C. What motivates health information exchange in social media? The roles of the social cognitive theory and perceived interactivity. *Inf. Manag.* **2018**, *55*, 771–780. [CrossRef]
65. Nazari, L.N.; Reisi, M.; Tahmasebi, R.; Javadzade, H. The effect of web-based educational intervention on physical activity-related energy expenditure among middle-aged women with overweight and obesity: An application of social cognitive theory. *Obes. Med.* **2020**, *18*, 100181. [CrossRef]
66. Thøgersen, J.; Grønhøj, A. Electricity saving in households—A social cognitive approach. *Energy Policy* **2010**, *38*, 7732–7743. [CrossRef]
67. Jani, A. Escalation of commitment in troubled IT projects: Influence of project risk factors and self-efficacy on the perception of risk and the commitment to a failing project. *Int. J. Proj. Manag.* **2011**, *29*, 934–945. [CrossRef]
68. Rana, N.P.; Dwivedi, Y.K. Citizen's adoption of an e-government system: Validating extended social cognitive theory (SCT). *Gov. Inf. Q.* **2015**, *32*, 172–181. [CrossRef]
69. Font, X.; Garay, L.; Jones, S. A social cognitive theory of sustainability empathy. *Ann. Tour. Res.* **2016**, *58*, 65–80. [CrossRef]
70. Boudreaux, C.J.; Nikolaev, B.N.; Klein, P. Socio-cognitive traits and entrepreneurship: The moderating role of economic institutions. *J. Bus. Ventur.* **2019**, *34*, 178–196. [CrossRef]
71. Boateng, H.; Adam, D.R.; Okoe, A.F.; Anning-Dorson, T. Assessing the determinants of internet banking adoption intentions: A social cognitive theory perspective. *Comput. Hum. Behav.* **2016**, *65*, 468–478. [CrossRef]
72. Wu, Z.; Yu, A.T.W.; Shen, L. Investigating the determinants of contractor's construction and demolition waste management behavior in China's mainland. *Waste Manag.* **2017**, *60*, 290–300. [CrossRef]
73. Abdelradi, F. Food waste behaviour at the household level: A conceptual framework. *Waste Manag.* **2018**, *71*, 485–493. [CrossRef]
74. Walker, B.J.; Wiersma, B.; Bailey, E. Community benefits, framing and the social acceptance of offshore wind farms: An experimental study in England. *Energy Res. Soc. Sci.* **2014**, *3*, 46–54. [CrossRef]
75. Tanaka, Y. Major psychological factors determining public acceptance of the siting of nuclear facilities. *J. Appl. Soc. Psychol.* **2004**, *34*, 1147–1165. [CrossRef]
76. Lindén, A.; Rapeli, L.; Brutemark, A. Community attachment and municipal economy: Public attitudes towards wind power in a local context. *Environ. Sci. Policy* **2015**, *54*, 10–14. [CrossRef]
77. Midden, C.J.; Huijts, N.M. The role of trust in the affective evaluation of novel risks: The case of CO_2 storage. *Risk Anal.* **2009**, *29*, 743–751. [CrossRef] [PubMed]
78. Slovic, P.; Finucane, M.L.; Peters, E.; MacGregor, D.G. The affect heuristic. *Eur. J. Oper. Res.* **2007**, *177*, 1333–1352. [CrossRef]
79. Ozer, E.M.; Bandura, A. Mechanisms governing empowerment effects: A self-efficacy analysis. *J. Personal. Soc. Psychol.* **1990**, *58*, 472. [CrossRef]
80. Arning, K.; Offermann-van Heek, J.; Sternberg, A.; Bardow, A.; Ziefle, M. Risk-benefit perceptions and public acceptance of Carbon Capture and Utilization. *Environ. Innov. Soc. Transit.* **2020**, *35*, 292–308. [CrossRef]
81. Chen, M.; Lin, Y.; Cheng, T. Public attitudes toward nanotechnology applications in Taiwan. *Technovation* **2013**, *33*, 88–96. [CrossRef]
82. Conway, M.; Connor, D.O. Social media, big data, and mental health: Current advances and ethical implications. *Curr. Opin. Psychol.* **2016**, *9*, 77–82. [CrossRef]
83. Bandura, A. Self-efficacy conception of anxiety. *Anxiety Res.* **1988**, *1*, 77–98. [CrossRef]
84. Bandura, A. Reflections on self-efficacy. *Adv. Behav. Res. Ther.* **1978**, *1*, 237–269. [CrossRef]
85. Johnson, T. The politics of waste incineration in Beijing: The limits of a top-down approach? *J. Environ. Policy Plan.* **2013**, *15*, 109–128. [CrossRef]
86. Inch, A. Creating 'a Generation of NIMBYs'? Interpreting the Role of the State in Managing the Politics of Urban Development. *Environ. Plan. C Gov. Policy* **2012**, *30*, 520–535. [CrossRef]
87. Mancini, E.; Raggi, A. Out of sight, out of mind? The importance of local context and trust in understanding the social acceptance of biogas projects: A global scale review. *Energy Res. Soc. Sci.* **2022**, *91*, 102697. [CrossRef]

88. Adamo, I.D.; Sassanelli, C. Biomethane Community: A Research Agenda towards Sustainability. *Sustainabilitiy* **2022**, *14*, 4735. [CrossRef]
89. Linnerud, K.; Toney, P.; Simonsen, M.; Holden, E. Does change in ownership affect community attitudes toward renewable energy projects? Evidence of a status quo bias. *Energy Policy* **2019**, *131*, 1–8. [CrossRef]

Disclaimer/Publisher's Note: The statements, opinions and data contained in all publications are solely those of the individual author(s) and contributor(s) and not of MDPI and/or the editor(s). MDPI and/or the editor(s) disclaim responsibility for any injury to people or property resulting from any ideas, methods, instructions or products referred to in the content.

sustainability

Article

Prediction of Battery Remaining Useful Life Using Machine Learning Algorithms

J. N. Chandra Sekhar [1,*], Bullarao Domathoti [2] and Ernesto D. R. Santibanez Gonzalez [3,*]

[1] Department of Electrical and Electronics Engineering, Sri Venkateswara University, Tirupati 517502, India
[2] Department of Computer Science and Engineering, Shree Institute of Technological Education, Tirupati 517501, India; bullaraodomathoti@gmail.com
[3] Department of Industrial Engineering and CES 4.0, Faculty of Engineering, University of Talca, Curico 3340000, Chile
* Correspondence: chandrasekhar.jn@svuniversity.edu.in (J.N.C.S.); santibanez.ernesto@gmail.com (E.D.R.S.G.)

Abstract: Electrified transportation systems are emerging quickly worldwide, helping to diminish carbon gas emissions and paving the way for the reduction of global warming possessions. Battery remaining useful life (RUL) prediction is gaining attention in real world applications to tone down maintenance expenses and improve system reliability and efficiency. RUL forms the prominent component of fault analysis forecast and health management when the equipment operation life cycle is considered. The uprightness of RUL prediction is vital in providing the effectiveness of electric batteries and reducing the chance of battery illness. In assessing battery performance, the existing prediction approaches are unsatisfactory even though the battery operational parameters are well tabulated. In addition, battery management has an important contribution to several sustainable development goals, such as Clean and Affordable Energy (SDG 7), and Climate Action (SDG 13). The current work attempts to increase the prediction accuracy and robustness with selected machine learning algorithms. A Real battery life cycle data set from the Hawaii National Energy Institute (HNEI) is used to evaluate accuracy estimation using selected machine learning algorithms and is validated in Google Co-laboratory using Python. Evaluated error metrics such as Mean Square Error (MSE), Root Mean Square Error (RMSE), Mean Absolute Error (MAE), R-Squared, and execution time are computed for different L methods and relevant inferences are presented which highlight the potential of battery RUL prediction close to the most accurate values.

Keywords: HNEI battery; machine learning algorithms; heat map; Mean Squared Error; Mean Absolute Error; Root Mean Squared Error; R-Squared Error

Citation: Sekhar, J.N.C.; Domathoti, B.; Santibanez Gonzalez, E.D.R. Prediction of Battery Remaining Useful Life Using Machine Learning Algorithms. *Sustainability* **2023**, *15*, 15283. https://doi.org/10.3390/su152115283

Academic Editor: Assunta Di Vaio

Received: 6 August 2023
Revised: 9 October 2023
Accepted: 10 October 2023
Published: 25 October 2023

Copyright: © 2023 by the authors. Licensee MDPI, Basel, Switzerland. This article is an open access article distributed under the terms and conditions of the Creative Commons Attribution (CC BY) license (https://creativecommons.org/licenses/by/4.0/).

1. Introduction

One of the major challenges facing society today is reducing greenhouse gas emissions Which has prompted the industry's top producers to promote the progress of electric vehicles (EVs). Since the majority of end user electronics are motorized by battery-like technology, and the use of renewable energy sources to generate electricity is expanding quickly, energy storage has emerged as one of the key sectors. Due to their high energy density and lengthy cyclical and calendrical lifetime, lithium-ion batteries (LIBs), which are efficient energy storage systems, have assumed a leading position in powering EVs. RUL is the required amount of time, which is commonly calculated using the numerical subsequent charge–discharge cycles, from the active profile point to the end of a battery's life. A battery's lifespan ends when its remaining capacity reaches 70–80% of its initial value, according to a widely accepted generalization. RUL is a well-liked research area in the realm of electric battery research because it has an obvious significance for the assessment of the safety of batteries. The length of time a machine or asset is liable to function before needing repair or substituent is known as the remaining useful life (RUL). In addition, it is important to highlight that good management and prediction of battery use can significantly contribute to improving the efficiency and sustainability of energy storage systems, promoting the adoption of clean energy, reducing emissions, and encouraging innovation and collaboration [1]. In this way, its contribution to several sustainable development goals is interesting, including, for example, Clean and Affordable Energy (SDG 7), and Climate Action (SDG 13) [2]. The Battery RUL problem is directly related to SDG 7 by supporting the development and deployment of affordable, reliable, and clean energy solutions. In systems relying on batteries, knowing the remaining useful life allows for efficient planning and operation. By optimizing the use of batteries and extending their useful life, we can enhance the sustainability of energy systems, promote renewable energy adoption, and work toward the achievement of SDG 7's targets for universal access to clean and affordable energy [3]. In the same vein, solving the Battery RUL problem contributes to SDG 13 by supporting efforts to combat climate change, promote renewable energy, reduce greenhouse gas emissions, and enhance resilience in the face of climate-related challenges. By extending the useful life of batteries and optimizing their use, we can minimize the environmental and economic impact of battery-related activities and contribute to a more sustainable and climate-resilient future [4].

The time period may be expressed in terms of days, miles, cycles, or any other number, depending on the system you are using. RUL prediction offers early failure alerts and has evolved into a crucial part of system prognostics and health management. Figure 1 shows deterioration of a machine over time.

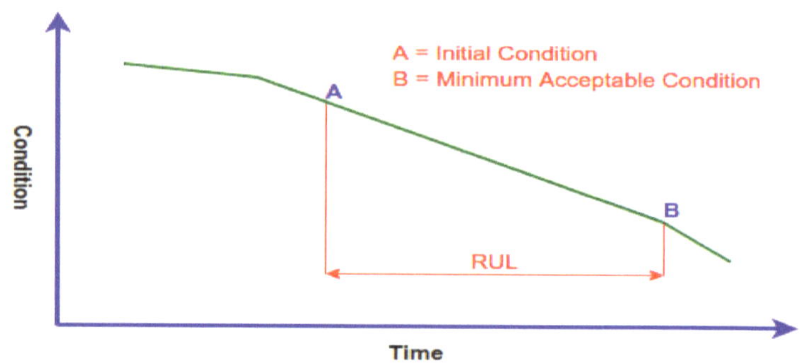

Figure 1. Prediction of RUL.

This picture illustrates how a machine degrades over time. Remaining useful life is calculated as the interval between these two points if A is the current condition and B is the minimal condition beyond which the machine will fail. RUL can be replaced or maintained to prevent unforeseen downtime and financial losses if it is forecast in advance. It is crucial to estimate RUL precisely because it is essential for operations and decision-making. The study of the most precise RUL calculation has recently gained popularity due to the swift advancement of situation and health tracking technology. In the most recent modelling trends for figuring out the RUL the examination places a strong emphasis on statistical methods that are fact-driven and use the finest readily available historical determined records and statistical models.

The different procedures that have been suggested and put into exercise to simulate the degradation of these complex systems have led to the evolution of two techniques: entirely model-based techniques and information-driven techniques. Due to the model's complexity, its universality is also limited and usually impossible to attain. As a result, the generalization impact cannot be determined. The kinetics of SEI generation and solvent transportation are considered in a one-dimensional model to examine possible decline in lithium-ion batteries. This work advises analyzing potential fade at various cyclic charge–discharge settings within a lithium-ion electrochemical pseudo-dimensional (P2D) battery version.

The model exhibits a stronger relationship between battery performance and electrolyte chemical characteristics by combining porosity exchange with electrolyte partial molar concentration [3]. The changing modern pricing and cut-off voltages are expected to cause fluctuations in the stable electrolyte interface (SEI) layer. The results show that the porosity version and convective heat switch coefficient have a substantial impact on the SEI layer development and battery life.

The semi-empirical cycle aging model for an LFP-based fully cylindrical cell presented here has been improved as a result of better knowledge of aging phenomena under the effect of DOD and Ah-throughput variables [4]. It is based on ability loss and is the primary LLI degradation technique since it takes into account both internal resistance aging and capacity loss. A Thevenin version has been described with a single RC branch and constant values for all model parameters. In this work, parameters of similar circuit models that account for battery aging and qualities for resistance aging, as well as several internal battery capabilities are included [5].

A substitute is used for random initialization, a prediction model with LSTM is initialized, and the system parameters are optimized based on test temperature and test dates, which presents an 89.18% average accuracy [6]. A comparison has been drawn between Naïve Bayes (NB) and Support Vector Machine (SVM): where NB is employed, the RUL of Li-ion batteries exhibit stable and better competitive prediction performance than SVM under constant discharge conditions [7].

A probabilistic-based adaptive estimator is used with caution for both SOE and SOC estimation using neural networks and information-pushed methodology, and accurate battery models are examined [8]. The voltage prediction effects that may be more accurate for the SOC calculation are provided by an electrochemical model and an n ordered RC equal circuit model. For SOE estimation, the relationship between the terminal voltage and the model input is often examined using a sliding window neural network model. The results demonstrate that the proposed voltage mistakes and records pushed neural network models can produce accurate version predictions.

A weighted least squares assist vector device is used for online estimation of the SOH of 2D-use lithium-ion batteries. The suggested method has been tested on a variety of 2D-use lithium-ion battery data units under various environmental conditions and battery types [9]. The results of the application show that, in comparison to the totally BPNN-based technique and the totally GPR-based approaches, the SOH accuracy anticipated via the totally WLSSVM-based approach is the best because the suggested approach has a higher ideal for nonlinear conditions with much less educational information. Battery terminal voltage characteristics during the charge process for RUL prediction has been considered Importance Sampling which has been employed in combination with a Feed Forward Neural Network [10].

In ref. [11] the author worked on a fusion of 3-D CNN and 2-D CNN which yielded a 1.1% test error for battery cycle life and a 36% test error for RUL prediction. The design and development of a lithium-ion cell version battery management system may also be used to execute state machines for charge and discharge current limit calculations, fault management, contactor management, and other duties [12]. The simulation model for designing and testing these methodologies includes Li-ion battery cell parameter estimation, battery packs with 6 and 96 cells, test documents with check cases to confirm state machine logic and linking requirements to models. Bayesian methodology is a probabilistic method for forecasting the RUL of a battery. When we initially offer a framework for feature extraction, we use the RUL estimation model to apply Bayesian inference of linear regression [13].

The lithium battery deterioration model and its RUL can both be accommodated by an integrated mastering approach that is entirely based on tracking records [14]. The relevance vector device (RVM), the random woodland (RF), the elastic net (EN), the autoregressive model (AR), and the lengthy quick-time period memory (LSTM) community are the five fundamental rookies used by the ensemble learning approach, which aims to improve prediction performance [15]. Remaining useful life (RUL) estimation of drilling pumps and critical components in fossil energy production is carried out with deep feature learning technique that uses a Convolutional Neural Network (CNN), Convolutional Block Attention Module (CBAM) [16].

Rechargeable battery systems with attractive potential for commercial applications are emerging as alternatives to traditional lithium-ion batteries (LIBs). They offer the promise of higher theoretical energy density and lower production costs compared to LIBs. However, despite their potential, these emerging lithium-sulfur Batteries (LSBs) continue to face persistent challenges that hinder their widespread industrial adoption. These issues include the inevitable dissolution of lithium polysulfide intermediates during electrochemical reactions and significant volume expansion, sometimes reaching up to 80%, upon the formation of Li_2S. These challenges result in severe limitations on battery lifespan and safety [17].

In this comprehensive review [18], an overview of the theoretical and experimental advancements in Ge-based monoelemental and binary two-dimensional (2D) materials is presented and encompasses various aspects, including crystal structures, as well as electronic, mechanical, thermal, optical, and photoelectric properties. Furthermore, we delve into the potential applications of these materials across a wide spectrum, including field-effect transistors, photodetectors, optical devices, catalysts, energy storage devices, solar cells, thermoelectric devices, sensors, biomedical materials, and spintronic devices.

Effective monomer design can lead to the creation of a sulfur-containing polymer with advantageous characteristics. These properties encompass ion and electron conductivity, high sulfur content, suitable viscosity, processability, and controllable morphology. These attributes hold significant promise for application in the cathodes of Li-S batteries, facilitating high capacity and consistent discharge even at elevated rates. This review provides a concise overview of the latest advancements in cathode materials based on sulfur-containing polymers [19].

The introduction of lithium iodide (LiI) as a Lewis basic salt serves to prevent the interaction between LiTFSI and the MoS_2 electrocatalyst, thereby inhibiting the formation of a surface gel layer. This electron-deficient center, induced by the trifluoromethanesulfonyl group in LiTFSI interacting with the Lewis acidic sites on MoS_2, triggers the cationic polymerization of the 1,3-dioxolane solvent and leads to a reduction in electrocatalytic activity. As a result of this modification, Li-S batteries incorporating the MoS_2 electrocatalyst and the LiI additive achieve an exceptionally high actual energy density of 416 W h kg^{-1} demonstrated at the pouch cell level [20].

Despite the notable theoretical energy density associated with lithium-sulfur (Li-S) batteries, they grapple with practical challenges, notably sluggish conversion kinetics and the notorious shuttle effect of polysulfides. In response to these challenges, a novel approach has been introduced involving a nitrogen-doped continuous porous carbon (CPC) host that firmly anchors monodispersed sub-10 nm FeS_2 nanoclusters, referred to as CPC@FeS_2. This innovative host exhibits a robust capability for adsorbing polysulfides, holding the promise of mitigating the polysulfide shuttle effect and enhancing the initial stages of the catalytic conversion process [21].

The other sections of this paper are prepared as follows: Section 2 explains the recent relevant works based on model-dependent and data-driven methods. Section 3 presents the pre-requisites of the data set, ML Algorithms, and Problem formulation. Section 4 presents Proposed Methodology of ML Methods for RUL Prediction and estimation of Performance error indices. Section 5 shows the results of our proposed method and discussion. Finally, Section 6 offers a conclusion and directions for future work.

2. Literature Review

To predict the RUL of lithium-ion batteries, completely version-based and statistics-driven techniques are frequently used. Primarily version-based procedures frequently make use of an extension of algebraic, differential, and empirical equations. Records-driven algorithms estimate battery RUL in the realm of mathematical fashions by utilizing latent mappings between inputs and outputs, hidden correlations between developments or features, hidden correlations between qualities and skills, and so on [22]. Several approaches for estimating lithium-ion battery RUL are currently linked. RUL prediction algorithms are classified into three categories: entirely version-based, facts-driven, and hybrid. Figure 2 depicts a thorough evaluation of the RUL prediction methods. The completely version-based approach makes use of modelling that is generally based on physics to enlarge the degradation and an empirical version to disclose how a linked item declines with time. Statistic-driven styles, on the other hand, embellish the degrading version to expand packages using ML approaches and historical statistics. In order to incorporate the respective advantages of the two models, hybrid-based solutions are finally developed.

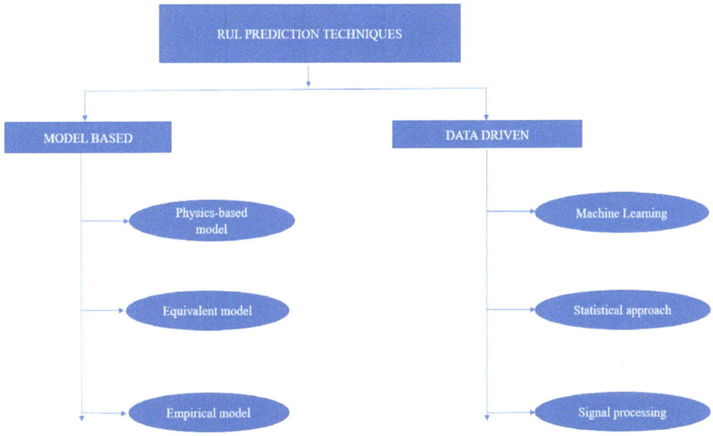

Figure 2. Classification of various RUL prediction methods for lithium-ion battery.

(A) Model Dependent Methods

Algebraic and empirical equations, linked elements, and mathematical models make up the model-based techniques. They also include a significant amount of empirical data gathering via experimentation. A physics or regression model can be used to create the battery's deterioration behavior, and the results can then be extrapolated to estimate battery performance.

(A.1) Physics-based model

The physical and electrochemical phenomena that take place inside the battery serve as the foundation of the physics model. The electrode porosity and reaction kinetics theories are used to create the physics model for RUL prediction. The physical and chemical processes, as well as calculating battery states like SOC and SOH, may also be better understood by using the physics model. The pros and cons of using dynamic models for the operation and diagnosis of Li-ion batteries are discussed in [23]. The summary of models and dynamics is succinct but clear, and it also looks at the necessary advancements for the field's future. The fundamental processes, including their dynamics, parameters, and time, must be understood in order to determine battery diagnostics and operation constants. Compared to other model-based methodologies, the accuracy of physics models is relatively poor [24]. The demand for batteries is rising as new energy electric vehicles and smart grids emerge quickly. An essential component of a system using batteries for energy storage is the battery management system (BMS). The state estimation approaches are evaluated in terms of ultimate potential and power estimation, strength functioning prediction, lifespan and health forecasting, and other important indicators in BMS.

(A.2) Equivalent circuit model

The equivalent circuit model (ECM), which stands based on knowledge of the physical and chemical processes involved in popular battery production, has been developed. It is entirely predicated on the idea that different electrical element combinations can result in outcomes that are electrically comparable to those of a battery. Consequently, a mathematical model can be developed using circuit assessment techniques. Additionally, the employment of ECM results in the simulation of several significant characteristics, such as the production of power, electricity, and heat under challenging operating conditions [25]. As the SEI movie layer matures, it has been seen that the battery's internal resistance continuously rises. Additionally, the electrochemical impedance spectroscopy (ECM) characteristics may be seen using EIS datasets, and RUL can be predicted using a regression model. One of the most important metrics for assessing the health of lithium-ion batteries is the electrochemical impedance spectrum (EIS). Ref. [26] outlines a method for online implementation that determines the EIS of lithium-ion batteries using a fractional-order equal circuit model (FOECM). The parameters of the fractional-order model are first generated based on measurements of current and voltage using a fractional-order kingdom variable clean out and the recursive least-squares approach. The parameters that were gathered are then used to construct the battery's expected EIS under various ageing scenarios. The regression version of the estimated EIS spectrum is then utilized to show how the battery is deteriorating in terms of the rise in internal resistance. Finally, compared to tests applicable to current FOECM-based regression models, the enhanced regression model is successfully applied in the context of particle filtering to predict the battery's final usable lifetime (RUL).

(A.3) Empirical model

The foundation of the empirical model is built by looking at the association between a wide range of factual aspects. Several battery degradation traits are included in the empirical model for RUL prediction. Empirical models can predict degradation trends using empirical formulas or exceptional regression models to represent degradation behavior. In [27], a framework for RUL prediction based on a logarithmic variant was established, proving its superiority over conventional empirical models. It is important to carefully choose version settings related to scalar quantities to prevent the introduction of complex values for expected ability. A two-term logarithmic variant was built to reflect the 2-segment nonlinear concave degradation of cylindrical-type Li-ion batteries. The proposed design accurately captured the inflection factor of the two-section concave deterioration. Reference [28] recommended a dual EKF (DEKF) based entirely on the Thevenin model for estimating battery kingdom. When a primary-order Taylor expression was employed for country estimate, the DEFK was used to convert a non-linear state and statement expression hooked on a linear feature. The empirical model remains a non-linear system, therefore conventional KF remains appropriate. The equivalent circuit version is constructed from basic circuit components, and the modelling process is challenging. The form acts as a real replacement for the outside function switch of battery charge and discharge. The comparable circuit version's parameter identification is rather straightforward; as a result, it is frequently used. As a result, we created the second-order Thevenin equal circuit version as well as the RC version.

(B) Data-driven-based Techniques for RUL Prediction

Utilizing past data, data-driven context predicts the rate of battery degeneration. At this time, it is still unclear how data-driven models function and propagate. Without using physics-based methods, the weight limitations in the data-driven mathematical context are determined using the training data. Recently, the flexibility and applicability of data driven models have attracted attention on a worldwide scale. Three categories can be made to classify the RUL estimation techniques created with data-driven methods: machine learning (ML), statistical methods, and signal processing.

(B.1) Machine learning (ML)

RUL can be predicted using model-based, statistics-driven, or hybrid strategies using historical data or direct statistics extraction. Developing version-based strategies in complex devices is particularly challenging, expensive, and time-consuming because it calls for a great deal of skill outside of the realm of machine learning, because data-driven tactics mainly rely on machine learning (ML) techniques.

The component's maximum useful life (MUL) is one of the most significant indicators of expected failure. The studies on RUL prediction using ML approaches are described in [29]. It is difficult, expensive, and time-consuming to develop totally version-based procedures in complex equipment since it requires extensive outside device expertise. Statistics-driven techniques had been heavily reliant on machine learning (ML) procedures.

Machine learning is employed in [30] to predict battery cycle life and early pleasant class throughout manufacture. Information-driven machine learning algorithms have been used to forecast and classify the early success of battery manufacture. The predictive power of linear regression models and artificial neural networks (ANNs) were investigated using a variety of datasets. However, the linear regression shows an equivalent look at errors of about 13% when only a few enter features are used. The focus of the RUL prediction investigation was on three top-ranked regions [31]. The first entails compiling battery data using publicly and commercially available Li-ion battery data sets. The second part of the section involves estimating battery statuses using a battery management system [32]. The symbolic regression machine learning algorithm was used to anticipate the final useful existence of these batteries. A single methodology for batch categorization, unit utilization, and battery recycling is provided by the information-pushed device mastery method used to anticipate strong state battery life. Offline training and online correction of data driven models for RUL prediction has been worked and the author inferred that a reduction of 40% in Root Mean Square Error and 34% in mean uncertainty calibration error is achieved with the augmented model [33].

(B.2) Statistical approach

The method of building a model for RUL prediction is statistical. A capacity deterioration model was created using data from earlier measurements. The application of statistical modelling is straightforward and precise. In the subsections below, a few of the statistical techniques utilized for RUL prediction are displayed.

(a) Autoregressive integrated moving average (ARIMA) technique

For battery RUL estimation, the ARIMA typical mostly substitutes the Autoregressive version due to the predominance of underfitting at some point in version schooling. To alleviate the problems associated with underfitting, ARIMA combines a shifting common approach with an autoregressive model. A novel hybrid Elman-LSTM approach for estimating the amount of usable existence that a battery has left was developed using Elman neural networks, long and short-term memory, and empirical model decomposition. This method is made available by way of [34]. The recorded battery potential versus cycle number data is broken down into several sub-layers using the empirical version decomposition set of rules. Then, utilizing Elman neural networks and recurrent long and short-term memory, high- and occasional-frequency sub-layer forecasting is performed. Large battery test datasets were used to parameterize the models and assess their performance. The appropriate HI for RUL prediction was determined using the capacity regeneration profile data from the state of health (SOH) estimation. When compared to the RVM, echo state network (ESN), and traditional ARIMA models, the determined RUL errors become insignificant. The performance of the model can be enhanced through PDF computation, as shown in [35]. A fractional-order equal circuit device is created using a combined Genetic Algorithm/Particle Swarm Optimization technique. The model is able to predict the voltage response using cycle-based comprehensive testing with a root-mean-squared error below many others. A dual fractional-order prolonged Kalman filter is suggested in order to obtain concurrent SOC and SOH estimates. Combining GM (1, 1), the ARIMA model, and the Variational mode decomposition (VMD) denoising scheme allowed us to forecast RUL, as is shown in Figure 3.

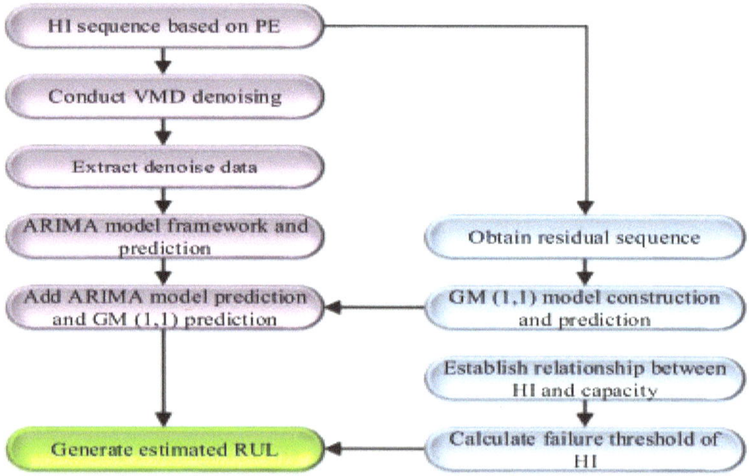

Figure 3. ARIMA based RUL prediction.

(b) Grey model technique

Enhancing the grey prediction classical means that we can use it to predict the number of users in a public road transportation system [36]. For evaluating the behavior of an unknown system, only small amounts of data are necessary. The forecasts are based on a small number of data points. Three different strategies are used in three different grey prediction models that have been presented to boost accuracy. To achieve the prediction of time series, these models convert discrete sequences into a continuous and dynamic differential equation. Each grey model is denoted by the notation GM (H, N), where H is the number of variables and N is the order of the differential equation. Small amounts of data are needed by the grey models to estimate the behavior of an unidentified system. The grey prediction model is described in greater detail in the ensuing subsections.

With the appropriate optimization of the extracted HI, the accuracy of the grey model's performance was increased [37]. The suggested model showed great accuracy, robustness, and dynamic flexibility. However, taking into account dynamic circumstances such as EV application, the provided results may vary. The RUL is predicted using an improved grey model, GM (1, 1), based on the HI that has been given. The experimental findings demonstrate the effectiveness and precision of the suggested strategy for battery degradation modelling and RUL prediction.

(B.3) Signal processing approach

With the help of the signal processing technique, significant data can be extracted. Data augmentation is one of the traits of signal processing methods. To estimate battery RUL, one can utilize a signal processing method. The DWT method is used to split a signal into its component parts. Although the DWT technique is rarely used for battery RUL prediction, it is commonly recognized for removing noise during the pre-processing of battery data. In ref. [38], lithium-ion battery remaining useful life prediction is performed using discrete wavelet transform. The remaining usable life and health are important factors in determining how safely and reliably an electric vehicle operates. Several methods have been used to estimate the RUL and SOH, with a focus on capacity loss, internal resistance growth, voltage drop, self-discharge, and cycle count. Most works only take batteries into account in specific circumstances. The model and identification techniques are the most crucial elements in figuring out how accurate the prediction results would be.

3. Problem Formulation

(A) Data set

The overall performance of a battery diminishes for a variety of reasons, and the reduction is not necessarily linear. Battery becoming older facts are required to develop an RUL prediction set of rules and compare its balance and correctness. In the early days of lithium-ion technology, battery life was limited. As a result, many ageing checks are required to swiftly acquire battery growing older facts. Numerous charge–discharge charges and ambient temperature are the most commonly utilized pressure parameters. However, apparent battery aging checks take time, and complex and expensive sign collection equipment is required to extract battery attributes from multiple aging cycles. Several researchers have used publicly available battery information to validate their RUL prediction algorithms. The Hawaii Natural Energy Institute evaluated 14 NMC-LCO 18,650 batteries with a nominal capacity of 2.8 Ah after being cycled over a thousand times at a temperature of 25 °C with a CC–CV rate of C/2 and a discharge price of 1.5C. The unique dataset is used to create features that emphasize the behavior of the voltage and current over the course of each cycle. These characteristics can be used to calculate the batteries' ultimate beneficial life (RUL). The dataset contains summaries for the 14 batteries. Finally, voltage, current, and time are used to extract seven functions from the original datasets. The goal is to forecast the battery's RUL using these features. The characteristics are listed in Figures 4–6:

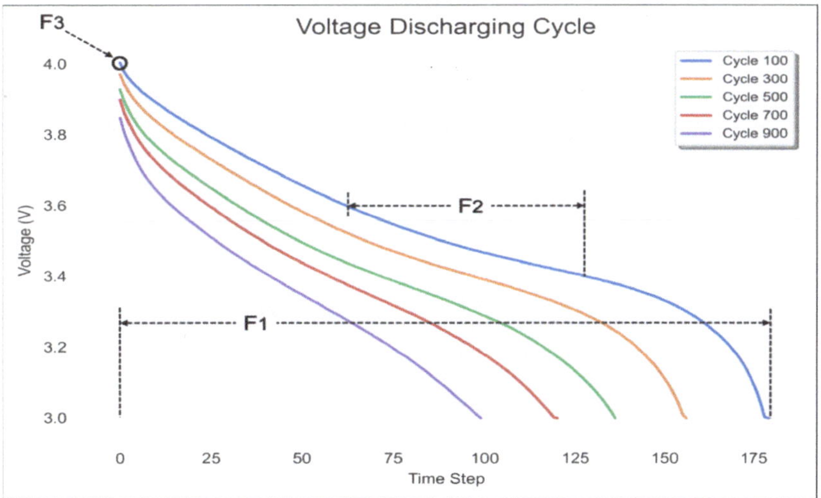

Figure 4. Profile of features F1, F2 and F3 in Voltage Discharge Cycle.

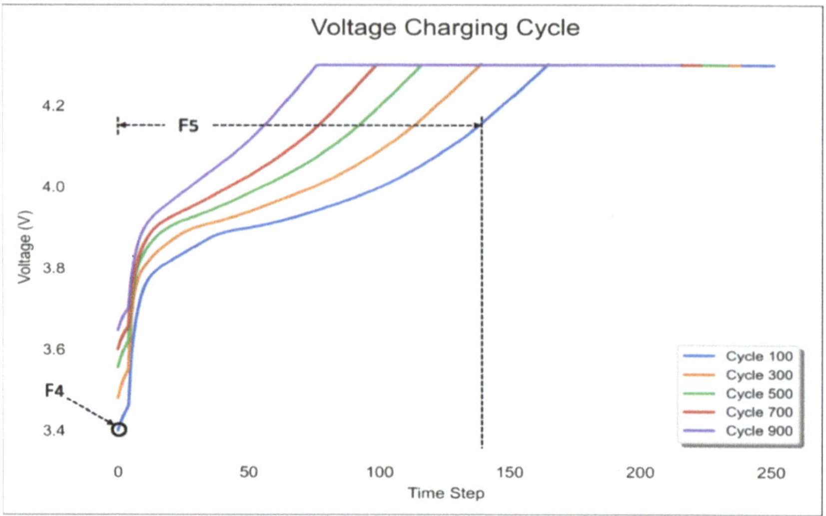

Figure 5. Profile of features F4 and F5 in Voltage charge Cycle.

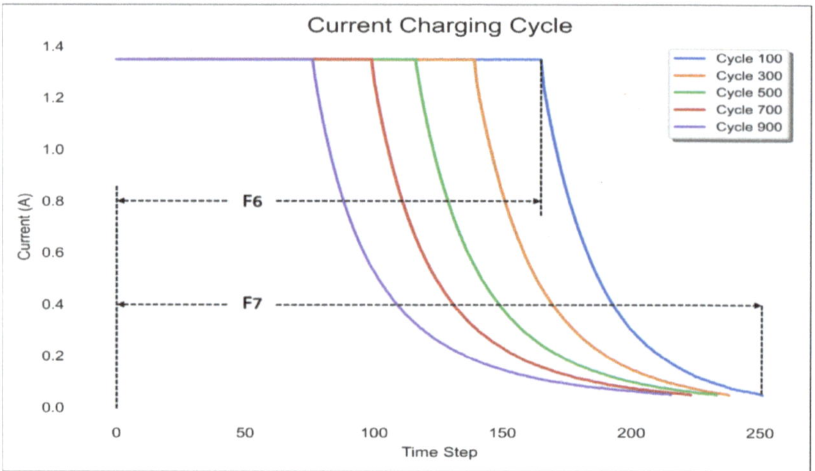

Figure 6. Profile of features F6 and F7 in Current charge Cycle.

Characteristic features:

- Cycle Index: number of cycle;
- F1: Discharge Time (s);
- F2: Time at 4.15 V (s);
- F3: Time Constant Current (s);
- F4: Decrement 3.6–3.4 V (s);
- F5: Max. Voltage Discharge (V);
- F6: Min. Voltage Charge (V);
- F7: Charging Time (s);
- Total time (s);
- RUL: target.

The battery life cycle databases that are available to the general public do not provide ready-to-use data for the assignment in [39]. Despite the fact that they incorporate various factors such as voltage (V), current (A), time (S), discharge and fee capability (Ah), and charge and discharge energy (Wh), not all of these factors can be employed for this enterprise. Voltage, cutting-edge, and time are the best inputs allowed. However, using such variables as inputs right away is impractical because they provide worthless information and are insufficient to construct a version. As a result, they must be used as a starting point and adjusted to produce new qualities, which the neural community can use to educate.

(B) Elucidation of ML Algorithms

Machine learning (ML), a completely AI-based technique, teaches a machine to mimic human behavior in order to carry out previously impossible tasks. It also encourages communication between individuals and ML systems in order to make ML outcomes intelligible to humans. Figure 7 depicts the extensive classification of gadget learning methods. The ML era has the potential to improve both prognostic accuracy and computing efficiency for predicting the RUL of Li-ion-based batteries. Machine learning aims to discover styles and laws mechanically, using data experiments or in-person interactions [40].

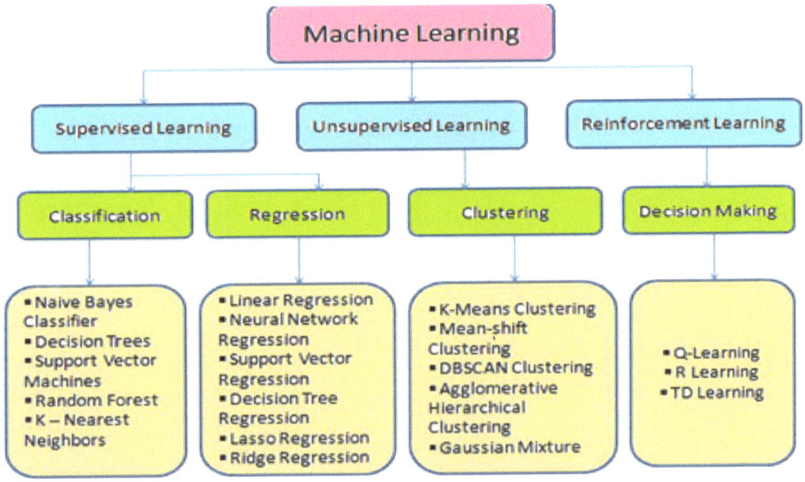

Figure 7. Classification of Machine learning algorithms.

This type of guide vector tool technique has grown in popularity during the last ten years. While using the technique as a regression strategy, its critical functions can be maintained. When records are regressed using the SVM, it is known as support vector regression (SVR). The RUL for Li-ion batteries can be predicted using the SVR technique. We decided to use a hybrid version of the SVR particle elimination technique [41]. For the terminal life cycle, an RUL prediction version that would offer the charge of RUL while updating the possibility distribution was provided. The results confirmed the efficacy of the cautioned RUL function prediction algorithms.

A simple and effective technique called KNN has been widely used in sample recognition for categorization. Using KNN and a specific number (k) of its closest neighbors in the feature area, a testing factor is categorized. These neighbors are chosen from a selection of educational websites with the right established categories. KNN regionally approximates the feature and waits to compute it because it is a lazy studying collection of rules [42].

Supervised Learning:

The supervised learning approach uses an enter-and-output pair-based learning dataset with the goal of implicitly establishing the mapping and intentional links between the inputs and outputs. The enter-and-output pairs' class output for the education dataset addresses a constrained set of discrete classes that need labelling. Information input and output are mapped as the basic goal of supervised learning. As it depends on supervision, supervised learning is analogous to when a student studies under a teacher's direction. Two further sorts of supervised learning issues can be identified:

(i) Classification;
(ii) Regression.

Unsupervised Learning:

Unsupervised learning is a sort of machine learning in which the computer is free to make decisions based solely on the information provided. The algorithm wishes to act on those records without supervision, thus the unsupervised models may be trained using the unlabeled dataset, which is neither categorized nor labelled. The version in unsupervised learning digs through the enormous number of facts in search of insightful data rather than providing a deliberate outcome. They are used to overcome the Association and Clustering difficulties As a result, they can be split into two groups:

(i) Clustering;
(ii) Association.

Reinforcement Learning:

When an agent uses reinforcement learning, it engages with its surroundings and learns from the feedback it receives. The agent receives feedback in the form of prizes; for instance, he gets a high compliment for every admirable activity and a bad reward for every distasteful conduct. The agent is not subject to any supervision. The decision-making process, which includes reinforcement algorithms, can be further divided into:

(i) Q-learning;
(ii) R-learning;
(iii) TD-learning.

Some of the popular machine learning algorithms which are used for prediction purpose are:

1. Linear Regression Algorithm;
2. Random Forest;
3. Decision Tree;
4. Support Vector Machine;
5. Naïve Bayes;
6. K-Nearest neighbor;
7. Gradient Boosting Regressor;
8. Deep Neural Network;
9. K-means Clustering;
10. Q learning.

(C) Work Objective and Problem Formulation

A twofold objective is proposed for the present work. The first step is to extract and build new features based on voltage, current, and time from the given source datasets. And the second is to predict a battery's RUL by working with machine learning methods using Python 3.6.8 in Google Co laboratory. The methods employed produce the outcome in terms of performance evaluation metrics like MSE, MAE, RMSE, and R2.

4. Proposed Methodology

(A) Framework for RUL Prediction

Although data-driven models can train quickly since they are based on historical data, choosing the right hyperparameters for effective model training requires human judgement. In this research, we suggest a novel RUL prediction framework. RUL prediction implementation is broken down into three stages, which are covered below.

Stage 1: Extraction of characteristics from considered dataset and subsequent pre-processing of data is performed.

Figure 8 shows the stage 1 flow chart. Here the battery dataset is obtained from Hawaii Natural Energy Institute (HNEI). By evaluating the charging profile data and capacity the input parameters voltage, current, and time are chosen, methodical sampling is employed. The capacity degradation data are attained from discharging cycles, and normalizations of data are done in this process.

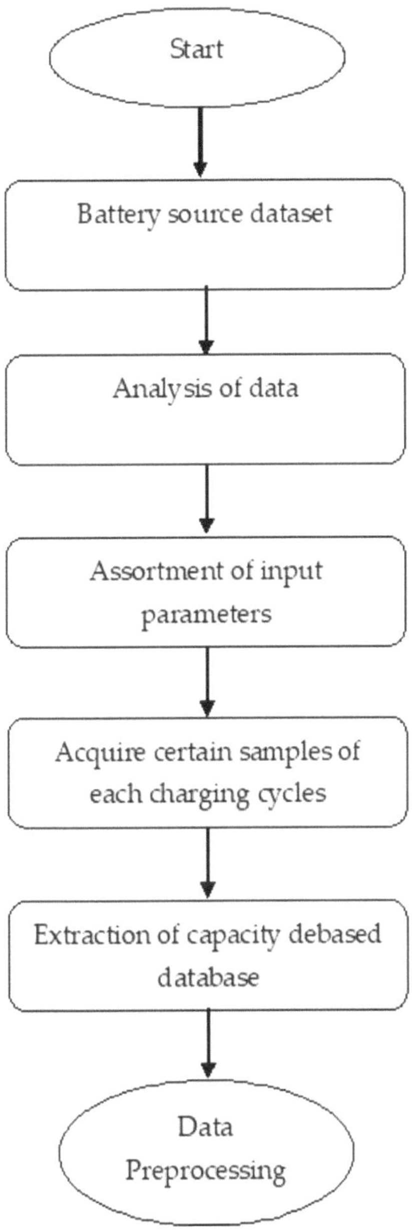

Figure 8. Extraction and Processing of data.

Stage 2: Data Division and Model Training

Figure 9 shows the flow chart of the stage 2 process. In the second stage, the data are divided into two components to facilitate model training, i.e., train and test the data. By using the hit and trail method for solution of focused parameters input features are selected and suitable algorithms are applied.

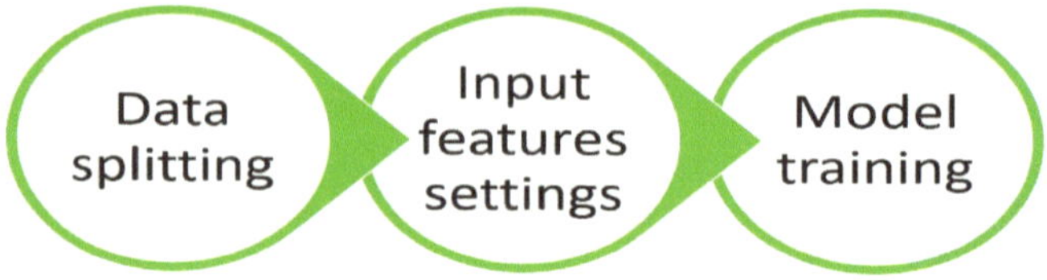

Figure 9. Data division and model training.

Stage 3: Prediction and Investigation

In the last stage, as displayed in Figure 10, RUL prediction and performance indices are calculated by Root Mean Square Error (RMSE), Mean Absolute Error (MAE), Mean Square Error (MSE), and Root Square (R^2), and various training dataset combinations are employed.

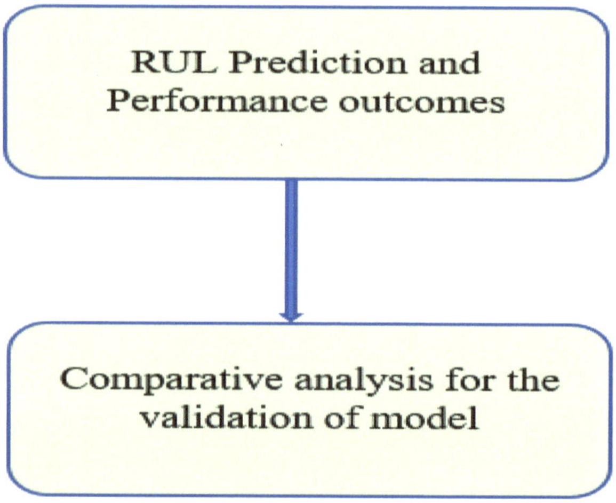

Figure 10. Prediction and Investigation of data.

(B) Selected ML Algorithms

The efficacy of a battery is determined by determining its remaining usable life. Through the identification of elements that will raise and improve their efficiency, it supports the development and testing of numerous EV variations. The method involves a number of parameters. Engineering difficulties with non-linearity and complexity can be modelled using machine learning approaches. ML approaches offer a non-invasive alternative with good accuracy and little processing, in contrast to the temporal constraints of battery degradation. The report provides a thorough and impartial assessment of the difficulties. Individual studies of numerous RUL evaluation methods are conducted. Finally, an overview that focuses on applications is provided, emphasizing the benefits in relations of accuracy and efficiency. The planned technique has six different algorithms that are used for obtaining the values of MSE, MAE, RMSE, and R Squared error. Figure 11 shows the flow chart for the proposed method.

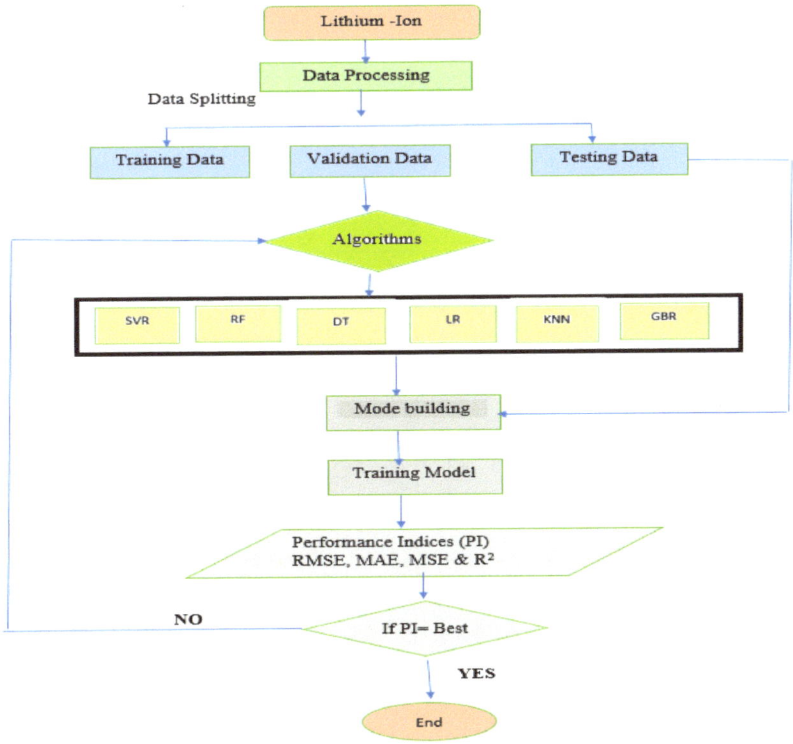

Figure 11. Flow chart for proposed method.

Random Forest Regression:

Random Forest, RF, may be applied for each type and regression, and it can manage a tremendous quantity of information. Many parallels exist between decision tree and random forest. The most basic distinction is that after the set of rules is run at some time in education, RF generates a larger number of DTs. The three primary benefits of using RF are that it eliminates the overfitting problem, has extremely high predictive accuracy, and is simple to use. The Gini index, which is frequently used, must be understood when using random forests that are entirely dependent on categorization data. The formula used to decide how nodes on a decision tree branch is given below.

$$\text{Gini} = 1 - \sum_{i=1}^{c} (p_i)^2 \tag{1}$$

The capacity to anticipate batteries for the purpose of maintaining a consistent supply of energy and the best possible use of that energy, remaining usable life (RUL), must be calculated beforehand. When it comes to accurately anticipating the battery management systems' state of charge, we decided to forecast RUL using a random forest model. With the help of this model, data may be gathered, pre-processed, and classified in order to forecast RUL. In terms of R2 and Root Mean Square Error (RMSE), the simulation is run. According to the findings, the ensemble random forest model has a greater level of prediction accuracy shows in Table 1. The random forest approaches utilized for RUL prediction are depicted in Figure 12.

Table 1. Features of random forest method.

Parameter	Value/Description
n_estimators	42
Max_features	09
Min_sample_leaf	10
Criterion	Gini Impurity
Max_leaf_nodes	10

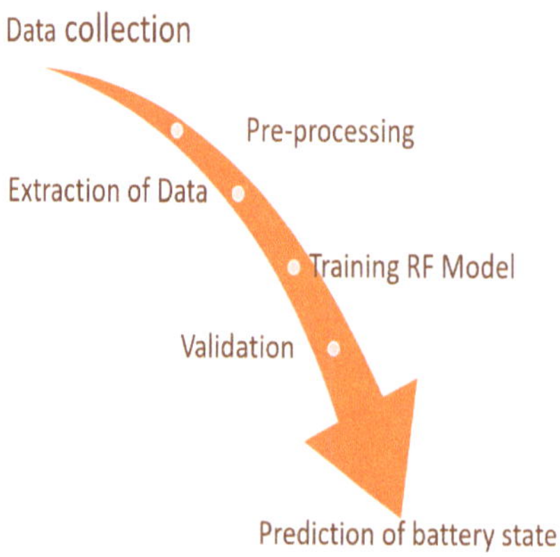

Figure 12. Flow chart for RF method.

Decision Tree Regression:

Although it can also be used to address regression problems, the decision tree approach to supervised learning is commonly used to solve classification problems. It is properly suited to each unique and continuous variable. Figure 13 depicts a tree-like structure with nodes and branches, beginning from the root node and progressing along extra branches to the leaf nodes. The central node represents the dataset's features, whereas branches indicate decision-making techniques, and leaf nodes represent results.

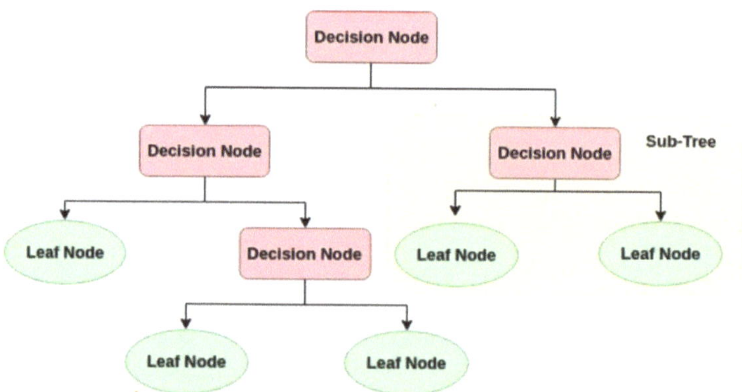

Figure 13. Decision tree Regression.

To determine which direction the node should head in, entropy analyzes the likelihood of a chosen outcome. It is more theoretically difficult to calculate than the Gini index since it uses a logarithmic feature. Entropy can also be used to analyze the branching patterns of nodes in a selection tree.

$$\text{Entropy} = \sum_{i=1}^{c} -P_i * \log_2 P_i \quad (2)$$

The DT classifier generates a final tree with leaves that reflect the final prediction and branches that represent the observations [43]. The dataset is split up into a number of subsets while the process is running, and these subsets are subsequently utilized to create the DT. The well-known classifier DT is applicable to both classification and regression. DT can therefore be applied to both numerical and categorical data. DTs are useful in a range of applications because they work well when the dataset contains missing values. Figure 14 shows the steps for predicting decision tree regression for RUL calculations shows in Table 2.

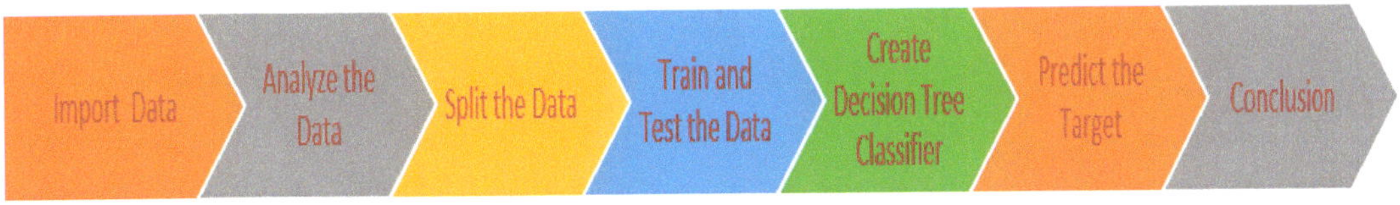

Figure 14. Decision tree regression for RUL.

Table 2. Features of decision tree method.

Parameter	Value/Description
Splitter	Random subset of features
Max_depth	-
Min_sample_split	Two
Criterion	Gini
Max_leaf_nodes	-
Min_sample_leaf	One
Max_features	Nine
Class_weight	-
Min_weight_fraction_leaf	-

Linear Regression (LR):

Linear regression is one of the most well-liked and straightforward device mastery procedures for predictive evaluation. Predictive analysis is used to explain what is predicted, and linear regression anticipates continuous values such as age, profitability, and other factors. In response to the unbiased variable (x), the dependent variable (y) adjusts, and the linear dating between the structured and independent variables is confirmed. The regression line is the path that aims to maximize the statistics among the dependent and independent variables. The equation for the regression line is:

$$y = a_0 + a^* x + b \quad (3)$$

where y = dependent variable; x = independent variable; and a_0 = Intercept of line.

Procedures for linear regression are applied if the output is a continuous variable. Instead, grading algorithms are utilized when output is classified into categories like pass/fail, good/average/bad, etc. [44] shows in Table 3. The LR method is the basic regression algorithm, but we also have other regression algorithms or classifying behaviors. Figure 15 displays the flow chart for the linear regression model.

Table 3. Features of linear regression.

Parameter	Value/Description
Linearity	Relationship between x and mean of y
Homoscedasticity	For any value of x, variance of residual is same
Observations	Independent
Normality	For any fixed x, y is normlly distributed

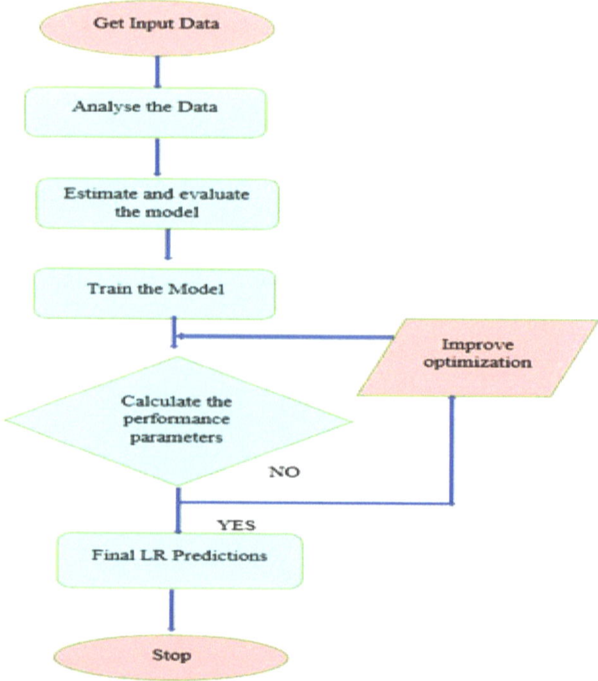

Figure 15. Flow chart for linear regression.

Gradient Boosting Regression:

An electric vehicle's battery management system must be reliable and durable in order to anticipate the state of charge. Lithium-ion batteries are heavily dependent on ageing information, which is occasionally pricey, or no longer accessible online. For the purpose of electrical car programs, it is strongly advised in [45] that the state of rate estimate of lithium-ion battery systems be carried out using a gradient boosting approach, which may also accumulate the nonlinear relationship model through offline training. The intensive gradient boosting methodology, which successfully works and accelerates, is used in the tree-based learning method.

To correct the flaws in the prior model, gradient boosting tree ensembles upload more timber at each stage. Those models are typically applied when the heterogeneity of the features is excessive. Unlike other boosting techniques like AdaBoost, this uses gradients to identify flaws. Additionally, the Gradient Boosting model is more resilient against outliers than the AdaBoost approach [46]. Evidently, gradient boosting is an expanded form of AdaBoost that is capable of handling unique loss features showed in Table 4. Typical loss features used in regression models include "huber" and "absolute" loss capabilities. More protection against outliers is provided by these loss features than by squared loss capabilities. We employed the Huber loss function as the loss function parameter in our gradient boosting regressor model. x Consider a training set (x_i, y_i) with $i = 1, 2, 3, \ldots,$

N and a differential loss function ($L(y_i, z)$, here referred to as "deviance"), where z is the predicted value. Set the model's parameters to:

$$f_0(x) = argmin_z \sum_{i=1}^{i=n} L(y_i, z). \quad (4)$$

Table 4. Features of gradient boosting regression method.

Parameter	Value/Description
n_estimators	500
Max_features	Nine
Learning rate	0.1
Random_state	100

For each observation in each tree calculate the gradient which is given by:

$$\theta_{jm} = argmin_z \sum L(y, z + f(x_i)). \quad (5)$$

Now, update the model of the succeeding tree with the gradient of the previous tree. This is given by:

$$f_m(x) = f_{m-1}(x) + \sum_{j=1}^{j=Jm} l_{jm} * I. \quad (6)$$

For each x which belongs to the region of that tree (i.e., R_{jm}), the overall output is given by $f_m(x)$.

The steps evolved for prediction of RUL using GBR are given below:

Step 1: Load the data.
Step 2: Define the target variables.
Step 3: Split the data into training and test sets.
Step 4: Initialize the GBR.
Step 5: Model the training data.
Step 6: Predict the target.
Step 7: Evaluate the performance.

Known as a Bayesian network, certain types of probabilistic graphics use Bayesian reasoning to calculate probabilities. Bayesian networks represent conditional dependency as edges in a directed graph with the objective of explaining conditional dependence and, by extension, causality. Using factors, these correlations can readily be exploited to extract data about the random variables in the graph [47].

To fully comprehend a Bayesian network, it is necessary to first grasp probability theory. It is vital to remember that the joint probability distribution of the random variables A_0, A_1, . . ., A_n, abbreviated as P(A_0, A_1, . . ., A_n), equals P(A_1 | A_2, . . ., A_n) * P(A_2 | A_3, . . ., A_n) *... * P(A_n). As it comprises N components, each with a localized probability, we can think of it as a factorized representation of the distribution.

$$P(\bigcap_{k=1}^{n} A_k) = \prod_{k=1}^{n} P(A_k | \bigcap_{j=1}^{k-1} A_J) \quad (7)$$

A and B are independent as long as C has a known and stable value. Remember that the conditional independence between the random variables A and B equals C * P B, assuming that C is the only other random variable. This can also be expressed in the way we will utilize it, which is C = A community software.

A Bayesian network's nodes and edges are directed acyclic graphs that represent conditional relationships and discrete random variables, respectively. Because the region (A, B) within the graph that implies A connects the random variables A and B, the joint opportunity distribution explicitly considers this. In order for it to execute inference, we must be aware of the P(B) possible values of A and B. Bayesian networks satisfy the nearby Markov asset, which asserts that a node is conditionally independent of its non-descendants given its parents. After simplification, In the Table 5 joint distribution of a Bayesian network is equal (mother and father(node)), as shown below:

$$P(X1....Xn) = \prod_{i=1}^{n} P(Xi|X1,....Xi-1) = \prod_{i=1}^{n} P(Xi|parents(Xi)) \tag{8}$$

Table 5. Features of Bayesian Network.

Parameter	Value/Description
Probability	Posterior—updated after evidence is considered
Conditional independence	Subset of variables
Directed acyclic graph	Continued valued variables

This feature allows us to drastically reduce the amount of processing required in larger networks because most nodes often have few parents relative to the size of the network as a whole.

(C) Performance Error Indices

To assess the model's accuracy and gauge how well it is functioning in its predictions, the process of developing system mastering models must include this step. The evaluation metrics vary depending on the problem type. The main issue with the regression problem—possibly the most famous example of this kind of issue—is that a dataset's objectives can only include real values [48]. The errors display how frequently the version makes erroneous predictions. The fundamental concept behind accuracy evaluation is to use a set of criteria to compare the original aim with the anticipated one. Following are the evaluation metrics:

C.1. MAE (Mean Absolute Error) is the difference between the original and predicted values calculated by averaging the absolute difference over the facts set. As the projected aim is the same, MAE is primarily a degree of accuracy for the labored model expressed on the same scale. If MAE is closer to zero, the more accurate the model is.

$$\text{Mean Absolute error} = \frac{1}{n}\sum\nolimits_{i=1}^{n}(y_i - \hat{y}) \tag{9}$$

C.2. MSE (Mean Squared Error) depicts the variance between the original and predicted values as determined by square rooting the mean variance throughout the data set.

$$\text{Mean Square Error} = \frac{1}{n}\sum\nolimits_{i=1}^{n}(y_i - \hat{y})^2 \tag{10}$$

C.3. The error rate by the square root of MSE is known as (Root Mean Squared Error) RMSE.

$$\text{Root Mean Square Error} = \sqrt{MSE} = \sqrt{\frac{1}{n}\sum\nolimits_{i=1}^{n}(y_i - \hat{y})^2} \tag{11}$$

C.4. R-Squared (Coefficient of determination) represents the degree to which the values fit the initial values. the proportions of values in the range of 0 and 1. The model improves as the value increases.

$$\text{R Squared} = 1 - \frac{\Sigma(y_i - \hat{y})^2}{\Sigma(y_i - \overline{y})^2} \tag{12}$$

5. Results Analysis

Heat Map:

To show how closely related various factors are, these coefficients are presented as a warmth map. It aids in locating trends that are useful for constructing device learning models. The warmth map converts the correlation matrix into a color-coding system. The correlation matrix displays how the variables are connected to one another on a scale of −1 to 1, with 1 signifying an extremely bad connection and −1 a flawless exceptional correlation. The main function of the Heatmap in Seaborn is to show a color-coded correlation matrix enabling easy visualization of the relationships between the statistical data.

From the heatmap in Figure 16, it can be inferred that:

- Features Time Constant Current' and Charging Time have a strong positive correlation with a value of 0.95.
- Features 'RUL' and 'Maximum Voltage Discharge' also have a strong positive correlation with a value of 0.78.
- Another positive correlation is between features 'Time at 4.15 V' and 'Discharge Time' with a value of 0.78.
- Feature 'Minimum Voltage Discharge' has a strong negative correlation with features 'RUL' and 'Maximum Voltage Discharge' with values −0.76 and −0.72, respectively.

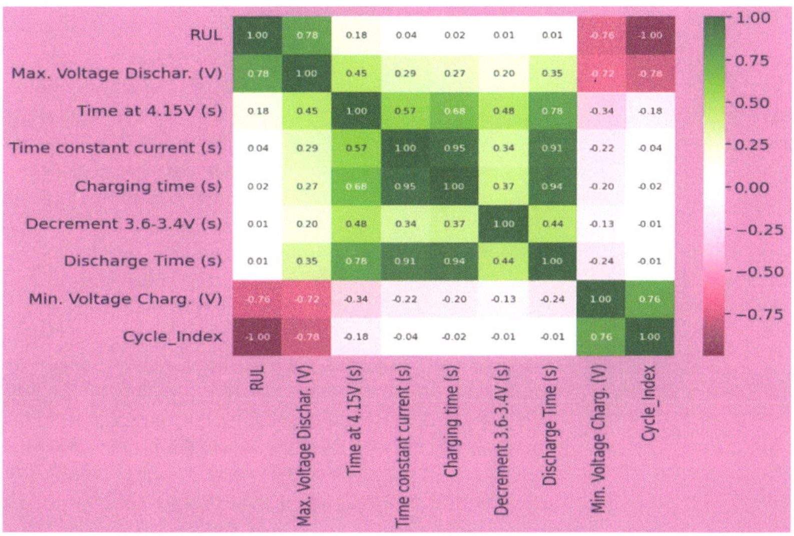

Figure 16. Heat map for the features of the source dataset.

Table 6 illustrates the numeric RUL prediction errors of the Hawaii Natural Energy Institute (HNEI) dataset's batteries. In addition, it compares ML strategies. We looked at five separate machine learning algorithms for predicting assessment metrics and the range of evaluation predictive accuracy. In this case, the records are educated, and the high-quality version forecasts are evaluated using the four outstanding overall performance error indices MSE, RMSE, MAE, and R-Squared. In this case, inputs are defined for a wide range of voltages, moderns, and temperatures for separate charging and discharging cycles. This provides a simple framework for including an expansion of input features for prediction, as well as general elapsed real time for training all algorithms. The battery RUL problem is crucial for the efficient and reliable operation of battery-dependent systems and is an active area of research and development, particularly in industries where batteries play a vital role in powering critical applications. Predicting RUL accurately can help extend the lifespan of batteries, reduce maintenance costs, and enhance overall system performance and safety [49].

Table 6. Comparison of ML methods with evaluation metrics.

ML ALGORITHMS	MSE	RMSE	MAE	R-Squared	TIME
Random Forest Regression	14.1186	3.7574	2.0930	0.9998	0.0651
Decision Tree Regression	26.8489	5.1816	2.2296	0.9997	0.1510
Linear Regression	54.5430	7.3853	4.6441	0.9994	0.0248
Bayesian network	54.5478	7.3856	4.6483	0.9993	1.2356
Gradient Boosting Regression	57.4476	7.5794	4.9842	0.9990	2.5640

In summary, the battery RUL problem is closely related to various Sustainable Development Goals, particularly those related to clean energy, sustainable infrastructure, climate action, responsible consumption, economic growth, and collaboration. Addressing the battery RUL problem through research and innovation can contribute significantly to achieving these global sustainability development goals:

Affordable and Clean Energy (SDG 7): one of the targets of SDG 7 aims to ensure access to affordable, reliable, sustainable, and modern energy for all. Batteries play a crucial role in energy storage and the transition to renewable energy sources such as solar and wind. Extending battery run-time can make these energy sources more reliable and accessible, helping to achieve the goal of clean and sustainable energy for everyone [50]. In systems relying on batteries, knowing the remaining useful life allows for efficient planning and operation. This information helps to ensure that energy storage systems and devices continue to function reliably until the end of their expected lifespan [51].

Climate Action (SDG 13): It's aims to combat climate change and its impacts. Batteries are essential for electric vehicles and renewable energy systems, both of which play a critical role in reducing greenhouse gas emissions. Improving battery run-time can accelerate the adoption of these technologies and contribute to mitigating climate change. Further regarding goal 13.1, one of the key targets is to strengthen resilience and adaptive capacity to climate-related hazards and natural disasters [52].

The comparison graph of RF, DT, LR, BN, and GBR are shown in Figure 17 and it is observed that the random forest regression (RF) method achieves the best value in MSE measurement with 14.1186 which is lower than all the other four methods. Therefore, this learning prediction method has higher robustness and better applicability. Accurate estimation of a battery's remaining useful life is crucial in various climate-related applications, such as renewable energy systems and electric vehicles. By optimizing battery usage and ensuring batteries operate at their full lifespan, we can reduce the need for premature replacements and the associated environmental impacts, such as the carbon footprint of manufacturing and disposing of batteries [53].

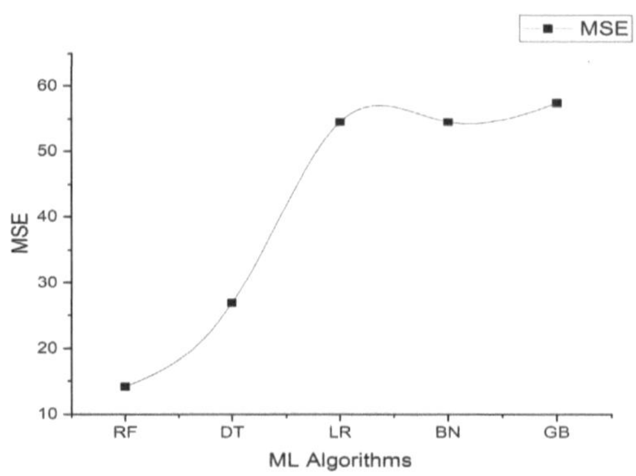

Figure 17. Mean Squared Error for different machine learning methods.

Figure 18 shows the training and prediction result for battery RUL prediction based on RMSE evaluation. It can also be seen that the random forest (RF) model stages the best performance for RUL prediction in terms of RMSE. This also verifies the advantage and superiority of all the five different machine learning algorithms.

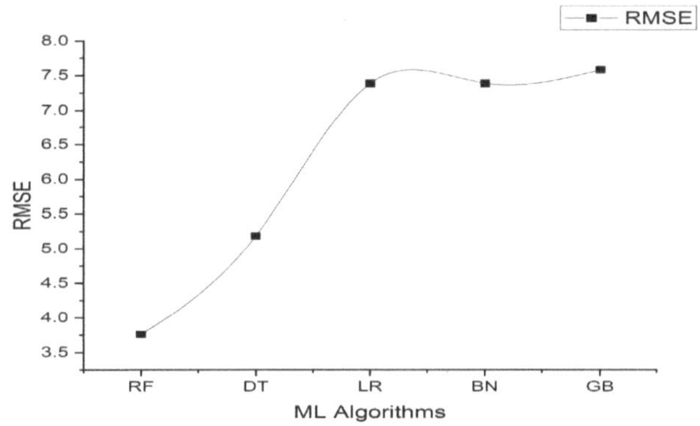

Figure 18. Root Mean Squared Error for different machine learning methods.

The MAE prediction of battery RUL is shown in Figure 19. This also verifies five different ML algorithms for prediction and determines that the random forest regression model has a good prediction capacity for determination of a battery's RUL. This also validates the effectiveness of the model for prediction.

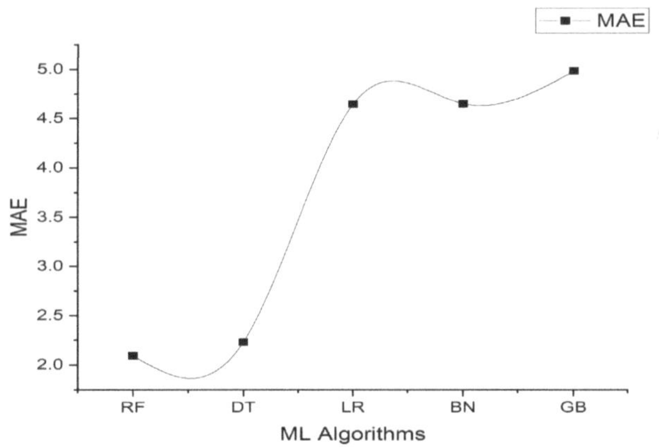

Figure 19. Mean Absolute Error for different machine learning methods.

The statistical indicator R-Squared is predicted for RF, DT, LR, BN, and GB as shown in Figure 20. This performance indicator describes the amount of uncertainty explained by an independent variable. From these performance error indices, it determines which approach achieves good efficiency with the other outperformed all the other methods.

The main contribution of this study is that it finds the optimized version and effectively captures characteristics of RUL model prediction with less execution time. Figure 21 shows the execution time obtained for all five of the ML algorithms to perform different performance error indices. This also shows the shorter training time and advantage features of all the models proposed for prediction of RUL.

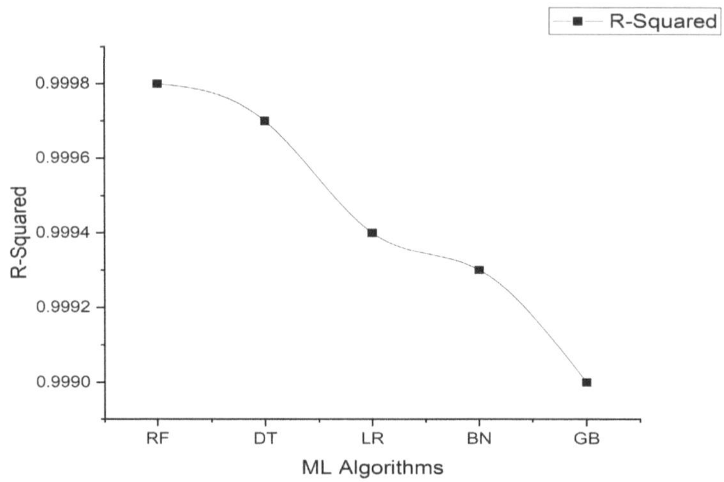

Figure 20. R-Squared Error for different machine learning methods.

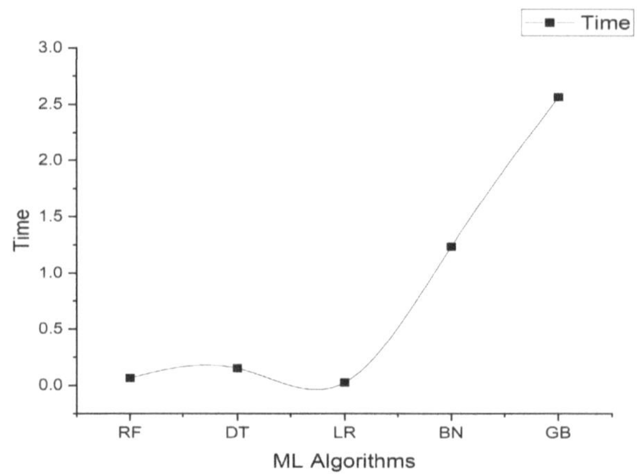

Figure 21. Time taken by different Machine Learning Methods.

In addition, the following two SDGs are also positively impacted by addressing battery run-time problem:

Responsible Consumption and Production (SDG 12): Sustainable batteries that last longer reduce the frequency at which batteries need to be replaced, resulting in less waste and fewer raw materials used in manufacturing. This aligns with SDG 12's target to ensure sustainable consumption and production patterns. Replacing batteries prematurely can be costly, especially in applications such as electric vehicles, where batteries are a significant part of the overall cost. Accurate RUL predictions help optimize the replacement schedule, reducing unnecessary expenses, waste, and raw materials used [54].

Industry, Innovation, and Infrastructure (SDG 9): SDG 9 emphasizes the development of resilient infrastructure and the promotion of sustainable industrialization. Longer battery run-time can contribute to the development of more efficient and sustainable transportation, communication, and industrial systems, reducing energy waste and promoting sustainable infrastructure [55].

The profile of the five performance error indices of the given data set using all five different machine learning algorithms is shown in Figure 22.

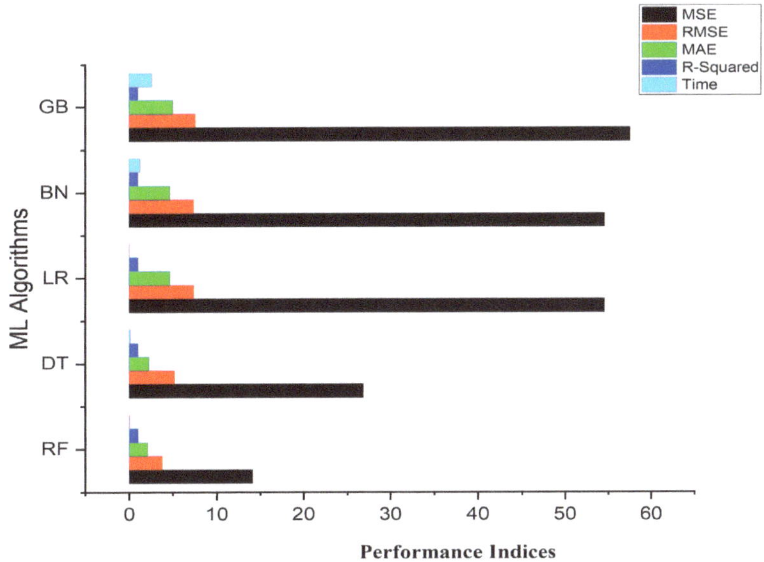

Figure 22. Error indices for different Machine Learning Methods.

6. Conclusions

This article studied the RUL prediction for the Hawaii Natural Energy Institute's (HNEI) real life battery dataset with various machine learning algorithms such as linear regression, gradient boosting, random forest, decision tree, and the Bayesian network. Subsequently the error metrics such as Mean Square Error (MSE), Root Mean Square Error (RMSE), Mean Absolute Error (MAE), R-Squared, and execution time were determined for the considered battery dataset. The proposed data driven model with all the ML algorithms was executed in python with Google Co-laboratory. Each algorithm exhibited a notable deviated relevancy with regard to the different system error indices. It is observed from the obtained results that the random forest method had better optimized evaluation metrics with less execution time and was also well suited for accurate RUL prediction. A heatmap in seaborn was also generated for the dataset where voltage, current, and time were considered to extract seven functions from the original datasets. A correlation matrix has also been computed to illustrate the association between the coefficients as presented in the warmth map.

In a global context, where energy sector emissions represent more than 75%, it is important to highlight that good management and prediction of battery use can significantly contribute to improving the efficiency and sustainability of energy storage systems. In addition to promoting the adoption of clean energy, good battery management can help reduce sector emissions and encourage innovation and collaboration between members of the supply chain. In this way, the proposed methodology is interesting for its contribution to several sustainable development goals, such as Clean and Affordable Energy (SDG 7), and Climate Action (SDG 13).

The research findings highlighted in this article may be useful for the researchers in predicting the battery RUL accuracy enhancing the overall battery efficiency. Future research can be focused on further improving the accuracy of prediction by employing Deep Learning, Hybrid Learning algorithms.

Author Contributions: Each author contributed equally in each part. All authors have read and agreed to the published version of the manuscript.

Funding: This research received no external funding.

Institutional Review Board Statement: This article does not contain any studies with human participants or animals performed by any of the authors.

Informed Consent Statement: Not applicable.

Data Availability Statement: The data used to support the findings of this study are available from the corresponding author upon request.

Acknowledgments: The authors thank to Sri Venkateswara University College of Engineering for providing characterization supports to complete this research work. The authors claim that none of the material in the paper has been published or is under consideration for publication elsewhere.

Conflicts of Interest: The authors declare that they have no conflict of interest.

Abbreviations

Abbreviation	Explanation
LIB	Lithium-ion battery
RUL	Remaining useful life
SOH	State of health
SOC	State of charge
ML	Machine learning
SVR	Support vector regression
KNN	K-Nearest neighbor
DT	Decision Tree
GBR	Gradient Boosting Regressor
DNN	Deep Neural Network
MAE	Mean Absolute error
MSE	Mean Square Error
RMSE	Root Mean Square Error
HNEI	Hawaii National Energy Institute

References

1. Si, X.-S.; Wang, W.; Hu, C.H.; Zhou, D.H. Remaining useful life estimation–a review on the statistical data driven approaches. *Eur. J. Oper. Res.* **2011**, *213*, 1–14. [CrossRef]
2. Sankarasubramanian, S.; Krishnamurthy, B. A capacity fade model for lithium-ion batteries including diffusion and kinetics. *Electrochim. Acta* **2012**, *70*, 248–254. [CrossRef]
3. Ashwin, T.R.; Chung, Y.M.; Wang, J. Capacity fade modelling of lithium-ion battery under cyclic loading conditions. *J. Power Sources* **2016**, *328*, 586–598. [CrossRef]
4. Sarasketa-Zabala, E.; Gandiaga, I.; Martinez-Laserna, E.; Rodriguez-Martinez, L.M.; Villarreal, I. Cycle ageing analysis of a lifepo4/graphite cell with dynamic model validations: Towards realistic lifetime predictions. *J. Power Sources* **2015**, *275*, 573–587. [CrossRef]
5. Liu, C.; Wang, Y.; Chen, Z. Degradation model and cycle life prediction for lithium-ion battery used in hybrid energy storge system. *Energy* **2019**, *166*, 796–806. [CrossRef]
6. Ma, J.; Shang, P.; Zou, X.; Ma, N.; Ding, Y.; Sun, J.; Cheng, Y.; Tao, L.; Lu, C.; Su, Y.; et al. A hybrid transfer learning scheme for remaining useful life prediction and cycle life test optimization of different formulation Li-ion power batteries. *Appl. Energy* **2021**, *282*, 116167. [CrossRef]
7. Ng, S.S.; Xing, Y.; Tsui, K.L. A Naïve Bayes Method for Robust Remaining Useful Life Prediction for Lithium-ion Battery. *Appl. Energy* **2014**, *118*, 114–123. [CrossRef]
8. Wang, Y.; Yang, D.; Zhang, X.; Chen, Z. Probability based remaining capacity estimation using data-driven and neural network model. *J. Power Sources* **2016**, *315*, 199–208. [CrossRef]
9. Feng, X.; Weng, C.; He, X.; Han, X.; Lu, L.; Ren, D. Online state-of-health estimation for Li-ion battery using partial charging segment based on support vector machine. *IEEE Trans. Veh. Technol.* **2019**, *68*, 8583–8592. [CrossRef]
10. Wu, J.; Zhang, C.; Chen, Z. An online method for lithium-ion battery remaining useful life estimation using importance sampling and neural networks. *Appl. Energy* **2016**, *173*, 134–140. [CrossRef]
11. Richardson, R.R.; Osborne, M.A.; Howey, D.A. Battery health prediction under generalized conditions using a Gaussian process transition model. *J. Energy Storage* **2019**, *23*, 320–328. [CrossRef]
12. Hitesh Chandra, S.; Chandra Sekhar, J.N. Battery Management Systems for Electric Vehicles using Lithium-Ion Batteries International. *J. Sci. Res.* **2022**, *11*, 1255–1261.

13. Yang, Y. A machine-learning prediction method of lithium-ion battery life based on charge process for different applications. *Appl. Energy* **2021**, *292*, 116897. [CrossRef]
14. Bhalaji, N. Rimi Chowdhury Remaining Useful Life (RUL) Estimation of Lead Acid Battery using Bayesian Approach. *J. Electr. Eng. Autom.* **2020**, *2*, 25–34.
15. Wua, J.; Kong, L. RUL Prediction for Lithium Batteries Using a Novel Ensemble Learning Method. *Energy Rep.* **2022**, *8*, 313–326. [CrossRef]
16. Ansari, S.; Ayob, A.; Lipu, M.H.; Hussain, A.; Saad, M.H.M. Remaining useful life prediction for lithium-ion battery storage system: A comprehensive review of methods, key factors, issues and future outlook. *Energy Rep.* **2022**, *8*, 12153–12185. [CrossRef]
17. Krewer, U.; Röder, F.; Harinath, E.; Braatz, R.D.; Bedürftig, B.; Findeisen, R. Review—Dynamic models of Li-ion batteries for diagnosis and operation: A review and perspective. *J. Electrochem. Soc.* **2018**, *165*, A3656–A3673. [CrossRef]
18. Wang, Y.; Tian, J.; Sun, Z.; Wang, L. A comprehensive review of battery modeling and state estimation approaches for advanced battery management systems. *Renew. Sustain. Energy Rev.* **2020**, *131*, 110015.
19. Laayouj, N.; Jamouli, H. Lithium-ion battery degradation assessment and remaining useful life estimation in hybrid electric vehicle. *Renew. Energy Sustain. Dev.* **2016**, *2*, 37–44. [CrossRef]
20. Guha, A.; Patra, A. Online estimation of the electrochemical impedance spectrum and remaining useful life of lithium-ion batteries. *IEEE Trans. Instrum. Meas.* **2018**, *67*, 1836–1849. [CrossRef]
21. Yang, F.; Wang, D.; Xing, Y.; Tsui, K.L. Prognostics of Li(NiMnCo)O2-based lithium-ion batteries using a novel battery degradation model. *Microelectron. Reliab.* **2017**, *70*, 70–78. [CrossRef]
22. Wang, A.; Chen, H.; Jin, P.; Huang, J.; Feng, D.; Zheng, M. RUL estimation of lithium-ion power battery based on DEKF algorithm. In Proceedings of the 2019 14th IEEE Conference on Industrial Electronics and Applications (ICIEA), Xi'an, China, 19–21 June 2019; pp. 1851–1856. [CrossRef]
23. Ferreira, C.; Gonçalves, G. Remaining Useful Life prediction and challenges: A literature review on the use of Machine Learning Methods. *J. Manuf. Syst.* **2022**, *63*, 550–562. [CrossRef]
24. Stock, S.; Pohlmann, S.; Guenter, F.J.; Hille, L.; Hagemeister, J.; Reinhart, G. Early Quality Classification and Prediction of Battery Cycle Life in Production Using Machine Learning. *J. Energy Storage* **2022**, *50*, 104144. [CrossRef]
25. Hasib, S.A.; Islam, S.; Chakrabortty, R.K.; Ryan, M.J.; Saha, D.K.; Ahamed, M.H.; Moyeen, S.I.; Das, S.K.; Ali, M.F.; Islam, M.R.; et al. A Comprehensive Review of Available Battery Datasets, RUL Prediction Approaches, and Advanced Battery Management. *IEEE Trans.* **2021**, *9*, 86166–86193. [CrossRef]
26. Cheng, D.; Sha, W.; Wang, L.; Tang, S.; Ma, A.; Chen, Y.; Wang, H.; Lou, P.; Lu, S.; Cao, Y.C. Solid-State Lithium Battery Cycle Life Prediction Using Machine Learning. *Appl. Sci.* **2021**, *11*, 4671. [CrossRef]
27. Thelen, A.; Li, M.; Hu, C.; Bekyarova, E.; Kalinin, S.; Sanghadasa, M. Augmented model-based framework for battery remaining useful life prediction. *Appl. Energy* **2022**, *324*, 119624. [CrossRef]
28. Li, X.; Zhang, L.; Wang, Z.; Dong, P. Remaining useful life prediction for lithium-ion batteries based on a hybrid model combining the long shortterm memory and elman neural networks. *J. Energy Storage* **2019**, *21*, 510–518. [CrossRef]
29. Chen, L.; Xu, L.; Zhou, Y. Novel Approach for Lithium-Ion Battery On-Line Remaining Useful Life Prediction Based on Permutation Entropy. *Energies* **2018**, *11*, 820. [CrossRef]
30. Balochian, S.; Baloochian, H. Improving grey prediction model and its application in predicting the number of users of a public road transportation system. *J. Intell. Syst.* **2020**, *30*, 104–114. [CrossRef]
31. Zhou, D.; Xue, L.; Song, Y.; Chen, J. On-line remaining useful life prediction of lithium-ion batteries based on the optimized gray model GM (1, 1). *Batteries* **2017**, *3*, 21. [CrossRef]
32. Wang, Y.; Pan, R.; Yang, D.; Tang, X.; Chen, Z. Remaining useful life prediction of lithium-ion battery based on discrete wavelet transform. *Energy Procedia* **2017**, *105*, 2053–2058. [CrossRef]
33. GitHub. How the Dataset Was Built. Available online: https://github.com/ignavinuales/Battery_RUL_Prediction (accessed on 1 March 2022).
34. Zhao, S.; Blaabjerg, F.; Wang, H. An Overview of Artificial Intelligence Applications for Power Electronics. *IEEE Trans. Power Electron.* **2020**, *36*, 4633–4658. [CrossRef]
35. Dong, H.; Jin, X.; Lou, Y.; Wang, C. Lithium-Ion Battery State of Health Monitoring and Remaining Useful Life Prediction Based on Support Vector Regression-Particle Filter. *J. Power Sources* **2014**, *271*, 114–123. [CrossRef]
36. Jayakumar, T.; Gowda, N.M.; Sujatha, R.; Bhukya, S.N.; Padmapriya, G.; Radhika, S.; Mohanavel, V.; Sudhakar, M.; Sathyamurthy, R. Machine Learning approach for Prediction of residual energy in batteries. *Energy Rep.* **2022**, *8*, 756–764. [CrossRef]
37. Chandran, V.; Patil, C.K.; Karthick, A.; Ganeshaperumal, D.; Rahim, R.; Ghosh, A. State of Charge Estimation of Lithium-Ion Battery for Electric Vehicles Using Machine Learning Algorithms. *World Electr. Veh. J.* **2021**, *12*, 38. [CrossRef]
38. Hu, C.; Jain, G.; Zhang, P.; Schmidt, C.; Gomadam, P.; Gorka, T. Data-driven method based on particle swarm optimization and k-nearest neighbor regression for estimating capacity of lithium-ion battery. *Appl. Energy* **2014**, *129*, 49–55. [CrossRef]
39. Jafari, S.; Shahbazi, Z.; Byun, Y.C.; Lee, S.J. Lithium-Ion Battery Estimation in Online Framework Using Extreme Gradient Boosting Machine Learning Approach. *Mathematics* **2022**, *10*, 888. [CrossRef]
40. Patil, S.; Patil, A.; Handikherkar, V.; Desai, S.; Phalle, V.M.; Kazi, F.S. Remaining Useful Life (Rul) Prediction of Rolling Element Bearing Using Random Forest and Gradient Boosting Technique. In Proceedings of the ASME 2018 International Mechanical Engineering Congress and Exposition, Pittsburgh, PA, USA, 9–15 November 2018.

41. Guo, J.; Wan, J.L.; Yang, Y.; Dai, L.; Tang, A.; Huang, B.; Zhang, F.; Li, H. A deep feature learning method for remaining useful life prediction of drilling pumps. *Energy* **2023**, *282*, 128442. [CrossRef]
42. Domathoti, B.; Ch, C.; Madala, S.; Berhanu, A.A.; Rao, Y.N. Simulation Analysis of 4G/5G OFDM Systems by Optimal Wavelets with BPSK Modulator. *J. Sens.* **2022**, *2022*, 8070428. [CrossRef]
43. Zhao, F.; Xue, J.; Shao, W.; Yu, H.; Huang, W.; Xiao, J. Toward high-sulfur-content, high-performance lithium-sulfur batteries: Review of materials and technologies. *J. Energy Chem.* **2023**, *80*, 625–657. [CrossRef]
44. Zhao, F.; Feng, Y.; Feng, W. Germanium-based monoelemental and binary two-dimensional materials: Theoretical and experimental investigations and promising applications. *InfoMat* **2022**, *4*, e12365. [CrossRef]
45. Zhao, F.; Li, Y.; Feng, W. Recent Advances in Applying Vulcanization/Inverse Vulcanization Methods to Achieve High-Performance Sulfur-Containing Polymer Cathode Materials for Li–S Batteries. *Small Methods* **2018**, *2*, 1800156. [CrossRef]
46. Li, X.Y.; Feng, S.; Zhao, C.X.; Cheng, Q.; Chen, Z.X.; Sun, S.Y.; Chen, X.; Zhang, X.Q.; Li, B.Q.; Huang, J.Q.; et al. Regulating Lithium Salt to Inhibit Surface Gelation on an Electrocatalyst for High-Energy-Density Lithium–Sulfur Batteries. *Am. Chem. Soc.* **2022**, *144*, 14638–14646. [CrossRef] [PubMed]
47. Sun, W.; Liu, S.; Li, Y.; Wang, D.; Guo, Q.; Hong, X.; Xie, K.; Ma, Z.; Zheng, C.; Xiong, S. Monodispersed FeS_2 Electrocatalyst Anchored to Nitrogen-Doped Carbon Host for Lithium–Sulfur Batteries. *Adv. Funct. Mater.* **2022**, *32*, 2205471. [CrossRef]
48. He, X.; Khan, S.; Ozturk, I.; Murshed, M. The role of renewable energy investment in tackling climate change concerns: En-vironmental policies for achieving SDG-13. *Sustain. Dev.* **2023**, *31*, 1888–1901. [CrossRef]
49. Ali, S.M.; Appolloni, A.; Cavallaro, F.; D'adamo, I.; Di Vaio, A.; Ferella, F.; Gastaldi, M.; Ikram, M.; Kumar, N.M.; Martin, M.A.; et al. Development Goals towards Sustainability. *Sustainability* **2023**, *15*, 9443. [CrossRef]
50. Türk, H. The Oil Crisis of 1973 as a Challenge to Multilateral Energy Cooperation among Western Industrialized Countries. *Hist. Soc. Res.* **2014**, *39*, 209–230. [CrossRef]
51. *India Becomes Associate Member of International Energy Agency*; The Economic Times: Mumbai, India, 2022.
52. Javaid, M.; Haleem, A.; Singh, R.P.; Suman, R.; Gonzalez, E.S. Understanding the adoption of Industry 4.0 technologies in improving environmental sustainability. *Sustain. Oper. Comput.* **2022**, *3*, 203–217. [CrossRef]
53. Yazdani, M.; Gonzalez, E.D.; Chatterjee, P. A multi-criteria decision-making framework for agriculture supply chain risk management under a circular economy context. *Manag. Decis.* **2021**, *59*, 1801–1826. [CrossRef]
54. Küfeoğlu, S. SDG-13: Climate Action. In *Emerging Technologies*; Sustainable Development Goals Series; Springer: Cham, Switzerland, 2022. [CrossRef]
55. Campbell, B.M.; Hansen, J.; Rioux, J.; Stirling, C.M.; Twomlow, S.; Wollenberg, E. Urgent action to combat climate change and its impacts (SDG 13): Transforming agriculture and food systems. *Curr. Opin. Environ. Sustain.* **2018**, *34*, 13–20. [CrossRef]

Disclaimer/Publisher's Note: The statements, opinions and data contained in all publications are solely those of the individual author(s) and contributor(s) and not of MDPI and/or the editor(s). MDPI and/or the editor(s) disclaim responsibility for any injury to people or property resulting from any ideas, methods, instructions or products referred to in the content.

sustainability

Article

Decent Work and Economic Growth in EU Countries—Static and Dynamic Analyses of Sustainable Development Goal 8

Beata Bieszk-Stolorz * and Krzysztof Dmytrów *

Institute of Economics and Finance, University of Szczecin, 71-101 Szczecin, Poland
* Correspondence: beata.bieszk-stolorz@usz.edu.pl (B.B.-S.); krzysztof.dmytrow@usz.edu.pl (K.D.)

Abstract: The goal of this research is a static geographic comparison of the degree of Sustainable Development Goal 8 (SDG8) implementation and to compare its dynamics for the European Union (EU) in the period of 2002–2021 with the consideration of the COVID-19 pandemic. The level of implementation of the 2030 Agenda for Sustainable Development objectives is monitored both jointly and individually. This research uses the COPRAS, the Dynamic Time Warping (DTW) and the hierarchical clustering methods. The study confirmed the geographical diversity in the degree of SDG8 implementation. Nordic countries have the highest degree of implementation, while Greece, Spain, Italy, Romania and Slovakia have the lowest. Some less-developed countries (Czechia, Poland and Malta) were in the cluster where the implementation improved, and some well-developed countries (France, Luxembourg and Portugal) were in the cluster in which the situation did not improve or improved at a low degree. It was not possible to identify geographical diversity in the dynamics of changes in the degree of SDG8 implementation. The COVID-19 pandemic caused the decline in the degree of SDG8 implementation. The added value of the study consists of comparing the dynamics of the degree of implementation of SDG8 using the DTW method. The obtained conclusions may help to create policies for EU countries in this regard.

Keywords: sustainable development; SDG8; COPRAS method; dynamic time warping; hierarchical clustering; COVID-19

Citation: Bieszk-Stolorz, B.; Dmytrów, K. Decent Work and Economic Growth in EU Countries—Static and Dynamic Analyses of Sustainable Development Goal 8. *Sustainability* **2023**, *15*, 13327. https://doi.org/10.3390/su151813327

Academic Editors: Idiano D'Adamo, Massimo Gastaldi and Manoj Kumar Nallapaneni

Received: 11 August 2023
Revised: 1 September 2023
Accepted: 4 September 2023
Published: 5 September 2023

Copyright: © 2023 by the authors. Licensee MDPI, Basel, Switzerland. This article is an open access article distributed under the terms and conditions of the Creative Commons Attribution (CC BY) license (https://creativecommons.org/licenses/by/4.0/).

1. Introduction

In 2015, the United Nations General Assembly approved the universal 2030 Agenda for Sustainable Development [1]. It includes 17 distinct Sustainable Development Goals (SDGs) and 169 individual targets under these goals. They were at first thought to be aspirational and global goals, and national governments were expected to decide how they should be integrated into national planning, policy and strategy processes. The goals were seen as milestones, setting the direction of changes but not defining implementation at the national level. The level of achievement of individual goals is monitored. Studies such as this one are carried out both for all the goals together and for individual goals. The conclusions of the evaluation activities are intended to assist in their implementation. Undoubtedly, disruptions in their implementation may be caused by crises of a local nature, i.e., affecting one country. However, due to the globalisation of socio-economic processes, the greatest disruptions to the implementation of the Sustainable Development Goals are brought about by crises of a global nature. Reporting on progress towards the Sustainable Development Goals is extremely important in terms of creating appropriate policy [2]. An analysis of trends in publications in the area of the Sustainable Development Goals shows that one of the less highlighted goals is Sustainable Development Goal 8 (SDG8—Decent Work and Economic Growth) [3].

The goal of this research is a static geographic comparison of the degree of Sustainable Development Goal 8 implementation and to compare its dynamics for the European Union (EU) in the period 2002–2021 with the consideration of the COVID-19 pandemic. The analysis was conducted in two planes. The first one was a static analysis in which the EU countries were divided into homogeneous clusters with regard to the degree of achievement of SDG8 in selected years: 2002 (the beginning of the research period), 2009 (the peak of the financial crisis) and 2021 (the health crisis resulting from the COVID-19 pandemic). The second aspect was a dynamic analysis that compared the dynamics of the degree of SDG8 implementation and the influence of the COVID-19 pandemic. The following research methods were applied: linear ordering using the COPRAS method, the Dynamic Time Warping method and a hierarchical cluster analysis with the application of Ward's method. The stated research objective is to identify the most divergent countries from those with the highest degree of SDG8 implementation (static analysis). The objective of the dynamic analysis was to identify groups of countries with a similar direction of change in the degree of SDG8 implementation. This may allow similar policies to be implemented to support sustainable development. The dynamics of the composite measure will identify which clusters of countries were most severely affected by the pandemic and which would require specific support. The use of less frequently used statistical and mathematical methods in the construction of composite measures can be considered an added value of the study [3].

This paper is organised as follows: Section 2 contains a discussion of the SDG8 targets and a literature review. In Section 3, the data used in the study and applied research methods are described. Section 4 contains a discussion of the results of the empirical study. Section 5 contains a discussion and Section 6 contains conclusions, directions for future research and limitations of the study.

2. Sustainable Development Goals

The Sustainable Development Goals were established by the United Nations General Assembly in 2015 (Resolution approved by the General Assembly) [1]. Progress towards the objectives is measured, monitored and evaluated using 17 goals. Sustainable Development Goal 8 (SDG8) is to "promote sustained, inclusive and sustainable economic growth, full and productive employment and decent work for all. SGD8 is the aspiration that the economic sector of each country should provide its citizens with the necessary needs for a good life, regardless of their origin, race or culture". SDG8 has a total of twelve targets. They are included in Figure 1.

Figure 1. Sustainable Development Goal 8 targets. Source: own elaboration on the basis of [4].

All Sustainable Development Goals are connected to each other. Research conducted by Fonseca et al. [5] indicates that SDG8 is positively correlated with most of the other SDGs. It is strongly and positively correlated with SDG2–SDG5, SDG7, SDG9, SDG11 and SDG16, and strongly negatively correlated with SDG12 (Responsible Consumption and Production). It should be noted here that SDG12 is negatively correlated with most of the other targets.

There is no doubt that crisis situations have an impact on the realisation of the Sustainable Development Goals. In recent years, the COVID-19 pandemic has been such an event. It has affected the socio-economic phenomena [6–10]. The Sustainable Development Goals Report 2021 [11] indicates that even before the pandemic outbreak, global economic growth had slowed. However, the beginning of the pandemic in 2020 severely disturbed global economic activity. The COVID-19-driven recession was assessed as the worst since the Great Depression.

As Kreinin and Aigner [12] highlight, SDG8 hinders society's ability to achieve other Sustainable Development Goals and prevent disastrous environmental crises. They point out that a focus on economic growth and a focus on an increase in employment lead to unsustainable and inequitable consequences. Currently, only sub-goals 8.4, 8.7 and 8.8 are adequate in achieving the overall SDGs and 2030 Agenda goals of ending poverty and ensuring prosperity and protection of the Earth. The authors proposed a redesigned SDG8 framework, together with new targets and a new set of indicators, which together are expected to contribute to harmonising SDG8 with the overall goals of the 2030 Agenda.

Projections for the SDG targets in Austria for 2020 were presented by Bilek-Steindl and Url [13]. They showed that the SDG8 targets are strongly linked to the underlying economic situation and that the outcomes of the COVID-19 pandemic are evident in the results for 2020. In particular, the evaluation of indicators involving the economic aspect has rapidly declined with the economic crisis. The authors point out that some of the data on the SDG8 target are insufficiently up to date. They present 'in-work at-risk-of-poverty rate' and the 'resource productivity' as examples. The lack of data published with sufficient frequency makes forecasting difficult.

Regional organisations play an important role in the implementation of the SDGs. Both the Association of Southeast Asian Nations (ASEAN) and the European Union are examining and experimenting with various institutional mechanisms to participate in the achievement of these goals. The result of these experiments is the creation of new institutional mechanisms and initiatives. These new institutional instruments and mechanisms are characterised by varying degrees of effectiveness [14].

One of the issues that inhibits efforts to predict progress in implementing the Sustainable Development Goals is the lack of data on crises and the poor quality of some data [15]. Despite this, numerous attempts are being made to assess the impact of the COVID-19 pandemic on the SDGs. The research conducted by Alibegovic et al. [16] presented the impact of the pandemic on the SDGs in Italy. They found that SDG1 (No Poverty), SDG4 (Quality Education) and SDG8 (Decent Work and Economic Growth) were the goals most affected by COVID-19. In contrast, research by Shulla et al. [17] indicates the impact of COVID-19 on stronger links between SDG4, SDG8, SDG3 (Health and Well-Being) and SDG13 (Climate Action) goals. In Portugal, implementation of SDG8 during the first wave of the pandemic was threatened by corporate bankruptcy and the increase in unemployment [18]. The research conducted by Lucas and Landman [19] consisted of an analysis of the impact of COVID-19 on target 8.7. They point out the major limitations of the study caused by limited data availability. As Ranjbari et al. [20] indicated, there was an urgent need for action to support the achievement of the SDGs, particularly in the following directions: (1) opportunities for transformative sustainable development following COVID-19 with a focus on SDG12 (Consumption and Production) and SDG9 (Industry, Innovation and Infrastructure), (2) innovative solutions for economic resilience in support of SDG1 (No Poverty), SDG8 and SDG17 (Partnerships for the Goals), and (3) an in-depth analysis of the long-term implications of COVID-19 for social sustainability to achieve SDG4,

SDG5 (Gender Equality) and SDG10 (Reducing Inequalities). The research conducted by Ranjbari et al. [21] attempted to identify priority areas for action after the COVID-19 crisis towards the implementation of the SDGs under the 2030 Agenda for Sustainable Development. Target 8.3 (development-oriented policies for supporting creativity and job creation) was identified as one of the highest priorities for action.

Most often, measures are taken to reduce the impact of crises negatively affecting the overall level of sustainable development. However, the impact of crises is not always negative. An example is the crisis resulting from the COVID-19 pandemic. The pandemic increased the demand for working from home, especially for parents with young children or school-age children. As a result, teleworking has gained attention and is estimated to continue in the future. Thus, a new form of employment in relation to the conventional workplace has become widespread [22,23]. Also, the acceleration of digitalisation may soon create more employment opportunities [24]. These factors have the potential to positively influence employment by using more effective approaches that can strengthen the progress towards the implementation of SDGS8 [25].

In their report, Sciarra et al. [26] ranked countries by the SDGs-GENEPY (GENeralised economic comPlexitY) framework [27] and SDG Index [28]. European and North American countries are at the top of these rankings. African and South Asian countries are in the worst situation. The analysis conducted by Pakkan et al. [29] points to correlations between the goals. SDG8 is often studied in conjunction with SDG9. In contrast, research conducted for Australia by Bandari et al. [30] indicates a strong link between SDG8 and SDG2 (Zero Hunger), and Rai et al. [31] argue that SDG8 focuses on Decent Work and Economic Growth insufficiently. In addition, they argue that SDG8 is at odds with SDG5, which calls for the value of domestic work and unpaid care to be recognised. On the other hand, the SDG8 implementation is positively influenced by the development of high-quality education, which is embedded within SDG4 [32].

The European Union is pursuing an accelerated transformation from the present linear economy to a circular one (CE). The CE is considered as a tool to achieve SDGs [33,34]. The research conducted by Rodríguez-Antón et al. [35] showed a significant positive correlation between the implementation of SDG8 and adherence to CE targets. This is particularly evident in such countries as Austria, Belgium, Denmark, France, Germany, Italy, the Netherlands and the United Kingdom.

Finally, Jianu et al. [36] investigated labour market inequalities in the EU, as expressed by specific indicators used for SDG8 of the 2030 Agenda for Sustainable Development on the basis of a cluster analysis for the 27 Member States. The first cluster comprised the majority of EU countries, geographically located in Eastern and Central Europe along with the Baltic countries, as well as two Mediterranean countries—Portugal and Greece. The principal characteristics of these countries were low values of the real GDP per capita, resource productivity and domestic material consumption and high values recorded for the NEET indicator for young people, the in-work at-risk-of-poverty rate the and long-term unemployment rate.

The analysis of the literature provided the basis for the following research hypotheses:

H1. *It is possible to identify regions of the European Union with a similar level of SDG8 implementation.*

H2. *It is possible to identify regions of the European Union with similar dynamics of change in the degree of SDG8 implementation.*

H3. *The COVID-19 pandemic caused the decline in the degree of SDG8 implementation.*

The problem addressed in H1 is reflected in the available literature. The contribution of this study is to present the changes in the geographical distribution of countries with a similar degree of SDG8 implementation over a long research horizon. The verification of H2 fills a research gap related to the assessment of the dynamics of change in the degree of SDG8 implementation and its comparison using the DTW method. Also, the study of the impact of the COVID-19 pandemic on the degree of SDG8 implementation using a synthetic measure created with the COPRAS method (verification of H3) is the added value of the research.

3. Data and Research Methodology

In this study, a pre-defined set of indicators included in the Eurostat database was used. There are criticisms and proposals for new indicators in the literature. In their study, Kreinin and Aigner [12] suggest a new SDG8 framework consistent with the strong sustainability concepts of 'Sustainable Work and Growth'. Similarly, Coscieme et al. [37] stress that pursuing unconditional GDP growth risks failing to achieve the Sustainable Development Goals. SDG8 implementation through continued GDP growth will make it more difficult to achieve environmental and inequality reduction goals. They propose guidelines for the selection of alternative indicators related to the achievement of SDG8. In addition, another problem arises. Many of the SDG indicators used to monitor implementation of the goals have long publication delays [38].

The Eurostat database collects data on sustainable development. These are grouped according to individual headline goals and include variables to examine these targets. The SDG8 data from period 2002–2021 were applied. There are eight variables related to the implementation of SDG8:

1. x_1—GDP per capita in constant prices from year 2021 (in Euro);
2. x_2—Investment share of GDP by institutional sectors (percentage of GDP);
3. x_3—Young people (aged 15–29 years) neither in employment nor in education and training by sex (NEET) (percentage of total population);
4. x_4—Employment rate (for persons aged 20–64 years) (percentage of total population);
5. x_5—Long-term unemployment rate (percentage of total population in the labour force);
6. x_6—In-work at-risk-of-poverty rate (percentage of total employed persons, aged 18 years and more);
7. x_7—Fatal accidents at work per 100,000 workers;
8. x_8—Inactive population due to caring responsibilities (percentage of population aged 20–64 outside the labour force and wanting to work).

The Eurostat data are public and freely available. They are collected in accordance with statistical reporting rules and have a high degree of reliability. These data do not require ethical considerations. The calculations were conducted in R language [39] with the use of the following libraries: `clusterSim` [40], `dtw` [41], `NbClust` [42] and `factoextra` [43].

The survey was conducted in several stages according to the following scheme (Figure 2).

In the first step of the static analysis, the pattern, i.e., best values of all variables, were determined (Table 1).

Table 1. Pattern values of indicators.

Specification	x_1	x_2	x_3	x_4	x_5	x_6	x_7	x_8
Value	EUR 86,550.00	53.59%	4.70%	81.80%	0.60%	2.70%	0.45	2.40%
Country/year	Luxembourg 2021	Ireland 2019	Denmark 2006	Sweden 2018	Czechia 2019	Finland 2017	Malta 2017	Sweden 2002

Source: own calculations on the basis of Eurostat data.

The best values of variables vary across the countries and years. For half of the variables, their best values were in the Nordic countries. This confirms the initial finding that these countries are the ones with generally the highest degree of SDG8 implementation.

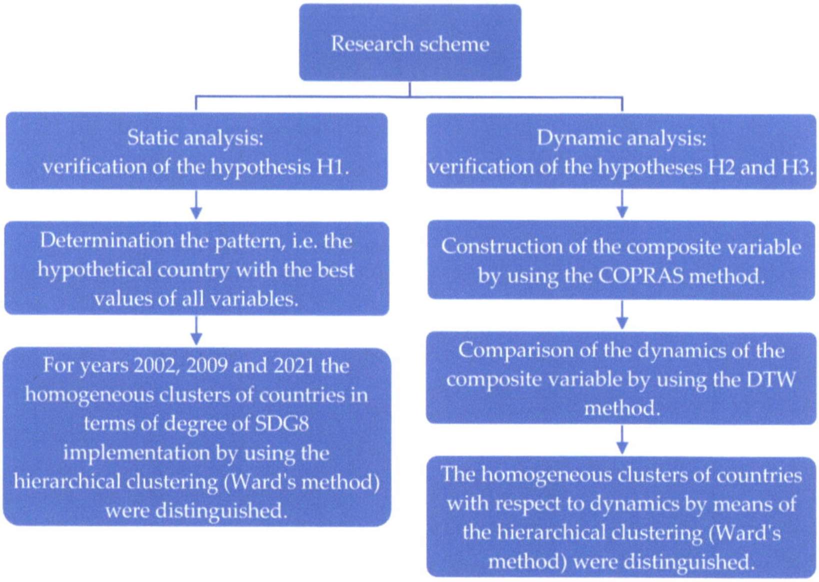

Figure 2. Research scheme. Source: own elaboration.

The static analysis (verification of H1) consisted of the selection of the homogeneous clusters of countries, with the artificial one being the pattern country (the hypothetical one with the best values of variables, as presented in Table 1). Although such an analysis can be performed every year, the size would exceed the capacity of this research. Therefore, three years were selected: 2002 (the beginning of the research period), 2009 (the peak of the financial crisis) and 2021 (the health crisis resulting from the COVID-19 pandemic). The agglomerative clustering method with Ward's method for minimising the total within-cluster variance was used [44]. The hierarchical clustering was based on the distance matrix. For the static analysis, the Euclidean distance matrixes between the countries were applied.

The dynamic analysis allowed for verification of H2 and H3. Its first stage was the assessment of the degree of SDG8 implementation every year by means of linear ordering. Linear ordering methods are types of multivariate statistical analysis. The idea of linear ordering methods is the creation of a composite measure from a set of a larger number of variables. There are many linear ordering methods, which can be divided into two main groups: the pattern and patternless methods. The pattern methods assess the position of every object by means of the weighed distance of them from the so-called pattern or anti-pattern, which is the hypothetical object with the best (worst) values in the dataset. The patternless methods are based on the weighed sums of normalised variables. One of the patternless methods—the COPRAS method—was selected. In the second stage of the dynamic analysis, we compared the time series created from the composite measures by means of the DTW method and used the results in hierarchical clustering to distinguish homogeneous clusters of EU countries with respect to the dynamics of the degree of SDG8 implementation (verification of H2). In the third stage of the dynamic analysis, we assessed the influence of COVID-19 on the degree of SDG8 implementation (verification of H3).

It is worth noting that the static and dynamic analyses are complimentary, and thus cannot be directly compared. The static analysis was conducted for every year separately, while the dynamic one was conducted for the whole period (2002–2021).

3.1. The COPRAS Method

The COPRAS (the COmplex PRoportional Assessment) method was invented by the Lithuanian researchers Zavadskas et al. [45]. It was created for the needs of the decision theory and belongs to the group of Multiple-Criteria Decision-Making (MCDM) methods. MCDM methods are used to analyse the Sustainable Development Goals [46]. There are no clearly defined ways of selecting methods. When it comes to ranking, different methods may give different results [47]. The COPRAS method is also applied in the analysis of socio-economic phenomena [48–51]. However, it can also be used in multivariate statistical analysis. There are five steps of the COPRAS method. In the first one, there is the observation matrix $\boldsymbol{X}$:

$$\mathbf{X} = \left[x_{ij}\right] = \begin{bmatrix} x_{11} & x_{12} & \cdots & x_{1m} \\ x_{21} & x_{22} & \cdots & x_{2m} \\ \vdots & \vdots & \ddots & \vdots \\ x_{n1} & x_{n2} & \cdots & x_{nm} \end{bmatrix} \text{ for } i = 1, \ldots, n; j = 1, \ldots, m \tag{1}$$

where the components are as follows:

x_{ij}—Value of j-th variable in the i-th object;
n—Number of objects;
m—Number of variables.

In the second step, the data are normalised in order to remove units and ensure that the variables have the same order of magnitude. In this research, one of the quotient inversions, given by the following formula, was selected:

$$z_{ij} = \frac{x_{ij}}{\sqrt{\sum_{i=1}^{n} x_{ij}^2}} \text{ for } i = 1, \ldots, n\ j = 1, \ldots, m \tag{2}$$

where z_{ij} is the value of the j-th variable in the i-th object after normalisation.

In the third step, the weighted normalised values of variables are calculated:

$$t_{ij} = z_{ij} w_j, i = 1, \ldots, n; j = 1, \ldots, m \tag{3}$$

where w_j is the weight of the j-th variable, satisfying the following conditions:

$$w_j \in \langle 0, 1 \rangle \tag{4}$$

$$\sum_{j=1}^{m} w_j = 1 \tag{5}$$

There are many methods used to assess the weights of variables. In this research, equal weights were assumed. The reason for this was that the weights allocated by statistical methods (on the basis of variation or correlation between variables) would vary from year to year. By contrast, using expert methods to assign weights would have resulted in a high level of subjectivity. As it is difficult to state whether some variables are of greater importance than others, equal weights are assumed. In the fourth step, the weighted sums for stimulants (S_i^+) and destimulants (S_i^-) for all objects are calculated:

$$S_i^+ = \sum_{j \in J^+} t_{ij} \text{ for } i = 1, \ldots, n \tag{6}$$

$$S_i^- = \sum_{j \in J^-} t_{ij} \text{ for } i = 1, \ldots, n \tag{7}$$

where the components are as follows:

J^+—Variables that are stimulants;
J^-—Variables that are destimulants.

Stimulant variables are the ones where the highest values are the most preferred, and destimulants are the opposite. In our case, the variables x_1, x_2 and x_4 are the stimulants. The remaining variables are the destimulants. In the fifth step, the value of the synthetic variable is obtained:

$$q_i = S_i^+ + \frac{\sum_{i=1}^{n} S_i^-}{S_i^- \sum_{i=1}^{n} \frac{1}{S_i^-}} \text{ for } i = 1, \ldots, n \tag{8}$$

The value of q_i is not normalised; $\max_i\{q_i\}$ is the best object and $\min_i\{q_i\}$ is the worst object. The composite measures for every country every year were calculated. These variables then created the time series of the degree of SDG8 implementation, which were compared with the use of the Dynamic Time Warping (DTW) method.

The advantage of the COPRAS method in this study over methods based on distances from the pattern and anti-pattern is due to the fact that their values also deteriorate during crises, which does not necessarily translate into a deterioration of the synthetic variable. Relying on the weighted sums of the normalised values alone (as in the COPRAS method) results in a decrease in the value of the synthetic measure when the indicators deteriorate and an increase if they improve. A disadvantage of the COPRAS method is that there must be at least one variable that is a destimulant. In this study, this assumption is satisfied.

3.2. Dynamic Time Warping (DTW)

DTW measures the similarity between time series. This method uses dynamic programming to find the optimal match between them with respect to a given score function. It was proposed by Bellman and Kalaba [52] and first used for speech recognition [53,54]. Its primary use has been in technical sciences, although more and more often its applications cover the areas of finance [55–57], labour markets [58,59] and energy markets [60–62].

There are four steps of the DTW method. Let $X = (x_1, x_2, \ldots, x_N)$ and $Y = (y, y_2, \ldots, y_M)$ be two time series. In the first step, the time series are normalised. The z-normalisation was selected. In the second step, the local measure of cost for two elements of X and Y is specified.

$$c(x_i, y_j) = |x_i - y_j| \text{ for } i = 1, 2, \ldots, N, j = 1, 2, \ldots, M \tag{9}$$

In the third step, the time warping path is constructed, which is the point-to-point match between X and Y by means of the following sequence:

$$p = (p_1, \ldots, p_L), p_l = (n_l, m_l) \in \{1, \ldots, N\} \times \{1, \ldots, M\} \tag{10}$$

for $l \in \{1, \ldots, L\} (L \in \{\max(N, M), \ldots, N + M - 1\})$.

This satisfies the boundary, monotonicity and step-size conditions. The first condition provides that the first and the last element of p are $p_1 = (1, 1)$ and $p_L = (N, M)$ (the first (last) index from the first sequence must be matched with the first (last) index from the second one). The monotonicity and step-size conditions provide that the path always goes up, right or up and right of the current position, i.e.,

$$p_{l+1} - p_l \in \{(1, 0), (0, 1), (1, 1)\} \text{for } i = 1, \ldots, L - 1 \tag{11}$$

In the fourth step, the optimal match is found. It satisfies all the aforementioned restrictions and minimises the total cost $c_p(X, Y)$ of a warping path p. This cost is calculated using the following equation:

$$c_p(X, Y) = \sum_{l=1}^{L} c(x_{n_l}, y_{m_l}) = \sum_{l=1}^{L} |x_{n_l} - y_{m_l}| \tag{12}$$

Therefore, the optimal alignment between X and Y is

$$DTW(X,Y) = c_{p*}(X,Y) = \min\{c_p(X,Y) \mid p \in P\} \quad (13)$$

where P is the set of all warping paths.

The DTW distance, i.e., a stretch-insensitive distance between the two time series, which is also the minimum distance between the series X and Y, is therefore defined as

$$DTW(X,Y) = D(N,M) \quad (14)$$

The degree of similarity between time series can also be tested using other distance measures, such as the Euclidean metric or other known distance measures. The first advantage of the DTW method in this respect is that the time series being compared do not have to be of equal lengths. The second advantage of this method is that it adjusts much better to differences in the cycles.

The calculated distances were applied in hierarchical clustering. The assessed similarities between the analysed time series by means of DTW were used in the hierarchical agglomerative clustering. A dendrogram was built with the use of the Ward's method.

4. Empirical Results and Discussion

The data used in the study covered the years 2002–2021. The static analysis was conducted for each year separately. The exact results of the cluster analysis for three years are presented: 2002 (the first analysed year), 2009 (the peak of the financial crisis) and 2021 (the COVID-19 pandemic). The countries were also joined by a hypothetical 'ideal' country with values for the variables that are the pattern of development. The variables were normalised by using the `data.Normalization` function from the `clusterSim` R package. In every analysed year, we specified the number of clusters by using the `NbClust` function from the `NbClust` R package and applying the Beale index [63]. The results are presented in Figures 3–5 (the dendrograms were drawn by using the `fviz_dend` function from the `factoextra` R package).

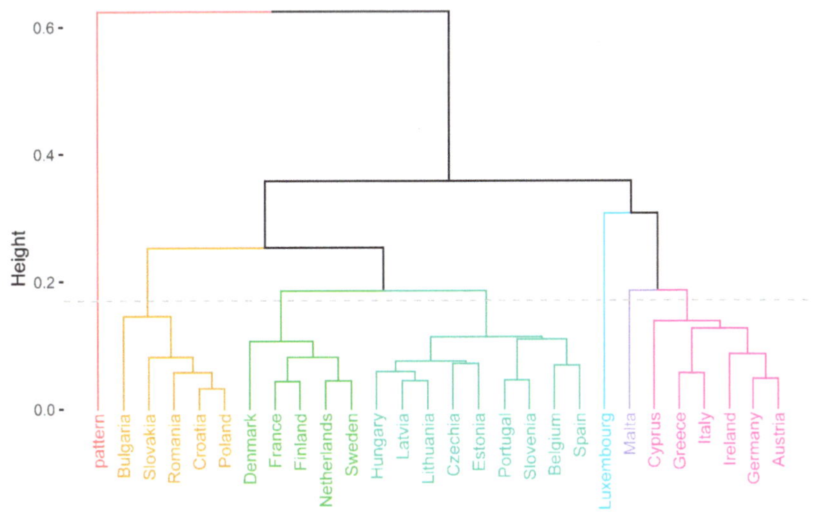

Figure 3. Homogeneous clusters of the European Union countries and the pattern—the degree of SDG8 implementation in 2002. Source: own calculations.

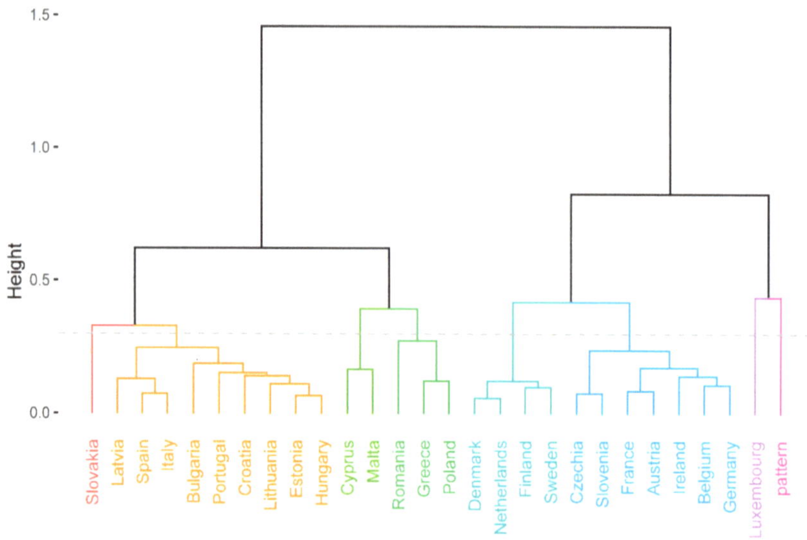

Figure 4. Homogeneous clusters of the European Union countries and the pattern—the degree of SDG8 implementation in 2009. Source: own calculations.

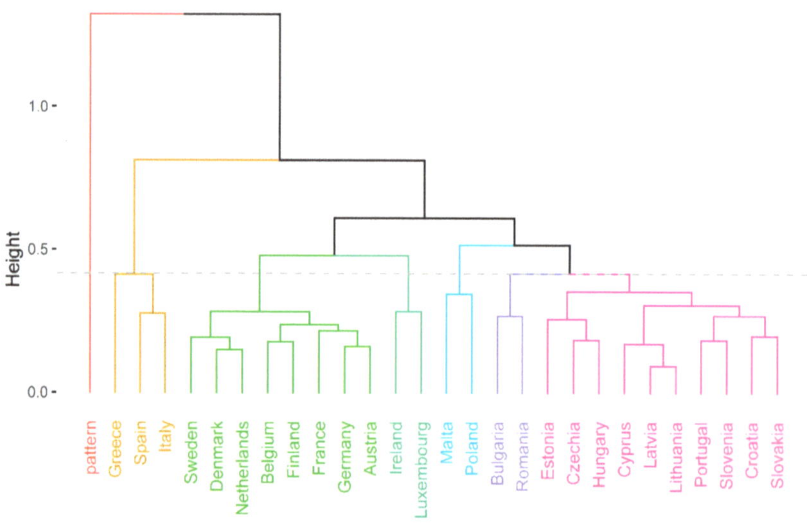

Figure 5. Homogeneous clusters of the European Union countries and the pattern—the degree of SDG8 implementation in 2021. Source: own calculations.

4.1. Results of the Static Analysis

The static analysis was used to verify H1.

In each of the three analysed years (2002, 2009 and 2021), the hypothetical pattern country always formed a separate cluster. This means that even the countries with the overall best situation with respect to the implementation of SDG8 had some weaker areas that did not allow them to be in the same cluster as the hypothetical pattern country. In 2009 only, Luxembourg had the most similar situation with respect to the degree of implementation of SDG8 as the pattern country (Figure 4). In each of the three analysed years, Sweden, Denmark, the Netherlands and Finland were in one cluster. They were the countries that, in the majority of the analysed years, were at the top of the degree of SDG8 implementation. Poland, Bulgaria, Romania, Croatia, Greece and Malta were, at the beginning of the period under analysis, in the group of countries with the lowest degree of SDG8 implementation (Figure 3). Over the years, Slovakia's position improved slightly, and Italy and Spain joined the cluster of countries with the lowest degree of SDG8 implementation. Germany, Belgium, Ireland and France were the countries forming clusters with a good (but not the highest) degree of SDG8 implementation throughout the analysed period. The remaining countries that joined the EU in 2004 were among those with an average degree of SDG8 implementation. The degree of SDG8 implementation in Ireland and Luxembourg was high, but their membership in the clusters was the most changing. The next two countries whose membership in the clusters was variable were Poland and Malta. They, in turn, are at the other pole of the degree of SDG8 implementation. Crisis periods have most severely deteriorated the degree of SDG8 implementation in Spain and Italy. This is also reflected in their membership in the clusters in 2009 and 2021 (Figures 4 and 5).

4.2. Results of the Dynamic Analysis

The dynamic analysis was used to verify H2 and H3. In the first stage of the dynamic analysis, the degree of SDG8 implementation was assessed by means of the COPRAS method. In each year, the synthetic measures were calculated for every country. Thus, the obtained time series of the synthetic measure for every country were compared pairwise by using the DTW method. In the second stage of the dynamic analysis, on the basis of these comparisons, the DTW distance matrix for the whole period by using the `dtw` function from the `dtw` R package was obtained. This matrix reflects the comparison of dynamics of the degree of SDG8 implementation and was the basis for the hierarchical clustering (Ward's method). Similarly, as in the static analysis, the Beale index in the `NbClust` function from the `NbClust` R package was applied to obtain the optimal number of clusters and the `fviz_dend` function from the `factoextra` R package to draw the dendrogram. The results of this analysis verify H2 and are presented in Figure 6.

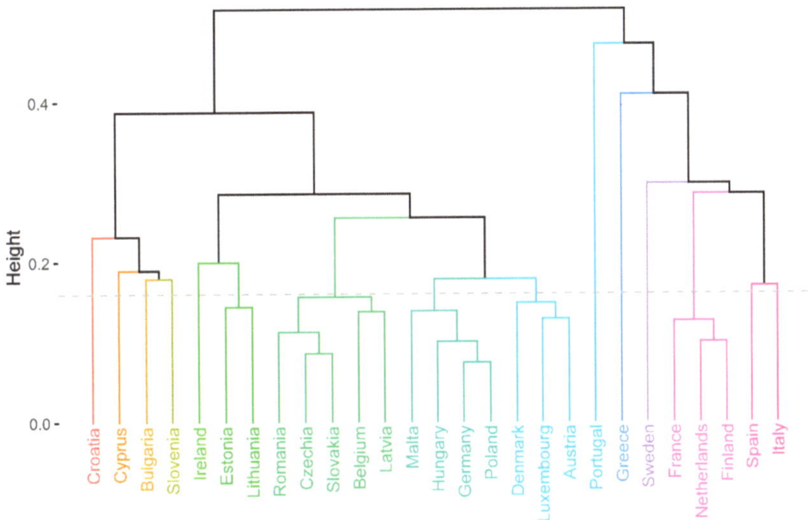

Figure 6. Homogeneous clusters of the European Union countries—dynamic analysis of the degree of SDG8 implementation. Source: own calculations.

A total of 15 clusters with respect to the similarity of dynamics of the degree of SDG8 implementation were obtained. Italy (further referred to as Cluster 1), Spain (Cluster 2), Sweden (Cluster 3), Greece (Cluster 4), Portugal (Cluster 5), Ireland (Cluster 6), Slovenia

(Cluster 7), Bulgaria (Cluster 8), Cyprus (Cluster 9) and Croatia (Cluster 10) formed separate single clusters. The dynamics of the synthetic variable formed clusters for the following countries:

- Estonia and Lithuania (Cluster 11);
- Romania, Czechia, Slovakia, Belgium and Latvia (Cluster 12);
- Malta, Hungary, Germany and Poland (Cluster 13);
- Denmark, Luxembourg and Austria (Cluster 14);
- France, the Netherlands and Finland (Cluster 15).

It can be found that the similarity of countries with the degree of SDG8 implementation (static analysis) does not have much in common with their dynamics. In general, it was hard to distinguish well-separated clusters with respect to dynamics of the SDG8 implementation. That is why the optimal number of them is that big. The results indicate that it is hard to find similarities between the dynamics of the degree of SDG8 implementation between the EU countries. The most similar countries with regard to the dynamics of the degree of SDG8 implementation were Poland and Germany, while Portugal was the most dissimilar with respect to all other EU countries. In the cases of Greece and Spain, the degree of SDG8 implementation at the end of the analysed period was lower than at the beginning. For Portugal (the most dissimilar country with comparison to other ones), it was the most steady (the degree of SDG8 implementation in Portugal had only small fluctuations). The highest increase was observed in the case of Ireland.

In the last stage of the dynamic analysis, H3 was verified. The average values of the COPRAS measure for each of the 15 clusters, obtained in the previous stage, are presented in Figure 7.

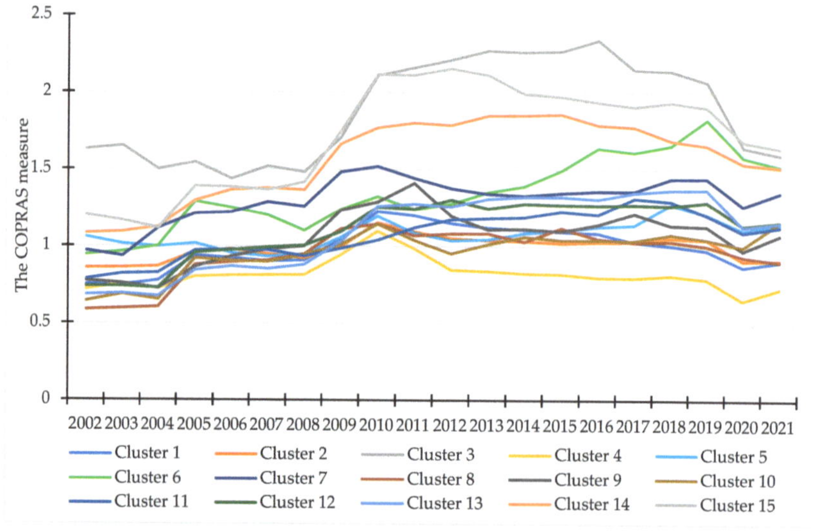

Figure 7. Dynamics of the degree of SDG8 implementation in the clusters of the EU countries in years 2002–2021. Source: own calculations.

The dynamics of all clusters varies, which confirms the findings from the second stage (Figure 6). However, some common regularities for each cluster can be seen. Firstly, the financial crisis of 2007–2008 did not cause a significant decline in the degree of SDG8 implementation. Some clusters (3 and 15) showed its high increase. The second common regularity for all clusters is the decline in the year 2020 (the first year of the COVID-19 pandemic). In 2021, the degree of SDG8 implementation increased in most clusters. Only Clusters 3, 6, 8, 14 and 15 reported a further decline in SDG8 implementation. Therefore, this confirms H3, that the COVID-19 pandemic caused deterioration in the degree of SDG8 implementation. However, high differences between the clusters' dynamics (especially since Clusters 1–10 consist of single countries) indicate that the directions of changes are different. This makes the implementation of a common policy with regard to the degree of SDG8 implementation difficult, if not impossible.

5. Discussion

The first research hypothesis concerned the geographical variation in the degree of SDG8 implementation. It was verified positively. The conducted study confirms the results of other studies. The results of the analysis by Rocchi et al. [64] showed that in 2019, the Nordic countries were the most advanced among European countries in terms of achieving the SDGs. These countries were at the top of the ranking for all dimensions, except for the environmental dimension in the case of Denmark. Moreover, Barbier and Burgess [65] explain such results by the institutional effectiveness and good governance associated with long-term development and success in sustainable development. This fact can explain the high position of the Nordic countries in terms of SDG implementation. The EU Baltic States and the post-communist countries performed the worst. The low position of the EU post-communist countries is due to their later accession to the EU. It should be noted that their accession had a positive impact on economic development, including the labour market [59]. However, the question of whether the economies of EU members are 'driving' the economies of EU candidates depends on a number of factors and is still open [66].

The high position of Sweden in terms of the implementation of all SDGs was pointed out by Stanujkic et al. [67]. And although this country was not leading in all the goals implemented, it ranked very high (e.g., SDG8) or high in all of them. Their research indicates a weak position of Romania, as well as Bulgaria. This is consistent with our analysis. Both countries are Eastern European ones and joined the EU in 2007. These countries have not included sustainable development as a demand in their policies yet. Interestingly, Romania ranks high in terms of implementing SDG7 [61]. Studies indicate an improvement in the energy use process in this country [68]. However, improvements are needed in the targets included in SDG8.

The targets and their constituting indicators of SDG8 in 2019 were assessed by Skvarciany and Astikė [69]. They showed that the most important indicator is the annual growth rate of the real GDP per employed person. In 2019, the countries that progressed the most in the implementation of SDG8 were Germany and France. Finland and Latvia made less progress in SDG8 implementation. The level of SDG8 implementation in EU countries between 2001 and 2020 was analysed by Grzebyk et al. [70]. Their results confirm that in most EU countries there have been positive changes in its implementation in the years studied. This is demonstrated by the increasing values of the synthetic measure calculated for each of the EU27 countries. According to their ranking in 2020, the Netherlands and Sweden were in the highest positions and Bulgaria and Romania were in the lowest positions. The latter two countries formed a separate cluster of countries in our 2021 study. In contrast, our analysis with respect to dynamics of the degree of SDG8 implementation indicated that both countries at the top of the ranking and at the bottom of the ranking were in separate clusters. This shows that changes in the implementation of the SDG8 targets in these countries have occurred in different ways. In contrast, in Carlsen's [71] ranking of the EU27 countries for 2019, the first two are Sweden and Austria and the last two are Greece and Bulgaria. In creating this ranking, the author considered the five main indicators that characterise SDG8. These were in relation to our variables: x_1, x_2, x_3, x_4 and x_5. The ranking created for 2020 by Kuc-Czarnecka et al. [72] showed that the best positions were held by the Netherlands and Finland, and the worst positions were held by Greece and Cyprus. In the latter study, the authors used different weights for all variables.

In the static analysis, Luxembourg and Ireland were often the outliers. This is confirmed by the study conducted by D'Adamo et al. [73]. As the authors point out, the distinguishing feature of these countries is a very attractive tax system, which appears to have succeeded in stimulating economic growth but has failed to reconcile it with the Sustainable Development Goals.

The verification of H2 was based on the DTW method, which is rarely used in studies on the SDGs. An example of such application was the study of SDG7 by Dmytrów et al. [61]. This study was able to identify homogeneous clusters of countries geographically. Unfortunately, it was not possible to extract such clusters in the current study on SDG8. A very large number of clusters were obtained, which were highly internally heterogeneous, as well as many clusters consisting of individual countries. This demonstrates the high diversity of EU countries according to the dynamics of the degree of SDG8 implementation. This may be related to the fact that EU countries are diverse in terms of their policies. There is a need to tailor programmes to the specifics of each region [74]. In addition, the main implications of the research by Anselmi et al. [75] are that Europe is moving towards the Sustainable Development Goals in an uneven manner, and so there is a need to define an agenda that allows for greater cooperation between several countries.

The third research hypothesis states that the COVID-19 pandemic caused a decline in the degree of SDG8 implementation. This is consistent with the research conducted by Grzebyk et al. [70]. In the last year of their study, 2020, the effect of the COVID-19 pandemic was observed. Also, other conducted studies are consistent with H3 [16–20]. Research conducted by Shuai et al. [76] identified progress towards the SDG targets after COVID-19 in the period 2020–2024 by means of predicted GDP and population growth and machine learning models. They justified that the total implementation of the SDGs declined by 7.7% in 2020 globally, with the performance of the 12 socio-economic SDGs (including SDG8) declining by 3.0–22.3% and the performance of the 4 environmental SDGs increasing by 1.6–9.2%. By 2024, progress on the 12 Sustainable Development Goals (including SDG8) will be delayed by one to eight years in comparison with their pre-COVID-19 trajectories, while the 4 environmental SDGs will gain additional time. Only the trajectory of SDG5 remained unchanged.

6. Conclusions

Decent Work and Economic Growth remains central to the prosperity of countries. Emerging crises cause countries around the world, including in the European Union, to face unexpected challenges.

It follows that the sustainable development of European countries is not uniform. It depends on their level of economic development, political and historical circumstances and the level of development of society. It will be interesting to see whether all countries will reach their targets in 2030. Therefore, research in this area is worth continuing.

A study of this type has some limitations. On the one hand, data for individual SDGs are available in the Eurostat database and grouped accordingly. On the other hand, the individual targets overlap. Hence, the question arises whether other variables should not also be taken into account when assessing the degree of achievement of a selected SDG. It is also interesting to note the proposals in the literature for other indicators. However, it happens that the data needed for the determination of other types of indicators are not collected with sufficient frequency and concern, for example, five-year periods. In addition, the data used for the classic indicators are provided with a delay of at least one year. Having more up-to-date data is essential in order to make effective and timely policy decisions. Another critical aspect is the fair assignment of weights. Researchers are also not consensual in assigning weights to individual indicators. Their inclusion may contribute to interesting new research results.

The conducted research allowed the first research hypothesis to be positively verified. The regions of the European Union with the highest degree of SDG8 implementation were the Nordic countries together with the Netherlands. On the other hand, the countries of Southern and Southeastern Europe were characterised by the lowest degree of SDG8 implementation. The second research hypothesis could not be positively verified. The dynamics of change in the degree of SDG8 implementation did not depend on the geographic location of a country, and no clearly differentiated clusters could be identified. The positive verification of the third hypothesis indicates a negative impact of the COVID-19 pandemic on the level of SDG8 implementation in EU countries.

This analysis may also encourage us to draw some policy recommendations. Countries with a low level of the degree of SDG8 implementation should follow the realisation of policy with respect to the Nordic countries, as they are the best in this regard. In the need for analysing the dynamics of the degree of SDG8 implementation, Ireland should be considered, as in the case of this country the increase in the indicators was the highest. The conclusions of this research may have a number of implications for governing bodies and policy makers in EU governments. When SDG8 implementation is low, it is necessary to change the governance model and take rigorous action. It is also important to establish cooperation between good and weak countries in order to equalise the level of SDG8 implementation. There is a need to adapt policies and programmes to the specific needs and characteristics of each region. There is unlikely to be a one-size-fits-all approach to guarantee success. The specific economic, social and environmental circumstances of the regions must always be taken into account.

Author Contributions: Conceptualization, B.B.-S. and K.D.; methodology, B.B.-S. and K.D.; software, B.B.-S. and K.D.; validation, B.B.-S.; formal analysis, K.D.; investigation, B.B.-S. and K.D.; resources, B.B.-S. and K.D.; data curation, B.B.-S. and K.D.; writing—original draft preparation, B.B.-S. and K.D.; writing—review and editing, B.B.-S. and K.D.; visualization, K.D.; supervision, B.B.-S. All authors have read and agreed to the published version of the manuscript.

Funding: This research received no external funding.

Institutional Review Board Statement: Not applicable.

Informed Consent Statement: Not applicable.

Data Availability Statement: All data are from https://ec.europa.eu/eurostat/web/main/data/database (accessed on 15 December 2022).

Conflicts of Interest: The authors declare no conflict of interest.

References

1. United Nations. Transforming Our World: The 2030 Agenda for Sustainable Development. UN General Assembley 70 Session. Available online: https://sustainabledevelopment.un.org/content/documents/21252030%20Agenda%20for%20Sustainable%20Development%20web.pdf (accessed on 15 September 2022).
2. Alsayegh, M.F.; Ditta, A.; Mahmood, Z.; Kouser, R. The Role of Sustainability Reporting and Governance in Achieving Sustainable Development Goals: An International Investigation. *Sustainability* **2023**, *15*, 3531. [CrossRef]
3. Ali, S.M.; Appolloni, A.; Cavallaro, F.; D'Adamo, I.; Di Vaio, A.; Ferella, F.; Gastaldi, M.; Ikram, M.; Kumar, N.M.; Martin, M.A.; et al. Development Goals towards Sustainability. *Sustainability* **2023**, *15*, 9443. [CrossRef]
4. United Nations. Goal 8 | Department of Economic and Social Affairs. Available online: https://sdgs.un.org/goals/goal8 (accessed on 23 August 2022).
5. Fonseca, L.M.; Domingues, J.P.; Dima, A.M. Mapping the sustainable development goals relationships. *Sustainability* **2020**, *12*, 3359. [CrossRef]
6. Fatih, C. What Can SVAR Models Tell us About the Impact of Public Expenditure Shocks on Macroeconomic Variables in Algeria? A Slight Hint to the COVID-19 Pandemic. *Folia Oecon. Stetin.* **2021**, *21*, 21–37. [CrossRef]
7. Bieszk-Stolorz, B.; Dmytrów, K. Assessment of the Similarity of the Situation in the EU Labour Markets and Their Changes in the Face of the COVID-19 Pandemic. *Sustainability* **2022**, *14*, 3646. [CrossRef]
8. Machová, R.; Korcsmáros, E.; Marča, R.; Esseová, M. An International Analysis of Consumers' Consciousness During the Covid-19 Pandemic in Slovakia and Hungary. *Folia Oecon. Stetin.* **2022**, *22*, 130–151. [CrossRef]
9. Musa, H.; Rech, F.; Yan, C.; Musova, Z. The Deterioration of Financial Ratios During the COVID-19 Pandemic: Does Corporate Governance Matter? *Folia Oecon. Stetin.* **2022**, *22*, 219–242. [CrossRef]
10. Streimikiene, D. Energy poverty and impact of Covid-19 pandemics in Visegrad (V4) countries. *J. Int. Stud.* **2022**, *15*, 9–25. [CrossRef]
11. The Sustainable Development Goals Report 2021. Available online: https://unstats.un.org/sdgs/report/2021/ (accessed on 31 August 2022).
12. Kreinin, H.; Aigner, E. From "Decent work and economic growth" to "Sustainable work and economic degrowth": A new framework for SDG 8. *Empirica* **2022**, *49*, 281–311. [CrossRef]

13. Bilek-Steindl, S.; Url, T. Nowcasting and monitoring SDG 8. *Empirica* **2022**, *49*, 313–345. [CrossRef] [PubMed]
14. Marx, A.; Pertiwi, S.B.; Depoorter, C.; Hoornick, M.; Mursitama, T.N.; Otteburn, K.; Arnakim, L.Y. What role for regional organizations in goal-setting global governance? An analysis of the role of the European Union and ASEAN in the Sustainable Development Goals. *GPPG* **2021**, *1*, 421–445. [CrossRef]
15. Marzouk, M.; Azab, S.; Elshaboury, N.; Megahed, A.; Metawie, M.; El-Hawary, M.; Ghaith, D.; Bayoumi, A.E.M. Modeling COVID-19 effects on SDGs using system dynamics in Egypt. *Environ. Sci. Pollut. Res.* **2022**, *29*, 59235–59246. [CrossRef] [PubMed]
16. Alibegovic, M.; Cavalli, L.; Lizzi, G.; Romani, I.; Vergalli, S. COVID-19 & SDGs: Does the current pandemic have an impact on the 17 Sustainable Development Goals? A qualitative analysis. *FEEM Policy Brief* **2020**, *7*, 1–18.
17. Shulla, K.; Voigt, B.F.; Cibian, S.; Scandone, G.; Martinez, E.; Nelkovski, F.; Salehi, P. Effects of COVID-19 on the Sustainable Development Goals (SDGs). *Discov. Sustain.* **2021**, *2*, 15. [CrossRef] [PubMed]
18. Borrego, A.C.; Carreira, F.A.; Pardal, P.; Abreu, R. Social Responsibility and SDG 8 during the First Wave of the COVID-19 Pandemic: The Role of Chartered Accountants in Portugal. *Sustainability* **2022**, *14*, 8625. [CrossRef]
19. Lucas, B.; Landman, T. Social listening, modern slavery, and COVID-19. *J. Risk Res.* **2021**, *24*, 314–334. [CrossRef]
20. Ranjbari, M.; Esfandabadi, Z.S.; Zanetti, M.C.; Scagnelli, S.D.; Siebers, P.-O.; Aghbashlo, M.; Peng, W.; Quatraro, F.; Tabatabaei, M. Three pillars of sustainability in the wake of COVID-19: A systematic review and future research agenda for sustainable development. *J. Clean. Prod.* **2021**, *297*, 126660. [CrossRef]
21. Ranjbari, M.; Esfandabadi, Z.S.; Scagnelli, S.D.; Siebers, P.-O.; Quatraro, F. Recovery agenda for sustainable development post COVID-19 at the country level: Developing a fuzzy action priority surface. *Environ. Dev. Sustain.* **2021**, *23*, 16646–16673. [CrossRef] [PubMed]
22. Djankov, S.; Panizza, U. Developing economies after COVID-19: An introduction. In *COVID-19 in Developing Economies*; Centre for Economic Policy Research: London, UK, 2020; Volume 8.
23. Ljungholm, D.P.; Olah, M.L. Regulating fake news content during COVID-19 pandemic: Evidence-based reality, trustworthy sources, and responsible media reporting. *Rev. Contemp. Philos.* **2020**, *19*, 43–49.
24. Avdiu, B.; Nayyar, G. When face-to-face interactions become an occupational hazard: Jobs in the time of COVID-19. *Econ. Lett.* **2020**, *197*, 109648. [CrossRef]
25. Clemente-Suárez, V.J.; Rodriguez-Besteiro, S.; Cabello-Eras, J.J.; Bustamante-Sanchez, A.; Navarro-Jiménez, E.; Donoso-Gonzalez, M.; Beltrán-Velasco, A.I.; Tornero-Aguilera, J.F. Sustainable Development Goals in the COVID-19 Pandemic: A Narrative Review. *Sustainability* **2022**, *14*, 7726. [CrossRef]
26. Sciarra, C.; Chiarotti, G.; Ridolfi, L.; Laio, F. A network approach to rank countries chasing sustainable development. *Sci. Rep.* **2021**, *11*, 15441. [CrossRef] [PubMed]
27. Sciarra, C.; Chiarotti, G.; Ridolf, L.; Laio, F. Reconciling contrasting views on economic complexity. *Nat. Commun.* **2020**, *11*, 3352. [CrossRef]
28. Sachs, J.D.; Schmidt-Traub, G.; Kroll, C.; Lafortune, G.; Fuller, G.; Woelm, F. *The Sustainable Development Goals and COVID-19. Sustainable Development Report 2020*; Cambridge University Press: Cambridge, UK, 2020.
29. Pakkan, S.; Sudhakar, C.; Tripathi, S.; Rao, M. A correlation study of sustainable development goal (SDG) interactions. *Qual. Quant.* **2023**, *57*, 1937–1956. [CrossRef] [PubMed]
30. Bandari, R.; Moallemi, E.A.; Lester, R.E.; Downie, D.; Bryan, B.A. Prioritising Sustainable Development Goals, characterising interactions, and identifying solutions for local sustainability. *Environ. Sci. Policy* **2022**, *127*, 325–336. [CrossRef]
31. Rai, S.M.; Brown, B.D.; Ruwanpura, K.N. SDG 8: Decent work and economic growth—A gendered analysis. *World Dev.* **2019**, *113*, 368–380. [CrossRef]
32. Sezgin, F.H.; Tekin Turhan, G.; Sart, G.; Danilina, M. Impact of Financial Development and Remittances on Educational Attainment within the Context of Sustainable Development: A Panel Evidence from Emerging Markets. *Sustainability* **2023**, *15*, 12322. [CrossRef]
33. Androniceanu, A.; Kinnunen, J.; Georgescu, I. Circular economy as a strategic option to promote sustainable economic growth and effective human development. *J. Int. Stud.* **2021**, *14*, 60–73. [CrossRef]
34. Skvarciany, V.; Lapinskaite, I.; Volskyte, G. Circular economy as assistance for sustainable development in OECD countries. *Oecon. Copernic.* **2021**, *12*, 11–34. [CrossRef]
35. Rodríguez-Antón, J.M.; Rubio-Andrada, L.; Celemín-Pedroche, M.S.; Ruíz-Peñalver, S.M. From the circular economy to the sustainable development goals in the European Union: An empirical comparison. *Int. Environ. Agreem.* **2022**, *22*, 67–95. [CrossRef] [PubMed]
36. Jianu, E.; Pîrvu, R.; Axinte, G.; Toma, O.; Cojocaru, A.V.; Murtaza, F. EU Labor Market Inequalities and Sustainable Development Goals. *Sustainability* **2021**, *13*, 2675. [CrossRef]
37. Coscieme, L.; Mortensen, L.F.; Anderson, S.; Ward, J.; Donohue, I.; Sutton, P.C. Going beyond Gross Domestic Product as an indicator to bring coherence to the Sustainable Development Goals. *J. Clean. Prod.* **2020**, *248*, 119232. [CrossRef]
38. Hopp, D.; Fu, E.; Peltola, A. Feasibility of nowcasting SDG indicators: A comprehensive survey. *Stat. J. IAOS* **2022**, *38*, 591–608. [CrossRef]
39. R Core Team. *R: A Language and Environment for Statistical Computing*; R Foundation for Statistical Computing: Vienna, Austria, 2020. Available online: https://www.R-project.org/ (accessed on 15 August 2022).

40. Walesiak, M.; Dudek, A. The Choice of Variable Normalization Method in Cluster Analysis. In *Education Excellence and Innovation Management: A 2025 Vision to Sustain Economic Development During Global Challenges*; Soliman, K.S., Ed.; International Business Information Management Association: King of Prussia, PA, USA, 2020; pp. 325–340.
41. Giorgino, T. Computing and Visualizing Dynamic Time Warping Alignments in R: The dtw Package. *J. Stat. Softw.* **2009**, *31*, 1–24. [CrossRef]
42. Charrad, M.; Ghazzali, N.; Boiteau, V.; Niknafs, A. NbClust: An R Package for Determining the Relevant Number of Clusters in a Data Set. *J. Stat. Softw.* **2014**, *61*, 1–36. [CrossRef]
43. Kassambara, A.; Mundt, F. factoextra: Extract and Visualize the Results of Multivariate Data Analyses. R Package Version 1.0.7. Available online: https://CRAN.R-project.org/package=factoextra (accessed on 15 August 2022).
44. Ward, J.H., Jr. Hierarchical Grouping to Optimize an Objective Function. *J. Am. Stat. Assoc.* **1963**, *58*, 236–244. [CrossRef]
45. Zavadskas, E.K.; Kaklauskas, A.; Sarka, V. The new method of multicriteria complex proportional assessment of projects. *Technol. Econ. Dev. Econ.* **1994**, *1*, 131–139.
46. Trane, M.; Marelli, L.; Siragusa, A.; Pollo, R.; Lombardi, P. Progress by Research to Achieve the Sustainable Development Goals in the EU: A Systematic Literature Review. *Sustainability* **2023**, *15*, 7055. [CrossRef]
47. Miola, A.; Schiltz, F. Measuring sustainable development goals performance: How to monitor policy action in the 2030 Agenda implementation? *Ecol. Econ.* **2019**, *164*, 106373. [CrossRef]
48. Tutak, M.; Brodny, J.; Bindzár, P. Assessing the Level of Energy and Climate Sustainability in the European Union Countries in the Context of the European Green Deal Strategy and Agenda 2030. *Energies* **2021**, *14*, 1767. [CrossRef]
49. Radulescu, C.Z.; Radulescu, M.; Boncea, R. A Multi-Criteria Decision Support and Application to the Evaluation of the Fourth Wave of COVID-19 Pandemic. *Entropy* **2022**, *24*, 642. [CrossRef]
50. Ünvan, Y.A.; Ergenç, C. Financial Performance Analysis with the Fuzzy COPRAS and Entropy-COPRAS Approaches. *Comput. Econ.* **2022**, *59*, 1577–1605. [CrossRef]
51. Xie, Z.; Tian, G.; Tao, Y. A Multi-Criteria Decision-Making Framework for Sustainable Supplier Selection in the Circular Economy and Industry 4.0 Era. *Sustainability* **2022**, *14*, 16809. [CrossRef]
52. Bellman, R.; Kalaba, R. On adaptive control processes. *IRE Trans. Automat. Control* **1959**, *4*, 1–9. [CrossRef]
53. Rabiner, L.; Rosenberg, A.; Levinson, S. Considerations in dynamic time warping algorithms for discrete word recognition. *IEEE Trans. Audio Speech Lang. Process.* **1978**, *26*, 575–582. [CrossRef]
54. Sakoe, H.; Chiba, S. Dynamic programming algorithm optimization for spoken word recognition. *IEEE Trans. Audio Speech Lang. Process.* **1978**, *26*, 43–49. [CrossRef]
55. Stübinger, J. Statistical arbitrage with optimal causal paths on high-frequency data of the S&P 500. *Quant. Financ.* **2019**, *19*, 921–935. [CrossRef]
56. Denkowska, A.; Wanat, S. Dynamic Time Warping Algorithm in Modeling Systemic Risk in the European Insurance Sector. *Entropy* **2021**, *23*, 1022. [CrossRef] [PubMed]
57. Šťastný, T.; Koudelka, J.; Bílková, D.; Marek, L. Clustering and Modelling of the Top 30 Cryptocurrency Prices Using Dynamic Time Warping and Machine Learning Methods. *Mathematics* **2022**, *10*, 3672. [CrossRef]
58. Dmytrów, K.; Bieszk-Stolorz, B. Mutual relationships between the unemployment rate and the unemployment duration in the Visegrad Group countries in years 2001–2017. *Equilib. Q. J. Econ. Econ. Policy* **2019**, *14*, 129–148. [CrossRef]
59. Dmytrów, K.; Bieszk-Stolorz, B. Comparison of changes in the labour markets of post-communist countries with other EU member states. *Equilib. Q. J. Econ. Econ. Policy* **2021**, *16*, 741–764. [CrossRef]
60. Dmytrów, K.; Landmesser, J.; Bieszk-Stolorz, B. The Connections between COVID-19 and the Energy Commodities Prices: Evidence through the Dynamic Time Warping Method. *Energies* **2021**, *14*, 4024. [CrossRef]
61. Dmytrów, K.; Bieszk-Stolorz, B.; Landmesser-Rusek, J. Sustainable Energy in European Countries: Analysis of Sustainable Development Goal 7 Using the Dynamic Time Warping Method. *Energies* **2022**, *15*, 7756. [CrossRef]
62. Cui, Y.; Zhu, Z.; Zhao, X.; Li, Z. Energy Schedule Setting Based on Clustering Algorithm and Pattern Recognition for Non-Residential Buildings Electricity Energy Consumption. *Sustainability* **2023**, *15*, 8750. [CrossRef]
63. Beale, E.M.L. *Cluster Analysis*; Scientific Control Systems: London, UK, 1969.
64. Rocchi, L.; Ricciolini, E.; Massei, G.; Paolotti, L.; Boggia, A. Towards the 2030 Agenda: Measuring the Progress of the European Union Countries through the SDGs Achievement Index. *Sustainability* **2022**, *14*, 3563. [CrossRef]
65. Barbier, E.B.; Burgess, J.C. Institutional Quality, Governance and Progress towards the SDGs. *Sustainability* **2021**, *13*, 11798. [CrossRef]
66. Radulović, M.; Kostić, M. Are EU Members' Economies an "Engine" of the EU Candidates' Economies? *Folia Oecon. Stetin.* **2021**, *21*, 97–117. [CrossRef]
67. Stanujkic, D.; Popovic, G.; Zavadskas, E.K.; Karabasevic, D.; Binkyte-Veliene, A. Assessment of Progress towards Achieving Sustainable Development Goals of the "Agenda 2030" by Using the CoCoSo and the Shannon Entropy Methods: The Case of the EU Countries. *Sustainability* **2020**, *12*, 5717. [CrossRef]
68. Bąk, I.; Tarczyńska-Łuniewska, M.; Barwińska-Małajowicz, A.; Hydzik, P.; Kusz, D. Is Energy Use in the EU Countries Moving toward Sustainable Development? *Energies* **2022**, *15*, 6009. [CrossRef]
69. Skvarciany, V.; Astikė, K. Decent work and economic growth: Case of EU. In Proceedings of the 12TH International Scientific Conference Business and Management 2022, Vilnius, Lithuania, 12–13 May 2022; pp. 184–190. [CrossRef]

70. Grzebyk, M.; Stec, M.; Hejdukova, P. Implementation of sustainable development goal 8 in European Union countries—A measurement concept and a multivariate comparative analysis. *Sustain. Dev.* **2023**, *31*, 2758–2769. [CrossRef]
71. Carlsen, L. Decent Work and Economic Growth in the European Union. A partial order analysis of Eurostat SDG 8 data. *Green Financ.* **2021**, *3*, 483–494. [CrossRef]
72. Kuc-Czarnecka, M.; Markowicz, I.; Sompolska-Rzechuła, A. SDGs implementation, their synergies, and trade-offs in EU countries–Sensitivity analysis-based approach. *Ecol. Indic.* **2023**, *146*, 109888. [CrossRef]
73. D'Adamo, I.; Gastaldi, M.; Morone, P. Economic sustainable development goals: Assessments and perspectives in Europe. *J. Clean. Prod.* **2022**, *354*, 131730. [CrossRef]
74. Chamusca, P. Public Policies for Territorial Cohesion and Sustainability in Europe: An Overview. *Sustainability* **2023**, *15*, 6890. [CrossRef]
75. Anselmi, D.; D'Adamo, I.; Gastaldi, M.; Lombardi, G.V. A comparison of economic, environmental and social performance of European countries: A sustainable development goal index. *Environ. Dev. Sustain.* 2023, *in press*. [CrossRef]
76. Shuai, C.; Zhao, B.; Chen, X.; Liu, J.; Zheng, C.; Qu, S.; Zou, J.-P.; Xu, M. Quantifying the impacts of COVID-19 on Sustainable Development Goals using machine learning models. *Fundam. Res.* 2022, *in press*. [CrossRef]

Disclaimer/Publisher's Note: The statements, opinions and data contained in all publications are solely those of the individual author(s) and contributor(s) and not of MDPI and/or the editor(s). MDPI and/or the editor(s) disclaim responsibility for any injury to people or property resulting from any ideas, methods, instructions or products referred to in the content.

sustainability

Article

Chemical Engineering beyond Earth: Astrochemical Engineering in the Space Age

Vassilis J. Inglezakis [1,*], Donald Rapp [2], Panos Razis [3] and Antonis A. Zorpas [4,*]

1 Department of Chemical & Process Engineering, University of Strathclyde, Glasgow G1 1XQ, UK
2 Independent Researcher, South Pasadena, CA 91030, USA; drdrapp@earthlink.net
3 Faculty of Pure and Applied Science, Department of Physics, University of Cyprus, 1st Panepistimiou Avenue, 2109 Aglantzia, P.O. Box 20537, Nicosia 1678, Cyprus; razis@ucy.ac.cy
4 Laboratory of Chemical Engineering and Engineering Sustainability, Faculty of Pure and Applied Sciences, Open University of Cyprus, Giannou Kranitiodi 89, Nicosia 2236, Cyprus
* Correspondence: vasileios.inglezakis@strath.ac.uk (V.J.I.); antonis.zorpas@ouc.ac.cy or antoniszorpas@yahoo.com (A.A.Z.)

Abstract: The Space Race in the second half of the 20th century was primarily concerned with getting there and back. Gradually, technology and international collaboration opened new horizons, but human activity was mostly restricted around Earth's orbit, while robotic missions were sent to solar system planets and moons. Now, nations and companies claim extraterrestrial resources and plans are in place to send humans and build bases on the Moon and Mars. Exploration and discovery are likely to be followed by exploitation and settlement. History suggests that the next step is the development of space industry. The new industrial revolution will take place in space. Chemical engineers have been educated for more than a century on designing processes adapted to the Earth's conditions, involving a range of raw materials, atmospheric pressure, ambient temperature, solar radiation, and 1-g. In space, the raw materials differ, and the unique pressure, temperature and solar radiation conditions require new approaches and methods. In the era of space exploration, a new educational concept for chemical engineers is necessary to prepare them for playing key roles in space. To this end, we introduce Astrochemical Engineering as an advanced postgraduate course and we propose a 2-year 120 ECTS MEng curriculum with a brief description of the modules and learning outcomes. The first year includes topics such as low-gravity process engineering, cryogenics, and recycling systems. The second year includes the utilization of planetary resources and materials for space resources. The course culminates in an individual design project and comprises two specializations: Process Engineering and Space Science. The course will equip engineers and scientists with the necessary knowledge for the development of advanced processes and industrial ecologies based on closed self-sustained systems. These can be applied on Earth to help reinvent sustainability and mitigate the numerous challenges humanity faces.

Keywords: chemical engineering; astrochemical engineering; curriculum development; extraterrestrial environment; space engineering; sustainable education

Citation: Inglezakis, V.J.; Rapp, D.; Razis, P.; Zorpas, A.A. Chemical Engineering beyond Earth: Astrochemical Engineering in the Space Age. *Sustainability* **2023**, *15*, 13227. https://doi.org/10.3390/su151713227

Academic Editors: Idiano D'Adamo, Massimo Gastaldi, Manoj Kumar Nallapaneni and Marc A. Rosen

Received: 8 July 2023
Revised: 28 August 2023
Accepted: 29 August 2023
Published: 4 September 2023
Corrected: 20 June 2024

Copyright: © 2023 by the authors. Licensee MDPI, Basel, Switzerland. This article is an open access article distributed under the terms and conditions of the Creative Commons Attribution (CC BY) license (https://creativecommons.org/licenses/by/4.0/).

1. Introduction

Chemical Engineering (ChE) has evolved dramatically over the past thirty years, embracing advances in the broader areas of *nanotechnology*, *biotechnology*, and *computer science*. Bio-, info-, and nano- are not new concepts to ChE but the exponential development of these areas has had a profound effect on the evolution of the discipline. Nanotechnology was first introduced in 1959 with a lecture by the Nobel-prize-winning physicist Richard Feynman, entitled *There's Plenty of Room at the Bottom* (*Data Storage*), where he envisioned the possibility of directly manipulating individual atoms as a powerful form of synthetic chemistry [1]. ChE embraced nanotechnology in the 1990s; suitable modules were integrated into the curriculum and research has been evolving at an accelerating pace. In recognition of biology as an important science impacting ChE, a number of ChE departments changed their names to include some bio terms [1], such as Chemical and Biomolecular Engineering (e.g., Georgia Tech, Atlanta, GA, USA), Chemical and Biological Engineering (e.g., Princeton, NJ, USA), Chemical Engineering and Biotechnology (e.g., Cambridge, UK), and Chemical and Biomedical Engineering (e.g., Florida State University, Tallahassee, FL, USA). The new hybrid plays a key role in advancing bioprocesses, biomaterials, and biomedicine. In tandem, a revolution in process engineering was brought about by computers in the 1990s [2]. The increasingly powerful hardware and software created unimaginable opportunities. ChE is undergoing a transformation driven largely by the booming of data science [3]. A new term, i.e., Industry 4.0, has emerged to represent the fourth revolution that has occurred in manufacturing driven by digitalization. There is ample evidence that the next revolution in ChE will be data-driven and powered by artificial intelligence [3].

The potential human bases on the Moon and Mars are coming within the reach of our current technological capabilities and *Astropolitics* shapes national strategies worldwide [4]. Several disciplines have embraced the new opportunity; *Astrochemistry* deals with the chemical evolution occurring in space and it is a recognized field of research [5,6], *Space Chemistry* is defined as the performance of chemistry in space [6], *Astrobiology* is the study of life in the universe, *Space Biology* is aimed at addressing the basic questions regarding the extent to which gravity plays a role in the growth, morphology, and function of cells in the space environment [7], *Astropharmacy* and *Astromedicine* address the question of how human explorers can receive effective medical and pharmaceutical care in the context of space [8], and Civil Engineering is expanding towards the design and construction of lunar and Martian structures, habitats, and outposts [9]. ChE contributes to space industry as well in a number of ways; life support systems (LSSs), materials, energy systems, and propulsion are a few examples. However, as Lobmeyer and Meneghelli argued two decades ago, ChE has been laboring in the shadows of other disciplines, namely Physics and Mechanical Engineering [10]. The situation has not changed much since then. In *New Directions for Chemical Engineering* published by the U.S. National Academy of Sciences, bio and info are thoroughly discussed but space is absent [3]. In *Revisiting the Future of Chemical Engineering*, where some of the profession's thought leaders share their visions of the future of ChE, data science and biotechnology (among others) are identified as strong forces shaping the future of ChE; however, again, space is not considered [11].

Looking beyond the immediacy of current space projects, we envision a significant future expansion of space exploration and exploitation of space resources. We believe that the current ChE curriculum does not adequately prepare students for leadership in space engineering. We therefore created and present in this paper a new 2-year curriculum leading to an MSc in Astrochemical Engineering to fill this anticipated need. Section 2 discusses the role of ChE in space exploration. Section 3 discusses the differences between space and terrestrial processing, and interprets these as both challenges and opportunities. That is followed by Section 4, where the structure of the 2-year course is presented. The first year is on the theory and fundamentals of space engineering from the ChE perspective and the second year is dedicated to space exploration applications. The course comprises seven core modules, four modules for Process Engineering, and four for the Space Science specializations and a Design Project. An important part of space processing is so-called circular or cyclic systems, where spent outputs are reprocessed to produce new useful input materials. In typical terrestrial processes, such spent outputs are often discharged to the environment, where they typically pollute and require great effort to mitigate. Eventually, as these circular systems designed for space applications also become available for terrestrial applications, they will contribute to sustainability goals.

The 2030 Agenda of the United Nations (UN) revolves around the Sustainable Development Goals (SDGs), and to achieve the related objectives, scientific production should align with the SDGs' achievement [12]. Universities should try to make the most of the many opportunities that the SDGs offer, not only in the field of teaching and research, but also in their university extension activities [13]. This commitment in universities is advancing with the help of academics who individually include it in their disciplines and course design but consider that there is still a long way to go [14]. The innovative technologies built to push humanity into space can also support sustainable development on Earth. Many projects related to sustainable development across industries make use of space-based technologies and services to contribute to the goals. In particular, space technologies directly support Good health and well-being (SDG3), Affordable and clean energy (SDG7) and Industry, Innovation and infrastructure (SDG9) and Climate Action (SDG13). Also, according to UNESCO, five other SDGs have direct reference to Quality Education (SDG4), including SDG3 (Target 3.7) and SDG 13 (Target 13.3). Thus, by educating chemical engineers for space applications, the proposed curriculum is focused on fulfilling all these SDGs. The modules, and in particular the Circular Systems module of the proposed curriculum, bring together theoretical approaches and innovative knowledge by providing students with the skills and knowledge to apply sustainable development solutions in extraterrestrial environments.

Space was a central theme in science fiction for centuries, but it is only recently that fiction and reality have converged. Science fiction as an educational resource was recognized from the very beginnings of the genre, but its use in ChE education was first discussed in a paper published in *Education for Chemical Engineers* in 2011 [15]. In the science fiction book *Reach for the Stars* published in 2021 by John Wegener, Ethan Richards, a space traveler, was sponsored by NASA to study *Astrochemical Engineering* at UCLA; to the best of our knowledge, this is the first time the term was used. Hereby, we argue that it is time to make science fiction a reality and introduce *Astrochemical Engineering* into the ChE curriculum.

"There is a tide in the affairs of men
Which, taken at the flood, leads on to fortune;
Omitted, all the voyage of their life
Is bound in shallows and in miseries.
On such a full sea are we now afloat;
And we must take the current when it serves,
Or lose our ventures."

William Shakespeare, Julius Caesar (Act IV, Scene III, lines 218–224)

2. Space Exploration and Chemical Engineering

The contribution of ChE to space exploration is not new and can be traced in the literature back to 1979; Waldron et al. discussed the possibilities and conditions of constructing and operating processing plants in space using materials taken from the lunar surface [16]. They list a number of processes, such as the electrolysis of molten silica, carbothermic/silicothermic reduction, the carbon–chlorination process, NaOH basic-leach process, and HF acid leach. Life support systems, low-gravity processes, and the associated automated control systems have long been identified as areas where ChE can contribute [17]. In a conference paper published in 2001 [10], Lobmeyer and Meneghelli list cryogenics, ISRU, miniaturization, launchability, and power/process efficiencies as several areas where chemical engineers can provide support for the exploration of space. ChE played a central role in the development of propulsion systems and the latest major contribution is the development of solid oxide electrolysis of CO_2 to O_2 recently demonstrated on Mars [18,19]. Also, several ChE processes are used in the Environmental Control and Life Support System (ECLSS) developed by NASA. The ECLSS provides clean air and water to the International Space Station (ISS) crew and laboratory animals and it consists of two key components: the Water Recovery System (WRS) and the Oxygen Generation System (OGS) [20,21]. However, except for propulsion systems, fuels, and Life Support Systems, hardcore ChE such as chemical reactor engineering and separation processes has not been utilized in the core of space research. This is about to change as the advent of in situ resource utilization (ISRU) brings ChE to the forefront of space programs, and it is expected to grow significantly in importance over the coming decades.

The current cost of exploration and subsequent settlement in extraterrestrial environments is challenging but can be reduced by use of resources found in situ. ISRU is defined as the conversion of local resources at a space destination to provide useful infrastructure and commodities [19,22]. A seminal paper published in 1978 by Ash et al. presented a revolutionary detailed analysis of Mars ISRU and its benefits [23]. The current ISRU interests on the Moon and Mars concentrate on building shelters and the harvesting of subsoil water and atmospheric CO_2 for the production of methane, hydrogen, and oxygen [24]. However, there is untapped potential as the Moon and Mars hold a number of other useful materials such as oxides of silicon, aluminum, iron, and titanium; salts; hydrated minerals; and atmospheric N_2 and Ar [25,26]. Minerals that exist on Earth in very limited quantities are abundant on some meteorites and asteroids [19,27]. The new generation of chemical engineers should be able to capture and/or recover such critical minerals from extraterrestrial environments as these are essential to develop smart technologies. For materials that are not readily available, chemical engineers would have to recover them from waste or end-of-life equipment applying circular economy strategies [28]. Clearly, ISRU is incomplete without the development of processes that transform raw materials into useful products—this is what chemical engineers carry out on Earth. Indeed, ChE is recognized as one of the key disciplines involved in ISRU technologies [29]. Reverse water–gas shift and Sabatier reactions are prime examples of reactions utilizing CO_2, the main component of the Martian atmosphere [25]. These reactions are the basis of the era of small-molecule activation [30]. Another resource is the waste produced by humans, notably urine. The closed-loop nutrient cycle from human urine has attracted interest in the last decade and the number of papers published on bioregenerative life support systems is growing [31]. However, Earth technologies cannot simply be implemented on other planets and need to be adapted to each location. There is a gap both in terms of technology and finances stemming from the major differences on other planets regarding available resources and the physical environment in terms of gravity, temperature, pressure, and radiation conditions. As one recent example, consider the Mars Oxygen In Situ Experiment ("MOXIE") that has been converting CO_2 to O_2 on Mars for the past two years. This operates with a continuous flow reactor system, where a compressor pulls the Martian atmosphere through a filter to a higher pressure, and the flow passes through the cathode of a high-temperature electrolysis stack, where a portion of Martian CO_2 is converted to CO while a commensurate flow of oxygen occurs in the anode. However, an undesirable side reaction can also occur that would produce carbon that would clog up the cells and essentially destroy the electrolysis stack. The minimum voltage required for the side reaction is greater than the minimum voltage for oxygen production, so the voltages on the cells in the stack must be maintained above the minimum for oxygen production, yet below that for carbon formation. The essential technical background for MOXIE involves thermodynamics, electrochemistry, gas flow in various regimes, the filtration of dust particles, and thermal control. All of this must be carried out within the context of space constraints on mass, volume, power, and reliability. Most of the work was beyond ordinary ChE education and required special study to work effectively on the project. ChE education must evolve and equip engineers for the Space Age.

3. Challenges and Opportunities

Throughout history, scientists and engineers have been educated to think in terms of a 1 g environment, traditional training that must be updated to account for low-gravity environments [32]. Unit operations in microgravity are not a new topic in ChE; for instance, they were discussed in a paper published by Allen and Pettit in 1987 [33]. The authors view microgravity as an advantage in crystallization processes. However, microgravity complicates fluid handling and processes governed by density differences, such as the stratification and separation of gases and liquids, due to the absence of buoyancy [34,35].

In the absence of gravity, convection, sedimentation, and buoyancy become irrelevant and mixing is not spontaneous. Diffusion is the only way that molecular heat and matter can be transported, and microgravity conditions provide a unique opportunity to study processes decoupled from sedimentation [36]. Microgravity benefits the formation of alloys as the process is diffusion-controlled rather than gravity-controlled, encouraging molecules to be distributed evenly in the material, resulting in a more uniform structure. On the other hand, chemistry under microgravity cannot be performed in flasks because the solutions would not mix well and reactions would not be reproducible [37]. The predominance of diffusion is a challenge for heterogeneous catalytic processes as the absence of density-driven convection hinders phase separation [37]. Distillation, arguably the most important separation process in ChE, may be useless in space since microgravity removes two of the variables in boiling: convection and buoyancy. This means that the vapor phase of a boiling liquid does not rise and the usual model of convection currents that distributes heat in the liquid phase that we know on Earth is no longer valid. Two-phase boiling experiments have shown that reduced gravity considerably alters the flow patterns and heat transfer as compared to 1 g conditions [38]. Another example is electrolysis, where under 1 g, buoyancy leads to the detachment of gas bubbles from the electrode surface and a separation of oxygen and hydrogen gas bubbles from the liquid electrolyte, but under reduced-gravity electrolysis systems, this is hindered, resulting in a lower efficiency [21]. Due to the effect of microgravity on mixing in batch systems, continuous flow is considered the best operational mode in space. Flow chemistry studies reactions taking place in continuous flow in tubes rather than in a flask [39]. The first journal publications regarding the overall concept and promises of flow chemistry for microgravity applications were published in 2017 [37]. Although these effects are more intense in zero gravity and microgravity, they are important in reduced-gravity conditions found on terrestrial planetary bodies. Moreover, process integration to small compact systems is compulsory in space [36]. At the plant level, the integration of extraterrestrial industrial systems needs to be complete, and process intensification, a relatively new concept in ChE, represents a way to achieve more with less. This is relevant to the miniaturization of processes, e.g., microreactors—in other words, to put a plant on the scale of a small laboratory setup [40].

Beyond the profound effect of gravity, as Allen and Pettit argued in 1987 [33], low temperature and high vacuum are advantages that space offers rather than challenges. Indeed, space offers an infinitely available vacuum and extremely low temperatures, which are expensive and difficult to achieve on Earth. Vacuum technology is ubiquitous and typical applications include distillation, drying, sublimation, and filtration [41]. Low temperature is used to produce and transport immense quantities of gas mixtures [42]. Also, we know that some reactions can be accelerated by operating at temperatures below −150 °C and some can even occur at temperatures below 10 K [43,44]. A challenge is the exposure of chemical reactions to the entire electromagnetic spectrum outside the Earth's atmosphere, which is likely to influence photocatalytic reactions and the stability of molecules due to higher UV radiation [37].

Sustainable development, as defined in the World Commission on Environment and Development's 1987 Brundtland report, should follow humans in the journey to space, considering circular economy principles [45]. A major challenge is to extract and transform the available in situ resources sufficiently in circular systems. The new concept of circular chemical processes is gaining traction [36]. In extraterrestrial environments, the objective is the development of an industrial ecology based on closed self-sustained systems with zero-discharge production processes. The knowledge acquired in extraterrestrial environments can be applied on Earth to help reinvent sustainability by developing new technologies, including smart ones that harness the power of chemistry to create sustainable products and processes. By studying the chemical properties of materials in space, scientists can identify new materials that are more resistant to environmental degradation. Furthermore, they can be used to create new processes for the production of energy, such as using solar power to create hydrogen fuel through a process known as artificial photosynthesis. Artificial photosynthesis can be an efficient alternative route to capture CO_2 and produce food and energy [46]. Hann et al. found a way to bypass the need for biological photosynthesis altogether and create food independent of sunlight by using artificial photosynthesis [47].

4. The Case for Astrochemical Engineering

In NASA's upcoming new mission "ARTEMIS II" [48], currently planned to be launched in November 2024, four astronauts will venture around the Moon, paving the way for future lunar missions. If we review the curriculum of the four astronauts, it will become apparent that nexus disciplines are needed for the next generations of engineers that intend to explore space. The commander spent hundreds of hours conducting valuable scientific research in areas such as Human Physiology, Medicine, Physical Science, Earth Science, and Astrophysics, and he studied Computer and Systems Engineering, with an MSc in Systems Engineering. The pilot holds a BSc in General Engineering, MSc in Flight Test Engineering, MSc in Systems Engineering, and MSc in Military Operational Art and Science. The other two members as specialists hold BSc degrees in Electrical Engineering and Physics and an MSc degree in Electrical Engineering and Space Science, as well as a PhD. The 17 members of the 2022 ESA astronaut class exhibit an amazing breadth of disciplines: Astronautical Engineering, Aerospace Engineering, Mechanical Engineering, Human Factors Engineering, Space Engineering, Physics, Astrophysics, Biotechnology, Neuroscience, Biomedical Engineering, Medicine, Transport Engineering, Military Technology, Electrical Engineering, and Astronomy. One of the class members studied Industrial Chemistry.

Beyond the standard Astronautical Engineering courses that focus on the design, development, and manufacturing of spacecrafts, there are a number of postgraduate courses related to space communications, data collection and processing, satellites, observation, and remote sensing. The *Technical University of Denmark* (Denmark) offers an MSc course in Earth and Space Physics and Engineering, with an astrophysics orientation and specialization primarily in the areas of instrumentation to observe the universe, data processing, and physical and mathematical modelling. The *University of Pisa* (Italy) offers a 2-year MSc of Science in Space Engineering course focused on aerospace engineering, instrumentation, and propulsion. Brno University of Technology (Czech Republic) offers a 2-year MSc course on Space Applications, which is an interdisciplinary association of electrical and mechanical engineering. *Gdansk University of Technology* (Poland) offers a 2-year MSc course on Engineering and Management of Space Systems and it is broader in scope, including robotics, gravity-related research, heat and mass transfer in zero gravity, space law, and several modules on applications of computer science. The ESPACE—Earth Oriented Space Science and Technology is a 2-year MSc course offered by the *Technical University of Munich* (Germany) positioned at the interface between space technology and the engineering and natural scientific use of satellite data. The 1-year Space Science MSc offered by *University College London* (UK) provides a broad understanding of all aspects of space science and space instrumentation, including modules on data, instrumentation, satellites and spacecrafts, planetary atmospheres, and astrophysics. The *University of Edinburgh* released a new master's course (1 year) on Astrobiology and Planetary Sciences targeting the development of knowledge on cometary and exoplanet science, investigating the origin, diversity, and behavior of planets, asteroids, and solar systems.

A course similar in spirit to the Astrochemical Engineering concept is the 4-year MEng course on Electronic Engineering with Space Science and Technology offered by the *University of Bath* (UK), focused on the design, operation, and building of electronic systems for the space environment. However, the modules offered do not seem to consider the environmental conditions found on terrestrial planets. In this paper, we propose a new ChE course, namely *Astrochemical Engineering* (120 ECTS), dedicated to enabling *space industry*, which is necessary for establishing self-sustaining communities on the Moon, Mars, and other solar system bodies. Astrochemical Engineering would have to answer the fundamental question of how terrestrial ChE processes can be optimized and/or transformed to adapt to, and benefit from, the conditions in space. In a paper entitled *Chemical Engineering Education in the Next Century,* Gillet argues that ChE has survived extremely well in a changing world by assimilating or developing new subjects into curricula, but there is a fear that this ability to adapt might go too far and might lead to its eventual demise [49]. We believe that ChE must adapt to the new trends without diluting its core and Astrochemical Engineering is seen as a stem of an evolving ChE discipline. Astrochemical Engineering (Figure 1) is a ChE course adapted to address extraterrestrial environments, typically characterized by high radiation and low gravity, pressure, and temperature.

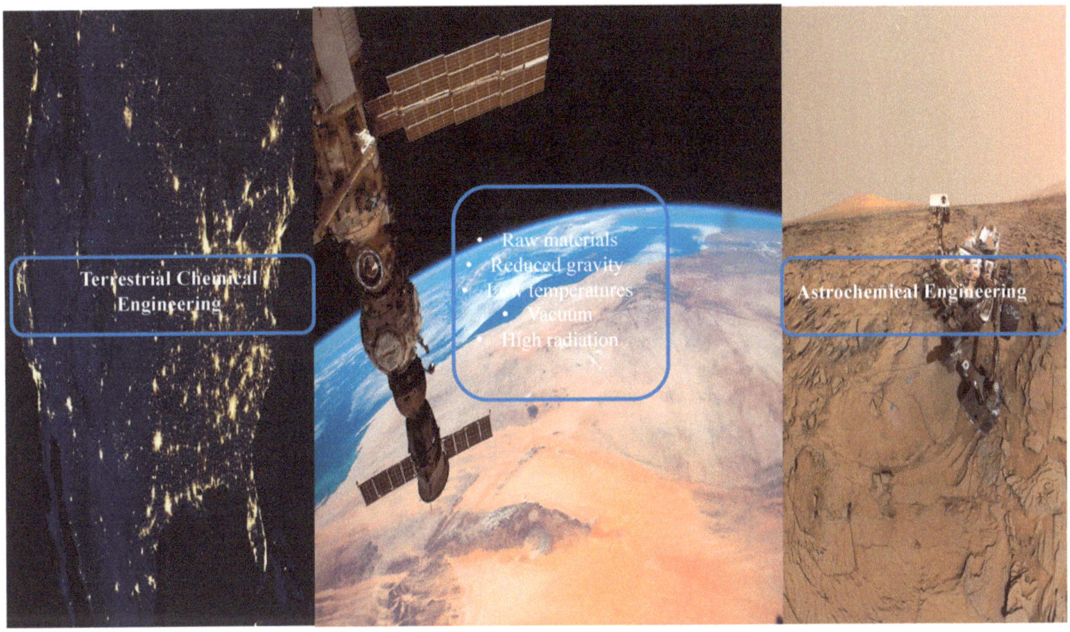

Figure 1. Astrochemical Engineering concept (images credit: NASA).

The 2-year MEng Astrochemical Engineering is a 120 ECTS course as presented below.

1st Year: Theory and fundamentals (60 ECTS)
Core modules (45 ECTS)

- Thermodynamics of terrestrial planets' and satellites' atmospheres and surfaces (7.5 ECTS): greenhouse effect, planetary atmosphere and climate, mass and heat flows, and case studies (Earth, Mars, Venus, Pluto, Titan, Triton, Enceladus, and Europa).
- Low-gravity process engineering (7.5 ECTS): the effect of gravity on transport phenomena and unit operations (boiling, flow, electrolysis, and crystallization).
- Principles of cryochemistry (7.5 ECTS): chemical reactions at low temperatures, in frozen systems, and at extremely low temperatures (<-150 °C).
- Programming (7.5 ECTS): numerical methods, open-source languages (Python), big data, machine learning, human–machine interfaces, the Metaverse environment, and gamification.
- Circular systems: circular economy, design for recycling, regenerative development remanufacturing, and reverse logistics for process optimization (7.5 ECTS).
- Design project I (7.5 ECTS).

Specialization: Process Engineering (15 ECTS)

- Fundamentals of cryogenic engineering (7.5 ECTS): the properties of materials and fluids in cryogenic conditions, cryogenic separation processes, the liquification of gases, cryogenic fluid storage and transfer, and cryogenic instrumentation.
- Process intensification (7.5 ECTS): equipment (microreactors, intensive mixing, and compact/microchannel heat exchangers) and methods (hybrid separations and reaction/separation integration).

Specialization: Space Science (15 ECTS)

- Energy production in space: solar energy, cosmic radiation, nuclear fission/fusion energy, artificial photosynthesis, energy production for life support systems, and carbon capture and utilization (7.5 ECTS).
- Space microbiology (7.5 ECTS): microbes in space, biochemical processes, synthetic biology and genetic engineering.

2nd Year: Space exploration applications (60 ECTS)
Core modules (45 ECTS)

- Martian in situ resource utilization (7.5 ECTS): atmospheric gas capture and purification, electrolysis (H_2O/CO_2), reactions for the production of useful substances (Sabatier, Fischer–Tropsch, reverse water–gas shift, and Haber–Bosch), and dust control and filtration systems.
- Materials for space applications: advanced composites (2D materials, graphene, carbon fiber, metal foams, aerogels, 3D-printed materials, and titanium composites), nanomaterials (carbon nanotubes; nanoscale sensors and actuators), and smart materials (shape memory alloys, self-healing materials, and bio-inspired materials) (7.5 ECTS).
- Design project II (30 ECTS).

Specialization: Process Engineering (15 ECTS)

- Lunar in situ resource utilization: polar ice, regolith, and fluid transfer under vacuum (pumps and compressors) (7.5 ECTS).
- Propulsion systems: methane–oxygen chemical propulsion, hydrogen storage and transport in space, solar electric propulsion, Xe, C60 and exotic propellants, nuclear thermal propulsion, nuclear electric propulsion, the storage and transport of cryogenic propellants on planetary surfaces, cryogenic propellants within aeroshells, and descent and ascent propulsion linkage (7.5 ECTS).

Specialization: Space Science (15 ECTS)

- Life support systems: human physiology and anatomy, life cycle assessment, instrumentation, and control systems (7.5 ECTS).
- Space agriculture (7.5 ECTS): plant science, farming technologies, the effect of reduced gravity, and genetic engineering.

The structure of the MSc provides the necessary transferable skills and ensures the learning outcomes (Figure 2) cover several pillars such as knowledge, comprehension, application, analysis, synthesis, and evaluation. The students will be equipped with solid theoretical knowledge and an understanding of applications, enabling them to operate in multiple contexts. The first year covers ChE fundamentals adapted to space conditions and applications, i.e., thermodynamics, transport phenomena, unit operations, and chemistry. The modules provide students with the tools needed to apply ChE in space applications. The circular systems module is of paramount importance in space and completes the set of core modules, ensuring that sustainability plays a central role in the curriculum [50]. In the second year, the approach is more specific and focuses on ISRU and materials needed for space applications. ISRU using the Mars paradigm as the planet offers a range of raw materials and appropriate, albeit challenging, conditions for the design of processes. Students can choose between two specializations in the first year, i.e., Process Engineering and Space Science, consisting of a total of four elective modules. The design project objective is the application of technical knowledge covered in the various teaching modules and it is divided into two phases: a group and an individual project. Bloom's taxonomy was used as a guide for the learning outcomes presented in Figure 2. As this is an advanced ChE course, it is best suited for students with ChE education. The course is designed to prepare a new generation of chemical engineers ready to work and innovate in the broader area of space industry.

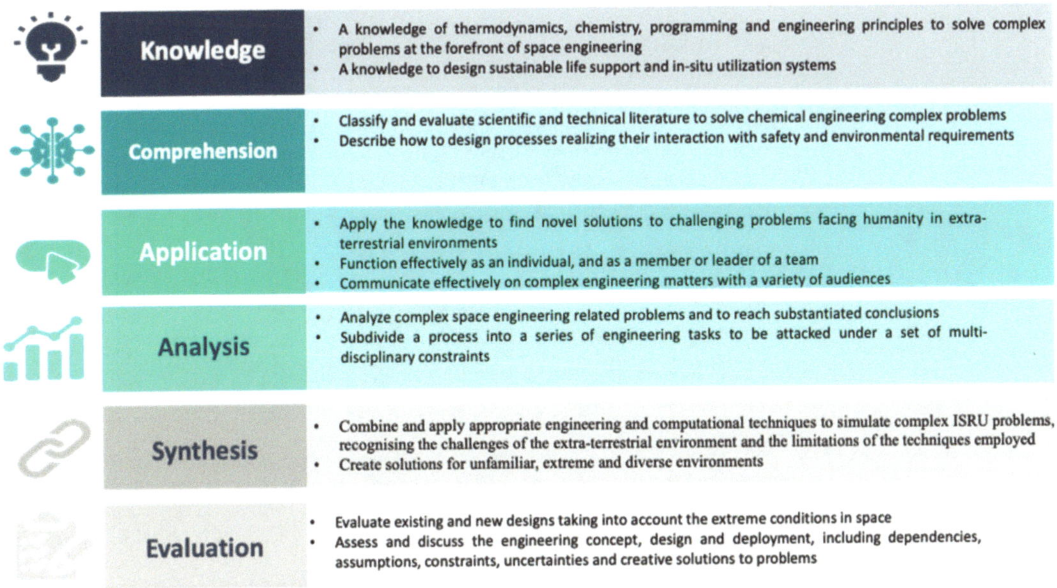

Figure 2. Learning outcomes of the course.

Education can be said to be part of an ongoing transition, which comprises opportunities and challenges. The proposed course competencies could vary depending on the institution and the evolving nature of the field. The course is multidisciplinary and concepts such as case-based learning (CBL) and problem-based learning (PBL), as a sustainable teaching practice, can be easily integrated [50]. Also, the proposed course contributes to SDG4 (Quality of Education) [51–53]. It links knowledge with real engineering problems through an interdisciplinary and cross-disciplinary approach to develop the competencies needed for building a sustainable future on Earth and beyond. It is relevant to mention that the notion of sustainability in Astrochemical Engineering is futuristic and one would have to look into the coming decades; it is difficult to make this specific in the year 2023.

5. Conclusions

Chemical Engineering education has passed through several significant stages of evolution, particularly in the incorporation of biological processes. We believe that Chemical Engineering education is poised for yet another evolution into preparation for operations in space, which we call Astrochemical Engineering. Chemical Engineering already plays a significant role in space technology via materials, environmental control, the reprocessing of waste materials, in situ resource utilization, and propulsion. These processes involve unique environmental conditions in gravity, radiation, and the need for recycling waste products under significant mass and power constraints. Individual chemical engineers have had to learn to adapt to these conditions ad hoc. We envision a significant future expansion in the global exploration and exploitation of space in which Chemical Engineering will play an increasing role, requiring targeted preparation at the university level. Looking beyond the immediacy of current space projects, we envision a significant future expansion of space exploration and the exploitation of space resources. We believe that the current Chemical Engineering curriculum does not adequately prepare students for leadership in space industry. We therefore created a new 2-year curriculum leading to an MS in Astrochemical Engineering to fill this anticipated need. Moreover, Astrochemical Engineering has much to offer to sustainability as space technologies can be used to solve problems on Earth. The proposed curriculum can support several SDGs, such as SDGs 3, 4, 7, 9, and 13.

Author Contributions: Conceptualization, V.J.I.; methodology, V.J.I.; software, V.J.I. and A.A.Z.; validation, V.J.I., D.R., P.R. and A.A.Z.; formal analysis, V.J.I., D.R., P.R. and A.A.Z.; investigation, V.J.I. and A.A.Z.; resources, V.J.I. and A.A.Z.; data curation, V.J.I. and A.A.Z.; writing—original draft preparation, V.J.I.; writing—review and editing, V.J.I., D.R., P.R. and A.A.Z.; visualization, V.J.I.; supervision, V.J.I.; project administration, V.J.I. and A.A.Z.; funding acquisition, V.J.I. and A.A.Z. All authors have read and agreed to the published version of the manuscript.

Funding: This research received no external funding.

Institutional Review Board Statement: Not applicable.

Informed Consent Statement: Not applicable.

Data Availability Statement: Not applicable.

Conflicts of Interest: The authors declare no conflict of interest.

References

1. Varma, A.; Grossmann, I.E. Evolving Trends in Chemical Engineering Education. *AIChE J.* **2014**, *60*, 3692–3700. [CrossRef]
2. Wei, J. Future Directions of Chemical Engineering. In *Advances in Chemical Engineering*; Academic Press: Cambridge, MA, USA, 1991; pp. 51–56.
3. Kaler, E. *New Directions for Chemical Engineering*; National Academies Press: Washington, DC, USA, 2022; ISBN 978-0-309-26842-4.
4. Marshal, T. *The Future of Geography*; Eliott and Thomson Limited: London, UK, 2023.
5. Puzzarini, C. Grand Challenges in Astrochemistry. *Front. Astron. Space Sci.* **2020**, *7*, 19. [CrossRef]
6. Sipos, G.; Bihari, T.; Milánkovich, D.; Darvas, F. Flow Chemistry in Space-A Unique Opportunity to Perform Extraterrestrial Research. *J. Flow Chem.* **2017**, *7*, 151–156. [CrossRef]
7. Clément, G. Introduction to Space Biology. In *Fundamentals of Space Biology*; Springer: New York, NY, USA, 2006; pp. 1–50.
8. Sawyers, L.; Anderson, C.; Boyd, M.J.; Hessel, V.; Wotring, V.; Williams, P.M.; Toh, L.S. Astropharmacy: Pushing the Boundaries of the Pharmacists' Role for Sustainable Space Exploration. *Res. Soc. Adm. Pharm.* **2022**, *18*, 3612–3621. [CrossRef]
9. Kalapodis, N.; Kampas, G.; Ktenidou, O.J. A Review towards the Design of Extraterrestrial Structures: From Regolith to Human Outposts. *Acta Astronaut.* **2020**, *175*, 540–569.
10. Lobmeyer, D.A.; Meneghelli, B. Chemical Engineering in Space. In Proceedings of the 6th World Congress of Chemical Engineering, Melbourne, Australia, 23–27 September 2001.
11. Westmoreland, P.R.; McCabe, C. Revisiting the Future of Chemical Engineering. *Chem. Eng. Prog.* **2018**, *114*, 26–38.
12. Sánchez-Roncero, A.; Garibo-i-Orts, Ò.; Conejero, J.A.; Eivazi, H.; Mallor, F.; Rosenberg, E.; Fuso-Nerini, F.; García-Martínez, J.; Vinuesa, R.; Hoyas, S. The Sustainable Development Goals and Aerospace Engineering: A Critical Note through Artificial Intelligence. *Results Eng.* **2023**, *17*, 100940. [CrossRef]
13. Leal Filho, W.; Shiel, C.; Paço, A.; Mifsud, M.; Ávila, L.V.; Brandli, L.L.; Molthan-Hill, P.; Pace, P.; Azeiteiro, U.M.; Vargas, V.R.; et al. Sustainable Development Goals and Sustainability Teaching at Universities: Falling behind or Getting Ahead of the Pack? *J. Clean. Prod.* **2019**, *232*, 285–294. [CrossRef]
14. Chaleta, E.; Saraiva, M.; Leal, F.; Fialho, I.; Borralho, A. Higher Education and Sustainable Development Goals (SDG)—Potential Contribution of the Undergraduate Courses of the School of Social Sciences of the University of Évora. *Sustainability* **2021**, *13*, 1828. [CrossRef]
15. Derjani-Bayeh, S.; Olivera-Fuentes, C. Winds Are from Venus, Mountains Are from Mars: Science Fiction in Chemical Engineering Education. *Educ. Chem. Eng.* **2011**, *6*, e103–e113. [CrossRef]
16. Waldron, R.D.; Criswell, D.R.; Erstfeld, T.E. Role of Chemical Engineering in Space Manufacturing. *Chem. Eng.* **1979**, *86*, 80–94.
17. Borman, S. Chemical Engineering: Ready for Space. *Chem. Eng. News Arch.* **1991**, *69*, 16–17. [CrossRef]
18. Hecht, M.; Hoffman, J.; Rapp, D.; McClean, J.; SooHoo, J.; Schaefer, R.; Aboobaker, A.; Mellstrom, J.; Hartvigsen, J.; Meyen, F.; et al. Mars Oxygen ISRU Experiment (MOXIE). *Space Sci. Rev.* **2021**, *217*, 9. [CrossRef]
19. Nasr, M.; Hoffman, J.; Masson-Zwaan, T.; Rapp, D.; Newman, D. A Policy Framework for Sustainable and Equitable Space Resource Utilization. *Soc. Sci. Res. Netw.* **2023**. [CrossRef]
20. Environmental Control and Life Support System, NASA. Available online: https://www.nasa.gov/centers/marshall/history/eclss.html (accessed on 15 April 2023).

21. Akay, Ö.; Bashkatov, A.; Coy, E.; Eckert, K.; Einarsrud, K.E.; Friedrich, A.; Kimmel, B.; Loos, S.; Mutschke, G.; Röntzsch, L.; et al. Electrolysis in Reduced Gravitational Environments: Current Research Perspectives and Future Applications. *NPJ Microgravity* **2022**, *8*, 56. [CrossRef]
22. Starr, S.O.; Muscatello, A.C. Mars in Situ Resource Utilization: A Review. *Planet. Space Sci.* **2020**, *182*, 104824.
23. Ash, R.L.; Dowler, W.L.; Varsi, G. Feasibility of Rocket Propellant Production on Mars. *Acta Astronaut.* **1978**, *5*, 705–724.
24. Bennett, N.J.; Ellender, D.; Dempster, A.G. Commercial Viability of Lunar In-Situ Resource Utilization (ISRU). *Planet. Space Sci.* **2020**, *182*, 104842. [CrossRef]
25. Rapp, D. *Use of Extraterrestrial Resources for Human Space Missions to Moon or Mars*, 2nd ed.; Springer: Berlin/Heidelberg, Germany, 2018.
26. Inglezakis, V.J. Extraterrestrial Environment. In *Environment and Development: Basic Principles, Human Activities, and Environmental Implications*; Elsevier: Amsterdam, The Netherlands, 2016; pp. 453–498, ISBN 9780444627339.
27. Petrovic, J.J. Review Mechanical Properties of Meteorites and Their Constituents. *J. Mater. Sci.* **2001**, *36*, 1579–1583. [CrossRef]
28. Zorpas, A.A. Strategy Development in the Framework of Waste Management. *Sci. Total Environ.* **2020**, *716*, 137088. [CrossRef]
29. Hadler, K.; Martin, D.J.P.; Carpenter, J.; Cilliers, J.J.; Morse, A.; Starr, S.; Rasera, J.N.; Seweryn, K.; Reiss, P.; Meurisse, A. A Universal Framework for Space Resource Utilisation (SRU). *Planet. Space Sci.* **2020**, *182*, 104811. [CrossRef]
30. Vogt, C.; Monai, M.; Kramer, G.J.; Weckhuysen, B.M. The Renaissance of the Sabatier Reaction and Its Applications on Earth and in Space. *Nat. Catal.* **2019**, *2*, 188–197. [CrossRef]
31. Maggi, F.; Tang, F.H.M.; Pallud, C.; Gu, C. A Urine-Fuelled Soil-Based Bioregenerative Life Support System for Long-Term and Long-Distance Manned Space Missions. *Life Sci. Space Res.* **2018**, *17*, 1–14. [CrossRef]
32. Sani, R.L.; Koster, J.N. (Eds.) *Low-Gravity Fluid Dynamics and Transport Phenomena*; American Institute of Aeronautics and Astronautics, Inc.: Reston, VA, USA, 1990.
33. Allen, D.T.; Pettit, D.R. Unit Operations in Microgravity. *Chem. Eng. Educ.* **1987**, *21*, 190–218.
34. Nijhuis, J.; Schmidt, S.; Tran, N.N.; Hessel, V. Microfluidics and Macrofluidics in Space: ISS-Proven Fluidic Transport and Handling Concepts. *Front. Space Technol.* **2022**, *2*, 779696. [CrossRef]
35. Wu, X.; Loraine, G.; Hsiao, C.-T.; Chahine, G.L. Development of a Passive Phase Separator for Space and Earth Applications. *Sep. Purif. Technol.* **2017**, *189*, 229–237. [CrossRef]
36. Hessel, V.; Sarafraz, M.M.; Tran, N.N. The Resource Gateway: Microfluidics and Requirements Engineering for Sustainable Space Systems. *Chem. Eng. Sci.* **2020**, *225*, 115774. [CrossRef]
37. Hessel, V.; Stoudemire, J.; Miyamoto, H.; Fisk, I.D. (Eds.) *In-Space Manufacturing and Resources*; Wiley: Hoboken, NJ, USA, 2022; ISBN 9783527348534.
38. Darr, S.; Dong, J.; Glikin, N.; Hartwig, J.; Majumdar, A.; Leclair, A.; Chung, J. The Effect of Reduced Gravity on Cryogenic Nitrogen Boiling and Pipe Chilldown. *NPJ Microgravity* **2016**, *2*, 16033. [CrossRef]
39. Plutschack, M.B.; Pieber, B.; Gilmore, K.; Seeberger, P.H. The Hitchhiker's Guide to Flow Chemistry. *Chem. Rev.* **2017**, *117*, 11796–11893. [CrossRef]
40. Dimian, A.C.; Bildea, C.S.; Kiss, A.A. *Integrated Design and Simulation of Chemical Processes*; Elsevier: Amsterdam, The Netherlands, 2014.
41. Jorisch, W. (Ed.) *Vacuum Technology in the Chemical Industry*; Wiley-VCH Verlag GmbH & Co. KGaA: Weinheim, Germany, 2014; ISBN 9783527653898.
42. Knapp, H. Chemical Engineering at Low Temperatures. *Int. J. Refrig.* **1988**, *11*, 352–355. [CrossRef]
43. Sergeev, G.B.; Batyuk, V.A. *Cryochemistry*, 2nd ed.; Mir Publishers: Moscow, Russia, 1986.
44. Widicus Weaver, S.L. Virtual Issue on Astrochemistry: From the Chemical Laboratory to the Stars. *J. Phys. Chem. A* **2019**, *123*. [CrossRef] [PubMed]
45. Papamichael, I.; Chatziparaskeva, G.; Pedreño, J.N.; Voukkali, I.; Almendro Candel, M.B.; Zorpas, A.A. Building a New Mind Set in Tomorrow Fashion Development through Circular Strategy Models in the Framework of Waste Management. *Curr. Opin. Green Sustain. Chem.* **2022**, *36*, 100638. [CrossRef]
46. Mahdi Najafpour, M. *Artificial Photosynthesis*; InTech: Rijeka, Croatia, 2012; ISBN 9789533079660.
47. Hann, E.C.; Overa, S.; Harland-Dunaway, M.; Narvaez, A.F.; Le, D.N.; Orozco-Cárdenas, M.L.; Jiao, F.; Jinkerson, R.E. A Hybrid Inorganic–Biological Artificial Photosynthesis System for Energy-Efficient Food Production. *Nat. Food* **2022**, *3*, 461–471. [CrossRef]
48. Our Artemis Crew, NASA. Available online: https://www.nasa.gov/specials/artemis-ii/ (accessed on 15 March 2023).
49. Gillett, J.E. Chemical Engineering Education in the next Century. *Chem. Eng. Technol.* **2001**, *24*, 561–570. [CrossRef]
50. D'Adamo, I.; Gastaldi, M. Perspectives and Challenges on Sustainability: Drivers, Opportunities and Policy Implications in Universities. *Sustainability* **2023**, *15*, 3564. [CrossRef]
51. Doukanari, E.; Ktoridou, D.; Efthymiou, L.; Epaminonda, E. The Quest for Sustainable Teaching Praxis: Opportunities and Challenges of Multidisciplinary and Multicultural Teamwork. *Sustainability* **2021**, *13*, 7210. [CrossRef]

52. Ali, S.M.; Appolloni, A.; Cavallaro, F.; D'Adamo, I.; Di Vaio, A.; Ferella, F.; Gastaldi, M.; Ikram, M.; Kumar, N.M.; Martin, M.A.; et al. Development Goals towards Sustainability. *Sustainability* **2023**, *15*, 9443. [CrossRef]
53. Efthymiou, L.; Kulshrestha, A.; Kulshrestha, S. A Study on Sustainability and ESG in the Service Sector in India: Benefits, Challenges, and Future Implications. *Adm. Sci.* **2023**, *13*, 165. [CrossRef]

Disclaimer/Publisher's Note: The statements, opinions and data contained in all publications are solely those of the individual author(s) and contributor(s) and not of MDPI and/or the editor(s). MDPI and/or the editor(s) disclaim responsibility for any injury to people or property resulting from any ideas, methods, instructions or products referred to in the content.

sustainability

MDPI

Article

Implications of Land Grabbing and Resource Curse for Sustainable Development Goal 2 in Africa: Can Globalization Be Blamed?

Samuel Chukwudi Agunyai * and Lere Amusan

Department of Political Studies and International Relations, Mafikeng Campus, North West University, Mmbatho 2790, South Africa; lereamusan@nwu.ac.za
* Correspondence: agunyais@oauife.edu.ng

Abstract: Globalization, as described by Joseph Stiglitz in his books *Globalization and its Discontents* and *Making Globalization Work*, draws on both pains and gains. These two seemingly incompatible positions, where globalization is used as a platform to partner or collaborate with other countries by grabbing lands for sustainable developmental initiatives such as the attainment of SDG 2, but ends up dispossessing the host communities of their lands, form the crux of this paper. Although not all land grabbing is illegal, especially if lands are leased within the confines of a country's land tenure laws, the reality in some African countries shows that lands are grabbed without following land tenure laws. This partly limits the capacity of African countries to effectively control and monitor the activities of foreign land grabbers or investors on leased lands. This loophole in the governance of arable lands in Africa has made many foreign partners use Africa's arable lands for their own benefit at the expense of Africa's food sovereignty initiative. It has partly made Africa appear to be a resource-cursed region, where it can hardly feed its population despite its global partnerships and huge land resources. Drawing on systematic desktop reviews of the literature, this study asks if globalization is contributing to Africa's hunger index and resource curse. The findings expand the discussion on how Africa is still not able to feed its population and end hunger, despite the potential offered by globalization. It suggests approaches through which Africa can optimize globalization in ways that support determined efforts at ending hunger in Africa.

Keywords: land grabbing; resource curse; SDGs; agroecology; complex interdependency theory

Citation: Agunyai, S.C.; Amusan, L. Implications of Land Grabbing and Resource Curse for Sustainable Development Goal 2 in Africa: Can Globalization Be Blamed?. *Sustainability* **2023**, *15*, 10845. https://doi.org/10.3390/su151410845

Academic Editors: Idiano D'Adamo, Massimo Gastaldi and Nallapaneni Manoj Kumar

Received: 1 May 2023
Revised: 7 July 2023
Accepted: 7 July 2023
Published: 11 July 2023

Copyright: © 2023 by the authors. Licensee MDPI, Basel, Switzerland. This article is an open access article distributed under the terms and conditions of the Creative Commons Attribution (CC BY) license (https://creativecommons.org/licenses/by/4.0/).

1. Introduction

Sustainable Development Goals (SDGs) are the greatest challenge of the current period and need to be pursued with all countries making a pragmatic contribution (Ali et al., 2023). Globalization, which is the interdependence of the world's populations, economies, or cultures, is driven by cross-border flows of people, investment, trade in goods and services, and information unquestionably has advantages and disadvantages [1]. While its benefits include trade liberalization, access to information, a vast flow of capital goods and services, technology transfer, and access to foreign aid, it is criticized for promoting unhealthy competition among states, exploitation of resources and labor, imbalanced trade, and the loss of domestic jobs [1]. Through globalization, nations have lost control of their resources, thus leading to what is known as the resource curse. It arises when a country or continent is not fully benefiting from its huge resource endowment [2]. Similarly, globalization stimulates land grabbing, which is the acquisition of large-scale land by an individual or entity, public or private, domestic or foreign, following laid-down land regulations. Land is usually allocated through ownership, lease, concession, quota, or general power. It could be legal or illegal for purposes of resource control, commodification, and speculation [3]. However, the motive behind international land grabbing has provoked studies that question the utility value of globalization in Africa. Land grabs, whether legal or illegal, somehow occur at the expense of agroecology, stewardship, subsistence farmers, food sovereignty, and human rights [4]. It has partly heightened poverty and hunger through the very thing (globalization, via global partnerships and collaborations) that the world believes will aid the achievement of the end to hunger. This is true even though the world has experienced tremendous improvement and civilization in technology, migration, productivity in food production, and innovations in global connectivity [5]. However, through globalization, developed countries such as China, the United States, the United Kingdom, France, and a host of others have maintained a stranglehold on Africa and its resources, including large-scale arable lands, as they promise to support the continent through partnerships, financial assistance, loans, humanitarian services, technology transfers, and other veiled collaborations [6]. Although these interventions are aided by the spirit of globalization, their consequences outweigh the benefits accrued by Africa [7]

Today, despite huge deposits of raw materials and natural resources in Africa as well as globalization spreading to the continent, Africa continues to struggle and lag behind other continents in critical indicators of development [8]. Africa still suffers greatly from poverty, hunger, and malnutrition. Sustainable Development Goal 2 (SDG 2) was proposed as part of globalization policy to end hunger by 2030 [9]. However, Africa seems to lack the capacity to achieve this goal, despite the description of Africa as a raw material continent [10,11]. The inability of Africa to feed its populations and end hunger despite the huge availability of arable landed resources, and the deliberate leasing or allocation of lands to foreign partners to boost food production in Africa raises some serious questions, such as: why is Africa still faced with the problem of hunger? Could Africa's hunger index rate be connected to the fallouts of globalization? Why is Africa not able to feed its population despite collaborations, collective action on climate change, and partnerships with developed countries? Can the resource curse in some African countries be linked to globalization? These questions are better addressed by focusing on international land grabbing, especially those specifically allocated for large-scale food production. Since the attainment of SDG 2 requires partnerships in terms of the transfer of technology that can combat harsh climatic threats or the leasing of lands to grow large-scale food production, this can enhance Africa's capacity to feed its population. It is therefore important to examine big international land grabbing drawn from different ethnic groups in host communities. One other reason for this choice is that globalization drives international land grabs, hence, the focus of this article.

All the foregoing questions are well documented in the SDGs, which are an element of globalization. Collective action on climate to boost food security and protect lives, the influx of multinational corporations (MNCs), and corporate social responsibility (CSR) to partner and engage in the transformation of African communities and assist in the best possible ways to boost food security in Africa, among others, are essential features of globalization [12,13]. However, the reality in most African countries shows that many developed countries, international organizations, and international non-governmental organizations, as well as global financial institutions or partners, have deployed the deceptive strategy of land grabbing, foreign aid, and assistance to impoverished Africa [14,15] The displacement of 15,000 Ugandans described in [16] to pave way for the planting of timber to mitigate climate change problems by New Forests, a British company, under the auspices of the United Nations Reduction of Emissions from Deforestation and Forest Degradation (REDD+) program is one amongst numerous examples of land grabs and displacement problems in Africa [11,17]. It raises the question as to whether the same dispossession of land and displacement can take place in the US or China [18], and the answer is that it could not happen, even though the US itself was created due to land dispossession [19].

It is the contention of this article that, similar to the era of colonialism, African countries are forcefully taken over or seized by foreign imperialists. Although not with force in this era, Africa's arable lands are subtly under the control of foreign investors or partners due to weak regulations on land tenure systems in some African states. It portrays an unlawful allocation of lands to foreign investors at the expense of local peasant farmers, human rights, and stewardship. In the name of global partnerships for agricultural development, food sovereignty in Africa is gradually waning due to the imposition of obnoxious food policies. These policies include genetically modified organisms (GMOs), mono-cropping as against the agro-ecological model, soulless capitalism, and the granting of low-interest credit or loans, among other unhealthy practices by Africa's global partners [20,21]. African lands have been deceptively used by these global giants to further under develop and annex Africa [22–24]. China's interest in Africa, especially in sectors such as agriculture, technology, infrastructure, and finance, aptly buttressed its resource control motive in some African countries [25,26]. Similarly, the West, especially the United States, through its dominance in organizations such as the UN, WHO, World Bank, FAO, IMF, and WTO, has imposed Western-oriented policies that have consistently made Africa's economy dependent on the West [10,27]. The policy of collective action on climate change may look good on paper, but many African countries lack the capacity to procure technology for the timely detection and prevention of climate change impacts, hence Africa's current dependent status [11,28]. The policy outcomes or goals of these Western-oriented policies do not always 'trickle down' to local communities in Africa [11,29]. For example, the erection of sophisticated machines to control, detect, and mitigate harsh climatic conditions often leads to the displacement of indigenous people and the dispossession of their lands with little or no compensation [30]. This does not only affect the capacity of peasant farmers to produce food but also creates tendencies toward hunger and poverty. Hunger is imminent when local farmers are dispossessed of their lands by foreign farmers in their host local farming communities. This is because it is on record that most global giants, having acquired land, used the same land to create more job opportunities and feed their own populations at the expense of indigenous African people [31].

The case of Africa as the most vulnerable continent, despite its lowest contribution to greenhouse emissions, heightens the already existing poverty and hunger index in Africa [32]. This is because, in addition to the inherent climatic problems suffered by Africa, the displacement of Africans from their lands in order to combat climate change has further impoverished them [33]. The large scale of arable lands, the inability of African farmers to predict uncertainties over future supplies of food, and increasing demand for food and biofuel drive foreign companies, land speculators or investors, and superpowers to Africa in search of lands, which some of them have subtly used to relegate Africa's food sovereignty and undermine peasant farmers [34,35]. Heightened international interest in the acquisition and leasing of land adversely affects food production at local and national levels [36]. Land grabs, especially those allocated without due process, often lead to the displacement of people and the dispossession of their lands. As people are displaced or dispossessed of their lands, it becomes very easy to grab land, which in turn limits the capacity of local peasant farmers to productively produce food [21,37]. How could local peasant farmers engage in farming after the dispossession of their lands? This question further points to the fact that land grabs are strongly related to the resource curse, which is the inability of Africa to fully benefit from the abundance of its land resources, raw materials, and global agricultural partnerships. Africa is currently not food sovereign and cannot adequately feed its populations despite being labeled the raw materials continent. Hunger still exists in Africa, despite the fact that it is home to valuable raw materials or natural resources [38]. This brings to the fore the importance of this article, which tries to examine how land grabs (legal acquisition of land for development, speculation, and commodification) and resource curses (i.e., hunger amid massive abundances of raw materials that can be used to grow food) undermine Africa's quest to achieve SDG 2. Zero hunger is the dream and goal of the world, including Africa. The extent to which this goal is achieved through effective utilization of the potential of globalization remains the focus of this article. Indeed, the positive and negative implications of globalization in Africa have received attention. However, empirical evidence is scarce on how globalization through land grabbing heightens hunger on an already hunger-ravaged continent. Research in this direction will guide the design of an Afrocentric policy intervention to resolve the problem of land grabbing and the resource curse, which, in the long run, will boost food security in Africa. The objective of this article is to assess this interdependent initiative by enhancing collaborative efforts and global partnerships to promote vibrant and mutually beneficial food security in Africa. The findings are significant for theory and policy. Theoretically, it expands the discussion on the dimensions of land grabs and resource curses and how they undermine the attainment of SDG 2. The findings also suggest a means by which African countries can optimize indigenous knowledge systems and technology to enhance strategic efforts at improving food security in Africa. More importantly, it expands knowledge on the relevance of integrated food security and sustainability in Africa.

Integrated food security requires an innovative multi-partner initiative to improve food security, nutrition analysis, and decision-making. This shows that the sustainability of food security in Africa depends largely on contributions from other countries [39]. The attainment of SDG 2, that is, zero hunger in Africa and other emerging economies, also depends on the ability of Africa to engage in multi-level partner initiatives that can promote its participation in the analysis and decision-making process to improve food security and nutrition [40]. One of Africa's important engagements in multi-partner initiatives, which further aids integrated food security, is that Africa, through these initiatives, can produce food that can sustain its teeming population [41]. In other words, this initiative can enhance the sustainability of Africa and other emerging economies towards ending hunger and malnutrition. Thus, Africa needs to be more proactive in innovative multi-partner decisions with other countries to boost its food security [42].

Following the introduction section, the article utilizes the social constructivist as a theoretical construct in explaining food security and SDGs in Africa. This article appraises the patterns and nature of land grabs, the resource curse, and their implications for food security in Africa. The next section discusses the impact of globalization on Africa's hunger index rate and the resource curse. It concludes with discussions on issues of indigenous knowledge systems and technology to boost food security in Africa. To achieve all these goals, the article utilizes a systematic research method that was sourced through extensive desktop reviews of extant studies that discussed issues in detail relating to the subject matter. The rationale for the choice of this method is based on the fact that issues of globalization, land grabbing, resource curse, and SDGs, especially those empirically analyzed, are readily available in the literature, hence the need for thorough systematic reviews. It is hoped that a systematic review method provides the opportunity for a detailed analysis, which is the focus of this paper.

2. Social Constructivism, Food Security, and SDG 2

Social constructivism advances the knowledge that theories are constructed due to the social development of an environment. Constructivism posits that the construction of knowledge is the product of social interaction, interpretation, and understanding. It posits that there is no perfect theory that is not fallible [43]. It argues that the social dimension of human relations is germane to the much-touted fourth industrial revolution (4IR) and food security in Africa. It allows for the training of people, especially women, who are known to be custodians of organic seeds for sustainable food production [44]. It is a variety of cognitive constructivism that emphasizes the collaborative nature of people in an environment. It argues that the development of society largely depends on social interactions and relations among those living in the environment. To interact in a locality, native intelligence or indigenous knowledge is required. It is needed to boost food security in Africa, and the functionality of the technology largely depends on human relations and interactions, which are largely based on local indigenous knowledge. All components of food security—availability, accessibility, affordability, and quality—are interdependent and can only be developed and sustained through healthy human relations and social interactions in an environment.

The theory provides alternative ways to conceptualize food security among states in the global system. It sees food security as a social construction, which makes it easier to understand how people are mobilized to promote food production, availability, accessibility, affordability, and utilization. In the analysis of food security, understanding 'reality' as a social construction means that identities, national interests, and security threats are socially constructed through interstate interaction and domestic political and strategic cultures [45]. Social constructivism has emphasized the role of international organizations as sites of social interaction that can produce changes in actors' identities and interests. Walters [45] specifies how norms, not just material power, provide the basis for an intersubjective understanding of social purpose among states. This understanding was shown to be critical to the formation and durability of the post-war international order. Social constructivism advocates for models and standards through mutual and cooperative interactions that can influence states' decisions to change their attitude against the use of force and coercion against other states in the international system. This explains why land grabs and resource curses are social constructions emanating from the environment. Social constructivism believes that these social constructs can be amicably addressed by socialization and diffusion, which are largely based on interactions and human relations.

Since the issue of international land grabbing is mostly perpetrated by actors such as international organizations and states, social constructivism advocates for authority, that is, "the ability of one actor to use institutional and discursive resources to induce deference from others" [46,47]. It underscores that international organizations (IOs) are not just servants of states; authority provides IOs with self-sufficiency regarding states and non-state actors. It is also opined that IOs sometimes accomplish more than states can accomplish; they adequately transmute broad mandates into specific ways of acting in the world that change the behavior of state and non-state actors and use knowledge and authority to regulate the social world [46]. Authority itself is understood to be a social construction that is deployed to check the excesses of some states or IOs in the international system.

This theory adequately explains the implications of land grabbing for zero hunger, as it stipulates that there should be mutual social interactions and human relations in the form of the sharing of ideas and knowledge on how to adequately tackle the issues of land grabbing and the resource curse that has been bedeviling the African continent. On the global scene, international organizations and other actors are expected to amicably discuss the issue of land grabbing, which somehow has given more foreign partners more control over resources and raw materials in Africa. This requires concerted efforts for sustainable solutions. Inclusion and consultation with the indigenous people over the use of their lands are essential for the attainment of these solutions. This might have led to the conclusion that human relations in terms of the sharing of indigenous knowledge and technology are required for sustainable food security in Africa [48,49]. Africa's quest for food sovereignty could be better achieved through global partnership and collaborative efforts aimed at considering the plight of peasant farmers, inclusiveness, and respect for the human rights of host communities.

The achievement of SDG 2 is well captured in the core principles of social constructivism. Zero hunger is a construct that is envisaged to be achieved through collaborations and mutual social interactions among citizens in society. Joint collaborations and interventions between the government and the people to address issues of poverty and hunger will serve as spurs for general development and food security [37]. Social interactions and the dynamism of constant change in the international system largely determine the quality of partnerships and collaborations among states toward the call for the eradication of hunger [48]. To end hunger, Africa's capacity to produce quality food depends on human relations and interactions. These social interactions should include women, who are more active in food production [49,50]. The empowerment of women is a strong requirement for food security in Africa. This is because women are more reliable custodians of organic seeds for sustainable food production [51]. Social interactions that include women, according to social constructivism, strengthen their productivity in food production. This is due to the fact that women are good managers of credit facilities to boost food security [52].

Thus, ensuring more access for women to relevant inputs will promote food security and empower the landless majority in the state with a greater focus on women and girls [53].

3. Methodology

A qualitative research method forms the basis of data collection in this article. It draws on systematic desktop reviews of key issues of land grabbing and resource curse in Africa. It has been argued that a systematic approach is appropriate for studies that uniquely provide detailed summaries of reviews in response to research questions [54]. Since this study draws on research questions (such as could globalization be responsible for Africa's high hunger index and could it be responsible for Africa's resource curse?), thus, it is more convenient to adopt a systematic review method. Its validation hinges on the fact that a similar approach has been used in studies and research conducted in education, engineering, humanities, social sciences, tropical medicine, and health [55–57]. Evidence shows most studies that used this approach provided answers to research questions in the form of findings [58]. Data were also drawn from empirical studies, verifiable government documents, online materials, periodicals, newspapers, journals, and policy notes for comprehensive reviews and discussions of findings or answers to the research questions. These data were analyzed using thematic descriptive content analysis.

4. Land Grabbing and Resource Curse: The Misery of Hunger in Africa

Land grabbing and the resource curse are not new in Africa. Africa has long continually suffered from these contemporary issues, which partly contribute to the misery of hunger in Africa [11]. Land grabbing is simply the practice of large-scale land acquisition by individuals, entities, governments, international organizations, and foreigners for speculation, investment, commodification, and resource control purposes [3]. It is the lawful acquisition of large-scale lands, especially those that follow the laid-down procedural means or approach of host countries. Evidence shows that some developed nations, foreign investors, and governments, in a bid to partner with Africa and end hunger or combat the impacts of climate change, have grabbed land for food production and biofuel needs [59]. Land grabbing as the right or power to land resources or large quantities of land acquired in a bid to respond to the food security crisis adequately, combat the impacts of climate change, and meet financial exigencies [60]. Although this is a global practice, the way it is acquired in Africa seems to be restrictive of Africa's quest for food sovereignty, agroecology, and human rights. This is because it has governance problems, such as deficient information or data on large-scale lands and the exclusion of host communities from the land allocation decision-making process, among others [21]. It tends to favor more foreign investors, governments, or states at the expense of stewardship and peasant farmers in home countries.

Peasant farmers, who largely depend on land as a means of livelihood, are more affected by unlawful land grabs. It also impacts herders, who are forest-dependent [61]. African governments with weak regulations on land tenure systems often violate existing communal and customary land tenure laws to re-allocate land to foreign governments, speculative investors, and organizations for agricultural development [11]. This inordinate practice by some African governments has partly contributed to the forceful acquisition or dispossession of land from the real owners under the guise that such land will be used for developmental purposes [21]. However, the reality in most African states reveals that land grabbing for developmental purposes hardly translates to the development of host communities where the lands were originally acquired [61]. It is the poor, displaced, and marginalized communities that lose land in Africa, mainly because they lack the power and influence to compete with foreign land speculators or investors. It is imperative to note that despite Africa's endowment of land resources and raw materials, hunger and malnutrition still persist on the continent. This is otherwise known as the "resource curse". A resource curse is an inability to fully benefit from the abundance of resources endowed or available in a country or state. Africa, despite its huge land resources and raw materials, as well as its initiative to lease land for developmental purposes, still cannot feed its populations [37,62]. It is a resource curse for Africa if, after partnering with foreign land speculators or investors through the leasing of its landed resources and huge raw materials, its people still suffer from hunger. One in five people, or 21% of the African population, still suffers from hunger, while 282 million are undernourished [63]. This negative outcome of Africa's land leasing or grabbing initiative has been blamed on a lack of accurate data on the scale, trends, geographical distribution of land, and actors in large-scale land deals [61].

Land deals, especially those in large quantities, are shrouded in secrecy and not transparent [21]. Variations in timescales, methodologies, and criteria for land deals make it very difficult to compare information about land grabbing in Africa [64]. From the foregoing, it can be deduced that the inability of African leaders to follow procedural and approved regulations on land tenure or the unlawful leasing of land to foreign partners is partly the reason for the resource curse in Africa. It shows that there is a strong relationship between land grabbing and the resource curse in Africa. African governments that jettisoned lawful allocation of land due to weak land tenure regulations or that allocated lands without due process because they wanted to balance their budgets would certainly not benefit fully from the proceeds of investment on the leased or allocated lands [21,61]. This is because once foreign land speculators or investors bribe their way to acquire land, as is currently performed in some African countries, it will be very difficult for these African governments to control and audit investment by these land speculators or grabbers [11]. This is because regulating agencies, at the point of land allocation, abandoned the land tenure laws [11,65]. Moreover, since one of the aims of land speculators or grabbers is to attain access to resources in host communities through collaborative efforts, they may likely benefit more from the proceeds of investment in the landed resources than African countries [21]. Thus, it can be interpreted to mean that while the abandonment of land tenure regulations by some African governments during land allocations has retarded Africa, it has also attracted more foreign land speculators to Africa [66]. Africa is home to uncultivated lands, and this could have informed the conclusion that uncultivated land with the prospect of boosting significant food outcomes attracts foreign investors [67]. Despite the fact that Africa accounts for about 60% of the world's arable land, most African countries hardly achieve 25% of their potential yield [68]. This may have partly contributed to an increase in China's investment in the agricultural sector in Africa. Land grabs raise concerns about corruption, large-scale resettlement of populations, and even the recolonization of Africa. It is imperative to note that some developed countries persistently hide under the pretense of climate mitigation action, corporate social responsibility (CSR), technology transfer, collaborative efforts in food security, foreign aid, and financial assistance to unduly exploit and occupy lands in Africa.

The case of Chinese loans and partnerships in agricultural food production in some African countries, where the latter default on loan repayment, often leads to the forfeiture of national assets. This may include the loss of cash crop plantations, where food meant for African host countries is shipped to China. This succinctly explains the problem of land grabbing in Africa [26,69,70]. It is the largest bilateral lender for public sector loans across the African continent [26,71,72]. Despite this large economic footprint, there is often very little information in the public domain on the specifics of its lending and investments [73,74]. Today, through its loan policy in Africa, China has been able to provide business and employment opportunities to Chinese citizens and contractors working in Africa [71]. Similarly, because China, on the one hand, often imposes most of these loans on infrastructural development in Africa, African governments, on the other hand, provide land for the construction of these infrastructure projects [26]. This imposition has benefited China more than host African countries because of her "going-out strategy", which tends to give more priority to Chinese companies as contractors or investors for any projects, including agricultural projects, financed by Chinese loans [26]. These companies, rather than giving job opportunities to citizens of host African countries, prefer giving job opportunities to Chinese citizens in host African countries. This describes a situation where

China gives out money with one hand and collects back all the proceeds of that money with another hand [26,75]. It is similar to a situation where China agreed to assist or partner with some African countries to boost food production through the recruitment of local peasant farmers but went behind the host nation's back by recruiting Chinese farmers and still transferred the harvest back to China [76]. This does not only undermine stewardship, peasant farmers, and human rights; it also raises the question of whether Africa wants to feed China or itself. This was well captured in the forms of large-scale investments, often involving transnational purchases of land to create jobs and bring new technology to the sector, and small-holder farming that neglects local rights, extracts short-term profits at the cost of long-term sustainability, ignores social standards, and fosters corruption on a large scale [68].

However, on the procurement of land for large investments in the creation of jobs, evidence such as that shows discrepancies between interest in large-scale investment in African agriculture and actual investment [10,68]. China has been criticized for grabbing land cheaply from Africa, using underpaid Africans as laborers, and producing food for its citizens back home at the expense of Africa [26,73,77]. This has implications for hunger as African countries, especially those that are politically sensitive and food insecure, will lose control over their own food supplies when they need them most. It also has severe implications for new colonialism, agrarian colonialism, or eco-colonialism. It is on record that countries such as Ghana, Ethiopia, Mali, Tanzania, Kenya, and Sudan have grossly engaged in the leasing of millions of hectares of their lands for biofuels and agricultural production; this could be similar to "new colonialism or agrarian colonialism" currently spreading to African countries [68,78]. Land acquisition in Africa has been unprecedented, and there is a huge gap between declaring an intention to lease land and the actual cultivation of the land for food production. Investment in African lands for food production that will benefit the population of the investing countries is not new; what is new is the rate of land deals that have been transacted. Land deals in Africa have been enormous, such that local farmers and investors hardly have the opportunity to procure land for food cultivation. This is even worse for women, who are major contributors to food production in Africa [9,37,79,80]. Another unprecedented thing is the issue of what the land is used for and the beneficiary. Africa's inability to address gaps in the effective management of leased or acquired lands for developmental purposes creates the tendency to have huge deposits of resources that cannot necessarily translate into wealth or freedom from hunger and poverty. This is similar to a resource curse because, despite Africa's huge land resources and abundance of raw materials, it can hardly boast of fully reaping positive outcomes from the resources or raw materials. This depicts a situation where those who own the resources (land) cheaply sell their land to foreigners or investors and work as laborers on their sold land to earn a livelihood. As foreign countries (US or China) smile to the bank to receive huge dividends from their investment [70,81]. This adequately describes the resource curse situation in Africa.

Similarly, multinational corporations, through CSR, have illegally grabbed lands for the benefit of their home states at the expense of the host countries. These corporations hide under the pretense of bringing socio-economic development to areas of their operations by illegally acquiring lands and assets belonging to host communities in Africa, as their assistance to developing African countries is embedded in an unequal exchange arrangement [10,82–84]. It has been argued that negotiation dexterity, diplomatic power, economic preponderance, and socio-cultural upper hands of multinational cooperations or states undermine African nations' bargaining prowess during negotiations over trade, including land; this has been found to be detrimental to the economic development of the continent [10]. MNCs have also been accused of preventing the free flow of indigenous knowledge among the local people in host countries. They have been accused of hijacking and converting high-income-producing indigenous knowledge to their own through patenting, despite the global geographical recognition of such knowledge belonging to host communities [20,37]. The promise to assist the people of San in processing the hoodia plant without infringing on their right of ownership, where there was a total disregard for the convention on biological diversity (CBD) by the foreign MNCs, aptly explained the domineering nature of the negotiation process by MNCs.

MNCs are opposed to local, vibrant initiatives that could have aided the development of host communities. They disarticulate the sources of economic stability in host communities as local people are forced to produce what they cannot eat and consume what they cannot produce. They are objects of de-capitalization and de-industrialization in Africa [10,27]. The claim that land is grabbed to create jobs in the agricultural sector is absolutely not correct and misleading, as most of these multinational cooperatives only cultivate the land for food production to feed their own populations and not those of Africa. They grabbed lands in host communities under the pretense of utilizing those lands for the improvement of infrastructure (roads, bridges, and schools, among others) and mitigation against climate change. In doing this, members of host communities are displaced and forcefully ejected from their lands, only to realize later that the main motive is to use the land to feed the population of the home countries [77,82].

Despite the endowment of Africa with raw materials and natural resources, it lacks the capacity to process these resources or raw materials into finished products. This has made Africa rely on developed countries for the processing of its resources. This reliance is partly due to a lack of sophisticated technology and technical know-how. However, empirical evidence shows that the refined or processed resources are sold back to Africa at a very high price, thus making it appear as a resource curse to Africa, which originally owned the resources [85]. The need for Africa to have some of its resources processed into finished products overseas makes her accept stringent conditions during loan negotiations with Bretton Woods institutions and other big powers (the IMF and World Bank, China, and the US) [10,26]. The imposition of mono-cropping against an agro-ecological model of sustainable development in food security in Africa is a sign and an indicator that Africa suffers from a resource curse. It literally means that Africa's resources cannot transform it from its poverty and hunger-striking state to a more desired and sustainable level.

This explains why Africa, despite her arable lands and her description as the world's capital of raw materials, is still largely poor and faces hunger, malnutrition, and inequalities. How could a continent so endowed with arable, fertile lands with huge raw materials be this poor and affected by hunger? This question confirms the fact that central Africa has the highest extreme poverty rate of 54.8%, followed by Southern Africa at 45.1%. Rates in Western and Eastern Africa are 36.8% and 33.8%, respectively [86]. The answer to this question is not far-fetched, as globalization not only triggered massive attraction to Africa for her unprocessed raw materials and uncultivated lands, but it also led to land grabbing for massive food production for commercial purposes rather than for humanitarian sustainability or purpose. This might have led to the conclusion that the description of Africa as the food basket and raw materials of the world has made many multinational corporations (MNCs) and states scramble for African countries' resources and, more recently, grab African arable lands to produce food for the North and Middle East markets [87]. Relatedly, Amusan (2018) argued that African resources are fast becoming a curse if multinational corporations (MNCs) continue to engage in capital flight and soulless capitalism, where monies realized from food production on the grabbed African land are sold outside African markets.

5. Mapping the Impact of Globalization on Global Partnerships, Hunger-Index, and Resource Curse in Africa

Although globalization often draws praise and recognition, it has also been blamed for its negative side effects across continents, including Africa [88]. Indeed, Africa has benefited and suffered greatly from globalization. While studies have extensively examined how Africa benefited from globalization, findings about Africa's pains, especially how they contribute to Africa's hunger index rate, are scarce in the literature. It is a fact that prior to the establishment of the SDGs, hunger already existed in Africa; SDG 2 was only a global response to end hunger in the world, including Africa [89]. To achieve this goal, all countries in the world are expected to take action against hunger [90]. However, while some countries seem to have the capacity in terms of technology, knowledge, and finance for massive food production that can solve their food crisis, others, especially their African counterparts, seem to be highly deficient in modern climate control, food production technologies, and finance to meet their food production needs [89,90]. In the same vein, while Africa has a large scale of arable lands, raw materials, and viable markets for agricultural harvests or products, their counterparts on other continents seem to have limited arable lands, markets, or raw materials required for their food production [11]. These shortfalls and limited capacities across continents form the basis of global partnerships, interdependency, and interconnectedness. With partnerships, countries across continents have been able to have access to agricultural funds, technology transfer, foreign aid, financial assistance, arable lands, raw materials, and markets to enhance their food production capacity [21]. From the foregoing, it is clear that partnerships can only thrive through flows of trade, people, communication, goods and services, ideas, information, and knowledge, which are essential features of globalization. It is expected that as countries engage in global partnerships or collaborative engagements, their capacity to achieve the SDGs, including SDG 2, will be enhanced, all things being equal [10,27]. Thus, Africa is expected to have the capacity to eradicate hunger following her global partnerships and collaborations with developed nations such as China, the US, Britain, and France, among others. In addition, its large scale of arable lands and endowment of raw materials are added advantages that ought to have improved Africa's capacity to produce food that can feed its populations [11,21]. However, despite her global partnerships, availability of arable lands, and raw materials, Africa still has a high hunger index rate, lacks the capacity to feed her populations and suffers from resource curse syndrome [91,92]. These deplorable conditions have been blamed on the fallout of globalization, which is manifested in the way global powers such as China have transgressed constituted authorities and undermined land allocation regulations in some African states with weak budgetary allocations to agricultural production [93]. It is a fact that the rising global demand for biofuel and food has shifted the focus of some developed nations to Africa, where raw materials are readily available [65]. As they scrambled for African raw materials and lands for food plantations, they tended to undermine the interests of peasant farmers and impugned human rights in Africa (Holmen, 2015). Although land grabbing is legal and commonly practiced across the world, its enforcement in some African countries limits Africa's food sovereignty initiative [56,94]. China's developmental drive in Africa indicates that, through globalization, most foreign land speculators or investors grab land in Africa. However, the motive for the acquisition of these lands appears to purportedly promote Africa's food production capacity, whereas, in reality, it is to provide job opportunities and promote the welfare of their citizens residing in host countries. China's "Going-out policy" tends to retain Chinese citizens as workers or contractors in any project financed by loans provided by China [26,73]. With this policy, peasant farmers in Africa would be denied the opportunity to grow food because the majority of them have been dispossessed of their lands [10]. Although these lands were allocated on the basis of positive agricultural outcomes such as the training of local farmers on how to operate agricultural technologies, high-yielding seedlings, irrigation, detecting a good and conducive planting season, climate adaptability, marketing, and storage of farm harvests, among others, that can be rewarding and beneficial to African countries food production capacity [11]. It is a fact that Africa is not fully benefiting from its allocated landed resources because the majority of its global partners often do not report all the gains or profits on agricultural investment. In the same vein, evidence indicated massive capital flights (transfers of profits on investment back home) overseas by these global partners [11,26,27]. In some severe cases, raw materials from Africa are transferred overseas by these global partners for the industrialization of their countries. In the colonial era, palm oil from southern Nigeria was used to fuel Britain's industrial revolution, just as French farmers resettled in the farmlands of local communities without any meaningful compensation in Algeria [11]. Today, Africa's natural resources, especially arable lands and other raw materials, have attracted global partners that have offered to assist Africa in ending hunger, but their assistance has been shrouded in secrecy and largely skewed to undermine the rights of peasant farmers, take control of resources in host communities, amass wealth at the expense of host communities, and limit food sovereignty in Africa [27,45]. These are the reasons why Africa still has a high hunger index rate, is unable to feed its populations, and suffers from a resource curse. The more peasant farmers are denied the opportunity to grow food, the higher the hunger index rate and the greater the inability to feed the population. This no doubt described Africa's resource curse situation, especially with the way Africa, despite her endowment of resources (large-scale arable lands and raw materials) and web of global partnerships, is still not fully benefiting from the proceeds of investment in its landed resources. This might have led to the conclusion that governance crises are the bane of underdevelopment in Africa [89]. Land grabbing in Africa has governance issues, such as a lack of information on leased lands and poor monitoring of the activities of land investors or foreign partners. These governance crises could limit the attainment of SDG 2 [10,27]. The practical effects or implications of these findings are that African governments can now understand where they get it wrong in terms of not following due process in land allocation and make swift adjustments against such practices. Similarly, the findings of this article advanced fresh knowledge on how African governments should now manage or handle their collaborative engagements or partnerships with global powers. Another practical effect of this article's findings is that they provided needed information on how African governments can optimize globalization to achieve positive SDG 2 outcomes. This will be accomplished by ensuring that all information about land grabbing is transparent, people-oriented, or focused, through effective land grab governance, periodic checks or reviews of the activities of foreign land investors or global partners, and by ensuring that Africa fully benefits from its landed resources.

6. Indigenous Knowledge Systems and Technology to Boost Food Security in Africa

Indigenous knowledge systems (IKS) are simply the application or use of native intelligence to foster or boost agricultural food production. Prior to globalization, indigenous knowledge systems were in existence and widely used by local farmers in Africa to produce mainly organic foods [95]. However, with the advent of globalization, the focus shifted from indigenous knowledge systems that specialize in mostly organic farming or food production to genetically modified organisms (GMOs) with the use of technology. The widespread nature of globalization, which eventually led to the adoption of technology in agricultural food production across the world, including Africa, is symptomatic of some factors. First, it attests to the expanding importance of technology in everyday human endeavors. Whereas technology can enhance and endanger food security, its gains offset the fallout. Second, it is reflective of an alternative method for addressing low food production that negatively impugns food security.

Although low food production is widespread on some continents, Africa's situation suffers from below-par and rushed attempts at increasing food production [21,49]. Studies on the use of crude farming equipment for food production in Africa show low food production outcomes [96]. The literature is replete with evidence of undesirable drifts in public perceptions of the capacity of crude technology in food security [97–100], indicating low food production, which could trigger food insecurity. The consequences of these actions have been stiffening food security challenges [101].

Thus, the advent of globalization, which comes with technology transfer and adoption, embodies opportunities for massive and fruitful food production that can meet the food demand of Africa. Africa's population is huge, and the indigenous technology systems cannot produce food that can meet the exponential population growth of Africa. Technology transfers and adoption or utilization are predominantly expedient in the context of Africa's exponential population growth. The ability of highly sophisticated technologies to produce the food requirements of Africa can enhance food availability, accessibility, and utilization and allow the people wider latitude to determine what to eat and produce what they can consume rather than receiving directives and impositions from the developed countries. Africa has, for a long time, been on the receiving end of out-of-use technologies from developed partners. No wonder it was concluded that inappropriate technologies are transferred to Africa, and even when sophisticated technologies are sent to Africa, local employees are hardly allowed to work on those machines [10,27]. It is also a fact that about 5% to 10% of the profit made on technology transfer to Africa is directed toward patenting payments [10]. This shows that technology transfer to Africa, which is purposely intended to boost food security, is meant to drain the pockets of Africa and enrich the MNCs and their states.

Importantly, as it was made clear, those technologies barely drive change without other factors; 'effectiveness arises from a combination of technology, organizational shifts, and policy reforms [102]'. Technologies need to be aggregated with other factors that are important for development [86]. The aspects of the reforms that could not be handled by technology require human judgment, intuition, and discretion [103]. Therefore, outcomes will be a function of the quality of decision-making resulting from the effective operation of farming technologies such as tractors, irrigation machines, food modification technologies or machines, robots, and agricultural drones, among others. For massive food production through technology to take firm root in Africa, 'it is vital that the training, mindset, and methods of food production and farming undergo fundamental change [104].' These underscore the human dimension. Local farmers need to be trained in the act of mechanization and the handling of sophisticated farming machines or technologies to boost food security in Africa.

Much progress in the use of technology platforms (such as tractors, climate detection devices, mobile phones, and drones) to boost food security depends on how farmers, especially women, can adapt them to daily operational workings such as the use of phones to alert customers or buyers of food harvests. In addition to what technology can add to food security, it is imperative that African farmers expand their infrastructure, especially in relation to the use of mobile phones and internet facilities, to showcase their harvests and attract customers. This is against a background of studies highlighting a shortage of basic technology used for food production, such as telephones, computers, and internet facilities, in rural local farming areas [75,82].

This limits the extent to which farmers are familiar with technology in their farming duties. Farmers need to deepen their use of technologies, as this can enable them to know the potency of the best technology platforms in addressing challenges, including seedlings, irrigation, planting season, prices of food, harvesting, and storage of food, and not just as platforms for increasing food production. The ability to institutionalize and strengthen the use of IKS and technology among farmers, especially women, will aid sustained efforts at boosting food security and ending hunger in Africa.

7. Limitations and Future Directions

One of the limitations of this paper was the difficulty in identifying the exact and global meaning of land grabbing. In the literature, while some scholars see it as a forceful dispossession of lands in foreign countries, especially in Africa, others see it as a legal and mutual allocation of land. In the long run, this article believes that land grabbing could be legal or illegal, especially when the due process or land tenure regulations of host African states are not followed in the process of leasing or allocating lands. Another limitation is that it places too much focus on developing countries in Africa. Other developing nations on other continents, where land grabbing is predominant, were excluded from the review. Importantly, the paper also suffers from the limitation of reviewing or examining only SDG 2. Its findings can only be applicable to countries in need of ways to improve their food production capacity to end hunger. Given these limitations, it is expected that more robust research should be conducted to compare land grabs and resource curses, with a specific focus on globalization in Africa and Asia. Similarly, research on this subject should cover more SDGs (specifically, SDGs 1, 13, and 17). Finally, research in this direction should strive to utilize mixed methods of quantitative and qualitative research design. It is hoped that in the future, the findings of the mixed-methods paper will provide a balanced view of how globalization impacts land grabbing and provokes resource curses.

8. Conclusions

This article has established that globalization in itself is not a bad thing. The real problem of globalization in Africa is the inability of some African governments to effectively optimize the potential of globalization for the development of Africa [105]. Land grabbing, if adequately enforced (in terms of following due process and good governance, which ensures that the local communities participate in land allocation decisions and have access to information about the land and transparency), has the potential to contribute to the attainment of SDG 2 [106]. Expectedly, with Africa's arable lands and lawful land grabs for food security purposes, the issue of hunger and mal-nutrition targeted by SDG 2 will be a thing of the past. The successful achievement of SDG 2 in Africa largely depends on transparent land grabs, sincere global partnerships, and effective monitoring of the activities of international land grabbers [107]. This, if adequately enforced, can improve Africa's capacity to feed its population and attain SDG 2. However, lands allocated to foreign partners have yielded little or no positive outcomes due to weak land tenure laws in Africa. This partly accounts for the increasing rate of dispossession of local farmers from their farmlands and the relegation of peasant farmers in host communities. The desire to amass wealth and earn foreign currencies from global partners is partly the reason why arable lands are allocated without due process or regulations [10,21,94]. This has implications for the loss of lands by local farmers; it also reduces their capacity to produce food and consequently aggravates Africa's hunger index rate. From the foregoing, it can be inferred that the more arable lands are unlawfully allocated at the expense of peasant farmers, human rights, and Africa's developmental motives, the lower the capacity to produce food and the greater the rate of hunger in Africa. It also has implications for Africa's inability to feed its populations. This is because most African governments fail to optimize and effectively utilize their collaborations with other developed states or global big powers (China, the US, Britain, and others) to advance sustainable development in their key sectors, including agriculture. Evidence shows that in most large-scale land deals, African governments with weak financial or budgetary allocations are usually under pressure to satisfy these big powers at the expense of their local farmers [26,27]. This satisfaction, if not properly checked, could lead these big powers to exert undue influence and control over resources in Africa. China's undue influence and control over gold mining and big farm lands and its ability to determine who works on these sites and farms succinctly explained why Africa's overdependency on foreign partners for technology, financial assistance, and foreign aid limits its capacity to feed its populations. It is also the reason why Africa is labeled a resource curse because Africa has yet to fully benefit from its landed resources and partnerships with big global powers.

To address this gap, the findings in this piece indicate the possibilities that IKS and technology provide for the effective cultivation of land and processing of resources. These are sacred to food security. Unfortunately, in Africa, governments have failed to optimize lawful land grabbing and foreign partnerships for the attainment of SDG 2 and, thus, food security. Uncultivated arable lands stimulate the massive attraction of foreign land investors and speculators to Africa, but the inordinate desire of some African governments, which made them allocate land without due process, caused many local peasant farmers to lose their lands, hence their limited capacity to produce food. Other restrictive challenges affecting food security, including the overbearing influence of MNCs, the imposition of food policies and methods, mono-cropping, a lack of technology, the marginalization of women farmers, and other related issues, as well as a lack of funds and the associated problems of GMOs, worsen food security in Africa.

Prospects provided by technology transfer are desirable for addressing some of the issues. It is established that a blend of IKS and technology used by farmers will help boost food production and, by extension, food security in Africa. Technologies and IKS enable deeper interactions, native intelligence, monitoring, and oversight leverage over massive food production. Gaps in food production and security in Africa can be alleviated by IKS, which relies on the low cost of technology to advance indigenous knowledge to boost food security efforts in Africa. Technology allows the joint production of indigenous knowledge, in which local farmers in host communities bring to the fore their native knowledge about problems affecting food production in their localities and support for ideas or efforts of partnering states or organizations, which are the determinants of food security. Continual cultivation of Africa's arable lands and processing of resources using a blend of IKS and technology will combine to boost food security in Africa.

Universities in Africa can also provide the expected change in the attainment of SDG 2, especially through the giving of assignments and research works to students, which specifically cover issues of global partnerships and food sovereignty in Africa. This can expose students to understanding how their work or assignments impact lives in society. Across continents, universities are thinking tanks that are strategically placed in society to conduct problem-solving research and cross-sectoral execution of the SDGs, providing an invaluable source of expertise in research and education on all sectors of the SDGs [108]. Universities, through the Higher Education Sustainability Initiative (HESI), can partner-with governments and communities in food production. This can be achieved by offering agricultural extension or educational services to local peasant farmers, creating awareness on how farmers can adapt and prevent harsh climatic impacts, and liaising with different sub-national governments to train and educate farmers in massive food production [109]. The departments of food technology, agricultural science, and extension services of various universities in Africa, through their community service functions, have rendered research expertise to governments and communities in the growing of cash crops that have yielded positive outcomes in food production [110]. This, if adequately and continually explored, could enhance the quick attainment of SDG 2 [111].

Conclusively, it can be deduced that while globalization stimulates Africa to partner with other nations or big powers to be able to attain SDG 2 (an end to hunger and malnutrition), globalization has also been disproportionately exploited to milk Africa of its resources, as Africa has benefited very little from allocating its arable lands to foreign land grabbers, who pretentiously cultivate Africa's lands to develop their home country and limit Africa's capacity to attain SDG 2, hence, the high hunger index in Africa.

Author Contributions: Conceptualization, S.C.A.; Methodology, S.C.A.; Software, L.A.; Validation, S.C.A.; Formal analysis, S.C.A.; Investigation, L.A.; Data curation, L.A.; Writing—original draft, S.C.A. All authors have read and agreed to the published version of the manuscript.

Funding: This research has no external funding. And the APC was funded by the University of North West, South Africa.

Institutional Review Board Statement: Not applicable.

Informed Consent Statement: Not applicable.

Data Availability Statement: Data Availability Statements are available in section "MDPI Research Data Policies" at https://www.mdpi.com/ethics.

Conflicts of Interest: The authors declare no conflict of interest.

References

1. Ali, S.M.; Appolloni, A.; Cavallaro, F.; D'Adamo, I.; Di Vaio, A.; Ferella, F.; Gastaldi, M.; Ikram, M.; Kumar, N.M.; Martin, M.A.; et al. Development Goals towards Sustainability. *Sustainability* **2023**, *15*, 9443. [CrossRef]
2. Peterson Institute for International Economics (PIIE). What Is Globalization? And How Has the Global Economy Shaped the United States? 2018. Available online: https://www.piie.com/microsites/globalization/what-is-globalization (accessed on 4 June 2023).
3. D'Adamo, I.; Gastaldi, M. Perspectives and Challenges on Sustainability: Drivers, Opportunities and Policy Implications in Universities. *Sustainability* **2023**, *15*, 3564. [CrossRef]

4. Hall, R.; Scoones, I.; Tsikata, D. *Africa's Land Rush: Rural Livelihoods and Agrarian Change*; James Currey: Woodbridge, UK, 2015.
5. He, X.; Khan, S.; Ozturk, I.; Murshed, M. The role of renewable energy investment in tackling climate change concerns: Environmental policies for achieving SDG-13. *Sustain. Dev.* **2023**, *31*, 1888–1901. [CrossRef]
6. Kumar, N.M.; Chopra, S.S. Leveraging Blockchain and Smart Contract Technologies to Overcome Circular Economy Implementation Challenges. *Sustainability* **2022**, *14*, 9492. [CrossRef]
7. Smol, M. Is the green deal a global strategy? Revision of the green deal definitions, strategies and importance in post-COVID recovery plans in various regions of the world. *Energy Policy* **2022**, *169*, 113152. [CrossRef]
8. Halkos, G.; Gkampoura, E.-C. Where do we stand on the 17 Sustainable Development Goals? An overview on progress. *Econ. Anal. Policy* **2021**, *70*, 94–122. [CrossRef]
9. Salahuddin, M.; Vink, N.; Ralph, N.; Gow, J. Globalisation, poverty and corruption: Retarding progress in South Africa. *Dev. South. Afr.* **2019**, *37*, 617–643. [CrossRef]
10. Ojo, A.S. Globalization and Agricultural Productivity Paradigm: The Nigeria Perspective. *Arch. Bus. Res.* **2018**, *6*, 94–104. [CrossRef]
11. Amusan, L. "Reap what you have not Sown", "Architect of Poverty": The Political Economy of Biodiversity Patenting in Africa. *Anthropologist* **2018**, *31*, 25–33. [CrossRef]
12. Batterbury, S.P.J.; Ndi, F. *Land grabbing in Africa. The Routledge Handbook of African Development*; Binns, J.A., Lynch, K., Nel, E., Eds.; Routledge: London, UK, 2018; pp. 573–582.
13. Gygli, S.; Haelg, F.; Potrafke, N.; Sturm, J.E. The KOF globalisation index–revisited. *Rev. Int. Organ.* **2019**, *14*, 543–574. [CrossRef]
14. Muthoka, S.; Muthuri, E.; Oginga, J. *Globalisation in Africa: An Overview*; MPRA Paper 65474; University Library of Munich: Munich, Germany, 2013.
15. Vibeke, B.; Henning, B.; Andre, F.; Rooyen, V. Why agricultural production in sub-Saharan Africa remains low compared to the rest of the world—A historical perspective. *Int. J. Water Resour. Dev.* **2020**, *36*, (Suppl. 1). S20–S53. [CrossRef]
16. Otsuka, K.; Muraoka, R. A green revolution for sub-Saharan Africa: Past failures and future prospects. *J. Afr. Econ.* **2017**, *26*, 173–198. [CrossRef]
17. Nel, A. *A Critical Reflection on Equity in Ugandan Carbon Forestry*; Social Justice and Environmental Governance; Routledge: London, UK, 2016.
18. Smith, W.; Jeppese, C. Britain, France and the Decolonization of Africa. 2017. Available online: www.ucl.ac.uk/ucl-press (accessed on 26 June 2023).
19. Bi, M.; Zhang, Z. Exploring the Path of Autonomous Development: The Development Dilemma and Coping Strategies of Sub-Saharan Africa in the Post-epidemic Era. *J. Knowl. Econ.* **2023**, *14*, 1–16. [CrossRef]
20. Carmody, P.; Taylor, D. Globalization, Land Grabbing, and the Present-Day Colonial State in Uganda: Ecolonization and Its Impacts. *J. Environ. Dev.* **2016**, *25*, 100–126. [CrossRef]
21. Amusan, L. Multinational Corporations' (MNCs) Engagement in Africa: Messiahs or Hypocrites? *J. Afr. Foreign Aff.* **2018**, *5*, 41–62. [CrossRef]
22. Holmén, H. Is Land Grabbing Always What It Is Supposed to Be? Large scale Land Investments in sub-Saharan Africa. *Dev. Policy Rev.* **2015**, *33*, 457–478. [CrossRef]
23. Yao, J.M.; Xu, M. Development dilemmas and prospects in Africa under the new crown pneumonia epidemic. *Contemp. World* **2020**, *9*, 64–71.
24. Zabella, A.A. The impact of the China-Africa cooperation Forum on the deepening of relations between China and Africa. *RUDN J. World Hist.* **2020**, *12*, 35–46. [CrossRef]
25. Zhou, J.; Jing, Z. Urban trap and breakthrough of Sub-Saharan Africa: Discussing the influence of China investment in Africa on urbanization. *Int. Urban. Plan.* **2018**, *33*, 29–38.
26. Zhou, J.Y. Why has poverty increased and life expectancy declined in sub-Saharan Africa? *Chin. Sci. Bull.* **2018**, *63*, 606–610. [CrossRef]
27. Gallous, A. The impact of Chinese Loans in Africa. *J. Am. Acad. Res.* **2022**, *10*, 1–26.
28. Amusan, L. SDGs 1, 2 and 5 Actualisation in the Age of Ultra-Capitalism: Likely Roles of State Intervention in South Africa. *Transylv. Rev.* **2020**, *28*, 12220–12226.
29. Wang, X.J. Africa's development situation and China-Africa joint construction of the Belt and Road. *Int. Stud.* **2019**, *2*, 35–48.
30. Wen, Z. The forum on China-Africa cooperation: An investment perspective. *China WTO Rev.* **2019**, *5*, 195–206. [CrossRef]
31. Yates, D.A. France and Africa. In *Africa and the World*; Palgrave Macmillan: Cham, Switzerland, 2018; pp. 95–118.
32. Zhao, Y.T. African development governance and China's role during the COVID-19 pandemic. *J. China-Afr. Stud.* **2021**, *2*, 24–43+155–156.
33. Quartey, S.H.; Oguntoye, O. Understanding and promoting industrial sustainability in Africa through the Triple Helix approach: A conceptual model and research propositions. *J. Knowl. Econ.* **2021**, *12*, 1100–1118. [CrossRef]
34. Rana, J.; Paul, J. Consumer behavior and purchase intention for organic food: A review and research agenda. *J. Retail. Consum. Serv.* **2017**, *38*, 157–165. [CrossRef]
35. Li, X.Y.; Li, J.Y.; Xu, J. Agriculture in Africa: Dilemma and prospects in the context of globalization. *Int. Econ. Rev.* **2020**, *5*, 5–6+61–75.

36. Liu, H.; Luo, J. Changes in the Geopolitics of Africa, and Issues Pertaining to Three-Way Cooperation among China, Africa and the West. In *Sino-African Development Cooperation*; Springer: Singapore, 2021; pp. 79–101.
37. Malah Kuete, Y.F.; Asongu, S.A. Infrastructure development as a prerequisite for structural change in Africa. *J. Knowl. Econ.* **2022**, *10*, 1–27. [CrossRef]
38. Amusan, L. Politics of Biopiracy: An Adventure into Hoodia/Xhoba Patenting in Southern Africa. *Afr. J. Tradit. Complement. Altern. Med.* **2017**, *14*, 103–109. [CrossRef]
39. Odinga, S.O. Looking for Leverage: Strategic Resources, Contentious Bargaining, and US-African Security Cooperation. CUNY Academic Works. 2016. Available online: https://academicworks.cuny.edu/gc_etds/1500 (accessed on 26 June 2023).
40. Prudence, A.; Wenche, B.E.; Kristin, R.K.; Per Ole, I.; Ane, C.W. Unlocking the potential for achievement of the UN Sustainable Development Goal 2—'Zero Hunger'—in Africa: Targets, strategies, synergies and challenges. *Food Nutr. Res.* **2021**, *65*, 7686.
41. Schöggl, J.-P.; Rusch, M.; Stumpf, L.; Baumgartner, R.J. Implementation of digital technologies for a circular economy and sustainability management in the manufacturing sector. *Sustain. Prod. Consum.* **2023**, *35*, 401–420. [CrossRef]
42. Alam, M.F.B.; Tushar, S.R.; Zaman, S.M.; Gonzalez, E.D.R.S.; Bari, A.B.M.M.; Karmaker, C.L. Analysis of the drivers of Agriculture 4.0 implementation in the emerging economies: Implications towards sustainability and food security. *Green. Technol. Sustain.* **2023**, *1*, 100021. [CrossRef]
43. Tseng, M.L.; Lim, M.K.; Ali, M.H.; Christianti, G.; Juladachah, P. Assessing the sustainable food system in Thailand under uncertainties: Governance, distribution and storage drive technological innovation. *J. Ind. Prod. Eng.* **2022**, *39*, 1–18. [CrossRef]
44. Fierke, K.M. Constructivism. In *International Relations Theories*; Dunne, T., Kurki, M., Smith, S., Eds.; Oxford University Press: Oxford, UK, 2016.
45. Amusan, L. The Plight of African Resources Patenting through the Lenses of the World Trade Organisation: An Assessment of South Africa's Rooibos Tea's Labyrinth Journey. *Afr. J. Tradit. Complement. Altern. Med.* **2017**, *5*, 41–47. [CrossRef]
46. Walters, R. *Eco Crime and Genetically Modified Food*; Routledge: New York, NY, USA, 2011.
47. Thun, E. The Globalization of Production. In *Global Political Economy*; Ravenhill, J., Ed.; Oxford University Press: Oxford, UK, 2017; pp. 174–196.
48. Wade, R.H. *Global Growth, Inequality, and Poverty*; The Globalization: London, UK, 2017.
49. Tickner, J.A.; Laura, S. Feminism. In *International Relations Theories*; Dunne, T., Kurki, M., Smith, S., Eds.; Oxford University Press: Oxford, UK, 2016.
50. Amusan, L.; Agunyai, S.C. The COVID-19 Pandemic and the crisis of Lockdowns in Nigeria: The food security perspective. *Afr. Public. Serv. Deliv. Perform. Rev.* **2021**, *9*, 78–98. [CrossRef]
51. Santos, M.P.; Brewer, J.D.; Lopez, M.A.; Paz-Soldan, V.A.; Chaparro, M.P. Determinants of food insecurity among households with children in Villa el Salvador, Lima, Peru: The role of gender and employment, a cross-sectional study. *BMC Public Health* **2022**, *22*, 717. [CrossRef]
52. Riley, L.; Caesar, M. Urban household food security in China and Mozambique: A gender-based comparative approach. *Dev. Pract.* **2018**, *28*, 1012–1021. [CrossRef]
53. Amusan, L.; Agunyai, S.C. Food Security without Women: A sine qua'non for Food Insecurity in Africa. In *Food Security and Food Sovereignty in Africa*; Amusan, L., Ed.; Ethics Press: Cambridge, UK, 2022; pp. 67–79.
54. Ehrlich, S.D. *The Politics of Fair Trade: Moving Beyond Free Trade & Protection*; Oxford University Press: Oxford, UK, 2018.
55. Owens, J.K. Systematic reviews: Brief overview of methods, limitations, and resources. *Nurse Author Ed.* **2021**, *31*, 69–72. [CrossRef]
56. Bedenlier, S.; Kondakci, Y.; Zawacki-Richter, O. Two decades of research into the internationalization of higher education: Major themes in the Journal of Studies in International Education (1997–2016). *J. Stud. Int. Educ.* **2018**, *22*, 108–135. [CrossRef]
57. Almudena, M.; Taísa, D.; Javier, M.S.; Raquel, G. Systematic Literature Reviews in Social Sciences and Humanities: A Case Study. *J. Inf. Technol. Res.* **2018**, *11*, 1–17.
58. Giang, H.T.N.; Banno, K.; Minh, L.H.N.; Trinh, L.T.; Loc, L.T.; Eltobgy, A.; Tai, L.L.T.; Khan, A.; Tuan, N.H.; Reda, Y.; et al. Dengue hemophagocytic syndrome: A systematic review and meta-analysis on epidemiology, clinical signs, outcomes, and risk factors. *Rev. Med. Virol.* **2018**, *28*, e2005. [CrossRef] [PubMed]
59. Tawfik, G.M.; Dila, K.A.S.; Mohamed, M.Y.F.; Tam, D.N.H.; Kien, N.D.; Ahmed, A.M.; Huy, N.T. A step by step guide for conducting a systematic review and meta-analysis with simulation data. *Trop. Med. Health* **2019**, *47*, 46. [CrossRef]
60. Hanson, N. Biofuels, land grabbing and food security in Africa. *Afr. Geogr. Rev.* **2013**, *32*, 190–192.
61. Ralph, T.; Fred, S. The Politics of Land Grabbing: State and corporate power and the (trans)nationalization of resistance in Cameroon. *J. Agrar. Chang.* **2018**, *19*, 41–63.
62. Cotula, L. *The Great African Land Grab?* Zed Books: London, UK, 2013.
63. Lisk, F. Land Grabbing or Harnessing of Development Potential in Agriculture? East Asia's land-based investments in Africa. *Pac. Rev.* **2013**, *26*, 563–587. [CrossRef]
64. Oxfam International. *Land and Power: The Growing Scandal Surrounding the New Wave Ofinvestments in Land*; Oxfam: Oxford, UK, 2011.
65. IIED. *The Global Land Rush: What the Evidence Reveals about Scale and Geography*; IIED Briefing; International Institute for Environment and Development: London, UK, 2012.

66. Hamouchene, H.; The Ouarzazate Solar Plant in Morocco: Triumphal 'Green' Capitalism and the Privatization of Nature. Adaliyya e-zine (Arab Studies Institute). Syndicated Article. 2016. Available online: http://www.jadaliyya.com/pages/index/24124/the-ouarzazate-solar-plant-in-morocco_triumphal-gr (accessed on 16 June 2023).
67. Tijo, S. Facing criticism: An analysis of (land-based) corporate responses to the large-scale land acquisition countermovement. *J. Peasant Stud.* **2018**, *46*, 1–18.
68. Amanor, K.S. Global Resource Grabs, Agribusiness Concentration and the Smallholder: Two West African Case Studies. *J. Peasant Stud.* **2012**, *39*, 731–749. [CrossRef]
69. Amadou, S. What Do We Know about the Chinese Land Grab in Africa? *Brookings*, 2015. Available online: https://www.brookings.edu/blog/africa-in-focus/2015/11/05/what-do-we-know-about-the-chinese-land-grab-in-africa/ (accessed on 20 June 2023).
70. Shaomin, X.; Jiang, L. The Emergence and Fallacy of 'China's Debt-Trap Diplomacy' Narrative. *China Intl. Stud.* **2020**, *81*, 69.
71. Gill, I.; Karakhula, K. Sounding the Alarm on Africa's Debt. *Brookings Institution Blog*, 6 April 2018. Available online: https://www.brookings.edu/blog/future-development/2018/04/06/sounding-the-alarm-on-africas-debt/ (accessed on 4 June 2023).
72. Bernstein, H. Agrarian Political Economy and Modern World Capitalism: The Contributions of Food Regime Analysis. *J. Peasant Stud.* **2016**, *43*, 611–647. [CrossRef]
73. Borras, S.M.; Franco, J.; Wang, C. The Challenge of Global Governance and Land Grabbing: Changing International Agricultural Context and Competing Political Views and Strategies. *Globalizations* **2013**, *10*, 161–179. [CrossRef]
74. Sun, Y. China's Aid to Africa: Monster or Messiah? Brookings Institution, 7 February 2014. Available online: https://www.brookings.edu/opinions/chinas-aid-to-africa-monster-or-messiah/ (accessed on 3 June 2023).
75. Were, A.; Debt Trap? Chinese Loans and Africa's Development Options. Africanportal, South African Institute of Internal Affiars, Policy Insights 66. 2018. Available online: https://www.africaportal.org/publications/debt-trap-chinese-loans-and-africas-development-options/ (accessed on 26 June 2023).
76. Su, X. Why Chinese Infrastructure Loans in Africa Represent a Brand-New Type of Neocolonialism. *The Diplomat*, 9 June 2017. Available online: https://thediplomat.com/2017/06/why-chinese-infrastructuralloans-in-africa-represent-a-brand-new-type-of-neocolonialism/ (accessed on 26 June 2023).
77. Scoones, I.; Hall, R.; Borras, S.M., Jr.; White, B.; Wolford, W. The Politics of Evidence: Methodologies for understanding the global land rush. *J. Peasant Stud.* **2013**, *40*, 469–483. [CrossRef]
78. Sassen, S. Land grabs today: Feeding the disassembling of national territory. *Globalizations* **2013**, *10*, 25–46. [CrossRef]
79. Rotberg, R. *Africa Emerges: Consummate Challenges, Abundant Opportunities*; Polity Press: Cambridge, UK, 2013.
80. Ndlovu-Gatsheni, S.J. Decolonising borders, decriminalising migration and rethinking citizenship. In *Crisis, Identity and Migration in Post-Colonial Southern Africa*; Springer: Cham, Switzerland, 2018; pp. 23–37.
81. Axford, B. *Theories of Globalization*; Polity Press: Cambridge, UK, 2015.
82. Dicken, P. *Global Shift: Mapping the Changing Contours of the World Economy*; Sage: London, UK, 2015.
83. Roberts, D.J. Technology is playing an expanding role in policing. Technology talk. *Police Chief* **2011**, *78*, 72–73.
84. Schwab, K. *The Fourth Industrial Revolution; World Economic Forum*; Geneva, Switzerland, 2016; 184p.
85. World Bank. *Poverty Rate in Africa: 2022*; World Bank: Geneva, Switzerland, 2022.
86. Pearce, F. *The Landgrabbers: The New Fight over Who Owns the Earth*; Eden Books Project: Croydon, UK, 2012.
87. Dwyer, M.B. Building the Politics Machine: Tools for 'Resolving' the Global Land Grab. *Dev. Chang.* **2013**, *44*, 309–333. [CrossRef]
88. United Nations General Assembly (UNGA). *World Economic and Social Survey*; United Nations: New York, NY, USA, 2015. Available online: https://www.un.org/en/desa/world-economic-and-social-survey-20142015 (accessed on 21 June 2023).
89. Kossi, A.; Wang, R.; Wang, W.; Stephanie, N.; Josphert, N.K.; Judith, M.P. The impact of globalization on African countries economic development. *Afr. J. Bus. Manag.* **2012**, *6*, 11057–11076.
90. Anseeuw, W. The rush for land in Africa: Resource grabbing or green revolution? *S. Afr. J. Int. Aff.* **2013**, *20*, 159–177. [CrossRef]
91. Zoomers, A.; Gekker, A.; Schäfer, M.T. Between two Hypes: Will 'Big Data' Help Unravel Blind Spots in Understanding the 'Global Land Rush'? Geoforum. *J. Phys. Hum. Reg. Geosci.* **2016**, *69*, 147–159. [CrossRef]
92. Liberti, S. *Land Grabbing: Journeys in the New Colonialism*; Verso: London, UK, 2013.
93. Nally, D. Governing precarious lives: Land grabs, geopolitics, and 'food security'. *Geogr. J.* **2015**, *181*, 340–349. [CrossRef]
94. Wolford, W.; Borras, S.M., Jr.; Hall, R.; Scoones, I.; White, B. Governing Global Land Deals: The Role of the State in the Rush for Land. *Dev. Chang.* **2013**, *44*, 189–210. [CrossRef]
95. Agunyai, S.C. Emerging Governance Crises in Twenty-First Century Nigeria. In *Africa Now: Emerging Issues and Alternative Perspective*; Adebusuyi, A., Ikuteijo, K., Eds.; Springer Publication, Palgrave Macmillan: London, UK, 2018; pp. 211–239.
96. Olaopa, O.; Ayodele, O.A. Building on the strengths of African indigenous knowledge and innovation (AIK&I) for sustainable development in Africa. *Afr. J. Sci. Technol. Innov. Dev.* **2021**, *14*, 1313–1326.
97. Baumüller, H. Mobile Technology Trends and Their Potential for Agricultural Development, 2013. ZEF Working Paper 123. Available online: https://ageconsearch.umn.edu/record/160565/?ln=en (accessed on 21 June 2023).
98. Chen, H. Applications of cyber-physical system: A literature review. *J. Ind. Integr. Manag.* **2017**, *2*, 1750012. [CrossRef]
99. Krell, N.T.; Giroux, S.A.; Guido, Z.; Hannah, C.; Lopus, S.E.; Caylor, K.K.; Evans, T.P. Smallholder farmers' use of mobile phone services in central Kenya. *Clim. Dev.* **2021**, *13*, 215–227. [CrossRef]

100. Scalzo, R.L.; Picchi, V.; Migliori, C.A.; Campanelli, G.; Leteo, F.; Ferrari, V.; Di Cesare, L.F. Variations in the phytochemical contents and antioxidant capacity of organically and conventionally grown Italian cauliflower (*Brassica oleracea* L. subsp. *botrytis*): *Results from a three-year field study. J. Agric. Food Chem.* **2013**, *61*, 10335–10344.
101. Raheem, D.; Shishaev, M.; Dikovitsky, V. Food System Digitalization as a Means to Promote Food and Nutrition Security in the Barents Region. *J. Agric.* **2019**, *9*, 168. [CrossRef]
102. Schwab, B.; China and Africa—Continuing an Unequal Relationship. A Fiedrich Ebert Stiftung Briefing Paper. 2016. Available online: http://www.fes-Tanzania.org/fileadmin/user_upload/pdf/FES_Briefing_Paper_China_and_Africa_-_Continuing_an_Unequal_Relationship__Benjamin_Schwab_.pdf (accessed on 2 June 2023).
103. West, D.M. *The Next Wave: Using Digital Technology to Further Social and Political Innovation*; The Brookings Institution: Washington, DC, USA, 2011.
104. World Bank. *Digital Dividends: World Development Report*; World Bank: Washington, DC, USA, 2016.
105. Nolte, K.; Väth, S.J. Interplay of Land Governance and Large-Scale Agricultural Investment: Evidence From Ghana and Kenya. *J. Mod. Afr. Stud.* **2015**, *53*, 69–92. [CrossRef]
106. Ishaku, B.L. Globalization and development. The impact on Africa; A political economy approach. *OIDA Int. J. Sustain. Dev.* **2014**, *7*, 153–162.
107. Odoh, S.I. The Challenges of Globalisation as Instrument of Imperialism in Africa. *South East J. Political Sci.* **2018**, *4*, 44–55.
108. Tewodros, K.; Yemiamrew, J. The Challenges of Globalization to Africa: Theoretical Reflections and Practical Assessments. *World J. Res. Rev.* **2020**, *11*, 21–38.
109. Schendel, R.; McCowan, T. Expanding higher education systems in low- and middle-income countries: The challenges of equity and quality. *High. Educ.* **2016**, *72*, 407–411. [CrossRef]
110. El-Jardali, F.; Ataya, N.; Fadlallah, R. Changing roles of universities in the era of SDGs: Rising up to the global challenge through institutionalizing partnerships with governments and communities. *Health Res. Policy Syst.* **2018**, *16*, 38. [CrossRef] [PubMed]
111. SDSN Australia/Pacific. *Getting Started with the SDGs in Universities. A Guide for Universities, Higher Education Institutions, and the Academic Sector. Australia, New Zealand and Pacific Edition*; Sustainable Development Solutions Network—Australia/Pacific: Melbourne, Australia, 2017.

Disclaimer/Publisher's Note: The statements, opinions and data contained in all publications are solely those of the individual author(s) and contributor(s) and not of MDPI and/or the editor(s). MDPI and/or the editor(s) disclaim responsibility for any injury to people or property resulting from any ideas, methods, instructions or products referred to in the content.

sustainability

MDPI

Review

Mapping Drivers, Barriers, and Trends in Renewable Energy Sources in Universities: A Connection Based on the SDGs

Vinicius dos Santos Skrzyzowski [1], Felipe Neves Farinhas [1], Maria Cecília Ferrari de Carvalho Teixeira [1], Murillo Vetroni Barros [1,*], Rodrigo Salvador [2], Sebastião Cavalcanti Neto [3] and Fernando Henrique Lermen [1,4]

1 Department of Industrial Engineering, State University of Paraná, Paranaguá 83203560, Brazil; viniciusdssnts78@gmail.com (V.d.S.S.); felipenfarinhas@gmail.com (F.N.F.); cecilia.ferraricarvalho@gmail.com (M.C.F.d.C.T.); fernando-lermen@hotmail.com (F.H.L.)

2 Department of Engineering Technology and Didactics, Technical University of Denmark, 2750 Ballerup, Denmark; rodsa@dtu.dk

3 Department of Management, State University of Paraná, Paranaguá 83203560, Brazil; sebastiao.cavalcanti@unespar.edu.br

4 Industrial Engineering Department, Universidad Tecnológica del Perú, Lima 15046, Peru

* Correspondence: murillo.vetroni@gmail.com; Tel.: +55-42-99142-9087

Abstract: Universities play a pivotal role in modern society and must lead the way in achieving energy efficiency, directly contributing to the Sustainable Development Goals (SDGs). Like small towns in resource consumption and population mobility, many universities and research centers face significant challenges transitioning to renewable electricity systems. This study aims to (i) map the current scientific literature on renewable energy sources used by universities; (ii) discuss the drivers, barriers, and trends of implementing renewable energy; and (iii) establish a connection with the SDGs. More specifically, the authors conducted a systematic literature review based on three stages: (i) data collection, (ii) bibliometric analysis, and (iii) content analysis. Forty-two articles were obtained and defined as the studied sample. The findings of this review illuminate critical research themes, leading countries in renewable energy adoption, and the prevalent electricity sources, shedding light on the primary authors shaping the discourse. Wind and solar energy exhibit a notable growth trajectory, offering environmentally friendly alternatives compared to conventional sources. Furthermore, it is essential to highlight that the distribution of research documents in the sample is uneven, with a predominant concentration in European countries. Additionally, the study identifies the field's key drivers, barriers, and emergent trends. The theoretical contributions encompass a comprehensive compilation of renewable energy sources, discernible research trajectories, and strategies to navigate obstacles. In practical terms, this work offers valuable insights for the selection of energy sources and stakeholder engagement, facilitating informed decision-making processes. This article's novelty lies in its holistic examination of renewable energy adoption in university settings, providing a comprehensive overview of the current landscape and actionable insights for stakeholders seeking sustainable energy solutions within these institutions. This aligns with multiple SDGs, including Goal 7 (Affordable and Clean Energy), Goal 11 (Sustainable Cities and Communities), and Goal 13 (Climate Action), underscoring the critical role of universities in driving sustainable development.

Keywords: renewable energy; sustainable energy; solar energy; university; review

Citation: Skrzyzowski, V.d.S.; Farinhas, F.N.; Teixeira, M.C.F.d.C.; Barros, M.V.; Salvador, R.; Neto, S.C.; Lermen, F.H. Mapping Drivers, Barriers, and Trends in Renewable Energy Sources in Universities: A Connection Based on the SDGs. *Sustainability* **2024**, *16*, 6583. https://doi.org/10.3390/su16156583

Academic Editors: Manoj Kumar Nallapaneni, Idiano D'Adamo and Massimo Gastaldi

Received: 4 June 2024
Revised: 28 July 2024
Accepted: 29 July 2024
Published: 1 August 2024

Copyright: © 2024 by the authors. Licensee MDPI, Basel, Switzerland. This article is an open access article distributed under the terms and conditions of the Creative Commons Attribution (CC BY) license (https://creativecommons.org/licenses/by/4.0/).

1. Introduction

Energy plays an essential role in any country's economic growth [1]. Finding alternative energy sources that meet the population's needs is necessary to meet its growing demand. Unfortunately, fossil fuels are a concerning problem with electricity generation [2], harming the environment, depleting natural resources, and causing pollution. In this sense, new energy sources, especially those of renewable origin, must be present in countries' energy matrices.

In residential and industrial settings, renewable energy has been pursuing the forefront on a large scale [3]. However, implementing renewable sources for electricity generation in higher education institutions (HEI) is still a topic for scientific research. Nevertheless, HEI have studied hybrid renewable energy adoption [4–6]. To Kolokotsa et al. [7], university campuses are like small-scale cities. In such a way, according to Avila et al. [8], the occupancy profiles of cities are vast, and the use of spaces is so diverse that a university campus resembles a community or a neighborhood in a city. Once universities can be characterized as small cities, they must embrace the implementation of renewable energy to set an example for the community on behalf of sustainable development. Based on this university–city relationship, it is possible to see the importance of expanding renewable and sustainable energy technologies on campuses.

Given universities' role in modern society, they must take the lead in analyzing energy efficiency, characterizing themselves as smart and sustainable universities [9–11]. The definition of a sustainable university is commonly associated with the three pillars of sustainability, as universities are responsible for mitigating environmental, economic, and social impacts; promoting health and well-bing; and disseminating these values globally [12–14]. Furthermore, Cortese [15] argued that teaching, research, operations, and extension are part of a sustainable university's integrated system; therefore, a sustainable university focuses on inseparability, considering the impacts it suffers.

Universities face barriers to joining a renewable energy system. For example, many public institutions are burdened with supplying and maintaining diesel generators [4]. Also, Avila et al. [8] suggested that bottom-up initiatives may fail due to a lack of financing and support from administrative boards. Further, it is notable that the barriers for implementing these systems in HEI are common across the board, including, e.g., lack of funding, lack of human and technological resources, lack of support from administration, resistance from collaborators, and lack of knowledge about the importance of renewable energy generation [16–20], with most of the impediments to the adoption of this type of energy system coming from a lack of governmental support for its adoption in universities.

Given this research gap, this study aims to (i) map the current scientific literature on renewable energy sources used by universities; (ii) discuss the drivers, barriers, and trends of implementing renewable energy; and (iii) to establish a connection with the Sustainable Development Goals. To do so, we used a systematic literature review in three stages: (i) data collection, (ii) bibliometric analysis, and (iii) content analysis. This manuscript presents a dyad of theoretical and practical contributions. A compilation of renewable energy sources, research trends, and strategies to mitigate barriers is given regarding the former. The latter is directed at the types of energy and stakeholders that support decision making and the list of management trends and opportunities that support clean energy implementation in universities. In addition, this study aligns with some of the Sustainable Development Goals (SDGs) established by the 2030 Agenda.

Therefore, this paper is organized as follows: The first section contextualizes the study's theme and aim. The following section shows the methodology used in the systematic literature review on the topic. Section 3 reports the results of the bibliometric analysis. Section 4 discusses the trends, barriers, and trends for further study. Finally, Section 5 presents the final remarks and limitations of the research.

2. Materials and Methods

A methodology was adopted for data collection, selection of relevant documents for analysis, and obtaining answers to the research questions: the systematic review, which was uniquely and methodically used on each selected paper to narrow down and summarize the content of the files to specific results consistent with the research [21–23]. In addition, the methods used enable the inclusion of an empirical analysis of the data obtained, commonly used to complement systematic reviews and show the practical and functional side of the research [19]. Based on the method of Denyer and Tranfield [20], it focuses on three stages: (i) data collection, (ii) bibliometric analysis, and (iii) content analysis.

2.1. Data Collection

The protocol used in this research was the Preferred Reporting Item for Systematic Reviews and Meta-Analysis (PRISMA), developed by Moher et al. [24]. It was established to assist in classifying and selecting the articles. This research used the following search string: ((University OR Universities) AND ("Renewable Energy" OR "Sustainable Energy") AND (Empirical OR "Case Study") AND (Strategy OR Strategies OR "Public Policy" OR "Public Policies")). Based on the string results, the entire gathered dataset was taken from the Web of Science database, a comprehensive and diverse platform regarding scientific content [25]. At the end of the searches, 2534 documents were found, as presented in Figure 1 through the PRISMA protocol.

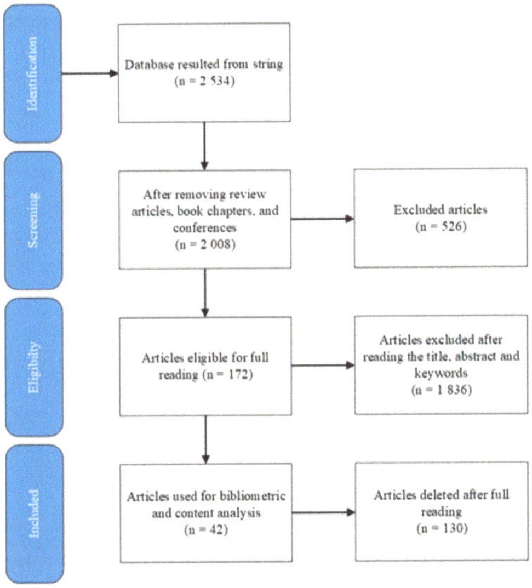

Figure 1. PRISMA protocol.

The use of PRISMA followed three filtering steps: (i) exclusion of review articles, book chapters, books, and conference articles, resulting in 2008 records; (ii) reading titles, abstracts, and keywords, resulting in 172 records; and (iii) complete reading of the sample articles, resulting in 42 documents. The Mendeley ® v1.19.8 software was used as a support tool for managing references [26]. The final sample of articles was used for bibliometric analysis (Section 2.2) and content analysis (Section 2.3).

2.2. Bibliometric Analysis

The Bibliometrix tool 4.1 was used for the research analysis and mapping. The tool developed by Aria and Cuccurullo [27] makes it possible to quantitatively analyze many documents based on the accumulated database examined by the package. It also helps in empirical studies since it is based on experiences that have occurred. As a form of modernization and adequacy, Bibliometrix used the RStudio® software 3.6.0 as a support tool for the graphics [27].

Thus, Bilbliometrix is a visualization tool for obtaining bibliometric data. Therefore, once the analysis method is defined, the platform needs a more apparent data classification to analyze each provided element accurately and concisely. Consequently, it was necessary to export the entire database in BibTex format. In this file model, it is possible to detail all the elements obtained from the database in text form [28]. Other information was considered in the bibliometric analysis because it is information referring to the articles in the sample studied: (i) research method employed, (ii) type of energy source used, and (iii) stakeholders.

2.3. Content Analysis

Lastly, content analysis was used to categorize all articles in the final sample according to the topics the research addresses [29]. This analysis followed the steps suggested by Elo and Kyngäs [30], such as open coding, categorization, and abstraction. Through these steps, we sought to identify relevant information through a deductive process while coding the studied documents. This information was analyzed in two ways: bibliometric analysis and content analysis. Finally, the abstraction step supported the discussions among the sample authors for the results presented in Sections 3 and 4. This was carried out by retrieving the information from the articles according to specified categories so that the content analysis could occur in an organized manner. Therefore, the following categories were used: (i) drivers, (ii) barriers, and (iii) trends. Thus, based on the established categories, a detailed reading of the documents in the final sample was initiated to identify the information related to each category.

3. Overview of the Studied Sample

This section discusses the bibliometric analyses carried out in this study. The obtained database allowed us to identify how the documents and authors position themselves and relate to the studied themes. Figure 2 shows the analysis of the articles concerning the year of publication, the journal of publication, the university of affiliation, and the global citations.

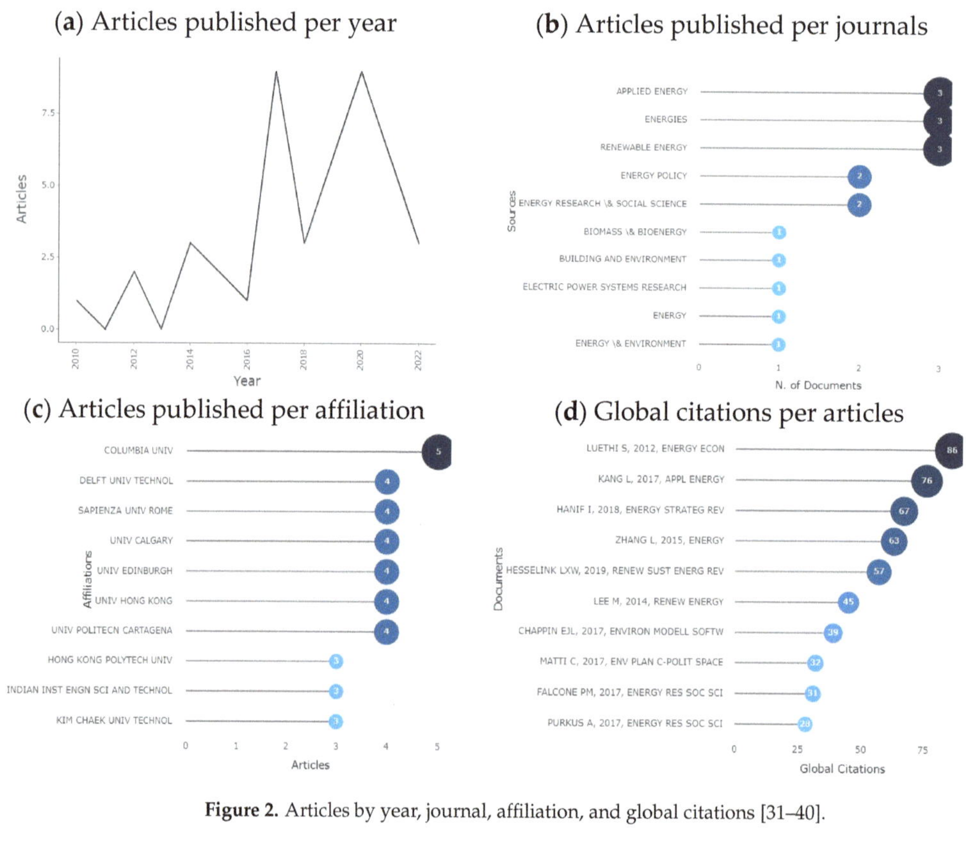

Figure 2. Articles by year, journal, affiliation, and global citations [31–40].

Figure 2a shows the publications on renewable energy and their applications in universities between 2010 and 2022, presenting a considerable increase starting in 2016. Furthermore, the volume of studies in this area declined between 2021 and 2022. Yet, they are still more frequent than in years before 2016, pointing to the topic's great importance in academia.

The manuscript was constructed from a systematic review of the literature, following methodological procedures for searching documents, filters, and selection. The literature does not provide a clear answer as to why the number of publications decreased in 2018 and 2022. However, given the general history, the topic is growing. One issue that might relate to the decline is that studies are more focused on empirical applications of renewable energy in other fields, such as manufacturing, agriculture, and cities.

Figure 2b exhibits an analysis of the journal's area of knowledge and the frequency of publication of these studies. The first three journals with the highest number of publications are *Applied Energy*, *Energy*, and *Renewable Energy*, which are part of the scope of the topic presented. Other journals cited were *Energy Policy* and *Energy Research & Social Science*; all five of these journals are recognized as top journals in energy research. Figure 2c shows which universities are affiliated with which articles in the study's portfolio. The leading university, having affiliation in five articles, is Columbia University/USA, followed by Delft University of Technology/The Netherlands, Sapienza University of Rome/Italy, University of Calgary/Canada, the University of Edinburgh/Scotland, and the University of Hong Kong/Japan, which have affiliations in four studies each. Most of the universities cited in this paper are related to the continents America, Europe, and Asia.

Figure 2d presents the number of citations of the analyzed articles globally. The most cited author is Luethi S., followed by Kang L. and Hanif I. They are also characterized as the three most cited authors in other articles. Another analysis was performed on the relationship among the main research themes. In this sense, Figure 3 shows the bibliometric analysis regarding the themes, their interconnections, and published studies by country.

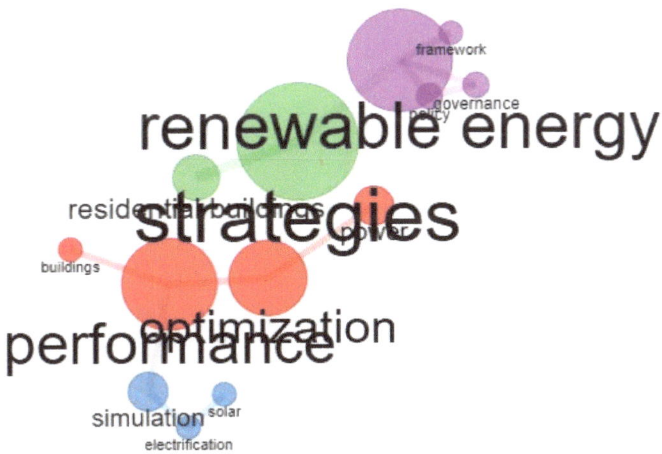

Figure 3. Main themes and their relationships.

The main themes in the articles highlighted in Figure 3 are renewable energy, strategies, and performance. Thus, it is notable that, within the selected articles, the main objects of study are the types of renewable energy, the possible strategies to be adopted, and the performance of tools within this theme. Secondary themes, still having significant incidence, are optimization, simulation, solar, buildings, and governance. Considering the occurrence of words, strategies were cited by 33 papers (78.54%), renewable energy by 22 papers (52.36%), and performance by 18 papers (42.84%). To complement the analysis, Figure 4 shows the map of countries with the most prominent number of analyzed publications.

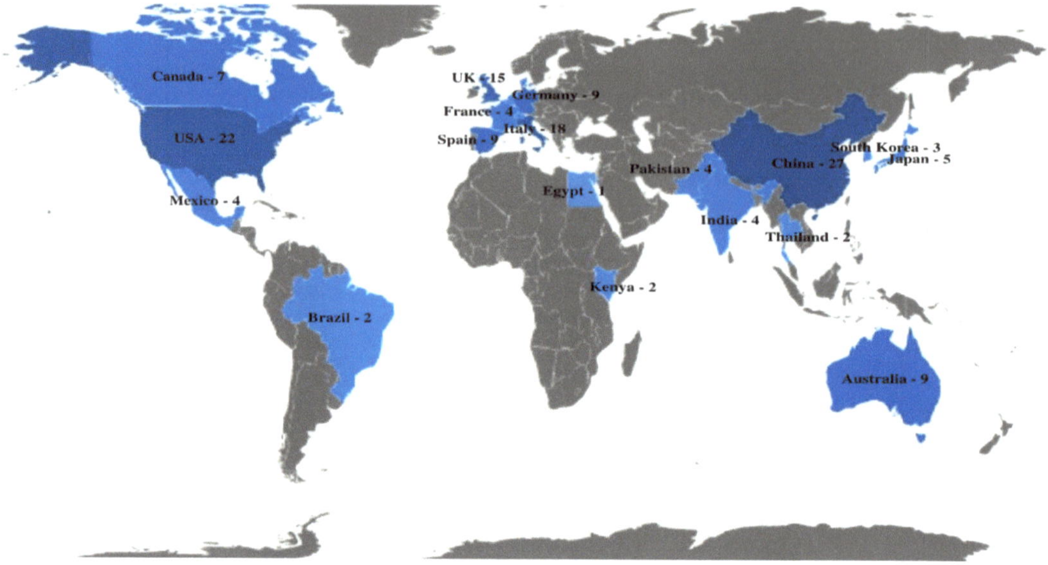

Figure 4. Published studies by authors in each country.

Figure 4 shows in more intense colors (darker shades of blue) where there is a higher incidence of articles being published. Considering occurrence per continent, Europe presented 55 authors (37.41%), Asia 45 authors (30.61%), America 35 authors (23.80%), Oceania 9 authors (6.12%), and Africa 3 authors (2.04%). China and the USA are the main countries in which this theme is addressed and have the highest number of publications. Subsequently, the incidence is also considerable in Italy, the UK, Germany, Spain, and Australia, with fewer publications. Therefore, there is a need to increase the discussion in different countries and continents on how to deal with the energy application of their universities to increase the number of studies published on the subject. Furthermore, in Figure 5, the correspondence analysis of the studied theme is shown based on interrelated studies.

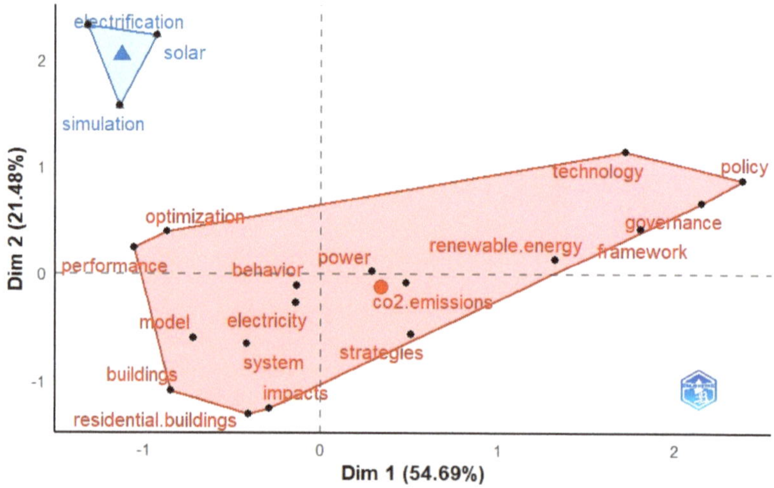

Figure 5. Correspondence analysis of the studied sample. The dotted line and red dot show where the zero is located, that is, it is possible to identify the four quadrants of the image.

In Figure 5, there are two groups with correlated words. The first group (the smaller one—in blue) presents the concepts of electrification, solar, and simulation. This indicates that these three themes intersect in similar works and are directly linked. The second group (in red) has a broader range of concepts that can be related more frequently by proximity, with the main topics being CO_2 emissions, policy, and electricity. It can be noticed from Figure 5 that the closer to a positive value in the axis, the greater the incidence of the word in the articles in the sample; in addition, the percentage shown in the dimensions demonstrates the number of studies that belong to it.

Furthermore, Table 1 allows one to identify the research method employed, energy type, and stakeholders present in the analyzed articles as well as the frequency or incidence with which they appear in the studies.

Table 1. Methods, energy source types, and stakeholders in the sample.

Classification		Articles (n)	References	%
Research method employed	Mixed method (qualitative and quantitative)	25	[31–35,41–60]	59.5%
	Qualitative	8	[33,36–39,61–63]	19.0%
	Quantitative	9	[40,55,64–70]	21.4%
Type of energy source	Solar	20	[31–33,35,37,40,44,46,52,55,58,62–64,67–69]	47.6%
	Wind	7	[38,47,56,60,61,66,70]	16.7%
	Geothermal	4	[41,48,53,71]	9.5%
	Hybrid	3	[43,50,65]	7.1%
	Biomass	3	[34,39,45]	7.1%
	Tidal wave	2	[54,59]	4.8%
	Nuclear	1	[51]	2.4%
	Hydropower	1	[69]	2.4%
	Fossil fuels	1	[36]	2.4%
Stakeholders	Community	4	[31,37,43,65]	9.5%
	Entrepreneurs	2	[41,55]	4.8%
	University	8	[35,40,44,49,60,62,64,66]	19.0%
	NGOs	1	[61]	2.4%
	Project developers	1	[32]	2.4%
	Farmers	1	[45]	2.4%
	Countries	21	[33,34,38,46,48,50–59,63,68–70]	50.0%
	Politicians	1	[39]	2.4%
	Entrepreneurs, community, and government	2	[47,71]	4.8%
	University and community	1	[67]	2.4%

The research methods employed by the sample articles are divided into qualitative, quantitative, and mixed methods (qualitative and quantitative). Most studies use mixed methods to measure their results, totaling 59.5% of the sample. This provides information stating that studies are not being conducted solely in diagnostic form but rather with empirical and longitudinal applications for implementing renewable energy in universities. Additionally, it is possible to analyze the types of energy used during the studies: geothermal, solar, wind, hybrid, fossil fuels, nuclear, tide wave, hydropower, and biomass. In this analysis, it was noted that some types are more prevalent than others within the articles. An example is photovoltaic and wind energy, which account for 45.2% and 16.7% of the sample. Geothermal and hybrid energy were characterized as 7.1% of the analyzed studies.

In fact, solar energy has a large and positive representation in the global context for potential installation in universities for a few reasons. First, it is clean, renewable energy with low-carbon prospects and follows global movements towards a lower-carbon economy. Secondly, installation and maintenance costs have decreased considerably in recent years, increasing consumer potential.

The stakeholders of each study were defined, with 10 of them based on reading the articles. Among them are the countries involved in half of the analyzed articles, while the university is present in 19.0%. The community and entrepreneurs are also present in 9.5% and 4.8% of the sample. Besides these stakeholders acting alone in the studies, a group is also involved in some studies, such as the entrepreneurs allied to the government and the community, representing 4.8% of the analyzed articles. Also, the university allied with the community in 2.4% of the sample.

4. Drivers, Barriers, and Trends of Renewable Energy Sources in Universities

This section addresses the main results of this article's content analysis, identifying drivers, barriers, and trends for future research.

4.1. Drivers

After analyzing the 42 documents in the final sample, the importance of using renewable energy was made explicit both concerning the efficiency of sustainable energy models and the benefits to the environment and the health of the population, as shown in Table 2.

Table 2. Drivers to implement renewable energy in universities.

Cluster	Type of Energy Source	Reference	Main Drivers
Sustainable and renewable energy employed by universities	Solar energy	[44]	Dealing with uncertain links and disturbances that are present in microgrid systems in some universities.
		[67]	Performing future evaluations and analyses of its potential is key to its consolidation and dissemination.
		[32]	Validating the data with larger samples through future research. Investigating whether renewable energy policy risk assessment differs by type of investors (e.g., large vs. small companies).
		[55]	Engaging in future research to understand the psychological benefits regarding the intention to install solar energy, alongside the mediation of an eco-friendly lifestyle.
		[52]	Standardizing ecosystem monetary methods.
		[58]	Exploring the integration of additional upgrading measures. Explore land and locations that are being used in large university campuses, with potential for installing solar energy
	Other (wind, hydroelectric, tidal, geothermal)	[70]	Identifying the best sustainable conditions for powering consumer.
		[48]	Conducting further analysis concerning large university buildings through future work.
		[51]	Studying the impacts of introducing taxes on carbon generation.
	Multiple energy types	[46]	Performing tests through case studies of policies for different issues, varied sectors, and delimitations.
		[33]	Extending the investigation to include environmental variables and conducting comparative analyses across sectors for a sustainable transition.
		[34]	Further studying the relative share of each alternative clean energy resource.

Table 2. *Cont.*

Cluster	Type of Energy Source	Reference	Main Drivers
Public policy and external projects to implement sustainable and renewable energy	Public policies	[37]	Relating the modeling of different energy technologies to the adoption barriers beyond those already known.
		[63]	Developing ways to engage the dynamic consciousness of historical relationships about the causes of contradictions in a proactive way to support systemic transformations based on future research.
		[36]	Enhancing the influence of energy management policies to ensure sustainable development in the region by integrating civil society into fundamental development and environmental policies.
	External projects	[41]	Initiating a multiple-case-study project could do more to suggest a general model of the process previously identified, in addition to conducting a survey for technology entrepreneurs about their experiences with the early stages of the commercialization process.
		[65]	Seeking to understand exactly how the physical and socioeconomic conditions of the landscape influence the development, success, and possible scaling up of sustainable energy initiatives.
		[71]	Conducting investigations across sectors and beyond the European continent.
		[43]	Using techniques considering the future uncertainty of electricity-type growth.

Through the analyses and after tabulating the data, it was possible to observe that most authors carried out research focused on solar energy. At least fifteen of them discussed their articles in various ways while commenting on the use of solar energy. In addition, some researchers showed how the transition from using energy with high carbon emission rates to renewable energy from the sun happens. Secondly, the type of energy they depicted the most is wind energy, which was demonstrated in some instances through calculations of the energy efficiency coming from atmospheric air currents. Furthermore, the authors referred to other renewable energy types, such as biomass, hydroelectric, and geothermal.

However, the data obtained note the importance of using renewable resources to capture energy. Due to increasing globalization, the rate of CO_2 emissions has also increased dramatically. Therefore, systems that avoid the emission of undesired gasses into the ecosystem must be sought at all costs.

Solar energy is the most convenient and straightforward technology to obtain today in terms of size and complexity, and it has become the most coherent technology when considering its application anywhere. After all, it requires roof space and a place for energy storage. According to [67], in addition to the great potential that exists in the production of energy from photovoltaic cells, there are incentives for those who have a network to generate energy from renewable sources, as the implementation of this type of energy demands a high investment, which must be considered in a compensation calculation.

Moreover, underdeveloped countries do not use this type of energy in ample supply [67] but only a tiny portion compared to the production and use of other methods. Therefore, aside from the potential environmental benefits of solar energy generation, there are other incentives to start this process, at least to some degree. Following Lee et al. [31], implementing a power distribution network based on photovoltaic cells presents many expenses due to the plate number, amount of energy stored, distance, and terrain. Considering the installed equipment, this distribution system offers one of the simplest and most affordable ways to spread the adoption of this type of energy because of its easy maintenance and modular components; i.e., its products have great versatility.

Bearing in mind that acquiring this system is very costly, Lüthi and Wustenhagen [32] commented that governments are creating financial incentives for projects with photovoltaic technology. Similarly, several countries are encouraging the adaptation and usage of renewable energy sources, considering the climate problems the planet has been suffering. In summary, microgrids were created to study energy generation capacity, in which reliability and compensation calculations are made for possible more significant investments.

Photovoltaic solar energy is one of the many renewable energy options available worldwide. It is a solution in several cases that require an energy source transition. Lakhani et al. [52] stated that the two most recurrent installations of this type of energy are the systems implemented on the roof and the photovoltaic panels installed on the ground. Therefore, regarding the implementation of photovoltaic technology, a precise analysis is necessary to decide where to place it. If the impact of land use for installing the system directly on the ground is too high, the solution is to implement it on roofs. It can be stated that for the direct implementation of this energy in a university, its impact must be analyzed and calculated in a general way, both concerning the structure of the building and the influence on the ecosystem. Li et al. [69] stated that several ways to implement a solar system in a building exist. They can be installed separately and added to the building after its construction, or they can even be used to replace some elements of the original building.

Wind energy is a renewable source from nature with infinite capacity, for it renews itself endlessly. In this sense, the wind is one of the forces of nature with an inexhaustible capacity, as it is a mass of air that varies in direction, intensity, and speed. The capture of wind energy has been happening for a long time, constituting a considerable portion of the world's energy supply. These data are remarkable if we analyze the number and extent of wind farms scattered around the globe. In agreement with Choe et al. [70], the capture occurs through wind turbines positioned in strategic locations with a high wind-incidence rate, which moves the propellers of the equipment, moving the turbine that generates energy. There are different sizes of turbines, and the energy production is proportional to them. Thus, the biggest problem for implementing a wind turbine at a university is the area to install the equipment, as the space limits its size, consequently limiting energy generation. Despite that, the turbines can be operated parallel to conventional methods, whereby storing the energy generated and directing it to more basic purposes is possible. Considering that this type of energy production depends on the constant occurrence of wind to move the turbines and that the storage generators are usually diesel-powered, this system would help to keep carbon emissions on the planet to a minimum and not stop the energy supply.

Tidal Energy is one of several ways to obtain renewable energy. The capture of energy through tidal power is performed by converting the waves' kinetic energy into electricity for the population or nearby industries. According to Gray et al. [59], this relatively new method of energy production has both positive and negative sides, the two most prominent being that the energy is clean, and its implementation cost is high. However, considering that some universities have their facilities in coastal cities, constructing a small machine for extracting kinetic energy is possible due to the advantages of the geographical location.

Many papers deal with energy transitions, i.e., the change of generating energy from conventional non-renewable means with high rates of carbon emissions that are incredibly harmful to people and the environment to a means of sustainable energy production from renewable sources. According to Falcone et al. [33], energy transition happens for different reasons, such as energy crises, an aggravating factor that generates the need for alternative energy production. However, many factors amidst crises should be considered to analyze the best alternative energy. Moreover, among the positive aspects, this transition generates many jobs for developing and implementing renewable sources, hence having a positive financial impact. Based on this principle, the application of a renewable energy source in a university should occur after the analysis of numerous factors, such as the energy consumption values of the location, labor costs, the prices of the equipment to be implemented, and the social acceptance of the people who circulate in the environment, alongside political factors that involve the energy application.

Public policies are an aggravating and necessary factor for implementing a renewable energy source. They will determine the bids needed for projects and the amount of energy produced, which, in some cases, should be able to meet local demand. Further, spare parts that are not stored can be destined for the population, determining sustainability laws concerning energy production and some other minor details. Moreover, since many energy generation processes require much capital, there is the possibility of acquiring investment from the public to assist in creating new renewable energy sources.

Finally, Thoyre [64] reported that companies that generate energy using fossil fuels have constantly damaged their investments because they still use methods that emit too much carbon, harming the environment. To change companies' methods of obtaining energy, laws to encourage the use of renewable sources to generate energy have been created, through which places like universities would benefit from receiving monetary aid to develop and assist studies on the energy efficiency of renewable sources.

4.2. Barriers and Strategies

The barriers refer directly to the difficulties identified regarding implementing a particular type of renewable energy, whether in a conventional residence, building, or university. These difficulties can be identified by analyzing all the aspects involving energy and their crucial factors, whether environmental, socioeconomic, or political.

It is possible to point out that solar energy is currently one of the cleanest and most widely used energy types in the world, but this does not mean there are no difficulties when implementing a project in a specific location. As exemplified by Sarkar et al. [49], one of the main barriers regarding this type of energy would be the variation of solar incidence in the region, which would negatively affect the energy production of the photovoltaic panels and make them not as effective as they should be.

Furthermore, Lakhani et al. [52] reported different ways to install photovoltaic panels to meet specific energy demands without harming the environment, such as installing these panels on the roof of the building in question without harming untouched land. Another option would be implementing them on the ground, as described by the authors. However, it is notable that, given their proportions, both ways have installation difficulties. Taking a university as an example, it is necessary to analyze the roof of the building to identify whether it is appropriate to receive the panels in a way that does not damage its structure and to study the possibility of an energy transition of the structure to an alternative type of energy. Additionally, a prior study should be performed on the possibility of such an installation, as its implementation on the ground harms the land it will occupy. There may not be available land around the university in question.

Wind power is a solution for certain cases since its implementation directly depends on ample available territorial space in which there is also an incidence of wind for power generation. According to Yuan et al. [56], this results in other barriers, such as the fact that, in most cases, the availability of these lands only happens in remote areas; in other words, there is difficulty in taking the energy produced to the final consumer across these great distances. Moreover, when it comes to implementing wind energy at a university, it can be said that, except in some specific cases, there is usually not enough space to install it within the university itself, further limiting the use of this technology. From this, Matti et al. [38] also highlighted the obligation to analyze the region's governmental policies regarding properly implementing this type of energy since it cannot be applied in any location or situation. Thus, it might not be a viable solution to meet the demand for sustainable energy.

Regarding tidal energy, which comes from the force exerted by tidal phenomena and the transformation of this kinetic energy into electrical energy, its application strongly depends on the sea's presence in the implementation region, thus prioritizing coastal regions. Regarding such locations, it is indispensable that tidal energy plays an increasingly active role in meeting the demand for energy production [54] since the most common types are solar, wind, and hydroelectric power. From this, Gray et al. [59] demonstrated that, compared to the types of energy mentioned above, tidal power has a much higher application cost due to its specificities and evident territorial limitations.

As for energy transition, the application of one or more sources of renewable energy is directly influenced by local laws and public policies. In this sense, Tanaka et al. [46] showed that, in Japan, electric utilities significantly impact the decisions regarding energy policy, which results in significant difficulties when the government implements legislation regarding energy generation from renewable sources. Therefore, it is essential to consider all the possibilities and feasibility of implementing these energies [34]; this becomes critical when deciding whether to create or continue a sustainable power generation project within universities.

Therefore, most of the barriers to implementing a renewable energy system are limited to the current country's public policies, which results in fewer possibilities for applications within universities or any other type of structure. As highlighted by some authors, scarce funding for renewable energy projects [43], barriers to energy-efficiency adoption [37], and the complexity of carrying out the energy transition process due to the extended timeframe [65] make the political and socioeconomic environment two of the biggest causes of difficulties when implementing sustainable energy generation and upgrading the energy system.

The innovative barriers and strategies are (i) the energy transition from fossil sources to renewable sources, (ii) installation and use of alternative sources such as wind and solar, and (iii) promoting research and development in the university environment for clean, safe, and affordable energy.

Towards climate change mitigation, the transition from fossil energy to renewable energy is an impetus for modern society, including universities. Given this, some energy sources have positive characteristics in terms of the low environmental impact generated, the efficiency in electricity production, and the reduction in implementation costs (compared to other sources). These characteristics are based on solar, wind, and biogas sources. Furthermore, the best environment to foster the development of new technologies and practical applications is the university.

4.3. Trends by SDGs

Studies on the 17 SDGs developed by the UN are widely reported in the scientific literature, such as in the area of indicators associated with equitable and sustainable well-being (BES) [72], pragmatic sustainability [73], digital transformation and Industry 5.0 [74], companies headquartered in the European Union [75], monitoring the outcomes of the SDGs by evaluating the Italian regions [76], and many others.

This section addresses questions that researchers and supporters can use to develop new empirical studies related to renewable and sustainable energy at universities and what their public policies are, as presented in Table 3.

Based on the list of trends (Table 3), it is suggested that authors in the area answer the questions raised, which are of paramount importance for the beginning of new research, in addition to identifying the best strategies and types of energy to be utilized. In a specific case, such as the implementation of renewable energy at a university, that list is essential in identifying the authors who relate to the methods of application and their connection to the geographical territory and its climate. Regarding the bureaucracy involved in the application, one must analyze which laws make the use of renewable energy viable at a university and how the creation of public policies can assist in this process.

Based on all the above, we highlight that this study aligns with some of the SDGs established by the 2030 Agenda [77].

Table 3. Trends to support researchers.

Cluster	Question	References
Sustainable and renewable energy employed by universities	What is the energy efficiency regarding the location of the application of a photovoltaic energy grid between the ground, roof, and curtains of capture?	[31,32,52,67]
	What is the feasibility and efficiency of using batteries for electrical energy storage in conjunction with wind systems?	[43,60,66]
	In what ways can tidal energy be further implemented despite its high cost?	[54,59]
	What are the alternatives for assisting the supply of energy in periods when no generation occurs due to the lack of incidence of both sun and wind?	[49,66]
	Considering the application of an energy grid at a university, what would be the advantage of applying a system based on photovoltaic cells?	[31,32,67]
	Considering the high incidence of tropical winds in Brazil, what are the advantages of creating mini wind grids in universities?	[66,70]
Public policies and external projects to implement sustainable and renewable energy	What are the implementation laws, and how do they influence the adoption of renewable energy in universities?	[38,63,64]
	What are the incentive policies for creating new renewable energy generators, and how do they apply to universities?	[37,63,64]
	What are the monetary aid policies used for renewable energy implementation at universities?	[35,64]
	If the study site has some energy supply, what are the public policies for this energy transition?	[36,65]
	How do social beliefs and interests impact the implementation of a renewable energy source?	[46,61,63]
	What is the influence of public policies to ensure sustainable development in the region by applying renewable energy?	[36,63]

Goal 4 refers to quality education. Some specific goals in higher education can contribute to achieving this SDG. Ensuring access to inclusive, quality, and equitable education and promoting lifelong learning opportunities for all needs to be on the agenda of many schools, research institutes, and universities [78,79]. In addition, a sustainable structure, in terms of renewable electricity, can be a viable solution in this field. Collaborating with other institutions, governments, and organizations to promote sustainable educational practices and sharing best practices is part of sustainable development in higher education. Furthermore, it should be ensured that the institution's physical infrastructure is adequate to support effective and modern learning, preparing students to face the challenges of the 21st century.

Goal 7 dwells on clean and affordable energy. Access to reliable, sustainable, modern energy sources in the teaching environment is important. Industries must produce and consume clean energy, and educational buildings (schools and universities) must also act. A higher education institution can adopt practices and technologies that promote the efficient use of energy and the adoption of renewable sources [80]. This includes installing solar, wind, or other renewable energy systems to reduce dependence on fossil fuels. Additionally, universities often conduct advanced research into energy technologies, including energy storage, energy efficiency, and new energy sources. This environment is extremely rich in possibilities for developing new technologies and patents in the electrical energy sector.

Goal 11 covers sustainable cities and communities. This study focuses on making communities and educational environments more inclusive, safe, resilient, and sustainable. Universities must produce knowledge; operate in inclusive, resilient systems; and, above all, be sustainable. Universities are most often established in central regions of the city or close to it. In few cases is a university located very far from the city. The reason for its location is the ease of transporting students, teachers, staff, and supplies. Therefore, working with sustainable mobility and using bikes, scooters, public transport, and walking can make sense. Additionally, universities often have large campuses that can be thought of as small urban communities. A higher education institution can adopt sustainable urban planning principles in its physical development, promoting the efficient use of space, the preservation of green areas, waste management, and the reduction of the carbon footprint [81].

Finally, strengthening the means of implementation and revitalizing the global partnership for sustainable development is an impetus for a more sustainable future in universities.

5. Conclusions

It is possible to identify different ways of capturing energy from renewable sources, the most cited being electrical grids based on photovoltaic cells and energy from wind turbines. When considering the application of an electric grid based on renewable sources in a university, several options are available regarding which source type to employ. This choice is usually based on the terrain, i.e., analyzing the incidence of wind, the amount of sunlight received, or the possibility of using the ocean to capture kinetic energy.

When choosing the renewable energy method or source, we must investigate the political and socioeconomic environment, identifying public policies that assist in implementing the energy network and facilitate the installation. In other words, one must consider the high costs of the installation and check the policies for financial aid for the implementation of the project. In addition, promoting studies concerning energy efficiency at the university and improving sustainability using low-carbon emission energy align directly with multiple SDGs.

Besides the difficulties due to current policies, environmental and socioeconomic factors are still directly related to their implementation. The incidence of sunlight for capturing energy through photovoltaic panels, the recurrence of wind force to make the turbines work, and the presence of waves and tides in coastal regions for kinetic energy generation are clear examples of the barriers encountered when dealing with these types of renewable energy. However, it is worth pointing out the advantages for the environment and its use in general through the decrease in harmful non-renewable sources, which contributes to sustainability and prolongs energy generation due to its inexhaustible source.

Finally, the best choice for implementing energy from a renewable source in a university would be based on photovoltaic cell networks. It presents a high degree of versatility. Besides the conventional panels deployed on the roofs or fixed on the ground, curtains can still capture the sun's rays. In addition, this type of energy generation uses a modular system in its panels, making the maintenance and exchange of equipment parts straightforward and practical. In this way, the energy transition from conventional carbon-based means to technological energy generation systems through renewable sources becomes feasible.

The study is not free from limitations. Some documents may have been left out of this analysis due to the set of keywords and the databases used. In further research, we highlight the need to explore other less prominent renewable energy sources that might be more readily available at the university level, such as biomass, which can be derived from food waste and might be a cheaper alternative in terms of infrastructure while also helping to manage existing waste and environmental influence on the campus. Other sources also deserve exploration. As a suggestion for future work, a case study reporting descriptions of sustainable universities implementing renewable sources, pointing out the technical and economic viability, can be developed. Furthermore, daily energy consumption in universities, different building types, seasonal variations, and external parameters can be considered. Additionally, the alignment of these initiatives with SDGs such as Goal 7 (Affordable and Clean Energy), Goal 11 (Sustainable Cities and Communities), and Goal 13 (Climate Action) should be emphasized to highlight their broader impact on sustainable development.

Author Contributions: Conceptualization, F.H.L., V.d.S.S., F.N.F. and M.C.F.d.C.T.; methodology, F.H.L.; software, M.C.F.d.C.T.; validation, M.V.B. and R.S.; formal analysis, V.d.S.S. and F.N.F.; investigation, V.d.S.S., F.N.F. and M.C.F.d.C.T.; resources, F.H.L.; data curation, V.d.S.S. and F.N.F.; writing—original draft preparation, F.H.L., V.d.S.S., F.N.F. and M.C.F.d.C.T.; writing—review and editing, M.V.B. and R.S.; visualization, S.C.N.; supervision, F.H.L.; project administration, F.H.L. and

M.V.B.; funding acquisition, S.C.N. All authors have read and agreed to the published version of the manuscript.

Funding: The authors appreciate the support of the Araucaria Foundation and the National Council for Scientific and Technological Development (CNPq).

Institutional Review Board Statement: Not applicable.

Informed Consent Statement: Not applicable.

Data Availability Statement: The data will be provided if requested from the corresponding author.

Acknowledgments: We would like to thank the State University of Paraná—UNESPAR's Academic Writing Center (Centro Acadêmico de Letramento e Escrita—UNESPAR—https://eri.unespar.edu.br/menu-geral/centro-de-escrita-academica-da-unespar, accessed on 30 July 2024) for assistance with the English language editing. The authors would also like to express their gratitude to Idiano D'Adamo of the Special Issue "Sustainable Development Goals: A Pragmatic Approach", and the reviewers for substantially improving this work.

Conflicts of Interest: The authors declare no conflicts of interest.

References

1. Bin Amin, S.; Chang, Y.; Khan, F.; Taghizadeh-Hesary, F. Energy security and sustainable energy policy in Bangladesh: From the lens of 4As framework. *Energy Policy* **2022**, *161*, 112719. [CrossRef]
2. Poudyal, R.; Loskot, P.; Nepal, R.; Parajuli, R.; Khadka, S.K. Mitigating the current energy crisis in Nepal with renewable energy sources. *Renew. Sustain. Energy Rev.* **2019**, *116*, 109388. [CrossRef]
3. Nezhnikova, E.; Papelniuk, O.; Dudin, M. Developing renewable and alternative energy sources to improve the efficiency of housing construction and management. *Int. J. Energy Econ. Policy* **2019**, *9*, 172–178. [CrossRef]
4. Karabulut, A.; Gedik, E.; Keçebaş, A.; Alkan, M.A. An investigation on renewable energy education at the university level in Turkey. *Renew. Energy* **2011**, *36*, 1293–1297. [CrossRef]
5. Akindeji, K.T.; Tiako, R.; Davidson, I.E. Use of Renewable Energy Sources in University Campus Microgrid—A Review. IEEE Conference Publication. 2019. IEEE Xplore n.d. Available online: https://ieeexplore.ieee.org/abstract/document/8734352 (accessed on 8 May 2023).
6. Babatunde, O.; Munda, J.; Hamam, Y. Off-grid hybrid photovoltaic—Micro wind turbine renewable energy system with hydrogen and battery storage: Effects of sun tracking technologies. *Energy Convers. Manag.* **2022**, *255*, 115335. [CrossRef]
7. Kolokotsa, D.; Gobakis, K.; Papantoniou, S.; Georgatou, C.; Kampelis, N.; Kalaitzakis, K.; Vasilakopoulou, K.; Santamouris, M. Development of a web based energy management system for University Campuses: The CAMP-IT platform. *Energy Build.* **2016**, *123*, 119–135. [CrossRef]
8. Ávila, L.V.; Beuron, T.A.; Brandli, L.L.; Damke, L.I.; Pereira, R.S.; Klein, L.L. Barriers to innovation and sustainability in universities: An international comparison. *Int. J. Sustain. High. Educ.* **2019**, *20*, 805–821. [CrossRef]
9. Rinaldi, C.; Cavicchi, A.; Spigarelli, F.; Lacchè, L.; Rubens, A. Universities and smart specialisation strategy: From third mission to sustainable development co-creation. *Int. J. Sustain. High. Educ.* **2018**, *19*, 67–84. [CrossRef]
10. Pérez, F.M.; Martínez, J.V.B.; Fonseca, I.L. Modelling and Implementing Smart Universities: An IT Conceptual Framework. *Sustainability* **2021**, *13*, 3397. [CrossRef]
11. Valdés, R.M.A.; Comendador, V.F.G. European Universities Initiative: How Universities May Contribute to a More Sustainable Society. *Sustainability* **2022**, *14*, 471. [CrossRef]
12. Alshuwaikhat, H.M.; Abubakar, I. An integrated approach to achieving campus sustainability: Assessment of the current campus environmental management practices. *J. Clean. Prod.* **2008**, *16*, 1777–1785. [CrossRef]
13. Cole, L. *Assessing Sustainability on Canadian University Campuses: Development of a Campus Sustainability Assessment Framework*; Royal Roads University: Victoria, BA, Canada, 2003.
14. Velazquez, L.; Munguia, N.; Platt, A.; Taddei, J. Sustainable university: What can be the matter? *J. Clean. Prod.* **2006**, *14*, 810–819. [CrossRef]
15. Cortese, A.D. The critical role of higher education in creating a sustainable future. *Plan. High. Educ.* **2003**, *31*, 15–22.
16. Babatunde, O.; Denwigwe, I.; Oyebode, O.; Ighravwe, D.; Ohiaeri, A.; Babatunde, D. Assessing the use of hybrid renewable energy system with battery storage for power generation in a University in Nigeria. *Environ. Sci. Pollut. Res.* **2022**, *29*, 4291–4310. [CrossRef] [PubMed]
17. Ávila, L.V.; Filho, W.L.; Brandli, L.; Macgregor, C.J.; Molthan-Hill, P.; Özuyar, P.G.; Moreira, R.M. Barriers to innovation and sustainability at universities around the world. *J. Clean. Prod.* **2017**, *164*, 1268–1278. [CrossRef]
18. Amaral, A.R.; Rodrigues, E.; Gaspar, A.R.; Gomes, Á. A review of empirical data of sustainability initiatives in university campus operations. *J. Clean. Prod.* **2020**, *250*, 119558. [CrossRef]

19. Qazi, A.; Hussain, F.; Rahim, N.A.; Hardaker, G.; Alghazzawi, D.; Shaban, K.; Haruna, K. Towards Sustainable Energy: A Systematic Review of Renewable Energy Sources, Technologies, and Public Opinions. *IEEE Access* **2019**, *7*, 63837–63851. [CrossRef]
20. Denyer, D.; Tranfield, D. Producing a systematic review. In *The Sage Handbook of Organizational Research Methods*; Buchanan, D.A., Bryman, A., Eds.; Sage Publications Ltd.: Thousand Oaks, CA, USA, 2009; pp. 671–689.
21. Graciano, P.; Lermen, F.H.; Reichert, F.M.; Padula, A.D. The impact of risk-taking and creativity stimuli in education towards innovation: A systematic review and research agenda. *Think. Ski. Creat.* **2022**, *47*, 101220. [CrossRef]
22. Cordeiro, E.R.; Lermen, F.H.; Mello, C.M.; Ferraris, A.; Valaskova, K. Knowledge management in small and medium enterprises: A systematic literature review, bibliometric analysis, and research agenda. *J. Knowl. Manag.* **2023**, *28*, 590–612. [CrossRef]
23. Kuakoski, H.S.; Lermen, F.H.; Graciano, P.; Lam, J.S.L.; Mazzuchetti, R.N. Marketing, entrepreneurship, and innovation in port management: Trends, barriers, and research agenda. *Marit. Policy Manag.* **2023**, 1–18. [CrossRef]
24. Moher, D.; Liberati, A.; Tetzlaff, J.; Altman, D.G. Preferred reporting items for systematic reviews and meta-analyses: The PRISMA statement. *PLoS Med.* **2009**, *6*, e1000097-269. [CrossRef] [PubMed]
25. Mongeon, P.; Paul-Hus, A. The journal coverage of Web of Science and Scopus: A comparative analysis. *Scientometrics* **2016**, *106*, 213–228. [CrossRef]
26. Thelwall, M. Are Mendeley reader counts useful impact indicators in all fields? *Scientometrics* **2017**, *113*, 1721–1731. [CrossRef]
27. Aria, M.; Cuccurullo, C. bibliometrix: An R-tool for comprehensive science mapping analysis. *J. Informetr.* **2017**, *11*, 959–975. [CrossRef]
28. Derviş, H. Bibliometric analysis using bibliometrix an R Package. *J. Sci. Res.* **2019**, *8*, 156–160. [CrossRef]
29. Spens, K.M.; Kovács, G. A content analysis of research approaches in logistics research. *Int. J. Phys. Distrib. Logist. Manag.* **2006**, *36*, 374–390. [CrossRef]
30. Elo, S.; Kyngäs, H. The qualitative content analysis process. *J. Adv. Nurs.* **2008**, *62*, 107–115. [CrossRef] [PubMed]
31. Lee, M.; Soto, D.; Modi, V. Cost versus reliability sizing strategy for isolated photovoltaic micro-grids in the developing world. *Renew. Energy* **2014**, *69*, 16–24. [CrossRef]
32. Lüthi, S.; Wüstenhagen, R. The price of policy risk—Empirical insights from choice experiments with European photovoltaic project developers. *Energy Econ.* **2012**, *34*, 1001–1011. [CrossRef]
33. Falcone, P.M.; Lopolito, A.; Sica, E. Policy mixes towards sustainability transition in the Italian biofuel sector: Dealing with alternative crisis scenarios. *Energy Res. Soc. Sci.* **2017**, *33*, 105–114. [CrossRef]
34. Zhang, L.; Zhou, P.; Newton, S.; Fang, J.-X.; Zhou, D.-Q.; Zhang, L.-P. Evaluating clean energy alternatives for Jiangsu, China: An improved multi-criteria decision making method. *Energy* **2015**, *90*, 953–964. [CrossRef]
35. Chappin, E.J.; de Vries, L.J.; Richstein, J.C.; Bhagwat, P.; Iychettira, K.; Khan, S. Simulating climate and energy policy with agent-based modelling: The Energy Modelling Laboratory (EMLab). *Environ. Model. Softw.* **2017**, *96*, 421–431. [CrossRef]
36. Hanif, I. Impact of fossil fuels energy consumption, energy policies, and urban sprawl on carbon emissions in East Asia and the Pacific: A panel investigation. *Energy Strat. Rev.* **2018**, *21*, 16–24. [CrossRef]
37. Hesselink, L.X.; Chappin, E.J. Adoption of energy efficient technologies by households—Barriers, policies and agent-based modelling studies. *Renew. Sustain. Energy Rev.* **2019**, *99*, 29–41. [CrossRef]
38. Matti, C.; Consoli, D.; Uyarra, E. Multi level policy mixes and industry emergence: The case of wind energy in Spain. *Environ. Plan. C Politics Space* **2017**, *35*, 661–683. [CrossRef]
39. Purkus, A.; Gawel, E.; Thrän, D. Addressing uncertainty in decarbonisation policy mixes—Lessons learned from German and European bioenergy policy. *Energy Res. Soc. Sci.* **2017**, *33*, 82–94. [CrossRef]
40. Kang, L.; Yang, J.; An, Q.; Deng, S.; Zhao, J.; Wang, H.; Li, Z. Effects of load following operational strategy on CCHP system with an auxiliary ground source heat pump considering carbon tax and electricity feed in tariff. *Appl. Energy* **2017**, *194*, 454–466. [CrossRef]
41. Earle, A.G.; Merenda, M.J.; Davis, J.M. Strategy-as-Process in a Technology Venture: A Case Study of Pivots, Pauses, Partners, and Progress. *Technol. Innov. Manag. Rev.* **2019**, *9*, 8–18. [CrossRef]
42. Gasbarro, F.; Rizzi, F.; Frey, M. Sustainable institutional entrepreneurship in practice: Insights from SMEs in the clean energy sector in Tuscany (Italy). *Int. J. Entrep. Behav. Res.* **2018**, *24*, 476–498. [CrossRef]
43. Fioriti, D.; Giglioli, R.; Poli, D.; Lutzemberger, G.; Micangeli, A.; Del Citto, R.; Perez-Arriaga, I.; Duenas-Martinez, P. Stochastic sizing of isolated rural mini-grids, including effects of fuel procurement and operational strategies. *Energy Convers Manag.* **2020**, *213*, 112824. [CrossRef]
44. Ndwali, P.K.; Njiri, J.G.; Wanjiru, E.M. Optimal Operation Control of Microgrid Connected Photovoltaic-Diesel Generator Backup System Under Time of Use Tariff. *J. Control. Autom. Electr. Syst.* **2020**, *31*, 1001–1014. [CrossRef]
45. Carrosio, G. Energy production from biogas in the Italian countryside: Modernization vs. repeasantization. *Biomass- Bioenergy* **2014**, *70*, 141–148. [CrossRef]
46. Tanaka, Y.; Chapman, A.; Tezuka, T.; Sakurai, S. Multiple Streams and Power Sector Policy Change: Evidence from the Feed-in Tariff Policy Process in Japan. *Politics Policy* **2020**, *48*, 464–489. [CrossRef]
47. Hale, T.; Urpelainen, J. When and how can unilateral policies promote the international diffusion of environmental policies and clean technology? *J. Theor. Politics* **2015**, *27*, 177–205. [CrossRef]

48. Ramos-Escudero, A.; Gil-García, I.C.; García-Cascales, M.S.; Molina-Garcia, A. Energy, economic and environmental GIS–based analysis of shallow geothermal potential in urban areas—A Spanish case example. *Sustain. Cities Soc.* **2021**, *75*, 103267. [CrossRef]
49. Sarkar, A.; Velasco, L.; Wang, D.; Wang, Q.; Talasila, G.; de Biasi, L.; Kübel, C.; Brezesinski, T.; Bhattacharya, S.S.; Hahn, H.; et al. High entropy oxides for reversible energy storage. *Nat. Commun.* **2018**, *9*, 3400. [CrossRef] [PubMed]
50. Maes, D.; Van Passel, S. Interference of regional support policies on the economic and environmental performance of a hybrid cogeneration-solar panel energy system. *Energy Policy* **2012**, *42*, 670–680. [CrossRef]
51. Percebois, J.; Pommeret, S. Efficiency and dependence in the European electricity transition. *Energy Policy* **2021**, *154*, 112300. [CrossRef]
52. Lakhani, R.; Doluweera, G.; Bergerson, J. Internalizing land use impacts for life cycle cost analysis of energy systems: A case of California's photovoltaic implementation. *Appl. Energy* **2014**, *116*, 253–259. [CrossRef]
53. Dadzie, J.; Runeson, G.; Ding, G. Assessing determinants of sustainable upgrade of existing buildings: The case of sustainable technologies for energy efficiency. *J. Eng. Des. Technol.* **2020**, *18*, 270–292. [CrossRef]
54. Harcourt, F.; Angeloudis, A.; Piggott, M.D. Utilising the flexible generation potential of tidal range power plants to optimise economic value. *Appl. Energy* **2019**, *237*, 873–884. [CrossRef]
55. Sun, P.-C.; Wang, H.-M.; Huang, H.-L.; Ho, C.-W. Consumer attitude and purchase intention toward rooftop photovoltaic installation: The roles of personal trait, psychological benefit, and government incentives. *Energy Environ.* **2020**, *31*, 21–39. [CrossRef]
56. Yuan, Q.; Zhou, K.; Yao, J. A new measure of wind power variability with implications for the optimal sizing of standalone wind power systems. *Renew. Energy* **2020**, *150*, 538–549. [CrossRef]
57. Li, C.; Lin, T.; Xu, Z. Impact of Hydropower on Air Pollution and Economic Growth in China. *Energies* **2021**, *14*, 2812. [CrossRef]
58. Abdallah, M.; El-Rayes, K. Multiobjective Optimization Model for Maximizing Sustainability of Existing Buildings. *J. Manag. Eng.* **2016**, *32*, 4. [CrossRef]
59. Gray, A.; Dickens, B.; Bruce, T.; Ashton, I.; Johanning, L. Reliability and O&M sensitivity analysis as a consequence of site specific characteristics for wave energy converters. *Ocean Eng.* **2017**, *141*, 493–511. [CrossRef]
60. Zhan, S.; Hou, P.; Enevoldsen, P.; Yang, G.; Zhu, J.; Eichman, J.; Jacobson, M.Z. Co-optimized trading of hybrid wind power plant with retired EV batteries in energy and reserve markets under uncertainties. *Int. J. Electr. Power Energy Syst.* **2020**, *117*, 105631. [CrossRef]
61. Szarka, J. Bringing interests back in: Using coalition theories to explain European wind power policies. *J. Eur. Public Policy* **2010**, *17*, 836–853. [CrossRef]
62. Streitferdt, V.; Chirarattananon, S.; Du Pont, P. Lessons learned from studying public initiatives to support energy efficiency finance in Thailand from 1992 to 2014. *Energy Effic.* **2017**, *10*, 905–923. [CrossRef]
63. Novalia, W.; Rogers, B.C.; Bos, J.J. Incumbency and political compromises: Opportunity or threat to sustainability transitions? *Environ. Innov. Soc. Transitions* **2021**, *40*, 680–698. [CrossRef]
64. Thoyre, A. Neoliberalizing negawatts: Governance of energy efficiency as accumulation strategy. *Geoforum* **2021**, *118*, 140–149. [CrossRef]
65. De Boer, J.; Zuidema, C. Towards an integrated energy landscape. *Proc. Inst. Civ. Eng.-Urban Des. Plan.* **2015**, *168*, 231–240. [CrossRef]
66. Hauer, I.; Balischewski, S.; Ziegler, C. Design and operation strategy for multi-use application of battery energy storage in wind farms. *J. Energy Storage* **2020**, *31*, 101572. [CrossRef]
67. Salvador, D.S.; Toboso-Chavero, S.; Nadal, A.; Gabarrell, X.; Rieradevall, J.; da Silva, R.S. Potential of technology parks to implement Roof Mosaic in Brazil. *J. Clean. Prod.* **2019**, *235*, 166–177. [CrossRef]
68. Arif, A.; Rizwan, M.; Elkamel, A.; Hakeem, L.; Zaman, M. Optimal Selection of Integrated Electricity Generation Systems for the Power Sector with Low Greenhouse Gas (GHG) Emissions. *Energies* **2020**, *13*, 4571. [CrossRef]
69. Li, L.; Qu, M.; Peng, S. Performance evaluation of building integrated solar thermal shading system: Active solar energy usage. *Renew. Energy* **2017**, *109*, 576–585. [CrossRef]
70. Choe, K.Y.; Kim, H.; Li, J.U.; Hyon, C.I.; Kang, I.Y. New architecture and SCADA for stand-alone hybrid (medium-sized asynchronous wind turbine + UPS with battery + photovoltaic array) power system without diesel generator. *Wind. Energy* **2019**, *22*, 959–974. [CrossRef]
71. Gasbarro, F.; Annunziata, E.; Rizzi, F.; Frey, M. The Interplay Between Sustainable Entrepreneurs and Public Authorities: Evidence from Sustainable Energy Transitions. *Organ. Environ.* **2017**, *30*, 226–252. [CrossRef]
72. D'adamo, I.; Di Carlo, C.; Gastaldi, M.; Rossi, E.N.; Uricchio, A.F. Economic Performance, Environmental Protection and Social Progress: A Cluster Analysis Comparison towards Sustainable Development. *Sustainability* **2024**, *16*, 5049. [CrossRef]
73. D'adamo, I.; Gastaldi, M.; Nallapaneni, M.K. Europe Moves toward Pragmatic Sustainability: A More Human and Fraternal Approach. *Sustainability* **2024**, *16*, 6161. [CrossRef]
74. Osorio, A.M.; Úsuga, L.F.; Restrepo-Carmona, J.A.; Rendón, I.; Sierra-Pérez, J.; Vásquez, R.E. Methodology for Stakeholder Prioritization in the Context of Digital Transformation and Society 5.0. *Sustainability* **2024**, *16*, 5317. [CrossRef]
75. Perevoznic, F.M.; Dragomir, V.D. Achieving the 2030 Agenda: Mapping the Landscape of Corporate Sustainability Goals and Policies in the European Union. *Sustainability* **2024**, *16*, 2971. [CrossRef]

76. D'adamo, I.; Gastaldi, M. Monitoring the Performance of Sustainable Development Goals in the Italian Regions. *Sustainability* **2023**, *15*, 14094. [CrossRef]
77. UN. United Nations. *Sustainable Development Goals.* 2015. Available online: https://sdgs.un.org/goals (accessed on 3 July 2024.).
78. Gamage, K.A.A.; Munguia, N.; Velazquez, L. Happy sustainability: A future quest for more sustainable universities. *Soc. Sci.* **2022**, *11*, 24. [CrossRef]
79. Zenchanka, S.; Gorbatchev, N.; Zagoumennov, I.; Frankenberger, F. Sustainability of university campus and SDG 2030: Social aspects. In *Handbook of Best Practices in Sustainable Development at University Level 2022*; Springer International Publishing: Cham, The Netherlands, 2022; pp. 337–352. [CrossRef]
80. Silva-Da-Nóbrega, P.I.; Chim-Miki, A.F.; Castillo-Palacio, M. A smart campus framework: Challenges and opportunities for education based on the sustainable development goals. *Sustainability* **2022**, *14*, 9640. [CrossRef]
81. Alawneh, R.; Jannoud, I.; Rabayah, H.; Ali, H. Developing a novel index for assessing and managing the contribution of sustainable campuses to achieve UN SDGs. *Sustainability* **2021**, *13*, 11770. [CrossRef]

Disclaimer/Publisher's Note: The statements, opinions and data contained in all publications are solely those of the individual author(s) and contributor(s) and not of MDPI and/or the editor(s). MDPI and/or the editor(s) disclaim responsibility for any injury to people or property resulting from any ideas, methods, instructions or products referred to in the content.

sustainability

Review

Decoding the Nexus: A Bibliometric Review on Sustainability, Circular Economy, and Consumer Food Waste †

Miray Baybars [1], Fatma Demircan Keskin [1], İpek Kazançoğlu [1] and Yiğit Kazançoğlu [2,*]

1 Department of Business Administration, Ege University, İzmir 35040, Turkey; miray.baybars@ege.edu.tr (M.B.); fatma.demircan.keskin@ege.edu.tr (F.D.K.); ipek.savasci@ege.edu.tr (İ.K.)
2 Department of Logistics Management, Yaşar University, İzmir 35040, Turkey
* Correspondence: yigit.kazancoglu@yasar.edu.tr
† This study is an extended and revised version of the paper entitled "Mapping the intersections of sustainability, circular economy, and consumer behavior: a bibliometric review on food waste" presented at the 3rd International Conference on Sustainable, Circular Management and Environmental Engineering (ISCMEE 2023) held in İzmir, Turkey on 12 July 2023.

Abstract: One of the issues that has gained importance within the scope of the United Nations Sustainable Development Goals (SDGs) is the issue of food waste. These goals, which represent very important and urgent problems to be solved at the global level, are extremely critical in terms of sustainability. Food waste, by its nature at the intersection of economic, social, and environmental sustainability goals, has become a global problem linked to key challenges in the global environment in terms of food security, climate change, malnutrition, and economic sustainability. The fact that consumers are one of the most important factors affecting food waste in the transition to a circular economy increases the importance of this study once again. Due to the lack of systematic, chronological studies showing how food waste develops over time, this study will examine the development and evolution of food waste research using a bibliometric analysis. In this way, it aims to gain a comprehensive insight into the field's current state and shed light on this highly important area of study. In addition to informing policymakers, practitioners, and consumers with the results of this research, it is also aimed to support all relevant individuals, institutions, and organizations in the efforts to combat food waste. One of the main objectives of this study is to contribute to the achievement of the United Nations Sustainable Development Goals (SDGs). For this reason, it can be stated that the research has objectives in line with SDG 12: Responsible Consumption and Production and SDG 13: Climate Action.

Keywords: circular economy; sustainability; food waste; bibliometric analysis

Citation: Baybars, M.; Demircan Keskin, F.; Kazançoğlu, İ.; Kazançoğlu, Y. Decoding the Nexus: A Bibliometric Review on Sustainability, Circular Economy, and Consumer Food Waste. *Sustainability* **2024**, *16*, 4408. https://doi.org/10.3390/su16114408

Academic Editors: Idiano D'Adamo, Massimo Gastaldi and Manoj Kumar Nallapaneni

Received: 7 December 2023
Revised: 1 February 2024
Accepted: 7 March 2024
Published: 23 May 2024

Copyright: © 2024 by the authors. Licensee MDPI, Basel, Switzerland. This article is an open access article distributed under the terms and conditions of the Creative Commons Attribution (CC BY) license (https://creativecommons.org/licenses/by/4.0/).

1. Introduction

Sustainability is an understanding that invites people to think beyond individual needs and motivates them to consider the long-term consequences of their current actions. One of the main axes of sustainable thinking is the effective and efficient use of limited resources to meet current needs and consider future generations' needs. This approach, which should not be limited to the sustainability of natural resources, is a holistic understanding that needs to be reflected in environmental, economic, and social spheres. While environmental sustainability emphasizes the importance and protection of the natural environment, economic sustainability underlines the need for the efficient, responsible, and effective use of economy-related scarce resources. Finally, social sustainability refers to equity and inclusiveness within society.

In recent years, there has been an increase in individual and academic interest in sustainability. Sustainability-oriented thinking and practices, which impact macro-scale practices at the level of institutions or enterprises and individual practices, have increased with increasing awareness on various occasions [1,2]. On the other hand, academic studies in this field have also risen significantly with increasing interest, awareness, and thus increasing examples of practice at institutional and individual levels [3,4].

The increasing interest in this field at both individual and academic levels has led to an intensification of studies in the field, which has necessitated a comprehensive evaluation of the existing studies. The field of sustainability, which is located at the intersection of many fields, is an interdisciplinary field of study with a wide range of evaluation with studies involving many sub-fields, from economics to sociology and public administration to consumer behavior. For this reason, this study aims to provide a valuable and structured framework by examining the extensive body of work in the field.

The United Nations (UN) Sustainable Development Goals Report data distinguished between food waste and food loss, highlighting 13% of food loss in production chains as well as 17% of food waste at the consumer level [5]. Food waste is a critical issue to prevent, not only in terms of economic sustainability but also in terms of the sustainability of natural resources and social justice and therefore social sustainability. Therefore, it is required to be examined in depth.

Understanding and gaining insights into the causes of food waste, which originates from consumers, is one of the most important steps to prevent it. This study, which aims to develop an in-depth understanding of the field to develop effective strategies to reduce this waste, is an important effort for a more sustainable universe by shedding light on the intersection of sustainability, circular economy, and food waste.

The bibliometric analysis method, which will be used in this study, is one of the leading tools to examine and synthesize comprehensive information in a field. Especially in intersectional fields such as this particular research study area, where many fields intersect and have rich literature, a bibliometric analysis emerges as a method that enables valuable results to be obtained. A bibliometric analysis, which differs from a traditional literature review in several respects, was chosen because of its potential for quantitative insights, its competence in visual representation, and its superiority in terms of enabling the identification of objective trends and patterns in the literature. Through this method, this study addressed the following research questions (RQs) by conducting descriptive analyses and network analyses using the R Bibliometrix package and VOSviewer, examining publications at the intersection of sustainability, circular economy, and food waste in the Web of Science (WoS) database until the end of August 2023.

RQ1. What are the variations in annual publications and citation trends within the research field?
RQ2. What keywords have shown the highest frequency of use and the most frequent co-occurrence within publications in the research field?
RQ3. Which countries demonstrate the highest research productivity, and what are the collaboration patterns among these countries within the research field?
RQ4. Who are the most prolific authors, what co-authorship patterns emerge among them, and which sources publish the most studies within the research field?

2. Food Waste, Sustainability, and Circular Economy

In line with the 2030 Agenda for Sustainable Development announced by the United Nations, our study focuses on key topics, such as responsible consumption and production (SDG 12) and climate action (SDG 13). The growing interest in these Sustainable Development Goals (SDGs) is also reflected in the literature [6]. While both are important for sustainability, food waste and food loss are distinguished from each other by definition by the UN [5]. Food loss is defined as food becoming unusable due to inefficiencies that occur during production, post-harvest processing, transportation, etc., before the product reaches the final consumer; food waste, on the other hand, is explained as the disposal of edible food as a result of consumer-induced practices such as overbuying, improper storage, etc., or personal preferences [7–9]. It has been stated that nations' economic development is directly related to the food waste or loss they cause. It has been emphasized that developed countries are more responsible for food waste due to reasons such as more effortless access to food.

In contrast, developing countries are more responsible for food loss due to ineffective management of inefficiency in the process [10–14]. Although food waste production varies according to countries' development levels, the problem is characterized as a universal sustainability issue affecting the whole world [15]. The economic cost of food waste from consumers is higher than food loss due to the product's added value until it reaches the consumer [16]. Therefore, understanding food waste at the consumer level is critical for achieving sustainability goals.

Food waste can be expressed as a holistic process that cannot be explained only by individual attitudes, intentions, and behaviors [17]. It is essential to understand its multifaceted nature to reduce food waste, which is influenced by social norms, cultural practices, and structural variables. It has also been stated that efforts to raise awareness and consciousness only at the individual level do not have a significant effect if other structural factors are ignored [9,17]. Despite the existence of individual awareness or positive attitudes or intentions to avoid food waste [18,19], the inability to prevent food waste suggests the importance of focusing on structural solutions with a macro approach. It is clear that studies that point to interdisciplinary collaboration and guide policymakers toward systemic change will contribute to a more sustainable food ecosystem.

Once it is accepted that individual factors do not fully explain food waste behavior, systemic influences must be closely examined. Culture is a phenomenon that should not be ignored when discussing food waste. For this very reason, studies indicate that taking steps to prevent food waste by considering local dynamics and using culturally adapted language that avoids giving generalized messages yields more meaningful results in the fight against food waste [20]. It is known that food culture patterns that differ from region to region, sometimes even within the same country, will lead to diversity in approaches to food waste and related behaviors [21,22]. Within the scope of individual practices related to food waste, contextual factors such as family composition, income level, and individual differences need to be considered alongside structural influences, such as culture [21,23,24]. Even time-based differences such as holidays and celebrations have the potential to create variability in wastage due to shopping and preparations [25,26]. Therefore, delivering a standard message about food waste to all consumers, even if they are located in the same geographical region, would reduce effectiveness, so it is crucial to provide consumers with customized messages that are appropriate for their segment [27].

Studies on food waste can be quite complex as they adopt a wide range of methodological approaches, focusing on various stages of waste and using measurement methods without adhering to a standard [28]. As a result of this situation, it can be stated that although there are a large number of studies in the field, the studies are disconnected from each other. Therefore, the synergistic potential of the research in the field is not realized. Because of the importance of analyzing this extensive literature in a structured way and with a common framework, it is crucial to organize the research in such a way that a greater synergy can be achieved. The present study aims to provide a clear view to researchers, practitioners, and policymakers working in this field by structuring the various studies on food waste around emerging themes, trends, and concepts.

Since existing studies on sustainability and consumer behavior are based either on traditional literature analysis [9,17] or consumer data [7,8,13], the current study addresses the lack of a comprehensive bibliometric analysis focusing on quantitative insights in the field.

3. Materials and Methods

This study employed a systematic approach to address the research questions posed. Initially, each set's search sets and relevant keywords were defined, and searches were conducted within WoS, one of the most widely utilized databases in the literature. The retrieved studies were subsequently selected in accordance with the research area, and datasets were prepared for analysis through the cleaning and standardization processes. Following this, bibliometric analyses were conducted using the R Bibliometrix package and VOSviewer.

3.1. Data Selection and Preparation for the Analysis

To systematically cover the breadth of academic research conducted within the scope of this research, three specific search categories were identified: "Circular Economy and Sustainability", "Food Waste", and "Consumer". On 21 August 2023, scientific studies that included at least one of the predefined keywords from these three different categories were found, as summarized in Table 1. The data were imported from the WoS database.

Table 1. Search sets.

Sets	Search Terms
Circular economy and sustainability	"Circular econom*" OR "Circular business model*" OR "Circular supply chain*" OR "Circular design*" OR "Closed-loop system*" OR "Resource efficien*" OR "Waste manag *" OR "Sustainable product*" OR "Industrial ecolog*" OR "Resource recover*" OR "Extended producer responsibilit*" OR "Remanufactur*" OR "Sustainable consumpt*" OR "Ecological footprint*" OR"Green procurement*"
Food waste	"Food wast*" OR "Food loss*" OR "Food dispos*" OR "Food recov*" OR "Food supply*" OR "Food consump*" OR "Food recycl*" OR "Food redistribut*" OR "Sustainable food*" OR "Food packaging wast*"
Consumer	"consum*"

The search was refined within the WoS categories, excluding research areas outside the scope of this study. Within the obtained dataset, duplicated studies were eliminated, and the expressions of terms were standardized. For instance, terms such as "LCA" and "life cycle assessment (LCA)" were changed to "life cycle management", while "consumer behavior" was changed to "consumer behavior". Following the editing process, a dataset comprising 1448 studies, spanning 1994 to 2023, was compiled. The main details about the studies encompassed within the analysis are presented in Table 2.

Table 2. Summary of the analysis data.

Main Information about Data	
Timespan	1994:2023
Sources (Journals, Books, etc.)	440
Documents	1448
Document Types	
Article	1147
Book Chapter	19
Data Paper	1

Table 2. *Cont.*

Main Information about Data	
Early Access	42
Proceeding Paper	71
Book Review	1
Correction	2
Editorial Material	5
Meeting Abstract	3
Review	157
Authors	
Authors	4799
Author Appearances	5637
Authors of Single-authored Documents	101
Authors of Multi-authored Documents	4698
Authors Collaboration	
Single-authored Documents	103
Co-Authors per Documents	3.89
Collaboration Index	3.49

The dataset mainly comprises articles, accounting for approximately 79% (1147), followed by reviews at 11% (157), Proceeding Papers at 5% (71), and Early Accesses constituting 3% (42). The studies within the dataset were published in 440 different sources and authored by 4799 authors. Among these studies, 103 were single-authored, while 1345 were collaboratively authored. One of the indicators that illustrates the extent of collaboration among authors in the research field is the average number of co-authors per study, which is 3.89. Meanwhile, the collaboration index indicator, calculated by dividing the total number of authors in multi-authored studies by the total number of multi-authored studies [29], yields a value of 3.49.

3.2. Bibliometric Analysis

The bibliometric methodology, which involves applying various quantitative techniques to bibliometric data, has seen a significant rise in popularity, especially in recent years. This increase can be attributed to several factors, including the availability of large-scale databases like WoS and Scopus that provide bibliometric data. Moreover, numerous software tools such as R version 3.6.3, VOSviewer version 1.6.19, and Gephi (accessed date: 21 August 2023) enable the analysis of bibliometric data to be more accessible, further fueling the growing utilization of bibliometric analysis [30]. Bibliometric analyses encompass the descriptive analysis of bibliometric data and examining networks encompassing keywords, authors, citations, countries, and their connections [31]. In this study, various aspects of research within the field have been investigated, including how the annual scientific production within the research field evolved over the years, the annual citation patterns within the research field, the most frequently used keywords and their co-occurrence patterns, the most prolific countries and authors in the research field, collaboration patterns among countries and authors within the research field, and the leading sources within the research field.

4. Results

As a result of the bibliometric analyses applied to the dataset within the scope of this research, various insights have been obtained on the research field. These insights include the year-by-year evolution of publications in the field up to the end of August 2023, annual citation patterns within the research area, the most frequently used keywords and their co-occurrence patterns, the most productive countries and authors within the research domain, collaboration patterns among countries and authors in the research area, and the primary sources contributing to the field. This section presents all of these findings.

4.1. Annual Scientific Production

The yearly publication figures within the research field are illustrated in Figure 1. The number of publications gradually increased, especially in the last ten years [14,32]. Between 1994 and 2008, the yearly number of studies within the research field remained below 10. However, starting from 2009, the annual number of studies in the field began to surpass 10, and by 2018, it had exceeded 100. In the research field, where the number of studies has been on the rise since 2014, an analysis of the last ten years shows that the most substantial increases compared to the previous year took place in 2015 and 2021, respectively. This increasing trend in annual study counts continued, and as of August 2023, the number of publications in 2023 reached 67% of the total number of publications produced in 2022.

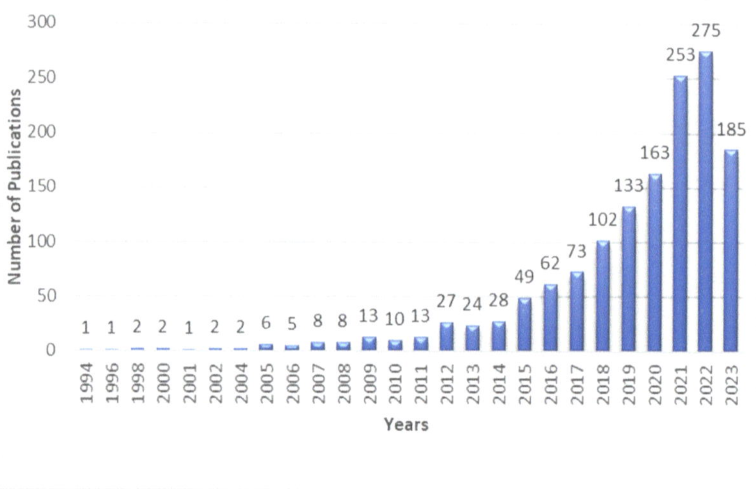

Figure 1. Annual scientific production.

4.2. Annual Citations

The yearly citation patterns within the research field are analyzed through the indicators of total publications (TPs), total citations (TCs), and Citations per Paper per year (CPY = Citations per Paper/Citable years). Citations per Paper are calculated by dividing the TCs by the TPs. In Figure 2, the annual variations in the TCs and TPs are depicted, while in Figure 3, the annual changes in the CPY and TCs are illustrated.

In the field of research, the number of publications has consistently increased each year since 2014, while the citation numbers have exhibited a fluctuating trend. Decreases in citation counts were observed in 2019, 2021, and 2022 compared to the previous year. The most significant decline occurred in 2022.

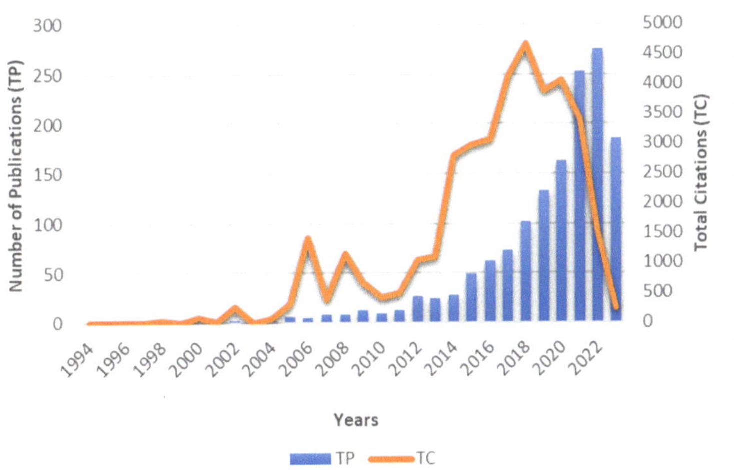

Figure 2. Annual publications–annual citations.

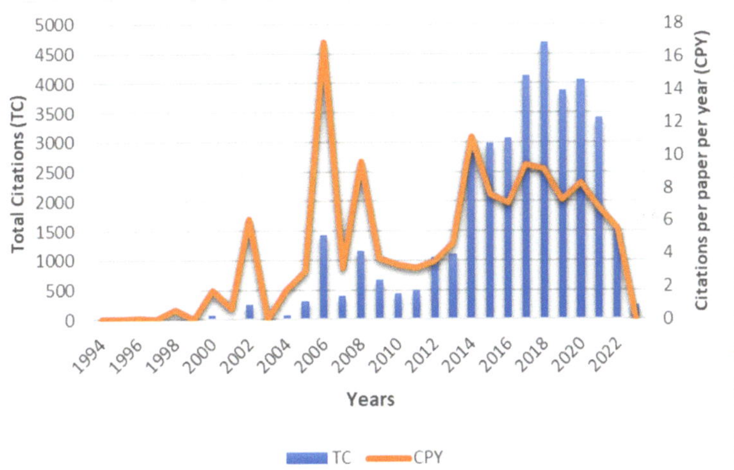

Figure 3. Annual citations–citations per paper per year.

Figure 3 indicates that although TCs have generally followed a similar path to the CPY, TCs increased in 2015, 2016, and 2018, while the CPY declined and TCs decreased in 2020, and the CPY increased. The CPY indicator, which provides a comprehensive viewpoint by considering the number of studies and the years cited together, reveals that 2006 was the most productive year. The article "Sustainable food consumption: Exploring the consumer' attitude–behavioral intention' gap", authored by [33] and published in the *Journal of Agricultural and Environmental Ethics*, has received 1258 citations in the WoS database. This study, which has the highest citation count among the publications covered in the research, played a crucial role in achieving the top position within the citation indicators for the year 2006.

Based on the CPY indicator, 2006 is followed by 2014 (CPY = 11.1) and 2008 (CPY = 9.64), respectively. In terms of the total citation count, the studies of [34–36] push these years to the top positions.

4.3. Co-Occurrence Analysis of Keywords

The most frequently used keywords were determined, and a keyword co-occurrence analysis was conducted to uncover recurring keywords and research patterns at the intersections of sustainability, circular economy, and food waste. The most frequently encountered terms include "food waste", "sustainability", "circular economy", "sustainable consumption", and "food", all of which are part of the search sets. Beyond the terms within these sets, words such as "life cycle assessment", "ecological footprint", "food supply chain", "environment", "organic food", "sustainable food", and "food security" have also demonstrated noteworthy usage.

Figure 4 displays the co-occurrence network, which encompasses keywords mentioned at least ten times within the Author Keywords sections of all the documents in the dataset. The co-occurrence analysis of these keywords resulted in seven primary clusters. Figure 5 presents the clusters and corresponding keywords within the clusters with their frequencies.

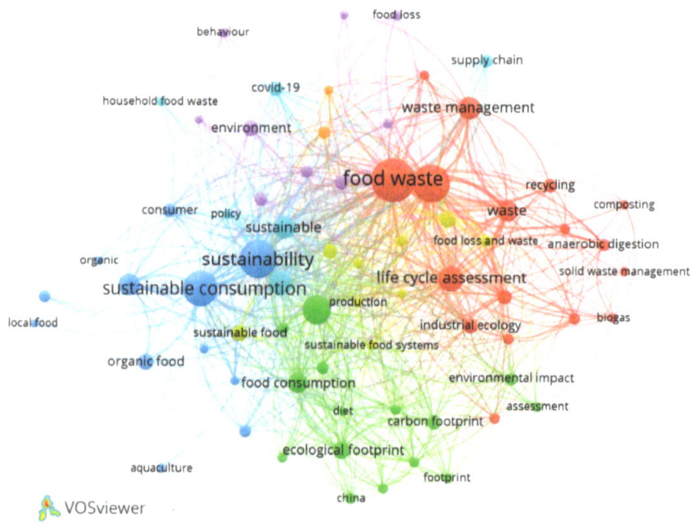

Figure 4. Co-occurrence network of most used keywords.

Cluster 1 (red)

- food waste (238)
- circular economy (168)
- life cycle assessment (76)
- waste management (68)
- waste (56)
- industrial ecology (26)
- management (26)
- recycling (22)
- anaerobic digestion (19)
- material flow analysis (17)
- food waste management (15)
- resource recovery (14)
- energy (13)
- bioeconomy (12)
- biogas (12)
- composting (11)
- solid waste management (11)

Cluster 2 (green)

- food (105)
- food consumption (51)
- ecological footprint (39)
- carbon footprint (26)
- agriculture (21)
- environmental impact (21)
- nutrition (17)
- china (16)
- footprint (15)
- environmental impacts (14)
- water (14)
- assessment (13)
- sustainable diets (13)
- water footprint (13)
- diet (10)

Cluster 3 (dark blue)

- sustainability (181)
- sustainable consumption (156)
- consumer behaviour (57)
- organic food (34)
- consumer (23)
- sustainable food consumption (21)
- climate change (19)
- meat (12)
- aquaculture (11)
- organic (11)
- survey (11)
- local food (10)

Figure 5. *Cont.*

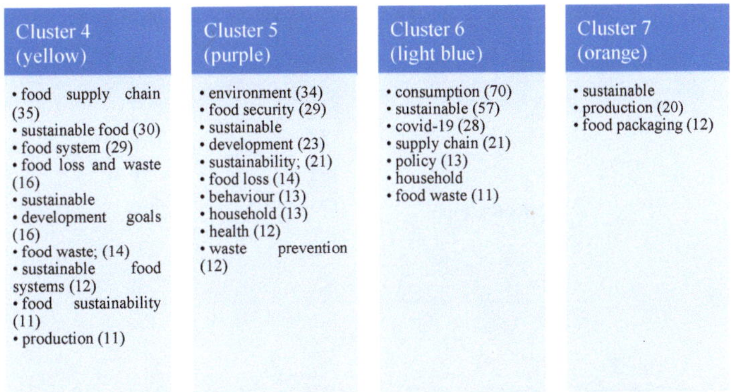

Figure 5. Clusters and corresponding keywords within clusters (frequencies).

The first cluster, named Waste and Sustainability, centers around food waste, circular economy, life cycle assessment, and waste management. The keywords co-occurring in this cluster implied a holistic exploration of waste management, circular economy principles, and the potential for value extraction from waste. The second cluster, Food and Consumption, focuses on food, incorporating food consumption and diet. The cluster collectively emphasizes evaluating the overall sustainability of food choices, providing a concise exploration of intersections between food, environmental impact, and sustainability.

Consumer and Sustainability, the third cluster, focuses intensely on sustainability and addresses sustainability in consumer behavior, emphasizing food choices, organic practices, and their broader implications for climate and local economies. The cluster that concentrates on the food supply chain, emphasizing sustainability and key themes within food systems, is cluster four, named Food Chain and Goals. This cluster comprehensively examines sustainable practices within the food supply chain, encompassing waste reduction, adherence to development goals, and promoting resilient food systems.

The fifth cluster, Environment and Security, delves into ecological considerations with the environment at its core. It centers on the intersection of environmental concerns, food security, and sustainable development, exploring various sustainability dimensions. The sixth cluster, Consumption and Policy, reflects the effects of the COVID-19 pandemic on consumption dynamics. It provides a perspective on the interplay between consumption, sustainability, the COVID-19 pandemic, and policy influences. The last cluster, Production and Packaging, focuses on sustainable production, particularly addressing the environmental impact of food packaging.

4.4. Co-Authorship Analysis: Countries

Patterns of collaboration, indicative of a more intense form of interaction, can be explored by creating co-authorship networks at the level of authors, organizations, or countries, thus facilitating the study of interactions within the collaborative realm [37,38]. The country analysis and the dynamics of collaborations are very important in terms of showing which countries are most interested in the field of research and shaping the dissemination of knowledge worldwide.

The analysis is based on co-authorship among countries considered to have participated in at least ten studies within the research field. A total of 44 countries met this criterion. Table 3 presents information about the top 10 countries with the most publications in the research field. It includes the TPs, TCs, Citations per Paper (CPs), and Total Link Strength (TLS) values. The TLS represents the collective intensity of the co-authorship connections between a specific country and others [39].

Table 3. Publication, citation, and collaboration indicators for top 10 productive countries.

Country	TP	TC	CP	TLS
Italy	211	6529	30.94	187
England	172	6036	35.09	198
USA	162	4805	29.66	187
Peoples R China	160	3205	20.03	158
Germany	102	2488	24.39	117
Spain	92	2565	27.88	128
India	81	1154	14.25	70
Netherlands	75	3770	50.27	70
Sweden	72	2042	28.36	75
Australia	67	1791	26.73	75

Italy, England, and the USA have the highest TPs and TCs in the research field. However, a different result emerges among the top three countries when considering the CP metric. In this case, Belgium, Austria, and Finland, which fall within the range of 20 to 30 in the TP-based ranking, are leading in CPs. Even though these countries have not contributed to as many collaborative studies as the countries at the top of the list, the research produced by the authors from these countries has received many citations. Of the top 10 countries based on TPs, India, China, and Germany have the lowest CP scores. This result suggests that despite these countries producing a substantial number of studies, the average number of citations per publication is comparatively lower.

Country co-authorship networks are highly relevant as they visually represent global research collaboration for understanding countries' efforts in a field that transcends geographical boundaries. This information is particularly valuable for researchers, institutions, or policymakers to identify potential collaborations. Furthermore, to build broader research relationships in the future, it is critical to recognize the existing strong links between countries. Through analyzing collaborative authorship relationships between countries, five distinct clusters were identified, each containing countries frequently participating in collaborative research. The visual representation of these clusters is shown in Figure 6. In the network depiction, countries with the highest productivity are shown with larger nodes. According to the TLC metric, which shows the strength of the co-authoring links between authors from different countries, the top three countries with the strongest co-authorship relationships are England, Italy, and the USA, in descending order.

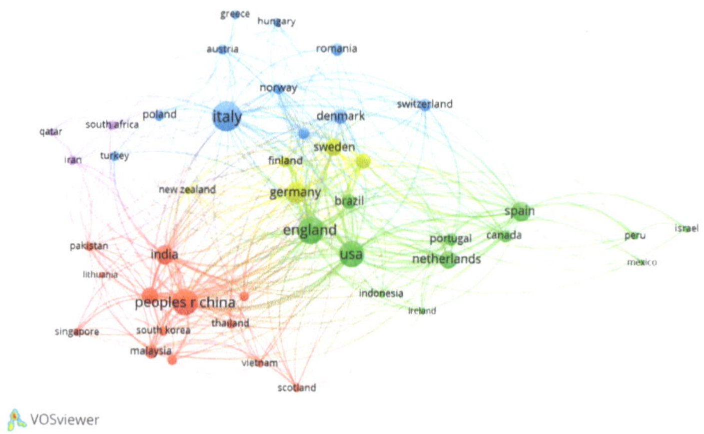

Figure 6. Co-authorship network of countries.

4.5. Co-Authorship Analysis: Authors

A co-authorship analysis was undertaken to investigate the collaborative patterns observed among authors who have contributed significantly to the research field under consideration. A total of 4797 authors were identified as contributors, of whom 57 individuals had participated in a minimum of four studies. Notably, the authors with the highest TPs are Filimonau, Viachaslau; Laso, Jara; Margallo, Maria; and Sala, Serenella, as presented in Table 4. It is important to recognize the most influential names in the field and identify individuals who have made highly influential contributions to enhance the collaborative nature of the discipline. These authors have significantly impacted the field through their publications and citations. Therefore, the values associated with the citations are also of great importance.

Table 4. Publication, citation, and collaboration indicators of the most productive authors.

Author	TP	TC	CP	TLS
Filimonau, Viachaslau	14	402	28.71	2
Laso, Jara	13	375	28.85	39
Margallo, Maria	10	356	35.60	35
Sala, Serenella	10	1360	136.00	0
Hoehn, Daniel	9	304	33.78	34
Aldaco, Ruben	8	361	45.13	28
Vazquez-Rowe, Ian	8	159	19.88	21
El Bilali, Hamid	6	134	22.33	1
Galli, Alessandro	6	272	45.33	1
Mangla, Sachin Kumar	6	157	26.17	0
Martens, Pim	6	50	8.33	7
Rousta, Kamran	6	218	36.33	5
Woolley, Elliot	6	131	21.83	0

Furthermore, it is worth highlighting that Verbeke, Wim (TC = 1977, CP = 395.4) and Vermeir, Iris (TC = 1969, CP = 393.8) emerged as the authors with the highest TCs and CPs, although they are not among the authors listed with the highest TPs. Following them in terms of TCs is Sala, Serenella (TC = 1360, CP = 136), who also has a position within the top three authors based on TPs. Sala, Serenella's position in TCs can be attributed to her significant role as a co-author in several highly cited works, most notably the studies of [35] with 595 citations and [40] with 338 citations.

Figure 7 visually represents the co-authorship density among authors involved in at least four studies. In this depiction, authors with strong co-authorship connections are represented by dark yellow shading, while dark blue shading represents those with weaker co-authorship links. The TLS metric, which signifies the strength of the co-authorship connections, highlights Laso, Jara; Margallo, Maria; and Hoehn, Daniel as the top three contributors.

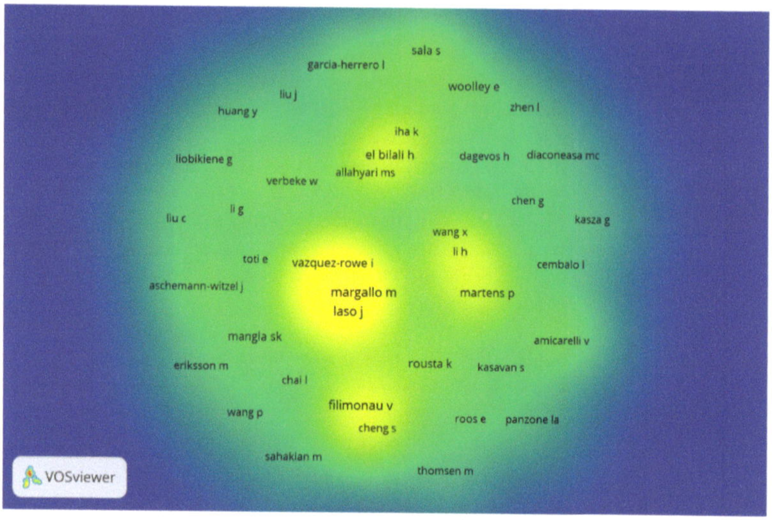

Figure 7. Co-authorship density visualization of authors.

4.6. Leading Sources

There have been 440 sources, including journals, books, conference proceedings, and reports, in which the studies within the context of this study were published. Among these sources, 44 published at least five studies in the research field. The leading ten sources with the highest publications are presented in Table 5, with the TC and CP indicators.

Table 5. Publication and citation indicators of 10 sources with the highest TPs.

Source	TP	TC	CP
Sustainability	178	2643	14.85
Journal of Cleaner Production	149	7663	51.43
British Food Journal	45	654	14.53
Foods	41	768	18.73
Science of The Total Environment	39	1577	40.44
Waste Management	37	2287	61.81
Resources Conservation and Recycling	36	1426	39.61
Sustainable Production and Consumption	29	430	14.83
Journal of Industrial Ecology	26	860	33.08
Frontiers in Sustainable Food Systems	23	105	4.57

All the sources in the top ten list are journals. *Sustainability*, the *Journal of Cleaner Production*, and the *British Food Journal* are the top three journals with the highest TPs, respectively. On the TCs side, the *Journal of Cleaner Production*, *Sustainability*, and *Waste Management* are in the top three. The CP metric resulted in a different picture. The journals *Ecological Economics* (TP = 14, CP = 124.50), *International Journal of Production Economics* (TP = 5, CP = 18.40), and *Food Policy* (TP = 11, CP = 76.18), while not having positions within the TP-based top 10 rankings, emerged as the top three in the CP list. This finding highlights the fine distinction in evaluating journals, pointing to the complex interplay between productivity and impact. Within the subset of journals ranked in the top 10 based on TPs, an internal CP ranking reveals that *Waste Management* (CP = 61.81), the *Journal of*

Cleaner Production (CP = 51.43), and *Science of The Total Environment* (CP = 40.44) take the top three positions.

The number of articles published by these ten journals in the last ten years on an annual basis is depicted in Figure 8. Until 2019, the journal with the highest number of publications in the intersection of sustainability, circular economy, and food waste was the *Journal of Cleaner Production*. However, starting from 2020, *Sustainability* took the lead. In 2022, both journals saw a decrease in published studies in the research area. The publication-based rankings of other journals have shown variability.

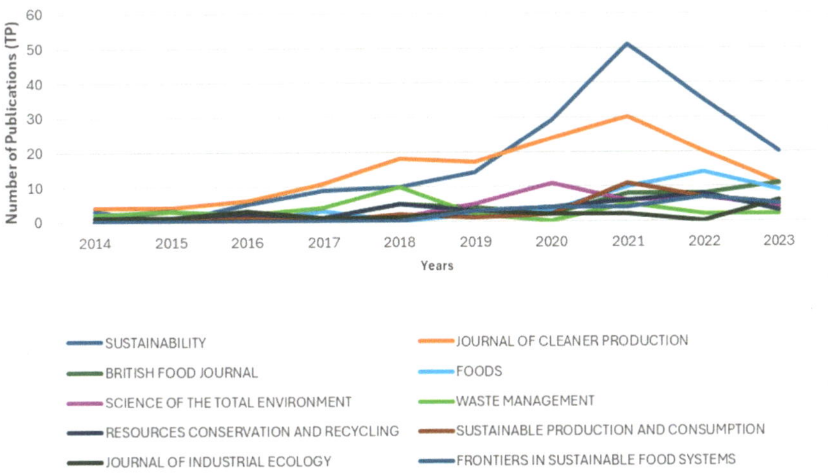

Figure 8. Annual publications of the top journals.

These journals demonstrate distinct scopes based on the most frequently occurring keywords and their co-occurrence. *Sustainability* has a broad focus encompassing circular economy, food waste, sustainable consumption, and other related topics. In contrast, the *Journal of Cleaner Production* emphasizes food waste, sustainable consumption, life cycle assessment, circular economy, and waste management. The *Science of The Total Environment Journal* focuses on the circular economy, life cycle assessment, food waste, and related environmental aspects. *Foods* covers sustainable consumption, food waste, circular economy, food security, and various food production and consumption aspects. Similarly, the *British Food Journal* shares similar focus areas, emphasizing sustainable consumption, food waste, circular economy, and consumer behavior, including meat consumption. The comparison highlights the subtle thematic orientations of these journals in the context of this research field.

5. Conclusions and Future Research

As mentioned in the introduction, the present study, which examines consumer food waste in the context of sustainability and circular economy, aims to provide a meaningful and structured framework by analyzing the extremely rich content of the field. The intersection of sustainability, circular economy, and food waste brings with it the advantage of the rich diversity of the field and the need to analyze and standardize this rich content that speaks different languages. Our study aims to provide a structured examination of the interaction of these fields by aiming to take advantage of this diversity.

With the current study focusing on the intersection of sustainability, circular economy, and consumer food waste, we prioritize the UN's goal of action to mitigate climate change (SDG 13) and responsible consumption and production (SDG 12).

Beyond the traditional literature review, employing a bibliometric analysis, which allows for quantitative insights into the broad and interactive nature of the field, this study prioritizes evaluating the findings from a unique perspective.

By identifying the publication trends and the most co-occurring keyword clusters, revealing co-authorship patterns, identifying prolific authors, and revealing the field's leading sources and their focal points, the current study's findings contribute to understanding the development of the field, identifying related concepts in the field, and articulating models of collaboration on a global scale.

Publication indicators reveal a continuous growth trend in the research field, consistent with the studies by [12,41]. Their findings indicate a surge in annual publications from 2017 to 2021, particularly in circular economy and food waste and losses. Publications through August 2023 suggest this upward trend will continue, underscoring the continued importance and expansion of research in these areas.

A co-authorship analysis of countries indicates that, in the field of research, Italy is the most productive country with the most publications, as stated in the study by [41]. Ref. [12] stated that India, one of the countries that stands out in collaborative publications in the research field, has publications in the research field. The most dominant journals in the research field with the most publication numbers are *Sustainability* and the *Journal of Cleaner Production*, in line with [41,42].

Also, by analyzing trend topics and frequently used words in recent research and examining recent studies conducted by authors with the highest collaboration in the research field, this study attempts to identify the research trends. One key trend centers around the imperative to ensure food security and enhance the sustainability of seafood production systems. Researchers are actively addressing the challenges arising from the escalating global demand for seafood, analyzing supply chains, and exploring ways to support sustainability merits. Another focal point is the comprehensive approach to addressing contamination issues in fish and aquaculture products, emphasizing the fishing sector's short- and medium-term sustainability and the critical goal of ensuring food safety. Additionally, there is a growing emphasis on understanding consumer attitudes and behaviors related to the environmental impact of seafood packaging, with a concerted effort to explore consumers' willingness to actively contribute to reducing this impact.

Furthermore, the field explores the nuanced interplay between objective and subjective knowledge and its impact on organic purchase intentions. It reflects a heightened interest in understanding consumer choices in sustainable and organic food preferences. Collectively, these research trends underscore a multifaceted approach aimed at fostering sustainability, environmental responsibility, and informed consumer decision-making in seafood and related industries.

However, the reliance on the WoS database represents a significant limitation of this study. Future studies are thought to improve their scope by including publications in databases, such as Scopus.

Funding: This research received no external funding.

Conflicts of Interest: The authors declare no conflicts of interest.

References

1. Cogut, G.; Webster, N.J.; Marans, R.W.; Callewaert, J. Links between sustainability-related awareness and behavior: The moderating role of engagement. *Int. J. Sustain. High.* **2019**, *20*, 1240–1257. [CrossRef]
2. Alsaati, T.; El-Nakla, S.; El-Nakla, D. Level of sustainability awareness among university students in the eastern province of Saudi Arabia. *Sustainability* **2020**, *12*, 3159. [CrossRef]
3. Cullen, J.G. Educating business students about sustainability: A bibliometric review of current trends and research needs. *J. Bus. Ethics* **2017**, *145*, 429–439. [CrossRef]
4. Kajikawa, Y.; Tacoa, F.; Yamaguchi, K. Sustainability science: The changing landscape of sustainability research. *Sustain. Sci.* **2014**, *9*, 431–438. [CrossRef]
5. United Nations. The Sustainable Development Goals Report 2022. 2022. Available online: https://unstats.un.org/sdgs/report/2022/The-Sustainable-Development-Goals-Report-2022.pdf (accessed on 1 February 2024).

6. Ali, S.M.; Appolloni, A.; Cavallaro, F.; D'Adamo, I.; Di Vaio, A.; Ferella, F.; Gastaldi, M.; Ikram, M.; Kumar, N.M.; Martin, M.A.; et al. Development Goals towards Sustainability. *Sustainability* **2023**, *15*, 9443. [CrossRef]
7. Dobernig, K.; Schanes, K. Domestic spaces and beyond: Consumer food waste in the context of shopping and storing routines. *Int. J. Consum. Stud.* **2019**, *43*, 480–489. [CrossRef]
8. Stancu, V.; Haugaard, P.; Lähteenmäki, L. Determinants of consumer food waste behaviour: Two routes to food waste. *Appetite* **2016**, *96*, 7–17. [CrossRef]
9. Dos Santos, J.I.A.S.; da Silveira, D.S.; da Costa, M.F.; Duarte, R.B. Consumer behaviour in relation to food waste: A systematic literature review. *Br. Food J.* **2022**, *124*, 4420–4439. [CrossRef]
10. Chauhan, Y. Food Waste Management with Technological Platforms: Evidence from Indian Food Supply Chains. *Sustainability* **2020**, *12*, 8162. [CrossRef]
11. Li, H.; Pan, P. Food waste in developed countries and cold chain logistics. *E3S Web Conf.* **2021**, *251*, 03001. [CrossRef]
12. De Oliveira, M.M.; Lago, A.; Dal'Magro, G.P. Food loss and waste in the context of the circular economy: A systematic review. *J. Clean. Prod.* **2021**, *294*, 126284. [CrossRef]
13. Gaiani, S.; Caldeira, S.; Adorno, V.; Segre, A.; Vittuari, M. Food wasters: Profiling consumers' attitude to waste food in Italy. *Waste Manag.* **2018**, *72*, 17–24. [CrossRef]
14. Tamasiga, P.; Miri, T.; Onyeaka, H.; Hart, A. Food waste and circular economy: Challenges and opportunities. *Sustainability* **2022**, *14*, 9896. [CrossRef]
15. United Nations. The Sustainable Development Goals Report 2023. 2023. Available online: https://unstats.un.org/sdgs/report/2023/The-Sustainable-Development-Goals-Report-2023.pdf (accessed on 4 February 2024).
16. Aktas, E.; Sahin, H.; Topaloglu, Z.; Oledinma, A.; Huda, A.K.S.; Irani, Z.; Sharif, A.M.; van't Wout, T.; Kamrava, M. A consumer behavioural approach to food waste. *J. Enterp. Inf. Manag.* **2018**, *31*, 658–673. [CrossRef]
17. Schanes, K.; Dobernig, K.; Gozet, B. Food waste matters—A systematic review of household food waste practices and their policy implications. *J. Clean. Prod.* **2018**, *182*, 978–991. [CrossRef]
18. Abeliotis, K.; Lasaridi, K.; Chroni, C. Attitudes and behaviour of Greek households regarding food waste prevention. *Waste Manag. Res.* **2014**, *32*, 237–240. [CrossRef]
19. Stefan, V.; van Herpen, E.; Tudoran, A.; Lahteenmaki, L. Avoiding food waste by Romanian consumers: The importance of planning and shopping routines. *Food Qual. Prefer.* **2013**, *28*, 375–381. [CrossRef]
20. Schmidt, K. Explaining and promoting household food waste-prevention by an environmental psychological based intervention study. *Resour. Conserv. Recycl.* **2016**, *111*, 53–66. [CrossRef]
21. Annunziata, A.; Agovino, M.; Ferraro, A.; Mariani, A. Household Food Waste: A Case Study in Southern Italy. *Sustainability* **2020**, *12*, 1495. [CrossRef]
22. Marangon, F.; Tiziano, T.; Troiano, S.; Daniel, V. Food waste, consumer attitudes and behaviour. A survey study in the North-Eastern part of Italy. In *CAP 2014–2020: Scenarios for European Agri-Food and Rural Systems*; Universitas Studiorum Srl: Mantova, Italy, 2015; pp. 67–72.
23. Ankiel, M.; Samotyja, U. Consumer opinions on the causes of food waste—Demographic and economic conditions. *Mark. Sci. Res. Organ.* **2021**, *42*, 75–96. [CrossRef]
24. Annunziata, A.; Agovino, M.; Ferraro, A.; Mariani, A. Food waste as a consequence of an inefficient consumer's choices: A microeconomic approach. *Appl. Econ.* **2021**, *53*, 6266–6285. [CrossRef]
25. Liao, C.H.; Qiao, L.G.; Wang, X.Z.; Lu, S.S. Exploring food waste prevention through advent food consumption: The role of perceived concern, consumer value, and impulse buying. *Front. Sustain. Food Syst.* **2022**, *6*, 988260. [CrossRef]
26. Elshaer, I.; Sobaih, A.E.E.; Alyahya, M.; Abu Elnasr, A. The impact of religiosity and food consumption culture on food waste intention in Saudi Arabia. *Sustainability* **2021**, *13*, 6473. [CrossRef]
27. Aschemann-Witzel, J. Helping You to Waste Less? Consumer Acceptance of Food Marketing Offers Targeted to Food-Related Lifestyle Segments of Consumers. *J. Food Prod. Mark.* **2018**, *24*, 522–538. [CrossRef]
28. Amicarelli, V.; Bux, C. Food waste measurement toward a fair, healthy and environmental-friendly food system: A critical review. *Br. Food J.* **2021**, *123*, 2907–2935. [CrossRef]
29. Aria, M.; Cuccurullo, C. A Brief Introduction to Bibliometrix. 2019. Available online: https://www.bibliometrix.org/vignettes/Introduction_to_bibliometrix.html (accessed on 26 August 2023).
30. Donthu, N.; Kumar, S.; Mukherjee, D.; Pandey, N.; Lim, W.M. How to conduct a bibliometric analysis: An overview and guidelines. *J. Bus. Res.* **2021**, *133*, 285–296. [CrossRef]
31. Liang, T.P.; Liu, Y.H. Research landscape of business intelligence and big data analytics: A bibliometrics study. *Expert Syst. Appl.* **2018**, *111*, 2–10. [CrossRef]
32. Kumar, D.; Choudhuri, S.; Shandilya, A.K.; Singh, R.; Tyagi, P.; Singh, A.K. Food Waste & Sustainability Through A Lens of Bibliometric Review: A Step Towards Achieving SDG 2030. In Proceedings of the 2022 International Conference on Innovations in Science and Technology for Sustainable Development (ICISTSD), Kollam, India, 25–26 August 2022; pp. 185–192.
33. Vermeir, I.; Verbeke, W. Sustainable food consumption: Exploring the consumer "attitude–behavioral intention" gap. *J. Agric. Environ. Ethics.* **2006**, *19*, 169–194. [CrossRef]
34. Papargyropoulou, E.; Lozano, R.; Steinberger, J.K.; Wright, N.; bin Ujang, Z. The food waste hierarchy as a framework for the management of food surplus and food waste. *J. Clean. Prod.* **2014**, *76*, 106–115. [CrossRef]

35. Mirabella, N.; Castellani, V.; Sala, S. Current options for the valorization of food manufacturing waste: A review. *J. Clean. Prod.* **2014**, *65*, 28–41. [CrossRef]
36. Vermeir, I.; Verbeke, W. Sustainable food consumption among young adults in Belgium: Theory of planned behaviour and the role of confidence and values. *Ecol Econ.* **2008**, *64*, 542–553. [CrossRef]
37. Melin, G.; Persson, O. Studying research collaboration using co-authorships. *Scientometrics* **1996**, *36*, 363–377. [CrossRef]
38. Kumar, S.; Jan, J.M. Mapping research collaborations in the business and management field in Malaysia, 1980–2010. *Scientometrics* **2013**, *97*, 491–517. [CrossRef]
39. Van Eck, N.J.; Waltman, L. *VOSviewer manual: Manual for VOSviewer Version 1*; CWTS: Leiden, The Netherlands, 2022.
40. Notarnicola, B.; Sala, S.; Anton, A.; McLaren, S.J.; Saouter, E.; Sonesson, U. The role of life cycle assessment in supporting sustainable agri-food systems: A review of the challenges. *J. Clean. Prod.* **2017**, *140*, 399–409. [CrossRef]
41. Del-Aguila-Arcentales, S.; Alvarez-Risco, A.; Carvache-Franco, M.; Rosen, M.A.; Yáñez, J.A. Bibliometric analysis of current status of circular economy during 2012–2021: Case of foods. *Processes* **2022**, *10*, 1810. [CrossRef]
42. Chauhan, C.; Dhir, A.; Akram, M.U.; Salo, J. Food loss and waste in food supply chains. A systematic literature review and framework development approach. *J. Clean. Prod.* **2021**, *295*, 126438. [CrossRef]

Disclaimer/Publisher's Note: The statements, opinions and data contained in all publications are solely those of the individual author(s) and contributor(s) and not of MDPI and/or the editor(s). MDPI and/or the editor(s) disclaim responsibility for any injury to people or property resulting from any ideas, methods, instructions or products referred to in the content.

sustainability

MDPI

Review

Exploring Circular Economy Practices in the Healthcare Sector: A Systematic Review and Bibliometric Analysis

Carlotta D'Alessandro [1,*], Katarzyna Szopik-Depczyńska [2], Małgorzata Tarczyńska-Łuniewska [3], Cecilia Silvestri [4] and Giuseppe Ioppolo [1]

1 Department of Economics, University of Messina, 98121 Messina, Italy; giuseppe.ioppolo@unime.it
2 Department of Business Management, Institute of Management, University of Szczecin, 71-004 Szczecin, Poland; katarzyna.szopik-depczynska@usz.edu.pl
3 Department of Econometrics and Statistics, Institute of Economics and Finance, University of Szczecin, Mickiewicza 64, 71-101 Szczecin, Poland; malgorzata.tarczynska-luniewska@usz.edu.pl
4 Department of Economics, Engineering, Society and Business Organization, University of "Tuscia" of Viterbo, 01100 Viterbo, Italy; c.silvestri@unitus.it
* Correspondence: carlotta.dalessandro@unime.it

Abstract: The healthcare sector produces 2 gigatons of CO_2. To address this impactful trend and contribute to the Sustainable Development Goals (SDGs), the adoption of circular economy (CE) practices could represent a strategic target. In this context, the present article provides a systematic and bibliometric literature review of CE practices applied in the healthcare sector by considering the collected case studies. This study aims to analyze the state of the art in CEs in the healthcare sector in order to identify CE practices in healthcare, examining how they contribute to sustainability goals and the critical issues in their implementation. A final selection of 36 articles from reputable databases, Web of Science and Scopus, was obtained and analyzed using VOSviewer. By systematically examining these papers, the study investigates the key CE practices implemented within the healthcare sector and their respective areas of application, which help the broader mission of achieving SDG 12, and also, to a lesser extent, SDG 9. Although the research criteria impose some limitations, this study offers a comprehensive review of successful circular practices adopted in the healthcare sector while shedding light on existing gaps and providing valuable insights for relevant stakeholders.

Keywords: healthcare sector; bibliometric analysis; systematic analysis; circular economy; CE practices

Citation: D'Alessandro, C.; Szopik-Depczyńska, K.; Tarczyńska-Łuniewska, M.; Silvestri, C.; Ioppolo, G. Exploring Circular Economy Practices in the Healthcare Sector: A Systematic Review and Bibliometric Analysis. *Sustainability* **2024**, *16*, 401. https://doi.org/10.3390/su16010401

Academic Editors: Andrea Appolloni and Steve W. Lyon

Received: 18 October 2023
Revised: 13 December 2023
Accepted: 19 December 2023
Published: 2 January 2024

Copyright: © 2024 by the authors. Licensee MDPI, Basel, Switzerland. This article is an open access article distributed under the terms and conditions of the Creative Commons Attribution (CC BY) license (https://creativecommons.org/licenses/by/4.0/).

1. Introduction

Nowadays, humanity is facing several challenges; among the many, climate change is one of the most pernicious, threatening clean air, safe drinking water, sufficient food, and secure shelter. As we look ahead to the period between 2030 and 2050, climate change is expected to have a grave impact, resulting in an estimated annual increase in deaths of approximately 250,000, predominantly caused by malnutrition, malaria, diarrhea, and heat stress [1]. It is worth noting that the health sector, which bears the responsibility of safeguarding human well-being, has also made substantial contributions to the climate crisis in recent decades [2]. According to the Healthcare Without Harm (HCWH) Annual Report in 2022, if the healthcare sector were a country, it would be the world's fifth largest producer of greenhouse gas (GHG) pollution [3]. In an international comparative analysis using analogous information taken from a selection of 36 OECD countries at various points in time, it was noted that in 2014, the healthcare sector was accountable for emitting 2 gigatons of CO_2, equivalent to 4.4% of the global ecological footprint [4]. Therefore, scientists have stressed that quality amelioration strategies are indispensable for sustainability [5]. Sustainability is perceived as an equilibrium between the social, environmental, and economic aspects of society and the planet as a whole [6].

To enhance environmental sustainability, solutions such as a complete life cycle inventory database for medical devices and drugs, reform of contagion check standards that guide the practical use of single-use disposable devices, implementation of consolidated sustainability operation measures at the medical level, and more national research financial support are needed [7]. Examples that boost social sustainability are domiciliary telemedicine; emerging technologies to ensure accessibility to and the availability of healthcare and ensure patient satisfaction; customized treatments exploiting 3D printing technology; financing expensive drugs for global use; sustainable health training as an approach to achieve sustainability; and improved personnel recruitment [8].

Finally, research, development, and reductions in costs are some of the strategies to achieve economic sustainability; furthermore, energy safeguarding, recycling, sourcing, training programs, and employability, despite covering environmental and social aspects, also contribute to the promotion of economic sustainability [8].

In this context, the United Nations General Assembly (UN) took a significant step in September 2015 by endorsing the 2030 Agenda for Sustainable Development, which encompasses 17 SDGs. The SDGs call upon nations to collaborate with the purpose of diminishing economic disparities and safeguarding the planet for peace and peoples' well-being [9]. Addressing the environmental impact of the healthcare sector and reorienting its practices towards long-term sustainability can play decisive roles in achieving the overarching goals of the 2030 Agenda for Sustainable Development, ultimately contributing to a high quality of life for both present and future generations.

Moreover, the European Union's (EU) effort to achieve circularity and become more sustainable led to the creation of the Green Deal, a policy developed to reach zero GHG emissions and carbon neutrality and protect human health by 2050 [10,11]. In pursuit of this goal, EU countries have collectively vowed to achieve a reduction in emissions of no less than 55% by 2030 in contrast to the emission levels recorded in 1990 [10]. In order to actively support the fulfillment of the SDGs and address the environmental challenges arising from the healthcare sector, the transition to a CE, considered an emerging and innovative paradigm, holds significant promise as a viable solution. In this economic framework, the central emphasis is on reducing resource consumption and managing environmental consequences [12]. Further, as reported in [13], the shift towards a CE is the requirement for reaching sustainability. A CE is an interdisciplinary topic that encompasses several fields of expertise, with the objective of banding together the enhancement of ecological well-being and financial growth for sustainable ecological development [14]. Indeed, the CE seems to encompass a diverse array of ideas and principles, making it more akin to an umbrella concept. The CE draws inspiration from a variety of sources, including concepts like Cradle to Cradle, Industrial Ecology, Biomimicry, Performance Economy, Blue Economy, Natural Capitalism, and Industrial Capitalism, among others [15]. From this perspective, the implementation of CE strategies in the healthcare sector could represent a response to the growing environmental threats. In fact, a CE primarily aims to reduce pollution and waste while simultaneously generating economic benefits [16], and is built on the suitable and ecological utilization of resources [17]. Moreover, a CE plays a pivotal role in guiding economic development towards sustainability, representing a sustainable profitable system in which the economy improves; it is disconnected from the consumption of resources thanks to the reduction in use and recycling of natural resources [18]. Reaching a CE predominantly means ensuring the reduction in the environmental effects of production and the efficient reuse and recycling of products, developing community interventions to adapt customer behavior to CE requirements [16]. For all the above reasons, the healthcare sector, and in general, all types of institutions, can support positive environmental policies [19].

Furthermore, researchers contend that the transition towards circularity is closely connected with the digitalization transformation [20]; Industry 5.0, conceived to use man's originality expertise in cooperation with intelligent and precise apparatus, is designed to enhance customer satisfaction [21]. In this sense, Industry 5.0 is facilitating the individualized tracking of essential health metrics, such as monitoring blood pressure and blood sugar levels, and offering customized medical care [21]. Likewise, green innovation or eco-innovation can be helpful in preserving environmental management, as it is a method focused on the development of novel advancements in production and technology, all while striving to mitigate environmental hazards, such as pollution and adverse impacts stemming from the exploitation of resources [22].

This method is commonly associated with the triple bottom line concept, which encompasses social, environmental, and economic operations [23]. Due to the huge environmental impacts of the healthcare sector, and the consequent necessity to achieve sustainable development, the purpose of this paper is to comprehend how the sector embraces circularity, identifying the most representative CE practices implemented in the sector.

Indeed, reference [24] demonstrates the existence of untapped sustainability opportunities that have not been thoroughly investigated yet.

In this context, a CE could be considered as a pathway to advance specific SDGs, for example, encouraging facilities to assume more responsible and sustainable practices (Goal 12) [25].

Up to now, the existing body of literature has been more focused on the depiction of singular practice, mainly implemented for healthcare waste management (HWM). While the circularity topic is currently researched within the sector, the categorization of existing practices is necessary and requires attention; none of the analyzed papers present a comprehensive overview of the above-mentioned practices. To achieve this goal, this paper addressed the following research questions:

RQ1: What is the state of the art in a CE implemented in the healthcare sector?

RQ2: What are the main areas of CE practices applied to the healthcare sector?

Following this Introduction, Section 2 presents a description of the literature review method applied, Section 3 underlines the main findings of the study, Section 4 critically examines the results, and finally, in the last section, the conclusions with the main outcomes are summarized, defining future outlooks.

2. Materials and Methods

The present section details the methodology employed for the literature review in this study, which involved both a systematic review and a bibliometric analysis. A systematic literature review involves "replicable, scientific, and transparent procedures to collect all related publications and documents that fit pre-defined inclusion criteria to answer a specific research question" [26]. Instead, a bibliometric analysis is the identification of emerging patterns in articles and a journal's impact, the examination of collaborative networks, and the investigation of the knowledge landscape within a particular field as documented in the existing literature [27]. A combination of both methodologies was proposed to obtain coherent, trustworthy, and robust research.

Sample selection was performed following the specifications suggested by the PRISMA protocol [28]. The systematic literature review was conducted to obtain a global view of the CE practices implemented in the healthcare sector. Figure 1 illustrates the comprehensive research method to provide transparency and allow readers to understand the methodology used to ensure the appropriateness and quality of the sources included in the study.

First, the study objectives, questions, keywords, inclusion and exclusion criteria, and databases were developed. Scopus and Web of Science (WoS) were used concurrently as main sources through the employment of chosen keywords in the title, abstract, and keywords of publications according to the Boolean operators "OR" and "AND". The preference for these two sources was to ensure the inclusion of peer-reviewed articles. These articles, found in reputable journals, are considered high-quality studies, and their relevance and significance were verified [29].

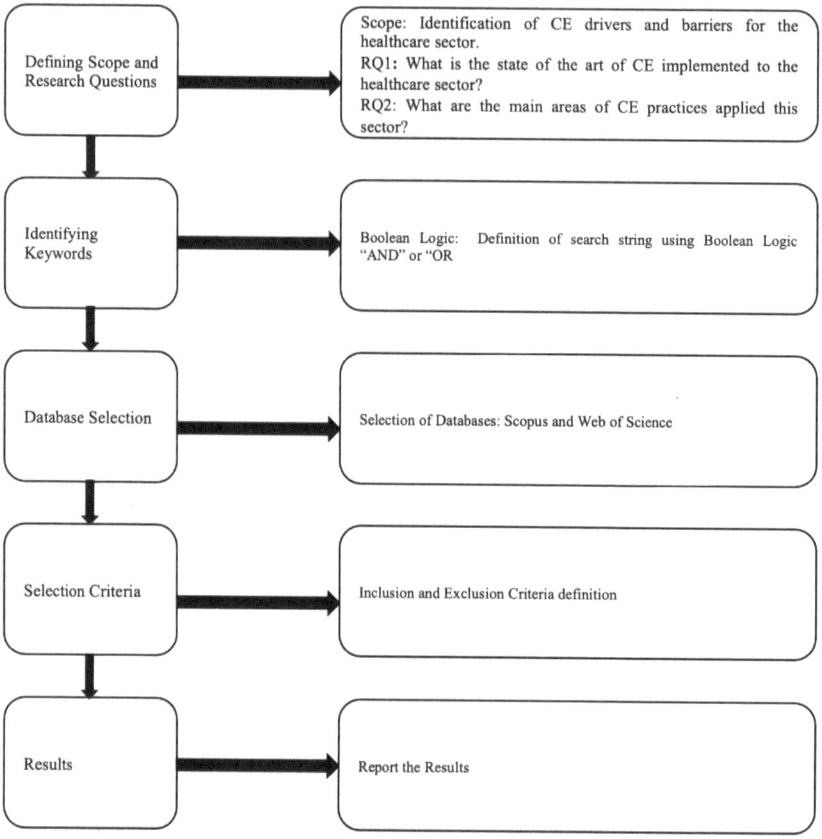

Figure 1. Methodological framework (Source: authors).

The search queries used to conduct the search are available in Table S5 in the Supplementary Materials.

Paper collection was conducted in January 2023; thus, potential forthcoming publications were not taken into account. Afterwards, inclusion (I) and exclusion (E) criteria were defined to evaluate the studies (see Table 1). In addition, articles presenting CE practices applied in multiple sectors were included if and only if one of the sectors analyzed is the healthcare sector, and only data regarding this were extracted.

Table 1. Selection criteria of the sources (Source: authors).

Inclusion Criteria	Exclusion Criteria
(a) Articles must be written in English	(a) Not aligned with the purpose of the study
(b) Accordance with the forward-looking perspective of the studies regarding circularity in the healthcare sector	(b) Inadequacy of information
(c) Main sector of application is healthcare	(c) If the document is a Conference paper; Conference Proceeding; Review; or Book Chapter
(d) Must present CE practices applied to the healthcare sector to face environmental challenges	

In the beginning, the above research strategy initially allowed the identification of a total of 324 articles: 203 from Scopus and 121 from WoS. The two results were merged in Microsoft Excel for Mac, Version 16.80. License Microsoft 365 to carry on the screening process (as shown in Figure 2). The PRISMA flowchart in Figure 2 enables readers to gain a comprehensive understanding of the systematic approach utilized for the identification and analysis of the literature, thereby reinforcing the credibility and reproducibility of the study's findings.

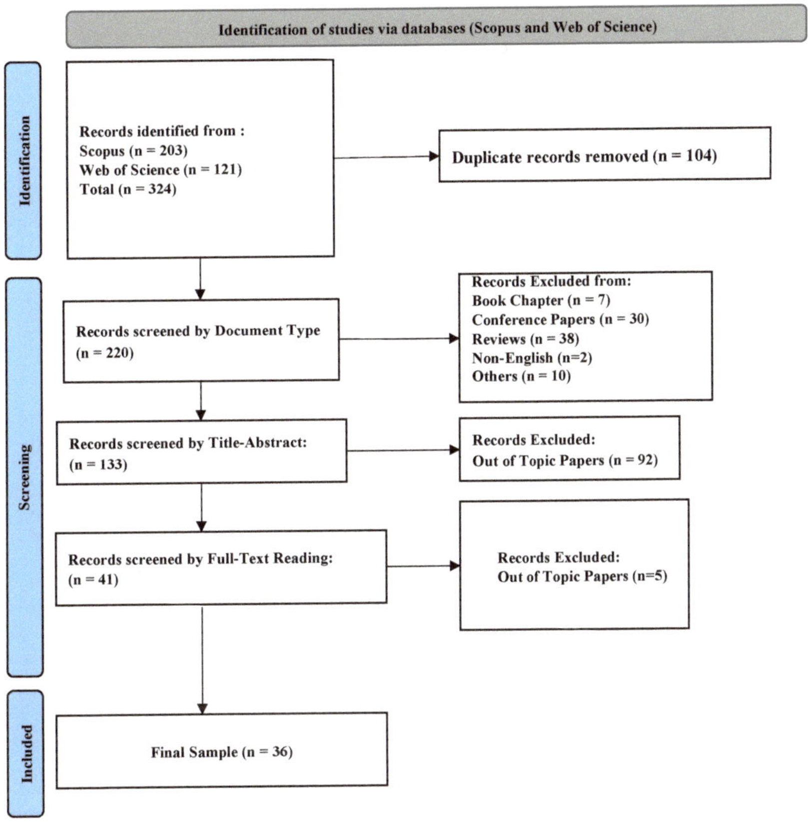

Figure 2. PRISMA flowchart.

The corresponding literature was analyzed to eliminate copies and all non-relevant documents, resulting in an initial sample of 220 articles. The residual articles were screened by considering the document type, excluding conference proceedings, book chapters, reviews, non-English language papers, and others. Thus, 133 articles were screened by taking into account titles and abstracts, not including papers that were off topic, reducing the number of articles to 41.

Later, the full text of these articles was downloaded in order to investigate it. This final screening process permitted the exclusion of five articles, leading to a total of 36 works.

While no temporal constraints were imposed in this analysis, papers deemed pertinent to the study's objectives could not be located preceding the year 2016, as delineated in Section 3.1.

From these 36 studies, the following types of data were extracted and analyzed: bibliometric data such as keywords, the journal and year of publication, subject areas, and geographical location of CE practices. Further, data concerning the practices (particularly their areas of application within the healthcare sector), benefits, and limitations were collected. Bibliometric data were represented in a network map, while the systematic review results were condensed into tables and charts.

Identification of CE Practices and Areas of Application

This section focuses on the practices gathered in the sample and emphasizes the specific areas in the healthcare sector where circular practices have been implemented. Table 2 provides an illustration of the categorization of these practices and their application areas.

Table 2. Classification of CE practices and areas of application (Source: authors).

CE Practice	Area of Application	Example
R-strategies: Analyze how to mitigate operating room (OR) waste through recycling and reuse possibilities. Reusing medical items, after thorough cleaning, decontamination, and sterilization, has shown potential in various areas, e.g., textiles (table covers, gowns, and facemasks), consumables (syringes and single-use plastic trays), and packaging. Recycling by optimizing the sorting process. Reuse options have emerged as the best solution [30].	HWM: It aims to identify and promote waste reduction and safe healthcare waste management, implement regulations to meet global standards, and raise awareness of safety practices [31].	Explore how waste can be minimized in large hospitals via eight observations and five expert interviews. Lowering medical waste cuts both financial and environmental costs [30].
Technology involvement: Clinical climate informatics can lead healthcare decarbonization efforts towards achieving net-zero emissions, minimizing electronic waste, advocating responsible resource management, and realizing environmental sustainability [32].	HSI: It develops awareness about CE strategies' importance to employees, patients, and all the actors involved in healthcare processes [33].	Involvement with relevant stakeholders to increase the awareness of environmental problems and foster change. Optimized algorithms, shared cloud computing resources, low-consumption CPUs, and telemedicine can be implemented to reduce energy consumption. Data analysis can be used to optimize work processes, procurement, and procedures, thus reducing supply waste [32].
Design opportunities: Tracks all materials entering and leaving intensive care through an MFA. The primary environmental footprint is from everyday materials instead of materials designated for specific therapies such as non-sterile gloves, isolation gowns, bed liners, surgical masks, and syringes, giving support to a shift to a circular system in intensive care [34].	Medical devices and supplies: It embraces a wide range of objects and substances, including instruments, apparatuses, implements, machines, materials, medical, or surgical items that are consumable, expendable, disposable, or non-durable for a medical purpose [35].	Application of a Material Flow Analysis (MFA) allows for an assessment of the environmental impacts of key product groups, including weight, carbon footprint, agricultural land occupation, and water usage [34].
Stakeholder involvement: Healthcare stakeholders demonstrate the capability to enact sustainable supply chain management practices and wield substantial influence in elevating the organization's sustainable performance and maintaining a heightened awareness of sustainability [36].	HSC: The healthcare supply chain sector, a significant contributor to worldwide greenhouse gas emissions, is linked to organizational factors, including forging partnerships, delineating roles and responsibilities, and coordinating and managing interface processes [37].	Examination of supply chain strategies aimed at achieving a circular economy within the Indian healthcare sector. Empirical research involving 145 healthcare organizations reveals the hidden connections of stakeholder involvement, sustainable supply chain practices, sustainable performance, and the circular economy in the sector [35].

Table 2. *Cont.*

CE Practice	Area of Application	Example
	HT: It incorporates actions to enhance consciousness about healthcare treatment to reduce carbon footprints [38].	LCA analysis to measure the carbon footprint of breast surgical treatment, revealing less environmental impacts for telehealth visits [38].
	HCP: It covers actions for the proper management of construction processes to ensure the control of CO_2 emissions [39].	Recommendations for sustainable materials in order to enhance environmental protection [39].

3. Results

In this section, we present the main outcomes of our academic literature review, discussing the data obtained from the bibliometric analysis and the relevant practices of circularity implemented in the healthcare sector.

3.1. Bibliometric Analysis

A bibliometric analysis was conducted using VOS viewer Version 1.6.18 to cluster emerging fields related to the chosen topic. The analysis also examined connections among publications and keywords to identify strengths and gaps in the topic and publications. In particular, in the 36 selected papers, the correlations between the keywords used by the top authors were determined through a co-occurrence network map (Figure 3).

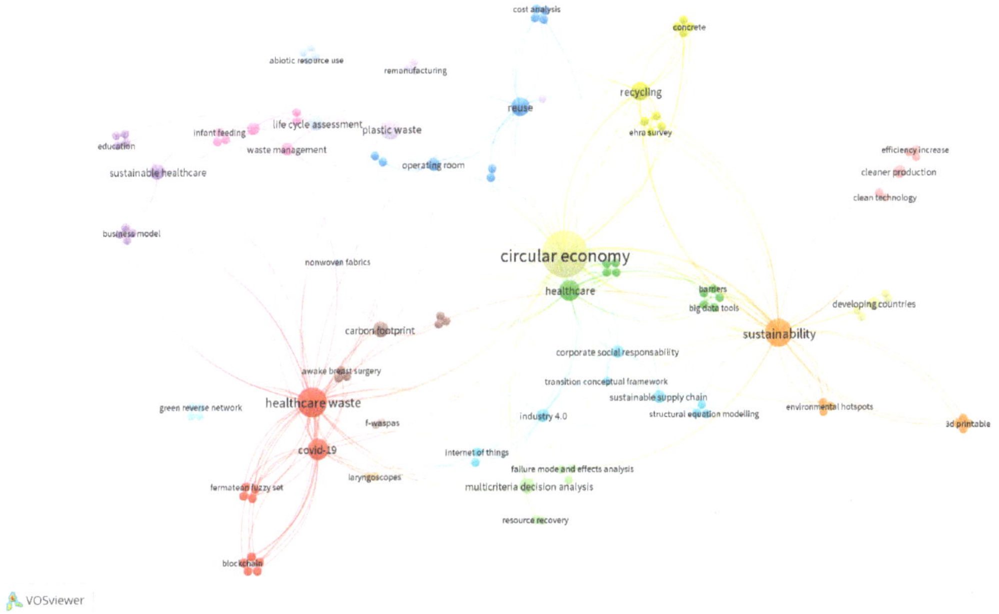

Figure 3. Co-occurrence network map (Source: authors).

The thickness of the lines in the map indicates the strength of the correlations between the nodes (keywords). This strength is calculated by tallying the number of publications where the two keywords appear together. A total of 122 fixed keywords were tested in order to standardize the topics. Of the 122 items in the network, 117 items were connected to each other. The resulting network graph presents clear connections among the investigated keywords, grouped in colored clusters. The most frequently occurring keywords in the study are "circular economy", "healthcare waste", "sustainability", and "plastic waste".

Specifically regarding "plastic waste" and "healthcare waste", they appeared connected with "life cycle assessment". Additionally, the network analysis revealed a connection between the issue of "plastic waste" and "remanufacturing", as they appeared in the same cluster (liliac). Furthermore, there is a correlation between "operating rooms" (ORs) and "reuse" (blue cluster). Similarly, the keyword "Industry 4.0" occurs many times and is linked to the repeated keyword "Internet of Things", and also connected to "sustainable supply chain", to which technologies and the keyword "stakeholder involvement" (light blue cluster) were interconnected, aiming at the achievement of circularity. Moreover, the keywords "waste management", "single-use plastic", and "decontamination" are consistently grouped together (purple cluster). Lastly, "recycling" and "reprocessing" practices are strictly interconnected with a CE (grouped in the dark yellow cluster). Further, to inspect the progress of implementation of CE practices in the healthcare sector, the final group of 36 papers was analyzed by year and considering the journal of publication, as shown in Figure 4. This provides a snapshot of the scholarly activity and the dissemination of knowledge regarding the subject matter.

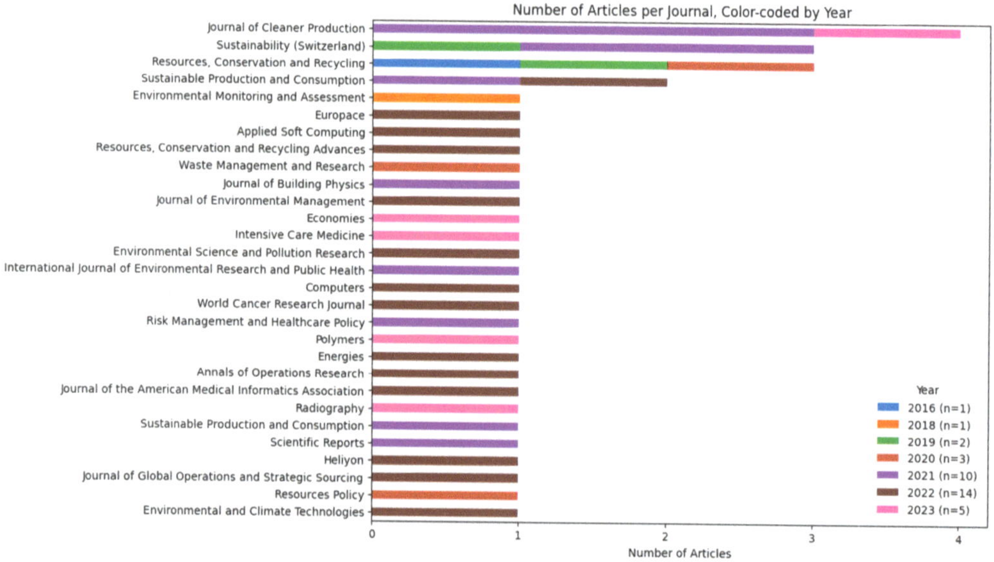

Figure 4. Count of articles distributed per year and by journal.

The first of these works was published in 2016 [40]; prior to 2021, there were only seven publications [40–46], but there was an explosive increase in that year [30,33,37,39,47–52]. Moreover, fourteen publications were recorded in 2022 [32,36,38,53–63], while in 2023, there were only five [34,64–67]. Additionally, the figure illustrates the distribution of articles among different scientific journals, considering the distribution by year as described. Importantly, this study was conducted in early 2023, which accounts for the relatively low number of articles retrieved for this year. The *Journal of Cleaner Production* (JCP) emerges with the highest publication count, with three articles in 2021 [33,47,52] and the most recent one in 2023 [66]. Starting with Resource, Conservation, and Recycling, the first publication on this subject was in 2016 [40]. Subsequently, two more articles were published in 2019 [45], and the most recent one was published in 2022 [56]. Moving on to Sustainability, it has received growing attention from 2019, as evidenced by reference [42]. However, in 2021, there was a decline in this interest, marked by the publication of the last two identified articles on the topic, which are [30,49]. Similarly, the *Journal of Sustainable Production and Consumption* received scholarly interest, as evidenced by the presence of two articles published in both 2021 [50] and 2022 [55], underscoring a commitment to this subject matter. Conversely, the remaining journals published only one paper delving into CE practices, with a particular focus on their implementation within the healthcare sector.

Furthermore, from an examination of research areas per journal, in addition to 15 subject areas of interest emerging, a clear focus on the environment in the subject area emerges, as "Environmental Sciences" is the prevailing area (27.40%), followed by "Engineering" (15.07%), "Medicine" (9.59%), "Economics, Econometrics and Finance" (8.22%), "Business, Management and Accounting", "Computer Science", and "Social Sciences" (6.85%). The other areas with minor relevance are displayed in the Supplementary Materials in Table S3. Finally, the analysis illustrates how publications are distributed geographically by country according to the context in which the studies were conducted.

The outcomes showed that significant contributions were made by the Netherlands (six publications) [33,34,50–52,62], followed by India (five works) [36,46,47,54,56], Spain [39,65], and Iran (two articles) [57,58]. Additionally, the analysis revealed that two articles did not specify the country (indicated as "Not Specified" countries) [32,48], while four articles were labeled as "Multiple countries", due to their reliance on surveys and questionnaires that were administered in more than one country [40,43,53,67]. The Supplementary Materials contains information on the other geographical areas with minor contributions in Table S1.

This examination demonstrates how Europe has made substantial contributions (58%), while Asia (22%) has been increasingly interested in transitioning from the conventional linear model to the implementation of CE practices in the healthcare sector. On the other hand, South America has made fewer contributions (6%), while 11% of contributions did not specify any continent or country.

3.2. *Classification of CE Practices*

This systematic analysis focuses on examining the well-studied CE practices within the healthcare sector, as documented in the literature, to evaluate their benefits and challenges. It involves classifying and analyzing the CE practices discussed in the collected articles (as shown in Figure 5).

To categorize CE practices in the healthcare sector, several practices were identified in the literature, including recycling, reuse, reprocessing, refurbishment, and recovery, which collectively constitute 47% of the strategies applied within the sector. These practices fall under the established categorization of "R-strategies". They are designed to mitigate the depletion of natural resources and reduce material consumption, all while actively attempting to minimize waste generation [49]; they include refusal, repair, remanufacture, reuse, repurpose, refurbishment, recycling, and recovery. In the hierarchy of circular strategies, "refusal" stands out as the most impactful, whereas "recovery" ranks as the least impactful [34]. Other practices designated as "technology involvement" underscore the role of technology in healthcare processes as a strategy to promote circularity (20%). The primary aim of healthcare waste technologies is to minimize the potential risks associated with waste. These technologies include thermal, chemical, irradiative, and biological treatment methods, alongside mechanical treatment technologies, and they serve as the primary methods of waste management [63]. The research has also identified other circular strategies, such as the redesign of products and processes to consider their end-of-life fate. These strategies, known as "design opportunities", account for 25% of the total practices identified.

Lastly, the familiarity and awareness of customers, particularly patients, and employees are considered potential drivers of CE adoption, representing a significant portion (8%), termed as "stakeholder involvement".

Categorization of Areas of Application

The analysis of the full text reveals several areas within the healthcare sector where CE practices have been applied, categorizing them by geographical areas. These findings are summarized and presented in Figure 6, providing an overview of the prevalence and distribution of CE practices across different categories within the healthcare sector and the geographical areas of their implementation.

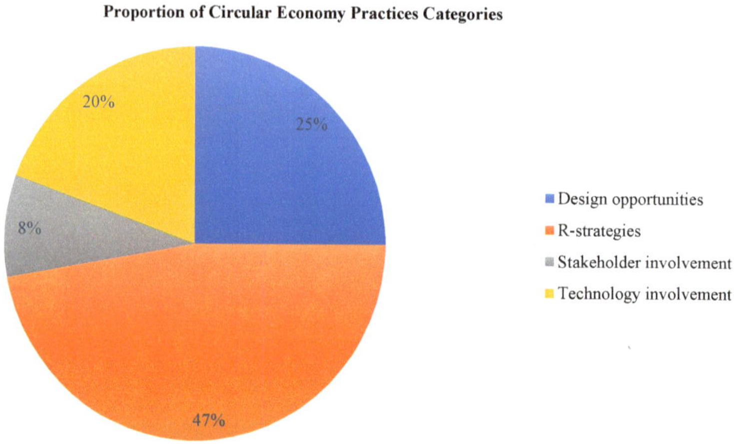

Figure 5. Classification and quantification of CE practices, according to the systematic analysis conducted by the authors.

Notably, HWM has received significant attention from scholars (39%). From this assessment, it emerges that the geographical area more interested in this category is Asia, with three articles originating from India [47,54,56], two from Iran [57,58], one from Pakistan [41], and another from Vietnam [59]. Following Asia, Europe also shows a significant interest in the category, with two articles on CE practice for waste management in the Netherlands [50,52], one in Belgium [30], one in Turkey [61], and another in Latvia [63].

Further, another important area of application has been identified, medical devices and supplies (39%), which encompasses a wide range of practices involving apparatus, tools, implants, in vitro reagents, and disposable or semi-disposable elements used individually or in combination for medical purposes. The focus on practices related to the circular use of devices and supplies is most pronounced in Europe, with three studies conducted in the Netherlands [34,51,62] and two articles spanning multiple countries, labeled as "Multiple Countries" [40,53]. Furthermore, individual studies in this category are located in various European countries, including the United Kingdom [55], the Czech Republic [64], Ireland [45], Spain [65], Denmark [66], and Germany [49]. Moreover, one work within this area does not specify a geographical region and has consequently been categorized as "Not Specified" [48].

Additionally, CE practices implemented in the supply chain have also been identified and are classified under the healthcare supply chain (HSC) category, although there are fewer publications than other areas (8%). The three publications focused on the HSC were conducted in Turkey [37], India [36], and Brazil [42]. Likewise, another category, given the healthcare stakeholder involvement (HSI) label, accounts for 8% of academic interest. These practices, related to the involvement of healthcare stakeholders, workers, and patients, are only addressed in a paper conducted in the Netherlands [33], in a study conducted in "Multiple Countries" [67] and in a study with no specified geographical area [32]. In conclusion, the final two areas in which CE practices were employed were categorized and designated as healthcare treatment (HT) and healthcare construction processes (HCPs). Academic attention was directed towards HT with a total share of 3%, with research conducted in Italy [38]. Meanwhile, HCPs, also representing 3% of the sample, are linked to a study conducted in Spain [39].

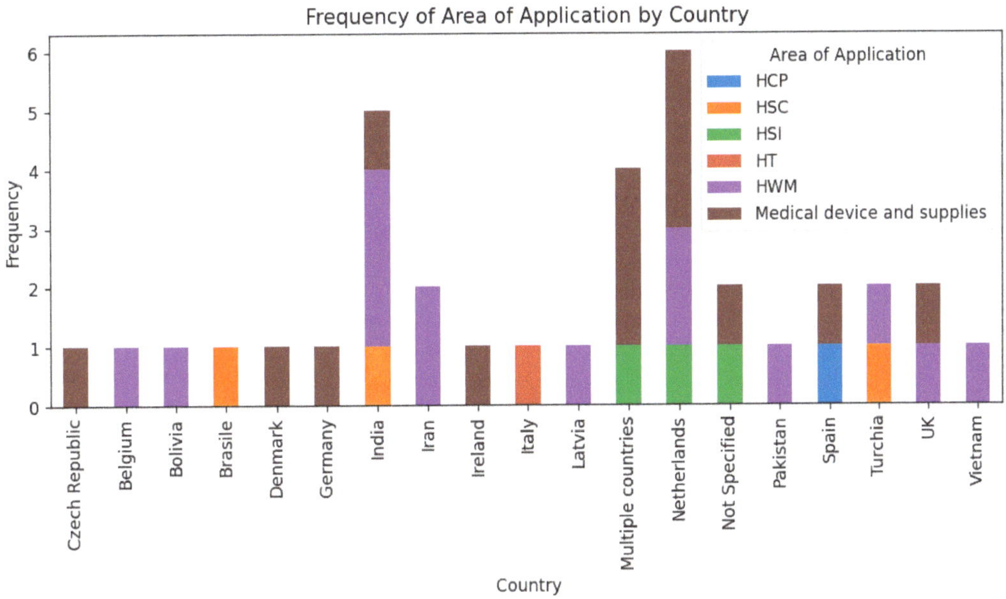

Figure 6. Frequency of CE practices distributed per category and country of application.

4. Discussion

The bibliometric analysis has outlined that the CE concept applied to the healthcare sector as a relatively recent development, as the first article in the collected sample was published in 2016; however, from 2021 to now, the increasing number of articles shows the growing interest of researchers on this topic, with a specific focus in Europe. In this regard, the analysis of keyword occurrence underscores the prevalence and interconnectedness of sustainability and healthcare waste within a CE. Remarkably, keywords such as "remanufacturing" are linked to the plastic waste issue, as it can be considered a solution to the plastic problem. Specifically, following an evaluation of the mechanical recycling of clinical plastic waste for secondary plastic recovery, it becomes evident that incorporating a design that considers the environment in the initial stages of plastic production is essential. This ensures the feasibility of post-use segregation and remanufacturing and enables the assembly and disassembly of material components as needed, thereby extending product life cycles indefinitely [60]. Further, it is evident that ORs are responsible for generating huge amounts of waste, suggesting that the implementation of reusable solutions could serve as a viable remedy [30]. Similarly, the connections between "waste management", "single-use plastic", and "decontamination" keywords can be attributed to the substantial usage of single-use plastic products in this sector, which ultimately leads to waste. Particularly, potential strategies for addressing this issue encompass several options, such as reducing the demand for single-use bottles, exploring alternative solutions for eliminating them, encouraging manufacturers to address product design [45], and motivating policymakers to implement consistent labeling systems for the recycling of materials and the enhancement of waste management and to implement collaborative efforts and assistance systems to achieve sustainable resource management [45].

Furthermore, the analysis of the subject area of the 28 journals in which articles were published clearly marked the close correlation with environmental science; in that light, the sector's profound link to the environment becomes evident, in addition to medicine, engineering, and business and accounting.

Moreover, in line with the co-occurrence analysis, the field of medical devices and supplies was highly prominent, together with the HWM area. Articles on the former were concentrated in Europe, and the latter in Asia.

Nevertheless, we aim to delve further into the assessment of CEs within the healthcare sector by comprehensively outlining all the practices that have been identified in this research. Particularly, the practices identified are categorized as an "R-strategy", considered a pivotal solution to mitigate environmental harm and its associated costs. These practices encompass the use of reusable medical devices [64]. For example, while reusable gynecological speculums have a negative environmental impact during the disinfection and sterilization phase due to the use of ethylene oxide or detergents, this impact is comparatively lower than that of disposable medical devices [64]. Despite this, a notable limitation arises from the preferences of physicians and patients, who often opt for disposable devices [64]. Further, an analysis has been conducted to assess the value of used laryngoscopes [40], with the aim of restoring the intrinsic value of materials. Achieving increased circularity in the management of used laryngoscopes lies in the sourcing of these devices; however, more efficient communication between relevant departments and staff involvement are necessary. One prominent application of reprocessing is the steam sterilization of medical equipment, particularly face masks. Notably, reprocessed masks exhibit a reduced carbon footprint compared to disposable masks, as indicated by the findings of a life cycle analysis (LCA) performed in [51]. A LCA is a methodological approach that evaluates the environmental impact of a product or service throughout its entire life cycle, from raw material extraction to disposal [68,69].

Moreover, the collected sample allowed the identification of the emergence of "design opportunities" practices. A prime example is presented by the development of 3D-printable bioresorbable materials for orthopedic implants (such as bioactive ceramics and bioinert ceramics), which not only reduce waste but also utilize less material during the manufacturing process. However, the limitation posed by 3D-printable bioresorbable materials lies in their sensitivity to repeated stress, which confines their use to small bones [46]. Another example involves designing a circular healthcare business model; gowns integrating nonwoven polyester exhibited worse environmental consequences in comparison to their counterparts composed of nonwoven polypropylene. A circular economy model centered on non-sterile polypropylene gowns holds the potential to slash carbon emissions given the increased usage of these gowns [65].

Further, "technological involvement" has been delineated as a new CE practice area. In fact, emerging technologies could play an important role in strengthening and accelerating the transition into a more circular and sustainable healthcare sector. Indeed, the introduction of technologies such as blockchain technology in [59], even if it is only the first trial of substituting traditional waste treatment processes with this technology, represents a potential solution to the production of waste in order to foster sustainable development. From this perspective, smart waste management is a strategic approach that leverages advanced technologies to guarantee a reduction in medical waste generation [59].

CE strategies identified in the sample also provide a clinical informatics framework designed to mitigate healthcare's contributions to environmental pollution and climate-related effects [32] and to adopt big data [37] to obtain social, economic, and environmental benefits. The framework proposed in [32] can play a fundamental role in promoting the contribution of health information technology (IT) in enhancing environmental sustainability and the betterment of planetary health in healthcare settings. However, the high costs for the implementation of these technologies [37] in this sector represent a major limitation, due to which their application is still a slow process. Finally, the last CE practice area identified was termed "stakeholder involvement", which foresees stakeholder engagement as essential for achieving the common goal of mitigating negative environmental impacts and lowering carbon emissions for the shift toward a CE [37].

Subsequently, in this study, we attempted to provide a categorization of the application areas of the aforementioned circular practices. "HWM" is the first identified area, which aims to promote the reduction in healthcare waste, with a focus on sustainable development as one of its guiding principles [31]. Studies concerning this aspect were concentrated in Asia. The majority of practices for HWM encompass recycling strategies; this offers clear benefits within the CE framework for the preservation of natural resources and reducing the need for extracting new materials. From this standpoint, reuse and recycling processes can contribute to reducing the costs caused by the extraction and processing of natural resources, minimizing healthcare waste. For example, as stated in [60], the reuse, or recycling, if the reuse process is not feasible, of plastic waste could ensure the transition toward a CE. Moreover, ORs are liable for 33% of the waste generated in a hospital [30].

In [30], reusing strategies were introduced to manage waste generated in ORs, registering benefits that extend beyond mere financial savings and leading to a decrease in the environmental impacts. Further, refurbishment and repair were identified as practices to enhance circularity. Refurbishment, for the revitalization of obsolete products and their transformation to align to contemporary standards, alongside the repair strategy, is an optimal solution to avoid waste and costs linked to its disposal [50]. However, ref. [50] evaluated the viability of implementing a circular approach for repurposing discarded medical instruments and stainless-steel waste within hospital settings, ultimately showing that repairing and refurbishing surgical instruments, rather than replacing them with new ones, hold the greatest potential for cost reduction and environmental benefits.

This study focuses on "medical devices and supplies", an area of significant scholarly interest, possibly driven by the positive impacts associated with their circular utilization. In particular, practices falling into this category are primarily classified as an "R-strategy", as possible solutions to manage environmental risks and their related expenses. To address these issues, it is imperative to enhance interdepartmental communication and increase staff involvement.

Thus, the area labeled "HSC" encompasses the reduction in overall resources needed to provide the required level of customer service by increasing product availability, decreasing the time taken for order processing, and simultaneously lowering costs. It is apparent that, for the overwhelming bulk of global greenhouse gas emissions within the HSC, the application of big data technologies is crucial for optimizing the healthcare supply chain [37].

Despite this, a lack of studies in this area was noted; however, the sustainable management of the supply chain could be useful to foster circularity and reduce environmental impacts and the creation of waste by enhancing collaborative partnerships with healthcare professionals.

Likewise, the HSI is another delineated area, including stakeholder, patient, and employee value and involvement, more focused on the social dimension of sustainability, which receives attention from European academics. Establishing familiarity and awareness among human resources engaged in healthcare processes could lead to circularity [37]. Ref. [67] emphasizes the importance of raising awareness among therapeutic radiographers/radiation therapists regarding the several facets of integrating a CE into healthcare, for example, "sustainable transportation", "eco-conscious procurement", "innovative hospital architecture", "efficient food processing", "water conservation", "energy sustainability", and "effective waste control". However, it could be interesting to consider the importance and also the point of view of other relevant actors involved in healthcare processes, like non-specialized figures such as manufacturers, managers, administrators, cooks, and cleaning attendants, to provide a more comprehensive perspective. Stakeholder engagement practices have been acknowledged as essential for achieving the common goal of becoming sustainable [66]; indeed, employees are also responsible for the consumption and separation of products, while producers could redesign medical devices and products to improve end-of-life solutions that can be reused [66]. Nevertheless, there is a lack of adequate training for patients and workers on circular practices. In this sense, a significant obstacle to the implementation of a CE within this sector arises from healthcare workers and professionals having a limited awareness of environmental issues due to insufficient human resources capabilities [37]. It is crucial to enhance their understanding to minimize the environmental consequences of their actions, while also considering the economic outcomes.

Finally, the last two areas identified in this study, which garnered less interest, are HT and HCPs. Although little attention from academics is focused on these subjects, these two areas represent fundamental aspects of the healthcare sector, as their proper management can ensure the reduction in or at least the control of CO_2 emissions [38,39]. To reduce environmental pollution due to treatment, the use of telemedicine is considered a CE solution [38]; instead, in [39], they show that concrete and steel are the most polluting materials. Therefore, by paying more attention to these identified practices, we can unlock the potential for positive environmental, economic, and social outcomes in the healthcare sector.

For example, in [49], the potential of reusing disposed medical devices and stainless-steel waste is demonstrated, while the possibility of cost-cutting in repair and recycling is considered, a new base for surgical waste management is produced, and long-term environmental benefits are actualized. The study indicates that the environmental benefits increase as the collection rates of catheters rise [49]. Further, in [56], it is shown that pharmaceutical blisters (PBs) ground into a powder form and incorporated into concrete, partially substituted with sand, with the aim of recycling PB waste and preserving natural aggregates, could represent a potential solution to fulfil sustainable development goals [56].

Another research work shows that reusable masks create 80–90% less waste than single-use face masks, and have up to 11-fold lower climate change impacts [55]. Moreover, the calculation of the ecocentric value of embodied energy in healthcare waste, performed in [41], proves the value of 100% waste recycling, which can help mitigate the costs of extracting virgin resources. An analysis of intelligent and sustainable technologies within healthcare facilities found that they can contribute to cost savings and enhance staff comfort [42]. Lastly, attention on the social dimension could lead to educational programs that can empower professionals to adjust their practices for greater environmental sustainability [67]. Additionally, the implementation of adaptive treatment methods, such as 3D printing technology [46] or telemedicine [38], contributes to enhancing social sustainability. While the majority of studies tend to emphasize environmental considerations, it is important to recognize that the implementation of circular practices in healthcare holds the potential to not only enhance economic sustainability but also yield positive social impacts. To summarize, the effective management of various categorized areas within healthcare organizations can lead to significant progress in building a more sustainable sector, reducing environmental impacts and fostering a healthier future for both people and the planet.

Hence, these CE practices are a cohesive framework for reorganization at the system level, and by using innovation and creativity, they can pave the way to a constructive and regenerative economy [70]. Consequently, by addressing barriers associated with sustainability in the healthcare sector, CE practices, promoting sustainable resource management and elimination of waste and pollution, hold a crucial role in meeting the SDGs [71]. Although there is not a distinct and specified elucidation of the realized benefits with regard to the SDGs, the identified practices appear to align to SDG 12: Responsible production and consumption. This involves achieving sustainable management and productive utilization of natural resources, as well as substantial waste reduction through prevention, reduction, recycling, and reuse. Additionally, some of these practices [32,37,38,59,63] could be in harmony with SDG 9: Industry, innovation, and infrastructure. SDG 9 emphasizes the modernization of infrastructure industries to make them sustainable, promoting a higher efficiency of resource use and encouraging the adoption of clean and environmentally friendly technologies and industrial processes, making them reliable, sustainable, and resilient.

Despite the growing interest in this topic, the absence of an established classification of CE practices hinders the development of circular healthcare businesses, as does the lack of appropriate behavioral, regulatory, and policy guidance. These recognized practices could be useful for the sector's relevant stakeholders. Thereby, it is recommended to implement managerial, policy, and theoretical measures that support sustainable development initiatives in the healthcare sector, ensuring the effective management of environmental, social, and economic outcomes throughout the process. In this sense, the primary objective of CE practices is to reduce resource consumption, waste production, emissions, and energy depletion [72], while simultaneously promoting social and economic growth; thus, the potential of employing a CE as a vehicle to advance certain SDGs has been suggested [73].

Further, the connection between Industrial Ecology (IE) and a CE is incontrovertible, since IE was denominated as a science of a CE, and central IE tools, such as LCAs or Material Flow Analyses (MFAs), have been gradually implemented within the sustainable CE framework [74]. However, a limited application of IE tools in contrast to other analyses found in the sample, for instance, Multi-Criteria Decision Analyses (MCDAs) and fuzzy theory, [58,61,63], was registered. LCAs were performed in five studies [39,49,51,55,65]; in contrast, only one study was identified that utilized an MFA to evaluate material flows within the healthcare sector [34]. An MFA focuses on managing and quantifying the stocks and flows of substances or materials within a specific system and serves to provide control and measurement over material movement across various stages [75].

Material Flow Analysis or Substance Flow Analysis could be a useful tool for sustainability assessment since it admits in parallel, evaluating environmental and socioeconomic subsystems, together with an analysis of resource utilization [76].

However, it is important to note that research in the field of circular practices within the healthcare sector is still in its early stages, and there are limited scientific studies on this topic.

This broader exploration not only enriches the understanding of circular practices in the healthcare sector but also bolsters commitment to achieving the SDGs, ultimately leading towards a more sustainable and patient-centered healthcare sector.

5. Conclusions

Through a systematic analysis and an in-depth examination of the relevant literature, this study has provided valuable insights into the implementation of CE practices in the healthcare sector, highlighting the pressing need to transform the sector into a more sustainable and circular one aligned with the SDGs. By presenting an overview of the current state of CE practices based on a bibliometric and systematic review, this research contributes to the advancement of sustainability in healthcare.

The increasing interest in CE practices within the healthcare sector reflects a growing focus on sustainability, particularly regarding waste management and resource consumption.

Specifically, the adoption of "R-strategies" addresses reducing waste, prolonging the usefulness of materials, and promoting circularity. Practices falling under "design opportunities" aim to redesign products and processes with attention on the final impact to reduce resource consumption and waste. "Technology involvement" practices are required to improve efficiency and innovation in the healthcare sector, sustaining economic growth and creating a sustainable infrastructure. Lastly, "stakeholder involvement" practices are fundamental for promoting CE adoption through collaboration. All these practices align to both Sustainable Development Goal 12: Responsible Production and Consumption and Sustainable Development Goal 9: Industry, Innovation, and Infrastructure. Although the specific benefits realized in relation to these SDGs are not explicitly outlined, some authors suggest that the proposed research could be a significant focal point in the Agenda for 2023.

Furthermore, these strategies have been applied in various healthcare areas, including HWM, medical devices and supplies, the HSC, HSI, HT, and HCPs.

The framework proposed in this article provides a comprehensive understanding of existing CE practices and offers valuable insights for future studies and applications.

It is important to acknowledge that this study has certain limitations. Firstly, the chosen keywords and database used may have resulted in the exclusion of relevant studies, as the literature on CEs within the healthcare sector is expanding rapidly. Nevertheless, this work represents an original contribution by the authors and represents progress in advancing circularity within the healthcare sector. These outcomes could be significant for the relevant stakeholders seeking to adopt the principles of a CE.

However, further research on CE practices in the healthcare sector should not be limited to qualitative approaches like this study. It is essential to integrate the findings of this review with quantitative IE methods in order to analyze and assess environmental risks more comprehensively. This will provide a more robust and global perspective for future research endeavors. In conclusion, the outcomes contribute to the existing body of knowledge on a CE in healthcare and pave the way for future studies that combine qualitative and quantitative approaches for a more holistic understanding of the environmental impacts and risks associated with healthcare practices.

Supplementary Materials: The following supporting information can be downloaded at https://www.mdpi.com/article/10.3390/su16010401/s1: Table S1: Dataset; Table S2: Scheme codes; Table S3: Subject Areas; Table S4: Keywords; Table S5: Search queries.

Author Contributions: Conceptualization, C.D.; methodology, C.D. and G.I.; data curation, G.I., C.D., K.S.-D., M.T.-Ł. and C.S.; writing—original draft preparation, C.D. and G.I.; writing—review and editing, M.T.-Ł., C.D. and C.S; visualization, G.I. and K.S.-D.; supervision, G.I. All authors have read and agreed to the published version of the manuscript.

Funding: This research received no external funding.

Institutional Review Board Statement: Not applicable.

Informed Consent Statement: Not applicable.

Data Availability Statement: Data is contained within the article and Supplementary Materials.

Conflicts of Interest: The authors declare no conflict of interest.

Abbreviations

The following abbreviations were used in this manuscript:

SDGs	Sustainable Development Goals
CE	Circular Economy
HWM	Healthcare Waste Management
HSC	Healthcare Supply Chain
HSI	Healthcare Stakeholder Involvement
HT	Healthcare Treatment
HCP	Healthcare Construction Process
OR	Operating Room
PB	Pharmaceutical Blister
IE	Industrial Ecology
LCA	Life Cycle Assessment
MFA	Material Flow Analysis

References

1. World Heath Organizations. Climate Change and Health. 2021. Available online: https://www.who.int/news-room/fact-sheets/detail/climate-change-and-health (accessed on 8 February 2023).
2. Karliner, J.; Slotterback, S.; Boyd, R.; Ashby, B.; Steele, K. Health Care's Climate Footprint: How the health sector contributes to the global climate crisis and opportunities for action. In *Health Care Without Harm and Arup*; 2019 Climate-Smart Health Care Series; Green Paper Number One; Ashland Creek Press: Ashland, OR, USA, 2019.
3. Health Care Without Harm. From Commitment to Action. 2022 Annual Report. Available online: https://noharm-global.org/sites/default/files/documents-files/7347/2022_AnnualReport_HCWH.pdf (accessed on 16 January 2023).
4. Pichler, P.-P.; Jaccard, I.S.; Weisz, U.; Weisz, H. International comparison of health care carbon footprints. *Environ. Res. Lett.* **2019**, *14*, 064004. [CrossRef]
5. Goh, C.Y.; Marimuthu, M. The path towards healthcare sustainability: The role of organisational commitment. *Procedia-Soc. Behav. Sci.* **2016**, *224*, 587–592. [CrossRef]

6. Eslami, Y.; Dassisti, M.; Lezoche, M.; Panetto, H. A survey on sustainability in manufacturing organisations: Dimensions and future insights. *Int. J. Prod. Res.* **2019**, *57*, 5194–5214. [CrossRef]
7. Sherman, J.D.; Thiel, C.; MacNeill, A.; Eckelman, M.J.; Dubrow, R.; Hopf, H.; Lagasse, R.; Bialowitz, J.; Costello, A.; Forbes, M.; et al. The green print: Advancement of environmental sustainability in healthcare. *Resour. Conserv. Recycl.* **2020**, *161*, 104882. [CrossRef]
8. Mehra, R.; Sharma, M.K. Measures of sustainability in healthcare. *Sustain. Anal. Model.* **2021**, *1*, 100001. [CrossRef]
9. Ali, S.M.; Appolloni, A.; Cavallaro, F.; D'Adamo, I.; Di Vaio, A.; Ferella, F.; Gastaldi, M.; Ikram, M.; Kumar, N.M.; Martin, M.A.; et al. Development Goals towards Sustainability. *Sustainability* **2023**, *15*, 9443. [CrossRef]
10. European Council. European Green Deal. 2019. Available online: https://www.consilium.europa.eu/en/policies/green-deal/#:~:text=The%20European%20Green%20Deal%20is%20a%20package%20of%20policy%20initiatives,a%20modern%20and%20competitive%20economy (accessed on 12 January 2023).
11. Cheba, K.; Bąk, I.; Szopik-Depczyńska, K.; Ioppolo, G. Directions of green transformation of the European Union countries. *Ecol. Indic.* **2022**, *136*, 108601. [CrossRef]
12. Oliveira, M.; Miguel, M.; van Langen, S.K.; Ncube, A.; Zucaro, A.; Fiorentino, G.; Passaro, R.; Santagata, R.; Coleman, N.; Lowe, B.H.; et al. Circular Economy and the Transition to a Sustainable Society: Integrated Assessment Methods for a New Paradigm. *Circ. Econ. Sust.* **2021**, *1*, 99–113. [CrossRef]
13. Geissdoerfer, M.; Savaget, P.; Bocken, N.M.; Hultink, E.J. The Circular Economy–A new sustainability paradigm? *J. Clean. Prod.* **2017**, *143*, 757–768. [CrossRef]
14. Schröder, P.; Lemille, A.; Desmond, P. Making the circular economy work for human development. *Resour. Conserv. Recycl.* **2020**, *156*, 104686. [CrossRef]
15. Voorter, J.; Iurascu, A.; Van Garsse, S. The Concept "Circular Economy": Towards a More Universal Definition; 2022. Available online: http://hdl.handle.net/1942/37690 (accessed on 12 January 2023).
16. Demirel, P.; Danisman, G.O. Eco-Innovation and Firm Growth in the Circular Economy: Evidence from European SMEs. *Electron. J.* **2019**, *23*, 1608–1618. [CrossRef]
17. Ghisellini, P.; Cialani, C.; Ulgiati, S. A review on circular economy: The expected transition to a balanced interplay of environmental and economic systems. *J. Clean. Prod.* **2016**, *114*, 11–32. [CrossRef]
18. Corona, B.; Shen, L.; Reike, D.; Carreón, J.R.; Worrell, E. Towards sustainable development through the circular economy—A review and critical assessment on current circularity metrics. *Resour. Conserv. Recycl.* **2019**, *151*, 104498. [CrossRef]
19. Ioppolo, G.; Cucurachi, S.; Salomone, R.; Saija, G.; Shi, L. Sustainable local development and environmental governance: A strategic planning experience. *Sustainability* **2016**, *8*, 180. [CrossRef]
20. Chauhan, C.; Parida, V.; Dhir, A. Linking circular economy and digitalisation technologies: A systematic literature review of past achievements and future promises. *Technol. Forecast. Soc. Change* **2022**, *177*, 121508. [CrossRef]
21. Maddikunta, P.K.R.; Pham, Q.-V.; B, P.; Deepa, N.; Dev, K.; Gadekallu, T.R.; Ruby, R.; Liyanage, M. Industry 5.0: A survey on enabling technologies and potential applications. *J. Ind. Inf. Integr.* **2022**, *26*, 100257. [CrossRef]
22. Takalo, S.K.; Tooranloo, H.S. Green innovation: A systematic literature review. *J. Clean. Prod.* **2021**, *279*, 122474. [CrossRef]
23. Szopik-Depczyńska, K.; Cheba, K.; Vikhasta, M.; Depczyński, R. New form of innovations related to the environment—A systematic review. *Procedia Comput. Sci.* **2021**, *192*, 5039–5049. [CrossRef]
24. D'Adamo, I.; Gastaldi, M. Sustainable Development Goals: A Regional Overview Based on Multi-Criteria Decision Analysis. *Sustainability* **2022**, *14*, 9779. [CrossRef]
25. Pasqualotto, C.; Callegaro-De-Menezes, D.; Schutte, C.S.L. An Overview and Categorization of the Drivers and Barriers to the Adoption of the Circular Economy: A Systematic Literature Review. *Sustainability* **2023**, *15*, 10532. [CrossRef]
26. Mengist, W.; Soromessa, T.; Legese, G. Method for conducting systematic literature review and meta-analysis for environmental science research. *MethodsX* **2020**, *7*, 100777. [CrossRef] [PubMed]
27. Donthu, N.; Kumar, S.; Mukherjee, D.; Pandey, N.; Lim, W.M. How to conduct a bibliometric analysis: An overview and guidelines. *J. Bus. Res.* **2021**, *133*, 285–296. [CrossRef]
28. Moher, D.; Shamseer, L.; Clarke, M.; Ghersi, D.; Liberati, A.; Petticrew, M.; Shekelle, P.; Stewart, L.A. Preferred reporting items for systematic review and meta-analysis protocols (PRISMA-P) 2015 statement. *Syst. Rev.* **2015**, *4*, 1. [CrossRef] [PubMed]
29. Kelly, J.; Sadeghieh, T.; Adeli, K. Peer Review in Scientific Publications: Benefits, Critiques, & A Survival Guide. *EJIFCC* **2014**, *25*, 227–243. [PubMed]
30. Harding, C.; Van Loon, J.; Moons, I.; De Win, G.; Du Bois, E. Design Opportunities to Reduce Waste in Operating Rooms. *Sustainability* **2021**, *13*, 2207. [CrossRef]
31. World Health Organization. Health-Care Waste. 2018. Available online: https://www.who.int/news-room/fact-sheets/detail/health-care-waste (accessed on 18 April 2023).
32. Sittig, D.F.; Sherman, J.D.; Eckelman, M.J.; Draper, A.; Singh, H. i-CLIMATE: A "clinical climate informatics" action framework to reduce environmental pollution from healthcare. *J. Am. Med. Inform. Assoc.* **2022**, *29*, 2153–2160. [CrossRef] [PubMed]
33. van Boerdonk, P.J.M.; Krikke, H.R.; Lambrechts, W. New business models in circular economy: A multiple case study into touch points creating customer values in health care. *J. Clean. Prod.* **2021**, *282*, 125375. [CrossRef]

34. Hunfeld, N.; Diehl, J.C.; Timmermann, M.; van Exter, P.; Bouwens, J.; Browne-Wilkinson, S.; de Planque, N.; Gommers, D. Circular material flow in the intensive care unit—Environmental effects and identification of hotspots. *Intensive Care Med.* **2023**, *49*, 65–74. [CrossRef]
35. World Health Organization. Medical Devices. 2019. Available online: https://www.who.int/medical_devices/full_defnition/en (accessed on 13 April 2023).
36. Vishwakarma, A.; Dangayach, G.S.; Meena, M.L.; Gupta, S.; Joshi, D.; Jagtap, S. Can circular healthcare economy be achieved through implementation of sustainable healthcare supply chain practices? Empirical evidence from Indian healthcare sector. *J. Glob. Oper. Strateg. Sourc.* **2022**. *ahead-of-print*. [CrossRef]
37. Kazançoğlu, Y.; Sağnak, M.; Lafcı, Ç.; Luthra, S.; Kumar, A.; Taçoğlu, C. Big Data-Enabled Solutions Framework to Overcoming the Barriers to Circular Economy Initiatives in Healthcare Sector. *Int. J. Environ. Res. Public Health* **2021**, *18*, 7513. [CrossRef]
38. Materazzo, M.; Facchini, A.; Garozzo, D.; Buonomo, C.; Pellicciaro, M.; Vanni, G. Maintaining good practice in breast cancer management and reducing the carbon footprint of care: Study protocol and preliminary results. *World Cancer Res. J.* **2022**, *9*, e2438. [CrossRef]
39. García-Sanz-Calcedo, J.; de Sousa Neves, N.; Fernandes, J.P.A. Assessment of the global warming potential associated with the construction process of healthcare centres. *J. Build. Phys.* **2021**, *44*, 309–325. [CrossRef]
40. Viani, C.; Vaccari, M.; Tudor, T. Recovering value from used medical instruments: A case study of laryngoscopes in England and Italy. *Resour. Conserv. Recycl.* **2016**, *111*, 1–9. [CrossRef]
41. Ali, M.; Geng, Y. Accounting embodied economic potential of healthcare waste recycling—A case study from Pakistan. *Environ. Monit. Assess.* **2018**, *190*, 678. [CrossRef] [PubMed]
42. Daú, G.; Scavarda, A.; Scavarda, L.F.; Portugal, V.J.T. The Healthcare Sustainable Supply Chain 4.0: The Circular Economy Transition Conceptual Framework with the Corporate Social Responsibility Mirror. *Sustainability* **2019**, *11*, 3259. [CrossRef]
43. Ertz, M.; Patrick, K. The future of sustainable healthcare: Extending product lifecycles. *Resour. Conserv. Recycl.* **2020**, *153*, 104589. [CrossRef]
44. Ferronato, N.; Ragazzi, M.; Torrez Elias, M.S.; Gorritty Portillo, M.A.; Guisbert Lizarazu, E.G.; Torretta, V. Application of healthcare waste indicators for assessing infectious waste management in Bolivia. *Waste Manag. Res.* **2020**, *38*, 4–18. [CrossRef]
45. Leissner, S.; Ryan-Fogarty, Y. Challenges and opportunities for reduction of single use plastics in healthcare: A case study of single use infant formula bottles in two Irish maternity hospitals. *Resour. Conserv. Recycl.* **2019**, *151*, 104462. [CrossRef]
46. Yadav, D.; Garg, R.K.; Ahlawat, A.; Chhabra, D. 3D printable biomaterials for orthopedic implants: Solution for sustainable and circular economy. *Resour. Policy* **2020**, *68*, 101767. [CrossRef]
47. Chauhan, A.; Jakhar, S.K.; Chauhan, C. The interplay of circular economy with industry 4.0 enabled smart city drivers of healthcare waste disposal. *J. Clean. Prod.* **2021**, *279*, 123854. [CrossRef]
48. Meissner, M.; Lichtnegger, S.; Gibson, S.; Saunders, R. Evaluating the Waste Prevention Potential of a Multi- versus Single-Use Surgical Stapler. *Risk Manag. Healthc. Policy* **2021**, *14*, 3911–3921. [CrossRef] [PubMed]
49. Schulte, A.; Maga, D.; Thonemann, N. Combining Life Cycle Assessment and Circularity Assessment to Analyze Environmental Impacts of the Medical Remanufacturing of Electrophysiology Catheters. *Sustainability* **2021**, *13*, 898. [CrossRef]
50. van Straten, B.; Dankelman, J.; Van der Eijk, A.; Horeman, T. A Circular Healthcare Economy; a feasibility study to reduce surgical stainless steel waste. *Sustain. Prod. Consum.* **2021**, *27*, 169–175. [CrossRef]
51. van Straten, B.; Ligtelijn, S.; Droog, L.; Putman, E.; Dankelman, J.; Weiland, N.H.S.; Horeman, T. A life cycle assessment of reprocessing face masks during the COVID-19 pandemic. *Sci. Rep.* **2021**, *11*, 17680. [CrossRef]
52. van Straten, B.; van der Heiden, D.R.; Robertson, D.; Riekwel, C.; Jansen, F.W.; Van der Elst, M.; Horeman, T. Surgical waste reprocessing: Injection molding using recycled blue wrapping paper from the operating room. *J. Clean. Prod.* **2021**, *322*, 129121. [CrossRef]
53. Boussuge-Roze, J.; Boveda, S.; Mahida, S.; Anic, A.; Conte, G.; Chun, J.K.R.; Marijon, E.; Sacher, F.; Jais, P. Current practices and expectations to reduce environmental impact of electrophysiology catheters: Results from an EHRA/LIRYC European physician survey. *Europacem* **2022**, *24*, 1300–1313. [CrossRef]
54. Chakraborty, S.; Saha, A.K. A framework of LR fuzzy AHP and fuzzy WASPAS for health care waste recycling technology. *Appl. Soft Comput.* **2022**, *127*, 109388. [CrossRef]
55. Chau, C.; Paulillo, A.; Ho, J.; Bowen, R.; La Porta, A.; Lettieri, P. The environmental impacts of different mask options for healthcare settings in the UK. *Sustain. Prod. Consum.* **2022**, *33*, 271–282. [CrossRef]
56. Dalal, S.P.; Dalal, P.; Motiani, R.; Solanki, V. Experimental investigation on recycling of waste pharmaceutical blister powder as partial replacement of fine aggregate in concrete. *Resour. Conserv. Recycl. Adv.* **2022**, *14*, 200076. [CrossRef]
57. Govindan, K.; Nosrati-Abarghooee, S.; Nasiri, M.M.; Jolai, F. Green reverse logistics network design for medical waste management: A circular economy transition through case approach. *J. Environ. Manag.* **2022**, *322*, 115888. [CrossRef]
58. Jafarzadeh Ghoushchi, S.; Memarpour Ghiaci, A.; Rahnamay Bonab, S.; Ranjbarzadeh, R. Barriers to circular economy implementation in designing of sustainable medical waste management systems using a new extended decision-making and FMEA models. *Environ. Sci. Pollut. Res.* **2022**, *29*, 79735–79753. [CrossRef] [PubMed]
59. Le, H.T.; Quoc, K.; Le Nguyen, T.A.; Dang, K.T.; Vo, H.K.; Luong, H.H.; Le Van, H.; Gia, K.H.; Cao Phu, L.; Van Nguyen Truong Quoc, D.; et al. Medical-Waste Chain: A Medical Waste Collection, Classification and Treatment Management by Blockchain Technology. *Computers* **2022**, *11*, 113. [CrossRef]

60. Sadhukhan, J.; Sekar, K. Economic Conditions to Circularize Clinical Plastics. *Energies* **2022**, *15*, 8974. [CrossRef]
61. Simic, V.; Ebadi Torkayesh, A.; Ijadi Maghsoodi, A. Locating a disinfection facility for hazardous healthcare waste in the COVID-19 era: A novel approach based on Fermatean fuzzy ITARA-MARCOS and random forest recursive feature elimination algorithm. *Ann. Oper. Res.* **2023**, *328*, 1105–1150. [CrossRef] [PubMed]
62. van Straten, B.; Tantuo, B.; Dankelman, J.; Weiland, N.H.S.; Boersma, B.J.; Horeman, T. Reprocessing Zamak laryngoscope blades into new instrument parts; an 'all-in-one'experimental study. *Heliyon* **2022**, *8*, E11711. [CrossRef] [PubMed]
63. Zlaugotne, B.; Pubule, J.; Gusca, J.; Kalnins, S.N. Quantitative and Qualitative Assessment of Healthcare Waste and Resource Potential Assessment. *Environ. Clim. Technol.* **2022**, *26*, 64–74. [CrossRef]
64. Hospodková, P.; Rogalewicz, V.; Králíčková, M. Gynecological Speculums in the Context of the Circular Economy. *Economies* **2023**, *11*, 70. [CrossRef]
65. Quintana-Gallardo, A.; del Rey, R.; González-Conca, S.; Guillén-Guillamón, I. The Environmental Impacts of Disposable Nonwoven Fabrics during the COVID-19 Pandemic: Case Study on the Francesc de Borja Hospital. *Polymers* **2023**, *15*, 1130. [CrossRef]
66. Ramos, T.; Christensen, T.B.; Oturai, N.; Syberg, K. Reducing plastic in the operating theatre: Towards a more circular economy for medical products and packaging. *J. Clean. Prod.* **2023**, *383*, 135379. [CrossRef]
67. Soares, A.L.; Buttigieg, S.C.; Couto, J.G.; Bak, B.; McFadden, S.; Hughes, C.; McClure, P.; Rodrigues, J.; Bravo, I. An evaluation of knowledge of circular economy among Therapeutic Radiographers/Radiation Therapists (TR/RTTs): Results of a European survey to inform curriculum design. *Radiography* **2023**, *29*, 274–283. [CrossRef]
68. *ISO 14040:2006*; Environmental Management—Life Cycle Assessment—Principles and Framework. International Organization for Standardization: Geneva, Switzerland, 2006.
69. *ISO 14044:2006*; Environmental Management—Life Cycle Assessment—Requirements and Guidelines. International Organization for Standardization: Geneva, Switzerland, 2006.
70. Ellen MacArthur Foundation. Towards the Circular Economy. Ellen MacArthur Foundation. 2012. Available online: https://www.ellenmacarthurfoundation.org/assets/downloads/publications/Ellen-MacArthur-Foundation-Towards-the-Circular-Economy-vol.1.pdf (accessed on 2 February 2023).
71. Economic, U.N.; Council, S. Circular Economy for the SDGs: From Concept to Practice. 2018. Available online: https://www.un.org/en/ga/second/73/jm_conceptnote.pdf (accessed on 2 February 2023).
72. Salvioni, D.M.; Bosetti, L.; Fornasari, T. Implementing and Monitoring Circular Business Models: An Analysis of Italian SMEs. *Sustainability* **2022**, *14*, 270. [CrossRef]
73. Romero-Perdomo, F.; Carvajalino-Umaña, J.D.; Moreno-Gallego, J.L.; Ardila, N.; González-Curbelo, M.Á. Research Trends on Climate Change and Circular Economy from a Knowledge Mapping Perspective. *Sustainability* **2022**, *14*, 521. [CrossRef]
74. Wiprächtiger, M.; Haupt, M.; Froemelt, A.; Klotz, M.; Beretta, C.; Osterwalder, D.; Burg, V.; Hellweg, S. Combining industrial ecology tools to assess potential greenhouse gas reductions of a circular economy: Method development and application to Switzerland. *J. Ind. Ecol.* **2023**, *27*, 254–271. [CrossRef]
75. Brunner, P.H.; Rechberger, H. *Handbook of Material Flow Analysis*; CRC Press: Boca Raton, FL, USA, 2016. [CrossRef]
76. Huang, C.L.; Vause, J.; Ma, H.W.; Yu, C.P. Using material/substance flow analysis to support sustainable development assessment: A literature review and outlook. *Resour. Conserv. Recycl.* **2012**, *68*, 104–116. [CrossRef]

Disclaimer/Publisher's Note: The statements, opinions and data contained in all publications are solely those of the individual author(s) and contributor(s) and not of MDPI and/or the editor(s). MDPI and/or the editor(s) disclaim responsibility for any injury to people or property resulting from any ideas, methods, instructions or products referred to in the content.

sustainability

MDPI

Review

Impact of Power Interruption on Buildings and Neighborhoods and Potential Technical and Design Adaptation Methods

Caroline Hachem-Vermette * and Somil Yadav

Department of Building, Civil & Environmental Engineering, Gina Cody School of Engineering and Computer Science, Concordia University, Montreal, QC H3G 1M8, Canada; somil.yadav@concordia.ca
* Correspondence: caroline.hachemvermette@concordia.ca

Abstract: This paper presents a state-of-the-art review of the impact of energy interruptions on people, buildings, and neighborhoods and discusses some technological and design strategies to mitigate some of these impacts. An exhaustive literature review was carried out employing keyword searches in the ScienceDirect and Scopus databases. The literature focuses mainly on 37 keywords, which occurred in more than two sources. Based on this literature survey, the paper highlights that, depending on their duration, power outages can have a severe impact on people, buildings, and neighborhoods. The lives of vulnerable individuals dependent on electrical medical devices can be threatened even in short-term power interruption scenarios. Longer-term power outages affect multiple aspects of daily life, such as communication, thermal comfort, life quality, transportation, health, and security, in addition to potential damage to buildings and their contents. The paper identifies and discusses various methods that can be implemented to reduce vulnerability and improve adaptation to climate-related power interruptions. These methods range from simple, low-tech solutions that enable users to temporarily cope with hours of interruption to more sophisticated methods requiring advanced planning. These adaptation and coping methods are classified according to various criteria, including their ease of implementation, accessibility, potential cost, ease of use by occupants, and their potential to address various needs. The paper finally discusses the impact of building and neighborhood design on improving adaptation to energy interruptions. High-performance building design can extend the time that a building can passively operate without reliance on mechanical systems for heating and for cooling. Building shape and geometry, as well as the spatial design of the neighborhood, can maximize solar access and therefore facilitate the implementation of PV and solar technologies. In addition, the design of mixed-use neighborhoods with access to various facilities and basic amenities assists in prolonging the self-reliance of the community as a whole. This work aligns with the vision of the Sustainable Development Goals: by identifying methods and technologies to reduce the impact of power interruptions and improve the energy resilience of urban areas around the globe, this work can contribute to the direct and indirect fulfillment of several Sustainable Development Goals (e.g., SDGs 7, 11, 13, and others). Although the work is performed in a North American context and specifically refers to the Canadian climate, the methodology can be implemented in other climatic and regional conditions.

Keywords: energy resilience; Sustainable Development Goals; climate crisis; energy interruption; renewable energy sources; building and neighborhood design

Citation: Hachem-Vermette, C.; Yadav, S. Impact of Power Interruption on Buildings and Neighborhoods and Potential Technical and Design Adaptation Methods. *Sustainability* **2023**, *15*, 15299. https://doi.org/10.3390/su152115299

Academic Editor: Davide Settembre-Blundo

Received: 22 September 2023
Revised: 18 October 2023
Accepted: 24 October 2023
Published: 26 October 2023

Copyright: © 2023 by the authors. Licensee MDPI, Basel, Switzerland. This article is an open access article distributed under the terms and conditions of the Creative Commons Attribution (CC BY) license (https://creativecommons.org/licenses/by/4.0/).

1. Introduction

Modern societies are significantly dependent on electric power to provide essential needs, including food and adequate living environment, as well as to power various activities including communication and transport [1]. The interruption of the electricity supply can cause significant damage to daily life activities and services, including household practices and basic needs, transportation systems, banking and financial systems, health services, and communication [2,3]. While maintaining a reliable and continuous electric supply might be challenging under regular conditions [4,5], anomalous conditions such as an extreme natural event can significantly increase the impact of this challenge [6].

The occurrence of extreme weather-related disasters, such as hurricanes, wildfires, and floods, has significantly increased during the last two decades. For example, the average number of disasters in the United States doubled in the period between 2014 and 2018, as compared to the average number of climate disasters that occurred between 1980 and 2018 [7,8].

A review of worldwide extreme weather events highlights that many of these events have a major effect on power grids, causing extensive power outages that in some cases may last for days, weeks, or even months (e.g., during Hurricane Sandy [9–11] and the ice storm in Canada [12,13], etc.), leading to prolonged electrical interruptions and major economic losses. Experiences with the energy infrastructure during the last two decades have demonstrated that the electricity network in European and North American countries is not as dependable and robust as was generally assumed and that it can be significantly affected by various weather events. Bushal et al. [7] provide a summary of major blackouts, technical problems, and cyber-attacks caused by weather events based on data collected from research papers and reports. To enhance communities' continuous functioning and sometimes survival during power interruption, there is a need to comprehensively understand the impact of power outages on various aspects of human life and the environment, to identify factors affecting the magnitude of these impacts, and to recognize existing and potential technologies that can mitigate these impacts, as well as their feasibility and limitations.

The existing literature on the impacts of power interruption is both generally scattered and does not clearly highlight the relation between the magnitude of these impacts and various factors related to humans and their environment. Although the role of technologies in enhancing energy resilience is increasingly investigated, a focused review of these technologies, their capabilities, and their limitations, with respect to feasibility criteria, has not yet been presented at the building and neighborhood scale.

The existing literature focuses mostly on specific issues during power outages and discusses specific solutions, such as thermal resilience, resilient cooling, and distributed energy generation, e.g., [14–17]. Although such issues and solutions are undeniably significant in improving the resilience of the built environment, it is vital to understand that the severity of the impact of power outages on households is affected by various factors. These factors include access to alternative methods of energy generation, the energy efficiency of buildings, other basic design features including building type, and the diversity of fuel types serving various functions (e.g., all-electric vs. hybrid for heating, cooling, etc.). In addition, the type and size of the neighborhood can play a significant role in the severity and duration of the disruption [18].

The severity of energy interruption goes beyond the energy systems themselves to affect other basic systems in the built environment, such as water, health, and economic systems. This is especially important due to the interaction of these systems leading to cascading effects during an unanticipated disruption or disturbance [19,20]. For instance, water systems rely on energy to bring water to buildings, and any prolonged energy disturbance can impact water supply. Various health problems can also result from large and prolonged energy interruptions [21,22].

Reducing vulnerabilities is a key concern of the Sustainable Development Goals (SDGs). For instance, the implementation of renewable energy technologies to support energy supply from local grids can lead to more sustainable and resilient cities (SDG 11), help mitigate the impact of climate change (SDG 13), and constitute a key aspect of affordable and clean energy access (SDG 7) [23]. Despite the importance of improving the energy resilience of communities, there are enormous gaps both in research and implementation [23]. For example, developing solutions that recognize the interconnection of critical infrastructure structures and their vulnerability to disruptions can induce events with major cascading effects on the social, economic, and environmental levels [23,24]. To meet some of the SDGs, it is increasingly important to multiply efforts toward the implementation of green technologies that can reduce the energy consumption of the built environment, reduce waste, and promote renewable energy generation and advanced energy management systems (e.g., community energy systems) while developing sustainable policies that support these efforts [23,25–27].

The paper aims to achieve the following: (1) present a state-of-the-art review of the impact of power interruptions on buildings and neighborhoods, (2) discuss the available technologies for reducing the impact of these interruptions, and (3) highlight design strategies for improved energy resilience. The paper provides an information-based tool to implement evidence-based energy-conscious design scenarios, thus assisting in designing resilient and sustainable urban developments. The merit of this focused review work lies in providing key insights that can drive future research, inform resilience policies, and assist in establishing practical procedures to reduce vulnerabilities. Moreover, identifying methods and technologies to reduce the vulnerability of urban areas to energy interruption contributes significantly to the Sustainable Development Goals, both directly and indirectly. The contribution of improved energy resilience in communities to the SDGs is discussed in the manuscript.

2. Methodology

Three main topics form the focus of this paper: (1) identifying the impacts of power outages on households' daily practices; (2) discussing various technologies for mitigating some of these impacts and reducing vulnerability during outages; (3) analyzing the role of building and neighborhood design in improving energy resilience. A neighborhood is considered in this work as a geographic area encompassing a number of buildings and their surroundings, featuring various land uses and associated infrastructure. A building is a single physical construction dedicated to a single or mixed use.

An extensive literature review was carried out in two major fields: the impacts of power interruptions on human survivability and their built environment, and potential technologies employed to mitigate, in different capacities, these impacts. Relevant literature (journal articles and reports) was identified through keyword searches in the ScienceDirect and Scopus databases, and a network visualization illustrating the relationship between keywords in the current literature review is presented in Figure 1. The literature focuses mainly on 37 keywords, occurring in more than two sources. A few keywords, such as critical infrastructure and solar irradiation, have been neglected due to weak link strength. This enhances the credibility of the review and is implemented in a number of current studies, e.g., [28,29]. The selected keywords were imported into VOS viewer 1.6.19 software to generate the visual maps. The size of a keyword's label and circle corresponds to its frequency in the cited literature, with more prominent items having larger labels and circles. These keywords are further clubbed together and presented by different colored clusters. The lines connecting clusters represent the connections between them. From the figure, it is evident that power outages are connected with natural disasters and climate change. Additionally, resilience in such events is associated with renewable energy sources, including solar energy and fuel cells, which are further linked to storage technologies. Smart grids and microgrids establish connections between storage technologies and demand-side management. All these facets are extensively discussed in the literature review. Figure 1 also illustrates the evolving research trends in these domains over time. The color bar in the right corner indicates the year of the cited research articles. It is evident from Figure 1 that the current emphasis is on studying power outage resilience.

In processing the various literature sources on outage impacts, topics need to be narrowed down to various specific criteria, such as the length of the power outage, health and safety impacts, communication, and other specific needs, to obtain meaningful results. The main observations are then grouped into tables to provide a holistic picture of the impact of power interruptions on various human needs.

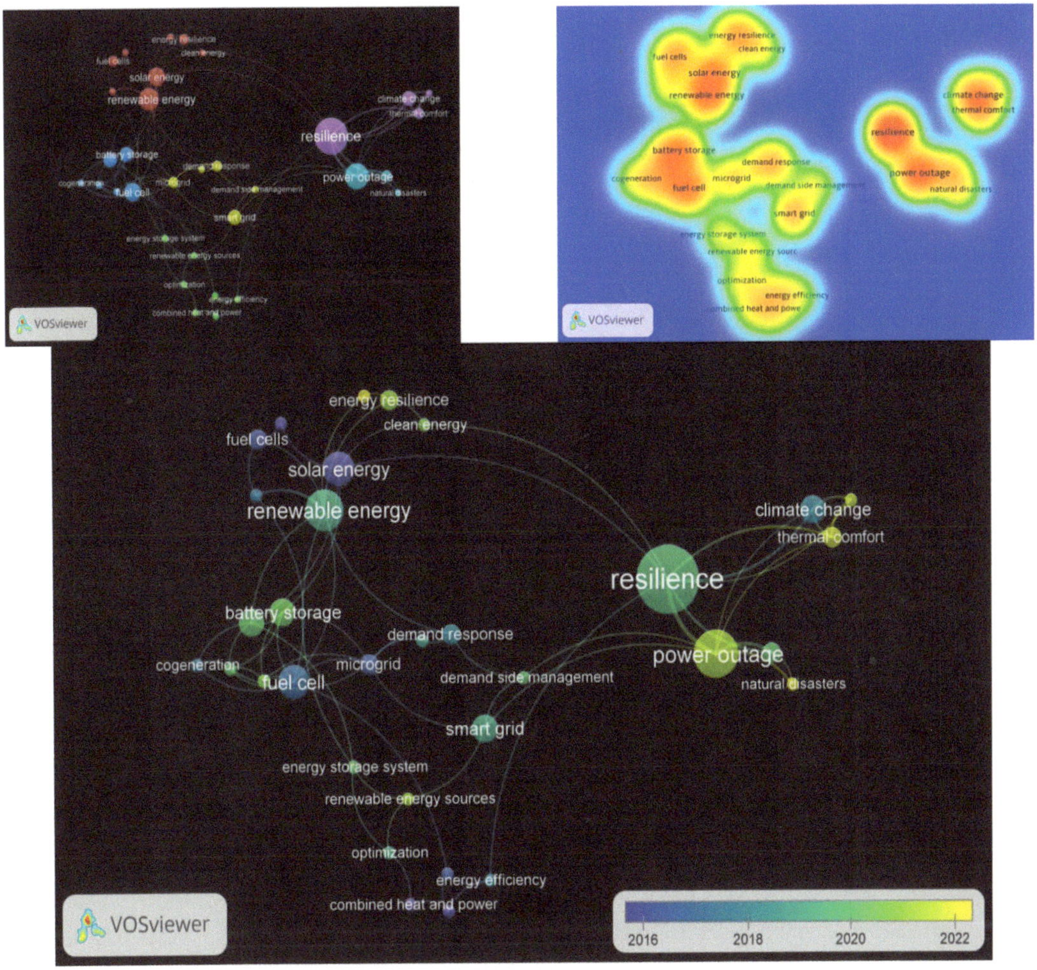

Figure 1. Visual representation of keywords in the current literature review and evolving research trends over time for the keywords in the current literature review.

The literature review on technologies employed to mitigate the impacts of power interruptions yields knowledge about technologies employed in specific circumstances (such as PV with batteries, combined heat and power generation, etc.) and highlights the role of distributed energy in general. A list of potential distributed energy technologies at the building and neighborhood level is then developed, and each of these is individually reviewed employing pertinent literature to identify their potential and their limitations.

The third part of this paper is based on the extensive research of the author related to the impact of building and neighborhood design parameters to allow energy efficiency and to optimize the implementation of various renewable and alternative energy resources, thus improving the overall resilience of a community. Figure 2 presents an illustration of the approach applied in this paper.

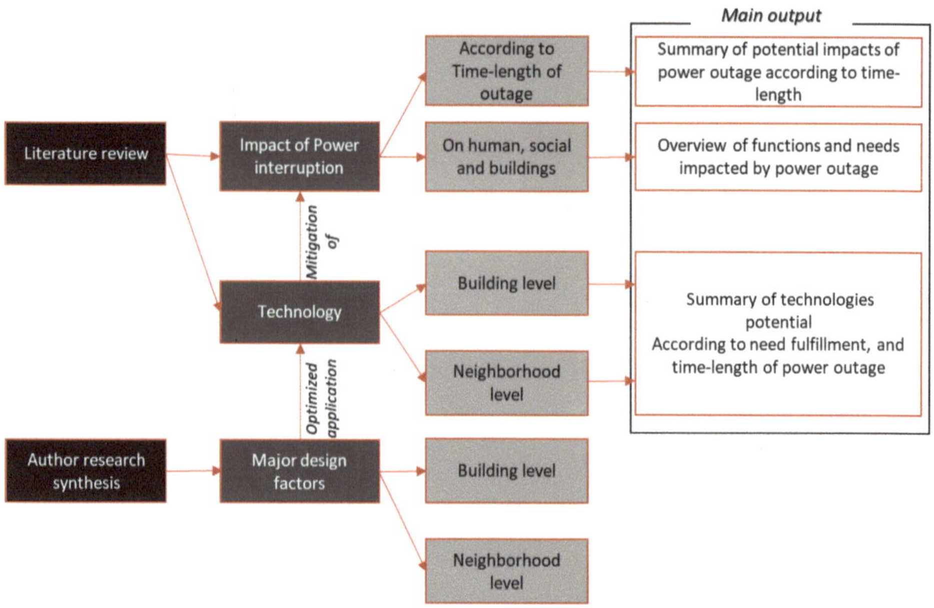

Figure 2. Illustration of the general approach employed in the paper.

3. Impacts of Power Outage

Power outages can have a range of direct and indirect impacts on humans and their environment [30,31]. The literature on technical issues related to power outages is increasing; however, studies that explore the psychological and emotional impacts of power outages are scarce to nonexistent [31].

This section assesses the potential impacts of power outage crises, which often result from an extreme climate crisis, on livelihood and survivability in residential buildings. The impacts are determined for varying outage durations, considered short-term (hours), medium-term (days), and long-term (weeks). The impacts are mostly classified in terms of communication, comfort (heating/cooling, warm water, electric lighting), food and water (e.g., nutrition, cooking, water supply), safety, security, damage to dwelling content, damage to dwelling itself, health concerns, and threat to life.

The impacts are discussed according to different types of residential buildings as well as the demographic of occupants (e.g., elderly and vulnerable individuals, families with vulnerable individuals, and young couples/single individuals). A literature survey was conducted to identify the common and specific impacts of power outages on residential buildings and their inhabitants.

The impacts can change dramatically according to the length of the outage. Some of the expected impacts are summarized below, according to the outage duration, and in Table 1.

- Short-term outage (<24 h):

Although a short-duration outage of less than 24 h has a limited impact on people and buildings, some serious damage can still occur. For instance, for vulnerable people who depend on electrical medical devices for survival (heart, breathing, etc.), even a short-term power outage may lead to life-threatening situations [32,33].

Other critical functions that can be affected by short-term outages include a potential interruption in communication, such as reliance on cellphones (with limited battery duration) [34,35]. In addition, thermal comfort can be compromised in low-performance buildings during extreme cold or heat waves [36,37]. Other critical impacts include access to high-rise residential buildings, especially for vulnerable people [38]. The various concerns are summarized in Table 1.

Table 1. Summary of potential impacts of power outage.

	12–24 h	1 Day–1 Week	>1 Week
Communication	• Cellphone interruption (battery running out). • Communications for police and fire departments can be compromised.	• All kinds of communication are severely impacted, including cellphones and internet.	• All kinds of communication are severely impacted, including cellphones and internet.
Comfort	• Heating in winter in electrically heated buildings. • Electric lighting.	• Heating in winter in all-electric buildings. • Electric lighting. • Overheating in summer can significantly affect survival, especially of vulnerable people.	• Heating in winter in all-electric buildings. • Electric lighting. • Overheating in summer can significantly affect survival, especially of vulnerable people.
Food	Minimal	• Refrigerated food is damaged. • Depending on availability of dry food, food demand can become an issue.	• Depending on availability of dry food, food demand can become an issue. • Potential disruption of large-scale food supplies and availability of supply.
Water	Minimal	• Potential water interruption. • Domestic hot water can be compromised if electricity is employed to heat water.	• Potential water interruption • Domestic hot water can be compromised if electricity is employed to heat water.
Transportation	Minimal	• Limited possibility of personal transport due to potential lack of fuel in motor vehicles, etc. • Interruption of electric transportation (trains, trams, etc.), charging electric vehicles, etc.	• Individual transportation is compromised due to lack of fuel, or disruption in operation of the fuel stations. • Interruption of electric transportation (trains, trams, etc.), charging electric vehicles, etc. • Potential disruptions to public transportation.
Safety	• Indoor air quality, especially in very tightly spaced buildings or apartment buildings. • CO_2 built up in case of using alternative heating. • Access, especially for multistorey buildings.	• Indoor air quality, especially with very tightly spaced buildings or apartments/condos. • CO_2 built up in parking lots (apartment buildings with underground parking). • Access, especially for multistorey buildings (for elderly or people with health issues).	• People with electrical medical devices (heart, breathing, etc.). • Risk of shortage of supply of medicaments. • Care for emerging sickness (especially for children and vulnerable people). • Mental health effects from traumatic or stressful experiences during outages. • Post-outage hazards from sheltering in place in unhealthy environments. • Potential exposure to hazards such as contaminated drinking water, contaminated floodwaters (if flooding occurs), and potential mold growth and moisture in housing. • Environmental hazards due to damage to sewage treatment. • Health and safety risks from clean-up and recovery activities.

Table 1. *Cont.*

	12–24 h	1 Day–1 Week	>1 Week
Security	Minimal	• Financial security: ATM not working. • Risk of looting.	• Financial security: ATM not working, cash availability. • Looting. • The risk of disturbance to public order and security.
Health concerns	• People with electrical medical devices (heart, breathing, etc.).	• People with electrical medical devices (heart, breathing, etc.). • Risk of shortage of supply of medicaments (for persons requiring supply for less than a week). • Water shortages, spoilage of foods.	• People with electrical medical devices (heart, breathing, etc.). • Risk of shortage of supply of medicaments. • Care for emerging sickness (especially for children and vulnerable people). • Mental health effects from traumatic or stressful experiences during outage. • Risks to health and safety emerging from certain clean-up and recovery activities. • Water shortages, spoilage of foods.
Threat to life	• People with electrical medical devices (heart, breathing, etc.).	• People with electrical medical devices (heart, breathing, etc.). • Potential compromise of medical and social care institutions, etc. • Increased traffic accidents.	• People with electrical medical devices (heart, breathing, etc.). • Potential compromise of medical and social care institutions, etc. • Potential risk to life and health due to the occurrence of secondary crisis. • Increased traffic accidents.
Damage to dwellings' contents	• Minimal	• Potential damage from power surge after an outage to computers, TVs, air conditioners, heaters, motors, and other HVAC components. • Appliances, including washers, dryers, and microwaves, are vulnerable to sudden and frequent changes in voltage.	• Potential damage from power surge after an outage to computers, TVs, air conditioners, heaters, motors, and other HVAC components. • Appliances, including washers, dryers, and microwaves, are vulnerable to sudden and frequent changes in voltage.
Damage to dwellings	• Minimal	• Frozen pipes- bursting (in cold period). • Water damage due to pipe burst. • Potential appearance of moisture and damp patches. • Potential basement flooding and consequent damage to equipment and furniture.	• Frozen pipes- bursting (in cold period). • Water damage due to pipe burst. • Potential appearance of moisture and damp patches. • Potential basement flooding and consequent damage to equipment and furniture.

- Medium-term outage (24 h–1 week):

Beyond 24 h, some functions and human needs become difficult to fulfill. For instance, communication can be severely disrupted, including communications for police and fire departments, compromising public safety [19]. Basic levels of thermal comfort through adequate heating and cooling may become unattainable, especially during severe climate events. This can significantly affect the well-being of occupants and, in some cases, lead to the buildings' evacuation. Domestic hot water can be compromised as well if electricity is employed to heat water [39]. Similarly to the short-duration outage, people with critical electric equipment can experience life-threatening situations. In addition, medical centers and the supply of medicines can be affected, even in the presence of some emergency power generation means [19].

Other crucial issues include damage to dwelling contents such as the refrigeration of food and damage to various electrical equipment appliances such as computers, TVs,

HVAC, and various electric appliances, which can be damaged by power surges [40]. Under extreme cold conditions, other significant issues can arise, such as the bursting of frozen pipes coupled with water damage [41,42].

At the safety level, indoor air quality can be significantly impacted, especially in tight-sealed buildings that rely on mechanical ventilation [43]. In addition, carbon dioxide can build up in underground spaces, such as in parking lots (particularly in apartment buildings), which could render these areas hazardous [44] necessitating, in some cases, the evacuation or partial evacuation of the buildings. Other issues relate to access to multistorey buildings due to the restricted function of elevators (especially for vulnerable people).

Additional issues include the limited possibility of transport due to the potential of fuel shortages or dysfunctional fuel stations, as well as the impact of power outages on electric-powered transport (e.g., trams, trains) and the charging of electric vehicles, which are becoming increasingly abundant. A summary of the main issues related to various criteria and outage periods is summarized in Table 1.

- Long-term outage (>one week):

Issues that occur during a single-week power outage continue and become more severe for longer-term periods of outage [45]. Additional potentially major issues (compared to the medium-term outage) include disruption of large-scale food and water supplies and sewage treatment. During long-period outages, serious damage to buildings can occur, potentially including pipes bursting (in cold periods, as discussed above), basement flooding, and consequent damage to equipment and furniture, as well as the appearance of moisture and damp patches [19].

Other concerns include a negative impact on living standards and social services, mental health issues due to traumatic experiences, and the potential occurrence of various emergencies during the outage that might endanger health and life [46]. Hazards from sheltering in inadequate housing conditions include exposure to various hazardous conditions such as contaminated water, mold, and moisture [46]. Population displacement leads to large-scale problems, such as housing shortages, accompanied by economic and social impacts. A detailed summary is presented in Table 1.

Table 2 graphically shows the various functions and needs that are impacted by power outages, according to the time length of the outage. The main impacts are indicated by triangles. Circles indicate that a combined impact of other factors may apply. These factors include residence types, location, and demographics, as well as the type of fuel employed to power various functions (e.g., heating, appliances, etc.). Residence types include detached houses, attached houses, low-rise apartments, mid-rise apartments, high-rise apartments, and apartments with or without underground indoor parking. Locations may include urban high-density areas, rural areas with moderate density, and farmland. The demographics of the population can play an important role in the resilience of a community; this includes the existence of elderly people (couples or singles), families with vulnerable individuals (the elderly, children, etc.), and young individuals (couples or singles). Other impactful factors relate to the design of the buildings and the neighborhoods and their energy systems. While the role of building and neighborhood design is discussed below (Section 4), the impact of the other factors mentioned above is beyond the scope of this paper. These factors (building types, location, and demographics) can have a significant effect on the overall resilience of a community to power outages and need to be thoroughly investigated.

Table 2. Functions and needs that are impacted by power outage.

Period	Communication	Comfort	Food	Water	Transportation	Safety	Security	Health Concerns	Threat to Life	Damage to Dwellings' Contents	Damage to Dwellings
12–24 hours	▲▲	●●	–	–	–	●	–	●●	●●	–	–
1 day–1 week	▲▲▲	●●●	▲▲	●●	▲▲	●●	▲	●●	●●	▲	▲▲
>1 week	▲▲▲	▲▲▲	▲▲▲	●●●	▲▲▲	●●●	▲▲	●●●	▲▲	▲	▲▲▲

▲, ● Indicate low impact, ▲▲, ●● indicate medium impact, ▲▲▲, ●●● indicate higher impact. All impacts are qualitatively assessed in relative terms.

4. Technologies for Improved Resilience

There are several strategies and technologies that can be used to improve resilience during power outages at the building and neighborhood levels. Some of these are very basic and address individual needs, while some other technologies can replace the interrupted power for a specific time span. For instance, the adoption of renewable and alternative energy sources presents a practical approach to adapting to power outages associated with climate crises and extreme events, allowing it to address not only potential power interruptions and energy deficits but also to alleviate climate change-associated disruptions. Neighborhoods and communities can become more energy resilient by diversifying their energy resources, including the implementation of reliable renewable energy systems [47].

In addition to improving distributed energy generation potential, energy efficiency in buildings constitutes a key aspect of energy resilience [48,49], assisting in reducing the energy consumption of urban areas and consequently the strain on local energy suppliers during an energy emergency. Enhancing passive strategies, including passive heating and cooling, can increase the preparedness of buildings and households to withstand disasters [16,18,50]. In addition to these measures, there are some commonly applied methods, including low-tech methods, that can assist in adapting, albeit temporarily, to a power outage crisis.

This section presents an overview of various methods and strategies that can mitigate the impact of power interruptions, as discussed above. These strategies are then preliminary classified according to various criteria. The work acknowledges, however, that there is a need to analyze in more depth the potential of various technologies and to quantify their impacts on improving energy resilience.

4.1. Building-Level Measures

Some methods are commonly employed in residential buildings to cope with power outages. Most of these methods can be helpful for short timespans and to address a specific need, including emergency lighting (e.g., emergency exits, floor lighting, and other important signage). Other methods relate to energy efficiency measures that allow a reduction in the electricity load and prolong the period of building autonomy without relying on mechanical systems. This has been investigated in a large number of studies, as presented in Section 4 (see below). Other advanced methods, including those focusing on energy generation, are presented below.

Low-tech and alternative methods

Some of the temporary coping methods include simple devices that can be easily acquired and utilized for restricted needs, such as flashlights and candles to provide lighting for limited visual tasks or portable gas stoves to allow food preparation. When the crisis lasts longer than a few hours, some basic needs start to be important to fulfill, including food preparation and thermal comfort, especially heating in cold weather. Alternative solutions such as a wood stove or wood fireplace can mitigate some of these critical issues for short- to medium-term outages [19,35]. However, these methods would not fulfill a wider range of needs and cannot be applied to all types of buildings (e.g., wood fireplaces).

For thermal comfort, an efficient building design, especially building envelopes, can assist in keeping the residence at acceptable conditions; however, these are also short- to medium-term solutions, depending on the exigency of the outdoor thermal conditions.

Energy efficiency measures

Energy efficiency plays a major role in reducing demand, especially during periods of extreme stress on the grid. For instance, extreme heat or cold events may increase in magnitude and frequency, leading to an increase in peak electricity demand [51]. Higher electricity demand can lead to prolonged power interruptions [52]. The energy efficiency of building equipment and appliances can help communities reduce their overall load and, as such, their demand on the local grid, which may assist in avoiding outage interruptions in critical periods.

Energy-efficient buildings with high-performance building envelopes allow the passive performance of buildings, leading to some level of inhabitability without the requirement of mechanical systems. For example, research carried out on some buildings in New York City to evaluate the impact of power outages on indoor temperatures in winter and summer demonstrates the better thermal resilience of high-energy-performance buildings. Such buildings that were designed with highly insulated and air-tight building envelopes maintained a comfortable indoor temperature (in the upper 50 s °F (10 s °C) in the winter during a theoretical weeklong power outage [22], while the indoor air temperature of older building types dropped to 40 s °F (4.5 °C) [53] in a period ranging from 1–3 days, causing health risks, especially to vulnerable inhabitants.

Various passive strategies can assist in reducing the impact of outdoor conditions, both high and low temperatures, on the indoor environment. The impact of building design, including the building envelope, is discussed more in the building design section (Section 4) below.

4.1.1. Energy Generation

Technologies that permit the use of renewable and alternative energy sources for power and heat production can offer resilient backups for communities [54,55]. These technologies comprise on-site generation using photovoltaic systems, including building-integrated photovoltaics (BIPVs), solar thermal collectors, geothermal heating, and micro wind turbines coupled with microgrids. The planning and implementation of such technologies can be beneficial for the whole community and particularly pertinent to critical facilities such as hospitals and buildings accommodating vulnerable people (e.g., retirement homes). Combined heat and power systems and microgrids can operate continuously (not only during emergencies) and, as such, they present reliable methods of energy supply.

PVs Coupled to Batteries

PV systems can withstand extreme weather events, providing backup power for buildings and critical facilities, thus enhancing the resilience of communities [29]. PV systems can be sized to provide the most critical functions to residences to survive power outages [56]. Buildings with hybrid photovoltaic–battery storage systems can provide a continuous electricity supply during power outages, depending on the weather conditions and the battery size as well as the building energy load [57]. Research highlights that a combination of PVs, battery storage, and grid connections is cost-effective and environmentally efficient [58].

The potential of PVs coupled to batteries has been investigated in a number of studies [16], including the capabilities of PVs and plug-in hybrid electric vehicles to enhance the resilience of households [59] and their return on investment [60]. The impact of a PVs–batteries combination on the resilience of whole communities was also studied, e.g., [61,62].

Micro Wind Turbines

Small-size urban wind turbines include building-integrated wind turbines or small stand-alone wind turbines [63]. Building-integrated wind turbines can be grid-tied or off-grid, requiring battery storage to store energy. The rotor diameter of residential wind turbines can range from 0.9 m to 7 m and require a height of 18 m to 30 m [64].

Wind turbines extract about 40–50% of the energy that passes through them [65]. A traditional single-home family would need one 10–20 KW turbine to produce sufficient energy to fulfill the total energy demand of a typical house (using approximately 10,000 kWh per year) [66].

The feasibility and efficiency of wind energy depend mainly on the location. Wind turbines are more suited for areas with reduced obstacles and with an average annual wind speed of at least 10 mph (16 km/h) [67].

Micro Combined Heat and Power

Micro combined heat and power (micro-CHP) is a heat and power cogeneration system designed for use at the building scale, such as single or multi-family houses, and can also serve small office buildings (up to 50 kW [68]). A micro-CHP system provides electric power while simultaneously generating thermal energy for space heating and hot water provisions for a building by recovering waste heat [69].

A micro-CHP system can be designed to follow the electricity or heat demand of a building, delivering heat or electricity as a by-product. This may result in the production of excess electricity or heat, which requires devising methods to manage the excess power, such as designing storage systems [70]. Excess electricity can be, under regular circumstances, fed to the grid.

The most common micro-CHP systems employ natural gas to cogenerate heat and power. Although natural gas CHP is responsible for GHG emissions, due to the effective efficiency of the CHP system, the produced emissions are less than those of other alternative systems for generating heat (e.g., a condensing boiler) [71].

Although connected to the electricity distribution network, CHP systems, including micro-CHPs, can be completely independent, allowing them to produce energy when it is required, thus improving the resilience of a building to power interruptions.

Fuel Cells

Fuel cell technologies present a promising option for clean power generation. This technology allows the generation of electrical energy while producing heat as a useful by-product. Fuel cells convert energy into electricity and heat through the chemical reaction of hydrogen and oxygen to produce water [72,73]. Although fossil fuel is generally used to produce hydrogen (the basic ingredient of fuel cells), fuel cells create significantly fewer emissions than most other fossil fuel generators [72].

Fuel cells are mainly employed in two major applications: powering vehicles and generating power for various types of buildings, utilities, and communities. Demonstrated applications of fuel cells include backup generation for hospitals, office buildings, and schools [74]. Some applications include remote villages and campgrounds [75]. Fuel cells can also be used to supply power for temporary needs, including shelters and construction sites. Fuel cells can play a significant role in enhancing distributed energy initiatives, providing electricity and heat, and reducing the vulnerability of central grid disruptions.

Integrated Micro-Generation Systems

Integrating different systems and methods to generate heat and power is continuing to attract attention to improve the efficiency of renewable and alternative energy resources. For instance, research on the integration of a fuel cell (FC) micro-cogeneration device, a heat pump (HP), and thermal storage highlights that such a combination presents an optimal solution to manage electrical and thermal storage while reducing energy consumption and evading energy excess production, e.g., [69,76]. An FC-based CHP system can be sized to produce sufficient energy for individual buildings.

Other research investigated the integration of heat pumps, PV systems, and the local grid connection, demonstrating the efficiency of such systems in multistorey buildings [77].

Portable Generators with Various Fuel Types

Generators are some of the most common systems to generate distributed energy. The fuel employed in generators ranges from high-GHG-producing, such as diesel generators, to natural gas (NG) [78]. Although NG generators are cleaner than diesel generators, they still have high GHG emissions. Other generators use gasoline and propane as fuel. Gasoline-fueled generators fall between diesel-fueled and natural gas generators in terms of GHG emissions, while propane generators are similar to natural gas generators (in terms of emissions).

Employing such generators can be useful during an emergency power interruption; however, they should be restricted to use as emergency back-up generation. Emission-control measures should be put in place when such power generation methods are employed to reduce their harmful environmental impacts. These measures comprise improvements in fuel and control technologies and enhanced efficiency.

4.2. Neighborhood-Level Energy Resilience

Most of the energy generation technologies suitable at the individual building level can be effective on a neighborhood level, with design modifications to suit larger-scale developments. For example, PV systems and wind turbines can be designed to be not only part of a building (integrated within the building envelope or add-on systems) but also part of the neighborhood outdoor surface (see Section 4). Similarly, wind turbines can be installed in various areas of the neighborhood. A CHP plant can be designed more efficiently at the urban scale, as discussed below. Additional technologies that can be applied only on the urban scale include district heating and cooling systems, microgrids, and smart grids. These are summarized below.

4.2.1. Combined Heat and Power

Urban- and community-level CHP systems serve the same function as a micro-CHP (described above) but are designed at a larger scale to serve multiple buildings within a neighborhood. Examples of CHP implementation and its performance and impact during power outages due to weather events are reported in the literature [79]. For instance, during Hurricane Sandy, CHP systems supplied heat and power (although sometimes in limited capacity) as well as other critical functions to multifamily buildings [54]. Maintaining critical services during a power outage is vital to improving the overall resilience of communities and can affect multiple households. This is because the risk to individual buildings and individuals can be significantly reduced by keeping vital services running (hospitals, water treatments, and others). On the other hand, designing such critical buildings and facilities with backup power can enable them to serve as temporary shelters for displaced residents [80], leading to increased social resilience and capacity to cope.

CHP systems can use various types of fuel to provide continuous operations. Most CHP systems are fueled by natural gas, which can be reliable during outages (as long as natural gas pipelines are not disrupted). CHP can be operated using waste generation biomass or biogas, which can be equally reliable in times of disaster [54]. In addition to providing emergency power, CHP systems are cost-effective and reduce overall net emissions [81].

4.2.2. District Energy

District energy systems can integrate several types of renewable energy sources to produce thermal and electrical energy [82,83] while reducing GHG emissions [84]. Some of the most common renewable sources incorporated in district energy systems include solar photovoltaics and stand-alone microgrids, as discussed below. Biomass is also used in some communities using local waste generated from tree trimming and other urban waste wood. These technologies are often coupled with electrical and thermal energy storage (e.g., batteries [85], fuel cells [86,87], and borehole geothermal energy storage [88]). The integration of different energy sources constitutes an effective method to enhance the efficiency of the entire system.

District energy systems can be designed to fulfill only one requirement, such as heating, cooling, combined heating and cooling needs, co-generation, or tri-generation [83]. District systems allow the shift in energy consumption from peak demand to off-peak periods, thus reducing the dependence on the electric grid.

4.2.3. Microgrids

The microgrid concept is explained employing a number of definitions [89–91]. Most commonly, a microgrid is defined as a mix of distributed energy resources and interconnected loads, constituting a single controllable entity that can be connected to or disconnected from the grid. When disconnected from the grid, the microgrid can operate in island mode [92].

Microgrids connect buildings and facilities within a neighborhood to various distributed energy resources, such as those described above (e.g., CHP, photovoltaic systems, wind turbines, district energy systems, and energy storage). The design of a microgrid system varies depending on the project specifications and requirements [89]. Microgrids can play a key role in enhancing the resilience of communities against power outages since they are able to maintain a reliable and continuous supply of power.

4.2.4. Smart Grid

Smart grids are advanced digital systems that enable a two-way power flow [93] and incorporate several technologies, comprising advanced metering and information and communication networks, which are integrated into power infrastructures [94]. Smart grids can use different energy resources, including intermittent solar- and wind-generated electricity [95], and can accommodate storage facilities. Smart grids can play a significant role in restricting the spread of power outages since they allow the identification of the impacted parts of the electricity system and can then isolate them, improving the resilience of a community to the power outage. Smart grids also allow users to preventively turn off the power in specific areas before extreme weather events to avoid system-wide damage. This capability assists in reducing the extent of power outages and shortening recovery times.

Overview of Technologies' Potential

This section gives an overview of the potential of the technologies presented above with respect to a number of criteria. These criteria include the current feasibility of these technologies and their environmental impact, including their efficiency (Table 3). The potential of these technologies is presented in a qualitative manner (Table 4), indicating the fulfillment of some of the criteria.

Table 3. Description of criteria.

Feasibility:	**Building types:** potential to accommodate variety of buildings. **Accessibility**: relates to various issues such as availability on the market, cost, etc. **Implementation:** relates to the ease of implementation in a building, ease of use, etc.
Impact	**Reduced Emissions**: potential to reduce GHG emissions (based on materials used or/and fuel consumed). **Efficiency**: increase in the value of reduced damage from power breakdown (fulfillment of the objectives) to investment in the technology.
Objectives	**Comfort:** achieving survivable and safe temperatures and acceptable overall comfort (e.g., lighting). **Indoor air quality:** potential improvement in indoor air quality. **Health and safety:** potential reduction in health and safety hazards. **Fulfilling specific need:** potential to fulfill various needs (more than one). **Continuous power generation:** potential to generate stable power.
Outage Period	Describes the suitability of the resilience measure to the outage time.

Table 4. Summary of technologies potential. ● indicates fulfillment, ▲ indicates unfulfillment, S: short-term, M: medium-term, L: long-term.

	Feasibility			Impact		Objectives					
	Accessibility	**Implementation**	**Building Types**	**Reduced Emissions**	**Efficiency**	**Thermal Comfort**	**Indoor Air Quality**	**Health and Safety**	**Fulfilling Other Needs**	**Continuous Power**	**Outage period**
	Low-tech and alternative measures										
Candles, flashlights, propane/kerosene lamps	●	●	●	●	▲	—	▲	▲	▲	—	S
Wood stove	▲	▲	▲	●	▲	▲	▲	▲	▲	—	M-L
Portable gas stove	●	●	●	▲	●	—	▲	▲	▲	—	S-M
Wood fireplace	▲	▲	▲	●	▲	▲	●	▲	▲	—	M-L
Batteries	▲	●	●	●	▲	—	—	—	●	●	S-M
	Efficiency measures										
High-performance building envelope	▲	▲	●	●	●	●	●	—	▲	—	S-M
Passive solar design	▲	▲	●	●	●	●	●	—	▲	—	S-M
Operable windows/natural ventilation	▲	▲	●	●	●	●	●	—	▲	—	S-M
High-efficiency appliances	▲	●	●	●	●	—	—	—	—	—	M-L
	Generation										
PV	▲	▲	●	●	▲	●	●	●	●	▲	S-L
Micro wind turbines	▲	▲	●	●	▲	●	●	●	●	▲	S-L
Micro combined heat and power	▲	▲	●	▲	●	●	●	▲	●	▲	S-L
Portable generators/propane/diesel	▲	▲	●	▲	●	●	●	▲	●	●	S-L
Fuel cells	▲	▲	●	●	●	●	●	●	●	●	S-L
	Neighborhood level										
PV	●	▲	●	●	▲	●	●	●	●	▲	S-L
Wind	●	▲	●	●	▲	●	●	●	●	▲	S-L
CHP	●	●	●	▲	●	●	●	●	●	●	S-L
District energy	▲	▲	●	●	●	●	●	●	●	▲	S-L
Microgrid	▲	▲	●	●	●	●	●	●	●	●	S-L
Smart grid	▲	▲	●	●	●	●	●	●	●	●	S-L

5. Impact of Building and Neighborhood Design

This section is based on the author's research and aims to illustrate the implementation of principles outlined in the literature review of Sections 2 and 3 in the design of buildings and neighborhoods and their energy systems to improve their overall energy resilience. The section highlights the impact of a holistic approach in the design of neighborhoods, their buildings, and the exterior areas surrounding the buildings in order to provide an adequate environment for implementing measures to enhance energy resilience and empower the neighborhoods' residents.

5.1. Building Design

The design of a building significantly affects its energy efficiency, its capability to withstand climate events, and its vulnerability to energy outages. Design considerations, such as the building type, form and configuration, and outer envelope, directly affect the building's energy demand [18,22,96] and its potential to incorporate solar technologies for sustainable energy production. Improving such capacity is, as mentioned above, a significant strategy to mitigate the impact of central power interruptions. Design aspects, such as being highly insulated, having an airtight building envelope, and having optimized window systems with an adequate window-to-wall ratio (WWR), can significantly impact the energy requirements of a building. The window-to-wall ratio (WWR) is a variable parameter, and a country's building and high-energy performance codes govern the appropriate WWR value. Some examples of high-performance windows are thermochromic glazing [97] and ventilated/non-ventilated PV-integrated windows [98]. These high-performance windows significantly diminish heating and cooling requirements while improving visual and thermal comfort. This is not only beneficial in reducing the size of mechanical equipment but also in increasing the duration of the self-sustainability of a building without relying on electric energy to provide these functions. The design of various other climate-responsive features and architectural passive design elements, including shading strategies, light shelves, and solar chimneys, can maximize the utilization of solar energy for heating, cooling, and daylighting, reducing dependence on local energy grids. Architectural passive design features can be implemented in different types of buildings, including residential, office, and commercial, allowing the enhancement of buildings' adaption to energy interruption.

Building envelope design for energy generation. The design of the building envelope plays a significant role in preparing buildings to integrate renewable energy generation technologies (e.g., PV and PV/thermal systems (PV/T)), thus improving their potential to withstand power outages. The shape of the building and of the envelope affects its solar exposure and thus its solar potential, which is highly dependent on the tilt and orientation angles of these surfaces (Figure 1 [99]). While integrating PV technologies in roofs is an optimal decision for low-rise buildings ($\leq$3 floors), multistorey buildings' facades can offer advantageous surfaces for the integration of PV systems due to their increased surface area from the increased height of multistorey buildings relative to the roofs of the same buildings (which remain unchanged [100]).

The geometric design of building envelope can be carefully designed to maximize solar radiation incident on these surfaces, consequently increasing the electric and thermal energy generated by solar collectors. Manipulating the tilt and orientation angles of individual building surfaces can lead to increased electricity generation and the extension of the timing peak generation. Multifaceted geometry (Figure 3d–f) can significantly increase the energy generation potential of roofs and particularly facades [100]. Such design considerations can provide the creative assimilation of PV systems, making them architecturally and esthetically pleasant.

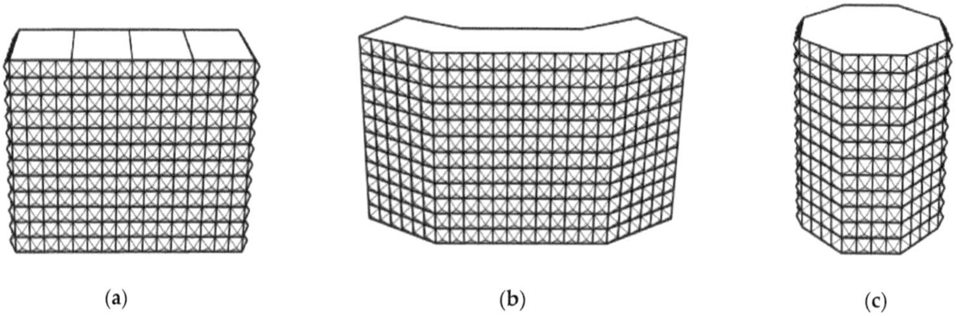

Figure 3. *Cont.*

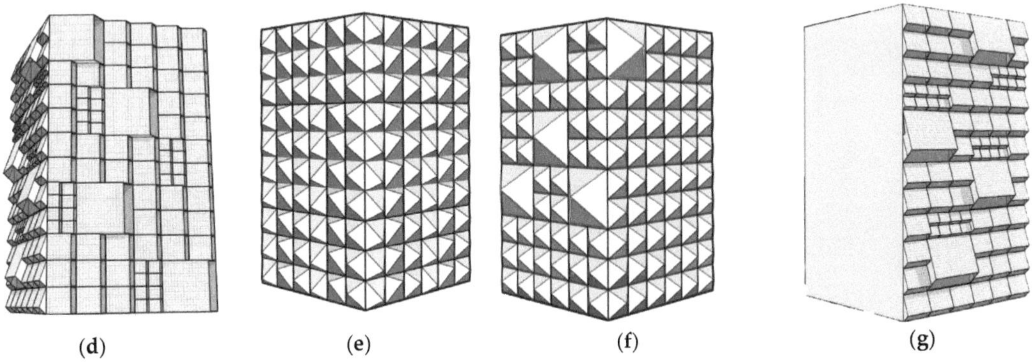

Figure 3. Design of buildings for enhanced solar potential: (a–c) building layout, (d–g) building facades.

5.2. Neighborhood Design

Neighborhood design has a significant impact on the energy consumption and GHG emissions of the overall neighborhood and enhances its resilience and capability to adjust to different stresses. Below is a summary of some factors that should be considered in the design of resilient communities and neighborhoods.

Type of neighborhood and building mix. A mixed-use neighborhood that encompasses diverse amenities within walking distance provides several possibilities for mitigating energy vulnerabilities. For instance, such neighborhoods can reduce the dependence on vehicle transport when fuel is restricted and can provide potential nearby temporary shelter in buildings that can be supplied with continuous alternative energy (e.g., schools [80]). Other opportunities offered by mixed-use neighborhoods include the prospective application of district energy, large-scale renewable energy, thermal storage, and energy sharing between buildings.

The variety of building types within a neighborhood can affect its energy consumption as well as its overall capacity to accommodate solar technologies within its buildings and neighborhood surfaces [101]. An optimal ratio of commercial to built land area ranges from 23% to 32%, allowing for a reduction in energy consumption and GHG emissions.

The mix of buildings also affects, as mentioned above, the potential of designating specific buildings as temporary shelters during power outages. Figure 4 presents an example neighborhood that includes various amenities and buildings that can serve as temporary shelters, which are strategically placed with respect to roads and residential buildings [80].

Designated temporary shelters can be selected, where possible, according to their energy intensity, prioritizing those that have a reduced overall energy intensity (energy per unit area) or emergency power generation capacity. In addition, such buildings should be easily accessible to the residents of different parts of the neighborhood.

Density. An increased built density, including a high ratio of built floor area in a neighborhood, is often associated with urban environmental sustainability [102]. To augment the potential of renewable solar energy generation, enhancing solar availability is a high priority. The solar access of buildings and the neighborhoods' outdoor areas can be negatively affected by high density, especially with the design of high-rise multistorey buildings. The impact of density on solar access can be offset through the thoughtful design of the urban layout, site coverage, and building heights [103,104]. Increasing spacing between buildings allows better solar access to buildings and increases solar availability at the ground level for the implementation of standalone PV systems for neighborhood-level electricity generation (see below). On the other hand, while low-density residential neighborhoods can achieve energy self-sufficiency from PVs integrated into buildings' surfaces, higher-density neighborhoods require diverse energy sources that are combined with energy storage [101]. A study that investigated and contrasted the energy resilience of low- and higher-density neighborhoods based on specific resilience indicators showed that higher-density, mixed-use neighborhoods are less energy resilient [80].

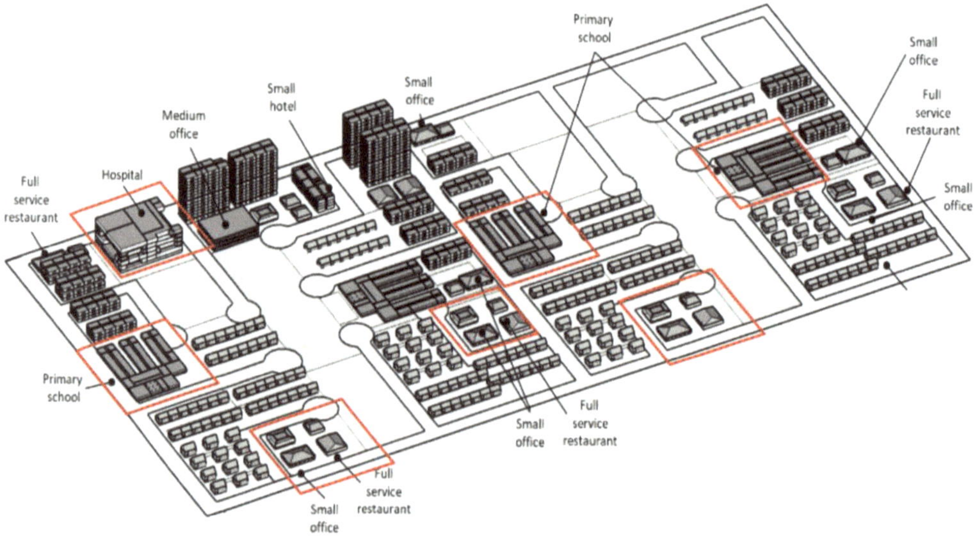

Figure 4. Example of mixed-use neighborhood with various amenities. Red lines indicate buildings that can serve as temporary shelters, and/or buildings that need to maintain continuous operations.

Green and spatial areas. Landscape and outdoor areas surrounding buildings can be designed to maximize the incorporation of PV and solar thermal collectors within neighborhoods (Figure 5a). For solar applications, the design of public green areas should address various issues such as avoiding shade from surrounding buildings.

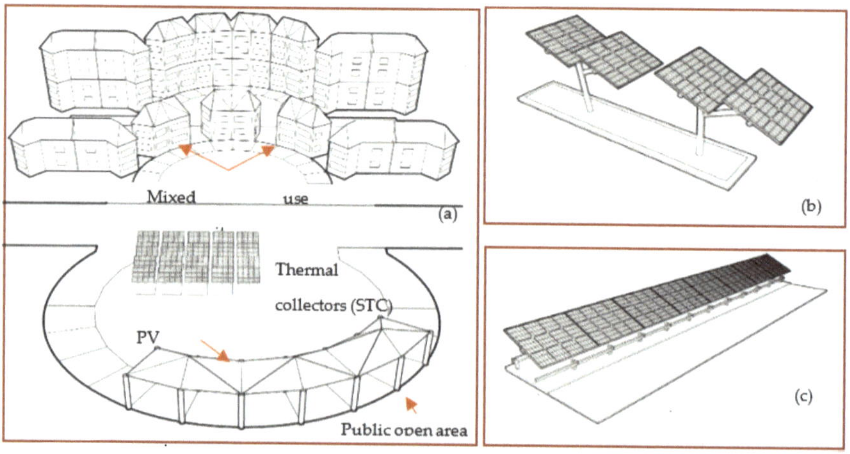

Figure 5. (**a**) PV and STC integration in public areas, (**b**) PV as parking structure, (**c**) PV on street borders.

The implementation of solar collector structures in the public landscape can be designed to enhance outdoor thermal comfort, creating attractive areas for social interaction and, as such, improving the quality of life in the community. For example, landscape PV structures can be employed as shading structures for thermal shelters during colder periods (Figure 5b,c).

Street design. The street layout, including shape and orientation, influences the design and orientation of the buildings on them. Street design therefore has an impact on the solar potential of the surrounding buildings.

Beyond their impact on the energy consumption and energy generation of the buildings and surrounding areas, the street's design has a significant impact on the choice of transport and resulting energy consumption and GHG emissions. Research highlights that streets with a higher density of intersections may decrease the reliance on individual cars for transportation [103].

The design of connection nodes and available routes affects the potential for evacuation during emergencies and therefore impacts the resilience of the whole community. A key criterion for creating resilient neighborhood layouts is to reduce dependence on major streets. This design consideration prevents the destabilization of the entire street network system when some central nodes are disabled due to power outages or climate-related events [100]. Potential shelter buildings should be located at street intersections to improve their access during evacuations [80]. Although analyzing the response to disasters is beyond the scope of this paper, the impact of neighborhood design on some responses (such as evacuation) is mentioned due to the interconnection of various design elements and their direct and indirect impacts on energy resilience.

5.3. Impact of Design on Energy Systems

Resilient neighborhood design should consider energy demand as well as local and distributed energy generation strategies. The impact of neighborhood design on urban energy systems, both on the demand and supply sides, is briefly discussed below.

Energy demand. Energy demand can be significantly reduced through the architectural design of buildings and proper spatial neighborhood design. Employing high-energy-performance mechanical systems and energy-efficient appliances can further reduce energy consumption in buildings. Efficient mechanical systems, including heat pump technologies, heat pumps coupled with PV systems, heat recovery systems, mechanical ventilation (e.g., displacement ventilation), and effective distribution and controls, are increasingly implemented to improve the energy performance of different building types. For instance, smart management systems can be implemented to preheat or precool buildings before peak hours [105,106]. The deliberate exploitation of thermal mass, together with energy-efficient building envelopes and mechanical ventilation, can assist in the strategic preheating and precooling of buildings [107].

Other promising systems consist of ground source heat pumps (GSHPs), which, using ground-extracted heat, are considerably more energy-efficient than conventional mechanical systems [108]. Coupling a GCHP system with solar thermal collectors offers a significant opportunity for energy savings, especially for heating-dominated neighborhoods [109]. Integrating energy efficiency measures with microgrid technology allows for a reduction in energy demand on the microgrid itself.

Energy supply and neighborhood planning. Diversification of energy resources constitutes a promising strategy to enhance the energy resilience of a community. PV technologies integrated in buildings and in neighborhood outdoor areas form a mature technology that is ready to be deployed at various scales and capacities [110], forming an important layer of energy resilience. Together with solar technologies (PV and STC), other energy resources, including combined heat and power utilizing various sources, such as waste to energy (WtE) or reduced impact fuel (see above) and small wind turbines, can be explored, especially when available surfaces for installing solar systems are restricted. To achieve self-sufficient neighborhoods, energy storage is required, together with an optimal mix of energy sources [101]. To be completely independent from the grid, during an energy crisis, electrical and thermal storage should be sized to provide the necessary demand of a neighborhood. It should be considered that increased levels of distributed energy resources may lead to issues in energy balance and congestion. These potential congestion issues should be addressed when designing urban energy systems, allowing for the proper management of energy production, utilization, control, and storage.

On the other hand, microgrids and smart metering can be beneficial in controlling the zones that can be supplied with energy at specific times. For instance, a study carried out by Singh and Hachem-Vermette [77] shows that, for long periods of power interruption, it might be beneficial to evacuate the population to designated temporary shelters (e.g., schools) and to prioritize these buildings when supplying local energy. Such a scenario will be less vulnerable than sheltering in place, as it is easier and more efficient to supply a few buildings with energy rather than a larger number of residences. In such cases, smart grids can be useful in controlling the energy supply to specific zones.

6. Discussion and Conclusions

This work presents an overview of the impact of power interruptions on buildings, buildings' occupants, and neighborhoods. It also discusses the existing technologies, ranging from simple to more advanced, to mitigate some of these impacts. The paper finally highlights the role of building and neighborhood design in resilient energy systems.

Depending on its length, a power interruption can severely impact people, buildings, and neighborhoods. These impacts may include threat to the lives of vulnerable populations, even under short-term power interruptions, while, at a longer term, other issues can arise, including displacement of the population, serious damage to buildings, and instability of neighborhood organization. The severity of the power interruption impact is enhanced by the interconnection of energy systems with many other basic systems, such as water, food, connectivity, and transportation systems. The gravity of power interruption impacts can be affected by various factors, including demographics, building type, building design, and location. Research focusing on highlighting the correlation between the severity of power interruptions and the factors mentioned above is still lacking.

There are various methods to cope with power interruptions at building and neighborhood levels. At a building level, some of these methods are simple and affordable, allowing the fulfillment of a specific need for a short time, while others are more sophisticated and need planning and a larger budget to be efficiently implemented. Such methods include PV-integrated systems with battery storage or integrated wind turbines. Several strategies need to be analyzed with respect to the building type to fully understand their potential and feasibility. For example, a wood stove cannot be easily installed in a multistorey building if it is not designed at an early stage. On the neighborhood level, the implementation of technologies requires more advanced planning, involving various stakeholders. Although some single technologies, such as neighborhood PV installations or combined heat and power (CHP), can mitigate power interruption, research shows that the integrated design of various renewable and alternative energy sources within smart microgrids presents more efficient and low-environmental impact solutions. Other employed and relatively easy-to-implement solutions such as generators, including portable ones, should be limited to emergency back-up generation. The utilization of such methods should be governed by emission-control measures to reduce harmful environmental impacts.

On the design side, a holistic approach should be applied in planning energy-resilient neighborhoods. This approach depends on the integration of building and urban design considerations, as well as on the interaction between these design considerations. Buildings and surrounding open public spaces need to be considered active elements of the energy network, contributing to production, storage, and supply. The various components of a neighborhood should be planned to maintain uninterrupted operation and to maximize their energy efficiency and potential contribution to the neighborhood energy system. For example, street layouts can be designed to allow the optimal orientation of the surrounding buildings while ensuring functionality during disruptions, thus enhancing the overall operation and resilience of the neighborhood. Additionally, the thoughtful consideration of building density assists in achieving a range of economic, social, and environmental benefits without compromising the solar potential of open public spaces and building surfaces.

A flow chart presenting the main research output is included below (Figure 6).

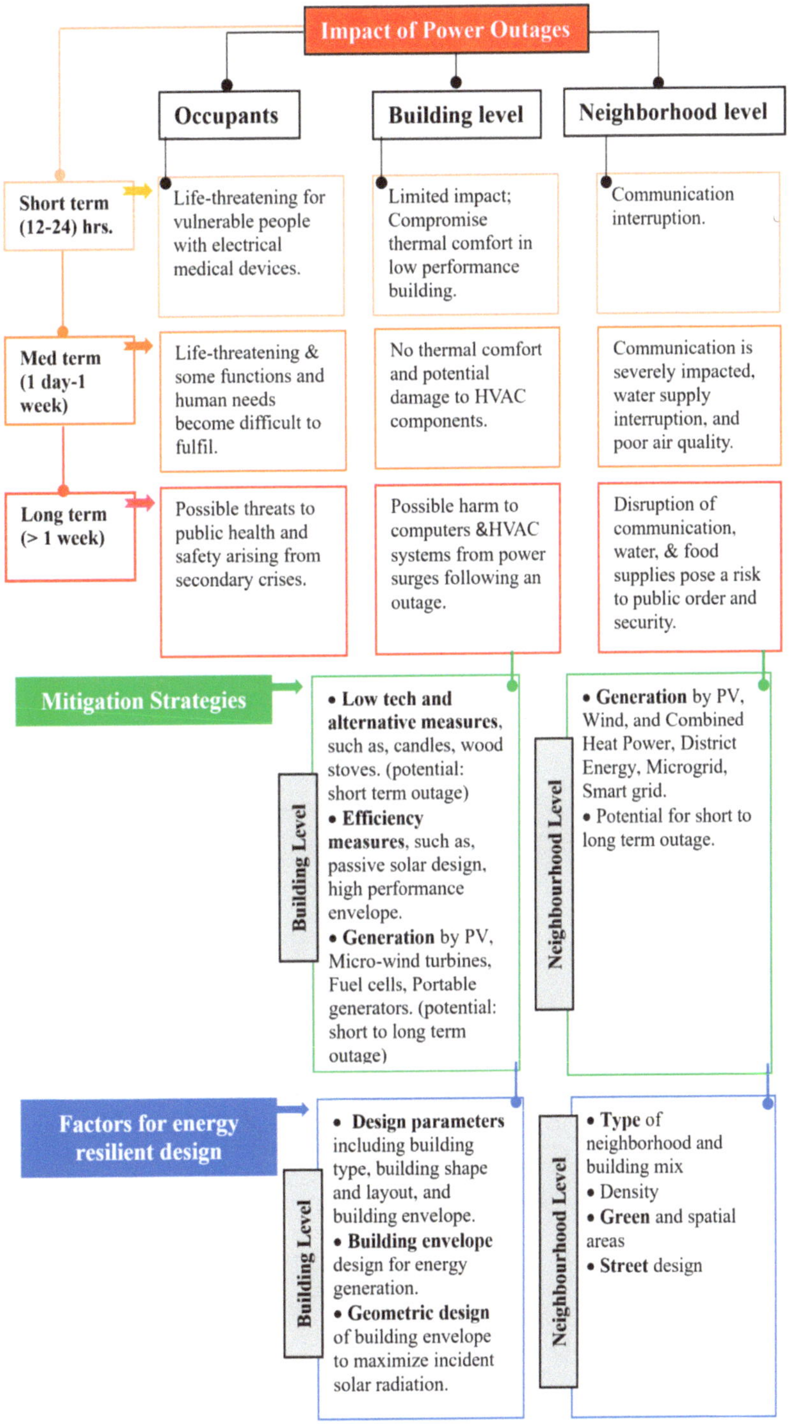

Figure 6. Flowchart of main research output.

The strategies and technologies discussed in this work can assist in minimizing the impact of power interruptions and ensuring a reliable energy supply. This can contribute to fulfilling several of the SDGs. For example, developing and implementing microgrids and promoting the use of renewable energy sources and energy storage technologies can reduce dependence on a single energy source (usually the local grid), contributing to SDG 7 (Affordable and Clean Energy), SDG 11 (Sustainable Cities and Communities), and SDG 13 (Climate Action). Reducing energy consumption through the implementation of various energy efficiency measures reduces resource consumption and waste generation, promoting responsible consumption (SGC 12). In addition, a resilient and uninterrupted energy supply is vital for vulnerable people (with health issues) and for healthcare facilities, ensuring continuous access to medical services during climate disasters (SDG 3: Good Health and Well-being). Other indirect contributions of this work to the SDGs include encouraging the development of innovative solutions to mitigate power interruptions and to improve the robustness and efficiency of infrastructure. This can foster economic growth and industrial development, thus supporting SDG 9 (Industry, Innovation, and Infrastructure).

Conclusions

This paper gives an insight into the impact of power interruptions and mitigation strategies and the role of neighborhood composition and building types in increasing the energy resilience of a specific neighborhood.

The review of potential issues caused by power interruptions highlights the need for substantial research to be conducted in various domains. This includes determining correlations between the severity of impacts and the parameters representative of the built environment and the population, as well as the assessment of various technologies with respect to useful criteria. In addition, investigating the impact of power interruptions and their length on the psychological performance of the population is still lacking. Such work can assist municipalities and governmental agencies in fully understanding the impact of power interruption in specific communities and, accordingly, in adopting feasible and effective mitigating strategies, as well as emergency response and management.

This review of the severe and diverse impacts of energy interruptions that span social, economic, and environmental domains supports the importance of a serious and urgent commitment to achieving the Sustainable Development Goals, as many of these goals have direct or indirect impacts on reducing the vulnerabilities of cities. Additionally, the discussed energy and design strategies can play a key role in achieving affordable and clean energy, sustainable cities and communities, climate action, and responsible consumption, among others, which are central to numerous SDGs.

Author Contributions: C.H.-V. is the first author of this work and has contributed to all aspects of its creation, research, and development. S.Y. contributed to the literature review and its visualization. All authors have read and agreed to the published version of the manuscript.

Funding: This research was funded by NSERC Discovery Grant held by the first author.

Institutional Review Board Statement: Not applicable.

Informed Consent Statement: Not applicable.

Data Availability Statement: The datasets generated and/or analyzed during the current study are not publicly available but are available from the corresponding author on reasonable request.

Conflicts of Interest: The authors declare no conflict of interest.

Nomenclature

ATM	Automated teller machine	NG	Natural gas
CHP	Combined heat power	PV	Photovoltaic
CO_2	Carbon dioxide	PV/T	Photovoltaics thermal
FC	Fuel cell	STC	Solar thermal collector
GHG	Greenhouse gases	SDG	Sustainable Development Goals
HP	Heat pump	WWR	Window–wall ratio
HVAC	Heating, ventilation, and air conditioning		

References

1. Kent, M.G.; Huynh, N.K.; Mishra, A.K.; Tartarini, F.; Lipczynska, A.; Li, J.; Sultan, Z.; Goh, E.; Karunagaran, G.; Natarajan, A.; et al. Energy savings and thermal comfort in a zero energy office building with fans in Singapore. *Build. Environ.* **2023**, *243*, 110674. [CrossRef]
2. Abi Ghanem, D. Energy, the city and everyday life: Living with power outages in post-war Lebanon. *Energy Res. Soc. Sci.* **2018**, *36*, 36–43. [CrossRef]
3. Hasselqvist, H.; Renström, S.; Strömberg, H.; Håkansson, M. Household energy resilience: Shifting perspectives to reveal opportunities for renewable energy futures in affluent contexts. *Energy Res. Soc. Sci.* **2022**, *88*, 102498. [CrossRef]
4. Chu, S.; Majumdar, A. Opportunities and challenges for a sustainable energy future. *Nature* **2012**, *488*, 294–303. [CrossRef]
5. Alhelou, H.; Hamedani-Golshan, M.E.; Njenda, T.C.; Siano, P. A survey on power system blackout and cascading events: Research motivations and challenges. *Energies* **2019**, *12*, 682. [CrossRef]
6. McLellan, B.; Zhang, Q.; Farzaneh, H.; Utama, N.A.; Ishihara, K.N. Resilience, sustainability and risk management: A focus on energy. *Challenges* **2012**, *3*, 153–182. [CrossRef]
7. Bhusal, N.; Abdelmalak, M.; Kamruzzaman, M.; Benidris, M. Power system resilience: Current practices, challenges, and future directions. *IEEE Access* **2020**, *8*, 18064–18086. [CrossRef]
8. Mora, C.; Spirandelli, D.; Franklin, E.C.; Lynham, J.; Kantar, M.B.; Miles, W.; Smith, C.Z.; Freel, K.; Moy, J.; Louis, L.V.; et al. Broad threat to humanity from cumulative climate hazards intensified by greenhouse gas emissions. *Nat. Clim. Chang.* **2018**, *8*, 1062–1071. [CrossRef]
9. Robinson, D.; Cruikshank, K. Hurricane Hazel: Disaster relief, politics, and society in Canada, 1954–55. *J. Can. Stud.* **2006**, *40*, 37–70. [CrossRef]
10. Subaiya, S.; Moussavi, C.; Velasquez, A.; Stillman, J. A rapid needs assessment of the Rockaway peninsula in New York City after hurricane Sandy and the relationship of socioeconomic status to recovery. *Am. J. Public Health* **2014**, *104*, 632–638. [CrossRef] [PubMed]
11. Comes, T.; Van de Walle, B.A. Measuring disaster resilience: The impact of hurricane sandy on critical infrastructure systems. *ISCRAM* **2014**, *11*, 195–204.
12. Lecomte, E.L.; Pang, A.W.; Russell, J.W. *Ice Storm '98*; Research Paper Series No. 1; Institute for Catastrophic Loss Reduction: Toronto, ON, Canada, 1998; p. 45.
13. Abbey, C.; Cornforth, D.; Hatziargyriou, N.; Hirose, K.; Kwasinski, A.; Kyriakides, E.; Platt, G.; Reyes, L.; Suryanarayanan, S. Powering through the storm: Microgrids operation for more efficient disaster recovery. *IEEE Power Energy Mag.* **2014**, *12*, 67–76. [CrossRef]
14. Attia, S.; Levinson, R.; Ndongo, E.; Holzer, P.; Kazanci, O.B.; Homaei, S.; Zhang, C.; Olesen, B.W.; Qi, D.; Hamdy, M.; et al. Resilient cooling of buildings to protect against heat waves and power outages: Key concepts and definition. *Energy Build.* **2021**, *239*, 110869. [CrossRef]
15. Amada, K.; Kim, J.; Inaba, M.; Akimoto, M.; Kashihara, S.; Tanabe, S.I. Feasibility of staying at home in a net-zero energy house during summer power outages. *Energy Build.* **2022**, *273*, 112352. [CrossRef]
16. Rahif, R.; Hamdy, M.; Homaei, S.; Zhang, C.; Holzer, P.; Attia, S. Simulation-based framework to evaluate resistivity of cooling strategies in buildings against overheating impact of climate change. *Build. Environ.* **2022**, *208*, 108599. [CrossRef]
17. Sun, K.; Specian, M.; Hong, T. Nexus of thermal resilience and energy efficiency in buildings: A case study of a nursing home. *Build. Environ.* **2020**, *177*, 106842. [CrossRef]
18. Kavan, Š.; Dvořáčková, O.; Pokorný, J.; Brumarová, L. Long-Term Power Outage and Preparedness of the Population of a Region in the Czech Republic—A Case Study. *Sustainability* **2021**, *13*, 13142. [CrossRef]
19. Zimmerman, R.; Restrepo, C.E. Analyzing cascading effects within infrastructure sectors for consequence reduction. In Proceedings of the 2009 IEEE Conference on Technologies for Homeland Security, Waltham, MA, USA, 11–12 May 2009; pp. 165–170.
20. Pescaroli, G.; Alexander, D. Critical infrastructure, panarchies and the vulnerability paths of cascading disasters. *Nat. Hazards* **2016**, *82*, 175–192. [CrossRef]
21. Ribeiro, D.; Mackres, E.; Baatz, B.; Cluett, R.; Jarrett, M.; Kelly, M.; Vaidyanathan, S. *Enhancing Community Resilience through Energy Efficiency*; American Council for an Energy-Efficient Economy: Washington, DC, USA, 2015.

22. Casey, J.A.; Fukurai, M.; Hernández, D.; Balsari, S.; Kiang, M.V. Power outages and community health: A narrative review. *Curr. Environ. Health Rep.* **2020**, *7*, 371–383. [CrossRef]
23. Ali, S.M.; Appolloni, A.; Cavallaro, F.; D'Adamo, I.; Di Vaio, A.; Ferella, F.; Gastaldi, M.; Ikram, M.; Kumar, N.M.; Martin, M.A.; et al. Development Goals towards Sustainability. *Sustainability* **2023**, *15*, 9443. [CrossRef]
24. Kumar, N.M.; Chopra, S.S. Leveraging blockchain and smart contract technologies to overcome circular economy implementation challenges. *Sustainability* **2022**, *14*, 9492. [CrossRef]
25. Di Vaio, A.; Zaffar, A.; Balsalobre-Lorente, D.; Garofalo, A. Decarbonization technology responsibility to gender equality in the shipping industry: A systematic literature review and new avenues ahead. *J. Shipp. Trade* **2023**, *8*, 9. [CrossRef]
26. Yeğin, T.; Ikram, M. Performance evaluation of green furniture brands in the marketing 4.0 period: An integrated MCDM approach. *Sustainability* **2022**, *14*, 10644. [CrossRef]
27. He, X.; Khan, S.; Ozturk, I.; Murshed, M. The role of renewable energy investment in tackling climate change concerns: Environmental policies for achieving SDG-13. *Sustain. Dev.* **2023**, *31*, 1888–1901. [CrossRef]
28. Omrany, H.; Chang, R.; Soebarto, V.; Zhang, Y.; Ghaffarianhoseini, A.; Zuo, J. A bibliometric review of net zero energy building research 1995–2022. *Energy Build.* **2022**, *262*, 111996. [CrossRef]
29. Agbodjan, Y.S.; Wang, J.; Cui, Y.; Liu, Z.; Luo, Z. Bibliometric analysis of zero energy building research, challenges and solutions. *Sol. Energy* **2022**, *244*, 414–433. [CrossRef]
30. Rubin, G.J.; Rogers, M.B. Behavioural and psychological responses of the public during a major power outage: A literature review. *Int. J. Disaster Risk Reduct.* **2019**, *38*, 101226. [CrossRef]
31. Experts Call for Home Battery Storage to Protect Vulnerable during Outages. Available online: https://www.publichealth.columbia.edu/news/experts-call-home-battery-storage-protect-vulnerable-during-outages (accessed on 21 September 2023).
32. Mango, M.; Casey, J.A.; Hernandez, D. Resilient Power: A home-based electricity generation and storage solution for the medically vulnerable during climate-induced power outages. *Futures* **2021**, *128*, 102707. [CrossRef]
33. Campbell, R.J.; Lowry, S. *Weather-Related Power Outages and Electric System Resiliency*; Congressional Research Service, Library of Congress: Washington, DC, USA, 2012.
34. Wethal, U. Practices, provision and protest: Power outages in rural Norwegian households. *Energy Res. Soc. Sci.* **2020**, *62*, 101388. [CrossRef]
35. Palmer, J.; Bennetts, H.; Chileshe, N.; Pullen, S.; Zuo, J.; Ma, T. Heat Wave Risks and Residential Buildings. Ph.D. Thesis, Griffith University, Brisbane, Australia, 2012.
36. Erba, S.; Barbieri, A. Retrofitting buildings into thermal batteries for demand-side flexibility and thermal safety during power outages in winter. *Energies* **2022**, *15*, 4405. [CrossRef]
37. Anderson, G.B.; Bell, M.L. Lights out: Impact of the August 2003 power outage on mortality in New York, NY. *Epidemiology* **2012**, *23*, 189. [CrossRef] [PubMed]
38. Meireles, I.; Sousa, V.; Bleys, B.; Poncelet, B. Domestic hot water consumption pattern: Relation with total water consumption and air temperature. *Renew. Sustain. Energy Rev.* **2022**, *157*, 112035. [CrossRef]
39. Liddiard, R.; Gowreesunker, B.L.; Spataru, C.; Tomei, J.; Huebner, G. The vulnerability of refrigerated food to unstable power supplies. *Energy Procedia* **2017**, *123*, 196–203. [CrossRef]
40. Bhattacharyya, S.; Myrzik, J.M.A.; Kling, W.L. Consequences of poor power quality-an overview. In Proceedings of the 2007 42nd International Universities Power Engineering Conference, Brighton, UK, 4–6 September 2007; pp. 651–656.
41. Gordon, J.R. *An Investigation into Freezing and Bursting Water Pipes in Residential Construction*; Building Research Council, School of Architecture, College of Fine and Applied Arts, University of Illinois at Urbana-Champaign: Champaign, IL, USA, 1996.
42. Chang, S.E.; McDaniels, T.L.; Mikawoz, J.; Peterson, K. Infrastructure failure interdependencies in extreme events: Power outage consequences in the 1998 Ice Storm. *Nat. Hazards* **2007**, *41*, 337–358. [CrossRef]
43. Babu, P.; Suthar, G. Indoor air quality and thermal comfort in green building: A study for measurement, problem and solution strategies. In *Indoor Environmental Quality*; Springer: Singapore, 2020; pp. 139–146.
44. Amnian, J.; Maerefat, M.; Heidarinejad, G. Offering a method for reducing pollution and criterion for evaluation of ventilation flow in multilevel enclosed parking lots. *Modares Mech. Eng.* **2016**, *16*, 285–296.
45. Dugan, J.; Byles, D.; Mohagheghi, S. Social Vulnerability to Long-Duration Power Outages. *Int. J. Disaster Risk Reduct.* **2023**, *85*, 103501. [CrossRef]
46. Lane, K.; Charles-Guzman, K.; Wheeler, K.; Abid, Z.; Graber, N.; Matte, T. Health effects of coastal storms and flooding in urban areas: A review and vulnerability assessment. *J. Environ. Public Health* **2013**, *2013*, 913064. [CrossRef]
47. Ribeiro, D.; Bailey, T. *Indicators for Local Energy Resilience*; American Council for an Energy-Efficient Economy: Washington, DC, USA, 2017.
48. Dodman, D.; Diep, L.; Colenbrander, S. Making the case for the nexus between resilience and resource efficiency at the city scale. *Int. J. Urban Sustain. Dev.* **2017**, *9*, 97–106. [CrossRef]
49. Economidou, M.; Todeschi, V.; Bertoldi, P.; D'Agostino, D.; Zangheri, P.; Castellazzi, L. Review of 50 years of EU energy efficiency policies for buildings. *Energy Build.* **2020**, *225*, 110322. [CrossRef]
50. Hong, T.; Chang, W.K.; Lin, H.W. A fresh look at weather impact on peak electricity demand and energy use of buildings using 30-year actual weather data. *Appl. Energy* **2013**, *111*, 333–350. [CrossRef]

51. Petermann, T.; Bradke, H.; Lüllmann, A.; Poetzsch, M.; Riehm, U. *What Happens during a Blackout: Consequences of a Prolonged and Wide-Ranging Power Outage*; Office of Technology Assessment at The German Bundestag: Berlin, Germany, 2014.
52. Leigh, R.; Kleinberg, J.; Scheib, C.; Unger, R.; Kienzl, N.; Esposito, M.; Hagen, E.; Tillou, M. Leaks and Lives: How Better Building Envelopes Make Blackouts Less Dangerous. In *2014 ACEEE Summer Study on Energy Efficiency in Buildings*; American Council for an Energy-Efficient Economy: Washington, DC, USA, 2014; pp. 17–22.
53. Bourgeois, T.; Gerow, J.; Litz, F.; Martin, N. *Community Microgrids: Smarter, Cleaner, Greener*; Pace Energy and Climate Center, Pace Law School: White Plains, NY, USA, 2013.
54. Tostado-Véliz, M.; Icaza-Alvarez, D.; Jurado, F. A novel methodology for optimal sizing photovoltaic-battery systems in smart homes considering grid outages and demand response. *Renew. Energy* **2021**, *170*, 884–896. [CrossRef]
55. Gong, H.; Ionel, D.M. Improving the Power Outage Resilience of Buildings with Solar PV through the Use of Battery Systems and EV Energy Storage. *Energies* **2021**, *14*, 5749. [CrossRef]
56. Yang, Y.; Wang, S. Resilient residential energy management with vehicle-to-home and photovoltaic uncertainty. *Int. J. Electr. Power Energy Syst.* **2021**, *132*, 107206. [CrossRef]
57. Chatterji, E.; Bazilian, M.D. Battery storage for resilient homes. *IEEE Access* **2020**, *8*, 184497–184511. [CrossRef]
58. Gupta, R.; Bruce-Konuah, A.; Howard, A. Achieving energy resilience through smart storage of solar electricity at dwelling and community level. *Energy Build.* **2019**, *195*, 1–15. [CrossRef]
59. Thompson, J.; Krarti, M. Resiliency Evaluation of Net-Zero Residential Communities. In Proceedings of the ASME 2021 15th International Conference on Energy Sustainability collocated with the ASME 2021 Heat Transfer Summer Conference. ASME 2021 15th International Conference on Energy Sustainability, Virtual, 16–18 June 2021; Volume 84881, p. V001T15A001.
60. Stankovic, S.; Campbell, N.; Harries, A. *Urban Wind Energy*; Routledge: London, UK, 2009.
61. Nelson, V. *Innovative Wind Turbines: An Illustrated Guidebook*; CRC Press: Boca Raton, FL, USA, 2019.
62. Blackwood, M. Maximum efficiency of a wind turbine. *Undergrad. J. Math. Model. One+ Two* **2016**, *6*, 2. [CrossRef]
63. Ko, D.H.; Jeong, S.T.; Kim, Y.C. Assessment of wind energy for small-scale wind power in Chuuk State, Micronesia. *Renew. Sustain. Energy Rev.* **2015**, *52*, 613–622. [CrossRef]
64. Thresher, R.; Robinsion, M.; Veers, P. *Wind Energy Technology: Current Status and R&D Future*; No. NREL/CP-500-43374; National Renewable Energy Lab. (NREL): Golden, CO, USA, 2008.
65. Hussain, F.; Ashfaq Ahmad, M.; Badshah, S.; Raza, R.; Ajmal Khan, M.; Mumtaz, S.; Dilshad, S.; Riaz, R.A.; Jafar Hussain, M.; Abbas, G. A modeling approach for low-temperature SOFC-based micro-combined heat and power systems. *Int. J. Mod. Phys. B* **2019**, *33*, 1950001. [CrossRef]
66. Sorace, M.; Gandiglio, M.; Santarelli, M. Modeling and techno-economic analysis of the integration of a FC-based micro-CHP system for residential application with a heat pump. *Energy* **2017**, *120*, 262–275. [CrossRef]
67. Bianchi, M.; De Pascale, A.; Spina, P.R. Guidelines for residential micro-CHP systems design. *Appl. Energy* **2012**, *97*, 673–685. [CrossRef]
68. Adam, A.; Fraga, E.S.; Brett, D.J. Options for residential building services design using fuel cell based micro-CHP and the potential for heat integration. *Appl. Energy* **2015**, *138*, 685–694. [CrossRef]
69. Stambouli, A.B. Fuel cells: The expectations for an environmental-friendly and sustainable source of energy. *Renew. Sustain. Energy Rev.* **2011**, *15*, 4507–4520. [CrossRef]
70. Edwards, P.P.; Kuznetsov, V.L.; David, W.I.; Brandon, N.P. Hydrogen and fuel cells: Towards a sustainable energy future. *Energy Policy* **2008**, *36*, 4356–4362. [CrossRef]
71. Garche, J.; Jürissen, L. Applications of fuel cell technology: Status and perspectives. *Electrochem. Soc. Interface* **2015**, *24*, 39. [CrossRef]
72. Wilberforce, T.; Alaswad, A.; Palumbo, A.; Dassisti, M.; Olabi, A.G. Advances in stationary and portable fuel cell applications. *Int. J. Hydrog. Energy* **2016**, *41*, 16509–16522. [CrossRef]
73. Vijay, A.; Hawkes, A. Demand side flexibility from residential heating to absorb surplus renewables in low carbon futures. *Renew. Energy* **2019**, *138*, 598–609. [CrossRef]
74. Singh, K.; Hachem-Vermette, C. Scheduling of Hybrid Heating and Cooling System Based on Energy Resources for Mixed-Use Multistorey Building (VC-20-C004). In Proceedings of the 2020 ASHRAE Virtual Conference, 29 June–2 July 2020.
75. Bohman, A.D.; Abdulla, A.; Morgan, M.G. Individual and collective strategies to limit the impacts of large power outages of long duration. *Risk Anal.* **2022**, *42*, 544–560. [CrossRef]
76. Wang, J.; You, S.; Zong, Y.; Træholt, C.; Dong, Z.Y.; Zhou, Y. Flexibility of combined heat and power plants: A review of technologies and operation strategies. *Appl. Energy* **2019**, *252*, 113445. [CrossRef]
77. Singh, K.; Hachem-Vermette, C. Techniques of Improving Infrastructure and Energy Resilience in Urban Setting. *Energies* **2022**, *15*, 6253. [CrossRef]
78. Hachem-Vermette, C.; Singh, K. Developing an optimization methodology for urban energy resources mix. *Appl. Energy* **2020**, *269*, 115066. [CrossRef]
79. Gilleo, A.; Chittum, A.; Farley, K.; Neubauer, M.; Nowak, S.; Ribeiro, D.; Vaidyanathan, S. *The 2015 State Energy Efficiency Scorecard*; American Council for an Energy-Efficient Economy: Washington, DC, USA, 2015.
80. Nguyen, T.; Gustavsson, L. Production of district heat, electricity and/or biomotor fuels in renewable-based energy systems. *Energy* **2020**, *202*, 117672. [CrossRef]

81. Mahmoud, M.; Ramadan, M.; Naher, S.; Pullen, K.; Baroutaji, A.; Olabi, A.G. Recent advances in district energy systems: A review. *Therm. Sci. Eng. Prog.* **2020**, *20*, 100678. [CrossRef]
82. Aste, N.; Caputo, P.; Del Pero, C.; Ferla, G.; Huerto-Cardenas, H.E.; Leonforte, F.; Miglioli, A. A renewable energy scenario for a new low carbon settlement in northern Italy: Biomass district heating coupled with heat pump and solar photovoltaic system. *Energy* **2020**, *206*, 118091. [CrossRef]
83. Murthy, S.S.; Dutta, P.; Sharma, R.; Rao, B.S. Parametric studies on a stand-alone polygeneration microgrid with battery storage. *Therm. Sci. Eng. Prog.* **2020**, *19*, 100608. [CrossRef]
84. Ijaodola, O.; Ogungbemi, E.; Khatib, F.N.; Wilberforce, T.; Ramadan, M.; El Hassan, Z.; Thompson, J.; Olabi, A.G. Evaluating the effect of metal bipolar plate coating on the performance of proton exchange membrane fuel cells. *Energies* **2018**, *11*, 3203. [CrossRef]
85. Baroutaji, A.; Wilberforce, T.; Ramadan, M.; Olabi, A.G. Comprehensive investigation on hydrogen and fuel cell technology in the aviation and aerospace sectors. *Renew. Sustain. Energy Rev.* **2019**, *106*, 31–40. [CrossRef]
86. Rosato, A.; Ciervo, A.; Ciampi, G.; Scorpio, M.; Guarino, F.; Sibilio, S. Impact of solar field design and back-up technology on dynamic performance of a solar hybrid heating network integrated with a seasonal borehole thermal energy storage serving a small-scale residential district including plug-in electric vehicles. *Renew. Energy* **2020**, *154*, 684–703. [CrossRef]
87. Hirsch, A.; Parag, Y.; Guerrero, J. Microgrids: A review of technologies, key drivers, and outstanding issues. *Renew. Sustain. Energy Rev.* **2018**, *90*, 402–411. [CrossRef]
88. Olivares, D.E.; Mehrizi-Sani, A.; Etemadi, A.H.; Cañizares, C.A.; Iravani, R.; Kazerani, M.; Hajimiragha, A.H.; Gomis-Bellmunt, O.; Saeedifard, M.; Palma-Behnke, R.; et al. Trends in microgrid control. *IEEE Trans. Smart Grid* **2014**, *5*, 1905–1919. [CrossRef]
89. Martin-Martínez, F.; Sánchez-Miralles, A.; Rivier, M. A literature review of Microgrids: A functional layer based classification. *Renew. Sustain. Energy Rev.* **2016**, *62*, 1133–1153. [CrossRef]
90. Ton, D.T.; Smith, M.A. The US department of energy's microgrid initiative. *Electr. J.* **2012**, *25*, 84–94. [CrossRef]
91. Dileep, G. A survey on smart grid technologies and applications. *Renew. Energy* **2020**, *146*, 2589–2625. [CrossRef]
92. Judge, M.A.; Khan, A.; Manzoor, A.; Khattak, H.A. Overview of smart grid implementation: Frameworks, impact, performance and challenges. *J. Energy Storage* **2022**, *49*, 104056. [CrossRef]
93. Khan, A.; Javaid, N. Jaya learning-based optimization for optimal sizing of stand-alone photovoltaic, wind turbine, and battery systems. *Engineering* **2020**, *6*, 812–826. [CrossRef]
94. Hachem, C.; Athienitis, A.; Fazio, P. Evaluation of energy supply and demand in solar neighborhood. *Energy Build.* **2012**, *49*, 335–347. [CrossRef]
95. Hachem, C.; Athienitis, A.; Fazio, P. Parametric investigation of geometric form effects on solar potential of housing units. *Sol. Energy* **2011**, *85*, 1864–1877. [CrossRef]
96. Hachem, C.; Athienitis, A.; Fazio, P. Energy performance enhancement in multistory residential buildings. *Appl. Energy* **2014**, *116*, 9–19. [CrossRef]
97. Liang, R.; Kent, M.; Wilson, R.; Wu, Y. The effect of thermochromic windows on visual performance and sustained attention. *Energy Build.* **2021**, *236*, 110778. [CrossRef]
98. Kim, S.; An, J.; Choi, H.; Hong, T. Assessment the technical and economic performance of a window-integrated PV system using third-generation PV panels. *Energy Build.* **2023**, *286*, 112978. [CrossRef]
99. Hachem-Vermette, C. Multistory building envelope: Creative design and enhanced performance. *Sol. Energy* **2018**, *159*, 710–721. [CrossRef]
100. Hachem-Vermette, C.; Grewal, K.S. Investigation of the impact of residential mixture on energy and environmental performance of mixed-use neighborhoods. *Appl. Energy* **2019**, *241*, 362–379. [CrossRef]
101. Hachem-Vermette, C.; Singh, K. Optimization of the mixture of building types in a neighborhood and their energy and environmental performance. *Energy Build.* **2019**, *204*, 109499. [CrossRef]
102. Jabareen, Y.R. Sustainable urban forms: Their typologies, models, and concepts. *J. Plan. Educ. Res.* **2006**, *26*, 38–52. [CrossRef]
103. Hachem, C. Impact of neighborhood design on energy performance and GHG emissions. *Appl. Energy* **2016**, *177*, 422–434. [CrossRef]
104. Lee, K.S.; Lee, J.W.; Lee, J.S. Feasibility study on the relation between housing density and solar accessibility and potential uses. *Renew. Energy* **2016**, *85*, 749–758. [CrossRef]
105. Frincu, M.; Draghici, R. Towards a scalable cloud enabled smart home automation architecture for demand response. In Proceedings of the 2016 IEEE PES Innovative Smart Grid Technologies Conference Europe (ISGT-Europe), Ljubljana, Slovenia, 9–12 October 2016; pp. 1–6.
106. Miller, W.; Ma, Y.; Susilawati, C.; Arlinkasari, F. *Residential Solar Pre-Cooling and Pre-Heating: H1 Opportunity Assessment*; RACE for Homes Program: Haymarket, Australia, 2021.
107. Ramos, J.S.; Moreno, M.P.; Delgado, M.G.; Domínguez, S.Á.; Cabeza, L.F. Potential of energy flexible buildings: Evaluation of DSM strategies using building thermal mass. *Energy Build.* **2019**, *203*, 109442. [CrossRef]
108. Lim, T.H.; De Kleine, R.D.; Keoleian, G.A. Energy use and carbon reduction potentials from residential ground source heat pumps considering spatial and economic barriers. *Energy Build.* **2016**, *128*, 287–304. [CrossRef]

109. Zhai, X.Q.; Qu, M.; Yu, X.; Yang, Y.; Wang, R.Z. A review for the applications and integrated approaches of ground-coupled heat pump systems. *Renew. Sustain. Energy Rev.* **2011**, *15*, 3133–3140. [CrossRef]
110. Victoria, M.; Haegel, N.; Peters, I.M.; Sinton, R.; Jäger-Waldau, A.; del Cañizo, C.; Breyer, C.; Stocks, M.; Blakers, A.; Kaizuka, I.; et al. Solar photovoltaics is ready to power a sustainable future. *Joule* **2021**, *5*, 1041–1056. [CrossRef]

Disclaimer/Publisher's Note: The statements, opinions and data contained in all publications are solely those of the individual author(s) and contributor(s) and not of MDPI and/or the editor(s). MDPI and/or the editor(s) disclaim responsibility for any injury to people or property resulting from any ideas, methods, instructions or products referred to in the content.

sustainability MDPI

Systematic Review

An Overview and Categorization of the Drivers and Barriers to the Adoption of the Circular Economy: A Systematic Literature Review

Carina Pasqualotto [1,*], Daniela Callegaro-De-Menezes [1] and Cornelius Stephanus Lodewyk Schutte [2,*]

[1] Agribusiness Research Center, Federal University of Rio Grande do Sul, Porto Alegre 91540-000, RS, Brazil; daniela.callegaro@ufrgs.br
[2] Department of Industrial Engineering, Stellenbosch University, Stellenbosch 7602, South Africa
* Correspondence: pasqualotto.carina@gmail.com (C.P.); corne@sun.ac.za (C.S.L.S.)

Abstract: The adoption of the circular economy (CE) can help to solve the dilemmas of food, economic and social crises, environmental pollution, and continuous decreases in non-renewable resources, caused by the continuous increase in the size of the global population. Identifying drivers of and barriers to the CE is important for the implementation of the CE. In this context, this study aims to identify and categorize the drivers of and barriers to the adoption of the CE through a systematic literature review. In doing this, ten categories of drivers and barriers were identified: environmental, supply chain, economic, information, legal, market, organizational, public, social, and technological. The results of this study may contribute to the development of circular processes, the promotion of sustainability, and may encourage the implementation of the CE in many areas. The CE's implementation can be a way to achieve some of the Sustainable Development Goals from the 2030 Agenda.

Keywords: circular economy; drivers; barriers; categories

Citation: Pasqualotto, C.; Callegaro-De-Menezes, D.; Schutte, C.S.L. An Overview and Categorization of the Drivers and Barriers to the Adoption of the Circular Economy: A Systematic Literature Review. *Sustainability* **2023**, *15*, 10532. https://doi.org/10.3390/su151310532

Academic Editors: Idiano D'Adamo, Massimo Gastaldi and Nallapaneni Manoj Kumar

Received: 12 April 2023
Revised: 23 June 2023
Accepted: 26 June 2023
Published: 4 July 2023

Copyright: © 2023 by the authors. Licensee MDPI, Basel, Switzerland. This article is an open access article distributed under the terms and conditions of the Creative Commons Attribution (CC BY) license (https://creativecommons.org/licenses/by/4.0/).

1. Introduction

We live in an era of food, economic, and social crises, environmental pollution, growing awareness of social responsibility, sustainability, and concern for the environment, and heightened growth in some economies, coupled with urbanization [1,2]. The modern economy threatens environmental protection, and this fact places pressure on environmental stakeholders, especially firms and policymakers [2]. Arising from the perception that current consumption patterns are at the root of the environmental crisis, criticism of consumerism came to be seen as a contribution to the development of sustainable societies [3]. In this sense, the adoption of the circular economy (CE) is seen as one of the ways that we might solve this dilemma. The CE paradigm aims to attain sustainability by preventing environmental degradation and ensuring the social and economic wellbeing of present and future generations [4]. CE has become a popular strategy for improving sustainability, and is a theme that has been extensively researched over the past five years [5]. Arthur et al. (2023) [2] assumed that some CE variables have a significant impact on the environment; variables such as the level of materials considered as input factors for economic production, the amount of waste generated because of the extraction and usage of these materials, and the rate of recycling of the generated waste.

From a different perspective, while the terms circular economy and sustainability are increasingly gaining traction within academia, industry, and for policymakers, and are often being used in similar contexts, the similarities and differences between these concepts have not been made explicit in the literature, and remain ambiguous [6]. However, Velenturf and Purnell (2021) [7] suggest every actor should do their very best to develop a more sustainable CE, which requires research and constant learning to ensure progress towards sustainability, even with imperfections.

The adoption of the CE fosters a reduction in the consumption of raw materials, improves brand image, encourages the emergence of new demands for services and new potential markets, reduces the costs and risks of emissions and waste, and increases the potential to attract new investors [8]. Therefore, the CE approach has attracted the attention of many firms, enabling them to make the production process more efficient, especially when material and energy inputs become more expensive [9]. Considering the importance of the topic, a growing number of authors have explored the theme of the CE, specifically its drivers and barriers. However, some studies have focused only on issues that facilitate the implementation of the CE (drivers) [10–12], while other studies have focused only on factors that hinder the implementation of the CE (barriers) [13–17]. Some studies deal with both drivers and barriers, but in specific contexts, e.g., the supply chain [18,19], the textile and apparel industry [20], small and medium enterprises [21], and the building and infrastructure sector [22], or in specific countries, e.g., Brazil [23], China [24], Taiwan [25] and Finland [26]. Additionally, there are some studies in the literature that have only categorized drivers of the CE in the leather industry [27] and in the Italian economy [10]; and barriers to the CE applied to the Danish economy [15], to the construction sector [28], and to five European regions [29], as well as both (drivers and barriers) applied together to the built environment sector [22] and to the manufacturing sector in the UK [30]. Moreover, Mishra et al. (2022) [31] developed, measured, and validated an instrument for barriers to the adoption of the CE practices in micro, small, and medium enterprises (MSMEs), determining seven dimensions; however, the drivers for the adoption of the CE were not considered in their study.

An in-depth and complete analysis joining the drivers of and barriers to the adoption of the CE is necessary to enable a general application. Elia et al. (2020) [32] analyzed the relationship between the level of supply chain integration and the adopted CE strategies from the industrial field, rather than specifically analyzing the drivers of and barriers to the adoption of the CE. Thus, notwithstanding the studies that have already been completed, in the literature, there is a lack of investigation into these drivers and barriers in a more detailed way that could be applied to multiple sectors [18,27], different markets [33], distinct economies [23], and different geographic contexts [19]. Additionally, Jia et al. (2020) [20] demand more databases to find relevant articles. In this sense, searching for a theoretical proposition that can help to attend to such demands, this study aims to identify and categorize the drivers of and barriers to the adoption of the CE for a general application.

The study contributes in different ways to research and practice in the CE field. It extends the body of knowledge on CE by assessing a significant number of papers that contain the drivers of and barriers to the adoption of the CE, equipping researchers and practitioners with prior information about the realities that will be faced. It also helps in planning for the transition from a linear economy to CE, making companies more efficient with their resources and advancing toward sustainable economies [14]. The results of D'Adamo and Gastaldi (2022) [34] in their study regarding the Sustainable Development Goals (SDGs) showed that many sustainability opportunities have not yet been well explored. In this sense, the adoption of the CE can be considered a way to achieve some of the SDGs, e.g., companies producing with more responsibility, encouraging them, especially large and transnational companies, to adopt sustainable practices (goal 12) and reducing the environmental impact of cities through waste management (goal 11). Additionally, Ali et al. (2023) [35] mention circular economy-centric education being the solution to the social, economic, and environmental problems stemming from climate change. Moreover, green technologies, through the optimization of the use of resources, the reduction of waste, and the reduction of demand for new resources, may promote the development of green products and services, thereby helping to reduce the environmental impact of consumption [35]. The study presents an extended literature review, enabling a broader view and a categorization of the drivers of and barriers to the adoption of the

CE. These contributions are important to help companies develop the CE and help the government obtain knowledge to work on public policies fostering the CE. It is important to have a clear understanding of the context of the CE in order to provide a common basis of assumptions and targets on which policymaking can be developed [36]. Additionally, it would facilitate practitioners to understand the drivers of and barriers to the adoption of the CE to handle them effectively.

To summarize, the present study addresses the following research question: What are the drivers of and barriers to the adoption of the CE, and how can we categorize them according to the literature?

The paper is structured as follows. Sections 2 and 3 present the theoretical basis and material and methods of the study, respectively. The fourth section presents the results and discussion of the paper. Finally, Section 5 concludes the paper by presenting the implications and limitations of the study, and suggestions for future studies.

2. Theoretical Basis

2.1. Circular Economy

The concept of the CE, which was created primarily by practitioners, the business community, and policymakers, is currently promoted by the European Union, several national governments, and various business organizations around the world [8]. CE is becoming part of popular discourse, especially in the government and corporate sector [37].

The CE focuses on the maintenance, reuse/redistribution/remanufacturing/recycling, circularity and optimization of resources, the use of clean energy, and processing efficiency, with zero waste as a basic premise [38]. According to Zhang et al. (2022) [4] (p. 656), the CE is perceived as a substitute for the take–make–waste linear economy.

The concept involves careful management of two types of material flows, as described by McDonough and Braungart (2010) [39]: biological nutrients, designed to re-enter the biosphere safely and build natural capital, and technical nutrients, designed to circulate in high quality without entering the biosphere. According to Sehnem and Pereira (2019) [38], the CE emphasizes the biological cycle and technical cycle of materials.

According to the Ellen MacArthur Foundation (2019) [40], the CE is based on three principles: (1) designing waste and pollution, (2) keeping products and materials in use, and (3) regenerating natural systems. It makes sense to extract resources from nature to transform them into a product or service that can be used not just once, but many times, thus reducing the need for virgin input extraction and waste production [8]. Designing waste, keeping products and materials in use, and regenerating natural systems creates vital opportunities for economic growth, thereby creating jobs and benefiting society [41]. Substantially reducing waste generation through prevention, reduction, recycling, and reuse, and achieving sustainable management and efficient use of natural resources are some goals of the 12th SDG, proposed in the 2030 Agenda (2015) [42].

While the CE is increasingly attracting attention in academia, industry, and with policymakers [6], Friant et al. (2020) [37] argue that the definition, objectives, and forms of implementation of the CE are still unclear, inconsistent, and contested. This is the case because different actors and sectors are articulating circular discourses which align with their own interests and which do not often sufficiently examine the ecological, social, and political implications of circularity [37]. In line with these authors, Corvellec et al. (2022) [43] addressed critiques of the CE in their study, considering that the CE has diffused limits, unclear theoretical grounds, and that its implementation faces structural obstacles. According to Velenturf and Purnell (2021) [7], every actor should do their very best to develop a more sustainable CE. Sustainable development is fraught with imperfection, and so is the circular economy, both requiring research and constant learning to ensure progress in pursuit of sustainability [7].

Joining the CE allows for a reduction in the consumption of raw materials [8] and gains in resource efficiency [19,44], thus promoting waste reduction [30], in addition to reducing a company's environmental impact [10].

The cost reductions arising from the implementation of the CE are one of the most frequently considered aspects in the literature [26,44]. The CE encourages the emergence of new demands for services and expansion into new markets, thus increasing a company's potential to attract new investors [8], and generating competitive advantages for circular companies [10,20].

In addition, the adoption of the CE makes it possible to improve the reputation and recognition of the brand [8], the relationship with customers [44], and to increase consumer satisfaction [10]. The adoption of the CE is also seen as a potential source of new jobs [33].

2.2. *Drivers and Barriers to the Circular Economy*

As the concept of the CE becomes more prevalent among the topics covered in the literature, more studies are focusing on the drivers of and barriers to the adoption of the CE. Many authors have worked with drivers of the adoption of the CE in their studies to encourage, motivate, and facilitate companies to adopt the CE [18,22,27,33], thus helping in the transition from a linear economy to the CE; it is thought that the CE is much more efficient in resources, and will generate greater competitiveness for the company and advancement towards a more sustainable economy [14].

Some of the main drivers addressed by the literature were concern for environmental impact and the environment [27]; increased transparency and engagement in the supply chain [26]; reduced costs [45,46]; the existence of laws and regulations regarding the CE [47]; awareness of environmental issues among consumers [18]; investment in science and technology [19,48]; and government support [21,49]. In this sense, Arthur et al. (2023) [2] concluded in their study that a blend of government policies is the most effective means of achieving a CE.

Drivers regarding companies, such as increasing the network and partnerships [50] and gains in market share and competitiveness [33], were also heavily addressed in the literature.

On the other hand, many studies have also pointed out the barriers to the adoption of the CE, hindering or inhibiting its implementation [30,33,51]. There is a lack of funding [19], financial resources [29], economies of scale [52], appropriate technology for the CE [47], information [36], and laws and rules supporting the CE [18]. Furthermore, the initial investment cost for companies is high [52].

Within the market, demand for circular products and processes is still restricted [29], and there is a lack of environmental awareness among consumers [53]. In companies, there is a lack of commitment at the management level [54], and a shortage of qualified personnel to work with CE [55]. There is also a lack of encouragement and support from the government [56].

The literature has categorized the presentation of drivers and barriers in different ways. Table 1 presents the categorization of only drivers, only barriers, and both drivers and barriers, from the literature.

Table 1. Categorization of drivers and barriers from the literature.

Author(s)	Drivers	Barriers
Moktadir et al. (2018) [27]	Knowledge about CE, consumer awareness, leadership, and commitment from top management and government support and legislation	
Gusmerotti et al. (2019) [10]	Economic drivers and resource risk drivers	

Table 1. *Cont.*

Author(s)	Drivers	Barriers
Ritzén and Sandström (2017) [57]		Technological, financial, and supply chain barriers
Campbell-Johnston et al. (2019) [58]		Soft factors (soft) and hard factors (hard)
Kirchherr et al. (2018) [53]		Cultural, market, regulatory, and technological barriers
Ormazabal et al. (2018) [46]		Hard barriers and human-based
Guldmann and Huulgaard (2020) [15]		Market and institutional, value chain, organizational, and employees
Nohra et al. (2020) [29]		Cultural, economic, informational, regulatory, political, institutional, technological, and environmental
Kazancoglu et al. (2020) [52]		Managerial and decision-making, work, design challenges, materials, rules and regulation, knowledge and awareness, integration and collaboration, costs and technological infrastructure
Masi et al. (2017) [19]		Financial, technological, social, informational, and institutional
Ababio and Lu (2023) [28]		Socio-cultural, technological, political and legislative, and financial and economic
Kumar et al. (2019) [30]		External, organizational, social, environmental, technological, and legal barriers
Jia et al. (2020) [20]	Organizational, consumer, and institutional	Organizational, financial, and political
Govindan and Hasanagic (2018) [18]	Governmental, corporate, consumer, organizational, and supplier perspectives	
Ranta et al. (2018) [51]	Regulatory, normative and cultural-cognitive	
Hart et al. (2019) [22]	Cultural, regulatory, financial, and sectoral	
Agyemang et al. (2019) [44]	Internal and external	
Tura et al. (2019) [26]	Environmental, economic, social, institutional, technological and informational factors, supply channels, and organizational	
Geng and Doberstein (2008) [48]	Policy, technology, and public participation	
De Jesus and Mendonça (2018) [45]	Soft factors and hard factors	

3. Materials and Methods

In order to identify and categorize drivers and barriers to the adoption of the CE, a systematic literature review was used as a method, as suggested by Snyder (2019) [59]. The method used consists of a content analysis of selected studies based on specific criteria defined by the authors. This study followed four stages, as suggested by Wolfswinkel et al. (2013) [60] and Flores and Jansson (2022) [61].

3.1. Stage 1—Selection of Database

First, the database to be used was identified. Following Paul and Criado (2020) [62], we decided to use Scopus, as it captures more articles than Web of Science and includes the main journals, thus providing a more comprehensive and relevant set of articles that could potentially be relevant, even considering that this decision may have resulted in the unintentional exclusion of other pertinent papers listed in other databases. Scopus is a consolidated database that is widely used in systematic review studies [18,19,63].

3.2. Stage 2—Selection of Keywords and Search for Studies According to Clear Criteria

After selecting the database, we needed to determine the keywords for searching relevant papers. The expressions used when searching the title, abstract, and keyword fields were "circular economy" AND "drivers" OR "barriers". We have not included synonyms, as the selected keywords are well-established terms used in academia. We considered papers and articles published in English, limiting the results to academic/scientific journals and conference proceedings. Only full papers were included, and book chapters, reviews, and books were excluded. This produced a list of 532 papers for further analysis.

3.3. Stage 3—Selection of Articles

To select the articles to be reviewed and included in our paper, we applied the following criteria: (1) the abstracts of the 532 selected articles were analyzed. As inclusion criteria, they had to be theoretical or empirical articles that presented, as a result, a list of drivers and/or barriers. Based on this criterion, 435 articles were excluded, and 97 articles were selected for inclusion. (2) These 97 articles were then read in their entirety to verify that the lists of drivers and/or barriers were related to the adoption of the circular economy, and not to adjacent fields that were not of interest to our study, such as recycling, sustainability, and green marketing. Thus, 44 articles were excluded, and 53 articles were included.

3.4. Stage 4—Analysis through Data Coding and Structuring of Findings

A spreadsheet was created for the analysis of the 53 selected articles. The information from these articles was released in the form of an Excel spreadsheet. The articles were tabulated under title, year and place of publication, area, sector, or geographic context in which the study was carried out, objectives, methodological approach, and the main conclusions, as well as the list of drivers of and barriers to the adoption of the CE found in the articles. Data were analyzed using the content analysis technique [64]. To ensure the quality of the interpretation, the drivers and barriers emanating from the literature were systematically organized according to Wolfswinkel et al. (2013) [60], Xiao and Watson (2019) [65], and Flores and Jansson (2022) [61]. In cases in which there were doubts regarding the organization of the drivers and barriers among the categories defined by the authors of this study, a discussion took place between the authors until a consensus was reached.

The flowchart in Figure 1 presents the research process.

It was observed that some studies categorized the presentation of drivers and barriers in their research (Table 1). So, based initially on the literature review categorization of drivers and barriers, presented in Table 1, the list of drivers of and barriers to the adoption of the CE was grouped, and ten categories were created for the purpose of the final presentation of the study results.

The first category identified for the present study was the environmental category. Nohra et al. (2020) [29] and Kumar et al. (2019) [30] used the environmental category to present barriers to the CE, and Tura et al. (2019) [26] used the environmental category to present drivers of and barriers to the CE. In this way, issues related to sustainability, the environment, waste management, recycling, and the scarcity of resources were allocated to the environmental category.

The second category identified in the study was the supply chain category. Ritzén and Sandström (2017) [57] used the supply chain category to present barriers to the CE, and Tura et al. (2019) [26] used the supply chain category to present the drivers of and barriers to the CE. In this sense, aspects from distribution channels, logistics, reverse logistics, and the potential to reduce channel dependence were allocated to the supply chain category.

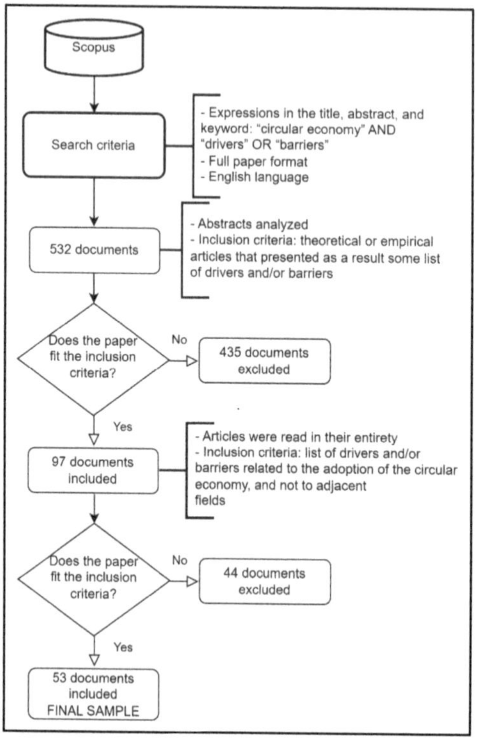

Figure 1. Flowchart of the research design.

The third category identified was the economic category. Gusmerotti et al. (2019) [10], Nohra et al. (2020) [29], Ababio and Lu (2023) [28] and Tura et al. (2019) [26] used the economic category to present only drivers of, only barriers to, and both (drivers and barriers), concerning the CE. In this way, aspects involving finance, sales, profitability, revenues, earnings, costs, accounting, raw material costs and prices, and the regulatory costs of environmental pollution and waste were allocated to the economic category.

The fourth category identified in our study was the information category. Nohra et al. (2020) [29] and Masi et al. (2017) [19] used the information category to present barriers to the CE, and Tura et al. (2019) [26] used the informational factors category to present the drivers of and barriers to the CE. In this way, aspects such as knowledge, information sharing, learning, training, and experiences were allocated to the information category.

The fifth category identified in the study was the legal category. Kumar et al. (2019) [30] and Ababio and Lu (2023) [28] used this category to present barriers to the CE in their article. Issues related to normativity, regulations, and legislation were allocated to the legal category.

The market was the sixth category identified in the article. Kirchherr et al. (2018) [53] and Guldmann and Huulgaard (2020) [15] used the market category to present barriers to the adoption of the CE. Aspects embracing the external aspects of the organization, for instance, the environmental awareness of consumers, consumer preferences, market demands, and market trends were allocated to the market category.

The seventh category identified was the organizational category. Several authors used this category in their studies; Guldmann and Huulgaard (2020) [15] and Kumar et al. (2019) [30] used it to present barriers to the adoption of the CE, Jia et al. (2020) [20] and Tura et al. (2019) [26] used it to present drivers of and barriers to the adoption of the CE, and Govindan and Hasanagic (2018) [18] used it to present the drivers of and barriers to the adoption of the CE. Internal aspects related to companies and commercial institutions, such as competitiveness, performance indicators, organizational culture, company policy, human resources, the value and quality of products, raw materials and components, suppliers, partnerships, customer satisfaction, customer relationship, branding, and company image were allocated to the organizational category.

The public category was identified as the eighth category in this study. Geng and Doberstein (2008) [48] used public participation as a category to present the drivers of and barriers to the adoption of the CE. All issues related to the government, states, and municipalities, for instance, support, incentives, financial assistance, and public policies were allocated to the public category.

The ninth category identified in this study was the social category. Masi et al. (2017) [19] and Kumar et al. (2019) [30] and Ababio and Lu (2023) [28] used the social category to present barriers to the CE, and Tura et al. (2019) [26] used it to present the drivers of and barriers to the adoption of the CE. Aspects of society and community, involving job creation and reduction of the unemployment rate, population size, public health, safety, hygiene, social responsibility, social projects, public awareness, social recognition, and stakeholders were allocated to the social category.

The tenth category identified in this study was the technological category. Ritzén and Sandström (2017) [57], Kirchherr et al. (2018) [53], Nohra et al. (2020) [29], Masi et al. (2017) [19], Kumar et al. (2019) [30], and Ababio and Lu (2023) [28] used the technological category to present the barriers to the CE, and Tura et al. (2019) [26] used it to present the drivers of and barriers to the CE. Geng and Doberstein (2008) [48] used the technology category to present the drivers of and barriers to the adoption of the CE. Aspects related to science, technology, and innovation were allocated to the technological category.

Table 2 summarizes the authors who motivated the choice of each of the ten categories for the purpose of the final presentation of the study results. Table 2 also shows the authors that used other categories to work with the drivers of and barriers to the CE, and the authors who did not use any category in their studies to present the drivers of and barriers to the CE.

Table 2. Motivation for choosing the categories of drivers and barriers.

Categories	Authors
Environmental	Kumar et al. (2019) [30], Tura et al. (2019) [26] and Nohra et al. (2020) [29].
Supply Chain	Ritzén and Sandström (2017) [57] and Tura et al. (2019) [26].
Economic	Gusmerotti et al. (2019) [10], Tura et al. (2019) [26], Ababio and Lu (2023) [28] and Nohra et al. (2020) [29].
Information	Masi et al. (2017) [19], Tura et al. (2019) [26] and Nohra et al. (2020) [29].
Legal	Kumar et al. (2019) [30] and Ababio and Lu (2023) [28]
Market	Kirchherr et al. (2018) [53] and Guldmann and Huulgaard (2020) [15].
Organizational	Govindan and Hasanagic (2018) [18], Kumar et al. (2019) [30], Tura et al. (2019) [26], Guldmann and Huulgaard (2020) [15] and Jia et al. (2020) [20].
Public	Geng and Doberstein (2008) [48]
Social	Masi et al. (2017) [19], Kumar et al. (2019) [30], Ababio and Lu (2023) [28], and Tura et al. (2019) [26].
Technological	Geng and Doberstein (2008) [48], Masi et al. (2017) [19], Ritzén and Sandström (2017) [57], Kirchherr et al. (2018) [53], Kumar et al. (2019) [30], Tura et al. (2019) [26], Ababio and Lu (2023) [28] and Nohra et al. (2020) [29].
Other categories use	De Jesus and Mendonça (2018) [45], Moktadir et al. (2018) [27], Ormazabal et al. (2018) [46], Ranta et al. (2018) [51], Agyemang et al. (2019) [44], Campbell-Johnston et al. (2019) [58], Hart et al. (2019) [22], and Kazancoglu et al. (2020) [52].
Authors that did not use categories in their studies to present drivers and barriers to CE	Xue et al. (2010) [24], Ilić and Nikolić (2016) [49], De Mattos and De Albuquerque (2018) [50], Ghisellini et al. (2018) [66], Mahpour (2018) [67], Mangla et al. (2018) [54], Masi et al. (2017) [19], Barbaritano et al. (2019) [33], Bolger and Doyon (2019) [68], Camacho-Otero et al. (2019) [69], Chang and Hsieh (2019) [25], Farooque et al. (2019) [70], Garcés-Ayerbe et al. (2019) [71], Gue et al. (2019) [72], Milios et al. (2019) [55], Piyathanavong et al. (2019) [73], Rajput and Singh (2019) [74], Scarpellini el al. (2019) [75], Šebo et al. (2019) [47], Singh and Giacosa (2019) [76], Tseng et al. (2019) [77], Dieckmann et al. (2020) [13], García-Quevedo et al. (2020) [14], Galvão et al. (2020) [63], Hartley et al. (2020) [11], Jabbour et al. (2020) [23], Jaeger and Upadhyay (2020) [78], Kanters (2020) [16], Mura et al. (2020) [21], Ozkan-Ozen et al. (2020) [79], Robaina et al. (2020) [12], Shao et al. (2020) [56], and Werning and Spinler (2020) [17].

4. Results and Discussion

In this chapter, the results of the analysis of the 53 articles on the drivers of and barriers to the CE that were analyzed for this study will be described and analyzed.

Of the total of 53 articles analyzed, 27 articles addressed drivers of the CE, and 47 articles addressed barriers to the CE. However, only 6 articles dealt only with drivers (11%), 26 articles dealt only with barriers (49%), and 21 articles dealt with both drivers and barriers (40%) (Figure 2).

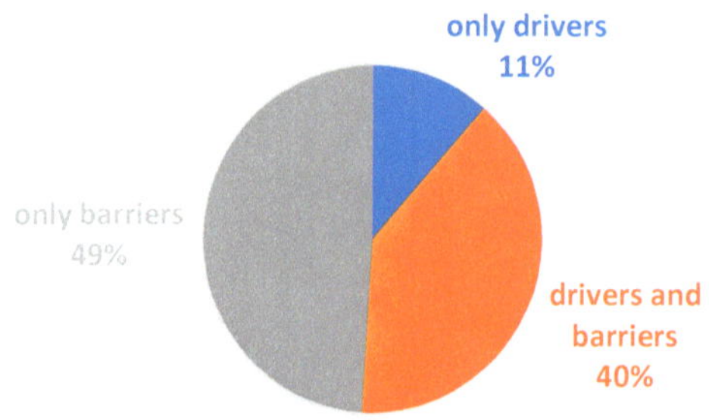

Figure 2. Drivers and barriers article topic analysis.

The drivers and barriers extracted from the analyzed articles were categorized based on the categorization of drivers and barriers from the literature shown in Table 1. The ten categories identified and used to present the drivers that can help the implementation of the CE and the barriers disrupting its adoption are as follows:

- Environmental, which involves aspects related to sustainability, environment, waste management, recycling, and the scarcity of resources;
- Supply chain, which covers aspects involving the supply chain, distribution channels, logistics and reverse logistics, as well as, according to Tura et al. (2019) [26], drivers related to the potential to reduce channel dependence;
- Economic, which includes financial aspects, sales, profitability, revenues, earnings, costs, accounting, raw material costs and prices, and regulatory costs of environmental pollution and waste;
- Information, which involves aspects related to information, knowledge about the CE, information sharing, learning, training, and experience;
- Legal, which encompasses normative, regulatory, and legislative aspects, as well as the costs arising from these aspects;
- Market, which involves aspects of the market, that is, aspects external to the organization, for example, the environmental awareness of consumers, consumer preferences, market demands and market trends;
- Organizational, which involves managerial aspects and aspects related to companies and commercial institutions, that is, the internal aspects of the company/organization such as competition and competitiveness, performance indicators, organizational culture, company policy, environmental aspects (such as environmental collaboration of customers and suppliers, reduction of the environmental impact of the company and processes), aspects regarding ownership, aspects of management and personnel department (such as leadership, employees, workers, and shareholders), aspects regarding the product (product value and quality), raw materials and components, suppliers, partnerships, customers (customer satisfaction and customer relationship), branding, and company image;
- Public, which encompasses aspects regarding the government, states, and municipalities, such as, for example, their support, incentive, financial assistance, and public policies;

- Social, which encompasses aspects of society and the community, involving job creation and reduction of the unemployment rate, population size, public health, safety, hygiene, social responsibility, social projects, public awareness, social recognition, and stakeholders;
- Technological, which involves aspects related to science and technology, technological innovation, and technical aspects, as well as the costs arising from these technologies.

It should be noted that the dividing line between some categories is very tenuous. When this occurred, the authors used the conceptual definition of the category as a criterion.

4.1. Drivers to the Circular Economy

A number of studies have pointed out drivers of the adoption of the CE in order to encourage, motivate, facilitate, and drive companies to adopt CE, and different approaches were used. In their study, Govindan and Hasanagic (2018) [18] examined the drivers in order to understand the motivational factors for implementing the CE in the supply chain. Moktadir et al. (2018) [27] and Hart et al. (2019) [22] refer to a driver as a facilitator in their studies. Motivations for CE practices and facilitating factors for the implementation of circular practices are raised in the study by Barbaritano et al. (2019) [33]. In the study by Jabbour et al. (2020) [23], CE motivators and CE drivers were considered to be synonymous. Finally, in their study, Xue et al. (2010) [24] address methods that drive the development of the CE, and Piyathanavong et al. (2019) [73] presented reasons to implement CE.

Based on this, it was observed that there is no clear-cut definition of drivers in the literature. What is known is that all expressions and nomenclatures used in the literature when it comes to drivers express driving forces leading companies to adopt the CE. The boundaries between the definitions are not clear; there are overlapping areas between the concepts. For the purposes of this study, drivers are therefore defined as forces that motivate or encourage companies to adopt CE.

There were several areas in which CE drivers were studied in the literature. Table 3 presents the contexts and sectors in which the papers were developed.

Table 3. Contexts of application of studies of the drivers of the CE.

Context/sector	Authors
Supply chain	Masi et al. (2017) [19] and Govindan and Hasanagic (2018) [18]
Textile industry	Jia et al. (2020) [20]
Automotive sector	Agyemang et al. (2019) [44]
Leather industry	Moktadir et al. (2018) [27]
Manufacturing industry	Gusmerotti et al. (2019) [10], Kumar et al. (2019) [30], Piyathanavong et al. (2019) [73], and Šebo et al. (2019) [47]
Business	Gue et al. (2019) [72]
Luxury furniture industry	Barbaritano et al. (2019) [33]
Waste management	Ilić and Nikolić (2016) [49]
Construction sector	Chang and Hsieh (2019) [25] and Hart et al. (2019) [22]
Small and medium-sized companies	Ormazabal et al. (2018) [46] and Mura et al. (2020) [21]

In addition, the studies' geographic contexts cover a significant range of countries around the world. Continents, countries, and authors are presented in Table 4.

Table 4. Countries and regions from which articles on CE drivers and barriers came.

	Countries	Author(s) Drivers	Author(s) Barriers
America	Brazil	De Mattos and De Albuquerque, 2018 [50]	
	Brazil	Jabbour et al., 2020 [23]	Jabbour et al., 2020 [23]
	USA	Ranta et al., 2018 [51]	Ranta et al., 2018 [51]
Asia	Bangladesh	Moktadir et al., 2018 [27]	
	China	Geng and Doberstein, 2008 [48]	Geng and Doberstein, 2008 [48]
	China	Xue et al., 2010 [24]	Xue et al., 2010 [24]
	China	Ranta et al., 2018 [51]	Ranta et al., 2018 [51]
	China		Shao et al., 2020 [56]
	China		Farooque et al., 2019 [70]
	India		Mangla et al., 2018 [54]
	Pakistan	Agyemang et al., 2019 [44]	Agyemang et al., 2019 [44]
	Philippines	Gue et al., 2019 [72]	
	Taiwan	Chang and Hsieh, 2019 [25]	Chang and Hsieh, 2019 [25]
	Taiwan	Piyathanavong et al., 2019 [73]	Piyathanavong et al., 2019 [73]
Europe	Belgium, Bulgaria, Czech Republic, Denmark, Germany, Estonia, Ireland, Greece, Spain, Italy, France, Cyprus, Latvia, Lithuania, Luxembourg, Hungary, The Netherlands, Austria, Poland, Portugal, Romania, Slovenia, Slovakia, Finland, Sweden, and United Kingdom	Robaina et al., 2020 [12]	
	Belgium, Denmark, The Netherlands, and United Kingdom		Kanters, 2020 [16]
	Belgium, Germany, The Netherlands, Portugal, Sweden, and United Kingdom		Kirchherr et al., 2018 [53]
	Denmark		Guldmann and Huulgaard, 2020 [15]
	Finland	Tura et al., 2019 [26]	Tura et al., 2019 [26]
	France, Belgium, The Netherlands, Germany, Italy, Luxembourg, Denmark, Ireland, United Kingdom, Greece, Spain, Portugal, Finland, Sweden, Austria, Cyprus (Republic), Czech Republic, Estonia, Hungary, Latvia, Lithuania, Malta, Poland, Slovakia, Slovenia, Bulgaria, Romania, and Croatia		Garcés-Ayerbe et al., 2019 [71]
	France, Cyprus, Belgium, Czech Republic, The Netherlands, Estonia, Germany, Hungary, Italy, Latvia, Luxembourg, Lithuania, Denmark, Malta, Ireland, Poland, United Kingdom, Slovakia, Greece, Slovenia, Spain, Bulgaria, Portugal, Romania, Finland, Croatia, Sweden, and Austria		García-Quevedo et al., 2020 [14]
	Italy	Barbaritano et al., 2019 [33]	Barbaritano et al., 2019 [33]
	Italy	Mura et al., 2020 [21]	Mura et al., 2020 [21]
	Italy	Gusmerotti et al., 2019 [10]	
	Italy, Spain, France, Romania, and Slovenia		Nohra et al., 2020 [29]
	The Netherlands	Campbell-Johnston et al., 2019 [58]	Campbell-Johnston et al., 2019 [58]
	The Netherlands, United Kingdom, Denmark, Belgium, Portugal, and Serbia	Hartley et al., 2020 [11]	
	not mentioned	Ranta et al., 2018 [51]	Ranta et al., 2018 [51]

Table 4. *Cont.*

	Countries	Author(s) Drivers	Author(s) Barriers
	not mentioned	Kumar et al., 2019 [30]	Kumar et al., 2019 [30]
	Scandinavia		Milios et al., 2019 [55]
	Serbia	Ilić and Nikolić, 2016 [49]	Ilić and Nikolić, 2016 [49]
	Slovakia	Šebo et al., 2019 [47]	Šebo et al., 2019 [47]
	Spain	Ormazabal et al., 2018 [46]	Ormazabal et al., 2018 [46]
	Spain		Scarpellini el al., 2019 [75]
	Sweden	Bolger and Doyon, 2019 [68]	
	United Kingdom		Dieckmann et al., 2020 [13]
	United Kingdom (England, Scotland, Northern Ireland, and Wales)	Kumar et al., 2019 [30]	Kumar et al., 2019 [30]
Oceania	Australia	Bolger and Doyon, 2019 [68]	Bolger and Doyon, 2019 [68]

Based on the analysis of data from articles that contained drivers of the adoption of the CE, the drivers were categorized into the ten categories proposed for this study. As a result, a list of 160 drivers was obtained and categorized. Some of these drivers will be presented below.

(a) Environmental drivers support the adoption of the CE, as there is great concern about environmental impacts and the state of the environment [20], the scarcity of resources [26], and global warming and climate change [18]. The CE can act as a solution to these problems, helping to minimize environmental impact [73], reduce waste [30], and develop sustainability [22].

(b) For supply chains, the adoption of the CE brings improvements to the entire chain [27]. For example, it may improve material efficiency and energy use [18], and increase chain transparency [26] and chain engagement [22].

(c) Considering the economic category, one of the main drivers identified in the literature that motivates companies to adopt the CE was cost reduction [45]. By adopting the CE, gains in resource efficiency are made [12], and new value streams are created using byproducts and waste, thus giving companies a new source of revenue and minimizing costs related to the treatment and disposal of waste [19]. In addition, the regulatory costs of environmental pollution and waste are avoided [19]. Finally, the CE generates economic growth for companies [49].

(d) Information drivers help the implementation of the CE, providing information [20], training and education [27], and knowledge exchange [58].

(e) Considering the category of legal drivers, the existence of laws and regulations regarding CE [23] was a very important aspect found in the literature that helps companies to adopt CE.

(f) As market drivers, awareness of environmental issues among consumers [73], customer awareness of green initiatives [20], and the preference and demand for circular products [23] drive the adoption of the CE.

(g) There are several organizational drivers identified in the literature that drive companies to adopt the CE: gains in market share and competitiveness [10,20], environmental collaboration with customers and suppliers [27], companies' willingness to adopt circulars [72], employee engagement and motivation [20], increased product value and quality [44], improving relationships with customers, building loyalty, and increasing their satisfaction [10], the promotion of the company's reputation, its brand and improving the corporate image [19], the collaboration between organizations and the enlarging of the network and partnerships [50], and the stability of the company [44]. All of these aspects motivate companies to adopt the CE.

(h) Public drivers motivate the adoption of the CE due to the support of the government and public institutions, whether in the form of financial, tax, or fiscal support, the waste collection system, or public policies [21,49].

(i) Social drivers support the adoption of the CE due to the possibility of generating jobs [33], concern for public health [30], social awareness [45], community pressure to adopt the CE [27], and stakeholder pressure for sustainable consumption [51].
(j) In the category of technological drivers, investment in science and technology for the CE's implementation is considered a very important aspect [48], in addition to the use of new and state-of-the-art technologies [23].

The categories of drivers for the adoption of the CE and the authors can be seen in Appendix A.

4.2. Barriers to the Circular Economy

Although on the one hand, a series of studies have pointed out drivers for the adoption of the CE to encourage, motivate, facilitate, and encourage companies to adopt a circular process, a number of barriers to the adoption of the CE were also found in the literature, expressing forces opposing the CE's implementation, inhibiting or barring the CE's development. Different approaches were used by authors in the literature regarding the barriers to the adoption of the CE. According to Barbaritano et al. (2019) [33], barriers are factors that hinder the implementation of the CE's practices. Ranta et al. (2018) [51] consider CE barriers to be difficulties face in the CE's implementation. Kumar et al. (2019) [30] and Masi et al. (2017) [19] treat barriers as inhibitors to the CE's implementation, whereas Rajput and Singh (2019) [74] address challenges involved in the implementation of the CE.

Based on this, it is observed that there is also no clear-cut definition of barriers in the literature. However, all expressions and nomenclatures used in the literature when it comes to barriers express forces that oppose the adoption of the CE. The boundaries between the definitions are not clear; there are overlapping areas between the concepts. For the purposes of this study, barriers are considered to be obstacles that hinder, or even prevent, the implementation of the CE.

Several sectors were the focus of the studies of the barriers to the adoption of the CE in the literature. Based on the literature review of the articles that deal with barriers, we observed the application of studies in the contexts and sectors presented in Table 5.

Table 5. Contexts of application of studies of barriers to the CE.

Context/sector	Authors
Supply chain	Masi et al. (2017) [19], Govindan and Hasanagic (2018) [18], Farooque et al. (2019) [70], Mangla et al. (2018) [54], Kazancoglu et al. (2020) [52], and Ozkan-Ozen et al. (2020) [79]
Food supply chain	Farooque et al. (2019) [70]
Food system	Tseng et al. (2019) [77]
Poultry industry	Dieckmann et al. (2020) [13]
Textile industry	Jia et al. (2020) [20], and Kazancoglu et al. (2020) [52]
Electronic industry	Werning and Spinler (2020) [17]
Services	Ritzén and Sandström (2017) [57]
Maritime	Milios et al. (2019) [55]
Fashion	Camacho-Otero et al. (2019) [69]
Automotive sector	Agyemang et al. (2019) [44], and Shao et al. (2020) [56]
Manufacturing industry	Kumar et al. (2019) [30], Piyathanavong et al. (2019) [73], Šebo et al. (2019) [47], and Jaeger and Upadhyay (2020) [78]
Luxury furniture industry	Barbaritano et al. (2019) [33]
Waste management	Ilić and Nikolić (2016) [49]
Construction	Chang and Hsieh (2019) [25], Hart et al. (2019) [22], Ababio and Lu (2023) [28], and Kanters (2020) [16]
Construction and demolition	Ghisellini et al. (2018) [66], and Mahpour (2018) [67]
Small and medium-sized companies	Ormazabal et al. (2018) [46], Garcés-Ayerbe et al. (2019) [71], García-Quevedo et al. (2020) [14], and Mura et al., 2020 [21]

It was also observed that studies related to barriers were applied in a significant number of countries around the world. Table 4 presents these continents, countries, and authors.

From the analysis of data from articles that contained barriers to the adoption of the CE, barriers were grouped according to the ten categories proposed for this study. As a result, a list of 430 barriers was obtained. Some of these barriers will be presented below.

(a) Environmental barriers can hamper the adoption of the CE due to the difficulty of validating, verifying, and predicting environmental pollution and all environmental effects and impacts [13]; the lack of benefits in relation to environmental sustainability, and uncertainty about potential environmental benefits [70]; the low level of reuse, recycling, and recovery of waste [52]; inefficiency in waste management [67]; the limited availability and quality of recycling materials [19]; and the underdeveloped waste infrastructure, which is to take components back for reuse [55]. There is also an informal sector tradition that collects recyclables and food waste (China), and textile waste [52].

(b) Within supply chains, the adoption of the CE may be hampered due to costs, lack of priority, lack of employee skills, lack of enthusiasm and leadership from managers [18], lack of customer awareness [54], the fragile return and collection system [17], the lack of reverse logistics infrastructure [52], the lack of collaboration with stakeholders in the supply chain in CE initiatives [54], and the lack of suitable partners in supply chains [30].

(c) Considering the economic category, some aspects can obstruct the implementation of the CE, such as high initial investment costs [22,28,70]; high production costs, management costs, and planning costs [29]; the low price of virgin raw materials compared to recycled/reused materials [52]; the uncertain profits and returns [28,45], and uncertain economic and fiscal benefits [71]; the lack of funding [15,28]; the lack of working capital [45]; the lack of financial resources [73]; financial risk [16]; and the lack of economies of scale [70].

(d) Some aspects of the information category can obstruct the implementation of the CE, given that there is a lack of training and education of the people involved in the chain [48], a lack of knowledge and skills about CE [15], an abundance of poor data and a lack of data quality [29], and a lack of information [20].

(e) Regarding the legal category, the lack of clarity or lack of support for the CE in the form of laws, norms, and rules becomes a major obstacle to the implementation of the CE. For example, there is a lack of legislation for recycling [51], waste management [55], the environment [48], green production [48], and laws that are specific to the CE's implementation [18]. There is also low regulatory support for increasing reuse activities [51], and bureaucratic difficulty in enforcing legislation on the sustainability (e.g., waste, water) of companies [21]. Furthermore, there are still laws and regulations that are opposed to CE solutions [13].

(f) Market barriers can hamper the CE's implementation, mainly considering two aspects: demand and consumers. The demand for circular products and processes is still restricted [29] and unclear [15]. In addition, there is a lack of demand for remanufactured products due to their appearance [30]. Regarding the consumer, there is a lack of interest in circular processes and products [53] and a lack of environmental awareness [73]. In addition, customers have a negative perception of the quality of remanufactured products [56], preferring new products and materials over reused or remanufactured ones [67].

(g) With regard to organizational barriers, it was identified that the lack of metrics, measurements, and a system with indicators and a method of performance evaluation can hinder the implementation of the CE [52], in addition to the fear of possible risks arising from the implementation of the CE [76] and the lack of collaboration between business functions and departments [17]. There is a business culture that operates a linear system [53] and maintains conservatism in current practices [22] and resistance to change [70], thus causing cultural conflicts [26]. Companies' own cultures do not favor the adoption of the CE [53], and their organizational structures make it difficult to implement the CE [18]. Regarding the managerial level of the organization, there is a lack of commitment [70], will [30], environmental awareness [19], support, and collaboration on the part of management [73], making the adoption of the CE difficult. Regarding the raw material, there is volatility in terms of quantity and availability [17] and a low quality of recovered used parts [56] negatively impacting the adoption of the CE. The difficulty of establishing partnerships is also an aspect that can hinder the implementation of the CE; there may be, for example, lack of suitable partners [30], difficulty in business-to-business (B2B) cooperation [78], and lack of a shared vision and willingness to collaborate with chain partners [52]. It takes time to build new partnerships and mutual trust [15]. Organizations are also faced with the challenge of a lack of qualified personnel to work with the CE and related areas (environmental, remanufacturing, reuse of products and components) [44,46]. Regarding the company's products, it is difficult to maintain quality throughout the product's life cycle, because returned materials may cause low quality in recycled products [52]. It is also difficult for companies to manage products made from reclaimed materials [18].

(h) Regarding the public category, the lack of incentives and support (industrial and financial) from the government is one of the main obstacles to the implementation of the CE identified in the literature [24,63]. In addition, existing recycling policies are ineffective in achieving high-quality recycling, thus hampering the development of the CE [18]. Moreover, there is a lack of public awareness about the CE [24], and public incentives regarding the CE are misaligned, complex, and confusing [45].

(i) The category of social barriers can hamper the adoption of the CE due to the low involvement of society in circular actions and practices [67]; the lack of a global consensus on CE [53]; the lack of awareness in society about circularity [76]; the linear mindset rooted in society [58]; the values, norms, lifestyles, and current social practices; a lack of cultural diversity, and social ignorance about the resource cycle [29].

(j) Technological barriers can impede the development of the CE due to the lack of technological systems [20]; lack of technology transfer [54]; lack of appropriate technologies for the CE [46]; lack of technical support [45,70]; insufficient technical resources [46]; weak demand and acceptance of environmental technologies [54], and the lack of skills and technical capacities of workers [78]. In addition, as there is a need for investment in technology for the adoption of the CE, the costs arising from this investment may also become an obstacle to the CE's implementation [74].

The categories of barriers to the adoption of the CE and the authors can be seen in Appendix A.

5. Conclusions

5.1. Contributions and Implications

This study aimed to identify and categorize drivers of and barriers to the adoption of the CE through a systematic literature review. The study's conclusion will be presented based on 53 analyzed articles. The results of the study show that there are more articles in the literature reflecting barriers to the adoption of the CE than those reflecting drivers of its adoption. Consequently, the study indicates that in the literature, there are more barriers to than drivers of the adoption of the CE. Thus, it was observed that the literature demonstrates a greater concern with what bars, prevents, or hinders companies and society implementing circular behavior; this circular behavior is necessary, considering the current increasing concern about the scarcity of resources in the environment.

Furthermore, it was observed that there are different contexts and sectors in which the analyzed studies were applied, and a significant range of countries and regions around the world were involved. The region in which the largest number of searches was observed was Europe. Corroborating this, in the last 50 years, there has been an intense debate on energy policies and issues related to the environment among the countries of the European

Union [80], culminating in 2019 with the affirmation of commitments to face the challenges of sustainability by adopting the European Green Deal (EGD), a set of initiatives for environmental protection whose main objective is to promote sustainable development strategies focused on energy emissions and the mitigation of climate change [81]. Furthermore, these European Green Deal policies can be incorporated into economic models that promote sustainable development, such as the CE [82].

Finally, the authors of 19 of the articles analyzed in this paper used some categorization to present the drivers and barriers in their studies. Following this idea, the drivers and barriers listed in the literature review were grouped according to ten categories identified and proposed for this study. The ten categories of drivers that can help the implementation of the CE and the barriers disrupting its adoption are presented as follows.

The first category defined in the study, the environmental category, looks at environmental aspects related to sustainability, waste management, recycling, and the scarcity of resources [26]. Concern regarding its environmental impact and concern for environmental sustainability are subjects discussed in the literature [1,20].

The second category identified was the supply chain category, which involves the aspects of the supply chain, distribution channels, logistics, and reverse logistics [57]. It is important to pay close attention to the supply chain as a whole in order to succeed in the CE's development. CE strategies are crucial to restructure the take-make-discard model through the active participation of all actors in the supply chain [83].

The economic category was the third category defined in the study, and covers financial aspects, sales, profitability, revenues, earnings, costs, accounting, the costs of raw materials, and the regulatory costs of environmental pollution and waste [29]. Lack of financial resources is a major limitation for companies in adopting the CE [73]. This could be dealt with by government support and public policies to achieve the CE, as suggested by Arthur et al. (2023) [2].

The fourth category identified was the information category, which considers aspects related to information, CE knowledge, information sharing, learning, training, and experience [19]. Everyone involved in the circular chain must have the necessary information to develop the CE successfully. One of the goals of the 12th SDG of the 2030 Agenda (2015) [42] is to ensure that people have relevant information and awareness about sustainable development.

The fifth category identified was the legal category, looking at normative and regulatory aspects, and their costs [30]. The lack of laws supporting CE practices is one of the major obstacles to the implementation of the CE [51,55]. Policymakers should attend to developing laws that could incentivize companies to adopt the practices of the CE.

The market category was the sixth category defined for the present study. External aspects of the organization, such as consumer environmental awareness, consumer preference, market demands, and trends, constitute this category [15]. There has been an increase in environmental awareness and concern for the environment on the part of consumers, thus generating an increase in demand for circular products. While joining the CE encourages the emergence of new demands for services and new potential markets [8], companies' interest in adopting the CE is increasing.

The seventh category identified in this study is the organizational category, which considers aspects related to companies and commercial institutions, i.e., internal company/organizational features such as competition and competitiveness, performance indicators, organizational culture, company policy, the environmental aspects of the company, aspects related to property, the management and personnel department, products, raw materials, components, suppliers, partnerships, customers, branding, and company image [18]. The benefits arising from the implementation of the CE for companies are numerous, but on the other hand, the challenges are also great. The adoption of the CE generates competitive advantages for circular companies [10,20]; thus, the CE has attracted interest from the business community wanting to work on sustainable development [8].

The public category was the eighth category defined in the study, involving aspects from the government, states, and municipalities, via the government's support, encouragement, financial aid, and public policies [48]. The lack of incentives and industrial and financial support from the government is one of the main obstacles to the implementation of the CE identified in the literature [29,56,73]. It is thus observed that public involvement is fundamental for the development of the CE. Financial resources are scarce, making implementing CE unfeasible for many companies [46,70], especially small and medium-sized companies. Through financial, tax, and fiscal support, and public policy, the adoption of the CE can be encouraged. In accordance with this, Arthur et al. (2023) [2] mentioned that a blend of government policies is an effective means of achieving a CE.

The ninth category identified in the study is the social category. This category involves society and community aspects, such as job creation and the reduction of the unemployment rate, population size, public health, safety, hygiene, social responsibility, social projects, and public awareness [30]. Increased awareness of social responsibility [1] should be harnessed as a driver for new circular business opportunities and the CE's development.

Finally, the tenth category defined in the study is the technological category, which considers science, technology, and technical aspects [53]. Considering that we live in a technological era of digitization and great technological developments, companies could embrace these aspects to help them in the development of the CE. It is suggested that the development of specific technologies will be necessary for the development of the CE. In this sense, the 2030 Agenda (2015) [42] points to the goal of the 12th SDG, supporting developing countries to strengthen their scientific and technological capacities in pursuit of more sustainable patterns of production and consumption.

A summary of the drivers of and barriers to the adoption of the CE is presented in Figure 3.

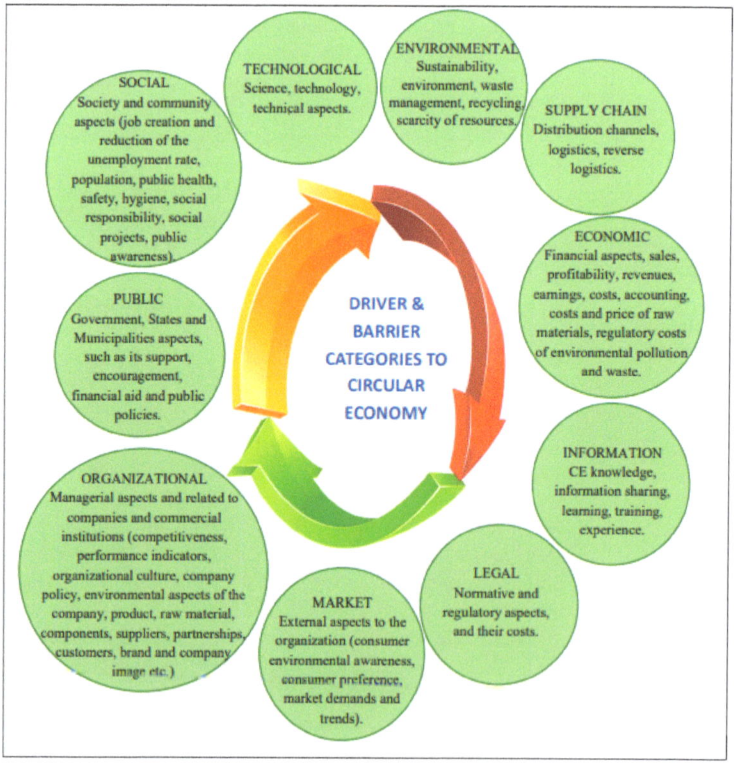

Figure 3. Summary of drivers of and barriers to the adoption of the circular economy.

It is important to highlight the multilevel approach within which these categories were constructed. Even though it was not the focus of this study, allocation at the micro, meso and macro levels is evident, following the findings of Ababio and Lu (2023) [28], which reinforce the current consensus that the CE should be discussed using a multilevel approach.

First, this study contributes by extending the body of knowledge on the CE, helping to integrate the existing literature and develop a comprehensive theoretical framework for guiding future research, as mentioned by Zhang et al. (2022) [4]. Due to the assessment of an extensive number of articles, a list of 160 drivers of and 430 barriers to the adoption of the CE was obtained. Having these drivers of and barriers to the adoption of the CE listed in the literature, it is possible to gain an in-depth understanding of the CE's context, thus providing researchers and practitioners with prior information about the realities that they will face. This will also allow companies and governments to work toward the CE's implementation, helping the transition from a linear economy to a CE, which is more efficient in terms of resources and will help us to advance toward sustainable economies [14]. The transition from a linear economy to a CE is a matter of extreme relevance in the pursuit of more sustainable development [84].

In accordance with this, the second contribution of this study is to bring about sustainable benefits for companies and society, by making them aware of aspects that foster the CE and aspects that make the implementation of the CE difficult. Substantially reducing waste generation through prevention, reduction, recycling, and reuse, and achieving sustainable management and efficient use of natural resources are some of the targets of the 12th SDG addressed in the 2030 Agenda (2015) [42], ensuring standards of sustainable production and consumption by the year 2030 (2030 Agenda, 2015). In this sense, the CE can be considered a promising concept for sustainable development [8]. Additionally, Ali et al. (2023) [35] mention circular economy-centric education being the solution to the social, economic, and environmental problems stemming from climate change. Moreover, green technologies, through the optimization of the use of resources, reduction of waste, and reduction of the demand for new resources, promote the development of green products and services, helping to reduce the environmental impact of consumption [35].

Third, the presentation of an extensive literature review enabled a broader view and a categorization of the drivers of and barriers to the adoption of the CE. These contributions are important to help companies develop a CE and encourage them to implement it. Additionally, the literature review contributes to helping governments to gain knowledge, in order to work on public policy implementation and actions to foster the CE's adoption. According to Schraven et al. (2019) [85], government incentives, such as research funds or stimulating legislation, should be created. Yazdanpanah et al. (2019) [86] support policy-making and fine-tuning the regulations that will foster the transition to a CE; for instance, due to a lack of regulations, firms may face no prohibition of the disposal of some particular (hazardous) wastes, and may have no incentives in the case of substituting some of their raw materials with reusable waste inputs. In this sense, the authors suggest that policy-makers could introduce monetary incentives to foster such practices [86]. Furthermore, the study results would facilitate practitioners' understanding of the drivers of and barriers to the adoption of the CE, that they might handle them effectively.

5.2. Limitations and Future Research

The main limitations of this research refer to the number of databases used. Only the Scopus database was used to collect data. We suggest that future studies also use other databases.

Even though this review was quite comprehensive, the search strategy used only the most consolidated terms in the literature, and may have left out some articles that used other nomenclatures. Thus, it is suggested that future studies use not only consolidated terms (for example, "circular economy"), but also different combinations of keywords that may be synonymous, such as, for example, "circular practices", "circularity", and "circular model".

Although some studies have discussed the circular economy in specific regions, e.g., in the Baltic region [87] and Central and Eastern European countries [88], the present study did not aim to discuss drivers and barriers in different regions or countries. However, due to the importance of countries having different behaviors, it is suggested that future studies take these local idiosyncrasies into account, testing the current categorization.

The analysis showed that there are some aspects of categories such as social, for example, hygiene, public health, and safety, that are not explored in depth in the literature. Thus, future research could identify the less explored categories in the literature and explore them, thereby leading to new frontiers.

This research is part of a research project focused on understanding how to promote an ideal structure of the CE in the organic products sector. The next step of this research is to validate the drivers and barriers that apply to producers and consumers in this sector. At the same time, it will be possible to verify if this general proposal can be applied to different sectors and markets.

Author Contributions: Conceptualization, C.P. and D.C.-D.-M.; methodology, C.P., D.C.-D.-M. and C.S.L.S.; validation, C.P., D.C.-D.-M. and C.S.L.S.; formal analysis, C.P.; investigation, C.P.; writing—original draft preparation, C.P.; writing—review and editing, C.P., D.C-D-M. and C.S.L.S.; supervision, C.P.; funding acquisition, C.S.L.S. All authors have read and agreed to the published version of the manuscript.

Funding: This study was financed in part by the Coordenação de Aperfeiçoamento de Pessoal de Nível Superior—Brazil (CAPES)—Finance Code 001.

Institutional Review Board Statement: Not applicable.

Informed Consent Statement: Not applicable.

Data Availability Statement: Data are available upon request from the corresponding author.

Conflicts of Interest: The authors declare no conflict of interest.

Appendix A

Table A1. Categories of drivers of and barriers to the adoption of the CE and the authors.

Barriers Authors	Category	Drivers Authors
Geng and Doberstein (2008) [48]; Ilić and Nikolić (2016) [49]; Mahpour (2018) [67]; Masi et al. (2017) [19]; Ranta et al. (2018) [51]; Barbaritano et al. (2019) [33]; Farooque et al. (2019) [70]; Kumar et al. (2019) [30]; Milios et al. (2019) [55]; Piyathanavong et al. (2019) [73]; Tura et al. (2019) [26]; Dieckmann et al. (2020) [13]; Guldmann and Huulgaard (2020) [15]; Kazancoglu et al. (2020) [52]; Nohra et al. (2020) [29].	ENVIRONMENTAL	Ilić and Nikolić (2016) [49]; Govindan and Hasanagic (2018) [18]; Moktadir et al. (2018) [27]; Ormazabal et al. (2018) [46]; Ranta et al. (2018) [51]; Hart et al. (2019) [22]; Kumar et al. (2019) [30]; Piyathanavong et al. (2019) [73]; Tura et al. (2019) [26]; Hartley et al. (2020) [11]; Jia et al. (2020) [20].
Ritzén and Sandström (2017) [57]; Govindan and Hasanagic (2018) [18]; Mangla et al. (2018) [54]; Masi et al. (2017) [19]; Agyemang et al. (2019) [44]; Farooque et al. (2019) [70]; Kumar et al. (2019) [30]; Tura et al. (2019) [26]; Dieckmann et al. (2020) [13]; Jaeger and Upadhyay (2020) [78]; Kazancoglu et al. (2020) [52]; Nohra et al. (2020) [29]; Werning and Spinler (2020) [17].	SUPPLY CHAIN	Govindan and Hasanagic (2018) [18]; Moktadir et al. (2018) [27]; Hart et al. (2019) [22]; Tura et al. (2019) [26]; Jia et al. (2020) [20].

Table A1. *Cont.*

Barriers Authors	Category	Drivers Authors
Geng and Doberstein (2008) [48]; Masi et al. (2017) [19]; Ritzén and Sandström (2017) [57]; De Jesus and Mendonça (2018) [45]; Ghisellini et al. (2018) [66]; Govindan and Hasanagic (2018) [18]; Kirchherr et al. (2018) [53]; Mahpour (2018) [67]; Mangla et al. (2018) [54]; Ormazabal et al. (2018) [46]; Agyemang et al. (2019) [44]; Barbaritano et al. (2019) [33]; Campbell-Johnston et al. (2019) [58]; Farooque et al. (2019) [70]; Garcés-Ayerband et al. (2019) [67]; Hart et al. (2019) [22]; Kumar et al. (2019) [30]; Milios et al. (2019) [55]; Piyathanavong et al. (2019) [73]; Scarpellini el al. (2019) [75]; Šebo et al. (2019) [47]; Tura et al. (2019) [26]; Dieckmann et al. (2020) [13]; Guldmann and Huulgaard (2020) [15]; Jabbour et al. (2020) [23]; Jaeger and Upadhyay (2020) [78]; Jia et al. (2020) [20]; Kanters (2020) [16]; Kazancoglu et al. (2020) [52]; Nohra et al. (2020) [29].	ECONOMIC	Ilić and Nikolić (2016) [49]; Masi et al. (2017) [19]; De Jesus and Mendonça (2018) [45]; Govindan and Hasanagic (2018) [18]; Moktadir et al. (2018) [27]; Ormazabal et al. (2018) [46]; Agyemang et al. (2019) [44], Barbaritano et al. (2019) [33]; Gue et al. (2019) [72]; Gusmerotti et al. (2019) [10]; Piyathanavong et al. (2019) [73]; Šebo et al. (2019) [47]; Tura et al. (2019) [26]; Hartley et al. (2020) [11]; Jia et al. (2020) [20]; Robaina et al. (2020) [12]; Ababio and Lu (2023) [28].
Geng and Doberstein (2008) [48]; Masi et al. (2017) [19]; De Jesus and Mendonça (2018) [45]; Govindan and Hasanagic (2018) [18]; Mangla et al. (2018) [54]; Ormazabal et al. (2018) [46]; Agyemang et al. (2019) [44]; Bolger and Doyon (2019) [68]; Campbell-Johnston et al. (2019) [58]; Farooque et al. (2019) [70]; Hart et al. (2019) [22]; Rajput and Singh (2019) [74]; Galvão et al. (2020) [63]; Guldmann and Huulgaard (2020) [15]; Jaeger and Upadhyay (2020) [78]; Jia et al. (2020) [20]; Kazancoglu et al. (2020) [52]; Kirchherr et al. (2018) [53]; Nohra et al. (2020) [29]; Ozkan-Ozen et al. (2020) [79]; Piyathanavong et al. (2019) [73].	INFORMATION	Geng and Doberstein (2008) [48]; Masi et al. (2017) [19]; De Jesus and Mendonça (2018) [45]; Moktadir et al. (2018) [27]; Campbell-Johnston et al. (2019) [58]; Gue et al. (2019) [72]; Hart et al. (2019) [22]; Jia et al. (2020) [20].
Geng and Doberstein (2008) [48]; Xue et al. (2010) [24]; Masi et al. (2017) [19]; De Jesus and Mendonça (2018) [45]; Govindan and Hasanagic (2018) [18]; Kirchherr et al. (2018) [53]; Mangla et al. (2018) [54]; Ranta et al. (2018) [51]; Campbell-Johnston et al. (2019) [58]; Farooque et al. (2019) [70]; Garcés-Ayerband et al. (2019) [67]; Hart et al. (2019) [22]; Kumar et al. (2019) [30]; Milios et al. (2019) [55]; Piyathanavong et al. (2019) [73]; Rajput and Singh (2019) [74]; Šebo et al. (2019) [47]; Singh and Giacosa (2019) [76]; Tseng et al. (2019) [77]; Tura et al. (2019) [26]; Dieckmann et al. (2020) [13]; García-Quevedo et al. (2020) [14]; Guldmann and Huulgaard (2020) [15]; Jia et al. (2020) [20]; Kanters (2020) [16]; Kazancoglu et al. (2020) [52]; Mura et al. (2020) [21]; Nohra et al. (2020) [29]; Shao et al. (2020) [56]; Werning and Spinler (2020) [17].	LEGAL	Xue et al. (2010) [24]; Govindan and Hasanagic (2018) [18]; De Jesus and Mendonça (2018) [45]; De Mattos and De Albuquerque (2018) [50]; Moktadir et al. (2018) [27]; Ranta et al. (2018) [51]; Agyemang et al. (2019) [44]; Gue et al. (2019) [72]; Hart et al. (2019) [22]; Piyathanavong et al. (2019) [73]; Šebo et al. (2019) [47]; Tura et al. (2019) [26]; Hartley et al. (2020) [11]; Jabbour et al. (2020) [23]; Jia et al. (2020) [20]; Ababio and Lu (2023) [28].
Masi et al. (2017) [19]; De Jesus and Mendonça (2018) [45]; Govindan and Hasanagic (2018) [18]; Kirchherr et al. (2018) [53]; Mahpour (2018) [67]; Ormazabal et al. (2018) [46]; Ranta et al. (2018) [51]; Agyemang et al. (2019) [44]; Camacho-Otero (2019) [69]; Campbell-Johnston et al. (2019) [58]; Farooque et al. (2019) [70]; Kumar et al. (2019) [30]; Milios et al. (2019) [55]; Piyathanavong et al. (2019) [73]; Scarpellini el al. (2019) [75]; Šebo et al. (2019) [47]; Singh and Giacosa (2019) [76]; Tura et al. (2019) [26]; Guldmann and Huulgaard (2020) [15]; Jabbour et al. (2020) [23]; Kanters (2020) [16]; Mura et al. (2020) [21]; Nohra et al. (2020) [29]; Shao et al. (2020) [56]; Werning and Spinler (2020) [17].	MARKET	De Jesus and Mendonça (2018) [45]; Govindan and Hasanagic (2018) [18]; Moktadir et al. (2018) [27]; Ranta et al. (2018) [51]; Barbaritano et al. (2019) [33]; Gue et al. (2019) [72]; Hart et al. (2019) [22]; Piyathanavong et al. (2019) [73]; Tura et al. (2019) [26]; Jabbour et al. (2020) [23]; Jia et al. (2020) [20].
Geng and Doberstein (2008) [48]; Masi et al. (2017) [19]; Ritzén and Sandström (2017) [57]; De Jesus and Mendonça (2018) [45]; Ghisellini et al. (2018) [66]; Govindan and Hasanagic (2018) [18]; Kirchherr et al. (2018) [53]; Mahpour (2018) [67]; Mangla et al. (2018) [54]; Ormazabal et al. (2018) [46]; Agyemang et al. (2019) [44]; Barbaritano et al. (2019) [33]; Campbell-Johnston et al. (2019) [58]; Chang and Hsieh (2019) [25]; Farooque et al. (2019) [70]; Garcés-Ayerband et al. (2019) [67]; Hart et al. (2019) [22]; Kumar et al. (2019) [30]; Milios et al. (2019) [55]; Piyathanavong et al. (2019) [73]; Rajput and Singh (2019) [74]; Scarpellini el al. (2019) [75]; Šebo et al. (2019) [47]; Singh and Giacosa (2019) [76]; Tura et al. (2019) [26]; Dieckmann et al. (2020) [13]; Galvão et al. (2020) [63]; García-Quevedo et al. (2020) [14]; Guldmann and Huulgaard (2020) [15]; Jabbour et al. (2020) [23]; Jaeger and Upadhyay (2020) [78]; Jia et al. (2020) [20]; Kanters (2020) [16]; Kazancoglu et al. (2020) [52]; Mura et al. (2020) [21]; Ozkan-Ozen et al. (2020) [79]; Shao et al. (2020) [56]; Werning and Spinler (2020) [17].	ORGANIZATIONAL	Ilić and Nikolić (2016) [49]; Masi et al. (2017) [19]; De Mattos and De Albuquerque (2018) [50]; Govindan and Hasanagic (2018) [18]; Moktadir et al. (2018) [27]; Ormazabal et al. (2018) [46]; Ranta et al. (2018) [51]; Agyemang et al. (2019) [44]; Barbaritano et al. (2019) [33]; Gue et al. (2019) [72]; Gusmerotti et al. (2019) [10]; Hart et al. (2019) [22]; Kumar et al. (2019) [30]; Piyathanavong et al. (2019) [73]; Šebo et al. (2019) [47]; Tura et al. (2019) [26]; Jia et al. (2020) [20]; Hartley et al. (2020) [11]; Mura et al. (2020) [21].

Table A1. *Cont.*

Barriers Authors	Category	Drivers Authors
Geng and Doberstein (2008) [48]; Xue et al. (2010) [24]; Ilić and Nikolić (2016) [49]; Masi et al. (2017) [19]; De Jesus and Mendonça (2018) [45]; Ghisellini et al. (2018) [66]; Govindan and Hasanagic (2018) [18]; Mahpour (2018) [67]; Mangla et al. (2018) [54]; Ormazabal et al. (2018) [46]; Agyemang et al. (2019) [44]; Bolger and Doyon (2019) [68]; Hart et al. (2019) [22]; Kumar et al. (2019) [30]; Piyathanavong et al. (2019) [73]; Tura et al. (2019) [26]; Dieckmann et al. (2020) [13]; Galvão et al. (2020) [63]; Guldmann and Huulgaard (2020) [15]; Jia et al. (2020) [20]; Mura et al. (2020) [21]; Nohra et al. (2020) [29]; Shao et al. (2020) [56].	PUBLIC	Geng and Doberstein (2008) [48]; Xue et al. (2010) [24]; Ilić and Nikolić (2016) [49]; Masi et al. (2017) [19]; De Mattos and De Albuquerque (2018) [50]; Moktadir et al. (2018) [27]; Barbaritano et al. (2019) [33]; Bolger and Doyon (2019) [68]; Campbell-Johnston et al. (2019) [58]; Chang and Hsieh (2019) [25]; Hart et al. (2019) [22]; Piyathanavong et al. (2019) [73]; Jabbour et al. (2020) [23]; Jia et al. (2020) [20]; Mura et al. (2020) [21]; Robaina et al. (2020) [12].
Geng and Doberstein (2008) [48]; Masi et al. (2017) [19]; Galvão et al. (2018) [63]; Kirchherr et al. (2018) [53]; Mahpour (2018) [67]; Mangla et al. (2018) [54]; Ranta et al. (2018) [51]; Campbell-Johnston et al. (2019) [58]; Chang and Hsieh (2019) [25]; Scarpellini el al. (2019) [75]; Singh and Giacosa (2019) [76]; Tseng et al. (2019) [77]; Tura et al. (2019) [26]; Dieckmann et al. (2020) [13]; Guldmann and Huulgaard (2020) [15]; Jabbour et al. (2020) [23]; Nohra et al. (2020) [29].	SOCIAL	Xue et al. (2010) [24]; Ilić and Nikolić (2016) [49]; Masi et al. (2017) [19]; De Jesus and Mendonça (2018) [45]; Govindan and Hasanagic (2018) [18]; Moktadir et al. (2018) [27]; Ranta et al. (2018) [51]; Agyemang et al. (2019) [44]; Barbaritano et al. (2019) [33]; Bolger and Doyon (2019) [68]; Campbell-Johnston et al. (2019) [58]; Gue et al. (2019) [72]; Kumar et al. (2019) [30]; Tura et al. (2019) [26]; Jabbour et al. (2020) [23]; Jia et al. (2020) [20]; Ababio and Lu (2023) [28].
Geng and Doberstein (2008) [48]; Xue et al. (2010) [24]; Ilić and Nikolić (2016) [49]; Masi et al. (2017) [19]; De Jesus and Mendonça (2018) [45]; Govindan and Hasanagic (2018) [18]; Mangla et al. (2018) [54]; Ormazabal et al. (2018) [46]; Agyemang et al. (2019) [44]; Campbell-Johnston et al. (2019) [58]; Farooque et al. (2019) [70]; Hart et al. (2019) [22]; Kumar et al. (2019) [30]; Rajput and Singh (2019) [74]; Šebo et al. (2019) [47]; Tura et al. (2019) [26]; Dieckmann et al. (2020) [13]; Galvão et al. (2020) [63]; Jabbour et al. (2020) [23]; Jaeger and Upadhyay (2020) [78]; Jia et al. (2020) [20]; Kazancoglu et al. (2020) [52]; Nohra et al. (2020) [29]; Ozkan-Ozen et al. (2020) [79].	TECHNOLOGICAL	Geng and Doberstein (2008) [48]; Masi et al. (2017) [19]; De Jesus and Mendonça (2018) [45]; Moktadir et al. (2018) [27]; Agyemang et al. (2019) [44]; Barbaritano et al. (2019) [33]; Chang and Hsieh (2019) [25]; Hart et al. (2019) [22]; Kumar et al. (2019) [30]; Tura et al. (2019) [26]; Hartley et al. (2020) [11]; Jabbour et al. (2020) [23]; Ababio and Lu (2023) [28].

References

1. Flores, P. Latin America. In *The World of Organic Agriculture Statistics and Emerging Trends 2022*; Willer, H., Trávnícek, J., Meier, C., Eds.; Fibl & Ifoam—Organics International: Bonn, Germany, 2022; pp. 272–274. Available online: https://www.organic-world.net/yearbook/yearbook-2022.html (accessed on 6 March 2023).
2. Arthur, E.E.; Gyamfi, S.; Gerstlberger, W.; Stejskal, J.; Prokop, V. Towards Circular Economy: Unveiling Heterogeneous Effects of Government Policy Stringency, Environmentally Related Innovation, and Human Capital within OECD Countries. *Sustainability* **2023**, *15*, 4959. [CrossRef]
3. Portilho, F. *Sustentabilidade Ambiental, Consumo e Cidadania*; Cortez: São Paulo, Brazil, 2010.
4. Zhang, Q.; Dhir, A.; Kaur, P. Circular economy and the food sector: A systematic literature review. *Sustain. Prod. Consum.* **2022**, *32*, 655–668. [CrossRef]
5. Hojnik, J.; Ruzzier, M.; Ruzzier, M.K.; Sučić, B.; Soltwisch, B. Challenges of demographic changes and digitalization on eco-innovation and the circular economy: Qualitative insights from companies. *J. Clean. Prod.* **2023**, *396*, 136439. [CrossRef]
6. Geissdoerfer, M.; Savaget, P.; Bocken, N.M.P.; Hultink, E.J. The circular economy—A new sustainability paradigm? *J. Clean. Prod.* **2017**, *143*, 757–768. [CrossRef]
7. Velenturf, A.P.; Purnell, P. Principles for a sustainable circular economy. *Sustain. Prod. Consum.* **2021**, *27*, 1437–1457. [CrossRef]
8. Korhonen, J.; Honkasalo, A.; Seppälä, J. Circular Economy: The Concept and its Limitations. *Ecol. Econ.* **2018**, *143*, 37–46. [CrossRef]
9. Horbach, J.; Rammer, C. Circular economy innovations, growth and employment at the firm level: Empirical evidence from Germany. *J. Ind. Ecol.* **2020**, *24*, 615–625. [CrossRef]
10. Gusmerotti, N.M.; Testa, F.; Corsini, F.; Pretner, G.; Iraldo, F. Drivers and approaches to the circular economy in manufacturing firms. *J. Clean. Prod.* **2019**, *230*, 314–327. [CrossRef]
11. Hartley, K.; van Santen, R.; Kirchherr, J. Policies for transitioning towards a circular economy: Expectations from the European Union (EU). *Resour. Conserv. Recycl.* **2020**, *155*, 104634. [CrossRef]
12. Robaina, M.; Villar, J.; Pereira, E.T. The determinants for a circular economy in Europe. *Environ. Sci. Pollut. Res.* **2020**, *27*, 12566–12578. [CrossRef]
13. Dieckmann, E.; Sheldrick, L.; Tennant, M.; Myers, R.; Cheeseman, C. Analysis of Barriers to Transitioning from a Linear to a Circular Economy for End of Life Materials: A Case Study for Waste Feathers. *Sustainability* **2020**, *12*, 1725. [CrossRef]
14. García-Quevedo, J.; Jové-Llopis, E.; Martínez-Ros, E. Barriers to the circular economy in European small and medium-sized firms. *Bus. Strat. Environ.* **2020**, *29*, 2450–2464. [CrossRef]
15. Guldmann, E.; Huulgaard, R.D. Barriers to circular business model innovation: A multiple-case study. *J. Clean. Prod.* **2020**, *243*, 118160. [CrossRef]

16. Kanters, J. Circular Building Design: An Analysis of Barriers and Drivers for a Circular Building Sector. *Buildings* **2020**, *10*, 77. [CrossRef]
17. Werning, J.P.; Spinler, S. Transition to circular economy on firm level: Barrier identification and prioritization along the value chain. *J. Clean. Prod.* **2020**, *245*, 118609. [CrossRef]
18. Govindan, K.; Hasanagic, M. A systematic review on drivers, barriers, and practices towards circular economy: A supply chain perspective. *Int. J. Prod. Res.* **2018**, *56*, 278–311. [CrossRef]
19. Masi, D.; Day, S.; Godsell, J. Supply Chain Configurations in the Circular Economy: A Systematic Literature Review. *Sustainability* **2017**, *9*, 1602. [CrossRef]
20. Jia, F.; Yin, S.; Chen, L.; Chen, X. The circular economy in textile and apparel industry: A systematic literature review. *J. Clean. Prod.* **2020**, *259*, 120728. [CrossRef]
21. Mura, M.; Longo, M.; Zanni, S. Circular economy in Italian SMEs: A multi-method study. *J. Clean. Prod.* **2020**, *245*, 118821. [CrossRef]
22. Hart, J.; Adams, K.; Giesekam, J.; Tingley, D.D.; Pomponi, F. Barriers and drivers in a circular economy: The case of the built environment. *Procedia CIRP* **2019**, *80*, 619–624. [CrossRef]
23. Jabbour, C.J.C.; Seuring, S.; de Sousa Jabbour, A.B.L.; Jugend, D.; Fiorini, P.D.C.; Latan, H.; Izeppi, W.C. Stakeholders, innovative business models for the circular economy and sustainable performance of firms in an emerging economy facing institutional voids. *J. Environ. Manag.* **2020**, *264*, 110416. [CrossRef] [PubMed]
24. Xue, B.; Chen, X.-P.; Geng, Y.; Guo, X.-J.; Lu, C.-P.; Zhang, Z.-L.; Lu, C.-Y. Survey of officials' awareness on circular economy development in China: Based on municipal and county level. *Resour. Conserv. Recycl.* **2010**, *54*, 1296–1302. [CrossRef]
25. Chang, Y.-T.; Hsieh, S.-H. A Preliminary Case Study on Circular Economy in Taiwan's Construction. *IOP Conf. Ser. Earth Environ. Sci.* **2019**, *225*, 012069. [CrossRef]
26. Tura, N.; Hanski, J.; Ahola, T.; Ståhle, M.; Piiparinen, S.; Valkokari, P. Unlocking circular business: A framework of barriers and drivers. *J. Clean. Prod.* **2019**, *212*, 90–98. [CrossRef]
27. Moktadir, M.A.; Rahman, T.; Rahman, M.H.; Ali, S.M.; Paul, S.K. Drivers to sustainable manufacturing practices and circular economy: A perspective of leather industries in Bangladesh. *J. Clean. Prod.* **2018**, *174*, 1366–1380. [CrossRef]
28. Ababio, B.K.; Lu, W. Barriers and enablers of circular economy in construction: A multi-system perspective towards the development of a practical framework. *Constr. Manag. Econ.* **2023**, *41*, 3–21. [CrossRef]
29. Nohra, C.G.; Pereno, A.; Barbero, S. Systemic Design for Policy-Making: Towards the Next Circular Regions. *Sustainability* **2020**, *12*, 4494. [CrossRef]
30. Kumar, V.; Sezersan, I.; Garza-Reyes, J.A.; Gonzalez, E.D.; Al-Shboul, M.A. Circular economy in the manufacturing sector: Benefits, opportunities and barriers. *Manag. Decis.* **2019**, *57*, 1067–1086. [CrossRef]
31. Mishra, R.; Singh, R.K.; Govindan, K. Barriers to the adoption of circular economy practices in Micro, Small and Medium Enterprises: Instrument development, measurement and validation. *J. Clean. Prod.* **2022**, *351*, 131389. [CrossRef]
32. Elia, V.; Gnoni, M.G.; Tornese, F. Evaluating the adoption of circular economy practices in industrial supply chains: An empirical analysis. *J. Clean. Prod.* **2020**, *273*, 122966. [CrossRef]
33. Barbaritano, M.; Bravi, L.; Savelli, E. Sustainability and Quality Management in the Italian Luxury Furniture Sector: A Circular Economy Perspective. *Sustainability* **2019**, *11*, 3089. [CrossRef]
34. D'adamo, I.; Gastaldi, M. Sustainable Development Goals: A Regional Overview Based on Multi-Criteria Decision Analysis. *Sustainability* **2022**, *14*, 9779. [CrossRef]
35. Ali, S.M.; Appolloni, A.; Cavallaro, F.; D'adamo, I.; Di Vaio, A.; Ferella, F.; Gastaldi, M.; Ikram, M.; Kumar, N.M.; Martin, M.A.; et al. Development Goals towards Sustainability. *Sustainability* **2023**, *15*, 9443. [CrossRef]
36. Kirchherr, J.; Yang, N.-H.N.; Schulze-Spüntrup, F.; Heerink, M.J.; Hartley, K. Conceptualizing the Circular Economy (Revisited): An Analysis of 221 Definitions. *Resour. Conserv. Recycl.* **2023**, *194*, 107001. [CrossRef]
37. Friant, M.C.; Vermeulen, W.J.; Salomone, R. A typology of circular economy discourses: Navigating the diverse visions of a contested paradigm. *Resour. Conserv. Recycl.* **2020**, *161*, 104917. [CrossRef]
38. Sehnem, S.; Pereira, S.C.F. Rumo à Economia Circular: Sinergia Existente entre as Definições Conceituais Correlatas e Apropriação para a Literatura Brasileira. *Rev. Eletrônica Ciência Adm.* **2019**, *18*, 35–62. [CrossRef]
39. McDonough, W.; Braungart, M. *Cradle to Cradle: Remaking the Way We Make Things*; North Point Press: New York, NY, USA, 2010.
40. Ellen MacArthur Foundation. Cities and Circular Economy for Food. 2019. Available online: ellenmacarthurfoundation.org (accessed on 18 April 2022).
41. Ellen MacArthur Foundation. A Solution to Build Back Better: The Circular Economy. 2020. Available online: https://www.ellenmacarthurfoundation.org/assets/downloads/emf-jointstatement.pdf (accessed on 8 December 2020).
42. Agenda 2030. 2015. Agenda 2023 Para o Desenvolvimento Sustentável. Available online: https://brasil.un.org/pt-br/91863-agenda-2030-para-o-desenvolvimento-sustentavel (accessed on 14 May 2022).
43. Corvellec, H.; Stowell, A.F.; Johansson, N. Critiques of the circular economy. *J. Ind. Ecol.* **2022**, *26*, 421–432. [CrossRef]
44. Agyemang, M.; Kusi-Sarpong, S.; Khan, S.A.; Mani, V.; Rehman, S.T.; Kusi-Sarpong, H. Drivers and barriers to circular economy implementation. *Manag. Decis.* **2019**, *57*, 971–994. [CrossRef]
45. De Jesus, A.; Mendonça, S. Lost in transition? drivers and barriers in the eco-innovation road to the circular economy. *Ecol. Econ.* **2018**, *145*, 75–89. [CrossRef]

46. Ormazabal, M.; Prieto-Sandoval, V.; Puga-Leal, R.; Jaca, C. Circular Economy in Spanish SMEs: Challenges and opportunities. *J. Clean. Prod.* **2018**, *185*, 157–167. [CrossRef]
47. Šebo, J.; Kádárová, J.; Malega, P. Barriers and motives experienced by manufacturing companies in implementing circular economy initiatives: The case of manufacturing industry in Slovakia. In Proceedings of the 2019 International Council on Technologies of Environmental Protection (ICTEP), Starý Smokovec, Slovakia, 23–25 October 2019; IEEE: New York, NY, USA, 2019; pp. 226–229. [CrossRef]
48. Geng, Y.; Doberstein, B. Developing the circular economy in China: Challenges and opportunities for achieving 'leapfrog development'. *Int. J. Sustain. Dev. World Ecol.* **2008**, *15*, 231–239. [CrossRef]
49. Ilić, M.; Nikolić, M. Drivers for development of circular economy—A case study of Serbia. *Habitat Int.* **2016**, *56*, 191–200. [CrossRef]
50. De Mattos, C.A.; De Albuquerque, T.L.M. Enabling factors and strategies for the transition toward a circular economy (CE). *Sustainability* **2018**, *10*, 4628. [CrossRef]
51. Ranta, V.; Aarikka-Stenroos, L.; Ritala, P.; Mäkinen, S.J. Exploring institutional drivers and barriers of the circular economy: A cross-regional comparison of China, the US, and Europe. *Resour. Conserv. Recycl.* **2018**, *135*, 70–82. [CrossRef]
52. Kazancoglu, I.; Kazancoglu, Y.; Yarimoglu, E.; Kahraman, A. A conceptual framework for barriers of circular supply chains for sustainability in the textile industry. *Sustain. Dev.* **2020**, *28*, 1477–1492. [CrossRef]
53. Kirchherr, J.; Piscicelli, L.; Bour, R.; Kostense-Smit, E.; Muller, J.; Huibrechtse-Truijens, A.; Hekkert, M. Barriers to the Circular Economy: Evidence From the European Union (EU). *Ecol. Econ.* **2018**, *150*, 264–272. [CrossRef]
54. Mangla, S.K.; Luthra, S.; Mishra, N.; Singh, A.; Rana, N.P.; Dora, M.; Dwivedi, Y. Barriers to effective circular supply chain management in a developing country context. *Prod. Plan. Control* **2018**, *29*, 551–569. [CrossRef]
55. Milios, L.; Beqiri, B.; Whalen, K.A.; Jelonek, S.H. Sailing towards a circular economy: Conditions for increased reuse and remanufacturing in the Scandinavian maritime sector. *J. Clean. Prod.* **2019**, *225*, 227–235. [CrossRef]
56. Shao, J.; Huang, S.; Lemus-Aguilar, I.; Ünal, E. Circular business models generation for automobile remanufacturing industry in China. *J. Manuf. Technol. Manag.* **2020**, *31*, 542–571. [CrossRef]
57. Ritzén, S.; Sandström, G. Barriers to the Circular Economy—Integration of Perspectives and Domains. *Procedia CIRP* **2017**, *64*, 7–12. [CrossRef]
58. Campbell-Johnston, K.; Cate, J.T.; Elfering-Petrovic, M.; Gupta, J. City level circular transitions: Barriers and limits in Amsterdam, Utrecht and The Hague. *J. Clean. Prod.* **2019**, *235*, 1232–1239. [CrossRef]
59. Snyder, H. Literature review as a research methodology: An overview and guidelines. *J. Bus. Res.* **2019**, *104*, 333–339. [CrossRef]
60. Wolfswinkel, J.F.; Furtmueller, E.; Wilderom, C.P.M. Using grounded theory as a method for rigorously reviewing literature. *Eur. J. Inf. Syst.* **2013**, *22*, 45–55. [CrossRef]
61. Flores, P.J.; Jansson, J. SPICe—Determinants of consumer green innovation adoption across domains: A systematic review of marketing journals and suggestions for a research agenda. *Int. J. Consum. Stud.* **2022**, *46*, 1761–1784. [CrossRef]
62. Paul, J.; Criado, A.R. The art of writing literature review: What do we know and what do we need to know? *Int. Bus. Rev.* **2020**, *29*, 101717. [CrossRef]
63. Galvão, G.D.A.; Homrich, A.S.; Geissdoerfer, M.; Evans, S.; Scoleze Ferrer, P.S.; Carvalho, M.M. Towards a value stream perspective of circular business models. *Resour. Conserv. Recycl.* **2020**, *162*, 105060. [CrossRef]
64. Bardin, L. *Análise de Conteúdo*; Edições 70: São Paulo, Brazil, 2016.
65. Xiao, Y.; Watson, M. Guidance on Conducting a Systematic Literature Review. *J. Plan. Educ. Res.* **2019**, *39*, 93–112. [CrossRef]
66. Ghisellini, P.; Ripa, M.; Ulgiati, S. Exploring environmental and economic costs and benefits of a circular economy approach to the construction and demolition sector. A literature review. *J. Clean. Prod.* **2018**, *178*, 618–643. [CrossRef]
67. Mahpour, A. Prioritizing barriers to adopt circular economy in construction and demolition waste management. *Resour. Conserv. Recycl.* **2018**, *134*, 216–227. [CrossRef]
68. Bolger, K.; Doyon, A. Circular cities: Exploring local government strategies to facilitate a circular economy. *Eur. Plan. Stud.* **2019**, *27*, 2184–2205. [CrossRef]
69. Camacho-Otero, J.; Boks, C.; Pettersen, I.N. User acceptance and adoption of circular offerings in the fashion sector: Insights from user-generated online reviews. *J. Clean. Prod.* **2019**, *231*, 928–939. [CrossRef]
70. Farooque, M.; Zhang, A.; Liu, Y. Barriers to circular food supply chains in China. *Supply Chain Manag. Int. J.* **2019**, *24*, 677–696. [CrossRef]
71. Garcés-Ayerbe, C.; Rivera-Torres, P.; Suárez-Perales, I.; Leyva-De La Hiz, D.I. Is it possible to change from a linear to a circular economy? an overview of opportunities and barriers for european small and medium-sized enterprise companies. *Int. J. Environ. Res. Public Health* **2019**, *16*, 851. [CrossRef]
72. Gue IH, V.; Ubando, A.T.; Promentilla MA, B.; Tan, R.R. Determining the Causality between Drivers of Circular Economy using the DEMATEL Framework. *Chem. Eng. Trans.* **2019**, *76*, 121–126. [CrossRef]
73. Piyathanavong, V.; Garza-Reyes, J.A.; Kumar, V.; Maldonado-Guzmán, G.; Mangla, S.K. The adoption of operational environmental sustainability approaches in the Thai manufacturing sector. *J. Clean. Prod.* **2019**, *220*, 507–528. [CrossRef]
74. Rajput, S.; Singh, S.P. Industry 4.0—Challenges to implement circular economy. *Benchmarking Int. J.* **2019**, *28*, 1717–1739. [CrossRef]
75. Scarpellini, S.; Portillo-Tarragona, P.; Aranda-Usón, A.; Llena-Macarulla, F. Definition and measurement of the circular economy's regional impact. *J. Environ. Plan. Manag.* **2019**, *62*, 2211–2237. [CrossRef]

76. Singh, P.; Giacosa, E. Cognitive biases of consumers as barriers in transition towards circular economy. *Manag. Decis.* **2019**, *57*, 921–936. [CrossRef]
77. Tseng, M.-L.; Chiu, A.S.; Chien, C.-F.; Tan, R.R. Pathways and barriers to circularity in food systems. *Resour. Conserv. Recycl.* **2019**, *143*, 236–237. [CrossRef]
78. Jaeger, B.; Upadhyay, A. Understanding barriers to circular economy: Cases from the manufacturing industry. *J. Enterp. Inf. Manag.* **2020**, *33*, 729–745. [CrossRef]
79. Ozkan-Ozen, Y.D.; Kazancoglu, Y.; Mangla, S.K. Synchronizes Barriers for Circular Supply Chains in Industry 3.5/Industry 4.0 Transition for Sustainable Resource Management. *Resour. Conserv. Recycl.* **2020**, *161*, 104986. [CrossRef]
80. Panarello, D.; Gatto, A. Decarbonising Europe—EU citizens' perception of renewable energy transition amidst the European Green Deal. *Energy Policy* **2023**, *172*, 113272. [CrossRef]
81. Cuadros-Casanova, I.; Cristiano, A.; Biancolini, D.; Cimatti, M.; Sessa, A.A.; Angarita, V.Y.M.; Dragonetti, C.; Pacifici, M.; Rondinini, C.; Di Marco, M. Opportunities and challenges for Common Agricultural Policy reform to support the European Green Deal. *Conserv. Biol.* **2023**, *37*, e14052. [CrossRef] [PubMed]
82. Camilleri, M.A. European environment policy for the circular economy: Implications for business and industry stakeholders. *Sustain. Dev.* **2020**, *28*, 1804–1812. [CrossRef]
83. Borrello, M.; Caracciolo, F.; Lombardi, A.; Pascucci, S.; Cembalo, L. Consumers' Perspective on Circular Economy Strategy for Reducing Food Waste. *Sustainability* **2017**, *9*, 141. [CrossRef]
84. Testa, F.; Iovino, R.; Iraldo, F. The circular economy and consumer behaviour: The mediating role of information seeking in buying circular packaging. *Bus. Strat. Environ.* **2020**, *29*, 3435–3448. [CrossRef]
85. Schraven, D.; Bukvić, U.; Di Maio, F.; Hertogh, M. Circular transition: Changes and responsibilities in the Dutch stony material supply chain. *Resour. Conserv. Recycl.* **2019**, *150*, 104359. [CrossRef]
86. Yazdanpanah, V.; Yazan, D.M.; Zijm, W.H.M. FISOF: A formal industrial symbiosis opportunity filtering method. *Eng. Appl. Artif. Intell.* **2019**, *81*, 247–259. [CrossRef]
87. Ahmadov, T.; Gerstlberger, W.; Prause, G.K. Fiscal Incentives for Circular Economy: Insights from the Baltic States. In *Business Models for the Circular Economy*; Sustainability and Innovation; Prokop, V., Stejskal, J., Horbach, J., Gerstlberger, W., Eds.; Springer: Cham, Switzerland, 2022. [CrossRef]
88. Prokop, V.; Gerstlberger, W.; Zapletal, D.; Striteska, M.K. The double-edged role of firm environmental behaviour in the creation of product innovation in Central and Eastern European countries. *J. Clean. Prod.* **2022**, *331*, 129989. [CrossRef]

Disclaimer/Publisher's Note: The statements, opinions and data contained in all publications are solely those of the individual author(s) and contributor(s) and not of MDPI and/or the editor(s). MDPI and/or the editor(s) disclaim responsibility for any injury to people or property resulting from any ideas, methods, instructions or products referred to in the content.

MDPI AG
Grosspeteranlage 5
4052 Basel
Switzerland
Tel.: +41 61 683 77 34

Sustainability Editorial Office
E-mail: sustainability@mdpi.com
www.mdpi.com/journal/sustainability

Disclaimer/Publisher's Note: The statements, opinions and data contained in all publications are solely those of the individual author(s) and contributor(s) and not of MDPI and/or the editor(s). MDPI and/or the editor(s) disclaim responsibility for any injury to people or property resulting from any ideas, methods, instructions or products referred to in the content.

www.ingramcontent.com/pod-product-compliance
Lightning Source LLC
Chambersburg PA
CBHW042014260926
51850CB00020B/3403
9783725825790